College Algebra

Third Edition

eText Reference

Kirk Trigsted
University of Idaho

PEARSON

Boston Columbus Indianapolis New York San Francisco Upper Saddle River
Amsterdam Cape Town Dubai London Madrid Milan Munich Paris Montreal Toronto
Delhi Mexico São Paulo Sydney Hong Kong Seoul Singapore Taipei Tokyo

Editor in Chief: Anne Kelly
Acquisitions Editor: Dawn Murrin
Editorial Assistant: Joseph Colella
Design Manager: Andrea Nix
Art Director: Heather Scott
Project Manager: Ron Hampton
Senior Math Media Producer: Tracy Menoza
Project Manager, MathXL: Kristina Evans
Marketing Manager: Peggy Lucas
Marketing Assistant: Justine Goulart
Manufacturing Buyer: Debbie Rossi

**For my wife, Wendy, and our children—
Benjamin, Emily, Gabrielle, and Isabelle.**

6

ISBN-13: 978-0-321-86934-0
ISBN-10:   0-321-86934-6

Contents

Preface

Introduction

This *eText Reference* contains the pages of Kirk Trigsted's *College Algebra* in a portable, bound format. The structure of this *eText Reference* helps student organize their notes by providing them with space to summarize the videos and animations. Students can also use the *eText Reference* to review the eText material anytime, anywhere.

A Note to Students

This eText was created for you and was specifically designed to be read online. Unlike a traditional textbook, I have created content that gives you, the reader, the ability to be an active participant in your learning. The eText pages have large, readable fonts and were designed to avoid scrolling. Throughout the material, I have carefully integrated hundreds of hyperlinks to definitions, previous chapters, interactive videos, animations, and other important content. Many of the videos and animations require you to actively participate. Take some time to "click around" and get comfortable with the navigation of the eText and explore its many features. To log into your ecourse powered by MyMathLab, you must first register at www.mymathlab.com. This eText Reference is a bound, printed version of the eText that provides a place for you to do practice work and summarize key concepts from the online videos and animations.

Before you attempt each homework assignment, read the appropriate section(s) of the eText. At the beginning of each section (starting in Chapter One), you will encounter a feature called Things to Know. This feature includes all the prerequisite objectives from previous sections that you will need to successfully learn the material presented in the new section. If you do not think that you have a basic understanding of any of these objectives, click on any of the hyperlinks and rework those objectives, taking advantage of the videos or animations.

Try testing yourself by working through the corresponding You Try It exercises. Remember, you learn math by *doing* math! The more time you spend working through the videos, animations, and exercises, the more you will understand. If your instructor assigns homework in MyMathLab or MathXL, rework the exercises until you get them right. Be sure to go back and read the eText at anytime while you are working on your homework. This eText is catered to your educational needs, and I hope you enjoy the experience.

A Note to Instructors

I have taught with MyMathLab for many years and have experienced first-hand how fewer and fewer students are using their traditional textbooks. As the use of technology plays an ever increasing role in how we are teaching our students, it is only natural to have a textbook that mirrors the way our students are learning. I am excited to have written an eText from the ground up to be used as an online, interactive tool for students to read while working in MyMathLab. I wrote this eText entirely from an online perspective, keeping MyMathLab and its existing functionality specifically in mind. Every hyperlink, video, and animation was strategically integrated within the context of each page to maximize the student learning experience. All of the interactive media was designed so students could actively participate while they learn math.

I am a proponent of students learning terms and definitions. Therefore, I have created hyperlinks throughout the text to the definitions of important mathematical terms. I have also inserted a significant amount of just-in-time review throughout the text by creating links to prerequisite topics. Students have the ability to reference these review materials with just a click of the mouse.

Each section has five reading assessment questions for those instructors who like to assign reading. These questions are conceptual in nature and were designed to test students on their reading comprehension. Each question was specifically designed to give specific feedback that directs the back to the appropriate pages of the eText. Note that these questions are static. Students are given two chances to obtain the correct answer. Students will not be able to regenerate a similar exercise. The reading assessment questions can be identified in the Homework/Test Manager by the code "RA."

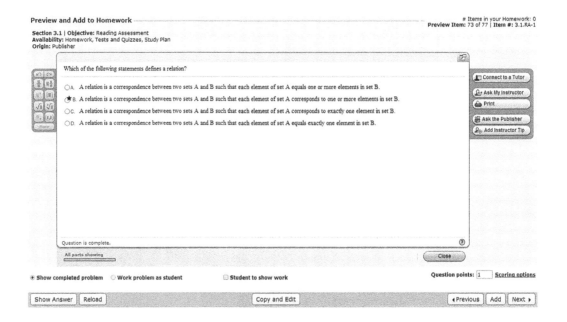

The first edition of *College Algebra* included several multipart exercises. I received feedback from many users who liked to use these multipart exercises for homework but often wished that they could assign only one of the parts for testing. This led to the creation of a new type of exercise called brief exercises. The multipart exercises are now called step-by-step exercises.

Step-by-Step Exercises

The step-by-step exercises were designed to use the power of MathXL to systematically walk the student through some of the more complex, conceptual topics. For example, instead of simply asking for the graph of quadratic function, the step-by-step exercise walks the student through the entire graphing process by asking for all of the important aspects of a parabola. The step-by-step exercises can be readily identified in the Homework/Test Manager by the code "SbS" that precedes the exercise number. An example of such an exercise follows.

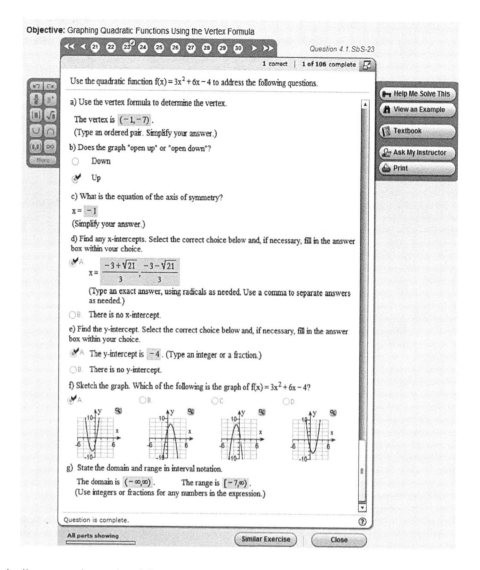

Objective: Graphing Quadratic Functions Using the Vertex Formula

Use the quadratic function $f(x) = 3x^2 + 6x - 4$ to address the following questions.

a) Use the vertex formula to determine the vertex.

The vertex is $(-1, -7)$.
(Type an ordered pair. Simplify your answer.)

b) Does the graph "open up" or "open down"?
- ○ Down
- ◉ Up

c) What is the equation of the axis of symmetry?
$x = -1$
(Simplify your answer.)

d) Find any x-intercepts. Select the correct choice below and, if necessary, fill in the answer box within your choice.
- ◉ A. $x = \dfrac{-3 + \sqrt{21}}{3}, \dfrac{-3 - \sqrt{21}}{3}$
  (Type an exact answer, using radicals as needed. Use a comma to separate answers as needed.)
- ○ B. There is no x-intercept.

e) Find the y-intercept. Select the correct choice below and, if necessary, fill in the answer box within your choice.
- ◉ A. The y-intercept is -4. (Type an integer or a fraction.)
- ○ B. There is no y-intercept.

f) Sketch the graph. Which of the following is the graph of $f(x) = 3x^2 + 6x - 4$?

g) State the domain and range in interval notation.
The domain is $(-\infty, \infty)$.   The range is $[-7, \infty)$.
(Use integers or fractions for any numbers in the expression.)

In the exercise sets at the end of each section of this eText Reference, all step-by-step exercises are labelled with an SbS icon.

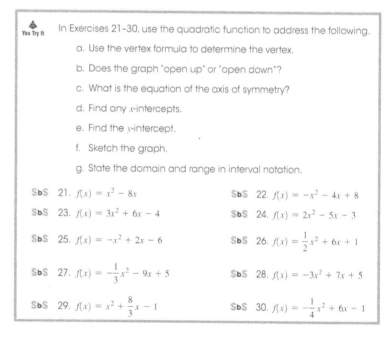

You Try It

In Exercises 21–30, use the quadratic function to address the following.

a. Use the vertex formula to determine the vertex.

b. Does the graph "open up" or "open down"?

c. What is the equation of the axis of symmetry?

d. Find any x-intercepts.

e. Find the y-intercept.

f. Sketch the graph.

g. State the domain and range in interval notation.

SbS 21. $f(x) = x^2 - 8x$

SbS 22. $f(x) = -x^2 - 4x + 8$

SbS 23. $f(x) = 3x^2 + 6x - 4$

SbS 24. $f(x) = 2x^2 - 5x - 3$

SbS 25. $f(x) = -x^2 + 2x - 6$

SbS 26. $f(x) = \frac{1}{2}x^2 + 6x + 1$

SbS 27. $f(x) = -\frac{1}{3}x^2 - 9x + 5$

SbS 28. $f(x) = -3x^2 + 7x + 5$

SbS 29. $f(x) = x^2 + \frac{8}{3}x - 1$

SbS 30. $f(x) = -\frac{1}{4}x^2 + 6x - 1$

Brief Exercises

The brief exercises are copies of the step-by-step exercises and are designed to test one concept of a multistep problem and were designed for instructors who may want to pinpoint a specific skill for a quiz and testing purposes. The brief exercises can be identified in the Homework/Test Manager by the code "BE" that precedes the exercise number. Below is the brief exercise that corresponds to the step-by-step exercise seen on the previous page.

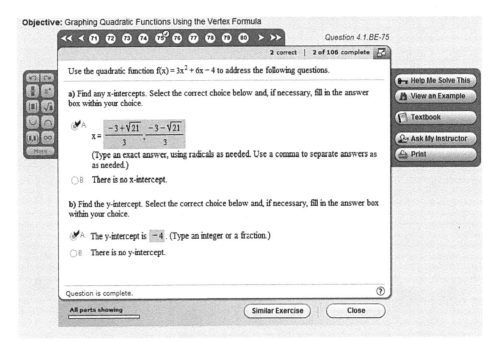

Skill Check Exercises

The Skill Check Exercises (SCE) were created based on comments submitted by users of the Trigsted precalculus series. Skill Check Exercises help students make algebra connections as they work through exercises that present new concepts. These exercises appear at the beginning of most exercises sets and are assignable in MyMathLab.

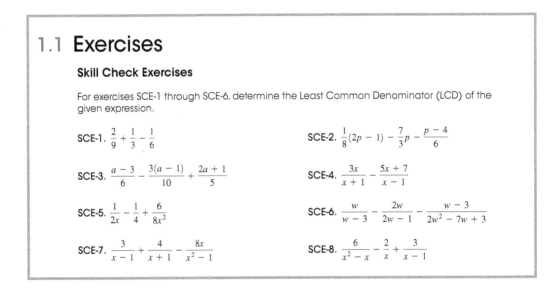

1.1 Exercises

Skill Check Exercises

For exercises SCE-1 through SCE-6, determine the Least Common Denominator (LCD) of the given expression.

SCE-1. $\dfrac{2}{9} + \dfrac{1}{3} - \dfrac{1}{6}$

SCE-2. $\dfrac{1}{8}(2p - 1) - \dfrac{7}{3}p - \dfrac{p - 4}{6}$

SCE-3. $\dfrac{a - 3}{6} - \dfrac{3(a - 1)}{10} + \dfrac{2a + 1}{5}$

SCE-4. $\dfrac{3x}{x + 1} - \dfrac{5x + 7}{x - 1}$

SCE-5. $\dfrac{1}{2x} - \dfrac{1}{4} + \dfrac{6}{8x^2}$

SCE-6. $\dfrac{w}{w - 3} - \dfrac{2w}{2w - 1} - \dfrac{w - 3}{2w^2 - 7w + 3}$

SCE-7. $\dfrac{3}{x - 1} + \dfrac{4}{x + 1} - \dfrac{8x}{x^2 - 1}$

SCE-8. $\dfrac{6}{x^2 - x} - \dfrac{2}{x} + \dfrac{3}{x - 1}$

All Skill Check Exercises will be labeled as SCE-1, SCE-2, SCE-3, and so forth in the Assignment Manager in MyMathLab and MathXL.

Resources for Success

MyMathLab® Online Course (access code required)

MyMathLab delivers **proven results** in helping individual students succeed. It provides **engaging experiences** that personalize, stimulate, and measure learning for each student. And, it comes from an **experienced partner** with educational expertise and an eye on the future. MyMathLab helps prepare students and gets them thinking more conceptually and visually through the following features:

Adaptive Study Plan

The Study Plan makes studying more efficient and effective for every student. Performance and activity are assessed continually in real time. The data and analytics are used to provide personalized content-reinforcing concepts that target each student's strengths and weaknesses.

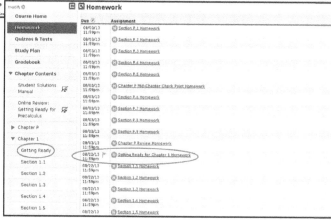

Getting Ready

Students refresh prerequisite topics through assignable skill review quizzes and personalized homework integrated in MyMathLab.

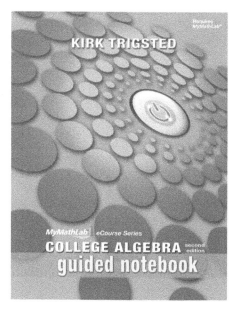

Guided Notebook

The *Guided Notebook* is a printed, interactive workbook created by the author to guide students through the eText. The workbook asks students to write down key definitions and work through important examples before they start their homework.

Skills for Success Modules

MyMathLab's new Skills for Success modules are designed to help students succeed in college and perpare for the future professions. Activities and assignments are availlable for topics such as "Time Management," "Stress Management," and "Financial Literacy."

Support and Resources

MyMathLab for College Algebra Student Access Kit, by Kirk Trigsted
0-321-93820-8/978-0-321-93820-6

Guided Notebook

0-321-92383-9/978-0-321-92383-7

The *Guided Notebook* is a printed, interactive workbook created by the author to guide students through the eText. The *Guided Notebook* asks students to write down key definitions and work through important examples before they start their homework. This resource is available in a loose-leaf, three-hole-punched format to provide the foundation for a personalized course notebook. Instructors can also customize the *Guided Notebook* files found within MyMathLab®.

Instructor Resources

PowerPoint® slides present key concepts and definitions from the text. Slides are available for download from within MyMathLab and from Pearson Education's online catalog.

TestGen® software enables instructors to build, edit, print, and administer tests using a computerized bank of questions. TestGen is algorithmically based, allowing instructors to create multiple but equivalent versions of the same question or test. Instructors can also modify test bank questions or add new questions. The software and test bank are available for download from Pearson Education's online catalog.

Acknowledgments

There are so many wonderful, dedicated people that I would like to thank. First of all, I must thank my beautiful wife, Wendy, for her continued loving support. I am also so grateful to my loving children, Benjamin, Emily, Gabrielle, and Isabelle for being so patient and understanding when daddy was locked in his basement office. Thank you for making daddy take breaks from time to time!

There are so many extraordinary and talented people who have contributed to this project. From Pearson, I would like to thank Anne Kelly, Dawn Murrin, and Joe Colella for their editorial superiority; Greg Tobin and Chris Hoag for their continued support; Eileen Moore, Kristina Evans, and Phil Oslin and the rest of the MathXL development team for a truly incredible job; Tracy Menoza and Ruth Berry for taking care of the eText details; the entire art team for their exceptional eye for detail; and Ron Hampton, our project manager, for all his support.

There are so many people from so many colleges and universities whose work on this project was immeasurable. I must personally thank my friend Phoebe Rouse from LSU whose contributions to this book were enormous—"Miss Phoebe" has officially become one of the family! I appreciate everyone else who made this book a reality: Tracy Duff from PreMedia Global helped me in so many ways; Alice Champlin and Anthony T. J. Kilian at Magnitude Entertainment for creating all of the awesome media assets; Pamela Trim for her continued dedication to detail; and Phil Veer and Bruce Yarbrough for sitting through the video shoots.

The list on the following pages includes all reviewers, focus group attendees, and class testers. Please accept my deepest apologies if I have inadvertently omitted anyone. I am humbled and so very grateful for all of your help and I thank you from the bottom of my heart.

—Kirk Trigsted

Teri Barnes, McLennan Community College
Linda Barton, Ball State University
Sam Bazzi, Henry Ford Community College
Molly Beauchman, Yavapai College
Brian Beaudrie, Northern Arizona University
Annette Benbow, Tarrant County College
Patricia Blus, National Louis University
Nina Bohrod, Anoka-Ramsey Community College
Barbara Boschmans, Northern Arizona University
David Bramlet, Jackson State University
Densie Brown, Collin County Community College
Connie Buller, Metropolitan Community College
Joe Castillo, Broward Community College
Mariana Coanda, Broward Community College
Alicia Collins, Mesa Community College
Earl W. Cook, Chattahoochee Valley Community
   College
Kemba C. Countryman, Chattahoochee Valley
   Community College
Douglas Culler, Midlands Technical College
Momoyo Dahle, Las Positas Community College
Diane Daniels, Mississippi State University
Emmett Dennis, Southern Connecticut State
   University
Donna Densmore, Bossier Parish Community
   College
Debbie Detrick, Kansas City Kansas Community
   College
Holly Dickin, University of Idaho
Timothy Doyle, University of Illinois at Chicago
Christina Dwyer, Manatee Community College
Stephanie Edgerton, Northern Arizona University
Jeanette Eggert, Concordia University
Brett Elliott, Southeastern Oklahoma State University
Amy Erickson, Georgia Gwinnett College
Nicki Feldman, Pulaski Technical College
Catherine Ferrer, Valencia Community College
Gerry Fitch, Louisiana State University
Cynthia Francisco, Oklahoma State University
Robert Frank, Westmoreland County Community
   College
Jim Frost, St. Louis Community College-Meramec
Angelito Garcia, Truman College
Lee Gibson, University of Louisville
Charles B. Green, University of North
   Carolina-Chapel Hill
Jeffrey Hakim, American University
Mike Hall, Arkansas State University
Melissa Hardeman, University of Arkansas, Little Rock
Celeste Hernandez, Richland College
Pamela Howard, Boise State University
Jeffrey Hughes, Hinds Community College
Eric Hutchinson, College of Southern Nevada
Robert Indrihovic, Florence-Darlington Technical
   College
Philip Kaatz, Mesalands Community College
Cheryl Kane, University of Nebraska-Lincoln

Robert Keller, Loras College
Mike Kirby, Tidewater Community College
Susan Knights, Boise State University
Marie Kohrmann, North Lake College
Debra Kopcso, Louisiana State University
Stephanie Kurtz, Louisiana State University
Jennifer LaRose, Henry Ford Community College
Jeff Laub, Central Virginia Community College
Jennifer Legrand, St. Charles Community College
Kurt Lewandowski, Clackamas Community College
Oscar Macedo, University of Texas at El Paso
Shanna Manny, Northern Arizona University
Pamela Spurlock Mills, University of Arkansas
Peggy L. Moch, Valdosta State University
Susan Moosai, Florida Atlantic University
Rebecca Morgan, Wayne State University
Rebecca Muller, Southeastern Louisiana University
Veronica Murphy, Saint Leo University
Charlie Naffziger, Central Oregon Community
   College
Prince Raphael A. Okojie, Fayetteville State
   University
Enyinda Onunwor, Stark State University
Shahla Peterman, University of Missouri-St. Louis
Sandra Poinsett, College of Southern Maryland
Steve Proietti, Northern Essex Community College
Ray Purdom, University of North Carolina at
   Greensboro
Nancy Ressler, Oakton Community College
Mary Revels, Southeast Community College
Joe Rody, Arizona State University
Cheryl Roddick, San Jose State University
Amy Rouse, Louisiana State University Laboratory
   School
Phoebe Rouse, Louisiana State University
Patricia Rowe, Columbus State Community
   College
Amy Rushall, Northern Arizona University
Chie Sakabe, University of Idaho
Jorge Sarmiento, County College of Morris
John Savage, Montana State University-College
   of Technology
Julie Sawyer, University of Idaho
Victoria Seals, Gwinnett Technical College
Susan P. Sherry, Northern Virginia Community
   College
Randell Simpson, Temple College
Rita Sowell, Volunteer State Community College
John Squires, Cleveland State Community College
Pam Stogsdill, Bossier Parish Community College
Eleanor Storey, Front Range Community College
Robert Strozak, Old Dominion University
Lalitha Subramanian, Potomac State College of
   West Virginia University
Mary Ann Teel, University of North Texas
Gwen Terwilliger, University of Toledo
Jamie Thomas, University of Wisconsin-Manitowoc

Keith B. Thompson, Davidson County Community College

Terry Tiballi, State University of New York at Oswego

Suzanne Topp, Salt Lake Community College

Diann Torrence, Delgado Community College

Philip Veer, Johnson County Community College

Marcia Vergo, Metropolitan Community College

Kimberly Walters, Mississippi State University

Aimee Welch, Louisiana State University Laboratory School

Jacci White, Saint Leo University

Ralph L. Wildy Jr., Georgia Military College

Sherri M. Wilson, Fort Lewis College

Xuezheng Wu, Madison College

Janet Wyatt, Metropolitan Community College–Longview

Ghidei Zedingle, Normandale Community College

Brian Zimmerman, University of Mississippi

Review Chapter

CHAPTER R CONTENTS

R.1 Real Numbers

OBJECTIVES

1 Understanding the Real Number System
2 Writing Sets Using Set-Builder Notation and Interval Notation
3 Determining the Intersection and Union of Sets and Intervals
4 Understanding Absolute Value

OBJECTIVE 1   UNDERSTANDING THE REAL NUMBER SYSTEM

A **set** is a collection of objects. Each object in the set is called an **element** or a **member** of the set. We typically use braces { } to enclose all elements of a set. Capital letters are sometimes used to name a set. For example, suppose we define sets A and B as

$$A = \{1, 2, 3, 4, 5, 6, 7, 8, 9, 10\} \quad \text{and} \quad B = \{2, 4, 6, 8, 10\}.$$

Notice that each element of set B is also an element of set A. We say that set B is a **subset** of set A and write $B \subset A$. The set of **real numbers** consists of several subsets of numbers. We now describe these subsets.

NATURAL NUMBERS

The set of natural numbers, sometimes referred to as the counting numbers, is defined as $\mathbb{N} = \{1, 2, 3, 4 \ldots\}$. The three dots (...), called an ellipsis, are used to indicate that the elements of the set follow the same pattern.

WHOLE NUMBERS

The set of whole numbers is composed of the natural numbers combined with the number 0. Thus, $\mathbb{W} = \{0, 1, 2, 3, 4 \ldots\}$. Note that $\mathbb{N} \subset \mathbb{W}$.

INTEGERS

Combining the negatives of the natural numbers with the whole numbers gives the set of integers. The symbol $\mathbb{Z}$ is often used to denote the set of integers; thus, $\mathbb{Z} = \{\ldots, -3, -2, -1, 0, 1, 2, 3, \ldots\}$. Note that $\mathbb{N} \subset \mathbb{W} \subset \mathbb{Z}$.

RATIONAL NUMBERS

A rational number is a number that can be expressed as a quotient of two integers. The symbol $\mathbb{Q}$ is often used to represent the set of rational numbers. We can use **set-builder notation** to define the set of rational numbers as $\mathbb{Q} = \left\{\dfrac{p}{q} \,\middle|\, p \text{ and } q \text{ are integers, } q \neq 0\right\}$, which is read as "the set of all elements $\dfrac{p}{q}$ such that p and q are integers, $q \neq 0$." We revisit set-builder notation later in this section. The set of rational numbers includes all repeating and terminating decimals. Examples of rational numbers include $7, \dfrac{1}{2}, -\dfrac{11}{43}, -0.56, .3333\ldots$, and $.\overline{1234}$. (The bar above the decimal .1234 represents a repeating decimal.) Note that $\mathbb{N} \subset \mathbb{W} \subset \mathbb{Z} \subset \mathbb{Q}$.

IRRATIONAL NUMBERS

Numbers that are not rational (numbers that cannot be expressed as a quotient of two integers) are called irrational numbers. We can define the set of irrational numbers as $\mathbb{I} = \{x \,|\, x \text{ is not a rational number}\}$. Examples of irrational numbers include $\sqrt{7}, \sqrt{17}, \sqrt[3]{29}$, and π.

The set of real numbers, denoted as $\mathbb{R}$, is the set of rational numbers joined with the set of irrational numbers. Figure 1 gives us a visual image of the set of real numbers.

Real Numbers $\mathbb{R}$

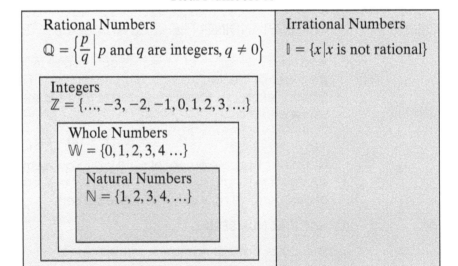

Figure 1 The real numbers

Real numbers can be represented by points located on a number line called the *real axis*. The point corresponding to 0 is called the *origin*. Figure 2 illustrates several real numbers plotted on a number line. In this text, we use a solid circle (•) to represent a point on a number line.

$$-\frac{7}{2} \qquad 0.\overline{4} \qquad \pi \quad \sqrt{17}$$

$$-5\ -4\ -3\ -2\ -1\quad 0\quad 1\quad 2\quad 3\quad 4\quad 5$$

origin

Figure 2

Example 1 Classify Real Numbers

Classify each number in the set $\left\{-5, -\frac{1}{3}, 0, \sqrt{3}, 1.\overline{4}, 2\pi, 11\right\}$ as a natural number, whole number, integer, rational number, irrational number, and/or real number.

Solution
The number -5 is an integer, rational number, and real number.

The number $-\frac{1}{3}$ is a rational number and a real number.

The number 0 is a whole number, integer, rational number, and real number.
The number $\sqrt{3}$ is an irrational number and a real number.
The number $1.\overline{4}$ is a rational number and a real number.
The number 2π is an irrational number and a real number.
The number 11 is a natural number, whole number, integer, rational number, and real number.

You Try It Work through the following You Try It problem.

Work Exercises 1–6 in this textbook or in the MyMathLab Study Plan.

OBJECTIVE 2 WRITING SETS USING SET-BUILDER NOTATION AND INTERVAL NOTATION

Set-builder notation can be a convenient way to describe certain sets of numbers. For example, we used set-builder notation to describe the set of rational numbers and the set of irrational numbers. Suppose that we want to describe the set of all real numbers less than 10. Using set-builder notation, we write this set as $\{x | x < 10\}$. This set is read as "the set of all x such that x is less than 10." Figure 3 illustrates how we can represent this set on a number line.

$$10$$

Figure 3 The set $\{x | x < 10\}$ represented on a number line.

The open circle (○) at the number 10 in Figure 3 represents that 10 is *not* included in the set. If we wanted to include the number 10, we would have used a solid circle (•). The arrow extending to the left indicates that the set goes on indefinitely toward negative infinity. In this text, we often use **interval notation** to describe sets of numbers such as the set displayed in Figure 3. The interval that describes this set is called an open infinite interval and is written as $(-\infty, 10)$. It is extremely important to be able to write intervals of numbers in both set-builder notation and interval notation. Table 1 illustrates the different types of intervals and the corresponding set-builder notation.

Table 1

Type of Interval and Graph	Interval Notation	Set-Builder Notation
Open interval	(a, b)	$\{x\|a < x < b\}$
Closed interval	$[a, b]$	$\{x\|a \leq x \leq b\}$
Half-open intervals	$(a, b]$ $[a, b)$	$\{x\|a < x \leq b\}$ $\{x\|a \leq x < b\}$
Open infinite intervals	(a, ∞) $(-\infty, b)$	$\{x\|x > a\}$ $\{x\|x < b\}$
Closed infinite intervals	$[a, \infty)$ $(-\infty, b]$	$\{x\|x \geq a\}$ $\{x\|x \leq b\}$

Example 2 Write a Set Using Set-Builder Notation and Interval Notation

Given the set sketched on the following number line,

a. Identify the type of interval.

b. Write the set using set-builder notation.

c. Write the set using interval notation.

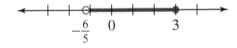

Solution

a. This interval is called a half-open interval. The solid circle at 3 indicates that the set includes the number 3. The open circle at $-\dfrac{6}{5}$ indicates that the set does not include $-\dfrac{6}{5}$.

b. This set can be described in set-builder notation as $\left\{ x \left| -\dfrac{6}{5} < x \le 3 \right. \right\}$.

c. We write this set in interval notation as $\left(-\dfrac{6}{5}, 3 \right]$.

My video summary ⊙ **Example 3** Write a Set Using Set-Builder Notation and Interval Notation

a. Write the set $\left[-\dfrac{1}{3}, \infty \right)$ in set-builder notation and graph the set on a number line.

b. Write the set $\left\{ x \left| -\dfrac{7}{2} < x \le \pi \right. \right\}$ in interval notation and graph the set on a number line.

Solution Watch the video to see the solution worked out in detail.

You Try It Work through the following You Try It problem.

Work Exercises 7–18 in this textbook or in the MyMathLab Study Plan.

OBJECTIVE 3 DETERMINING THE INTERSECTION AND UNION OF SETS AND INTERVALS

We are often interested in looking at two or more sets at a time. Joining all elements of a set A with all elements of a set B is known as the **union** of sets A and B and is denoted as $A \cup B$. The set of elements that are *common* to both sets A and B is known as the **intersection** of sets A and B and is written as $A \cap B$. The set containing no elements is called the empty set and is denoted $\varnothing$. Sometimes the intersection of two sets results in the empty set.

Example 4 Find the Intersection and Union of Sets

Let $A = \left\{ -5, 0, \dfrac{1}{3}, 11, 17 \right\}$, $B = \{ -6, -5, 4, 17 \}$, and $C = \left\{ -4, 0, \dfrac{1}{4} \right\}$.

a. Find $A \cup B$.   **b.** Find $A \cap B$.   **c.** Find $B \cap C$.

Solution

a. $A \cup B = \left\{ -6, -5, 0, \dfrac{1}{3}, 4, 11, 17 \right\}$ List all elements in set A or set B.

b. $A \cap B = \{ -5, 17 \}$ List all elements common to set A and set B.

c. $B \cap C = \varnothing$ There are no elements common to both B and C.

My video summary ⊙ **Example 5** Find the Intersection of Sets

Find the intersection of the following sets and graph the set on a number line.

a. $[0, \infty) \cap (-\infty, 5]$ **b.** $((-\infty, -2) \cup (-2, \infty)) \cap [-4, \infty)$

Solution Watch the video to verify the following solutions.

a. $[0, \infty) \cap (-\infty, 5] = [0, 5]$

b. $((-\infty, -2) \cup (-2, \infty)) \cap [-4, \infty) = [-4, -2) \cup (-2, \infty)$

You Try It Work through the following You Try It problem.

Work Exercises 19–27 in this textbook or in the My MathLab Study Plan.

OBJECTIVE 4   UNDERSTANDING ABSOLUTE VALUE

The absolute value of a real number a is defined as the distance between a and 0 on a number line and is denoted $|a|$. For example, $|-5| = 5$ because the number -5 is 5 units away from 0 on a number line. See Figure 4.

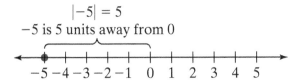

Figure 4 The absolute value of -5 is 5 because -5 is 5 units away from 0.

Definition   **Absolute Value**

The absolute value of a real number a is defined by

$$|a| = \begin{cases} a & \text{if } a \geq 0 \\ -a & \text{if } a < 0 \end{cases}$$

and represents the distance between a and 0 on a number line.

Because the absolute value of a number a represents the *distance* from a to 0 on a number line, it follows that $|a| \geq 0$ for any real number a. This property and several other properties of absolute value are stated as follows.

Properties of Absolute Value

For all real numbers a and b,

1. $|a| \geq 0$ 2. $|ab| = |a||b|$ 3. $\left|\dfrac{a}{b}\right| = \dfrac{|a|}{|b|}, b \neq 0$

4. $|a|^2 = a^2$ 5. $|a - b| = |b - a|$

Example 6 Evaluate Absolute Value Expressions

Evaluate the following expressions involving absolute value.

a. $\left| -\sqrt{2} \right|$ b. $\left| \dfrac{-4}{16} \right|$ c. $|2 - 5|$

Solution

a. $\left| -\sqrt{2} \right| = -(-\sqrt{2}) = \sqrt{2}$

b. $\left| \dfrac{-4}{16} \right| = \dfrac{|-4|}{|16|} = \dfrac{-(-4)}{16} = \dfrac{4}{16} = \dfrac{1}{4}$

c. $|2 - 5| = |-3| = -(-3) = 3$ or $|2 - 5| = |5 - 2| = |3| = 3$

Let's take a closer look at the absolute value expression $|2 - 5|$ from Example 6c and interpret its meaning, geometrically. The expression $|2 - 5|$ represents the distance between the numbers 2 and 5 on a number line. Notice in Figure 5 that the distance between 2 and 5 on a number line is three units.

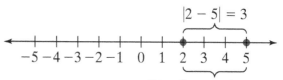

$|2 - 5| = 3$

The distance between
2 and 5 is 3 units.

Figure 5

Because the distance between 2 and 5 is the same as the distance between 5 and 2, we can conclude that $|2 - 5| = |5 - 2|$. Thus, for any real numbers a and b, $|a - b| = |b - a|$, which is precisely property 5 of absolute values. We now restate property 5 as the distance between two real numbers on a number line.

Distance between Two Real Numbers on a Number Line

For any real numbers a and b, the distance between a and b on a number line is defined by $|a - b|$ or $|b - a|$.

Example 7 Find the Distance between Two Real Numbers

Find the distance between the numbers -5 and 3 using absolute value.

Solution Let $a = -5$ and $b = 3$. The distance between a and b is given by $|a - b|$ or $|b - a|$. $|a - b| = |-5 - 3| = |-8| = 8$. So, the distance between -5 and 3 is 8 units.

You Try It Work through the following You Try It problem.

Work Exercises 28–35 in this textbook or in the MyMathLab Study Plan.

R.1 Exercises

In Exercises 1–4, classify each number as a natural number, whole number, integer, rational number, irrational number, and/or real number. Each number may belong to more than one set.

1. 259

2. $-11,401$

3. $-\dfrac{43}{7}$

4. $-\sqrt{17}$

In Exercises 5 and 6, given each set of real numbers, list the numbers that are a) natural numbers, b) whole numbers, c) integers, d) rational numbers, and/or e) irrational numbers.

5. $\left\{-17, -\sqrt{5}, -\dfrac{25}{19}, 0, 0.331, 1, \dfrac{\pi}{2}\right\}$

6. $\left\{-11, -3.\overline{2135}, -\dfrac{3}{9}, \dfrac{\sqrt{7}}{2}, 21.1\right\}$

In Exercises 7–10, given the set sketched on the number line, a) identify the type of interval, b) write the set using set-builder notation, and c) write the set using interval notation.

7.

8.

9.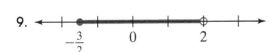

10.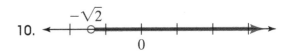

In Exercises 11–14, write each interval in set-builder notation and graph the set on a number line.

11. $\left[-\dfrac{1}{2}, 5\right]$

12. $\left(0, \dfrac{5}{2}\right)$

13. $(-\infty, 3]$

14. $(-1, \infty)$

In Exercises 15–18, write the given set in interval notation and graph the set on a number line.

15. $\left\{x \,\middle|\, -\dfrac{5}{2} \le x \le 1\right\}$

16. $\{x \mid 0 \le x < 3\}$

17. $\left\{x \,\middle|\, x \ge \dfrac{3}{4}\right\}$

18. $\{x \mid x > -4\}$

In Exercises 19–21, sets A and B are given. Find a) $A \cup B$ and b) $A \cap B$.

19. $A = \{-1, 0, 1, 2\}$ and $B = \{0, 2, 4\}$

20. $A = \left\{-\dfrac{3}{4}, -\dfrac{1}{4}, \dfrac{1}{4}, \dfrac{3}{4}\right\}$ and $B = \left\{-1, \dfrac{1}{4}, 1\right\}$

21. $A = \{1, 2, 3, 4\}$ and $B = \{0, 5, 6\}$

In Exercises 22–27, find the intersection of the given intervals.

22. $(-\infty, \infty) \cap (-\infty, 3)$

23. $(-8, 5] \cap (-12, 3)$

24. $[0, \infty) \cap (-\infty, 4]$

25. $((-\infty, 1) \cup (1, \infty)) \cap (-\infty, \infty)$

26. $((-\infty, 0) \cup (0, \infty)) \cap [-3, \infty)$

27. $(-\infty, 10) \cap (-5, \infty) \cap (-\infty, 1]$

In Exercises 28–32, evaluate the absolute value expression.

28. $|286|$ **29.** $|-2.2|$ **30.** $|-2 \cdot 3|$ **31.** $-\left|\dfrac{-2 \cdot 6}{4}\right|$ **32.** $|2 - \pi|$

In Exercises 33–35, find the distance between the given two numbers using absolute a value.

33. 3 and 14   **34.** -5 and 7   **35.** -24 and -1.5

R.2 The Order of Operations and Algebraic Expressions

OBJECTIVES

1 Understanding the Properties of Real Numbers

2 Using Exponential Notation

3 Using the Order of Operations to Simplify Numeric and Algebraic Expressions

Introduction to Section R.2

In this section, we learn how to simplify **algebraic expressions**. An algebraic expression consists of one or more terms that include **variables**, constants, and operating symbols such as $+$ or $-$. For example, $4x^3y^4 - 1$ is an algebraic expression consisting of two terms. The letters x and y used in this expression are called **variables** and are used to represent any number. Before we work with algebraic expressions, it is important to understand some basic properties of real numbers. Our understanding of these properties will enhance our ability to simplify numeric and algebraic expressions.

OBJECTIVE 1   UNDERSTANDING THE PROPERTIES OF REAL NUMBERS

We all know that when adding two real numbers or multiplying two numbers, the order in which we perform the addition or multiplication does not affect the result. For example, $3 + 5 = 5 + 3$ and $(-4)(6) = (6)(-4)$. The operations of addition and multiplication are said to be commutative. The commutative property of addition and multiplication are stated here along with two other important properties of real numbers.

Commutative Property of Addition

If a and b are real numbers, then $a + b = b + a$.

Example: $3 + 5 = 8$ and $5 + 3 = 8$

Commutative Property of Multiplication

If a and b are real numbers, then $ab = ba$.

Example: $(-4)(6) = -24$ and $(6)(-4) = -24$

Associative Property of Addition

If a, b, and c are real numbers, then $(a + b) + c = a + (b + c)$.

Example: $(-7 + 11) + 6 = 4 + 6 = 10$ and $-7 + (11 + 6) = -7 + 17 = 10$

Associative Property of Multiplication

If a, b, and c are real numbers, then $(ab)c = a(bc)$.

Example: $(-8 \cdot -2) \cdot 3 = 16 \cdot 3 = 48$ and $-8 \cdot (-2 \cdot 3) = -8 \cdot (-6) = 48$

Distributive Property

If a, b, and c are real numbers, then $a(b + c) = ab + ac$.

Example: $3(6 + 4) = 3(6) + 3(4)$
$$= 18 + 12$$
$$= 30$$

Example 1 Use the Properties of Real Numbers to Rewrite Expressions

a. Use the commutative property of addition to rewrite $11 + y$ as an equivalent expression.

b. Use the commutative property of multiplication to rewrite $7(y + 4)$ as an equivalent expression.

c. Use the associative property of multiplication to rewrite $8(pq)$ as an equivalent expression.

d. Use the distributive property to multiply $11(a - 4)$.

e. Use the distributive property to rewrite $aw + az$ as an equivalent expression.

Solution

a. By the commutative property of addition, the algebraic expression $11 + y$ is equivalent to $y + 11$.

b. By the commutative property of multiplication, the algebraic expression $7(y + 4)$ is equivalent to $(y + 4) \cdot 7$.

c. By the associative property of multiplication, the algebraic expression $8(pq)$ is equivalent to $(8p)q$.

d. To use the distributive property, we multiply each variable or number inside the parentheses by the factor outside the parentheses.

$$11(a - 4) = 11 \cdot a - 11 \cdot 4 = 11a - 44$$

e. By the distributive property, $aw + az = a(w + z)$.

You Try It Work through the following You Try It problem.

Work Exercises 1–18 in this textbook or in the MyMathLab Study Plan.

OBJECTIVE 2 USING EXPONENTIAL NOTATION

Numeric and algebraic expressions often have terms involving exponents such as x^3, a^4, or $(-2)^5$. It is important to understand how to simplify expressions involving exponents. We now define exponential notation.

Exponential Notation

If b is a real number and n is a positive integer, then b^n is equivalent to $\underbrace{b \cdot b \cdot b \cdots b.}$

n factors of b

Note: b is called the base of the exponential expression and n is called the exponent.

In Section R.3, we establish the laws of exponents.

Example 2 Using Exponential Notation

a. Rewrite the expression $5 \cdot 5 \cdot 5$ using exponential notation. Then identify the base and the exponent.

b. Identify the base and the exponent of the expression $(-3)^4$, and then evaluate.

c. Identify the base and the exponent of the expression -2^5, and then evaluate.

Solution

a. We can rewrite $5 \cdot 5 \cdot 5$ as 5^3. The base is 5, and the exponent is 3.

b. The base of the expression $(-3)^4$ is -3, and the exponent is 4. Therefore, $(-3)^4 = (-3) \cdot (-3) \cdot (-3) \cdot (-3) = 81$.

c. The expression -2^5 can be rewritten as $-1 \cdot 2^5$. The base of the exponential expression is 2, and the exponent is 5. Thus, $-2^5 = -1 \cdot 2^5 = -1 \cdot 2 \cdot 2 \cdot 2 \cdot 2 \cdot 2 = -1 \cdot 32 = -32$.

You Try It Work through the following You Try It problem.

Work Exercises 19–25 in this textbook or in the MyMathLab Study Plan.

OBJECTIVE 3   USING THE ORDER OF OPERATIONS TO SIMPLIFY NUMERIC
AND ALGEBRAIC EXPRESSIONS

When we encounter an expression with more than one operating symbol, it is crucial that we simplify the expression in the correct order. For example, suppose we are given the expression $5 - 3 \cdot 7$. In this expression, there are two operating symbols, symbolizing subtraction and multiplication. If we choose to subtract first and then multiply, we get

$$5 - 3 \cdot 7 = 2 \cdot 7 = 14.$$

If we multiply first and then subtract the result from 5, we get

$$5 - 3 \cdot 7 = 5 - 21 = -16.$$

As you can see, depending on the order in which we perform the operations of subtraction and multiplication, we obtain different results. The correct way to simplify the expression $5 - 3 \cdot 7$ is to perform the multiplication before the subtraction. Therefore, $5 - 3 \cdot 7 = -16$.

When simplifying an expression having more than one operating symbol, we use the following order of operations.

ORDER OF OPERATIONS

1. Always start within the innermost grouping symbols. This includes parentheses ( ), brackets [ ], braces { }, and absolute value bars | |.

2. Simplify exponential expressions.

3. Perform multiplication and division working from left to right.

4. Perform addition and subtraction working from left to right.

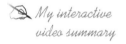 *My interactive video summary*

⊙ **Example 3** Use the Order of Operations to Simplify a Numeric Expression

Simplify each expression.

a. $-2^3 + [3 - 5 \cdot (1 - 3)]$.

b. $\dfrac{|2 - 3^3| + 5}{5^2 - 4^2}$

Solution Watch the interactive video to verify that

a. $-2^3 + [3 - 5 \cdot (1 - 3)] = 5$ and   b. $\dfrac{|2 - 3^3| + 5}{5^2 - 4^2} = \dfrac{10}{3}$

 You Try It   Work through the following You Try It problem.

Work Exercises 26–33 in this textbook or in the MyMathLab **Study Plan.**

 *My video summary*

⊙ **Example 4** Use the Order of Operations to Evaluate an Algebraic Expression

Evaluate the algebraic expression $-x^3 - 4x$ for $x = -2$.

Solution

$-x^3 - 4x$	Write the original algebraic expression.
$= -(-2)^3 - 4(-2)$	Substitute -2 for x.
$= -(-8) - 4(-2)$	$(-2)^3 = -8$.
$= 8 + 8$	Simplify.
$= 16$	

You Try It   Work through the following You Try It problem.

Work Exercises 34–41 in this textbook or in the MyMathLab Study Plan.

When simplifying algebraic expressions, we must remove all grouping symbols and combine **like terms**. Like terms are two or more terms consisting of the same variable (or variables) raised to the same power. For example, $3x^2$ and $5x^2$ are like terms because each term contains the same variable, x, raised to the same power, 2.

My video summary ⊙ **Example 5** Simplify an Algebraic Expression

Simplify each algebraic expression.

a. $3x^2 - 2 + 5x^2$ **b.** $-3(4a - 7) + 5(3 - 5a)$

Solution

a.

$3x^2 - 2 + 5x^2$	Write the original algebraic expression.
$= 3x^2 + 5x^2 - 2$	Use the commutative property of addition to rearrange the terms.
$= (3 + 5)x^2 - 2$	Use the distributive property $ab + ac = (b + c)a$ to combine like terms.
$= 8x^2 - 2$	Simplify.

b.

$-3(4a - 7) + 5(3 - 5a)$	Write the original algebraic expression.
$= -3(4a) - 3(-7) + 5(3) + 5(-5a)$	Use the distributive property.
$= -12a + 21 + 15 - 25a$	Simplify.
$= -12a - 25a + 21 + 15$	Use the commutative property of addition to rearrange the terms.
$= (-12 - 25)a + 21 + 15$	Use the distributive property $ab + ac = (b + c)a$ to combine like terms.
$= -37a + 36$	Simplify.

You Try It   Work through the following You Try It problem.

Work Exercises 42–51 in this textbook or in the MyMathLab Study Plan.

R.2 Exercises

In Exercises 1–3, use the commutative property of addition to rewrite each expression as an equivalent expression.

1. $13 + m$

2. $ab + c$

3. $11w + 10t$

In Exercises 4–6, use the commutative property of multiplication to rewrite each expression as an equivalent expression.

4. $3 \cdot z$

5. mn

6. $6(v + 4)$

In Exercises 7–9, use the associative property of addition to rewrite each expression as an equivalent expression.

7. $4 + (a + 11)$

8. $(d + a) + 30$

9. $(zw + a) + y$

In Exercises 10 and 11, use the associative property of multiplication to rewrite the expression as an equivalent expression.

10. $45 \cdot (y \cdot z)$

11. $8(2(y + z))$

In Exercises 12–15, use the distributive property to multiply the given expressions.

12. $12(11a + 4)$

13. $(x + y)20$

14. $15(a - 10 + 6b)$

15. $(h + 2v - 6)5$

In Exercises 16–18, use the distributive property to rewrite each sum or difference as a product.

16. $20c + 12c$

17. $9a - 11a$

18. $zx - zy$

In Exercises 19–21, write each expression using exponential notation. Then identify the base and the exponent.

19. $4 \cdot 4 \cdot 4$

20. $w \cdot w \cdot w \cdot w$

21. $(-6y) \cdot (-6y) \cdot (-6y) \cdot (-6y) \cdot (-6y)$

In Exercises 22–25, identify the base and exponent of each expression, and then evaluate.

22. 8^3

23. -5^4

24. $(-4)^3$

25. $(2z)^5$

In Exercises 26–33, simplify each expression using the order of operations.

26. $7 - (4 \cdot 6 - 26)$

27. $2 + 8^3 \div (-56) - 7$

28. $21^3 \div \dfrac{3^2}{5 - 2} - (-1)$

29. $-\dfrac{7}{6}\left(\dfrac{1}{2}\right) + \dfrac{5}{9} \div \dfrac{7}{6}$

30. $10^2 + 20^2 \div 5^2$

31. $3 \cdot (18 + 3)^2 - 4 \cdot (8 - 3)^2$

32. $\dfrac{7 \cdot 3 - 2^2}{16 - 2^3}$

33. $\dfrac{\left| -5^2 + (-3)^2 \right| + 4 \cdot 5}{6 \div 2 - 3 \cdot 2^2}$

In Exercises 34–41, evaluate each algebraic expression for the indicated variable.

34. $4a \div 16a^2$ for $a = 4$

35. $30 \div 3x$ for $x = -2$

36. $-z^2 - 6z$ for $z = -3$

37. $\dfrac{6b - 9b^2}{b^2 - 6}$ for $b = 3$

38. $54 \div 3^2 \cdot n(n - 4)$ for $n = 7$

39. $\left|8x^2 - 5y\right|$ for $x = -1$ and $y = 3$

40. $b^2 - 4ac$ for $a = -4$, $b = 3$, and $c = -2$

41. $\dfrac{y_2 - y_1}{x_2 - x_1}$ for $x_1 = 4$, $x_2 = 6$, $y_1 = -2$, and $y_2 = 7$

In Exercises 42–51, simplify each algebraic expression.

42. $-7x - 10x$

43. $10z^2 - z^2$

44. $-7y^5 - 7y^5$

45. $7a - 20 - 17a + 50$

46. $3k + 3k^2 + 7k + 7k^2$

47. $5(3t - 5) - 6t$

48. $6 - 3[4 - (7r - 2)]$

49. $-5(3w - 9) + 2(2w + 7)$

50. $56p^2 - 36 - [4(p^2 - 9) + 1]$

51. $\dfrac{3}{4}x - \dfrac{4}{3} + \dfrac{7}{2}x + \dfrac{5}{6}$

R.3 The Laws of Exponents; Radicals

OBJECTIVES

1 Simplifying Exponential Expressions Involving Integer Exponents

2 Simplifying Radical Expressions

3 Simplifying Exponential Expressions Involving Rational Exponents

OBJECTIVE 1   SIMPLIFYING EXPONENTIAL EXPRESSIONS INVOLVING INTEGER EXPONENTS

In Section R.2, we introduced exponential notation. In this section, we learn how to simplify exponential expressions of the form $b^{m/n}$, where m/n is a rational number. We start by simplifying expressions involving *positive* integer exponents.

Properties of Positive Integer Exponents

Suppose m and n are positive integers and a and b are real numbers, then

1. $b^m b^n = b^{m+n}$

2. $\dfrac{b^m}{b^n} = b^{m-n}$, where $m > n$ and $b \neq 0$

   $\dfrac{b^m}{b^n} = 1$, where $m = n$

   $\dfrac{b^m}{b^n} = \dfrac{1}{b^{n-m}}$, where $m < n$ and $b \neq 0$

3. $(b^m)^n = b^{mn}$

4. $(ab)^n = a^n b^n$

5. $\left(\dfrac{a}{b}\right)^n = \dfrac{a^n}{b^n}$, where $b \neq 0$

Although these five properties are true for positive integer exponents, they are actually true for *all* integer exponents, including exponents that are negative integers or zero. By property (1), for positive integer exponents m and n, we know that $b^m b^n = b^{m+n}$. If we want this property to hold true for $n = 0$, then $b^m b^0 = b^{m+0} = b^m$ if and only if $b^0 = 1$. This suggests the following rule.

Zero Exponent Rule

If b is a real number such that $b \neq 0$, then $b^0 = 1$.

Similarly, if we let $n = -m$, then $b^m \cdot b^{-m} = b^{m+(-m)} = b^0 = 1$ if and only if b^{-m} is the reciprocal of b^m or $b^{-m} = \dfrac{1}{b^m}$. This is known as the reciprocal rule of exponents.

Reciprocal Rule of Exponents

If b is a real number such that $b \neq 0$ and m is a positive integer, then $b^{-m} = \dfrac{1}{b^m}$.

We now state the following laws of exponents, which hold true for *all* integer exponents.

Laws of Exponents

Suppose m and n are integers and a and b are real numbers, then

1. $b^m b^n = b^{m+n}$ Product rule of exponents

2. $\dfrac{b^m}{b^n} = b^{m-n}$, where $b \neq 0$ Quotient rule of exponents

3. $(b^m)^n = b^{mn}$ Power rule of exponents

4. $(ab)^n = a^n b^n$ Product to power rule of exponents

(continued)

5. $\left(\dfrac{a}{b}\right)^n = \dfrac{a^n}{b^n}$, where $b \neq 0$ Quotient to power rule of exponents

6. $b^0 = 1$ Zero exponent rule

7. $b^{-m} = \dfrac{1}{b^m}$, where $b \neq 0$ Reciprocal rule of exponents

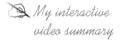
My interactive video summary

> **Example 1** Simplify Exponential Expressions Using the Laws of Exponents

Simplify each exponential expression. Write your answers using positive exponents. Assume all variables represent positive real numbers.

a. $-3^{-4} \cdot 3^4$ b. $5z^4 \cdot (-9z^{-5})$ c. $\left(\dfrac{a^{-3}b^4c^{-6}}{2a^5b^{-4}c}\right)^{-3}$

Solution

a. $-3^{-4} \cdot 3^4 = -3^{-4+4}$
$= -3^0$
$= -1$

b. $5z^4 \cdot (-9z^{-5}) = 5 \cdot (-9) \cdot z^4 \cdot z^{-5}$
$= -45 \cdot z^{4-5}$
$= -45 \cdot z^{-1}$
$= \dfrac{-45}{z}$

c. $\left(\dfrac{a^{-3}b^4c^{-6}}{2a^5b^{-4}c}\right)^{-3} = \left(\dfrac{b^4 b^4}{2a^5 a^3 c c^6}\right)^{-3}$
$= \left(\dfrac{b^8}{2a^8 c^7}\right)^{-3}$
$= \dfrac{b^{-24}}{2^{-3}a^{-24}c^{-21}}$
$= \dfrac{2^3 a^{24} c^{21}}{b^{24}}$
$= \dfrac{8a^{24}c^{21}}{b^{24}}$

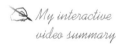
My interactive video summary

> You may want to watch the interactive video to see these three solutions worked out in detail.

You Try It   Work through the following You Try It problem.

Work Exercises 1–30 in this textbook or in the MyMathLab Study Plan.

OBJECTIVE 2   SIMPLIFYING RADICAL EXPRESSIONS

Before we discuss exponential expressions with rational exponents, we must first define a radical.

A number a is said to be an nth root of b if $a^n = b$. For example, 4 is a square root (2nd root) of 16 because $(4)^2 = 16$. The number -4 is also a square root of 16 because $(-4)^2 = 16$. Similarly, 3 and -3 are 4th roots of 81 because $(3)^4 = 81$ and $(-3)^4 = 81$. It follows that every positive real number has two real nth roots when n is an even integer. This is not the case if n is an odd integer. Every real number has exactly one nth root when n is an odd integer. For example, the only real cube root (3rd root) of -8 is -2 because $(-2)^3 = -8$. We use the symbol $\sqrt[n]{b}$ to mean the *positive* nth root of b, also called the **principal nth root** of a number b. The symbol $\sqrt[n]{}$ is called a **radical sign**. The integer n is called the **index**. The complete expression $\sqrt[n]{b}$ is called a **radical**. We now define the principal nth root.

Definition   Principal *n*th Root

For any integer $n \geq 2$, the **principal *n*th root** of a number b is defined as

$$\sqrt[n]{b} = a \quad \text{if and only if} \quad a^n = b.$$

If n is even and $b \geq 0$, then $a \geq 0$.

If n is even and $b < 0$, then a is not a real number.

If n is odd and $b \geq 0$, then $a \geq 0$.

If n is odd and $b < 0$, then $a < 0$.

For a real principal *n*th root to exist when *n* is a positive integer, the radicand must be greater than or equal to zero. Otherwise, the *n*th root is not a real number.

Example 2 Simplify Radical Expressions

Simplify each radical expression.

a. $\sqrt{144}$ b. $\sqrt[3]{\dfrac{1}{27}}$ c. $\sqrt[5]{-32a^5b^5}$ d. $\sqrt[6]{-64}$

Solution

a. $\sqrt{144} = 12$ because $12^2 = 144$.

b. $\sqrt[3]{\dfrac{1}{27}} = \dfrac{1}{3}$ because $\left(\dfrac{1}{3}\right)^3 = \dfrac{1}{27}$.

c. $\sqrt[5]{-32a^5b^5} = -2ab$ because $(-2ab)^5 = -32a^5b^5$.

d. The radical expression $\sqrt[6]{-64}$ has no real principal 6th root.

You Try It   Work through the following You Try It problem.

Work Exercises 31–36 in this textbook or in the MyMathLab Study Plan.

Note that the numeric part of each of the radicands from Examples 2a, 2b, and 2c are perfect *n*th powers. To simplify radicals in which the radicand is not a perfect *n*th power, we must first state some properties of radicals.

Properties of Radicals

If a and b are real numbers and $n \geq 2$ and $m \geq 2$ are integers such that $\sqrt[n]{a}$ and $\sqrt[n]{b}$ are real numbers*, then

1. $\sqrt[n]{ab} = \sqrt[n]{a}\sqrt[n]{b}$ 2. $\sqrt[n]{\dfrac{a}{b}} = \dfrac{\sqrt[n]{a}}{\sqrt[n]{b}}$ 3. $\sqrt[n]{a^n} = a$ if n is odd

4. $\sqrt[n]{a^n} = |a|$ if n is even

To simplify radicals in which the radicand is not a perfect *n*th power, we first factor all perfect power factors of the radicand and then use the properties of radicals to simplify.

Example 3 Simplify Radical Expressions

Simplify each radical expression.

a. $\sqrt{108}$ **b.** $\sqrt[3]{-40x^5y^3}$ **c.** $\sqrt[4]{16x^4y^5}$

Solution

a. $\sqrt{108}$ Write the original expression.

$\quad = \sqrt{36\cdot3}$ Factor out 36, the largest perfect square factor of 108.

$\quad = \sqrt{36}\sqrt{3}$ Rewrite using radical property (1).

$\quad = 6\sqrt{3}$ The principal square root of 36 is 6.

b. $\sqrt[3]{-40x^5y^3}$ Write the original expression.

$\quad = \sqrt[3]{-8\cdot5\cdot x^3\cdot x^2\cdot y^3}$ Factor out all perfect cubes.

$\quad = \sqrt[3]{-8}\sqrt[3]{x^3}\sqrt[3]{y^3}\sqrt[3]{5x^2}$ Rewrite using radical property (1).

$\quad = -2xy\sqrt[3]{5x^2}$ $\sqrt[3]{-8} = -2$, $\sqrt[3]{x^3} = x$, and $\sqrt[3]{y^3} = y$.

c. $\sqrt[4]{16x^4y^5}$ Write the original expression.

$\quad = \sqrt[4]{16\cdot x^4\cdot y^4\cdot y}$ Factor out the perfect 4th powers.

$\quad = \sqrt[4]{16}\sqrt[4]{x^4}\sqrt[4]{y^4}\sqrt[4]{y}$ Rewrite using radical property (1).

$\quad = 2|x||y|\sqrt[4]{y}$ $\sqrt[4]{16} = 2$, $\sqrt[4]{x^4} = |x|$ and $\sqrt[4]{y^4} = |y|$.

$\quad = 2|xy|\sqrt[4]{y}$ Use the absolute value property $|a||b| = |ab|$. ⬤

We can add or subtract radicals only if the index and the radicand of each expression are the same.

Example 4 Combine Radical Expressions

Simplify the radical expression. Assume all variables represent positive real numbers.

$$5\sqrt[3]{16x} + 3\sqrt[3]{2x} - \sqrt[3]{2x^4}$$

Solution

$5\sqrt[3]{16x} + 3\sqrt[3]{2x} - \sqrt[3]{2x^4}$

$\quad = 5\sqrt[3]{8}\sqrt[3]{2x} + 3\sqrt[3]{2x} - \sqrt[3]{x^3}\sqrt[3]{2x}$ Factor out all perfect cubes.

$\quad = 5(2)\sqrt[3]{2x} + 3\sqrt[3]{2x} - x\sqrt[3]{2x}$ $\sqrt[3]{8} = 2$ and $\sqrt[3]{x^3} = x$.

$\quad = 10\sqrt[3]{2x} + 3\sqrt[3]{2x} - x\sqrt[3]{2x}$ Simplify.

$\quad = 13\sqrt[3]{2x} - x\sqrt[3]{2x}$ Combine radicals.

$\quad = (13 - x)\sqrt[3]{2x}$ Use the distributive property. ⬤

Determine whether you can simplify the radical expressions from Example 5. Watch the interactive video to see each solution worked out in detail.

My interactive video summary

Example 5 Simplify Radical Expressions

Simplify each radical expression. Assume all variables represent positive real numbers.

a. $\sqrt{192}$

b. $(3\sqrt{2} + \sqrt{5})(2\sqrt{2} - 4\sqrt{5})$

c. $\sqrt[5]{2x^4y^4}\sqrt[5]{48xy^2}$

d. $8\sqrt{2x^2} - 3\sqrt{28x} - 2\sqrt{8x^2}$

Solution

Watch the interactive video to see each solution worked out in detail.

You Try It Work through the following You Try It problem.

Work Exercises 37–50 in this textbook or in the My MathLab **Study Plan.**

When a radical appears in the denominator of an algebraic expression, we typically multiply the numerator and denominator by a suitable form of 1 in order to remove the radical from the denominator. This process is called **rationalizing the denominator.** Example 6 illustrates this process.

My interactive video summary

Example 6 Rationalize the Denominator of a Rational Expression

Rationalize the denominator of each expression.

a. $\sqrt{\dfrac{3}{2}}$

b. $\dfrac{\sqrt[3]{375}}{\sqrt[3]{4}}$

c. $\dfrac{7}{\sqrt{5} + \sqrt{2}}$

Solution

a. $\sqrt{\dfrac{3}{2}}$ Write the original expression.

$= \dfrac{\sqrt{3}}{\sqrt{2}}$ Rewrite using radical property (2).

$= \dfrac{\sqrt{3}}{\sqrt{2}} \cdot \dfrac{\sqrt{2}}{\sqrt{2}}$ Multiply the expression by $\dfrac{\sqrt{2}}{\sqrt{2}} = 1$.

$= \dfrac{\sqrt{6}}{2}$ Simplify. The denominator is now rationalized.

b. $\dfrac{\sqrt[3]{375}}{\sqrt[3]{4}}$ Write the original expression.

$= \dfrac{\sqrt[3]{125}\sqrt[3]{3}}{\sqrt[3]{4}}$ Rewrite the numerator using radical property (1).

$= \dfrac{5\sqrt[3]{3}}{\sqrt[3]{4}}$ Simplify $\sqrt[3]{125} = 5$.

$= \dfrac{5\sqrt[3]{3}}{\sqrt[3]{4}} \cdot \dfrac{\sqrt[3]{2}}{\sqrt[3]{2}}$ Multiply the expression by $\dfrac{\sqrt[3]{2}}{\sqrt[3]{2}} = 1$.

$$= \frac{5\sqrt[3]{6}}{\sqrt[3]{8}}$$ Multiply using radical property (1).

$$= \frac{5\sqrt[3]{6}}{2}$$ Simplify $\sqrt[3]{8} = 2$. The denominator is now rationalized.

c. $\dfrac{7}{\sqrt{5} + \sqrt{2}}$ Write the original expression.

$$= \frac{7}{\sqrt{5} + \sqrt{2}} \cdot \frac{\sqrt{5} - \sqrt{2}}{\sqrt{5} - \sqrt{2}}$$ Multiply the expression by $\dfrac{\sqrt{5} - \sqrt{2}}{\sqrt{5} - \sqrt{2}} = 1$. (The expression $\sqrt{5} - \sqrt{2}$ is called the **radical conjugate** of the expression $\sqrt{5} + \sqrt{2}$.)

$$= \frac{7\sqrt{5} - 7\sqrt{2}}{3}$$ Simplify.

You Try It Work through the following You Try It problem.

Work Exercises 51–57 in this textbook or in the MyMathLab Study Plan.

OBJECTIVE 3 SIMPLIFYING EXPONENTIAL EXPRESSIONS INVOLVING RATIONAL EXPONENTS

We can use our knowledge of radicals to simplify expressions involving rational exponents. We start by defining the expression $b^{1/n}$.

Definition $b^{1/n}$

For any integer $n \geq 2$ and any real number b, we define the expression $b^{1/n}$ as

$$b^{1/n} = \sqrt[n]{b}$$

provided that the expression $\sqrt[n]{b}$ is a real number.

Example 7 Evaluate Expressions Involving Rational Exponents
Simplify each expression.

a. $4^{1/2}$ **b.** $(-27)^{1/3}$ **c.** $\left(\dfrac{1}{64}\right)^{1/6}$

Solution

a. $4^{1/2} = \sqrt{4} = 2$ **b.** $(-27)^{1/3} = \sqrt[3]{-27} = -3$ **c.** $\left(\dfrac{1}{64}\right)^{1/6} = \sqrt[6]{\dfrac{1}{64}} = \dfrac{1}{2}$

Each exponent from Example 7 was a rational number of the form $\dfrac{1}{n}$. The following definition provides a way to deal with expressions involving rational exponents of the form $\dfrac{m}{n}$.

Definition $b^{m/n}$

If $\dfrac{m}{n}$ is a rational number with $n \geq 2$ and b is a real number, then we define the expression $b^{m/n}$ as

$$b^{m/n} = \sqrt[n]{b^m} = \left(\sqrt[n]{b}\right)^m .$$

provided that the expression $\sqrt[n]{b}$ is a real number.

The laws of exponents that were stated previously for integer exponents hold true for rational exponents as well and are worth stating again.

Laws of Exponents

Suppose s and t are rational numbers and a and b are real numbers, then

1. $b^s b^t = b^{s+t}$ — Product rule of exponents

2. $\dfrac{b^s}{b^t} = b^{s-t}$, where $b \neq 0$ — Quotient rule of exponents

3. $(b^s)^t = b^{st}$ — Power rule of exponents

4. $(ab)^s = a^s b^s$ — Product to power rule of exponents

5. $\left(\dfrac{a}{b}\right)^s = \dfrac{a^s}{b^s}$, where $b \neq 0$ — Quotient to power rule of exponents

6. $b^0 = 1$ — Zero exponent rule

7. $b^{-s} = \dfrac{1}{b^s}$, where $b \neq 0$ — Reciprocal rule of exponents

---

When we simplify expressions involving rational exponents, we always simplify until all exponents are positive and until each base appears only once in the expression.

My interactive video summary

▶ **Example 8** Simplify Expressions Involving Rational Exponents

Simplify each expression using positive exponents. Assume all variables represent positive real numbers.

a. $(9)^{3/2}$ **b.** $(-32)^{-3/5}$ **c.** $\dfrac{(125x^4 y^{-1/4})^{2/3}}{(x^2 y)^{1/3}}$

Solution

a. $(9)^{3/2} = \sqrt{9^3} = (\sqrt{9})^3 = 3^3 = 27$

b. $(-32)^{-3/5} = \sqrt[5]{(-32)^3} = (\sqrt[5]{-32})^3 = (-2)^3 = -8$

c. $\dfrac{(125x^4 y^{-1/4})^{2/3}}{(x^2 y)^{1/3}} = \dfrac{125^{2/3} x^{8/3} y^{-1/6}}{x^{2/3} y^{1/3}} = \dfrac{(\sqrt[3]{125})^2 x^{8/3 - 2/3}}{y^{1/3 + 1/6}} = \dfrac{25x^2}{y^{1/2}}$

Watch the interactive video to see each solution worked out in detail. ●

You Try It   Work through the following You Try It problem.

Work Exercises 58–70 in this textbook or in the MyMathLab Study Plan.

R.3 Exercises

In Exercises 1–30, use the laws of exponents to simplify each expression using positive exponents only. Assume all variables represent nonzero real numbers.

1. $3^2 \cdot 3^3$

2. $k^4 \cdot k^5$

3. $x^5 \cdot x^9 \cdot x^{12}$

4. $(a^5 b^4)(a^7 b^2)$

5. $\dfrac{t^9}{t^2}$

6. $(r^5)^{-7}$

7. $(w^{-8})^{-6}$

8. $\dfrac{n}{n^{-7}}$

9. $\dfrac{m^{-6}}{m^{-9}}$

10. $(-5y)^0$

11. -6^0

12. 2^{-2}

13. -4^{-2}

14. n^{-3}

15. $5x^{-6}$

16. $-7y^{-2}$

17. $\dfrac{1}{t^{-2}}$

18. $\dfrac{5}{y^{-9}}$

19. $\dfrac{a^{-11}}{b^{-4}}$

20. $\dfrac{-5p^{-7}q^4}{w^5}$

21. $b^{11}b^{-4}$

22. $a^{-11}a^{-3}$

23. $3^{-4} \cdot 3$

24. $\dfrac{z^{14}}{z^{-4}}$

25. $(-5x^2 y^5)^2$

26. $\dfrac{24x^3 y^5}{32x^8 y^{-7}}$

27. $\dfrac{(a^{-3}b)^{-2}}{(a^3 b^{-1})^2}$

28. $(16a^{-3}bc^{-6})(2ab)^{-5}$

29. $\left(\dfrac{q^3 p^4 w^5}{q^{-3} p^{-4} w^{-5}}\right)^{-2}$

30. $\dfrac{(3^{-1}x^{-1}y^{-1})^{-2}(3x^{-4}y^3)(9x^{-2}y^3)^0}{(3x^{-3}y^{-5})^2}$

In Exercises 31–50, simplify each radical expression. Assume all variables represent positive real numbers.

31. $\sqrt{169}$

32. $\sqrt{\dfrac{1}{25}}$

33. $\sqrt[3]{216}$

34. $\sqrt[3]{-343}$

35. $\sqrt{49w^2}$

36. $\sqrt[4]{(x+1)^4}$

37. $\sqrt{160}$

38. $\sqrt{363x^2}$

39. $\sqrt[3]{250}$

40. $\sqrt[3]{-27x^4}$

41. $\sqrt[4]{32z^5}$

42. $\sqrt{180x^{12}y^{27}}$

43. $\sqrt[5]{\dfrac{x^{24}y^{13}}{x^9 y^8}}$

44. $\sqrt{6}\left(\sqrt{5} - \sqrt{11}\right)$

45. $\left(3\sqrt{2} + 3\sqrt{7}\right)\left(\sqrt{2} - 2\sqrt{7}\right)$

46. $\left(4 + \sqrt{3}\right)^2$

47. $\sqrt[3]{x^2}\left(8\sqrt[3]{4x} - 5\sqrt[3]{x^5}\right)$

48. $11\sqrt{5} + 3\sqrt{5}$

49. $\sqrt{48x} - \sqrt{12x}$

50. $\sqrt[3]{875xy^3} - 3\sqrt[3]{56xy^3} + 2y\sqrt[3]{448x}$

In Exercises 51–57, rationalize the denominator. Assume all variables represent positive real numbers.

51. $\dfrac{\sqrt{5}}{\sqrt{6}}$ **52.** $\dfrac{1}{\sqrt{7}}$ **53.** $\dfrac{\sqrt[3]{7}}{\sqrt[3]{2}}$ **54.** $\dfrac{\sqrt[3]{2y^5}}{\sqrt[3]{5x}}$

55. $\dfrac{9}{5-\sqrt{2}}$ **56.** $\dfrac{\sqrt{5}-\sqrt{3}}{\sqrt{5}+\sqrt{3}}$ **57.** $\dfrac{\sqrt{a}}{\sqrt{a}+\sqrt{b}}$

In Exercises 58–70, simplify each exponential expression. Write any negative exponents as positive exponents. Assume all variables represent positive real numbers.

58. $25^{1/2}$ **59.** $\left(\dfrac{1}{16}\right)^{1/4}$ **60.** $-81^{1/4}$ **61.** $16^{3/4}$

62. $8^{-4/3}$ **63.** $x^{-1/3}$ **64.** $x^{-3/4} \cdot x^{7/4}$ **65.** $x^{1/3} \cdot x^{1/4} \cdot x^{-1/2}$

66. $(x^5 y^{10})^{1/5}$ **67.** $\dfrac{(x^2 y^4)^{4/7}(xy^2)^{1/7}}{x^{4/7} y^{4/7}}$ **68.** $(9x^2 y^{-5/2})^{1/2}$

69. $(32x^{-5} y^{10})^{-1/5}(xy^{1/5})$ **70.** $\left(\dfrac{x^{-3/4} y^{1/4}}{x^{-1/4}}\right)^{-8}$

R.4 Polynomials

OBJECTIVES

1 Understanding the Definition of a Polynomial
2 Adding and Subtracting Polynomials
3 Multiplying Polynomials
4 Dividing Polynomials Using Long Division

OBJECTIVE 1   UNDERSTANDING THE DEFINITION OF A POLYNOMIAL

In Section R.2, we introduced algebraic expressions. An algebraic expression with one term in which the exponents of all variable factors are nonnegative integers is called a **monomial**. Examples include 7, x^5, $-\sqrt{2}x^2 y^7$, and πxyz. The **coefficient** of a monomial is the numeric factor. The **degree** of a monomial is the sum of the variable exponents.

Monomial	Degree	Coefficient
7	0	7
x^5	5	1
$-\sqrt{2}x^2 y^7$	9	$-\sqrt{2}$
πxyz	3	π

⚸
You Try It Work through the following You Try It problem.

Work Exercises 1–5 in this textbook or in the My Math Lab **Study Plan.**

A **polynomial** is the sum or difference of one or more monomials. A polynomial with two terms is called a **binomial**. A polynomial with three terms is called a **trinomial**. The **degree** of a polynomial is the degree of the highest monomial term.

Example 1 Identify Polynomials

Determine whether each algebraic expression is a polynomial. If the expression is a polynomial, state the degree.

a. $8x^3y^2 - 2$ **b.** $5w^5 - 3y^2 - \dfrac{6}{w}$ **c.** $7p^8m^4 - 5p^5m^3 + 2pm$

Solution

a. The expression $8x^3y^2 - 2$ is a polynomial. Because this polynomial has two terms, we say that $8x^3y^2 - 2$ is a binomial. The degree of this binomial is 5.

b. The expression $5w^5 - 3y^2 - \dfrac{6}{w}$ is not a polynomial because the term $\dfrac{6}{w}$ is equivalent to $6w^{-1}$. For an expression to be a polynomial, all exponents must be nonnegative integers.

c. The expression $7p^8m^4 - 5p^5m^3 + 2pm$ is a polynomial. Because this polynomial has three terms, we say that $7p^8m^4 - 5p^5m^3 + 2pm$ is a trinomial. The degree of this trinomial is 12.

⚸
You Try It Work through the following You Try It problem.

Work Exercises 6–10 in this textbook or in the My Math Lab **Study Plan.**

For the remainder of this section, we learn the arithmetic of polynomials. That is, we learn how to add, subtract, multiply, and divide polynomials. Our discussion focuses mainly on polynomials in one variable.

Definition Polynomial in One Variable

The algebraic expression $a_nx^n + a_{n-1}x^{n-1} + a_{n-2}x^{n-2} + \cdots + a_1x + a_0$ is a polynomial in one variable of degree n, where n is a nonnegative integer. The numbers $a_0, a_1, a_2, \ldots, a_n$ are called the **coefficients** of the polynomial.

We typically write polynomials in descending order, meaning that we first list the monomial with the highest degree, then the monomial with the next highest degree, and so on. Polynomials that are written in descending order are said to be written in **standard form**.

OBJECTIVE 2   ADDING AND SUBTRACTING POLYNOMIALS

To add or subtract polynomials, we simply add or subtract the coefficients of the corresponding terms with like exponents. We always write the final sum or difference in standard form.

My video summary ⊘ **Example 2** Add and Subtract Polynomials

Add or subtract as indicated and express your answer in standard form.

a. $(-x^5 + 3x^4 - 9x^2 + 7x + 9) + (x^4 - 5x^5 + x^3 - 3x^2 - 8)$

b. $(-x^5 + 3x^4 - 9x^2 + 7x + 9) - (x^4 - 5x^5 + x^3 - 3x^2 - 8)$

Solution

a. To add polynomials, we remove the parentheses then combine the like terms.

$(-x^5 + 3x^4 - 9x^2 + 7x + 9) + (x^4 - 5x^5 + x^3 - 3x^2 - 8)$	Write the original expression.
$= -x^5 + 3x^4 - 9x^2 + 7x + 9 + x^4 - 5x^5 + x^3 - 3x^2 - 8$	Remove the parentheses.
$= (-x^5 - 5x^5) + (3x^4 + x^4) + x^3 + (-9x^2 - 3x^2) + 7x + (9 - 8)$	Collect like terms.
$= -6x^5 + 4x^4 + x^3 - 12x^2 + 7x + 1$	Simplify and write in standard form.

b. To subtract polynomials, we remove the parentheses, making sure to change the sign of each term of the second polynomial and then combine the like terms.

$(-x^5 + 3x^4 - 9x^2 + 7x + 9) - (x^4 - 5x^5 + x^3 - 3x^2 - 8)$	Write the original expression.
$= -x^5 + 3x^4 - 9x^2 + 7x + 9 - x^4 + 5x^5 - x^3 + 3x^2 + 8$	Remove the parentheses and change the sign of each term of the second polynomial.
$= (-x^5 + 5x^5) + (3x^4 - x^4) - x^3 + (-9x^2 + 3x^2) + 7x + (9 + 8)$	Collect like terms.
$= 4x^5 + 2x^4 - x^3 - 6x^2 + 7x + 17$	Simplify and write in standard form.

Watch the video to see each solution worked out in detail.

You Try It Work through the following You Try It problem.

Work Exercises 11–15 in this textbook or in the MyMathLab Study Plan.

OBJECTIVE 3 MULTIPLYING POLYNOMIALS

To find the product of two monomials in one variable, we multiply the coefficients together and use the product rule of exponents to add any exponents. For example, $(8x^4)(-3x^7) = (8)(-3) \cdot x^4 \cdot x^7 = -24x^{11}$. To multiply a monomial by a polynomial with more than one term, we need to use the distributive property multiple times. Thus, if m is a monomial and $p_1, p_2, p_3, \ldots p_n$ are the terms of a polynomial, then

$$m(p_1 + p_2 + p_3 + \cdots + p_n) = mp_1 + mp_2 + mp_3 + \cdots + mp_n.$$

Example 3 Multiply a Polynomial by a Monomial

Multiply $(-2x^5)(3x^3 - 5x^2 + 1)$.

Solution

$$(-2x^5)(3x^3 - 5x^2 + 1) = -2x^5(3x^3) - 2x^5(-5x^2) - 2x^5(1)$$
$$= -6x^8 + 10x^7 - 2x^5$$

When we find the product of any two polynomials, we use the **distributive property** to multiply each term of the first polynomial by each term of the second polynomial and then simplify.

 *My video summary*

⊘ **Example 4** Multiply Polynomials

Find the product and express your answer in standard form.

$$(2x^2 - 3)(3x^3 - x^2 + 1)$$

Solution

Watch the video to verify that $(2x^2 - 3)(3x^3 - x^2 + 1) = 6x^5 - 2x^4 - 9x^3 + 5x^2 - 3$. ●

You Try It Work through the following You Try It problem.

Work Exercises 16–18 in this textbook or in the MyMathLab Study Plan.

THE FOIL METHOD

When we find the product of two binomials, we can use a technique known as the FOIL method. FOIL is an acronym that stands for First, Outside, Inside, and Last.

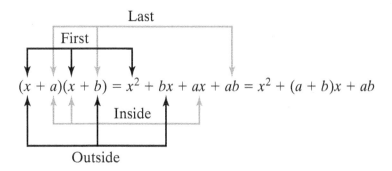

We illustrate the FOIL method by finding the product of the binomials $3x + 5$ and $2x + 7$.

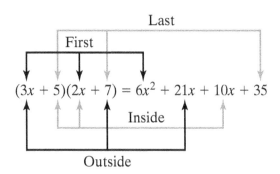

The product of the **First** terms:	$(3x)(2x) = 6x^2$
The product of the **Outside** terms:	$(3x)(7) = 21x$
The product of the **Inside** terms:	$(5)(2x) = 10x$
The product of the **Last** terms:	$(5)(7) = 35$

We now simplify by combining like terms:

$$6x^2 + 21x + 10x + 35 = 6x^2 + 31x + 35$$

Therefore, $(3x + 5)(2x + 7) = 6x^2 + 31x + 35$.

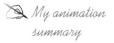 *My animation summary*

⊡ Watch this animation to see how the FOIL method works.

🔺

You Try It Work through the following You Try It problem.

Work Exercises 19–21 in this textbook or in the MyMathLab Study Plan.

Example 5 Multiply Binomials Using the FOIL Method

Find the product: $(x + 4)(x - 4)$.

Solution

$$(x + 4)(x - 4)$$ Write the original expression.

$$= x^2 - 4x + 4x - 16$$ Use the FOIL method.

$$= x^2 - 16$$ Combine like terms. ●

Notice that the product of the two binomials in Example 5 is also a binomial. (There is no x-term.) The product in Example 5 is a special product known as the **difference of two squares** because the resulting binomial is the difference of two perfect square terms. In general, $(a + b)(a - b) = a^2 - b^2$. This formula is worth remembering. There are other special products that are worth remembering as well. These products include the **square of a binomial** and the **cube of a binomial**. We now list some special products.

Special Products

Let a and b represent any real number or algebraic expression, then

$$(a + b)(a - b) = a^2 - b^2$$ Difference of two squares

$$(a + b)^2 = a^2 + 2ab + b^2$$ Squaring a binomial

$$(a - b)^2 = a^2 - 2ab + b^2$$

$$(a + b)^3 = a^3 + 3a^2b + 3ab^2 + b^3$$ Cubing a binomial

$$(a - b)^3 = a^3 - 3a^2b + 3ab^2 - b^3$$

Determine whether you can use these special product formulas to find the products in Example 6.

Example 6 Use Special Product Formulas

Find each product using a special product formula.

a. $(2x + 3)(2x - 3)$ b. $(3x + 2y)^2$ c. $(5 - 3z)^3$

Solution

a. We can use the difference of two squares formula $(a + b)(a - b) = a^2 - b^2$ with $a = 2x$ and $b = 3$ to get $(2x + 3)(2x - 3) = (2x)^2 - (3)^2 = 4x^2 - 9$.

b. We can use the squaring a binomial formula
$(a + b)^2 = a^2 + 2ab + b^2$ with $a = 3x$ and $b = 2y$ to get
$(3x + 2y)^2 = (3x)^2 + 2(3x)(2y) + (2y)^2 = 9x^2 + 12xy + 4y^2$.

c. Using the cubing a binomial formula $(a - b)^3 = a^3 - 3a^2b + 3ab^2 - b^3$ with $a = 5$ and $b = 3z$ we get

$$(5 - 3z)^3 = (5)^3 - 3(5)^2(3z) + 3(5)(3z)^2 - (3z)^3$$
$$= 125 - 225z + 135z^2 - 27z^3$$

You Try It Work through the following You Try It problem.

Work Exercises 22–28 in this textbook or in the MyMathLab Study Plan.

OBJECTIVE 4 DIVIDING POLYNOMIALS USING LONG DIVISION

We can long divide two polynomials in much the same way as the long division of two integers.

For example, suppose we want to divide the polynomial $2x^2 - 7x + 4$ by the polynomial $x + 8$. The polynomial $2x^2 - 7x + 4$ is the **dividend** and $x + 8$ is the **divisor**. We set up this problem using the familiar symbol used for long division with the dividend on the inside of the symbol and the divisor on the outside.

$$x + 8 \overline{)2x^2 - 7x + 4}$$

We now illustrate the long division process.

Step 1. Divide the first term of the dividend, $2x^2$, by the first term of the divisor, x. Write the result, $2x$, above the first term of the dividend.

$$\begin{array}{r} 2x \phantom{{}- 7x + 4} \\ x + 8 \overline{)2x^2 - 7x + 4} \end{array} \qquad \frac{2x^2}{x} = 2x, \text{ or equivalently, } 2x \cdot x = 2x^2.$$

Step 2. Multiply $2x$ by $x + 8$ and line up like terms vertically.

$$\begin{array}{r} 2x \phantom{{}- 7x + 4} \\ x + 8 \overline{)2x^2 - 7x + 4} \\ 2x^2 + 16x \phantom{{}+ 4} \end{array} \qquad 2x(x + 8) = 2x^2 + 16x.$$

Step 3. Subtract $2x^2 + 16x$ from $2x^2 - 7x$. We perform the subtraction by **changing the sign** of each term of the polynomial being subtracted and then combining like terms.

$$\begin{array}{r} 2x \phantom{{}- 7x + 4} \\ x + 8 \overline{)\,2x^2 - 7x + 4} \\ -2x^2 - 16x \phantom{{}+ 4} \\ \hline -23x \phantom{{}+ 4} \end{array} \qquad \text{Change the signs: } -(2x^2 + 16x) = -2x^2 - 16x.$$

Step 4. Bring down the constant 4.

$$\begin{array}{r} 2x \phantom{{}- 7x + 4} \\ x + 8 \overline{)2x^2 - 7x + 4} \\ -2x^2 - 16x \phantom{{}+ 4} \\ \hline -23x + 4 \end{array}$$

Step 5. Divide $-23x$ by the first term of the divisor, x. Write the result, -23, above the second term of the dividend.

$$\begin{array}{r} 2x - 23 \\ x + 8 \overline{)2x^2 - 7x + 4} \\ -2x^2 - 16x \phantom{{}+ 4} \\ \hline -23x + 4 \end{array} \qquad \frac{-23x}{x} = -23, \text{ or equivalently, } -23 \cdot x = -23x.$$

Step 6. Multiply -23 by $x + 8$ and line up like terms vertically.

$$\begin{array}{r}
2x - 23 \\
x + 8\overline{)2x^2 - 7x + 4} \\
-2x^2 - 16x \\
\hline
-23x + 4 \\
-23x - 184
\end{array}$$

$$-23(x + 8) = -23x - 184.$$

Step 7. Subtract $-23x - 184$ from $-23x + 4$.

$$\begin{array}{r}
2x - 23 \\
x + 8\overline{)2x^2 - 7x + 4} \\
-2x^2 - 16x \\
\hline
-23x + 4 \\
+23x + 184 \\
\hline
188
\end{array}$$

Change the signs. $-(23x - 184) = 23x + 184$

We know that we are done because 188 is of lower degree (degree zero) than the degree of the divisor (degree one). The polynomial $2x - 23$ is called the **quotient**, and the constant 188 is the **remainder**. We can always check our long division process by verifying that the dividend is equal to the product of the quotient and divisor plus the remainder.

$$\text{dividend} = (\text{divisor})(\text{quotient}) + \text{remainder}$$

$$2x^2 - 7x + 4 \overset{?}{=} (x + 8)(2x - 23) + 188$$

$$2x^2 - 7x + 4 \overset{?}{=} 2x^2 - 7x - 184 + 188$$

$$2x^2 - 7x + 4 = 2x^2 - 7x + 4 \checkmark$$

My video summary ▶ **Example 7** Divide Polynomials

Find the quotient and remainder when $3x^4 + x^3 + 7x + 4$ is divided by $x^2 - 1$.

Solution Using the long division symbol, place the dividend on the inside of the symbol and the divisor on the outside of the symbol. We write both polynomials in standard form, inserting coefficients of 0 for any missing terms.

$$x^2 + 0x - 1\overline{)3x^4 + x^3 + 0x^2 + 7x + 4}$$

The long division process is shown here. Watch the video to see every step of this process.

$$\begin{array}{r}
3x^2 + x + 3 \\
x^2 + 0x - 1\overline{)3x^4 + x^3 + 0x^2 + 7x + 4} \\
-(3x^4 + 0x^3 - 3x^2) \\
\hline
x^3 + 3x^2 + 7x \\
-(x^3 + 0x^2 - x) \\
\hline
3x^2 + 8x + 4 \\
-(3x^2 + 0x - 3) \\
\hline
8x + 7
\end{array}$$

The quotient is $3x^2 + x + 3$, and the remainder is $8x + 7$.

You Try It Work through the following You Try It problem.

Work Exercises 29–32 in this textbook or in the MyMathLab Study Plan.

R.4 Exercises

In Exercises 1–5, determine whether the algebraic expression is a monomial. If the expression is a monomial, state the degree and coefficient.

1. $5x^4$ **2.** $\dfrac{13}{x^2}$ **3.** $-x^4 y^8$ **4.** $2x^2 - 3y^5$ **5.** 0

In Exercises 6–10, determine whether the algebraic expression is a polynomial. If the expression is a polynomial, state the degree.

6. $5x^3 - \dfrac{13}{x^2}$ **7.** π **8.** $3wx^2 - \sqrt{5}$ **9.** $\dfrac{2x^2 - 3y^5}{x^5 - 1}$

10. $2p^3 m^4 - 5p^2 m^3 + 2p^4 m - 4pm^3 + 2p^4 m^3$

In Exercises 11–15, perform the indicated operations. Write each answer in standard form.

11. $(4x + 5) + (-5x + 9)$

12. $(x^2 + 11x + 7) + (3x - 13)$

13. $(9x^3 - 12x^2 + 11x + 5) - (3x^2 - 5x + 11)$

14. $(4x^5 + 8x^3 + 9x) + (7x^4 - 2x^3 + 8x^2)$

15. $(4x^2 - 4x + 4) + (3x^2 - 5x + 3) - (7x^2 + 8)$

In Exercises 16–18, find the indicated product. Write each answer in standard form.

16. $x(x^3 - x + 7)$ **17.** $-5x^4(7x^6 - 2)$ **18.** $(x + 3)(x^2 + 7x - 6)$

In Exercises 19–21, find the indicated product using the FOIL method. Write each answer in standard form.

19. $(x + 9)(x - 4)$ **20.** $(3x + 2)(x + 5)$ **21.** $(-x - 8)(-3x - 7)$

In Exercises 22–28, find each product using a special product formula. Write each answer in standard form.

22. $(6x + 2)(6x - 2)$ **23.** $(x + 5)^2$ **24.** $(4x - 9)^2$

25. $(8x + y)(8x - y)$ **26.** $(x - 6)^3$ **27.** $(3x + 2)^3$

28. $(x - 4)(x + 4)(x^2 - 16)$

In Exercises 29–32, find the quotient and remainder.

29. $7x^3 - 5x^2 + x + 4$ is divided by $x + 3$

30. $9x^4 - 2x^2 + 2x + 9$ is divided by $x^2 + 5$

31. $6x^5 - 5x^2 + 7x + 6$ is divided by $2x^3 - 1$

32. $7 - x^2 + x^4$ is divided by $x^2 + x + 9$

R.5 Factoring Polynomials

OBJECTIVES

1 Factoring Out a Greatest Common Factor

2 Factoring by Grouping

3 Factoring Trinomials with a Leading Coefficient Equal to One

4 Factoring Trinomials with a Leading Coefficient Not Equal to One

5 Factoring Using Special Factoring Formulas

Introduction to Section R.5

In this section, we learn techniques used to factor polynomial expressions. Being able to factor polynomials (and other algebraic expressions) is an essential algebraic skill and is used throughout this text to solve equations. Factoring a polynomial is basically the reverse of multiplying polynomials. The first factoring technique that we discuss is to factor out a greatest common factor.

OBJECTIVE 1 FACTORING OUT A GREATEST COMMON FACTOR

 *My video summary*

Factoring out a greatest common factor is essentially the distributive property. Note that the expression $AB + AC$ has two terms, AB and AC. The term AB has two factors, A and B. The term AC also has two factors, A and C. Notice that the terms AB and AC share a common factor of A. This common factor is called the **greatest common factor (GCF)** because it is the largest factor common to both terms. We can thus rewrite the expression $AB + AC$ by "pulling out" (factoring) this greatest common factor of A to get $AB + AC = A(B + C)$. When determining the GCF of a polynomial written as the sum and/or difference of monomials, we must determine the GCF of the numeric factors and then determine the GCF of the variable factors. For example, the polynomial $4x^2 - 6x^3 + 2x$ is composed of three monomials. The number 2 is the largest integer numeric factor common to each monomial term.

The largest variable factor common to each monomial is x. Therefore, the GCF is $2x$. Using the distributive property and the GCF of $2x$, we get the following:

$$4x^2 - 6x^3 + 2x$$
$$= 2x \cdot 2x - 2x \cdot 3x^2 + 2x \cdot 1$$
$$= 2x(2x - 3x^2 + 1)$$

Note We can always verify whether we have factored correctly by multiplying the factors together and checking to see if the product is equal to the original polynomial.

Example 1 Factor Out a Greatest Common Factor

Factor each polynomial by factoring out the GCF.

a. $5w^2y^3 - 15w^4y + 20w^3y^2$

b. $x(x^2 + 6) - 5(x^2 + 6)$

Solution

a. The numeric GCF is 5, and the variable GCF is w^2y. Thus, the GCF is $5w^2y$.

$$5w^2y^3 - 15w^4y + 20w^3y^2 \qquad \text{Write the original polynomial.}$$
$$= 5w^2y \cdot y^2 - 5w^2y \cdot 3w^2 + 5w^2y \cdot 4wy \qquad \text{The GCF is } 5w^2y.$$
$$= 5w^2y(y^2 - 3w^2 + 4wy) \qquad \text{Factor out the GCF.}$$

b. GCF is the binomial factor $(x^2 + 6)$.

$$x(x^2 + 6) - 5(x^2 + 6) = (x^2 + 6)(x - 5)$$

You Try It   Work through the following You Try It problem.

Work Exercises 1–8 in this textbook or in the My MathLab Study Plan.

OBJECTIVE 2   FACTORING BY GROUPING

 *My video summary*   ◉ We often encounter polynomials that, at first, appear to be difficult or impossible to factor. For example, consider the polynomial $x(x^2 + 6) - 5(x^2 + 6)$ seen in Example 1. Using the distributive property, this polynomial can be rewritten as $x^3 + 6x - 5x^2 - 30$. At first glance, it appears that we cannot factor this polynomial because the GCF is 1. However, if we group the terms in pairs and factor the GCF from each pair of terms, we see that we can indeed factor the polynomial.

$$(x^3 + 6x) + (-5x^2 - 30) \qquad \text{Group the terms in pairs.}$$
$$= x(x^2 + 6) - 5(x^2 + 6) \qquad \text{Factor the GCF out of each pair.}$$
$$= (x^2 + 6)(x - 5) \qquad \text{Factor out the GCF of } (x^2 + 6).$$

This technique is called **factoring by grouping** and often works when a polynomial has an even number of terms.

Example 2 Factor Polynomials by Grouping

Factor each polynomial by grouping.

a. $6x^2 - 2x + 9x - 3$

b. $2x^2 + 6xw - xy - 3wy$

Solution

a. $6x^2 - 2x + 9x - 3$

$$= (6x^2 - 2x) + (9x - 3) \qquad \text{Group the terms in pairs.}$$
$$= 2x(3x - 1) + 3(3x - 1) \qquad \text{Factor the GCF out of each pair.}$$
$$= (3x - 1)(2x + 3) \qquad \text{Factor out the GCF of } (3x - 1).$$

b. $2x^2 + 6xw - xy - 3wy$

$= (2x^2 + 6xw) + (-xy - 3wy)$ Group the terms in pairs.

$= 2x(x + 3w) - y(x + 3w)$ Factor the GCF out of each pair.

$= (x + 3w)(2x - y)$ Factor out the GCF of $(x + 3w)$.   ●

You Try It   Work through the following You Try It problem.

Work Exercises 9–16 in this textbook or in the My Math Lab Study Plan.

OBJECTIVE 3   FACTORING TRINOMIALS WITH A LEADING COEFFICIENT EQUAL TO ONE

My video summary   ⊙ In Section R.4, the FOIL method for multiplying two binomials was introduced. For example, we can use the FOIL method to multiply the two binomials $(x + a)$ and $(x + b)$.

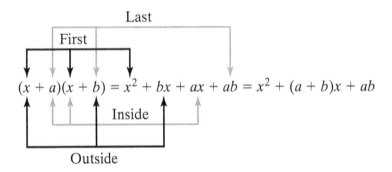

The binomial product $(x + a)(x + b)$ is said to be a factored form of the trinomial $x^2 + (a + b)x + ab$. Our goal is to start with a trinomial and try to find two binomials whose product is equal to the given trinomial. If we carefully look at the last two terms of the trinomial $x^2 + (a + b)x + ab$, we see that the middle term, $(a + b)x$, is the *sum* of the products of the inside and outside terms of the two binomials. The third term, ab, is the *product* of the last two terms of each binomial. Thus, when factoring trinomials of the form $x^2 + bx + c$, we must find two numbers whose product is c and whose sum is b.

Example 3   Factor Trinomials with a Leading Coefficient of One

Factor each trinomial.

a. $x^2 + 8x + 15$ **b.** $x^2 - 2x - 24$ **c.** $x^2 - 12x + 32$ **d.** $x^2 - 2x - 6$

Solution

a. We first note that the constant, 15, is positive and the coefficient of the x-term, 8, is positive. The only way for the product *and* the sum of two real numbers to be positive is if *both* numbers are positive. The only two pairs of positive integer factors of 15 are $1 \cdot 15$ and $3 \cdot 5$. Because $3 + 5 = 8$, we conclude that $x^2 + 8x + 15 = (x + 3)(x + 5)$.

b. We are looking for two integers whose product is −24 and whose sum is −2. We list all possible factors until we find the two integers that we are seeking or until we have exhausted all possibilities.

Product	Sum
$(1) \cdot (-24) = -24$	$(1) + (-24) = -23$
$(2) \cdot (-12) = -24$	$(2) + (-12) = -10$
$(3) \cdot (-8) = -24$	$(3) + (-8) = -5$
$(4) \cdot (-6) = -24$	$(4) + (-6) = -2$

We see that 4 and −6 are the correct integers. Thus, $x^2 - 2x - 24 = (x + 4)(x - 6)$.

c. In this case, we are seeking two integers whose product is 32 and whose sum is −12. The only way for the product of two real numbers to be positive and for the sum to be negative is if *both* numbers are negative.

Product	Sum
$(-1) \cdot (-32) = 32$	$(-1) + (-32) = -33$
$(-2) \cdot (-16) = 32$	$(-2) + (-16) = -18$
$(-4) \cdot (-8) = 32$	$(-4) + (-8) = -12$

We see that −4 and −8 are the correct integers. Thus, $x^2 - 12x + 32 = (x - 4)(x - 8)$.

d. We are searching for two integers whose product is −6 and whose sum is −2. The possibilities are listed as follows:

Product	Sum
$(1) \cdot (-6) = -6$	$(1) + (-6) = -5$
$(2) \cdot (-3) = -6$	$(2) + (-3) = -1$
$(3) \cdot (-2) = -6$	$(3) + (-2) = 1$
$(6) \cdot (-1) = -6$	$(6) + (-1) = 5$

We see that there are no two integers whose product is −6 and whose sum is −2. Therefore, we say that the trinomial $x^2 - 2x - 6$ is not factorable with integers. A polynomial that is not factorable with integers is called a prime polynomial*.

You Try It Work through the following You Try It problem.

Work Exercises 17–27 in this textbook or in the My Math**Lab Study Plan.**

In Example 3, all trinomials had a leading coefficient of one. We are now ready to explore techniques used to factor trinomials of the form $ax^2 + bx + c$, where $a \neq 1$.

OBJECTIVE 4 FACTORING TRINOMIALS WITH A LEADING COEFFICIENT NOT EQUAL TO ONE

We can try to factor trinomials of the form $ax^2 + bx + c$, where $a \neq 1$ using a **trial-and-error method**. That is, we basically "guess" until we stumble on the correct two binomial factors. For example, suppose that we want to factor the trinomial $6x^2 + 29x + 35$. The possible first terms of the two binomial factors are x

and $6x$ or $2x$ and $3x$. The only integer factors of the constant, 35, are $(1) \cdot (35)$ and $(5) \cdot (7)$. This means that there are eight possible combinations of potential pairs of binomial factors for this trinomial. The eight possibilities are as follows:

1. $(x + 1)(6x + 35)$ **2.** $(x + 35)(6x + 1)$ **3.** $(x + 7)(6x + 5)$ **4.** $(x + 5)(6x + 7)$

5. $(2x + 1)(3x + 35)$ **6.** $(2x + 35)(3x + 1)$ **7.** $(2x + 7)(3x + 5)$ **8.** $(2x + 5)(3x + 7)$

My animation summary

Try to determine which of the eight possibilities above are the correct factors of $6x^2 + 29x + 35$. Carefully work through this animation to see how to use the trial and error method to factor trinomials. Although this "guessing" method works, it can often be a long and arduous process, especially when the leading coefficient and constant coefficient have several factors. Fortunately, there is another more methodical approach to factoring trinomials. We now outline this five-step method using the same trinomial $6x^2 + 29x + 35$ as our example.

Factoring Trinomials of the Form $ax^2 + bx + c, a \neq 0, a \neq 1$

Example: $6x^2 + 29x + 35, a = 6, b = 29$, and $c = 35$

Step 1. Multiply $a \cdot c$.

$$(6) \cdot (35) = 210$$

Step 2. Find two integers, n_1 and n_2, whose product is ac and whose sum is b. So $n_1 \cdot n_2 = ac$ and $n_1 + n_2 = b$. If no such pair of integers exists, the trinomial is prime.
The two integers are $n_1 = 15$ and $n_2 = 14$ because $(15) \cdot (14) = 210$ and $15 + 14 = 29$.

Step 3. Rewrite the middle term as the sum of two terms using the integers found in step 2. So, $ax^2 + bx + c = ax^2 + n_1x + n_2x + c$.

$$6x^2 + \underbrace{15x + 14x}_{15x + 14x = 29x} + 35$$

Step 4. Factor by grouping.

$$6x^2 + 15x + 14x + 35 = (6x^2 + 15x) + (14x + 35)$$
$$= 3x(2x + 5) + 7(2x + 5)$$
$$= (2x + 5)(3x + 7)$$

Step 5. Check your answer by multiplying out the factored form.

$$(2x + 5)(3x + 7) = 6x^2 + 29x + 35$$

Example 4 Factor a Trinomial with a Leading Coefficient Not Equal to One

Factor each trinomial.

a. $4x^2 + 17x + 15$ **b.** $12x^2 - 25x + 12$

Solution

My animation summary

a. To see how to use the ac method to factor the trinomial $4x^2 + 17x + 15$, work through this animation.

b. To factor the trinomial $12x^2 - 25x + 12$, follow the five steps outlined previously.

Step 1. $a \cdot c = 12 \cdot 12 = 144$

Step 2. $n_1 = -16$ and $n_2 = -9$

$$(-16) \cdot (-9) = 144 \quad \text{and} \quad (-16) + (-9) = -25$$

Step 3. $12x^2 - 25x + 12 = 12x^2 - 16x - 9x + 12$

Step 4. $(12x^2 - 16x) + (-9x + 12)$

$$= 4x(3x - 4) - 3(3x - 4)$$

$$= (3x - 4)(4x - 3)$$

Step 5. $(3x - 4)(4x - 3) = 12x^2 - 25x + 12$

You Try It Work through the following You Try It problem.

Work Exercises 28–34 in this textbook or in the My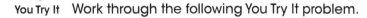Lab Study Plan.

OBJECTIVE 5 FACTORING USING SPECIAL FACTORING FORMULAS

There are several factoring formulas that are quite useful to remember and worth mentioning at this time. In Section R.4, we saw several special product formulas. We now revisit three of these formulas and restate them as factoring formulas.

> **Difference of Two Squares**
>
> Let a and b represent any real number or algebraic expression, then
>
> $$a^2 - b^2 = (a + b)(a - b).$$

Recognizing a polynomial that can be factored using the difference of two squares pattern is fairly straightforward because such a polynomial has two terms, each of which is a perfect square. Following are three examples:

$$z^2 - 4 = (z)^2 - (2)^2 = (z + 2)(z - 2)$$

$$36y^2 - 1 = (6y)^2 - (1)^2 = (6y + 1)(6y - 1)$$

$$z^4 - 16 = (z^2)^2 - (4)^2 = (z^2 + 4)(z^2 - 4) = (z^2 + 4)(z + 2)(z - 2)$$

Note The sum of two squares, $a^2 + b^2$, is not factorable using integers. A polynomial of this form is a prime polynomial.

In Section R.4, we introduced the following two special product formulas that were stated as the **square of a binomial**.

$$(a + b)^2 = a^2 + 2ab + b^2 \quad \text{and} \quad (a - b)^2 = a^2 - 2ab + b^2$$

The resulting trinomials are called *perfect square trinomials* because they can each be factored as a binomial squared.

Perfect Square Trinomials

Let a and b represent any real number or algebraic expression, then

$$a^2 + 2ab + b^2 = (a + b)^2$$
$$a^2 - 2ab + b^2 = (a - b)^2$$

Note that the first term of a perfect square trinomial is the square of some algebraic expression a. The last term is the square of some algebraic expression b, and the middle term is the sum or difference of twice the product of a and b. Following is an example of a perfect square trinomial:

$$4x^2 - 12xy + 9y^2 = \underbrace{(2x)^2}_{a^2} - \underbrace{2(2x)(3y)}_{2ab} + \underbrace{(3y)^2}_{b^2} = \underbrace{(2x - 3y)^2}_{(a - b)^2}$$

We stated previously that the sum of two squares cannot be factored with integers. However, this is not the case for the sum of two cubes. In fact, the difference of two cubes can be factored as well. We now state these final two special factoring formulas.

Sum and Difference of Two Cubes

Let a and b represent any real number or algebraic expression, then

$$a^3 + b^3 = (a + b)(a^2 - ab + b^2)$$
$$a^3 - b^3 = (a - b)(a^2 + ab + b^2).$$

Note that the sum of two cubes and the difference of two cubes have very similar factoring patterns. The only two differences are the signs of the binomial factor and the sign of the second term of the trinomial factor. Following is an example of a sum of two cubes:

$$8y^3 + 125 = \underbrace{(2y)^3}_{a^3} + \underbrace{(5)^3}_{b^3} = \underbrace{(2y + 5)}_{(a + b)}\underbrace{(4y^2 - 10y + 25)}_{(a^2 - ab + b^2)}$$

Determine whether you can use a special factoring formula to factor one or more of the polynomials stated in Example 5. Do not forget to first try to factor out a GCF if possible. Watch the interactive video to see if you are correct.

My interactive video summary

▶ **Example 5** Use Special Factoring Formulas to Factor Polynomials

Completely factor each expression.

a. $2x^3 + 5x^2 - 12x$ **b.** $5x^4 - 45x^2$ **c.** $8z^3x - 27y + 8z^3y - 27x$

Solution Watch the interactive video to see each solution worked out in detail.

You Try It Work through the following You Try It problem.

Work Exercises 35–50 in this textbook or in the MyMathLab Study Plan.

R.5 Exercises

In Exercises 1–8, factor each polynomial by factoring out the GCF.

1. $40x + 4$

2. $wx^4 + w$

3. $3y^7 + 6xy^8$

4. $4x^4 - 20x^3 + 16x^2$

5. $39w^9y^9 - 12w^2y^2 + 12wy + 15w^2y$

6. $11(x + 7) + 3a(x + 7)$

7. $8x(z + 1) + (z + 1)$

8. $9y(x^2 + 11) - 7(x^2 + 11)$

In Exercises 9–16, factor each polynomial by grouping.

9. $xy + 5x + 3y + 15$

10. $ab + 7a - 2b - 14$

11. $9xy + 12x + 6y + 8$

12. $3xy - 2x - 12y + 8$

13. $5x^2 + 4xy + 15x + 12y$

14. $3x^2 + 3xy - 2x - 2y$

15. $x^3 + 9x^2 + 4x + 36$

16. $x^3 - x^2 - 7x + 7$

In Exercises 17–50, completely factor each polynomial or state that the polynomial is prime.

17. $x^2 + 13x + 30$

18. $y^2 + 5y - 14$

19. $w^2 - 17w + 72$

20. $x^2 - 2x - 48$

21. $x^2 - 6x + 10$

22. $x^2 - 33x - 108$

23. $x^2 + 6xy + 8y^2$

24. $24 + 2x - x^2$

25. $5x^2 + 10x - 75$

26. $4x^2w + 20xw + 24w$

27. $5x^2 - 30x - 25$

28. $9x^2 + 18x + 5$

29. $12y^2 + 29y + 15$

30. $2x^2 + 5x - 25$

31. $8x^2 - 6x - 9$

32. $10y^2 - 23y + 12$

33. $16x^2 + 36x + 18$

34. $12 - x - 20x^2$

35. $x^2 - 81$

36. $z^2 + 10z + 25$

37. $x^3 + 64$

38. $2m^2 - 48m + 288$

39. $64x^2 - 49$

40. $x^3 - 343$

41. $108t^2 - 147$

42. $(x + 8y)^2 - 25$

43. $125x^3 + 216$

44. $(2x - 5)^3 - 8$

45. $(w + 5)^2 - 100$

46. $16x^2 + 40x + 25$

47. $x^{14} - x^{11}$

48. $6 - 149x^2 - 25x^4$

49. $x^3y^4 - 27y^4$

50. $x^2 + 14x + 49 - x^4$

R.6 Rational Expressions

OBJECTIVES

1 Simplifying Rational Expressions

2 Multiplying and Dividing Rational Expressions

3 Adding and Subtracting Rational Expressions

4 Simplifying Complex Rational Expressions

OBJECTIVE 1   SIMPLIFYING RATIONAL EXPRESSIONS

A rational number is the quotient of two integers. A rational expression is the quotient of two polynomial expressions.

Definition   Rational Expression

A rational expression is an algebraic expression of the form $\dfrac{P}{Q}$, where P and Q are polynomials such that $Q \neq 0$.

The following are examples of rational expressions:

$$\frac{x^2 - 4x}{2x - 8}, \qquad \frac{3x^2 - 2x - 8}{3x^2 - 14x - 24}, \qquad \frac{x^2 y^3 + 11x^5 y^4 - 7xy}{xy^5 - x^2 y}$$

Because division by zero is never allowed, we must only consider values of the variable for which the denominator does not equal zero. For example, the rational expression $\dfrac{x^2 - 4x}{2x - 8}$ is defined for all values of x with the exception of $x = 4$. The value of $x = 4$ produces zero in the denominator and thus cannot be considered as a possible value of the variable. To simplify a rational expression, try factoring the numerator and denominator and cancel any common factors. The restrictions placed on the original expression still hold for the simplified expression.

For example, $\dfrac{x^2 - 4x}{2x - 8} = \dfrac{x(x - 4)}{2(x - 4)} = \dfrac{x\cancel{(x - 4)}}{2\cancel{(x - 4)}} = \dfrac{x}{2}$. Therefore, the rational expression $\dfrac{x^2 - 4x}{2x - 8}$ is equivalent to $\dfrac{x}{2}$ for all values of x not equal to 4.

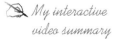 *My interactive video summary*

▶ **Example 1** Simplify Rational Expressions

Simplify each rational expression.

a. $\dfrac{x^2 + x - 12}{x^2 + 9x + 20}$
b. $\dfrac{x^3 + 1}{x + 1}$
c. $\dfrac{x^2 - x - 2}{2x - x^2}$

Solution

a. We first factor the rational expression. If you need a quick refresher of how to factor polynomials of the form $x^2 + bx + c$, watch the video that was first introduced in Section R.5 . The numerator and denominator of this rational expression factors as follows:

$$\frac{x^2 + x - 12}{x^2 + 9x + 20} = \frac{(x + 4)(x - 3)}{(x + 5)(x + 4)}$$

This rational expression is defined for all values of x such that $x \neq -5$ and $x \neq -4$. This restriction must also apply to any simplified version of the original rational expression.

(eText Screens R.6-1–R.6-22)

We now cancel all common factors within the numerator and denominator.

$$\frac{x^2 + x - 12}{x^2 + 9x + 20} = \frac{(x+4)(x-3)}{(x+5)(x+4)}$$ Cancel common factors.

Thus, $\frac{x^2 + x - 12}{x^2 + 9x + 20} = \frac{x-3}{x+5}$ for $x \neq -5$ and $x \neq -4$.

b. This rational expression is defined for all values of x not equal to -1. We can factor the numerator using the sum of two cubes factoring formula.

$$\frac{x^3+1}{x+1} = \frac{(x+1)(x^2 - x + 1)}{(x+1)}$$ Factor the numerator and cancel common factors.

$$= x^2 - x + 1 \text{ for } x \neq -1$$

c. Factor the numerator and denominator.

$$\frac{x^2 - x - 2}{2x - x^2} = \frac{(x-2)(x+1)}{x(2-x)}$$

We see that this rational expression is defined for all values of x such that $x \neq 0$ and $x \neq 2$. At first glance, it does not appear that the numerator and denominator share a common factor. However, note that if we were to multiply the factor $(2-x)$ by -1, we get $-1 \cdot (2-x) = -2 + x = x - 2$.

Therefore, multiply the numerator and denominator by 1 of the form $\frac{-1}{-1}$, and then cancel common factors.

$$\frac{x^2 - x - 2}{2x - x^2} = \frac{(x-2)(x+1)}{x(2-x)}$$

$$= \left(\frac{-1}{-1}\right)\frac{(x-2)(x+1)}{x(2-x)}$$ Multiply the rational expression by $\frac{-1}{-1}$.

$$= -\frac{(x-2)(x+1)}{x(x-2)}$$ $-1 \cdot (2-x) = (x-2)$.

$$= -\frac{(x-2)(x+1)}{x(x-2)}$$ Cancel common factors.

$$= -\frac{x+1}{x} \text{ for } x \neq 0 \text{ and } x \neq 2.$$

My interactive video summary ⊙ Watch the interactive video to see the solution to Example 1 worked out in detail.

You Try It Work through the following You Try It problem.

Work Exercises 1–10 in this textbook or in the MyMathLab Study Plan.

Note From this point forward, we do not make any restrictions on the variables. That is, we assume that all expressions in the denominators of all rational expressions represent nonzero real numbers.

OBJECTIVE 2 MULTIPLYING AND DIVIDING RATIONAL EXPRESSIONS

We multiply rational expressions similarly to the way we multiply rational numbers. If $\dfrac{A}{B}$ and $\dfrac{C}{D}$ are rational expressions, then $\dfrac{A}{B} \cdot \dfrac{C}{D} = \dfrac{AC}{BD}$.

Example 2 Multiply Rational Expressions

Multiply $\dfrac{2x^2 + 3x - 2}{3x^2 - 2x - 1} \cdot \dfrac{3x^2 + 4x + 1}{2x^2 + x - 1}$.

Solution Factor the numerator and denominator of each rational expression. Note that each numerator and denominator are polynomials of the form $ax^2 + bx + c$, where $a \neq 1$. If you need to review how to factor polynomials of this form, watch the video that was introduced in Section R.5.

$$\dfrac{2x^2 + 3x - 2}{3x^2 - 2x - 1} \cdot \dfrac{3x^2 + 4x + 1}{2x^2 + x - 1}$$

$$= \dfrac{(2x - 1)(x + 2)}{(x - 1)(3x + 1)} \cdot \dfrac{(3x + 1)(x + 1)}{(2x - 1)(x + 1)} \qquad \text{Factor the numerators and denominators.}$$

$$= \dfrac{(2x - 1)(x + 2)(3x + 1)(x + 1)}{(x - 1)(3x + 1)(2x - 1)(x + 1)} \qquad \text{Multiply numerators and denominators.}$$

$$= \dfrac{\cancel{(2x - 1)}(x + 2)\cancel{(3x + 1)}\cancel{(x + 1)}}{(x - 1)\cancel{(3x + 1)}\cancel{(2x - 1)}\cancel{(x + 1)}} \qquad \text{Cancel common factors.}$$

$$= \dfrac{x + 2}{x - 1}$$

We divide two rational expressions in exactly the same manner in which we divide two rational numbers. Recall that if $\dfrac{a}{b}$ and $\dfrac{c}{d}$ are rational numbers such that b, c, and d are nonzero, then $\dfrac{a}{b} \div \dfrac{c}{d} = \dfrac{a}{b} \cdot \dfrac{d}{c} = \dfrac{ad}{bc}$. The rational numbers $\dfrac{c}{d}$ and $\dfrac{d}{c}$ are called reciprocals. We follow this same pattern when dividing two rational expressions. To find the quotient of two rational expressions, we find the product of the first expression times the reciprocal of the second rational expression. Thus, if $\dfrac{A}{B}$ and $\dfrac{C}{D}$ are rational expressions, then $\dfrac{A}{B} \div \dfrac{C}{D} = \dfrac{A}{B} \cdot \dfrac{D}{C} = \dfrac{AD}{BC}$.

Example 3 Divide Rational Expressions

Divide and simplify $\dfrac{x^3 - 8}{2x^2 - x - 6} \div \dfrac{x^2 + 2x + 4}{6x^2 + 11x + 3}$.

Solution

$$\dfrac{x^3 - 8}{2x^2 - x - 6} \div \dfrac{x^2 + 2x + 4}{6x^2 + 11x + 3} \qquad \text{Write the original expression.}$$

$$= \dfrac{x^3 - 8}{2x^2 - x - 6} \cdot \dfrac{6x^2 + 11x + 3}{x^2 + 2x + 4} \qquad \text{Rewrite as a multiplication problem.}$$

$$= \frac{(x - 2)(x^2 + 2x + 4)}{(2x + 3)(x - 2)} \cdot \frac{(2x + 3)(3x + 1)}{x^2 + 2x + 4}$$

Factor the numerators and denominators. Use the difference of two cubes factoring formula to factor the first numerator.

$$= \frac{(x - 2)(x^2 + 2x + 4)(2x + 3)(3x + 1)}{(2x + 3)(x - 2)(x^2 + 2x + 4)}$$

Multiply.

$$= \frac{\cancel{(x - 2)}\cancel{(x^2 + 2x + 4)}\cancel{(2x + 3)}(3x + 1)}{\cancel{(2x + 3)}\cancel{(x - 2)}\cancel{(x^2 + 2x + 4)}}$$

Cancel common factors.

$$= 3x + 1$$

 *My video summary*   ⊙ **Example 4** Multiply and Divide Rational Expressions

Perform the indicated operations and simplify.

$$\frac{x^2 + x - 6}{x^2 + x - 42} \cdot \frac{x^2 + 12x + 35}{x^2 - x - 2} \div \frac{x + 7}{x^2 + 8x + 7}$$

Solution Watch the video to see that the expression simplifies to $\frac{(x + 3)(x + 5)}{x - 6}$.

You Try It   Work through the following You Try It problem.

Work Exercises 11–26 in this textbook or in the MyMathLab Study Plan.

OBJECTIVE 3   ADDING AND SUBTRACTING RATIONAL EXPRESSIONS

Recall that in order to add or subtract rational numbers, we must first rewrite both numbers in a different form using a common denominator. Once both rational numbers have a common denominator, we add or subtract the numerators, writing the result above the common denominator. For example, we subtract the rational number $\frac{1}{6}$ from $\frac{5}{8}$ as follows:

$$\frac{5}{8} - \frac{1}{6} = \left(\frac{5}{8}\right)\left(\frac{3}{3}\right) - \left(\frac{1}{6}\right)\left(\frac{4}{4}\right) = \ = \frac{15}{24} - \frac{4}{24} = \frac{15 - 4}{24} = \frac{11}{24}$$

We add and subtract rational expressions in much the same way. We start by determining the **least common denominator (LCD)**. The LCD is the smallest algebraic expression divisible by all denominators.

Example 5 Add and Subtract Rational Expressions

Perform the indicated operations and simplify.

a. $\dfrac{3}{x + 1} - \dfrac{2 - x}{x + 1}$

b. $\dfrac{3}{x^2 + 2x} + \dfrac{x - 2}{x^2 - x}$

Solution

a. Each rational expression has the same denominator, so we can combine by subtracting the numerators.

$$\frac{3}{x+1} - \frac{2-x}{x+1}$$ Write the original expression.

$$= \frac{3-(2-x)}{x+1}$$ Subtract numerators.

$$= \frac{3-2+x}{x+1}$$ Use the distributive property.

$$= \frac{x+1}{x+1}$$ Combine like terms.

$$= \frac{\cancel{x+1}}{\cancel{x+1}} = 1$$ Cancel common factors and simplify.

b. Factor the denominators to determine the LCD.

$$\left.\begin{array}{l} x^2 + 2x = x(x+2) \\ x^2 - x = x(x-1) \end{array}\right\} \quad \text{LCD} = x(x+2)(x-1).$$

$$\frac{3}{x(x+2)} + \frac{x-2}{x(x-1)}$$ Write the original expression with the denominators factored.

Multiply the first rational expression

$$= \frac{3}{x(x+2)}\frac{(x-1)}{(x-1)} + \frac{x-2}{x(x-1)}\frac{(x+2)}{(x+2)}$$ by $\frac{(x-1)}{(x-1)}$ and multiply the second

rational expression by $\frac{(x+2)}{(x+2)}$.

$$= \frac{3(x-1)+(x-2)(x+2)}{x(x+2)(x-1)}$$ Add the numerators.

$$= \frac{3x-3+x^2-4}{x(x+2)(x-1)}$$ Use the distributive property.

$$= \frac{x^2+3x-7}{x(x+2)(x-1)}$$ Combine like terms.

Try working through Example 6 on your own. Watch the interactive video when you are ready to see the solution.

My interactive video summary

▶ **Example 6** Add and Subtract Rational Expressions

Perform the indicated operations and simplify.

a. $\dfrac{3}{x-y} - \dfrac{x+5y}{x^2-y^2}$ **b.** $\dfrac{x+4}{3x^2+20x+25} + \dfrac{x}{3x^2+16x+5}$

Solution Watch the interactive video to see each solution worked out in detail.

You Try It Work through the following You Try It problem.

Work Exercises 27–44 in this textbook or in the MyMathLab Study Plan.

OBJECTIVE 4 SIMPLIFYING COMPLEX RATIONAL EXPRESSIONS

A complex rational expression (also called a complex fraction) is a fraction that contains a rational expression in the numerator and/or the denominator. The following are examples of complex rational expressions:

$$\frac{4 - \dfrac{5}{x-1}}{\dfrac{6}{x-1} - 7} \qquad\qquad \frac{\dfrac{1}{x} - \dfrac{3}{x+4}}{\dfrac{2}{x^2+4x} + \dfrac{2}{x}}$$

To simplify a complex rational expression, we want to rewrite the expression using only one fraction bar. There are two methods that can be used to simplify such expressions. We call these Method I and Method II. Method I involves the following three-step process.

METHOD I FOR SIMPLIFYING COMPLEX RATIONAL EXPRESSIONS

Step 1. Simplify the rational expression in the numerator.

Step 2. Simplify the rational expression in the denominator.

Step 3. Divide the expression in the numerator by the expression in the denominator. That is, multiply the expression in the numerator by the reciprocal of the expression in the denominator. Simplify if possible.

Example 7 Simplify a Complex Rational Expression Using Method I

Simplify using Method I: $\dfrac{4 - \dfrac{5}{x-1}}{\dfrac{6}{x-1} - 7}$

Solution

Step 1. Simplify the numerator.

$$4 - \frac{5}{x-1} = 4\frac{(x-1)}{(x-1)} - \frac{5}{x-1} = \frac{4(x-1)-5}{x-1}$$

$$= \frac{4x-4-5}{x-1} = \frac{4x-9}{x-1}$$

Step 2. Simplify the denominator.

$$\frac{6}{x-1} - 7 = \frac{6}{x-1} - 7\frac{(x-1)}{(x-1)} = \frac{6 - 7(x-1)}{x-1}$$

$$= \frac{6 - 7x + 7}{x-1} = \frac{13 - 7x}{x-1}$$

Step 3. Divide.

$$\frac{4 - \dfrac{5}{x-1}}{\dfrac{6}{x-1} - 7} = \frac{\dfrac{4x-9}{x-1}}{\dfrac{13-7x}{x-1}} = \frac{4x-9}{x-1} \cdot \frac{x-1}{13-7x}$$

$$= \frac{4x-9}{\cancel{x-1}} \cdot \frac{\cancel{x-1}}{13-7x} = \frac{4x-9}{13-7x}$$

As stated previously, there is another method that can be used to simplify a complex rational expression. Method II involves multiplying the numerator and denominator of the complex rational expression by the overall LCD. The overall LCD is the LCD of *all* fractions in both the numerator and the denominator.

METHOD II FOR SIMPLIFYING COMPLEX RATIONAL EXPRESSIONS

Step 1. Determine the overall LCD.

Step 2. Multiply the numerator and denominator of the complex rational expression by the overall LCD.

Step 3. Simplify.

We now use Method II to simplify the same complex rational expression from Example 7.

Example 8 Simplify a Complex Rational Expression Using Method II

Simplify using Method II: $\dfrac{4 - \dfrac{5}{x-1}}{\dfrac{6}{x-1} - 7}$

Solution

Step 1. The overall LCD is $x - 1$.

Step 2. Multiply the numerator and denominator by $x - 1$.

$$\frac{4 - \dfrac{5}{x-1}}{\dfrac{6}{x-1} - 7} \cdot \frac{x-1}{x-1}$$

Step 3. Simplify.

$$\frac{4 - \dfrac{5}{x-1}}{\dfrac{6}{x-1} - 7} \cdot \frac{x-1}{x-1} = \frac{4(x-1) - \dfrac{5(x-1)}{x-1}}{\dfrac{6(x-1)}{x-1} - 7(x-1)}$$

$$= \frac{4x - 4 - \dfrac{5(x-1)}{x-1}}{\dfrac{6(x-1)}{x-1} - 7x + 7}$$

$$= \frac{4x - 4 - 5}{6 - 7x + 7} = \frac{4x - 9}{13 - 7x}$$

As you can see from Examples 7 and 8, it does not matter which method we choose. Pick the method that you are most comfortable with. See if you can simplify the complex rational expression in Example 9. You can watch the interactive video and choose to see the solution using either method.

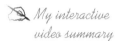 *My interactive video summary*

> **Example 9** Simplify a Complex Rational Expression

Simplify the complex rational expression using Method I or Method II.

$$\dfrac{\dfrac{1}{x} - \dfrac{3}{x+4}}{\dfrac{2}{x^2+4x} + \dfrac{2}{x}}$$

Solution To see the solution, watch the interactive video and click on Method I or Method II.

You Try It Work through the following You Try It problem.

Work Exercises 45–54 in this textbook or in the My**MathLab** Study Plan.

R.6 Exercises

In Exercises 1–10, simplify each rational expression.

1. $\dfrac{2x - 6x^2}{2x}$

2. $\dfrac{x^2 - 81}{9 + x}$

3. $\dfrac{9a - 27}{5a - 15}$

4. $\dfrac{x^2 + x - 6}{x + 3}$

5. $\dfrac{x - 13}{13 - x}$

6. $\dfrac{x^2 - 4}{-2 - x}$

7. $\dfrac{3x^2 - 2x - 8}{3x^2 - 14x - 24}$

8. $\dfrac{x^3 - 125}{3x - 15}$

9. $\dfrac{4x^2 - 7x - 2}{8x^3 + 2x^2 + 4x + 1}$

10. $\dfrac{36x^2 - 42x + 49}{216x^3 + 343}$

In Exercises 11–26, perform the indicated operations and simplify.

11. $\dfrac{18x + 18}{2x + 6} \cdot \dfrac{x + 3}{6x^2 - 6}$

12. $\dfrac{16x - 12}{35} \cdot \dfrac{10}{3 - 4x}$

13. $\dfrac{15a - 10a^2}{9a^2 + 6a + 1} \cdot \dfrac{6a^2 + 11a + 3}{4a^2 - 9}$

14. $\dfrac{4x^3 - 32}{2x^2 + 4x - 16} \cdot \dfrac{5x + 10}{4x^2 + 8x + 16}$

15. $\dfrac{a^3 + a^2b + a + b}{2a^3 + 2a} \cdot \dfrac{18a^2}{6a^2 - 6b^2}$

16. $\dfrac{x^2 - 6x - 7}{2x^2 - 98} \cdot \dfrac{x^2 + 14x + 49}{3x^2 + 24x + 21}$

17. $\dfrac{x^2 - 4}{9} \cdot \dfrac{x^2 - x - 2}{x^2 - 4x + 4}$

18. $\dfrac{4x}{9} \div \dfrac{8x + 16}{9x + 18}$

19. $\dfrac{a + b}{ab} \div \dfrac{a^2 - b^2}{7a^3b}$

20. $\dfrac{x^2 - 14x + 49}{x^2 - x - 42} \div \dfrac{x^2 - 49}{8}$

21. $\dfrac{x^2 - 6x - 16}{3x^2 - 192} \div \dfrac{x^2 + 10x + 16}{x^2 + 16x + 64}$

22. $\dfrac{3x - x^2}{x^3 - 27} \div \dfrac{x}{x^2 + 3x + 9}$

23. $\dfrac{\dfrac{5x^2 - 4x - 12}{3x^2 - 19x - 14}}{\dfrac{25x^2 + 35x + 6}{3x^2 + 17x + 10}}$

24. $\dfrac{8}{x} \div \dfrac{7xy}{x^4} \cdot \dfrac{5x^2}{x^7}$

25. $\dfrac{12x^2 + 8x + 1}{y^2 - 2y - 3} \cdot \dfrac{y^2 - 7y + 12}{6x^2 + 7x + 1} \div \dfrac{4x^2 - 4x - 3}{3x^2 - 2x - 5}$

26. $\dfrac{4a^2 - 64}{2a^2 - 4a} \div \dfrac{a^3 + 4a^2}{5a^2 - 10a} \cdot \dfrac{6a^3 + 4a^2}{4a^2 - 16a}$

27. $\dfrac{7}{x} + \dfrac{9}{x}$

28. $\dfrac{x + 3}{x - 4} + \dfrac{5x - 2}{x - 4}$

29. $\dfrac{x^2}{3x + 1} - \dfrac{4}{3x + 1}$

30. $\dfrac{6}{49x} + \dfrac{5}{7x^2}$

31. $\dfrac{9}{x - 3} + \dfrac{x}{3 - x}$

32. $\dfrac{8}{x - 1} - \dfrac{6}{x + 8}$

33. $\dfrac{x}{x + 3} + \dfrac{2x - 7}{x - 3}$

34. $\dfrac{x - 2}{x + 4} - \dfrac{x + 1}{x - 4}$

35. $\dfrac{x}{x^2 - 36} + \dfrac{4}{x}$

36. $\dfrac{7}{12x^2y} - \dfrac{13}{4xy}$

37. $\dfrac{10}{a + b} + \dfrac{10}{a - b}$

38. $\dfrac{z + 5}{z - 9} - \dfrac{z - 4}{z + 6}$

39. $\dfrac{9}{x - y} - \dfrac{x + 3y}{x^2 - y^2}$

40. $\dfrac{x - 8}{x^2 + 12x + 27} + \dfrac{x - 9}{x^2 - 9}$

41. $\dfrac{6x}{x^2 + 4x - 12} - \dfrac{x}{x^2 - 36}$

42. $\dfrac{y - 8}{2y^2 + 9y + 7} - \dfrac{y + 9}{2y^2 - 3y - 35}$

43. $\dfrac{5}{y} - \dfrac{5}{y - 3} + \dfrac{16}{(y - 3)^2}$

44. $\dfrac{x - 3}{x^2 + 12x + 36} + \dfrac{1}{x + 6} - \dfrac{2x + 3}{2x^2 + 3x - 54}$

In Exercises 45–54, simplify the complex rational expression using Method I or Method II.

45. $\dfrac{7 + \dfrac{1}{x}}{7 - \dfrac{1}{x}}$

46. $\dfrac{\dfrac{1}{x} + 3}{\dfrac{1}{x^2} - 9}$

47. $\dfrac{\dfrac{x + 2}{x - 6} - \dfrac{x + 12}{x + 5}}{x + 82}$

48. $\dfrac{\dfrac{6}{x} + \dfrac{5}{y}}{\dfrac{5}{x} - \dfrac{6}{y}}$

49. $\dfrac{\dfrac{1}{16} - \dfrac{1}{x^2}}{\dfrac{1}{4} + \dfrac{1}{x}}$

50. $\dfrac{\dfrac{8}{x + 4} - \dfrac{2}{x + 7}}{\dfrac{x + 8}{x + 4}}$

51. $\dfrac{\dfrac{2}{x + 13} + \dfrac{1}{x - 5}}{2 - \dfrac{x + 25}{x + 13}}$

52. $\dfrac{\dfrac{x - 5}{x + 5} + \dfrac{x - 5}{x - 9}}{1 + \dfrac{x + 5}{x - 9}}$

53. $\dfrac{1 + \dfrac{4}{x} - \dfrac{5}{x^2}}{1 - \dfrac{2}{x} - \dfrac{35}{x^2}}$

54. $\dfrac{6x^{-1} + 6y^{-1}}{xy^{-1} - x^{-1}y}$

Equations, Inequalities, and Applications

CHAPTER ONE CONTENTS

1.1   Linear Equations

THINGS TO KNOW

Before working through this section, be sure that you are familiar with the following concepts:

		VIDEO	ANIMATION	INTERACTIVE
You Try It	1. Factoring Trinomials with a Leading Coefficient Equal to 1 (Section R.5)	⊙		
You Try It	2. Factoring Trinomials with a Leading Coefficient Not Equal to 1 (Section R.5)	⊙	▭	

OBJECTIVES

1 Recognizing Linear Equations

2 Solving Linear Equations with Integer Coefficients

3 Solving Linear Equations Involving Fractions

4 Solving Linear Equations Involving Decimals

5 Solving Equations That Lead to Linear Equations

OBJECTIVE 1   RECOGNIZING LINEAR EQUATIONS

In the Review chapter, we manipulated algebraic expressions. Remember, algebraic expressions do *not* involve an equal sign. An **equation** indicates that two algebraic expressions are equal. For example, $2x + 6$ and $5x$ are two algebraic expressions. By equating these two expressions with an equal sign, we obtain the equation $2x + 6 = 5x$. This equation is an example of a **linear equation in one variable**. A linear equation in one variable involves only constants and variables that are raised to the first power. Examples of linear equations in one variable include the following:

$$7x - 4 = x - 5, \quad \frac{1}{3}y - \sqrt{2} = 11y, \quad \text{and} \quad 0.4a - 1 = 10$$

Definition   Linear Equation in One Variable

A linear equation in one variable is an equation that can be written in the form $ax + b = 0$ where a and b are real numbers and $a \neq 0$.

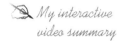
My interactive video summary

 Watch this interactive video to determine if equations are linear or nonlinear, or view this example of how to recognize linear and nonlinear equations.

You Try It   Work through this You Try It problem.

Work Exercises 1–7 in this textbook or in the MyMathLab Study Plan.

OBJECTIVE 2   SOLVING LINEAR EQUATIONS WITH INTEGER COEFFICIENTS

Let's now begin to solve linear equations. When we solve an equation for x, we are looking for all values of x that when substituted back into the original equation yield a true statement. For example, given the equation $2x - 1 = 5$, we see that if $x = 3$, the equation is a true statement. In other words, by substituting the number 3 into the equation for x, we get a true statement:

$$2x - 1 = 5 \qquad \text{Write the original equation.}$$
$$2(3) - 1 \overset{?}{=} 5 \qquad \text{Substitute 3 in for } x.$$
$$6 - 1 \overset{?}{=} 5 \qquad \text{Multiply.}$$
$$5 = 5 \qquad \text{True statement!}$$

We start by solving the most basic linear equations, which are linear equations with integer coefficients.

Example 1 Solve a Linear Equation with Integer Coefficients

Solve $5(x - 6) - 2x = 3 - (x + 1)$.

Solution   The goal here is to isolate the variable x on one side of the equation. First, use the distributive property to remove the parentheses on the left-hand side:

$$5(x - 6) - 2x = 3 - (x + 1) \qquad \text{Write the original equation.}$$
$$5x - 30 - 2x = 3 - x - 1 \qquad \text{Use the distributive property.}$$

$$3x - 30 = 2 - x \qquad \text{Combine like terms.}$$
$$3x - 30 + x = 2 - x + x \qquad \text{Add } x \text{ to both sides.}$$
$$4x - 30 = 2 \qquad \text{Simplify.}$$
$$4x - 30 + 30 = 2 + 30 \qquad \text{Add 30 to both sides.}$$
$$4x = 32 \qquad \text{Combine like terms.}$$
$$\frac{4x}{4} = \frac{32}{4} \qquad \text{Divide both sides by 4.}$$
$$x = 8 \qquad \text{Possible solution.}$$

You should now check by substituting $x = 8$ back into the original equation. Check your answer when finished.

Example 2 Solve a Linear Equation with Integer Coefficients

Solve $6 - 4(x + 4) = 8x - 2(3x + 5)$.

Solution To isolate the variable x, we first use the distributive property to remove all parentheses and then simplify:

$$6 - 4(x + 4) = 8x - 2(3x + 5) \qquad \text{Write the original equation.}$$
$$6 - 4x - 16 = 8x - 6x - 10 \qquad \text{Use the distributive property.}$$
$$-4x - 10 = 2x - 10 \qquad \text{Combine like terms.}$$
$$-4x - 10 + 4x = 2x - 10 + 4x \qquad \text{Add } 4x \text{ to both sides.}$$
$$-10 = 6x - 10 \qquad \text{Simplify.}$$
$$-10 + 10 = 6x - 10 + 10 \qquad \text{Add 10 to both sides.}$$
$$0 = 6x \qquad \text{Combine like terms.}$$
$$\frac{0}{6} = \frac{6x}{6} \qquad \text{Divide both sides by 6.}$$
$$0 = x \qquad \text{Possible solution.}$$

You should verify that $x = 0$ is the solution by substituting $x = 0$ back into the original equation.

It is important to point out that $x = 0$ can be a valid solution to an equation, as in Example 2. Once we encounter equations in which variables appear in one or more denominators, we have to be extremely careful when checking our answers because division by zero is never allowed! (See Example 6.)

You Try It Work through this You Try It problem.

Work Exercises 8–11 in this textbook or in the MyMathLab Study Plan.

OBJECTIVE 3 SOLVING LINEAR EQUATIONS INVOLVING FRACTIONS

To solve a linear equation involving fractions, we want to first transform the equation into a linear equation involving integer coefficients. We do this by multiplying both sides of the equation by the least common denominator (LCD).

My video summary ◎ **Example 3** Solve a Linear Equation Involving Fractions

Solve $\frac{1}{3}(1 - x) - \frac{x + 1}{2} = -2$.

Solution The first thing to do when solving equations involving fractions is to find the LCD. The LCD for this equation is **6**. We can eliminate the fractions by multiplying both sides of this equation by 6.

$$6\left(\frac{1}{3}(1 - x) - \frac{x + 1}{2}\right) = 6(-2) \qquad \text{Multiply both sides by 6.}$$

$$6\left(\frac{1}{3}(1 - x)\right) - 6\left(\frac{x + 1}{2}\right) = 6(-2) \qquad \text{Use the distributive property.}$$

$$2(1 - x) - 3(x + 1) = -12 \qquad \text{Multiply.}$$

Notice that the original equation containing fractions has now been transformed into a linear equation involving integer coefficients, as in Example 1. You should

My video summary ◎ solve this equation and/or watch the corresponding video to verify that the

solution set is $\left\{\dfrac{11}{5}\right\}$.　　　　　　　　　　　　　●

You Try It Work through this You Try It problem.

Work Exercises 12–18 in this textbook or in the MyMathLab **Study Plan**.

OBJECTIVE 4　SOLVING LINEAR EQUATIONS INVOLVING DECIMALS

The strategy for solving linear equations involving decimals is similar to the one used to solve linear equations involving fractions. We want to eliminate all decimals. We can eliminate decimals by multiplying both sides of the equation by the appropriate power of 10, such as $10^1 = 10$, $10^2 = 100$, $10^3 = 1,000$, etc. To determine the appropriate power of 10, look at the constants in the equation and choose the constant that has the greatest number of decimal places. Count those decimal places and then raise 10 to that power. Multiplying by that power of 10 will immediately eliminate all decimal places in the equation.

Example 4 Solve a Linear Equation Involving Decimals

Solve $.1(y - 2) + .03(y - 4) = .02(10)$.

My video summary ◎ **Solution** The constants .03 and .02 in the equation each have *two* decimal places. We can eliminate the decimals in this equation by multiplying both sides of the equation by $10^2 = 100$, thus moving each decimal two places to the right to obtain the new equation $10(y - 2) + 3(y - 4) = 2(10)$. Watch the video to see this problem solved in its entirety. You should verify that the solution set is $\{4\}$.　　●

You Try It Work through this You Try It problem.

Work Exercises 19–21 in this textbook or in the MyMathLab **Study Plan**.

OBJECTIVE 5   SOLVING EQUATIONS THAT LEAD TO LINEAR EQUATIONS

Some equations do not start out as linear equations but often lead to linear equations after simplification. Consider the following three examples.

Example 5 Solve an Equation That Leads to a Linear Equation

Solve $3a^2 - 1 = (a + 1)(3a + 2)$.

Solution The left side of this equation is not linear because the variable is raised to the second power. However, after multiplying the two binomials on the right-hand side and simplifying, we see that this equation does indeed simplify into a linear equation.

$$3a^2 - 1 = (a + 1)(3a + 2) \qquad \text{Write the original equation.}$$

$$3a^2 - 1 = 3a^2 + 5a + 2 \qquad \text{Multiply.}$$

$$3a^2 - 3a^2 - 1 = 3a^2 - 3a^2 + 5a + 2 \qquad \text{Subtract } 3a^2 \text{ from both sides.}$$

$$-1 = 5a + 2 \qquad \text{Combine like terms.}$$

$$-1 - 2 = 5a + 2 - 2 \qquad \text{Subtract 2 from both sides.}$$

$$-3 = 5a \qquad \text{Combine like terms.}$$

$$\frac{-3}{5} = \frac{5a}{5} \qquad \text{Divide both sides by 5.}$$

$$a = -\frac{3}{5} \qquad \text{Simplify.}$$

The solution set is $\left\{ -\dfrac{3}{5} \right\}$.

Example 6 Solve an Equation That Leads to a Linear Equation

Solve $\dfrac{2 - x}{x + 2} + 3 = \dfrac{4}{x + 2}$.

Solution This equation is nonlinear because there are variables in the denominator. This is an example of a *rational equation*. To solve this rational equation, first multiply both sides of the equation by the LCD, $x + 2$, to eliminate the denominators:

$$(x + 2)\left(\frac{2 - x}{x + 2} + 3 \right) = (x + 2)\left(\frac{4}{x + 2} \right) \qquad \text{Multiply both sides by } x + 2.$$

$$(x + 2)\left(\frac{2 - x}{x + 2} \right) + (x + 2)(3) = (x + 2)\left(\frac{4}{x + 2} \right) \qquad \text{Use the distributive property.}$$

$$2 - x + 3x + 6 = 4 \qquad \text{Multiply.}$$

$$2x + 8 = 4 \qquad \text{Combine like terms.}$$

$$2x = -4 \qquad \text{Subtract 8 from both sides.}$$

$$x = -2 \qquad \text{Divide both sides by 2.}$$

If we check by substituting $x = -2$ back into the original equation, we see that we obtain zeros in the denominators. Because division by zero is never permitted,

$x = -2$ is not a solution. The solution set to this equation is the empty set or $\varnothing$.
($x = -2$ is called an extraneous solution.)

> **Because rational equations often have extraneous solutions, *it is imperative to first determine all values that make any denominator equal to zero. Any solution that makes the denominator equal to zero must be discarded.***

 *My video summary*

⊙ **Example 7** Solve an Equation That Leads to a Linear Equation

Solve $\dfrac{12}{x^2 + x - 2} - \dfrac{x + 3}{x - 1} = \dfrac{1 - x}{x + 2}$.

Solution Watch and work through the video to see the solution.

You Try It Work through this You Try It problem.

Work Exercises 22–29 in this textbook or in the MyMathLab Study Plan.

1.1 Exercises

Skill Check Exercises

For exercises SCE-1 through SCE-6, determine the Least Common Denominator (LCD) of the given expression.

SCE-1. $\dfrac{2}{9} + \dfrac{1}{3} - \dfrac{1}{6}$

SCE-2. $\dfrac{1}{8}(2p - 1) - \dfrac{7}{3}p - \dfrac{p - 4}{6}$

SCE-3. $\dfrac{a - 3}{6} - \dfrac{3(a - 1)}{10} + \dfrac{2a + 1}{5}$

SCE-4. $\dfrac{3x}{x + 1} - \dfrac{5x + 7}{x - 1}$

SCE-5. $\dfrac{1}{2x} - \dfrac{1}{4} + \dfrac{6}{8x^2}$

SCE-6. $\dfrac{w}{w - 3} - \dfrac{2w}{2w - 1} - \dfrac{w - 3}{2w^2 - 7w + 3}$

SCE-7. $\dfrac{3}{x - 1} + \dfrac{4}{x + 1} - \dfrac{8x}{x^2 - 1}$

SCE-8. $\dfrac{6}{x^2 - x} - \dfrac{2}{x} + \dfrac{3}{x - 1}$

In Exercises 1–7, determine whether the given equation is linear or nonlinear.

1. $7x - \dfrac{1}{3} = 5$

2. $\dfrac{5}{x} + 4 = 10$

3. $\sqrt{2}x - 1 = 0$

4. $x^2 + x = 1$

5. $8x - 7 = \pi x - 3$

6. $\dfrac{5x - 1}{x + 2} = 3$

7. $5 - .2x = .1 - 4x$

In Exercises 8–29, solve each equation:

8. $5x - 8 = 3 - 7(x + 1)$

9. $4(2x + 5) = 10 - 2(x - 5)$

10. $3(3x + 4) = -(x - 3)$

11. $-3(2 - x) + 1 = 4 - (7x - 2)$

12. $\dfrac{1}{5} - \dfrac{1}{3}x = \dfrac{4}{3}$

13. $\dfrac{1}{2}x - 3 = 7 - \dfrac{3}{4}x$

14. $\dfrac{1}{2}y - \dfrac{1}{3}(y - 1) = 5y$

15. $\dfrac{1}{8}(2p - 1) = \dfrac{7}{3}p - \dfrac{p - 4}{6}$

16. $\dfrac{x + 5}{4} - \dfrac{x - 10}{5} = 2$

17. $\dfrac{x - 4}{2} - \dfrac{x + 1}{4} = \dfrac{2x - 3}{4}$

18. $\dfrac{a - 3}{6} - \dfrac{3(a - 1)}{10} = \dfrac{2a + 1}{5}$

19. $.12x + .3(x - 4) = .01(2x - 3)$

20. $.002(1 - k) + .01(k - 3) = 1$

21. $-.17x + .01(16x + 5) = .02(3 - x)$

22. $x(3x - 2) + x(5x - 1) = (4x + 1)(2x - 3)$

23. $(x + 3)^3 - 4 = x(x + 4)(x + 5) - 1$

24. $\dfrac{3x}{x - 4} + 1 = \dfrac{12}{x - 4}$

25. $\dfrac{w}{w - 3} - \dfrac{2w}{2w - 1} = \dfrac{w - 3}{2w^2 - 7w + 3}$

26. $\dfrac{3}{x - 1} + \dfrac{4}{x + 1} = \dfrac{8x}{x^2 - 1}$

27. $\dfrac{6}{x^2 - x} - \dfrac{2}{x} = \dfrac{3}{x - 1}$

28. $\dfrac{1}{x - 4} + \dfrac{2}{x - 2} = \dfrac{2}{x^2 - 6x + 8}$

29. $\dfrac{2x}{x^2 - 4} = \dfrac{4}{x^2 - 4} - \dfrac{1}{x + 2}$

1.2 Applications of Linear Equations

THINGS TO KNOW

Before working through this section, be sure that you are familiar with the following concepts:

VIDEO   ANIMATION   INTERACTIVE

You Try It

  1. Solving Linear Equations with Integer Coefficients (Section 1.1)

You Try It

  2. Solving Linear Equations Involving Fractions (Section 1.1)

You Try It

  3. Solving Linear Equations Involving Decimals (Section 1.1)

OBJECTIVES

1 Converting Verbal Statements into Mathematical Statements

2 Solving Applications Involving Unknown Numeric Quantities

3 Solving Applications Involving Decimal Equations (Money, Mixture, Interest)

4 Solving Applied Problems Involving Distance, Rate, and Time

5 Solving Applied Working Together Problems

Introduction to Section 1.2

We find algebraic applications in all sorts of subjects, including economics, physics, medicine, and countless other disciplines. In this section, we learn how to solve a variety of useful applications. Before we look at some specific examples, it is important to first practice translating or converting verbal statements into mathematical statements.

OBJECTIVE 1   CONVERTING VERBAL STATEMENTS INTO MATHEMATICAL STATEMENTS

When solving word problems, it is important to recognize key words and phrases that translate into algebraic expressions involving addition, subtraction, multiplication, and division.

Example 1 Convert Verbal Statements into Mathematical Statements

Rewrite each statement as an algebraic expression or equation:

a. 7 more than three times a number

b. 5 less than twice a number

c. Three times the quotient of a number and 11

d. The sum of a number and 9 is 1 less than half of the number.

e. The product of a number and 4 is 1 more than 8 times the difference of 10 and the number.

Solution

a. If x is the number, then three times a number is $3x$. Therefore, 7 more than three times a number is equivalent to $3x + 7$ (or $7 + 3x$).

b. Twice a number is equivalent to $2x$. So, 5 less than twice a number is written as $2x - 5$.

⚠ **The expression $2x - 5$ is not equivalent to $5 - 2x$ because subtraction is not commutative; that is, $a - b \neq b - a$.**

c. The quotient of a number and 11 is equivalent to $\dfrac{x}{11}$. So, three times the quotient of a number and 11 is equivalent to $3\left(\dfrac{x}{11}\right)$ or $\dfrac{3x}{11}$.

d. Notice the key word "*is.*" The word "*is*" often translates into an equal sign, thus indicating that we need to translate the phrase into an **equation**. In this case, the sum of a number and 9 translates to $x + 9$. The statement "1 less than half of the number" translates to $\frac{1}{2}x - 1$. Putting it all together, we get the following equation:

$$\underbrace{x + 9}_{\substack{\text{The sum of} \\ \text{a number} \\ \text{and } 9}} \;\; \underset{\text{is}}{=} \;\; \underbrace{\frac{1}{2}x - 1}_{\substack{\text{one less} \\ \text{than half} \\ \text{the number}}}$$

e. This statement translates into the following equation:

$$\underbrace{4x}_{\substack{\text{The product} \\ \text{of a number} \\ \text{and } 4}} \;\; \underset{\text{is}}{=} \;\; \underbrace{8(10 - x) + 1}_{\substack{\text{one more than} \\ \text{8 times the difference} \\ \text{of 10 and the number}}}$$

You Try It   Work through this You Try It problem.

Work Exercises 1–5 in this textbook or in the My MathLab **Study Plan.**

It is important to have a strategy in place before trying to solve an applied problem. George Polya, a Hungarian mathematician who taught at Stanford University for many years, was famous for his book *How to Solve It* (Princeton University Press). The Mathematics Lab at the University of Idaho was named The Polya Mathematics Learning Center in honor of George Polya. His problem-solving model can be summarized into four basic steps.

Polya's Guidelines for Problem Solving

1. Understand the problem.

2. Devise a plan.

3. Carry out the plan.

4. Look back.

Following is a four-step strategy for solving applied problems with Polya's problem-solving guidelines in mind.

FOUR-STEP STRATEGY FOR PROBLEM SOLVING

Step 1. Read the problem several times until you have an understanding of what is being asked. If possible, create diagrams, charts, or tables to assist you in your understanding.

⎫
⎬ Understand the
⎭ problem.

Step 2. Pick a variable that describes the unknown quantity that is to be found. All other quantities must be written in terms of that variable. Write an equation using the given information and the variable.

⎫
⎬ Devise a plan.
⎭

Step 3. Carefully solve the equation.

⎬ Carry out the plan.

Step 4. Make sure that you have answered the question
and then check all answers to make sure they } Look back.
make sense.

OBJECTIVE 2 SOLVING APPLICATIONS INVOLVING UNKNOWN NUMERIC QUANTITIES

Example 2 Number of Touchdowns Thrown

Roger Staubach and Terry Bradshaw were both quarterbacks in the National Football League. In 1973, Staubach threw three touchdown passes more than twice the number of touchdown passes thrown by Bradshaw. If the total number of touchdown passes between Staubach and Bradshaw was 33, how many touchdown passes did each player throw?

Solution

Step 1. After carefully reading the problem, we see that we are trying to figure out how many touchdown passes were thrown by each player.

Step 2. Let B be the number of touchdown passes thrown by Bradshaw. Because Staubach threw 3 more than twice the number of touchdowns thrown by Bradshaw, then $2B + 3$ represents the number of touchdown passes thrown by Staubach.

The sum of the number of touchdown passes is 33, so we can write the following equation:

$$\underbrace{\text{Bradshaw's touchdown passes}}_{B} + \underbrace{\text{Staubach's touchdown passes}}_{(2B\ +\ 3)} = \underbrace{\text{Total}}_{33}$$

Step 3. Solve:

$B + (2B + 3) = 33$	Write the equation.
$3B + 3 = 33$	Combine like terms.
$3B = 30$	Subtract 3 from both sides.
$B = 10$	Divide both sides by 3.

To answer the question, Bradshaw threw 10 touchdown passes in 1973, and Staubach threw $2(B) + 3 = 2(10) + 3 = 23$ touchdown passes in 1973.

Step 4. Check: We see that the total number of touchdown passes is $10 + 23 = 33$. The number of touchdown passes thrown by Staubach is 3 more than two times the number of touchdown passes thrown by Bradshaw.

You Try It Work through this You Try It problem.

Work Exercises 6–9 in this textbook or in the My MathLab **Study Plan.**

OBJECTIVE 3 SOLVING APPLICATIONS INVOLVING DECIMAL EQUATIONS (MONEY, MIXTURE, INTEREST)

In Example 2, the equation turned out to be a linear equation involving integer coefficients. In the next two examples, we see that the equations turn out to be linear equations involving decimals. You may want to review the technique used to solve linear equations involving decimals that was covered in Section 1.1.

 *My video summary*

⊗ Example 3 Money Application

Billy has $16.50 in his piggy bank, consisting of nickels, dimes, and quarters. Billy notices that he has 20 fewer quarters than dimes. If the number of nickels is equal to the number of quarters and dimes combined, how many of each coin does Billy have?

Solution

Step 1. We must find out how many nickels, dimes, and quarters Billy has.

Step 2. Let d = number of dimes.

The remaining quantities must also be expressed in terms of the variable d. Because Billy has 20 fewer quarters than dimes, we can express the number of quarters in terms of the number of dimes, or

$$d - 20 = \text{number of quarters.}$$

 The expression $d - 20$ is not equivalent to $20 - d$ because subtraction is not commutative, that is, $a - b \neq b - a$.

The number of nickels is equal to the sum of the number of dimes and quarters.

$$\underbrace{d}_{\substack{\text{number} \\ \text{of dimes}}} + \underbrace{(d - 20)}_{\substack{\text{number} \\ \text{of quarters}}} = \text{number of nickels}$$

Nickels are worth $.05, dimes are worth $.10, quarters are worth $.25, and the total amount is $16.50. So, we get the following equation:

$$\underbrace{.05(d + (d - 20))}_{\substack{\text{value of} \\ \text{the nickels}}} + \underbrace{.10d}_{\substack{\text{value of the} \\ \text{dimes}}} + \underbrace{.25(d - 20)}_{\substack{\text{value of} \\ \text{the quarters}}} = \underbrace{16.50}_{\text{Total}}$$

 *My video summary*

⊗ Watch the video to see the rest of the solution, or solve the equation yourself to verify that Billy has 80 nickels, 50 dimes, and 30 quarters. ●

You Try It Work through this **You Try It** problem.

Work Exercises 10 and 11 in this textbook or in the My MathLab **Study Plan.**

 *My video summary*

⊗ Example 4 Mixture Application

How many milliliters of a 70% acid solution must be mixed with 30 mL of a 40% acid solution to obtain a mixture that is 50% acid?

Solution

Step 1. We are mixing two solutions together to obtain a third solution.

Step 2. The unknown quantity in this problem is the amount (in mL) of a 70% acid solution. So, let x = amount of a 70% acid solution. A diagram may help set up the required equation. See Figure 1.

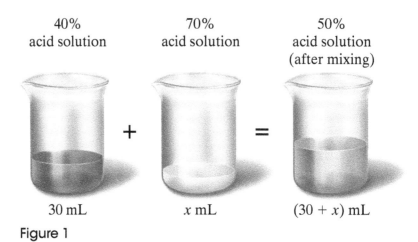

40%
acid solution

70%
acid solution

50%
acid solution
(after mixing)

30 mL

x mL

(30 + x) mL

Figure 1

Because we are mixing 30 mL of one solution with x mL of another, the resulting total quantity will be $(x + 30)$ mL, as in Figure 1. To set up an equation, notice that the number of milliliters of pure acid in the 40% solution plus the number of milliliters of pure acid in the 70% solution must equal the number of milliliters of pure acid in the 50% solution. The equation is as follows:

$$\underbrace{.40(30)}_{\substack{\text{amount of pure} \\ \text{acid in the 40\%} \\ \text{container}}} + \underbrace{.70x}_{\substack{\text{amount of pure} \\ \text{acid in the 70\%} \\ \text{container}}} = \underbrace{.50(30 + x)}_{\substack{\text{amount of pure} \\ \text{acid in the 50\%} \\ \text{container}}}$$

My video summary

Step 3. Solving this linear equation involving decimals, we get $x = 15$ mL. (Watch the video to see the problem solved in its entirety.) Therefore, we must mix 15 mL of 70% solution to reach the desired mixture.

You Try It   Work through this You Try It problem.

Work Exercises 12–15 in this textbook or in the MyMathLab Study Plan.

My video summary

Example 5 Interest Application

Kristen inherited $20,000 from her Aunt Dawn Ann, with the stipulation that she invest part of the money in an account paying 4.5% simple interest and the rest in an account paying 6% simple interest locked in for 3 years. If at the end of 1 year, the total interest earned was $982.50, how much was invested at each rate?

Solution Watch this video to see that Kristen invested $14,500 at 4.5% and $5,500 at 6%.

You Try It   Work through this You Try It problem.

Work Exercises 16 and 17 in this textbook or in the MyMathLab Study Plan.

OBJECTIVE 4   SOLVING APPLIED PROBLEMS INVOLVING DISTANCE, RATE, AND TIME

In Example 6, we use the formula **distance = rate × time**. It is important to draw a picture of the situation and then create a table to organize the information. Carefully work through Example 6 by watching the corresponding video.

My video summary

⊘ Example 6 Distance, Rate, and Time

Rick left his house on his scooter at 9:00 AM to go fishing. He rode his scooter at an average speed of 10 mph. At 9:15 AM, his girlfriend Deb (who did not find Rick at home) pedaled after Rick on her new 10-speed bicycle at a rate of 15 mph. If Deb caught up with Rick at precisely the time they both reached the fishing hole, how far is it from Rick's house to the fishing hole? At what time did Rick and Deb arrive at the fishing hole?

Solution It is 7.5 miles to the fishing hole from Rick's house. Deb and Rick arrived at the fishing hole at 9:45 AM. Watch the video to see this problem worked out in detail.

Example 7 Distance, Rate, and Time

An airplane that can maintain an average velocity of 320 mph in still air is transporting smokejumpers to a forest fire. On takeoff from the airport, it encounters a headwind and takes 34 minutes to reach the jump site. The return trip from the jump site takes 30 minutes. What is the speed of the wind? How far is it from the airport to the fire?

Solution

Step 1. We are asked to find the speed of the wind, and then to find the distance from the airport to the forest fire.

Step 2. Let w = speed of the wind.

The plane flies 320 mph in still air (with no wind). Therefore, when the plane encounters a headwind, the plane's rate will decrease by the amount of the wind. So, the net rate (speed) of the plane into a head-wind can be described as $320 - w$. Similarly, the net speed of the plane flying with the tailwind is $320 + w$. We know that the time, in hours, is $\frac{34}{60} = \frac{17}{30}$ hours into the headwind and $\frac{30}{60} = \frac{1}{2}$ hour with the tailwind. Knowing that distance = rate × time, we now create the following table.

	Rate	Time (hours)	Distance
Headwind	$320 - w$	$\dfrac{17}{30}$	$\dfrac{17}{30}(320 - w)$
Tailwind	$320 + w$	$\dfrac{1}{2}$	$\dfrac{1}{2}(320 + w)$

Because both distances are the same, we can equate the two distances:

$$\underbrace{\frac{17}{30}(320 - w)}_{\substack{\text{distance traveled} \\ \text{by the plane into} \\ \text{the headwind}}} = \underbrace{\frac{1}{2}(320 + w)}_{\substack{\text{distance traveled} \\ \text{by the plane with} \\ \text{the tailwind}}}$$

Step 3. Solve: $\dfrac{17}{30}(320 - w) = \dfrac{1}{2}(320 + w)$ Rewrite the equation.

$$17(320 - w) = 15(320 + w)$$ Multiply both sides by 30.

$$5440 - 17w = 4800 + 15w$$ Use the distributive property.

$$640 = 32w$$ Combine like terms.

$$20 = w$$ Divide by 32.

Step 4. To answer the question, the wind is blowing at a speed of 20 mph. We are also asked to find the distance from the airport to the fire. We can substitute $w = 20$ into either distance expression. Substituting $w = 20$ into the distance traveled with the tailwind expression, we see that the distance from the airport to the fire is as follows:

$$\frac{1}{2}(320 + 20) = \frac{1}{2}(340) = 170 \text{ mi}$$

You Try It   Work through this You Try It problem.

Work Exercises 18–21 in this textbook or in the MyMathLab Study Plan.

OBJECTIVE 5   SOLVING APPLIED WORKING TOGETHER PROBLEMS

Example 8 Working Together Application

Brad and Michelle decide to paint the entire upstairs of their new house. Brad can do the job by himself in 8 hours. If it took them 3 hours to paint the upstairs together, how long would it have taken Michelle to paint it by herself?

Solution

Step 1. We are asked to find the time that it would take Michelle to do the job by herself.

Step 2. Let $t =$ time it takes for Michelle to complete the job; thus, she can complete $\dfrac{1}{t}$ of the job in an hour. Brad can do this job in 8 hours by himself, so he can complete $\dfrac{1}{8}$ of the job per hour. Working together it took them 3 hours to complete the job. We can fill out the following table:

	Time needed to complete the job in hours	Portion of job completed in 1 hour (rate)
Brad	8	$\dfrac{1}{8}$
Michelle	t	$\dfrac{1}{t}$
Together	3	$\dfrac{1}{3}$

Step 3. There are two methods that we can use to solve this problem:

Method 1: From the table, we see that $\frac{1}{3}$ = portion of the job completed in 1 hour together. Adding Brad and Michelle's rate together, $\left(\frac{1}{8} + \frac{1}{t}\right)$, we get another expression describing the portion of the job completed together in 1 hour. Thus, equating these rates we obtain the following equation:

$$\left(\frac{1}{8} + \frac{1}{t}\right) = \frac{1}{3}$$

To solve for t, we first multiply by the LCD of $24t$ to clear the fractions:

$$\left(\frac{1}{8} + \frac{1}{t}\right) = \frac{1}{3} \qquad \text{Rewrite the equation.}$$

$$24t\left(\frac{1}{8} + \frac{1}{t}\right) = \left(\frac{1}{3}\right)24t \qquad \text{Multiply both sides by } 24t.$$

$$24t \cdot \frac{1}{8} + 24t \cdot \frac{1}{t} = \frac{1}{3} \cdot 24t \qquad \text{Use the distributive property.}$$

$$3t + 24 = 8t \qquad \text{Simplify.}$$

$$24 = 5t \qquad \text{Subtract } 3t \text{ from both sides.}$$

$$\frac{24}{5} = t \qquad \text{Divide both sides by 5.}$$

So, $t = \frac{24}{5}$ hours or $4\frac{4}{5}$ hours = 4 hours and 48 minutes $\left(\frac{4}{5}\text{ hour} \times \frac{60\text{ minutes}}{1\text{ hour}} = 48\text{ minutes}\right)$. Therefore, it would take Michelle 4 hours and 48 minutes to paint the upstairs by herself.

Method 2: Another method is based on the fact that *(time working on job)* × *(rate to complete job)* = *fraction of job completed*.

In this case, it took Brad and Michelle 3 hours to complete the entire job together. Thus, the fraction of the job completed is equal to 1. The rate to complete the job together was $\left(\frac{1}{8} + \frac{1}{t}\right)$.

Therefore, we can set up the following equation:

$$\underbrace{3}_{\substack{\text{time to}\\\text{complete}\\\text{the job}}} \cdot \underbrace{\left(\frac{1}{8} + \frac{1}{t}\right)}_{\substack{\text{rate to}\\\text{complete}\\\text{job}}} = \underbrace{1}_{\substack{\text{complete}\\\text{job}}}$$

You should verify that $t = \frac{24}{5}$.

Step 4. The solution makes sense because Michelle's time must be less than Brad's time. Also, we can substitute $t = \dfrac{24}{5}$ into the Method 1 or Method 2 equation to see that a true statement results.

You Try It   Work through this You Try It problem.

Work Exercises 22 and 23 in this textbook or in the MyMathLab Study Plan.

 *My video summary*   **⊙ Example 9** Working Together Application

Jim and Earl were replacing the transmission on Earl's old convertible. Earl could replace the transmission by himself in 8 hours, whereas it would take Jim 6 hours to do the same job. They worked together for 2 hours, but then Jim had to go to his job at the grocery store. How long did it take Earl to finish replacing the transmission by himself?

Solution Using Method 2 as in Example 8, it would take Earl 3 hours and 20 minutes to finish replacing the transmission. Watch this video to see the entire solution.

You Try It   Work through this You Try It problem.

Work Exercise 24 in this textbook or in the MyMathLab Study Plan.

1.2 Exercises

In Exercises 1–5, write the corresponding algebraic expression or equation for each verbal statement. Let n represent the unknown number.

1. 10 more than twice a number

2. 5 less than three times a number and 6

3. The quotient of three and four times a number

4. The product of a number and 2 is 3 less than the quotient of a number and 4.

5. Three less than four times the quotient of a number and 7 is equal to 8 less than twice the number.

In Exercises 6–24, solve the problems algebraically. Clearly define all variables, write an appropriate equation, and solve.

6. One number is 5 more than twice the other number. If the sum of the two numbers is 26, find the two numbers.

7. The perimeter of a rectangular garden is 32 feet. The length of the garden is 4 feet less than three times the width. Find the dimensions of the garden.

8. The sum of three consecutive even integers is 42. Find the integers.

9. Together, Steve and Tom sold 121 raffle tickets for their school. Steve sold 1 more than twice as many raffle tickets as Tom. How many raffle tickets did each boy sell?

10. Emily has $18.75, consisting of only dimes, quarters, and silver dollars. She has seven times as many dimes as silver dollars and 6 more quarters than dimes. How many of each coin does she have?

11. Shanika has invested in three different stocks. Stock A is currently worth $4.50 per share. Stock B is worth $8.00 per share, while stock C is worth $10.00 per share. Shanika purchased half as many shares of stock B as stock A and half as many shares of stock C as stock B. If her total investments are currently worth $704, how many shares of each stock does she own?

12. A coffee shop owner decides to blend a gourmet brand of coffee with a cheaper brand. The gourmet coffee usually sells for $9.00/lb. The cheaper brand sells for $5.00/lb. How much of each type should he mix in order to have 30 lb of coffee that is worth $6.50/lb?

13. How much of an 80% orange juice drink must be mixed with 20 gallons of a 20% orange juice drink to obtain a mixture that is 50% orange juice?

14. How many liters of pure water should be mixed with a 5-L solution of 80% acid to produce a mixture that is 70% water?

15. Ben recently moved from southern California to northern Alaska. After consulting with a local mechanic, Ben realized that his car radiator fluid should consist of 70% antifreeze. Currently, his 4.5-L radiator is full, with a concentration of 40% antifreeze. How much coolant should be drained and filled with pure antifreeze to reach the desired concentration?

16. A woman has $28,000 to invest in two investments that pay simple interest. One investment pays 4% simple annual interest, and the other pays 5% simple annual interest. How much would she have to invest in each investment if the total interest earned is to be $1,320?

17. Mark works strictly on commission of his gross sales from selling two different products for his company. Last month, his gross sales were $82,000. If he earns 6% commission on product A and 5% commission on product B, what were his gross sales for each product if he earned $4,310 in total commission?

18. A freight train leaves the train station 2 hours before a passenger train. The two trains are traveling in the same direction on parallel tracks. If the rate of the passenger train is 20 mph faster than the freight train, how fast is each train traveling if the passenger train passes the freight train in 3 hours?

19. Manuel traveled 31 hours nonstop to Mexico, a total of 2,300 miles. He took a train part of the way, which averaged 80 mph, and then took a bus the remaining distance, which averaged 60 mph. How long was Manuel on the train?

20. A homing pigeon left his coup to deliver a message. On takeoff, the pigeon encountered a tailwind of 10 mph and made the delivery in 15 minutes. The return trip into the wind took 30 minutes. How fast could the pigeon fly in still air?

21. A grain barge travels on a river from point A to point B loading and unloading grain. The barge travels at a rate of 5 mph relative to the water. The river flows downstream at a rate of 1 mph. If the trip upstream takes 2 hours longer than the trip downstream, how far is it from point A to point B?

22. Tommy can paint a fence in 5 hours. If his friend Huck helps, they can paint the fence in 1 hour. How fast could Huck paint the fence by himself?

23. It takes Joan three times longer than Jane to file the reports. Together, they can file the reports in 15 minutes. How long would it take each woman to file the reports by herself?

24. Two pumps were required to pump the water out of Lakeview in New Orleans after Hurricane Katrina. Pump A, the larger of the two pumps, can pump the water out in 24 hours, whereas it would take pump B 72 hours. Both pumps were working for the first 8 hours until pump A broke down. How long did it take pump B to pump out the remaining water?

1.3 Complex Numbers

THINGS TO KNOW

Before working through this section, be sure that you are familiar with the following concepts:

VIDEO   ANIMATION   INTERACTIVE

You Try It

1. Simplifying Radicals
   (Section R.3)

OBJECTIVES

1 Simplifying Powers of i

2 Adding and Subtracting Complex Numbers

3 Multiplying Complex Numbers

4 Finding the Quotient of Complex Numbers

5 Simplifying Radicals with Negative Radicands

....

Introduction to Section 1.3

 *My video summary*

⊘ **THE IMAGINARY UNIT**

It is easy to verify that $x = -1$ is the solution to the equation $x + 1 = 0$. Now consider the equation $x^2 + 1 = 0$. Is $x = -1$ a solution to this equation? To check, let's substitute $x = -1$ into the equation $x^2 + 1 = 0$ to see if we get a true statement:

$$x^2 + 1 = 0 \qquad \text{Write the original equation.}$$

$$(-1)^2 + 1 \overset{?}{=} 0 \qquad \text{Substitute } x = -1.$$

$$1 + 1 \neq 0 \qquad \text{The left-hand side does not equal the right-hand side.}$$

In fact, the equation $x^2 + 1 = 0$ has *no real solution* because any real number squared added to one is greater than zero. Does this equation have a solution? If so, what is the solution? To answer this question, suppose that the number i is a solution to the equation $x^2 + 1 = 0$. In other words, $i^2 + 1 = 0$ or $i^2 = -1$. This number i, which is a solution to the equation $x^2 + 1 = 0$, is known as the **imaginary unit**.

> **Definition** Imaginary Unit
>
> The imaginary unit, i, is defined by $i = \sqrt{-1}$, or equivalently, $i^2 = -1$.

OBJECTIVE 1   SIMPLIFYING POWERS OF i

Given the previous definition, we have the following:

$$i = \sqrt{-1} \qquad\qquad i^5 = i^4 \cdot i = (1) \cdot i = i$$
$$i^2 = (\sqrt{-1})^2 = -1 \qquad\qquad i^6 = i^4 \cdot i^2 = (1) \cdot (-1) = -1$$
$$i^3 = i^2 \cdot i = (-1) \cdot i = -i \qquad\qquad i^7 = i^4 \cdot i^3 = (1) \cdot (-i) = -i$$
$$i^4 = i^2 \cdot i^2 = (-1) \cdot (-1) = 1 \qquad\qquad i^8 = i^4 \cdot i^4 = (1) \cdot (1) = 1$$
$$\vdots$$

You can see the cyclic nature of i. After every fourth power, the cycle repeats itself. In fact, every integer power of i can be written as $i, -1, -i$, or 1.

Example 1 Simplify Powers of i

Simplify each of the following:

a. i^{43} b. i^{100} c. i^{-21}

Solution Because $i^4 = 1$, we factor out the largest fourth power of i.

a. $i^{43} = (i^4)^{10} \cdot i^3 = 1^{10} \cdot i^3 = i^3 = -i$

b. $i^{100} = (i^4)^{25} = 1^{25} = 1$

c. $i^{-21} = \dfrac{1}{i^{21}} = \dfrac{1}{(i^4)^5 \cdot i} = \dfrac{1}{(1)^5 \cdot i} = \dfrac{1}{i}$

It appears that we are done and $i^{-21} = \dfrac{1}{i}$. However, if we multiply this expression

by $\dfrac{i}{i}$, we get the following:

$$i^{-21} = \frac{1}{i} \cdot \frac{i}{i} = \frac{i}{i^2} = \frac{i}{-1} = -i$$

Therefore, $i^{-21} = -i$.

You Try It Work through this You Try It problem.

Work Exercises 1–5 in this textbook or in the MyMathLab Study Plan.

My video summary ⊙ **COMPLEX NUMBERS**

Now that we have defined the imaginary unit, it is time to introduce a new type of number known as a **complex number**.

> **Definition** Complex Number
>
> A complex number is a number that can be written in the form $a + bi$, where a and b are real numbers. We often use the variable z to denote a complex number.

If $z = a + bi$ is a complex number, then a is called the **real part**, and b is called the **imaginary part**. For example, $z = \sqrt{3} - 2i$ is a complex number, where $a = \sqrt{3}$ is the real part and $b = -2$ is the imaginary part. The number 5 is also a complex number because 5 can be written as $5 + 0i$. Therefore, every real number is also a complex number.

OBJECTIVE 2   ADDING AND SUBTRACTING COMPLEX NUMBERS

To add or subtract complex numbers, simply combine the real parts and combine the imaginary parts.

 *My video summary*

> **Example 2** Adding and Subtracting Complex Numbers

Perform the indicated operations:

a. $(7 - 5i) + (-2 + i)$ **b.** $(7 - 5i) - (-2 + i)$

Solution

a. $(7 - 5i) + (-2 + i)$ Write the original expression.

$= 7 - 5i - 2 + i$ Remove parentheses.

$= 5 - 4i$ Combine the real parts and the imaginary parts.

b. $(7 - 5i) - (-2 + i)$ Write the original expression.

$= 7 - 5i + 2 - i$ Remove the parentheses $[-(-2 + i) = 2 - i]$.

$= 9 - 6i$ Combine the real parts and the imaginary parts.

You Try It   Work through this You Try It problem.

Work Exercises 6–9 in this textbook or in the My MathLab **Study Plan.**

OBJECTIVE 3   MULTIPLYING COMPLEX NUMBERS

When multiplying two complex numbers, treat the problem as if it were the multiplication of two binomials. Just remember that $i^2 = -1$.

 *My video summary*

> **Example 3** Multiplying Complex Numbers

Multiply $(4 - 3i)(7 + 5i)$.

Solution Treating this multiplication as if it were the multiplication of two binomials, we get the following:

$(4 - 3i)(7 + 5i) = 4(7) + 4(5i) - (3i)(7) - (3i)(5i)$ Use the distributive property.

$= 28 + 20i - 21i - 15i^2$ Multiply.

$$= 28 - i - 15(-1)$$

$$= 28 - i + 15$$

$$= 43 - i$$

$20i - 21i = -i$ and $i^2 = -1$.

$-15(-1) = 15$.

Combine the real parts.

 *My video summary* ⊘ **Example 4** Squaring a Complex Number

Simplify $(\sqrt{3} - 5i)^2$.

Solution You should verify that the answer is $-22 - 10i\sqrt{3}$. Watch the video to see the worked out solution.

We now define the **complex conjugate**.

Definition   Complex Conjugate

The complex conjugate of a complex number $z = a + bi$ is denoted as $\bar{z} = \overline{a + bi} = a - bi$.

For example, the complex conjugate of $z = -2 - 7i$ is $\bar{z} = -2 - (-7i) = -2 + 7i$.

Example 5 Multiplying a Complex Number by Its Complex Conjugate

Multiply the complex number $z = -2 - 7i$ by its complex conjugate $\bar{z} = -2 + 7i$.

Solution

$$z\bar{z} = (-2 - 7i)(-2 + 7i) \qquad \text{Multiply.}$$

$$= 4 - 14i + 14i - 49i^2$$

$$= 4 - 49(-1) \qquad \text{Simplify and } i^2 = -1.$$

$$= 4 + 49 = 53 \qquad \text{Simplify.}$$

Thus, $(-2 - 7i)(-2 + 7i) = 53$.

Notice in Example 5 that the product of $z = -2 - 7i$ and its complex conjugate $\bar{z} = -2 + 7i$ is a real number. In fact, every complex number when multiplied by its complex conjugate will yield a real number. The following theorem is worth remembering:

Theorem

The product of a complex number $z = a + bi$ and its conjugate $\bar{z} = a - bi$ is the real number $a^2 + b^2$. In symbols, $z\bar{z} = a^2 + b^2$.

To see why this is true, read the proof of this theorem.

You Try It   Work through this You Try It problem.

Work Exercises 10–19 in this textbook or in the MyMathLab Study Plan.

OBJECTIVE 4   FINDING THE QUOTIENT OF COMPLEX NUMBERS

My video summary   ⊙ The goal when dividing two complex numbers is to eliminate the imaginary part from the denominator and express the quotient in the standard form of $a + bi$. We can do this by multiplying the numerator and denominator by the complex conjugate of the denominator.

Example 6 The Quotient of Complex Numbers

Write the quotient in the form $a + bi$:

$$\frac{1 - 3i}{5 - 2i}$$

Solution

Multiply numerator and denominator by the complex conjugate.

$$\frac{1 - 3i}{5 - 2i} \cdot \frac{5 + 2i}{5 + 2i}$$

Note: $\dfrac{5 + 2i}{5 + 2i} = 1$.

$$= \frac{5 - 13i - 6i^2}{(5)^2 + (2)^2}$$

Multiply the numerators and the denominators.
Note: $(a + bi)(a - bi) = a^2 + b^2$.

$$= \frac{5 - 13i - 6(-1)}{29}$$

Simplify denominator, and replace i^2 with -1.

$$= \frac{11 - 13i}{29}$$

Simplify.

$$= \frac{11}{29} - \frac{13}{29}i$$

Write in the form $a + bi$.

You Try It   Work through this You Try It problem.

Work Exercises 20–23 in this textbook or in the My MathLab **Study Plan.**

OBJECTIVE 5   SIMPLIFYING RADICALS WITH NEGATIVE RADICANDS

In Section 1.4, we solve equations that often lead to solutions with a negative number inside of a radical. Therefore, it is important to learn how to simplify a radical that contains a negative radicand such as $\sqrt{-49}$. We simplify such radicals using the following property, remembering that $\sqrt{-1} = i$.

Property

If M is a positive real number, then $\sqrt{-M} = \sqrt{-1} \cdot \sqrt{M} = i\sqrt{M}$.

Example 7 Simplify a Square Root with a Negative Radicand

Simplify $\sqrt{-108}$.

Solution

$$\sqrt{-108} = \sqrt{-1} \cdot \sqrt{108} = i\sqrt{108} = 6i\sqrt{3}$$

Thus, $\sqrt{-108} = 6i\sqrt{3}$. To review simplifying radicals, refer back to Section R.3.

 The property $\sqrt{a} \cdot \sqrt{b} = \sqrt{ab}$ is only true for real numbers $\sqrt{a}$ and $\sqrt{b}$. The expressions $\sqrt{a}$ and $\sqrt{b}$ are real numbers when both a and b are greater than or equal to zero. This property does not apply for nonreal numbers. Attempting to use this property with square roots of negative numbers will lead to false results:

$$\sqrt{-3} \cdot \sqrt{-12} = \sqrt{(-3) \cdot (-12)} = \sqrt{36} = 6 \qquad \text{False!}$$

$$\sqrt{-3} \cdot \sqrt{-12} = i\sqrt{3} \cdot i\sqrt{12} = i^2\sqrt{36} = -6 \qquad \text{True!}$$

Example 8 Simplifying Expressions That Contain Negative Radicands

Simplify the following expressions:

a. $\sqrt{-8} + \sqrt{-18}$

b. $\sqrt{-8} \cdot \sqrt{-18}$

c. $\dfrac{-6 + \sqrt{(-6)^2 - 4(2)(5)}}{2}$

d. $\dfrac{4 \pm \sqrt{-12}}{4}$

Solution

a. $\sqrt{-8} + \sqrt{-18} = 2i\sqrt{2} + 3i\sqrt{2} = 5i\sqrt{2}$

b. $\sqrt{-8} \cdot \sqrt{-18} = (2i\sqrt{2}) \cdot (3i\sqrt{2}) = 6i^2(2) = -12$

$$\text{or}$$

$$\sqrt{-8} \cdot \sqrt{-18} = (i\sqrt{8}) \cdot (i\sqrt{18}) = i^2\sqrt{144} = -12$$

c. $\dfrac{-6 + \sqrt{(-6)^2 - 4(2)(5)}}{2} = \dfrac{-6 + \sqrt{36 - 40}}{2} = \dfrac{-6 + \sqrt{-4}}{2}$

$$= \dfrac{-6 + 2i}{2} = \dfrac{\cancel{2}(-3 + i)}{\cancel{2}} = -3 + i$$

d. $\dfrac{4 \pm \sqrt{-12}}{4} = \dfrac{4 \pm 2i\sqrt{3}}{4} = \dfrac{2(2 \pm i\sqrt{3})}{4} = \dfrac{2(2 \pm i\sqrt{3})}{2 \cdot 2} = \dfrac{2 \pm i\sqrt{3}}{2}$

You Try It Work through this You Try It problem.

Work Exercises 24–31 in this textbook or in the MyMathLab Study Plan.

1.3 Exercises

Skill Check Exercises

For exercises SCE-1 through SCE-6, simplify the given expression.

SCE-1. $\left(\sqrt{2} + 3\right) + (5\sqrt{2} - 5)$ **SCE-2.** $\left(5\sqrt{3} - 9\right) - \left(-4\sqrt{3} - 13\right)$ **SCE-3.** $\sqrt{20}$

SCE-4. $\sqrt{108}$ **SCE-5.** $\left(\sqrt{2} - 5\right)^2$ **SCE-6.** $\left(3 + \sqrt{5}\right)^2$

In Exercises 1–5, write each power of i as i, -1, $-i$, or 1.

1. i^{17} **2.** $i^5 \cdot i^9$ **3.** i^{-6} **4.** i^{-59} **5.** $\dfrac{i^{24}}{i^{-23}}$

In Exercises 6–9, find the desired sum or difference. Write each answer in the form $a + bi$.

6. $(3 - 2i) + (-7 + 9i)$ **7.** $(3 - 2i) - (-7 + 9i)$

8. $i - (1 + i)$ **9.** $\left(\sqrt{2} - 3i\right) + \left(2\sqrt{2} + 3i\right)$

In Exercises 10–14, perform the desired operations. Write each answer in the form $a + bi$.

10. $2i(4 - 3i)$ **11.** $-i(1 - i)$ **12.** $(-2 - i)(3 - 4i)$

13. $(6 - 2i)^2$ **14.** $\left(\sqrt{2} - i\right)^2$

In Exercises 15–19, find the product of the given complex number and its conjugate. Write each answer in the form $a + bi$.

15. $5 - 2i$ **16.** $1 - i$ **17.** $\dfrac{1}{2} - 3i$ **18.** $\sqrt{5} + i$ **19.** $\dfrac{\sqrt{3}}{2} + \dfrac{1}{2}i$

In Exercises 20–23, write each quotient in the form $a + bi$.

20. $\dfrac{2 - i}{3 + 4i}$ **21.** $\dfrac{1}{2 - i}$ **22.** $\dfrac{3i}{2 + 2i}$ **23.** $\dfrac{5 + i}{5 - i}$

In Exercises 24–31, write each expression in the form $a + bi$.

24. $\sqrt{-36} - \sqrt{49}$ **25.** $\sqrt{-1} + 3 - \sqrt{-64}$ **26.** $\sqrt{-2} \cdot \sqrt{-18}$ **27.** $\left(\sqrt{-8}\right)^2$

28. $\left(i\sqrt{-4}\right)^2$ **29.** $\dfrac{-4 - \sqrt{-20}}{2}$ **30.** $\dfrac{-3 - \sqrt{-81}}{6}$ **31.** $\dfrac{4 + \sqrt{-8}}{4}$

1.4 Quadratic Equations

THINGS TO KNOW

Before working through this section, be sure you are familiar with the following concepts:

                                              VIDEO   ANIMATION   INTERACTIVE

You Try It
1. Simplifying Radicals (Section R.3)

You Try It
2. Simplifying Radicals with Negative Radicands (Section 1.3)

You Try It
3. Factoring Trinomials with a Leading Coefficient Equal to 1 (Section R.5)

You Try It
4. Factoring Trinomials with a Leading Coefficient Not Equal to 1 (Section R.5)   

OBJECTIVES

1 Solving Quadratic Equations by Factoring and the Zero Product Property

2 Solving Quadratic Equations Using the Square Root Property

3 Solving Quadratic Equations by Completing the Square

4 Solving Quadratic Equations Using the Quadratic Formula

5 Using the Discriminant to Determine the Type of Solutions of a Quadratic Equation

Introduction to Section 1.4

In Section 1.1, we studied linear equations of the form $ax + b = c, a \neq 0$. These equations are also known as first-order polynomial equations. In this section, we learn how to solve second-order polynomial equations. Second-order polynomial equations are called **quadratic equations**.

> **Definition**   Quadratic Equation in One Variable
>
> A **quadratic equation in one variable** is an equation that can be written in the form $ax^2 + bx + c = 0, a \neq 0$. Quadratic equations in this form are said to be in *standard form*.

OBJECTIVE 1   SOLVING QUADRATIC EQUATIONS BY FACTORING AND THE ZERO PRODUCT PROPERTY

 ⊙ Some quadratic equations can be easily solved by factoring and by using the following important property.

> **Zero Product Property**
> $$\text{If } AB = 0, \text{ then } A = 0 \text{ or } B = 0.$$

The zero product property says that if two factors multiplied together are equal to zero, then at least one of the factors must be zero. Example 1 shows how this property is used to solve certain quadratic equations.

 ⊙ **Example 1** Use Factoring and the Zero Product Property to Solve a Quadratic Equation

Solve $6x^2 - 17x = -12$.

Solution First, rewrite the equation in standard form by adding the constant 12 to both sides.

$$6x^2 - 17x = -12 \qquad \text{Write the original equation.}$$

$$6x^2 - 17x + 12 = 0 \qquad \text{Add 12 to both sides.}$$

$$(3x - 4)(2x - 3) = 0 \qquad \text{Factor the left-hand side.}$$

$$3x - 4 = 0 \quad \text{or} \quad 2x - 3 = 0 \qquad \text{Use the zero product property.}$$

$$3x = 4 \quad \text{or} \qquad 2x = 3$$

$$x = \frac{4}{3} \quad \text{or} \qquad x = \frac{3}{2} \qquad \text{Solve each equation for } x.$$

The solution is $\left\{\dfrac{4}{3}, \dfrac{3}{2}\right\}$.

 **You Try It** Work through this You Try It problem.

Work Exercises 1–10 in this textbook or in the MyMathLab Study Plan.

OBJECTIVE 2 SOLVING QUADRATIC EQUATIONS USING THE SQUARE ROOT PROPERTY

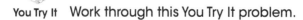 *My video summary* ⊘ Consider the quadratic equation $x^2 - 9 = 0$. The left-hand side is a difference of two squares, so we can easily solve by factoring.

$$x^2 - 9 = 0 \qquad \text{Write the original equation.}$$

$$(x - 3)(x + 3) = 0 \qquad \text{Factor using difference of squares formula.}$$

$$x - 3 = 0 \quad \text{or } x + 3 = 0 \qquad \text{Use the zero product property.}$$

$$x = 3 \quad \text{or} \quad x = -3 \qquad \text{Solve each equation for } x.$$

The solution is $x = \pm 3$. In fact, any quadratic equation of the form $x^2 - c = 0$ where $c > 0$ can be solved by factoring:

$$x^2 - c = 0$$

$$(x - \sqrt{c})(x + \sqrt{c}) = 0 \qquad \text{Factor using difference of squares formula.}$$

$$x - \sqrt{c} = 0 \quad \text{or} \quad x + \sqrt{c} = 0 \qquad \text{Use the zero product property.}$$

$$x = \sqrt{c} \quad \text{or} \quad x = -\sqrt{c}$$

Although quadratic equations of this form can be solved by factoring, they can be more readily solved by using the following square root property.

Square Root Property

The solution to the quadratic equation $x^2 - c = 0$, or equivalently $x^2 = c$, is $x = \pm\sqrt{c}$.

Example 2 Solve Quadratic Equations Using the Square Root Property

Solve each quadratic equation.

a. $x^2 - 16 = 0$ **b.** $2x^2 + 72 = 0$ **c.** $(x - 1)^2 = 7$

Solution

a. $x^2 - 16 = 0$ Write the original equation.

    $x^2 = 16$ Add 16 to both sides.

    $x = \pm 4$ Use the square root property.

b. $2x^2 + 72 = 0$ Write the original equation.

       $2x^2 = -72$ Subtract 72 from both sides.

         $x^2 = -36$ Divide by 2.

    $\sqrt{x^2} = \pm\sqrt{-36}$ Use the square root property.

          $x = \pm 6i$ Simplify the radical with a negative radicand.

c. $(x - 1)^2 = 7$

     $x - 1 = \pm\sqrt{7}$ Use the square root property.

        $x = 1 \pm \sqrt{7}$ Add 1 to both sides.

You Try It   Work through this You Try It problem.

Work Exercises 11–15 in this textbook or in the MyMathLab Study Plan.

In Example 2, we see that the quadratic equation $(x - 1)^2 = 7$ can be solved using the square root property. The left side of the equation is a perfect square. In fact, every quadratic equation can be written in the form $(x - h)^2 = k$ using a method known as **completing the square**.

OBJECTIVE 3   SOLVING QUADRATIC EQUATIONS BY COMPLETING THE SQUARE

Consider the following perfect square trinomials:

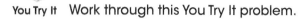

$$x^2 + 2x + 1 = (x + 1)^2 \qquad x^2 - 6x + 9 = (x - 3)^2 \qquad x^2 - 7x + \frac{49}{4} = \left(x - \frac{7}{2}\right)^2$$

$$\left(\frac{1}{2} \cdot 2\right)^2 = 1 \qquad\qquad \left(\frac{1}{2} \cdot (-6)\right)^2 = 9 \qquad\qquad \left(\frac{1}{2} \cdot (-7)\right)^2 = \frac{49}{4}$$

In each perfect square trinomial, notice the relationship between the coefficient of the linear term (x-term) and the constant term. The constant term of a perfect square trinomial is equal to the square of $\frac{1}{2}$ the linear coefficient.

Example 3   Completing the Square

What number must be added to each binomial to make it a perfect square trinomial?

a. $x^2 - 12x$ **b.** $x^2 + 5x$ **c.** $x^2 - \frac{3}{2}x$

Solution

a. The linear coefficient is -12, so we must add $\left(\frac{1}{2}(-12)\right)^2 = (-6)^2 = 36$ to complete the square. Thus, the expression $x^2 - 12x + 36$ is a perfect square trinomial and $x^2 - 12x + 36 = (x-6)^2$.

b. $\left(\frac{1}{2} \cdot 5\right)^2 = \left(\frac{5}{2}\right)^2 = \frac{25}{4}$ must be added to complete the square:
$$x^2 + 5x + \frac{25}{4} = \left(x + \frac{5}{2}\right)^2.$$

c. $\left(\frac{1}{2}\left(-\frac{3}{2}\right)\right)^2 = \left(-\frac{3}{4}\right)^2 = \frac{9}{16}$ must be added to complete the square:
$$x^2 - \frac{3}{2}x + \frac{9}{16} = \left(x - \frac{3}{4}\right)^2.$$

You Try It Work through this You Try It problem.

Work Exercises 16–19 in this textbook or in the MyMathLab Study Plan.

To solve a quadratic equation of the form $ax^2 + bx + c = 0, a \neq 0$, by completing the square, follow these steps:

Steps for Solving $ax^2 + bx + c = 0, a \neq 0$, by Completing the Square

1. If $a \neq 1$, divide the equation by a.

2. Move all constants to the right-hand side.

3. Take half the coefficient of the x-term, square it, and add it to both sides of the equation.

4. The left-hand side is now a perfect square. Rewrite it as a binomial squared.

5. Use the square root property to solve for x.

My video summary

> **Example 4** Solve by Completing the Square

Solve $3x^2 - 18x + 19 = 0$ by completing the square.

Solution

Step 1. Divide the equation by 3:
$$\frac{3x^2}{3} - \frac{18x}{3} + \frac{19}{3} = \frac{0}{3}$$
$$x^2 - 6x + \frac{19}{3} = 0$$

Step 2. Move all constants to the right-hand side:
$$x^2 - 6x = -\frac{19}{3}$$

Step 3. Take half the coefficient of the x-term, square it, and add it to both sides of the equation:

$$x^2 - 6x + 9 = -\frac{19}{3} + 9 \qquad \left(\frac{1}{2} \cdot (-6)\right)^2 = (-3)^2 = 9$$

$$x^2 - 6x + 9 = \frac{8}{3} \qquad\qquad \text{Simplify.}$$

Step 4. The left-hand side is now a perfect square. Rewrite it as a binomial squared.

$$(x - 3)^2 = \frac{8}{3}$$

Step 5. Use the square root property to solve for x.

$$x - 3 = \pm\sqrt{\frac{8}{3}} \qquad\qquad \text{Use the square root property.}$$

$$x - 3 = \pm\frac{\sqrt{8}}{\sqrt{3}} \qquad\qquad \sqrt{\frac{a}{b}} = \frac{\sqrt{a}}{\sqrt{b}}$$

$$x - 3 = \pm\frac{\sqrt{8}}{\sqrt{3}} \cdot \frac{\sqrt{3}}{\sqrt{3}} \qquad \text{Rationalize the denominator.}$$

$$x - 3 = \pm\frac{\sqrt{24}}{3} \qquad\qquad \text{Simplify.}$$

$$x - 3 = \pm\frac{2\sqrt{6}}{3} \qquad\qquad \sqrt{24} = 2\sqrt{6}$$

$$x = 3 \pm \frac{2\sqrt{6}}{3} \quad \text{or} \quad x = \frac{9 \pm 2\sqrt{6}}{3} \qquad \text{Add 3 to both sides and simplify.} \ \bullet$$

Example 5 Solve by Completing the Square

Solve $2x^2 - 10x - 6 = 0$ by completing the square.

Solution

Step 1. Divide the equation by 2:

$$\frac{2x^2}{2} - \frac{10x}{2} - \frac{6}{2} = \frac{0}{2}$$

$$x^2 - 5x - 3 = 0$$

Step 2. Move all constants to the right-hand side:

$$x^2 - 5x = 3$$

Step 3. Take half the coefficient of the x-term, square it, and add it to both sides of the equation:

$$x^2 - 5x + \frac{25}{4} = 3 + \frac{25}{4} \qquad \left(\frac{1}{2} \cdot (-5)\right)^2 = \left(-\frac{5}{2}\right)^2 = \frac{25}{4}$$

$$x^2 - 5x + \frac{25}{4} = \frac{37}{4} \qquad \text{Simplify.}$$

Step 4. The left-hand side is now a perfect square. Rewrite it as a binomial squared.

$$\left(x - \frac{5}{2}\right)^2 = \frac{37}{4}$$

Step 5. Use the square root property to solve for x.

$$x - \frac{5}{2} = \pm\sqrt{\frac{37}{4}} \qquad \text{Use the square root property.}$$

$$x - \frac{5}{2} = \pm\frac{\sqrt{37}}{\sqrt{4}} \qquad\qquad \sqrt{\frac{a}{b}} = \frac{\sqrt{a}}{\sqrt{b}}$$

$$x - \frac{5}{2} = \pm\frac{\sqrt{37}}{2} \qquad\qquad \text{Simplify.}$$

$$x = \frac{5}{2} \pm \frac{\sqrt{37}}{2} = \frac{5 \pm \sqrt{37}}{2} \qquad \text{Add } \frac{5}{2} \text{ to both sides and simplify.} \;\bullet$$

You Try It Work through this You Try It problem.

Work Exercises 20–24 in this textbook or in the MyMathLab Study Plan.

OBJECTIVE 4 SOLVING QUADRATIC EQUATIONS USING THE QUADRATIC FORMULA

In Example 5, we see how to solve a quadratic equation by completing the square. In fact, the method of completing the square can be used to solve *any* quadratic equation. However, this method is often long and tedious. If we solve the general quadratic equation $ax^2 + bx + c = 0, a \neq 0$, by the method of completing the square, we can obtain the following quadratic formula. The quadratic formula can be used to solve any quadratic equation and is often less tedious than the method of completing the square. Be sure to work through the derivation of the quadratic formula to see exactly how this formula is derived.

> **Quadratic Formula**
>
> The solution to the quadratic equation $ax^2 + bx + c = 0, a \neq 0$, is given by the following formula:
>
> $$x = \frac{-b \pm \sqrt{b^2 - 4ac}}{2a}$$

 *My video summary*

⊙ **Example 6** Solve a Quadratic Equation Using the Quadratic Formula

Solve $3x^2 + 2x - 2 = 0$ using the quadratic formula.

Solution This quadratic equation is in standard form with $a = 3, b = 2$, and $c = -2$. Substitute these values into the quadratic formula.

$$x = \frac{-2 \pm \sqrt{(2)^2 - 4(3)(-2)}}{2(3)}$$ Use the quadratic formula.

$$= \frac{-2 \pm \sqrt{28}}{6}$$ Simplify.

$$= \frac{-2 \pm 2\sqrt{7}}{6}$$ $\sqrt{28} = 2\sqrt{7}$

$$= \frac{2(-1 \pm \sqrt{7})}{6}$$ Factor.

$$= \frac{-1 \pm \sqrt{7}}{3}$$ Simplify.

My video summary **⊘ Example 7** Solve a Quadratic Equation Using the Quadratic Formula

Solve $4x^2 - x + 6 = 0$ using the quadratic formula.

Solution This quadratic equation is in standard form with $a = 4, b = -1$, and $c = 6$. Substitute these values into the quadratic formula.

$$x = \frac{-(-1) \pm \sqrt{(-1)^2 - 4(4)(6)}}{2(4)}$$ Use the quadratic formula.

$$= \frac{1 \pm \sqrt{-95}}{8}$$ Simplify.

$$= \frac{1 \pm i\sqrt{95}}{8}$$ Simplify the radical with a negative radicand.

You Try It Work through this You Try It problem.

Work Exercises 25–30 in this textbook or in the MyMathLab Study Plan.

OBJECTIVE 5   USING THE DISCRIMINANT TO DETERMINE THE TYPE OF SOLUTIONS OF A QUADRATIC EQUATION

My video summary   **⊘** In Example 7, the quadratic equation $4x^2 - x + 6 = 0$ had two nonreal solutions. The solutions were nonreal because the expression $b^2 - 4ac$ under the radical was a negative number. Given a quadratic equation of the form $ax^2 + bx + c = 0$, the expression $b^2 - 4ac$ is called the **discriminant**. Knowing the value of the discriminant can help us determine the number and nature of the solutions to a quadratic equation.

Definition   Discriminant

Given a quadratic equation $ax^2 + bx + c = 0, a \neq 0$, the expression $D = b^2 - 4ac$ is called the **discriminant**.

If $D > 0$, then the quadratic equation has two real solutions.
If $D < 0$, then the quadratic equation has two nonreal solutions.
If $D = 0$, then the quadratic equation has exactly one real solution.

Example 8 Use the Discriminant

Use the discriminant to determine the number and nature of the solutions to each of the following quadratic equations:

a. $3x^2 + 2x + 2 = 0$ 　　　　　　　**b.** $4x^2 + 1 = 4x$

Solution

a. The equation $3x^2 + 2x + 2 = 0$ is in standard form with $a = 3$, $b = 2$, and $c = 2$. $D = 2^2 - 4(3)(2) = 4 - 24 = -20 < 0$. Because the discriminant is less than zero, there are two nonreal solutions.

b. The equation $4x^2 + 1 = 4x$ is not in standard form. To get the equation into standard form, subtract $4x$ from both sides: $4x^2 - 4x + 1 = 0$. $D = (-4)^2 - 4(4)(1) = 16 - 16 = 0$. Because the discriminant is zero, there is exactly one real solution.

You Try It   Work through this You Try It problem.

Work Exercises 31–34 in this textbook or in the MyMathLab Study Plan.

1.4 Exercises

Skill Check Exercises

For exercises SCE-1 through SCE-8, simplify the given expression.

SCE-1. $-13 + \sqrt{529}$ 　　**SCE-2.** $-6 + \sqrt{44}$ 　　**SCE-3.** $4 + \sqrt{-196}$ 　　**SCE-4.** $6 - \sqrt{-116}$

SCE-5. $\dfrac{-13 + \sqrt{529}}{6}$ 　　**SCE-6.** $\dfrac{-6 + \sqrt{44}}{2}$ 　　**SCE-7.** $\dfrac{4 + \sqrt{-196}}{6}$ 　　**SCE-8.** $\dfrac{6 - \sqrt{-116}}{4}$

For exercises SCE-9 through SCE-13, factor each trinomial.

SCE-9. $x^2 + 10x + 24$ 　　**SCE-10.** $x^2 - 7x - 18$ 　　**SCE-11.** $x^2 - 10x + 21$

SCE-12. $2x^2 + 9x + 4$ 　　**SCE-13.** $6x^2 - x - 15$

In Exercises 1–10, solve each equation by factoring.

1. $x^2 - 8x = 0$ 　　**2.** $x^2 - x = 6$ 　　**3.** $x^2 + 9x + 20 = 0$

4. $x^2 - 10x + 24 = 0$ 　　**5.** $3x^2 + 8x - 3 = 0$ 　　**6.** $6x^2 - 37x - 35 = 0$

7. $12x^2 + 52x + 16 = 0$ 　　**8.** $8x^2 + 29x = 12$ 　　**9.** $8m^2 - 15 = 14m$

10. $28z^2 - 13z - 6 = 0$

In Exercises 11–15, solve each equation using the square root property.

11. $x^2 - 64 = 0$ 　　**12.** $x^2 + 64 = 0$ 　　**13.** $3x^2 = 72$

14. $(x + 2)^2 - 9 = 0$ 　　**15.** $(2x + 1)^2 + 4 = 0$

In Exercises 16–19, decide what number must be added to each binomial to make a perfect square trinomial.

16. $x^2 - 8x$ **17.** $x^2 + 10x$ **18.** $x^2 - 7x$ **19.** $x^2 + \dfrac{5}{3}x$

In Exercises 20–24, solve each quadratic equation by completing the square.

20. $x^2 - 8x - 2 = 0$ **21.** $x^2 + 7x + 14 = 0$ **22.** $2x^2 + 6x + 8 = 0$

23. $3x^2 + 24x - 7 = 0$ **24.** $3x^2 + 5x + 12 = 0$

In Exercises 25–30, solve each quadratic equation using the quadratic formula.

25. $3x^2 + 8x - 3 = 0$ **26.** $x^2 - 8x - 2 = 0$ **27.** $4x^2 - x + 8 = 0$

28. $3x^2 - 4x - 1 = 0$ **29.** $9x^2 - 6x = -1$ **30.** $5x^2 + 3x + 1 = 0$

In Exercises 31–34, use the discriminant to determine the number and nature of the solutions to each quadratic equation. Do not solve the equations.

31. $x^2 + 2x + 1 = 0$ **32.** $4x^2 + 4x + 1 = 0$

33. $2x^2 + x - 5 = 0$ **34.** $3x^2 + \sqrt{12}x + 4 = 0$

1.5 Applications of Quadratic Equations

THINGS TO KNOW

Before working through this section, be sure you are familiar with the following concepts:

		VIDEO	ANIMATION	INTERACTIVE

You Try It

1. Solving Applied Problems Involving Distance, Rate, and Time (Section 1.2)

You Try It

2. Solving Applied Working Together Problems (Section 1.2)

You Try It

3. Solving Quadratic Equations by Factoring and the Zero Product Property (Section 1.4)

You Try It

4. Solving Quadratic Equations by Completing the Square (Section 1.4)

You Try It

5. Solving Quadratic Equations Using the Quadratic Formula (Section 1.4)

OBJECTIVES

....

Introduction to Section 1.5

In Section 1.2, we learn how to solve applied problems involving linear equations. In this section, we learn to solve applications that involve quadratic equations. We follow the same four-step strategy for solving applied problems that is discussed in Section 1.2. The four steps are outlined as follows.

FOUR-STEP STRATEGY FOR PROBLEM SOLVING

Step 1. Read the problem several times until you have an understanding of what is being asked. If possible, create diagrams, charts, or tables to assist you in your understanding. } Understand the problem.

Step 2. Pick a variable that describes the unknown quantity that is to be found. All other quantities must be written in terms of that variable. Write an equation using the given information and the variable. } Devise a plan.

Step 3. Carefully solve the equation. } Carry out the plan.

Step 4. Make sure that you have answered the question and then check all answers to make sure they make sense. } Look back.

OBJECTIVE 1   SOLVING APPLICATIONS INVOLVING UNKNOWN NUMERIC QUANTITIES

Example 1 Find Two Numbers

The product of a number and 1 more than twice the number is 36. Find the two numbers.

Solution

Step 1. We are looking for two numbers.

Step 2. Let $x =$ the first number, and then $2x + 1 =$ the other number. Because their product is 36, we get the equation $x(2x + 1) = 36$, which simplifies to the quadratic equation $2x^2 + x - 36 = 0$.

Step 3. Solve:

$$2x^2 + x - 36 = 0 \qquad \text{Write the equation.}$$

$$(2x + 9)(x - 4) = 0 \qquad \text{Factor.}$$

$$2x + 9 = 0 \quad \text{or} \quad x - 4 = 0 \qquad \text{Use the zero product property.}$$

$$x = -\frac{9}{2} \quad \text{or} \qquad x = 4 \qquad \text{Solve.}$$

Step 4. If $x = -\frac{9}{2}$, then the other number is $2x + 1 = 2\left(-\frac{9}{2}\right) + 1 = -8$.

If $x = 4$, then the other number is $2x + 1 = 2(4) + 1 = 9$.

We can see that there are two possibilities for the two numbers, $-\frac{9}{2}$ and -8 or 4 and 9. In either case, the second number is one more than twice the first, and their products are 36.

You Try It   Work through this You Try It problem.

Work Exercises 1–4 in this textbook or in the My MathLab **Study Plan.**

OBJECTIVE 2   USING THE PROJECTILE MOTION MODEL

An object launched, thrown, or shot vertically into the air with an initial velocity of v_0 meters/second (m/s) from an initial height of h_0 meters above the ground can be modeled by the equation $h = -4.9t^2 + v_0 t + h_0$. The variable h is the height (in meters) of the projectile t seconds after its departure. This equation can be used to model the height of a toy rocket at any time after liftoff, as in Example 2.

Example 2 Launch a Toy Rocket

A toy rocket is launched at an initial velocity of 14.7 m/s from a 49-m-tall platform. The height h of the object at any time t seconds after launch is given by the equation $h = -4.9t^2 + 14.7t + 49$. When will the rocket hit the ground?

Solution

Step 1. We are looking for the time when h is equal to zero.

Step 2. Set h equal to 0, and solve for t: $0 = -4.9t^2 + 14.7t + 49$.

Step 3. Solve:

$$0 = -4.9t^2 + 14.7t + 49 \qquad \text{Write the original equation.}$$

$$0 = t^2 - 3t - 10 \qquad \text{Divide both sides by } -4.9.$$

$$0 = (t - 5)(t + 2) \qquad \text{Factor.}$$

$$t - 5 = 0 \quad \text{or} \quad t + 2 = 0 \qquad \text{Use the zero product property.}$$

$$t = 5 \quad \text{or} \qquad t = -2 \qquad \text{Solve.}$$

Step 4. Because t represents the time (in seconds) after launch, the value $t = -2$ seconds does not make sense. Therefore, the rocket will hit the ground 5 seconds after launch.

You Try It   Work through this You Try It problem.

Work Exercises 5 and 6 in this textbook or in the MyMathLab Study Plan.

OBJECTIVE 3   SOLVING GEOMETRIC APPLICATIONS

My interactive video summary

⊚ **Example 3** Find the Dimensions of a Rectangle

The length of a rectangle is 6 in. less than four times the width. Find the dimensions of the rectangle if the area of the rectangle is 54 in^2.

Solution Work through the interactive video to verify that the dimensions of the rectangle are 12 in. by 4.5 in.

You Try It   Work through this You Try It problem.

Work Exercises 7 and 8 in this textbook or in the MyMathLab Study Plan.

Example 4 Find the Width of a High-Definition Television

Jimmy bought a new 40-in. high-definition television. If the length of Jimmy's television is 8 in. longer than the width, find the width of the television.

Solution

Step 1. We are trying to find the width of the television.

Step 2. Let w = width of the television, then $w + 8$ = length of the television. The size of a television is the length of the diagonal, which in this case is 40 in. We can create an equation using the Pythagorean theorem $a^2 + b^2 = c^2$.

$$w^2 + (w + 8)^2 = 40^2 \qquad \text{Use the Pythagorean theorem.}$$

Step 3. Solve:

$$w^2 + w^2 + 16w + 64 = 1{,}600 \qquad \text{Square the binomial.}$$

$$2w^2 + 16w - 1{,}536 = 0 \qquad \text{Combine like terms.}$$

$$w^2 + 8w - 768 = 0 \qquad \text{Divide by 2.}$$

$$(w - 24)(w + 32) = 0 \qquad \text{Factor.}$$

$$w - 24 = 0 \quad \text{or} \quad w + 32 = 0 \qquad \text{Use the zero product property.}$$

$$w = 24 \quad \text{or} \qquad w = -32 \qquad \text{Solve for } w.$$

Step 4. Because w represents the width of the television, we can disregard the negative solution. Thus, the width of the television is 24 in.

You Try It   Work through this You Try It problem.

Work Exercise 9 in this textbook or in the MyMathLab Study Plan.

OBJECTIVE 4 SOLVING APPLICATIONS INVOLVING DISTANCE, RATE, AND TIME

 *My video summary*

ⓥ **Example 5** Find the Speed of an Airplane

Kevin flew his new Cessna O-2A airplane from Jonesburg to Mountainview, a distance of 2,560 miles. The average speed for the return trip was 64 mph faster than the average outbound speed. If the total flying time for the round trip was 18 hours, what was the plane's average speed on the outbound trip from Jonesburg to Mountainview?

Solution

Step 1. We are asked to find the average speed of the plane from Jonesburg to Mountainview.

Step 2. Let r = speed of plane on the outbound trip from Jonesburg to Mountainview, and then $r + 64$ = speed of plane on the return trip from Mountainview to Jonesburg.

Step 3. Because distance = rate × time, we know that $\text{time} = \dfrac{\text{distance}}{\text{rate}}$.

We can organize our data in the following table.

	Distance	Rate	Time = Distance/Rate
Outbound to Mountainview	2,560	r	$\dfrac{2{,}560}{r}$
Return to Jonesburg	2,560	$r + 64$	$\dfrac{2{,}560}{r + 64}$

The total flying time was 18 hours, so we can set up the following equation:

$$\underbrace{\frac{2{,}560}{r}}_{\substack{\text{Outbound} \\ \text{Time}}} + \underbrace{\frac{2{,}560}{r + 64}}_{\substack{\text{Return} \\ \text{Time}}} = \underbrace{18}_{\substack{\text{Total Flight} \\ \text{Time}}}$$

 *My video summary*

ⓥ Solving this equation for r, we get $r = 256$ or $r = -\dfrac{320}{9} \approx -35.56$. Watch the **video** to see the solution worked out in detail.

Step 4. Because the rate cannot be negative, the average outbound speed to Mountainview was 256 mph. ●

You Try It Work through this You Try It problem.

Work Exercises 10–12 in this textbook or in the MyMathLab **Study Plan**.

OBJECTIVE 5 SOLVING WORKING TOGETHER APPLICATIONS

My video summary

⊙ Example 6 Work Together to Create Monthly Sales Reports

Dawn can finish the monthly sales reports in 2 hours less time than it takes Adam. Working together, they were able to finish the sales reports in 8 hours. How long does it take each person to finish the monthly sales reports alone? (Round to the nearest minute.)

Solution

Step 1. We are asked to find the time that it takes each person to finish the monthly sales reports.

Step 2. Let t = Adam's time to do the job alone and $t - 2$ = Dawn's time to do the job alone. We can now set up a table with the given data.

	Time needed to complete the job, in hours	Portion of job completed in 1 hour (rate)
Adam	t	$\dfrac{1}{t}$
Dawn	$t - 2$	$\dfrac{1}{t - 2}$
Together	8	$\dfrac{1}{8}$

We can now use techniques discussed in **Section 1.2** to set up the following equation:

$$\underbrace{\frac{1}{t}}_{\substack{\text{Adam's} \\ \text{Rate}}} + \underbrace{\frac{1}{t - 2}}_{\substack{\text{Dawn's} \\ \text{Rate}}} = \underbrace{\frac{1}{8}}_{\substack{\text{Rate} \\ \text{Together}}}$$

Step 3. Solving for t, we see that Adam can do the job by himself in approximately 17 hours and 4 minutes, while it would take Dawn approximately 15 hours and 4 minutes to do the same job. To see this problem worked out in

My video summary ⊙ detail, watch this video.

Step 4. You should verify that Dawn's time is exactly 2 hours less than Adam's time and that the sum of their rates is equal to $\dfrac{1}{8}$. ●

You Try It Work through this You Try It problem.

Work Exercises 13–15 in this textbook or in the My MathLab **Study Plan.**

1.5 Exercises

1. The product of some negative number and 5 less than three times that number is 12. Find the number.

2. The square of a number plus the number is 132. What is the number?

3. The sum of the square of a number and the square of 5 more than the number is 169. What is the number?

4. Three consecutive odd integers are such that the square of the third integer is 15 more than the sum of the squares of the first two. Find the integers.

5. A toy rocket is launched from a 2.8-m-high platform in such a way that its height, h (in meters), after t seconds is given by the equation $h = -4.9t^2 + 18.9t + 2.8$. How long will it take for the rocket to hit the ground?

6. Benjamin threw a rock straight up from a cliff that was 24 ft above the water. If the height of the rock h, in feet, after t seconds is given by the equation $h = -16t^2 + 20t + 24$, how long will it take for the rock to hit the water?

7. The length of a rectangle is 1 cm less than three times the width. If the area of the rectangle is 30 cm^2, find the dimensions of the rectangle.

8. The length of a rectangle is 1 in. less than twice the width. If the diagonal is 2 in. more than the length, find the dimensions of the rectangle.

9. A 35-ft by 20-ft rectangular swimming pool is surrounded by a walkway of uniform width. If the total area of the walkway is 434 ft^2, how wide is the walkway?

10. A boat traveled downstream a distance of 45 mi and then came right back. If the speed of the current is 5 mph and if the total trip took 3 hours and 45 minutes, find the average speed of the boat relative to the water.

11. Meg rowed her boat upstream a distance of 9 mi and then rowed back to the starting point. The total time of the trip was 10 hours. If the rate of the current was 4 mph, find the average speed of the boat relative to the water.

12. Imogene's car traveled 280 miles. averaging a certain speed. If the car had gone 5 mph faster, the trip would have taken 1 hour less. Find the average speed.

13. Twin brothers, Billy and Bobby, can mow their grandparent's lawn together in 56 minutes. Billy could mow the lawn by himself in 15 minutes less time than it would take Bobby. How long would it take Bobby to mow the lawn by himself?

14. Jeff and Kirk can build a 75-ft. retaining wall together in 12 hours. Because Jeff has more experience, he could build the wall by himself 4 hours quicker than Kirk. How long would it take Kirk (to the nearest minute) to build the wall by himself?

15. Homer and Mike were replacing the boards on Mike's old deck. Mike can do the job alone in 1 hour less time than Homer. They worked together for 3 hours until Homer had to go home. Mike finished the job working by himself in an additional 4 hours. How long would it have taken Homer to fix the deck by himself? (*Hint:* Use Method 2, as discussed in Section 1.2.)

1.6 Other Types of Equations

THINGS TO KNOW

Before working through this section, be sure that you are familiar with the following concepts:

| | VIDEO | ANIMATION | INTERACTIVE |

You Try It 1. Factoring Trinomials with a Leading Coefficient Equal to 1 (Section R.5)

You Try It 2. Factoring Trinomials with a Leading Coefficient Not Equal to 1 (Section R.5)

You Try It 3. Factoring Polynomials by Grouping (Section R.5)

You Try It 4. Solving Quadratic Equations by Factoring and the Zero Product Property (Section 1.4)

You Try It 5. Solving Quadratic Equations Using the Quadratic Formula (Section 1.4)

OBJECTIVES

1 Solving Higher-Order Polynomial Equations

2 Solving Equations That Are Quadratic in Form (Disguised Quadratics)

3 Solving Equations Involving Radicals

OBJECTIVE 1 SOLVING HIGHER-ORDER POLYNOMIAL EQUATIONS

 *My video summary* So far in this text, we have learned methods for solving linear equations and quadratic equations. Linear equations and quadratic equations in one variable are both examples of polynomial equations of first and second degree, respectively. In this section, we first start by looking at certain higher-order polynomial equations that can be solved using special factoring techniques.

One useful technique is to factor out an expression included in each term.

Example 1 Remove the Common Factor
Find all solutions of the equation $3x^3 - 2x = -5x^2$.

Solution First, add $5x^2$ to both sides to set the equation equal to 0.

$$3x^3 + 5x^2 - 2x = 0$$

Now, try to factor the left-hand side.

$x(3x^2 + 5x - 2) = 0$	Factor out the greatest common factor, x.
$x(3x - 1)(x + 2) = 0$	Factor the quadratic.
$x = 0$ or $3x - 1 = 0$ or $x + 2 = 0$	Use the zero product property.
$x = 0$ or $x = \dfrac{1}{3}$ or $x = -2$	Solve for x.

The solution set is $\left\{ -2, 0, \dfrac{1}{3} \right\}$.

 In the equation $3x^3 + 5x^2 - 2x = 0$, do *not* divide both sides by x. This would produce the equation $3x^2 + 5x - 2 = 0$, which has only $\dfrac{1}{3}$ and -2 as solutions. The solution $x = 0$ would be "lost." In addition, because $x = 0$ is a solution of the original equation, dividing by x would mean dividing by 0, which of course is undefined and produces incorrect results.

Sometimes, polynomials can be solved by grouping terms and factoring (especially when the polynomial has four terms). You may want to review how to factor by grouping, which was discussed in Section R.5.

Example 2 Factor by Grouping to Solve a Polynomial Equation

Find all solutions of the equation $2x^3 - x^2 + 8x - 4 = 0$.

Solution Arrange the terms of the polynomial in descending order, and group the terms of the polynomial in pairs.

$2x^3 - x^2 + 8x - 4 = 0$	Write the original equation.
$(2x^3 - x^2) + (8x - 4) = 0$	Group the terms in pairs.
$x^2(2x - 1) + 4(2x - 1) = 0$	Factor the G.C.F from each pair.
$(2x - 1)(x^2 + 4) = 0$	Factor $2x - 1$ from the left-hand side.
$2x - 1 = 0$ or $x^2 + 4 = 0$	Use the zero product property.
$2x = 1$ or $x^2 = -4$	Solve for x.
$x = \dfrac{1}{2}$ or $x = \pm\sqrt{-4} = \pm 2i$	Simplify the radical with a negative radicand.

The solution set is $\left\{ -2i, 2i, \dfrac{1}{2} \right\}$.

 You Try It Work through this You Try It problem.

Work Exercises 1–6 in this textbook or in the MyMathLab Study Plan.

OBJECTIVE 2   SOLVING EQUATIONS THAT ARE QUADRATIC IN FORM (DISGUISED QUADRATICS)

Quadratic equations of the form $ax^2 + bx + c = 0, a \neq 0$, are relatively straightforward to solve because we know several methods for solving quadratics. Sometimes, equations that are not quadratic can be made into a quadratic equation by using a substitution. Equations of this type are said to be *quadratic in form* or *disguised quadratics*. These equations typically have the form $au^2 + bu + c = 0, a \neq 0$, after an appropriate substitution. Following are a few examples.

Original Equation	Make an Appropriate Substitution	New Equation Is a Quadratic
$2x^4 - 11x^2 + 12 = 0$	Determine the proper substitution $\longrightarrow$ Let $u = x^2$, then $u^2 = x^4$.	New equation is a quadratic $\longrightarrow$ $2u^2 - 11u + 12 = 0$
$\left(\dfrac{1}{x-2}\right)^2 + \dfrac{2}{x-2} - 15 = 0$	Determine the proper substitution $\longrightarrow$ Let $u = \dfrac{1}{x-2}$, then $u^2 = \left(\dfrac{1}{x-2}\right)^2$.	New equation is a quadratic $\longrightarrow$ $u^2 + 2u - 15 = 0$
$x^{\frac{2}{3}} - 9x^{\frac{1}{3}} + 8 = 0$	Determine the proper substitution $\longrightarrow$ Let $u = x^{\frac{1}{3}}$, then $u^2 = (x^{\frac{1}{3}})^2 = x^{\frac{2}{3}}$.	New equation is a quadratic $\longrightarrow$ $u^2 - 9u + 8 = 0$
$3x^{-2} - 5x^{-1} - 2 = 0$	Determine the proper substitution $\longrightarrow$ Let $u = x^{-1} = \dfrac{1}{x}$, then $u^2 = (x^{-1})^2 = x^{-2}$.	New equation is a quadratic $\longrightarrow$ $3u^2 - 5u - 2 = 0$

We now solve each of these equations listed in Example 3:

My interactive video summary

⊚ Example 3 Solve Equations That Are Quadratic in Form

Solve each of the following:

a. $2x^4 - 11x^2 + 12 = 0$ **b.** $\left(\dfrac{1}{x-2}\right)^2 + \dfrac{2}{x-2} - 15 = 0$

c. $x^{2/3} - 9x^{1/3} + 8 = 0$ **d.** $3x^{-2} - 5x^{-1} - 2 = 0$

Solution

a.

$2x^4 - 11x^2 + 12 = 0$	Write the original equation.
Let $u = x^2$ then $u^2 = x^4$.	Determine the proper substitution.
$2u^2 - 11u + 12 = 0$	Substitute u for x^2 and u^2 for x^4 into the original equation.
$(2u - 3)(u - 4) = 0$	Factor.
$2u - 3 = 0$ or $u - 4 = 0$	Use the zero product property.
$u = \dfrac{3}{2}$ or $u = 4$	Solve for u.

Don't stop yet! We want to solve for x. Because we know that $u = x^2$, we get

$$x^2 = \frac{3}{2} \quad \text{or} \quad x^2 = 4, \quad \text{so} \quad x = \pm\sqrt{\frac{3}{2}} = \pm\frac{\sqrt{6}}{2} \quad \text{or} \quad x = \pm 2.$$

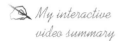

 Try to find the solutions to (b) through (d). Watch the interactive video to see each worked out solution.

You Try It Work through this You Try It problem.

Work Exercises 7–13 in this textbook or in the MyMathLab Study Plan.

OBJECTIVE 3 SOLVING EQUATIONS INVOLVING RADICALS

A radical equation is an equation that involves a variable inside a square root, cube root, or any higher root. To solve these equations, we must try to isolate the radical and then raise each side of the equation to the appropriate power to eliminate the radical. Consider the following examples:

Example 4 Solve an Equation Involving a Square Root

Solve $\sqrt{x-1} - 2 = x - 9$.

Solution First, we isolate the radical on the left side of the equation, adding 2 to both sides.

$$\sqrt{x-1} = x - 7 \qquad \text{Add 2 to both sides.}$$

To eliminate the radical, square both sides of the equation.

$\left(\sqrt{x-1}\right)^2 = (x-7)^2$ Square both sides of the equation.

$x - 1 = x^2 - 14x + 49$ Expand the right-hand side: $(x-7)^2 = (x-7)(x-7) = x^2 - 14x + 49$.

$0 = x^2 - 15x + 50$ Move all terms to the right-hand side.

$0 = (x-5)(x-10)$ Factor.

Setting each factor equal to zero gives us $x = 5$ or $x = 10$ as possible solutions. It is tempting to stop here. When solving equations involving radicals it is important to always check each potential solution by substituting back into the original equation.

Check $x = 10$:
$$\sqrt{x-1} - 2 = x - 9$$
$$\sqrt{10-1} - 2 \overset{?}{=} 10 - 9$$
$$\sqrt{9} - 2 \overset{?}{=} 1$$
$$3 - 2 \overset{?}{=} 1$$
$$1 = 1 \checkmark$$

Check $x = 5$:
$$\sqrt{x-1} - 2 = x - 9$$
$$\sqrt{5-1} - 2 \overset{?}{=} 5 - 9$$
$$\sqrt{4} - 2 \overset{?}{=} -4$$
$$2 - 2 \overset{?}{=} -4$$
$$0 \neq -4 \quad \times$$

We see that $x = 10$ checks, but $x = 5$ does not. Therefore, the solution set to this equation is $\{10\}$. $x = 5$ is called an extraneous solution.

 Because the squaring operation can make a false statement true ($-2 \neq 2$, but $(-2)^2 = (2)^2$), it is essential to always check your answers after solving an equation in which this operation was performed. The squaring operation is usually used when solving an equation involving square roots.

> ⚠ **Be careful when squaring an expression of the form** $(a + b)^2$ **or** $(a - b)^2$. **Remember,** $(a + b)^2 = (a + b)(a + b) = a^2 + 2ab + b^2$ **and** $(a - b)^2 = (a - b)(a - b) = a^2 - 2ab + b^2$.

My video summary ⊙ **Example 5** Solve an Equation Involving Two Square Roots

Solve $\sqrt{2x + 3} + \sqrt{x - 2} = 4$.

Solution When solving equations containing multiple radicals, first isolate one of the radicals and square both sides:

$$\sqrt{2x + 3} = 4 - \sqrt{x - 2}$$
$$(\sqrt{2x + 3})^2 = (4 - \sqrt{x - 2})^2$$

Watch the video to see how to obtain the solution of $x = 3$.

Example 6 Solve an Equation Involving a Cube Root

Solve $\sqrt[3]{1 - 4x} + 3 = 0$

Solution To eliminate the radical, we first isolate the radical and then cube both sides.

$$\sqrt[3]{1 - 4x} = -3 \qquad \text{Subtract 3 from both sides.}$$
$$(\sqrt[3]{1 - 4x})^3 = (-3)^3 \qquad \text{Cube both sides of the equation.}$$
$$1 - 4x = -27 \qquad \text{Simplify.}$$
$$-4x = -28 \qquad \text{Combine like terms.}$$
$$x = 7 \qquad \text{Divide both sides by } -4.$$

Check $\sqrt[3]{1 - 4(7)} + 3 = \sqrt[3]{-27} + 3 = -3 + 3 = 0$ ✓

You Try It Work through this You Try It problem.

Work Exercises 14–21 in this textbook or in the MyMathLab Study Plan.

1.6 Exercises

Skill Check Exercises

For exercises SCE-1 through SCE-6, perform the indicated operation and simplify.

SCE-1. $(\sqrt{2x + 4})^2$ **SCE-2.** $(\sqrt{18 - 7x})^2$ **SCE-3.** $(2x - 1)^2$

SCE-4. $(-4\sqrt{8x + 9})^2$ **SCE-5.** $(3 + \sqrt{y + 4})^2$ **SCE-6.** $(2 - \sqrt{4x + 5})^2$

In Exercises 1–6, find all solutions.

1. $x^3 + x = 0$ **2.** $2x^3 + 4x^2 - 30x = 0$ **3.** $x^2(x^2 - 1) - 9(x^2 - 1) = 0$

4. $x^3 - x^2 + 3x = 3$ **5.** $2x^3 + 3x^2 = 8x + 12$ **6.** $x^3 - x^2 - x = -1$

In Exercises 7–13, solve the equation after making an appropriate substitution.

SbS   **7.** $x^4 - 6x^2 + 8 = 0$ SbS   **8.** $(13x - 1)^2 - 2(13x - 1) - 3 = 0$

SbS   **9.** $2x^{2/3} - 5x^{1/3} + 2 = 0$ SbS   **10.** $x^6 - 7x^3 = 8$

SbS   **11.** $\sqrt{x} - 3\sqrt[4]{x} - 4 = 0$ SbS   **12.** $3\left(\dfrac{1}{x-1}\right)^2 - \dfrac{5}{x-1} - 2 = 0$

SbS   **13.** $2x^{-2} - 3x^{-1} - 2 = 0$

In Exercises 14–21, solve each radical equation.

14. $\sqrt{2x - 1} = 3$ **15.** $\sqrt{32 - 4x} - x = 0$ **16.** $x - 1 = \sqrt{2x + 1}$

17. $\sqrt{12 - 2x} = x + 6$ **18.** $\sqrt{3x + 1} - \sqrt{x + 4} = 1$ **19.** $\sqrt{6 + 5x} + \sqrt{3x + 4} = 2$

20. $\sqrt[3]{2x - 1} = -2$ **21.** $\sqrt[4]{x^2 - 6x} - 2 = 0$

Brief Exercises

In Exercises 22–28, solve the equation after making an appropriate substitution.

22. $x^4 - 6x^2 + 8 = 0$ **23.** $(13x - 1)^2 - 2(13x - 1) - 3 = 0$

24. $2x^{2/3} - 5x^{1/3} + 2 = 0$ **25.** $x^6 - 7x^3 = 8$

26. $\sqrt{x} - 3\sqrt[4]{x} - 4 = 0$ **27.** $3\left(\dfrac{1}{x-1}\right)^2 - \dfrac{5}{x-1} - 2 = 0$

28. $2x^{-2} - 3x^{-1} - 2 = 0$

1.7 Linear Inequalities

THINGS TO KNOW

Before working through this section, be sure that you are familiar with the following concepts:

                                                           VIDEO   ANIMATION   INTERACTIVE

You Try It   **1.** Describing Intervals of Real Numbers (Section R.1)

You Try It   **2.** Understanding the Intersection and Union of Sets (Section R.1)

OBJECTIVES

1 Solving Linear Inequalities

2 Solving Three-Part Inequalities

3 Solving Compound Inequalities

4 Solving Linear Inequality Word Problems

....

Introduction to Section 1.7

Unlike equations that usually have a finite number of solutions (or no solution at all), inequalities often have infinitely many solutions. For instance, the inequality $2x - 3 \leq 5$ has infinitely many solutions because there are infinite values of x for which the inequality is true. Take, for example, the numbers $-1, \frac{7}{2}$, and 4. Substituting each of these values into the inequality $2x - 3 \leq 5$ yields a true statement:

Inequality $2x - 3 \leq 5$

$x = -1 \qquad 2(-1) - 3 \overset{?}{\leq} 5$

$-5 \leq 5 \qquad$ True statement!

$x = \dfrac{7}{2} \qquad 2\left(\dfrac{7}{2}\right) - 3 \overset{?}{\leq} 5$

$4 \leq 5 \qquad$ True statement!

$x = 4 \qquad 2(4) - 3 \overset{?}{\leq} 5$

$5 \leq 5 \qquad$ True statement!

In fact, any real number x that is less than or equal to 4 makes this inequality a true statement. We certainly cannot list all solutions to this inequality individually. Thus, there are typically three methods used to describe the solution to an inequality:

1. Graph the solution on a number line.

2. Write the solution in set-builder notation.

3. Write the solution in interval notation.

Figure 2 shows the solution to the inequality in these three ways. In this section, you learn how to solve such inequalities. You may want to review the terms *set-builder notation* and *interval notation*, which are described in Section R.1.

Solution graphed on a number line:

Solution in set-builder notation: $\{x | x \leq 4\}$

Solution in interval notation: $(-\infty, 4]$

Figure 2 Solution to the inequality $2x - 3 \leq 5$

OBJECTIVE 1   SOLVING LINEAR INEQUALITIES

The inequality $2x - 3 \leq 5$ is called a **linear inequality**. In this section, you learn techniques necessary to solve linear inequalities.

Definition

A **linear inequality** is an inequality that can be written in the form $ax + b < c$, where a, b, and c are real numbers and $a \neq 0$.

Note The inequality symbol "$<$" in the previous definition and the following properties can be replaced with $>$, $\leq$, or $\geq$.

Before we solve a linear inequality, it is important to introduce some inequality properties:

PROPERTIES OF INEQUALITIES

Let a, b, and c be real numbers:

	Property	In Words	Example
1	If $a < b$, then $a + c < b + c$	The same number may be added to both sides of an inequality.	$-3 < 7$ $-3 + 4 < 7 + 4$ $1 < 11$
2	If $a < b$, then $a - c < b - c$	The same number may be subtracted from both sides of an inequality.	$9 \geq 2$ $9 - 6 \geq 2 - 6$ $3 \geq -4$
3	For $c > 0$, if $a < b$, then $ac < bc$	Multiplying both sides of an inequality by a *positive* number *does not reverse the direction* of the inequality.	$3 > 2$ $(3)(5) > (2)(5)$ $15 > 10$
4	For $c < 0$, if $a < b$, then $ac > bc$	Multiplying both sides of an inequality by a *negative* number *reverses the direction* of the inequality.	$3 > 2$ $(3)(-5) < (2)(-5)$ $-15 < -10$
5	For $c > 0$, if $a < b$, then $\dfrac{a}{c} < \dfrac{b}{c}$	Dividing both sides of an inequality by a *positive* number *does not reverse the direction* of the inequality.	$6 > 4$ $\dfrac{6}{2} > \dfrac{4}{2}$ $3 > 2$
6	For $c < 0$, if $a < b$, then $\dfrac{a}{c} > \dfrac{b}{c}$	Dividing both sides of an inequality by a *negative* number *reverses the direction* of the inequality.	$6 > 4$ $\dfrac{6}{-2} < \dfrac{4}{-2}$ $-3 < -2$

Example 1 Solve a Linear Inequality

Solve the inequality $-9x - 3 \geq 7 - 4x$. Graph the solution set on a number line, and express the answer in interval notation.

Solution The technique to use when solving linear inequalities is to isolate the variable on one side.

$$-9x - 3 \geq 7 - 4x$$

$$-9x + 4x - 3 \geq 7 - 4x + 4x \qquad \text{Add } 4x \text{ to both sides.}$$

$$-5x - 3 \geq 7 \qquad \text{Simplify.}$$

$$-5x - 3 + 3 \geq 7 + 3 \qquad \text{Add 3 to both sides.}$$

$$-5x \geq 10 \qquad \text{Simplify.}$$

$$\frac{-5x}{-5} \leq \frac{10}{-5} \qquad \text{Divide by } -5. \text{ Reverse the direction of the inequality!}$$

$$x \leq -2 \qquad \text{Simplify.}$$

The graph of the solution is depicted in Figure 3. The solution set in interval notation is $(-\infty, -2]$.

Solution graphed on number line:

Solution in interval notation: $(-\infty, -2]$

Figure 3 Solution to the inequality
$$-9x - 3 \geq 7 - 4x$$

 Remember to reverse the direction of the inequality when multiplying or dividing a linear inequality by a negative number.

 *My video summary*

⊙ Example 2 Solve a Linear Inequality

Solve the inequality $2 - 5(x - 2) < 4(3 - 2x) + 7$. Express the answer in set-builder notation.

Solution The solution in set-builder notation is $\left\{ x \mid x < \dfrac{7}{3} \right\}$. Watch the video to see the solution worked out in its entirety.

You Try It Work through this You Try It problem.

Work Exercises 1–8 in this textbook or in the MyMathLab Study Plan.

OBJECTIVE 2   SOLVING THREE-PART INEQUALITIES

The weight of a laptop computer depends on several factors including the size of the monitor, the thickness, the type of battery, and the type of internal hard drive. The weight of a typical 14-inch laptop varies between 1.2 lbs to 8.4 pounds. The weight can be expressed using the following *three-part inequality* $1.2 \leq W \leq 8.4$, where W is the weight of the laptop. Notice that the variable W is between the two inequality symbols. The technique to use when solving three-part inequalities is to simplify until the variable is "sandwiched" in the middle, as in our example, $1.2 \leq W \leq 8.4$.

Example 3 Solve a Three-Part Inequality

Solve the inequality $-2 \leq \dfrac{2 - 4x}{3} < 5$. Graph the solution set on a number line, and write the solution in set-builder notation.

Solution We start by multiplying all three parts by 3 to eliminate the fraction.

$$-2 \leq \frac{2 - 4x}{3} < 5$$ 　　　Write the original three-part inequality.

$$3 \cdot (-2) \leq 3 \cdot \frac{2 - 4x}{3} < 3 \cdot 5$$ 　　　Multiply by 3.

$$-6 \leq 2 - 4x < 15$$ 　　　Simplify.

$$-6 - 2 \leq 2 - 2 - 4x < 15 - 2$$ 　　　Subtract 2.

$$-8 \leq -4x < 13$$ 　　　Simplify.

$$\frac{-8}{-4} \geq \frac{-4x}{-4} > \frac{13}{-4}$$ 　　　Divide by −4, and reverse the direction of the inequalities.

$$2 \geq x > -\frac{13}{4}$$ 　　　Simplify.

Although this answer is adequate, it is good practice to rewrite the inequality so that the variable is in the middle and the smaller of the two outside numbers is on the left.

$$-\frac{13}{4} < x \leq 2$$ 　　　Rewrite the inequality.

We can write the solution in set-builder notation as $\left\{ x \,\middle|\, -\frac{13}{4} < x \leq 2 \right\}$. The graph of the solution is displayed in Figure 4.

Solution graphed on a number line:

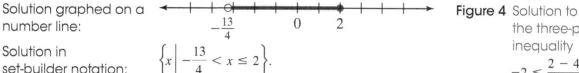

Solution in set-builder notation: $\left\{ x \,\middle|\, -\frac{13}{4} < x \leq 2 \right\}$.

Figure 4 Solution to the three-part inequality $-2 \leq \dfrac{2 - 4x}{3} < 5$ ●

 You Try It Work through this You Try It problem

Work Exercises 9–14 in this textbook or in the My MathLab **Study Plan.**

OBJECTIVE 3 SOLVING COMPOUND INEQUALITIES

Before we solve compound inequalities, it might be a good idea to review the concepts of intersection and union that were introduced in the Review chapter. To review the intersection and union of sets, refer back to Section R.1.

Compound inequalities are two inequalities that are joined together by the words *and* or *or*. A number is a solution to a compound inequality involving *and*, if it is a solution to *both* inequalities. A number is a solution to a compound inequality involving *or*, if it is a solution to *either* inequality.

Example 4 Solve a Compound Inequality with *and*

Solve $2x - 7 < -1$ and $3x + 5 \geq 3$. Graph the solution set, and write the solution in interval notation.

Solution To solve a compound inequality involving the word *and*, first solve each inequality independently and then find the intersection of the two solution sets.

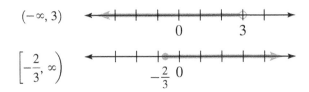

$$2x - 7 < -1 \quad \text{and} \quad 3x + 5 \geq 3$$
$$2x < 6 \quad \text{and} \quad 3x \geq -2$$
$$x < 3 \quad \text{and} \quad x \geq -\frac{2}{3}$$

We can now graph each solution on a number line:

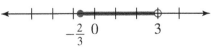

The solution to this compound inequality is the *intersection* of these two sets. In other words, we are looking for all values of x common to both solutions. The values in common are all points to the right of and including $-2/3$ and to the left of and not including 3, as depicted in Figure 5.

Solution graphed on a number line:

Figure 5 Solution to the compound inequality $2x - 7 < -1$ and $3x + 5 \geq 3$

Solution in interval notation: $\left[-\frac{2}{3}, 3\right)$

Example 5 Solve a Compound Inequality with *or*

Solve $1 - 3x \geq 7$ or $3x + 4 > 7$. Graph the solution set and write the solution in interval notation.

Solution Solving each inequality independently, we get $x \leq -2$ or $x > 1$. (To see how we arrive at this solution, read these steps.) We now graph each solution on a number line:

$(-\infty, -2]$

$(1, \infty)$

The solution to this compound inequality will be the *union* of these two sets. To express the solution in interval notation, we join the two intervals with the *union* symbol $\cup$ as shown in Figure 6.

Solution graphed on a number line:

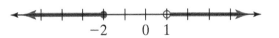

Figure 6 Solution to the compound inequality $1 - 3x \geq 7$ and $3x + 4 > 7$

Solution in interval notation: $(-\infty, -2] \cup (1, \infty)$

Example 6 Example of an Inequality with No Solution

Solve $3x - 1 < -7$ and $4x + 1 > 9$.

Solution We solve each inequality independently:

$$3x - 1 < -7 \quad \text{and} \quad 4x + 1 > 9$$
$$3x < -6 \quad \text{and} \quad 4x > 8$$
$$x < -2 \quad \text{and} \quad x > 2$$

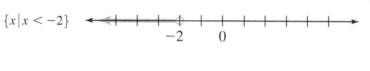

$\{x \mid x < -2\}$

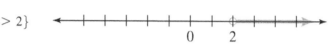

$\{x \mid x > 2\}$

These graphs do not intersect.

The solution is the intersection of all points to the left of −2 and all points to the right of 2. Because these two sets *do not intersect*, there is no solution to this inequality. We say that the solution set is empty, and use the symbol ∅ to represent the empty set. ●

You Try It Work through this You Try It problem.

Work Exercises 15–19 in this textbook or in the My MathLab **Study Plan.**

OBJECTIVE 4 SOLVING LINEAR INEQUALITY WORD PROBLEMS

Example 7 Solve a Linear Inequality Word Problem

Suppose you rented a forklift to move a pallet with 70-lb blocks stacked on it. The forklift can carry a maximum of 2,500 lb. If the pallet weighs 50 lb by itself with no blocks, how many blocks can be stacked on a pallet and lifted by the forklift?

Solution Let x be the number of blocks that can be stacked and lifted by the forklift. Each block weighs 70 lb, so $70x$ is the total weight in blocks that can be moved. The pallet by itself weighs 50 lb, so we can set up the following inequality:

$$\underbrace{50}_{\substack{\text{50 lb for} \\ \text{the pallet}}} + \underbrace{70x}_{\substack{\text{70 lb per} \\ \text{block}}} \underbrace{\leq}_{\substack{\text{is less than or} \\ \text{equal to}}} \underbrace{2{,}500}_{2{,}500 \text{ lb}}$$

Solving this inequality, we get $x \leq 35$. So, the forklift can lift no more than 35 blocks at a time. To see how we arrive at this solution, read these steps. ●

You Try It Work through this You Try It problem.

Work Exercises 20–23 in this textbook or in the My MathLab **Study Plan.**

Example 8 Solve a Three-Part Inequality Word Problem

The perimeter of a rectangular fence is to be at least 80 feet and no more than 140 feet. If the width of the fence is 12 feet, what is the range of values for the length of the fence?

Solution Let l be the length of the rectangle, as shown in Figure 7. The perimeter, P, of the rectangle is $P = 2(12) + 2l = 24 + 2l$.

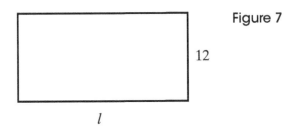

Figure 7

12

l

The perimeter must be greater than or equal to 80 and less than or equal to 140 or $80 \leq P \leq 140$. Because the perimeter is $24 + 2l$, we get the following three-part inequality: $80 \leq 24 + 2l \leq 140$.

Now, solve this three-part inequality:

$80 \leq 24 + 2l \leq 140$	Rewrite the three-part inequality.
$80 - 24 \leq 24 - 24 + 2l \leq 140 - 24$	Subtract 24.
$56 \leq 2l \leq 116$	Simplify.
$\dfrac{56}{2} \leq \dfrac{2l}{2} \leq \dfrac{116}{2}$	Divide by 2.
$28 \leq l \leq 58$	Simplify.

Therefore, the length of the fence must be between 28 feet and 58 feet, inclusive.

You Try It Work through this You Try It problem.

Work Exercises 24 and 25 in this textbook or in the MyMathLab Study Plan.

1.7 Exercises

In Exercises 1–4, solve each linear inequality. Express each solution in set-builder notation.

1. $4x + 2 > -8$

2. $2(y + 1) + 3 < -3(y - 2)$

3. $11 - 2a \leq 4a + 35$

4. $\dfrac{1}{2}w - 1 > 4 - \dfrac{2}{3}w$

In Exercises 5–8, solve each linear inequality. Graph the solution set on a number line, and express each solution using interval notation.

5. $-5x - 3 \leq 2x + 4$ **6.** $3(t - 1) - 5t \geq t + 6$ **7.** $-2a - 3 > 9 + 3a$ **8.** $\dfrac{4n - 1}{3} < 3 - n$

In Exercises 9–14, solve each three-part linear inequality. Graph the solution set on a number line, and express each solution using interval notation.

9. $-2 \leq 2m - 5 \leq 2$ **10.** $-3 < 1 - 2W < 4$ **11.** $\dfrac{1}{4} < \dfrac{t + 1}{12} \leq 1$

12. $-1 < \dfrac{3 - x}{2} < 5$ **13.** $-\dfrac{5}{6} \leq x + 1 < \dfrac{1}{2}$ **14.** $0 \leq -6 - (a + 3) \leq 1$

In Exercises 15–19, solve each compound inequality. Express each solution using interval notation or state that there is no solution.

15. $2x - 3 \leq 5$ and $5x + 1 > 6$ **16.** $4k + 2 \leq -10$ or $3k - 4 > 8$

17. $\dfrac{n - 1}{3} \geq 1$ and $\dfrac{4n - 2}{2} \leq 9$ **18.** $3b - 5 \leq 1$ and $1 - b < -3$

19. $2y - 1 \leq 7$ or $3y - 5 > 1$

In Exercises 20–25, set up the appropriate inequality and solve.

20. Derek can spend no more than \$300 to pay for a limousine. The limousine rental company charges \$200 to rent a limo, plus \$20 for each hour of use. What is the maximum amount of time that he can rent the limo?

21. Jason and Sadie went to Cancun for a week. They wanted to rent a car, so they looked at two different rental car companies. Company A charges \$40 per day plus 5 cents per mile. Company B charges \$200 per week plus 30 cents per mile. How many miles must they drive in order for the price of company A to be less than the price of company B?

22. Your scores on your first three algebra exams were 91, 92, and 85. What must you score on exam four so that the average of the four test scores is at least 90?

23. The heights of the two starting guards and two starting forwards on a basketball team are 6′2″, 5′11″, 6′7″, and 6′9″. How tall must the starting center be in order for the starting players to have an average height of at least 6′6″ (*Hint:* First convert all heights to inches.)

24. The perimeter of a rectangular fence is to be at least 100 feet and no more than 180 feet. If the width of the fence is 22 feet, what is the range of values for the length of the fence?

25. The perimeter of a rectangular fence is to be at least 120 feet and no more than 180 feet. If the length of the fence is to be twice the width, what is the range of values for the width of the fence?

1.8 Absolute Value Equations and Inequalities

THINGS TO KNOW

Before working through this section, be sure that you are familiar with the following concepts:

VIDEO   ANIMATION   INTERACTIVE

You Try It

1. Solving Three-Part Inequalities (Section 1.7)

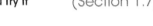
You Try It

2. Solving Compound Inequalities (Section 1.7)

OBJECTIVES

1 Solving an Absolute Value Equation

2 Solving an Absolute Value "Less Than" Inequality

3 Solving an Absolute Value "Greater Than" Inequality

Introduction to Section 1.8

My video summary ⊙ In Section R.1, we see that the absolute value of a number a, written as $|a|$, represents the **distance** from a number a to zero on the number line. Consider the equation $|x| = 5$. To solve for x, we must find all values of x that are five units away from zero on the number line. The two numbers that are five units away from zero on the number line are $x = -5$ and $x = 5$, as shown in Figure 8.

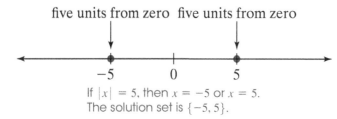

five units from zero five units from zero

If $|x| = 5$, then $x = -5$ or $x = 5$.
The solution set is $\{-5, 5\}$.

Figure 8 Solution to $|x| = 5$

The solution to the inequality $|x| < 5$ consists of all values of x whose distance from zero is less than five units on the number line. See Figure 9.

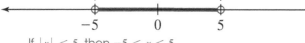

These values are all less
than five units from zero.

If $|x| < 5$, then $-5 < x < 5$.
The solution set is $\{x | -5 < x < 5\}$ in set-builder notation or $(-5, 5)$ in interval notation.

Figure 9 Solution to $|x| < 5$

Figure 10 shows the solution to the inequality $|x| > 5$. Notice that we are now looking for all values of x that are more than five units away from zero. The solution is the set of all values of x greater than 5 combined with the set of all values of x less than -5.

These values are more than five units from zero.   These values are more than five units from zero.

If $|x| > 5$, then $x < -5$ or $x > 5$.
The solution set is $\{x \mid x < -5 \text{ or } x > 5\}$ in set-builder notation or $(-\infty, -5) \cup (5, \infty)$ in interval notation.

Figure 10 Solution to $|x| < 5$

Table 1 illustrates how to solve absolute value equations and inequalities. Each absolute value equation and inequality in Table 1 is in standard form. When solving an absolute value equation or inequality, it is necessary to first rewrite it in standard form.

Table 1 Absolute Value Equations and Inequality Properties

Let u be an algebraic expression and c be a real number such that $c > 0$, then

1. $|u| = c$ is equivalent to $u = -c$ or $u = c$.

2. $|u| < c$ is equivalent to $-c < u < c$.

3. $|u| > c$ is equivalent to $u < -c$ or $u > c$.

OBJECTIVE 1   SOLVING AN ABSOLUTE VALUE EQUATION

My video summary   ⊙ **Example 1** Solve an Absolute Value Equation

Solve $|1 - 3x| = 4$.

Solution

This equation is in standard form. Therefore, by property 1, the equation $|1 - 3x| = 4$ is equivalent to the following two equations:

$$1 - 3x = 4 \quad \text{or} \quad 1 - 3x = -4 \qquad \text{Use property 1.}$$

$$-3x = 3 \quad \text{or} \quad -3x = -5 \qquad \text{Subtract 1 from both sides.}$$

$$x = -1 \quad \text{or} \quad x = \frac{5}{3} \qquad \text{Divide both sides by } -3.$$

The solution set is $\left\{-1, \dfrac{5}{3}\right\}$. See Figure 11.

Figure 11 Graph of the solution to $|1 - 3x| = 4$

You Try It   Work through this You Try It problem.

Work Exercises 1–6 in this textbook or in the MyMathLab Study Plan.

OBJECTIVE 2   SOLVING AN ABSOLUTE VALUE "LESS THAN" INEQUALITY

My video summary   ⊙ **Example 2** Solve an Absolute Value *Less Than* Inequality

Solve $|4x - 3| + 2 \le 7$.

Solution This inequality is not in standard form. Rewrite this inequality in standard form and then use property 2 to form a three-part inequality.

$|4x - 3| + 2 \le 7$	Write the original inequality.
$|4x - 3| \le 5$	Subtract 2 from both sides.
$-5 \le 4x - 3 \le 5$	Use property 2, and rewrite as a three-part inequality.
$-2 \le 4x \le 8$	Add 3 to all three parts and simplify.
$-\dfrac{1}{2} \le x \le 2$	Divide all three parts by 4 and simplify.

The solution is $\left\{ x \,\middle|\, -\dfrac{1}{2} \le x \le 2 \right\}$ in set-builder notation or $\left[-\dfrac{1}{2}, 2 \right]$ in interval notation. See Figure 12.

Figure 12 Graph of the solution to $|4x - 3| + 2 \le 7$

You Try It   Work through this You Try It problem.

Work Exercises 7–12 in this textbook or in the My Math Lab **Study Plan.**

OBJECTIVE 3   SOLVING AN ABSOLUTE VALUE "GREATER THAN" INEQUALITY

My video summary   ⊙ **Example 3** Solve an Absolute Value *Greater Than* Inequality

Solve $|5x + 1| > 3$.

Solution This inequality is in standard form. Therefore, we can use property 3 to rewrite the inequality as a compound inequality and solve.

$|5x + 1| > 3$	Write the original inequality.
$5x + 1 < -3 \quad \text{or} \quad 5x + 1 > 3$	Use property 3, and rewrite as compound inequality.
$5x < -4 \quad \text{or} \qquad 5x > 2$	Subtract 1 from both sides.
$x < -\dfrac{4}{5} \quad \text{or} \qquad x > \dfrac{2}{5}$	Divide both sides by 5.

The solution is $\left\{ x \,\middle|\, x < -\dfrac{4}{5} \text{ or } x > \dfrac{2}{5} \right\}$ in set-builder notation or $\left(-\infty, -\dfrac{4}{5} \right) \cup \left(\dfrac{2}{5}, \infty \right)$ in interval notation. See Figure 13.

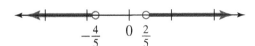

Figure 13 Graph of the solution to $|5x + 1| > 3$

In Example 3, $|5x + 1| > 3$ **is *not* equivalent to** $-3 > 5x + 1 > 3$. **In addition, a common error in this type of problem is to write** $5x + 1 > -3$ **for the first inequality, instead of** $5x + 1 < -3$. **Think carefully about the meaning of the inequality before writing it.**

You Try It Work through this You Try It problem.

Work Exercises 13–18 in this textbook or in the My MathLab Study Plan.

In Examples 1–3 (and in Table 1), the absolute value equation or inequality is written in standard form, and the constant on the right-hand side is *positive*. Example 4 illustrates how to solve absolute value equations and inequalities written in standard form when the right-hand side is negative or zero.

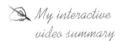
My interactive video summary

⊘ **Example 4** Absolute Value Equations and Inequalities Involving Zero or Negative Constants

Solve each of the following:

a. $|3x - 2| = 0$ b. $|x + 6| = -4$ c. $|7x + 5| \le 0$

d. $|3 - 4x| < -6$ e. $|8x - 3| > 0$ f. $|1 - 9x| \ge -5$

Solution

a. The expression $|3x - 2|$ will equal zero only when $3x - 2 = 0$ or when $x = \dfrac{2}{3}$. The solution set is $\left\{ \dfrac{2}{3} \right\}$.

b. The absolute value of an expression can never be less than zero. Therefore, there is no solution to this equation. The solution set is $\varnothing$.

c. The expression $|7x + 5|$ can never be less than zero. However, $|7x + 5|$ is equal to zero when $7x + 5 = 0$ or when $x = -\dfrac{5}{7}$. The solution set is $\left\{ -\dfrac{5}{7} \right\}$.

d. The absolute value of an expression can never be less than zero. Therefore, there is no solution to this inequality. The solution set is $\varnothing$.

e. The expression $|8x - 3|$ is equal to zero when $x = \dfrac{3}{8}$ and is greater than zero for all other values of x. Therefore, the solution set in set-builder notation is $\left\{ x \mid x \ne \dfrac{3}{8} \right\}$ or $\left(-\infty, \dfrac{3}{8} \right) \cup \left(\dfrac{3}{8}, \infty \right)$.

f. The absolute value of an expression is always nonnegative. Therefore, $|1 - 9x| \ge -5$ is true for all values of x. The solution is all real numbers or $(-\infty, \infty)$.

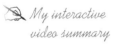
My interactive video summary

⊘ For the complete solution to this entire example, work through this interactive video.

1.8 Exercises

In Exercises 1–6, solve each equation.

1. $|x + 1| = 1$

2. $|6x + 7| - 8 = 3$

3. $\left|\dfrac{3x + 1}{5}\right| = \dfrac{2}{3}$

4. $|x^2 - 2x| = 1$

5. $|7x + 3| = 0$

6. $|4x - 1| = -5$

In Exercises 7–12, solve each inequality. Express each solution using interval notation.

7. $|x - 2| < 3$

8. $|5x - 1| + 7 \le 9$

9. $|3 - 4x| < 11$

10. $\left|\dfrac{2x - 5}{3}\right| \le \dfrac{1}{5}$

11. $|3x - 7| \le 0$

12. $|x - 1| < -2$

In Exercises 13–18, solve each inequality. Express each solution using interval notation.

13. $|x + 4| > 5$

14. $|7x - 1| - 5 \ge 8$

15. $\left|\dfrac{2x - 6}{7}\right| \ge \dfrac{3}{14}$

16. $|3 - 5x| \ge 2$

17. $|7x - 5| > 0$

18. $|x + 4| \ge -1$

1.9 Polynomial and Rational Inequalities

THINGS TO KNOW

Before working through this section, be sure you are familiar with the following concepts:

VIDEO   ANIMATION   INTERACTIVE

 You Try It
1. Factoring Trinomials with a Leading Coefficient Equal to 1 (Section R.5)

 You Try It
2. Factoring Trinomials with a Leading Coefficient Not Equal to 1 (Section R.5)

You Try It
3. Factoring Polynomials by Grouping (Section R.5)

You Try It
4. Solving Higher-Order Polynomial Equations (Section 1.6)

OBJECTIVES

1 Solving Polynomial Inequalities

2 Solving Rational Inequalities

OBJECTIVE 1 SOLVING POLYNOMIAL INEQUALITIES

In Section 1.7, we learned how to solve linear inequalities. In this section we will learn how to solve polynomial inequalities and rational inequalities. We start by solving a polynomial inequality in Example 1.

 *My video summary*

▶ Example 1 Solve a Polynomial Inequality

Solve $x^3 - 3x^2 + 2x \geq 0$.

Solution First, factor the left-hand side to get $x(x - 1)(x - 2) \geq 0$. Second, find all real values of x that make the expression on the left *equal to* 0. These values are called **boundary points**. To find the boundary points, set the factored polynomial on the left equal to 0 and solve for x.

$x(x - 1)(x - 2) = 0$ Set the factored polynomial equal to 0.

$x = 0$ or $x - 1 = 0$ or $x - 2 = 0$ Use the zero product property.

The boundary points are $x = 0, x = 1$, and $x = 2$.

Now plot each boundary point on a number line. Because the expression $x(x - 1)(x - 2)$ is equal to 0 at our three boundary points, we use a solid circle ● at each of the boundary points to indicate that the inequality $x(x - 1)(x - 2) \geq 0$ is satisfied at these points. (*Note:* If the inequality is a strict inequality such as > or <, then we represent the fact that the boundary points are not part of the solution by using an open circle, ○. See Example 2.)

Notice in Figure 14 that we have naturally divided the number line into four intervals.

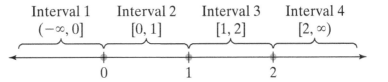

Figure 14

The expression $x(x - 1)(x - 2)$ is equal to zero *only* at the three boundary points 0, 1, and 2. This means that in any of the four intervals shown in Figure 14, the expression $x(x - 1)(x - 2)$ must be either *positive* or *negative* throughout the entire interval. To check whether this expression is positive or negative on each interval, pick a number from each interval called a **test value**. Substitute this test value into the expression $x(x - 1)(x - 2)$, and check to see if it yields a positive or negative value. Possible test values are plotted in red in Figure 15.

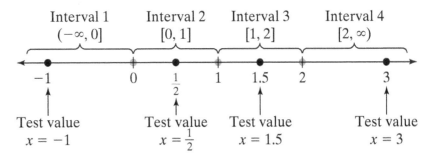

Figure 15

Interval	Test Value	Substitute Test Value into $x(x-1)(x+2)$	Comment
1. $(-\infty, 0]$	$x = -1$	$(-1)(-1-1)(-1-2) \Rightarrow (-)(-)(-) = -$	Expression is negative on $(-\infty, 0]$
2. $[0, 1]$	$x = \dfrac{1}{2}$	$\left(\dfrac{1}{2}\right)\left(\dfrac{1}{2}-1\right)\left(\dfrac{1}{2}-2\right) \Rightarrow (+)(-)(-) = +$	Expression is positive on $[0, 1]$
3. $[1, 2]$	$x = 1.5$	$(1.5)(1.5-1)(1.5-2) \Rightarrow (+)(+)(-) = -$	Expression is negative on $[1, 2]$
4. $[2, \infty)$	$x = 3$	$(3)(3-1)(3-2) \Rightarrow (+)(+)(+) = +$	Expression is positive on $[2, \infty)$

If the expression $x(x-1)(x-2)$ is positive on an interval, we place a "+" above the interval on the number line. If the expression $x(x-1)(x-2)$ is negative, we place a "−" above the interval. See Figure 16.

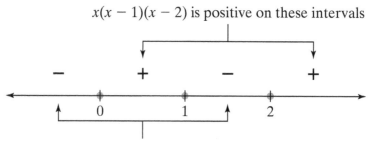

$x(x-1)(x-2)$ is positive on these intervals

$x(x-1)(x-2)$ is negative on these intervals

Figure 16

My video summary

From the number line in Figure 16, we see that $x(x-1)(x-2)$ is greater than or equal to zero on the interval $[0, 1] \cup [2, \infty)$ which is precisely the solution to the original inequality $x^3 - 3x^2 + 2x \geq 0$. Watch this video to see the solution to this polynomial inequaltiy.

Example 1 illustrates the following steps for solving a polynomial inequality that can be factored.

Steps for Solving Polynomial Inequalities

Step 1. Move all terms to one side of the inequality leaving zero on the other side.

Step 2. Factor the nonzero side of the inequality.

Step 3. Find all boundary points by setting the factored polynomial equal to zero.

Step 4. Plot the boundary points on a number line. If the inequality is ≤ or ≥, then use a solid circle ●. If the inequality is < or >, then use an open circle ○.

Step 5. Now that the number line is divided into intervals, pick a test value from each interval.

Step 6. Substitute the test value into the polynomial, and determine whether the expression is positive or negative on the interval.

Step 7. Determine the intervals that satisfy the inequality.

Example 2 Solve a Polynomial Inequality

Solve $x^2 + 5x < 3 - x^2$.

Solution

Step 1. Move all terms to one side of the inequality leaving zero on the other side.

$$2x^2 + 5x - 3 < 0$$

Step 2. Factor the nonzero side of the inequality.

$$(2x - 1)(x + 3) < 0$$

Step 3. Find all boundary points by setting the factored polynomial equal to zero.

The boundary points are $x = -3$ and $x = \dfrac{1}{2}$ because these are the values that are solutions to $(2x - 1)(x + 3) = 0$.

Step 4. Plot the boundary points on a number line.

Note that we use open circles on our boundary points because these points are *not* part of the solution. (We are looking for values that make the expression strictly less than zero.)

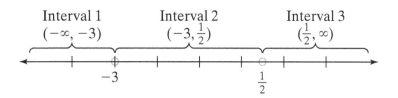

Step 5. Now that the number line is divided into three intervals, pick a test value from each interval.

Interval 1: Test value $x = -4$

Interval 2: Test value $x = 0$

Interval 3: Test value $x = 1$

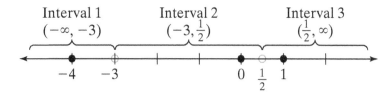

Step 6. Substitute the test value into the polynomial, and determine whether the expression is positive or negative on the interval.

$x = -4$: $(2(-4) - 1)((-4) + 3) \Rightarrow (-)(-) = +$ Expression is positive on $(-\infty, -3)$.

$x = 0$: $(2(0) - 1)((0) + 3) \Rightarrow (-)(+) = -$ Expression is negative on $\left(-3, \dfrac{1}{2}\right)$.

$x = 1$: $(2(1) - 1)((1) + 3) \Rightarrow (+)(+) = +$ Expression is positive on $\left(\dfrac{1}{2}, \infty\right)$.

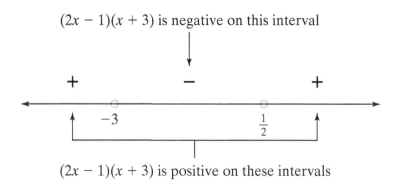

$(2x - 1)(x + 3)$ is negative on this interval

$(2x - 1)(x + 3)$ is positive on these intervals

Step 7. Determine the intervals that satisfy the inequality.

Because we are looking for values of x that are less than zero (negative values), the solution must be the interval $\left(-3, \dfrac{1}{2}\right)$.

You Try It   Work through this You Try It problem.

Work Exercise-s 1–10 in this textbook or in the MyMathLab Study Plan.

OBJECTIVE 2   SOLVING RATIONAL INEQUALITIES

 *My video summary*

A **rational inequality** can be solved using a technique similar to the one used in Examples 1 and 2, except that the boundary points are found by setting both the polynomial in the numerator and the denominator equal to zero. We will follow the seven steps outlined below.

Steps for Solving Rational Inequalities

Step 1. Move all terms to one side of the inequality leaving zero on the other side.

Step 2. Factor the numerator and denominator of the nonzero side of the inequality and cancel any common factors.

Step 3. Find all boundary points by setting the factored polynomials in the numerator and the denominator equal to zero.

Step 4. Plot the boundary points on a number line.
- For the boundary points found by setting the numerator equal to zero:

  If the inequality is $\leq$ or $\geq$, then use a solid circle ●.
  If the inequality is $<$ or $>$, then use an open circle ○.

- Use an open circle ○ to represent all boundary points found by setting the denominator equal to zero regardless of the inequality symbol that is used.

Step 5. Now that the number line is divided into intervals, pick a test value from each interval.

Step 6. Substitute the test value into the polynomial, and determine whether the expression is positive or negative on the interval.

Step 7. Determine the intervals that satisfy the inequality.

Example 3 Solve a Rational Inequality

Solve $\dfrac{x-4}{x+1} \geq 0$.

Solution Because the inequality is already in completely factored form, we can skip steps 1 and 2 and go right to step 3.

Step 3. Find all boundary points by setting the factored polynomial in the numerator and the denominator equal to zero.

Numerator: $x - 4 = 0$, so $x = 4$ is a boundary point.

Denominator: $x + 1 = 0$, so $x = -1$ is a boundary point.

Step 4. Plot the boundary points on a number line.

The inequality is a greater than *or equal to* inequality; thus, the boundary point $x = 4$ is represented by a closed circle. Because division by zero is *never* permitted, the boundary point, $x = -1$, that was found by setting the polynomial in the denominator equal to zero, must always be represented by an open circle on the number line.

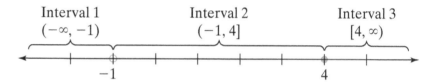

Step 5. Now that the number line is divided into three intervals, pick a test value from each interval.

Interval 1: Test value $x = -2$
Interval 2: Test value $x = 0$
Interval 3: Test value $x = 5$

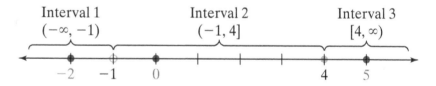

Step 6. Substitute each test value into the rational expression and determine whether the expression is positive or negative on the interval.

$x = -2$: $\dfrac{(-2-4)}{(-2+1))} \Rightarrow \dfrac{(-)}{(-)} = +$ Expression is positive on the interval $(-\infty, -1)$.

$x = 0$: $\dfrac{(0-4)}{(0+1)} \Rightarrow \dfrac{(-)}{(+)} = -$ Expression is negative on the interval $(-1, 4]$.

$x = 5$: $\dfrac{(5-4)}{(5+1)} \Rightarrow \dfrac{(+)}{(+)} = +$ Expression is positive on the interval $[4, \infty)$.

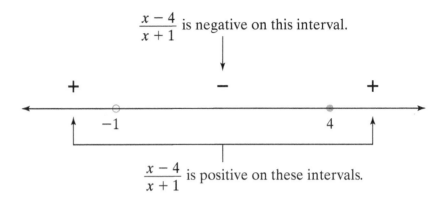

$\dfrac{x-4}{x+1}$ is negative on this interval.

$\dfrac{x-4}{x+1}$ is positive on these intervals.

Step 7. Determine the intervals that satisfy the inequality.

Finally, because we are looking for values of x for which the rational expression is greater than or equal to zero, looking at the previous number line we see that the solution to the inequality is $(-\infty, -1) \cup [4, \infty)$.

Note that we include 4 as a solution but not -1 because $x = -1$ makes the denominator equal to 0.

 My video summary ▶ **Example 4** Solve a Rational Inequality with Nonzero Factors on Both Sides of the Inequality

Solve $x > \dfrac{3}{x-2}$

Solution

Step 1. Move all terms to one side of the inequality, leaving zero on the other side.

$$x - \frac{3}{x-2} > 0$$

⚠ **You cannot multiply both sides of the inequality by $x - 2$ to eliminate the fraction. This is because we do not know whether $x - 2$ is negative or positive; therefore, we do not know whether we would need to reverse the direction of the inequality.**

Now that all nonzero terms are on the left-hand side, we must combine the terms by getting a common denominator:

$$x - \frac{3}{x-2} > 0 \qquad \text{Rewrite the inequality.}$$

$$\frac{x(x-2)}{x-2} - \frac{3}{x-2} > 0 \qquad \text{Rewrite using a common denominator of } x - 2.$$

$$\frac{x^2 - 2x - 3}{x-2} > 0 \qquad \text{Combine terms.}$$

Step 2. Factor the numerator and denominator of the nonzero side of the inequality and cancel any common factors.

$$\frac{(x-3)(x+1)}{x-2} > 0$$

 Finish steps 3–7 on your own, and see if you can come up with the correct solution of $(-1, 2) \cup (3, \infty)$. Watch this video to see the entire solution.

You Try It Work through this You Try It problem.

Work Exercises 11–22 in this textbook or in the MyMathLab Study Plan.

1.9 Exercises

In Exercises 1–10, determine all boundary points and solve the polynomial inequality. Express each solution using interval notation.

SbS **1.** $(x - 1)(x + 3) \geq 0$ SbS **2.** $x(3x + 2) \leq 0$ SbS **3.** $(x - 1)(x + 4)(x - 3) \geq 0$

SbS **4.** $2x^2 - 4x > 0$ SbS **5.** $x^2 + 3x - 21 < 0$ SbS **6.** $x^2 \leq 1$

SbS **7.** $3x^2 + x < 3x - 1$ SbS **8.** $2x^3 > 24x - 2x^2$ SbS **9.** $x^3 + x^2 - x \leq 1$

SbS **10.** $x^3 \geq -2x^2 - x$

In Exercises 11–22, determine all boundary points and solve the rational inequality. Express each solution using interval notation.

SbS **11.** $\dfrac{x + 3}{x - 1} \leq 0$ SbS **12.** $\dfrac{2 - x}{3x + 9} \geq 0$ SbS **13.** $\dfrac{x}{x - 1} > 0$

SbS **14.** $\dfrac{x^2 - 9}{x + 2} \leq 0$ SbS **15.** $\dfrac{x - 2}{x^2 + 5x - 24} \leq 0$ SbS **16.** $\dfrac{x + 2}{x^2 - x - 6} \leq 0$

SbS **17.** $\dfrac{4}{x + 1} \geq 2$ SbS **18.** $\dfrac{x + 5}{2x - 3} > 1$ SbS **19.** $\dfrac{x^2 - 8}{x + 4} \geq x$

SbS **20.** $\dfrac{x}{2 - x} \leq \dfrac{1}{x}$ SbS **21.** $\dfrac{x - 1}{x - 2} \geq \dfrac{x + 2}{x + 3}$ SbS **22.** $\dfrac{x - 1}{x + 1} + \dfrac{x + 1}{x - 1} \leq \dfrac{x + 5}{x^2 - 1}$

Brief Exercises

In Exercises 23-32, solve each polynomial inequality. Express each solution using interval notation.

23. $(x - 1)(x + 3) \geq 0$ **24.** $x(3x + 2) \leq 0$ **25.** $(x - 1)(x + 4)(x - 3) \geq 0$

26. $2x^2 - 4x > 0$ **27.** $x^2 + 3x - 21 < 0$ **28.** $x^2 \leq 1$

29. $3x^2 + x < 3x - 1$ **30.** $2x^3 > 24x - 2x^2$ **31.** $x^3 + x^2 - x \leq 1$

32. $x^3 \geq -2x^2 - x$

In Exercises 33-42, solve each rational inequality. Express each solution using interval notation.

33. $\dfrac{x+3}{x-1} \le 0$
34. $\dfrac{2-x}{3x+9} \ge 0$
35. $\dfrac{x}{x-1} > 0$
36. $\dfrac{x^2-9}{x+2} \le 0$

37. $\dfrac{x-2}{x^2+5x-24} \le 0$
38. $\dfrac{x+2}{x^2-x-6} \le 0$
39. $\dfrac{4}{x+1} \ge 2$
40. $\dfrac{x+5}{2x-3} > 1$

41. $\dfrac{x^2-8}{x+4} \ge x$
42. $\dfrac{x}{2-x} \le \dfrac{1}{x}$
43. $\dfrac{x-1}{x-2} \ge \dfrac{x+2}{x+3}$
44. $\dfrac{x-1}{x+1} + \dfrac{x+1}{x-1} \le \dfrac{x+5}{x^2-1}$

The Rectangular Coordinate System, Lines, and Circles

CHAPTER TWO CONTENTS

2.1 The Rectangular Coordinate System

THINGS TO KNOW

Before working through this section, be sure that you are familiar with the following concepts:

VIDEO   ANIMATION   INTERACTIVE

You Try It

1. Simplifying Radicals
   (Section R.3)

OBJECTIVES

1 Plotting Ordered Pairs

2 Graphing Equations by Plotting Points

3 Finding the Midpoint of a Line Segment Using the Midpoint Formula

4 Finding the Distance between Two Points Using the Distance Formula

Introduction to Section 2.1

Throughout chapter one, we solved several types of equations, including linear equations, quadratic equations, and rational equations. Each equation had something in common. They were all examples of **equations in one variable**. In this chapter, we study **equations involving two variables**. To motivate the discussion, consider the two-variable equation $x - 2y = -8$. What does it mean for this equation to have a solution? A solution to this equation consists of a *pair of numbers*, that is, an x-value and a y-value for which the equation is true.

For example, the pair of numbers $x = 2$ and $y = 5$ is a solution to this equation because substituting these two values into the equation produces a true statement:

$$x - 2y = -8 \qquad \text{Write the two-variable equation.}$$
$$(2) - 2(5) \overset{?}{=} -8 \qquad \text{Substitute } x = 2 \text{ and } y = 5.$$
$$-8 = -8 \checkmark \qquad \text{The statement is true.}$$

The pair of numbers, $x = -2$ and $y = 3$ is also the solution to this equation:

$$x - 2y = -8 \qquad \text{Write the two-variable equation.}$$
$$(-2) - 2(3) \overset{?}{=} -8 \qquad \text{Substitute } x = -2 \text{ and } y = 3.$$
$$-8 = -8 \checkmark \qquad \text{The statement is true.}$$

In fact, there are infinitely many pairs of numbers that satisfy this equation. Table 1 illustrates a few solutions to this equation.

Table 1

x	y	$x - 2y = -8$
-2	3	$-2 - 2(3) = -8$
0	4	$0 - 2(4) = -8$
2	5	$2 - 2(5) = -8$
4	6	$4 - 2(6) = -8$

Each pair of values is called an **ordered pair** because you can see that the order does matter. We use the notation (x, y) to represent an ordered pair. Notice that the x-coordinate or **abscissa** is first, followed by the y-coordinate or **ordinate** listed second. To represent an ordered pair graphically, we use the rectangular coordinate system, also called the Cartesian coordinate system, named after the French mathematician René Descartes. The plane used in this system is called the **coordinate plane** or **Cartesian plane**. The horizontal axis (x-axis) and the vertical axis (y-axis) intersect at the **origin** and naturally divide the plane into four quadrants, labeled quadrants I, II, III, and IV. See Figure 1.

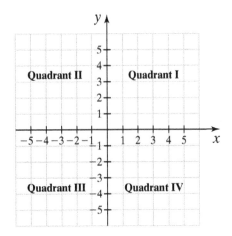

Figure 1 Rectangular coordinate system

OBJECTIVE 1   PLOTTING ORDERED PAIRS

My video summary   ⊙ **Example 1** Plot Ordered Pairs

Plot the ordered pairs $(-2, 3)$, $(0, 4)$, $(2, 5)$, and $(4, 6)$, and state in which quadrant or on which axis each pair lies.

Solution To plot the point $(-2, 3)$, go two units to the left of the origin on the x-axis, and then move three units up, parallel to the y-axis. The point corresponding to the ordered pair $(-2, 3)$ is labeled A in Figure 2 and is located in Quadrant II. To plot the point $(0, 4)$, start at the origin and move up four units. The point corresponding to the ordered pair $(0, 4)$ is labeled B in Figure 2 and lies on the y-axis. The points $(2, 5)$ and $(4, 6)$ are labeled C and D and are both located in Quadrant I.

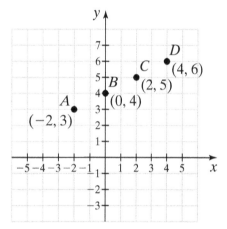

Figure 2

The four ordered pairs plotted in Figure 2 were all solutions to the equation $x - 2y = -8$. Connecting these points forms the graph of the equation $x - 2y = -8$. See Figure 3. The graph of the equation $x - 2y = -8$ is a straight line. We study the equations of lines extensively in Section 2.3.

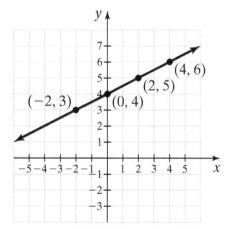

Figure 3 Graph of $x - 2y = -8$ ●

You Try It   Work through this You Try It problem.

Work Exercises 1 and 2 in this textbook or in the MyMathLab **Study Plan**.

OBJECTIVE 2 GRAPHING EQUATIONS BY PLOTTING POINTS

Not every equation in two variables has a graph that is a straight line. One way to sketch the graph of an equation is to find several ordered pairs that satisfy the equation, plot those ordered pairs, and then connect the points with a smooth curve. We choose arbitrary values for one of the coordinates and then solve the equation for the other coordinate.

Example 2 Graph an Equation by Plotting Points

Sketch the graph of $y = x^2$.

Solution Arbitrarily choose several values of x and then find the corresponding values for y:

Let $x = -2$, then $y = (-2)^2 = 4$, and thus the ordered pair $(-2, 4)$ satisfies the equation $y = x^2$.

Let $x = -1$, then $y = (-1)^2 = 1$, and thus the ordered pair $(-1, 1)$ satisfies the equation $y = x^2$.

Let $x = 0$, then $y = (0)^2 = 0$, and thus the ordered pair $(0, 0)$ satisfies the equation $y = x^2$.

Table 2 shows several more ordered pairs that are solutions to the equation $y = x^2$. The graph of $y = x^2$ is called a parabola and is shown in Figure 4.

Table 2

x	$y = x^2$	Ordered pair that lies on the graph of $y = x^2$
-2	4	$(-2, 4)$
-1	1	$(-1, 1)$
0	0	$(0, 0)$
1	1	$(1, 1)$
2	4	$(2, 4)$

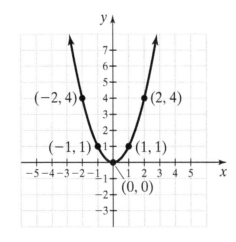

Figure 4 The graph of $y = x^2$

You Try It Work through this You Try It problem.

Work Exercises 3–6 in this textbook or in the MyMathLab Study Plan.

My video summary ⊙ **Example 3** Determine Whether a Point Lies on the Graph of an Equation

Determine whether the following ordered pairs lie on the graph of the equation $x^2 + y^2 = 1$.

a. $(0, -1)$ b. $(1, 0)$ c. $\left(\dfrac{1}{3}, \dfrac{2}{3}\right)$ d. $\left(-\dfrac{\sqrt{2}}{2}, \dfrac{\sqrt{2}}{2}\right)$

Solution Watch the video to verify that the points $(0, -1)$, $(1, 0)$, and $\left(-\dfrac{\sqrt{2}}{2}, \dfrac{\sqrt{2}}{2}\right)$ lie on the graph of $x^2 + y^2 = 1$, and the point $\left(\dfrac{1}{3}, \dfrac{2}{3}\right)$ does not lie on the graph. ●

You Try It Work through this You Try It problem.

Work Exercises 7–9 in this textbook or in the My MathLab **Study Plan**.

OBJECTIVE 3 FINDING THE MIDPOINT OF A LINE SEGMENT USING THE MIDPOINT FORMULA

Suppose we want to find the midpoint $M(x, y)$ of the line segment between the points $A(x_1, y_1)$ and $B(x_2, y_2)$. See Figure 5. To find this midpoint, we simply average the x- and y-coordinates, respectively. In other words, the x-coordinate of the midpoint is $\dfrac{x_1 + x_2}{2}$, whereas the y-coordinate of the midpoint is $\dfrac{y_1 + y_2}{2}$.

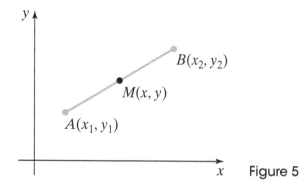

Figure 5

Midpoint Formula

The midpoint of the line segment from $A(x_1, y_1)$ to $B(x_2, y_2)$ is

$$\left(\frac{x_1 + x_2}{2}, \frac{y_1 + y_2}{2}\right).$$

Example 4 Find the Midpoint of a Line Segment

Find the midpoint of the segment whose endpoints are $(-3, 2)$ and $(4, 6)$.

Solution The midpoint of this line segment is $\left(\dfrac{-3 + 4}{2}, \dfrac{2 + 6}{2}\right) = \left(\dfrac{1}{2}, 4\right)$.
See Figure 6.

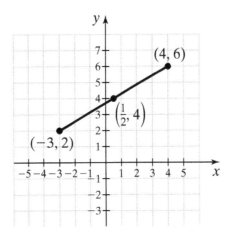

Figure 6

 **You Try It** Work through this You Try It problem.

Work Exercises 10–14 in this textbook or in the MyMathLab Study Plan.

My video summary ⊙ **Example 5** Application of the Midpoint Formula

In geometry, it can be shown that four points in a plane form a **parallelogram** if the two diagonals of the quadrilateral formed by the four points **bisect** each other. Do the points $A(0, 4)$, $B(3, 0)$, $C(9, 1)$, and $D(6, 5)$ form a parallelogram?

Solution To prove that the two diagonals bisect each other, we can show that they share the same midpoint. We can see in Figure 7 that the diagonals are the segments $\overline{AC}$ and $\overline{DB}$.

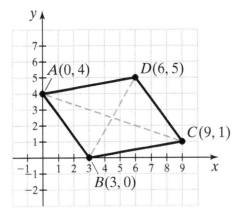

Figure 7

The midpoint of diagonal $\overline{AC}$ is $\left(\dfrac{0 + 9}{2}, \dfrac{4 + 1}{2} \right) = \left(\dfrac{9}{2}, \dfrac{5}{2} \right)$.

The midpoint of diagonal $\overline{DB}$ is $\left(\dfrac{6 + 3}{2}, \dfrac{5 + 0}{2} \right) = \left(\dfrac{9}{2}, \dfrac{5}{2} \right)$.

My video summary ⊙ Because the diagonals share the same midpoint, it is implied that they bisect each other. Therefore, the quadrilateral is a parallelogram. Watch this video for a detailed solution.

 **You Try It** Work through this You Try It problem.

Work Exercises 15–17 in this textbook or in the MyMathLab Study Plan.

OBJECTIVE 4 FINDING THE DISTANCE BETWEEN TWO POINTS USING THE DISTANCE FORMULA

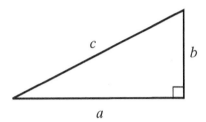 *My video summary*

⊘ Recall that the Pythagorean theorem states that the sum of the squares of the two sides of a right triangle is equal to the square of the hypotenuse or $a^2 + b^2 = c^2$. See Figure 8.

Figure 8 Pythagorean theorem: $a^2 + b^2 = c^2$

We can use the Pythagorean theorem to find the distance between any two points in a plane. For example, suppose we want to find the distance, $d(A, B)$, between the points $A(x_1, y_1)$ and $B(x_2, y_2)$. To find the length $d(A, B)$ of the line segment AB, consider the point $C(x_1, y_2)$ in **Figure 9**, which is on the same vertical line segment as point A and the same horizontal line segment as point B. The triangle formed by points A, B, and C is a right triangle whose hypotenuse has length $d(A, B)$. The horizontal leg of the triangle has length $a = |x_2 - x_1|$, whereas the vertical leg of the triangle has length $b = |y_2 - y_1|$.

Therefore, by the Pythagorean theorem, $\underbrace{|x_2 - x_1|^2}_{a^2} + \underbrace{|y_2 - y_1|^2}_{b^2} = \underbrace{[d(A, B)]^2}_{c^2}$.

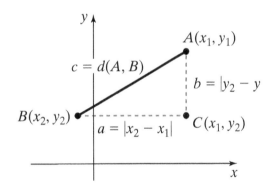

Figure 9

$$[d(A, B)]^2 = |x_2 - x_1|^2 + |y_2 - y_1|^2 \qquad \text{Use the Pythagorean theorem.}$$
$$d(A, B) = \pm\sqrt{|x_2 - x_1|^2 + |y_2 - y_1|^2} \qquad \text{Use the square root property.}$$

However, because distance cannot be negative, exclude $-\sqrt{|x_2 - x_1|^2 + |y_2 - y_1|^2}$.

$$d(A, B) = \sqrt{|x_2 - x_1|^2 + |y_2 - y_1|^2} \qquad \text{Use the positive square root only.}$$
$$d(A, B) = \sqrt{(x_2 - x_1)^2 + (y_2 - y_1)^2} \qquad \text{For any real number } a, |a|^2 = a^2.$$

This formula is known as the distance formula.

Distance Formula

The distance between any two points $A(x_1, y_1)$ and $B(x_2, y_2)$ is given by the formula

$$d(A, B) = \sqrt{(x_2 - x_1)^2 + (y_2 - y_1)^2}.$$

2.1 The Rectangular Coordinate System   **2-7**

Example 6 Use the Distance Formula to Find the Distance between Two Points

Find the distance, $d(A, B)$, between the points A and B.

$A(-1, 5); B(4, -5)$

Solution

$$d(A, B) = \sqrt{(x_2 - x_1)^2 + (y_2 - y_1)^2} \qquad \text{Use the distance formula.}$$
$$= \sqrt{(4 - (-1))^2 + (-5 - 5)^2} \qquad \text{Substitute in the distance formula.}$$
$$= \sqrt{5^2 + (-10)^2} \qquad \text{Combine terms.}$$
$$= \sqrt{25 + 100} \qquad \text{Simplify.}$$
$$= \sqrt{125} \qquad \text{Add.}$$
$$= 5\sqrt{5} \qquad \text{Simplify the radical.}$$

The distance between the two given points is $d(A, B) = 5\sqrt{5}$ units. See Figure 10.

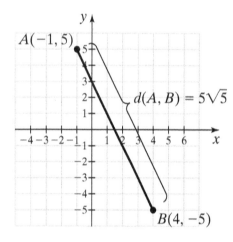

Figure 10 Distance from A to B is $d(A, B) = 5\sqrt{5}$.

You Try It Work through this You Try It problem.

Work Exercises 18–21 in this textbook or in the MyMathLab Study Plan.

My video summary ⊙ **Example 7** Application of the Distance Formula

Verify that the points $A(3, -5)$, $B(0, 6)$, and $C(5, 5)$ form a right triangle.

Solution The three points do form a right triangle. Watch the video to see the solution to this example.

You Try It Work through this You Try It problem.

Work Exercises 22–25 in this textbook or in the MyMathLab Study Plan.

2.1 Exercises

Skill Check Exercises

For exercises SCE-1 through SCE-3, find the average of the two indicated real numbers.

SCE-1. $-3, 8$ **SCE-2.** $\dfrac{1}{3}, -2$ **SCE-3.** $-\dfrac{2}{5}, \dfrac{3}{7}$

For exercises SCE-4 through SCE-7, simplify the radical.

SCE-4. $\sqrt{108}$ **SCE-5.** $\sqrt{(-7)^2 + (24)^2}$ **SCE-6.** $\sqrt{(-9)^2 + (-3)^2}$

SCE-7. $\sqrt{\left(-\dfrac{\sqrt{29}}{2}\right)^2 + \left(\dfrac{\sqrt{71}}{2}\right)^2}$

In Exercises 1 and 2, plot each ordered pair in the Cartesian plane, and state in which quadrant or on which axis it lies.

1. a. $A(-1, 4)$ **b.** $B(-2, -2)$ **c.** $C(0, -3)$ **d.** $D\left(1, \dfrac{1}{2}\right)$ **e.** $E(2, 0)$

2. a. $A(2, -4)$ **b.** $B\left(-\dfrac{5}{2}, 1\right)$ **c.** $C(\sqrt{2}, \pi)$ **d.** $D\left(0, -\dfrac{9}{2}\right)$ **e.** $E(0, 0)$

In Exercises 3–6, sketch the graph for each equation by plotting points.

3. $y = 2x + 1$ **4.** $2y = x^3$ **5.** $y^2 + 3x = 0$ **6.** $y = 3x^2$

In Exercises 7–9, determine whether the indicated ordered pairs lie on the graph of the given equation.

7. $y = 2x^2 + 1$ **8.** $y = \sqrt{x} - 1$ **9.** $y = |x|$

  a. $(0, 1)$ **a.** $(4, 1)$ **a.** $(-2, 2)$

  b. $(-1, 1)$ **b.** $(1, 4)$ **b.** $(-9, 9)$

  c. $(1, 3)$ **c.** $(0, 1)$ **c.** $(3, 3)$

In Exercises 10–13, find the midpoint of the line segment joining points A and B.

10. $A(-1, 4)$; $B(-2, -2)$ **11.** $A(2, -5)$; $B(4, 1)$

12. $A(0, 1)$; $B\left(-3, \dfrac{1}{2}\right)$ **13.** $A(a, b)$; $B(c, d)$

14. A line segment has a midpoint of $\left(1, \dfrac{1}{2}\right)$. If one endpoint is $(7, 3)$, what is the other endpoint?

In Exercises 15–17, determine whether the points A, B, C, and D form a parallelogram.

15. $A(1, 4)$; $B(-2, -1)$; $C(4, 2)$; $D(7, 7)$

16. $A(-2, 4)$; $B(-1, 2)$; $C(1, 1)$; $D(3, 3)$

17. $A(-2, -1)$; $B(2, 0)$; $C(3, 3)$; $D(-1, 2)$

In Exercises 18–21, find the distance, $d(A, B)$ between points A and B.

18. $A(1, 5)$; $B(-2, 1)$

19. $A(3, 5)$; $B(-2, -2)$

20. $A(2, -4)$; $B(-3, 6)$

21. $A\left(-\dfrac{\sqrt{3}}{2}, \dfrac{\sqrt{17}}{2}\right)$; $B(0, 0)$

In Exercises 22–24, determine whether the points A, B, and C form a right triangle.

22. $A(-3, -3)$; $B(3, -1)$; $C(2, 2)$

23. $A(1, 4)$; $B(-2, -1)$; $C(4, 2)$

24. $A(1, 2)$; $B(2, 6)$; $C(9, 0)$

25. Find the y-coordinates of the points that are 5 units away from the point $(-1, 3)$ that have an x-coordinate of 3.

2.2 Circles

THINGS TO KNOW

Before working through this section, be sure that you are familiar with the following concepts:

VIDEO   ANIMATION   INTERACTIVE

You Try It

1. Solving Quadratic Equations by Completing the Square (Section 1.4)

You Try It

2. Solving Quadratic Equations Using the Square Root Property (Section 1.4)

You Try It

3. Finding the Midpoint of a Line Segment Using the Midpoint Formula (Section 2.1)

You Try It

4. Finding the Distance between Two Points Using the Distance Formula (Section 2.1)

OBJECTIVES

1 Writing the Standard Form of an Equation of a Circle

2 Sketching the Graph of a Circle

3 Converting the General Form of a Circle into Standard Form

Introduction to Section 2.2

A **circle** is the set of all points (x, y) in the Cartesian plane that are a fixed distance, r, from a fixed point, (h, k). The fixed distance, r, is called the **radius** of the circle, and the fixed point (h, k), is called the **center** of the circle. See Figure 11. To derive the equation of a circle, we use the distance formula that was discussed in Section 2.1.

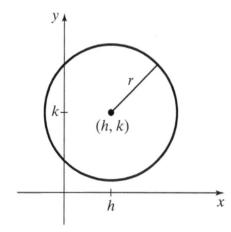

Figure 11 Circle with center, (h, k) and radius, r

My animation summary

▶ Watch this animation to see how the standard form of the equation of a circle is derived.

The **standard form of an equation of a circle** with center, (h, k), and radius, r, is

$$(x - h)^2 + (y - k)^2 = r^2.$$

The standard form of an equation of a circle centered at the origin with radius, r, is

$$x^2 + y^2 = r^2.$$

OBJECTIVE 1 WRITING THE STANDARD FORM OF AN EQUATION OF A CIRCLE

My video summary

⊙ **Example 1 Find the Standard Form of an Equation of a Circle**

Find the standard form of the equation of the circle whose center is $(-2, 3)$ and with radius 6.

Solution We know that the standard form of the equation of a circle is $(x - h)^2 + (y - k)^2 = r^2$. We are given $h = -2, k = 3$, and $r = 6$. Substituting these values into the equation, we get

$$(x - (-2))^2 + (y - 3)^2 = 6^2 \quad \text{or} \quad (x + 2)^2 + (y - 3)^2 = 36.$$

You Try It Work through this You Try It problem.

Work Exercises 1–4 in this textbook or in the MyMathLab Study Plan.

Example 2 Find the Equation of a Circle Given the Center and a Point on the Circle

Find the standard form of the equation of the circle whose center is $(0, 6)$ and that passes through the point $(4, 2)$.

Solution The graph of this circle is sketched in Figure 12. The center is labeled $A(0, 6)$, and the point on the circle is labeled $B(4, 2)$. To find the equation of this circle, we must calculate the radius. The radius can be found using the distance formula.

$r = d(A, B) = \sqrt{(x_2 - x_1)^2 + (y_2 - y_1)^2}$ Write the distance formula.

$r = d(A, B) = \sqrt{(4 - 0)^2 + (2 - 6)^2}$ Substitute into the distance formula.

$r = d(A, B) = \sqrt{(4)^2 + (-4)^2}$ Combine terms.

$r = d(A, B) = \sqrt{32}$ Add.

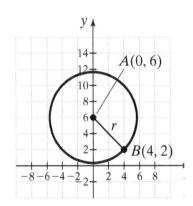

Figure 12

Now that we know the center and the radius, we can determine the equation of this circle.

$(x - h)^2 + (y - k)^2 = r^2$ Write the standard equation of a circle.

$(x - 0)^2 + (y - 6)^2 = \sqrt{32}^2$ Substitute values for the center and the radius.

$x^2 + (y - 6)^2 = 32$ Simplify.

The equation of this circle in standard form is $x^2 + (y - 6)^2 = 32$.

 *My video summary*

▷ Example 3 Find the Equation of a Circle Given the Endpoints of a Diameter

Find the standard form of the equation of the circle that contains endpoints of a diameter at $(-4, -3)$ and $(2, -1)$.

Solution We can use the midpoint formula to determine the center of this circle. The radius can be found using the distance formula. Watch this video to verify that the equation of this circle is $(x + 1)^2 + (y + 2)^2 = 10$.

You Try It   Work through this You Try It problem.

Work Exercises 5–10 in this textbook or in the MyMathLab Study Plan.

OBJECTIVE 2   SKETCHING THE GRAPH OF A CIRCLE

Once you know the center and the radius of a circle, it is fairly straightforward to graph the circle.

   ▷ **Example 4** Sketch the Graph of a Circle

Find the center and the radius, and sketch the graph of the circle $(x - 1)^2 + (y + 2)^2 = 9$. Also find any intercepts.

Solution The circle $(x - 1)^2 + (y + 2)^2 = 9$ has center $(h, k) = (1, -2)$ and $r = 3$.

⚠ **Note that the y-coordinate of the center** $(k = -2)$ **is negative because** $(y + 2)^2 = (y - (-2))^2$.

To sketch the graph of this circle, plot the center, and then locate a few points on the circle. The points that are easiest to locate are the points that are three units left and right of the center and three units up and down from the center. Complete the graph by drawing the circle through these four points, as in Figure 13.

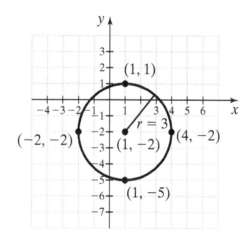

Figure 13 Graph of $(x - 1)^2 + (y + 2)^2 = 9$

To find the x-intercepts, set $y = 0$ and solve for x.

$$(x - 1)^2 + (0 + 2)^2 = 9 \qquad \text{Set } y = 0.$$
$$(x - 1)^2 + 4 = 9 \qquad \text{Simplify.}$$
$$(x - 1)^2 = 5 \qquad \text{Subtract 4 from both sides.}$$
$$x - 1 = \pm\sqrt{5} \qquad \text{Use the square root property.}$$
$$x = 1 \pm \sqrt{5} \qquad \text{Add 1 to both sides.}$$

The two x-intercepts are $x \approx 3.24$ and $x \approx -1.24$.

To find any y-intercepts, set $x = 0$ and solve for y. You should verify that the y-intercepts are $y = -2 + 2\sqrt{2} \approx .83$ and $y = -2 - 2\sqrt{2} \approx -4.83$. Watch the

   ▷ video to see this example worked out in its entirety.

You Try It   Work through this You Try It problem.

Work Exercises 11–17 in this textbook or in the MyMathLab Study Plan.

OBJECTIVE 3   CONVERTING THE GENERAL FORM OF A CIRCLE INTO STANDARD FORM

As you can see from Example 4, if the equation of a circle is in standard form, it is not too difficult to determine the center and the radius. For example, the center of the circle $(x + 3)^2 + (y - 1)^2 = 49$ is at $(-3, 1)$, and the radius is 7. Let's now take this same equation, eliminate both sets of parentheses, and simplify:

$(x + 3)^2 + (y - 1)^2 = 49$	Write the original equation in standard form.
$x^2 + 6x + 9 + y^2 - 2y + 1 = 49$	Remove parentheses.
$x^2 + 6x + y^2 - 2y + 10 = 49$	Combine like terms.
$x^2 + y^2 + 6x - 2y - 39 = 0$	Subtract 49 from both sides and rearrange terms.

We see that the equation $(x + 3)^2 + (y - 1)^2 = 49$ is equivalent to the equation $x^2 + y^2 + 6x - 2y - 39 = 0$; however, determining the center and the radius of this equation is less obvious. The equation $x^2 + y^2 + 6x - 2y - 39 = 0$ is said to be in **general form**.

> The **general form of the** equation of a circle* is $Ax^2 + By^2 + Cx + Dy + E = 0$, where A, B, C, D, and E are real numbers such that $A \neq 0$, $B \neq 0$, and $A = B$.

Before working through Example 5, it may be beneficial to review how to complete the square, which was discussed in Section 1.4.

My video summary

> **Example 5** Convert the General Form of a Circle into Standard Form

Write the equation $x^2 + y^2 - 8x + 6y + 16 = 0$ in standard form; find the center, radius, and intercepts, and sketch the graph.

Solution   Because we are going to complete the square on both x and y, we first rearrange the terms, leaving some room to complete the square, and move any constants to the right-hand side.

$x^2 - 8x \qquad + y^2 + 6y \qquad = -16$	Rearrange the terms.
$x^2 - 8x + 16 + y^2 + 6y + 9 = -16 + 16 + 9$	Complete the square on x and y. Remember to add 16 and 9 to both sides.

$$\left(\frac{1}{2} \cdot (-8)\right)^2 = 16 \qquad \left(\frac{1}{2} \cdot 6\right)^2 = 9$$

$(x - 4)^2 + (y + 3)^2 = 9$	Factor the left side.

We have now converted the general form of the circle into standard form. The center is at $(4, -3)$, and the radius is $r = 3$. Watch the video to verify that there are no y-intercepts. The only x-intercept is $x = 4$. The graph of the circle is shown in Figure 14.

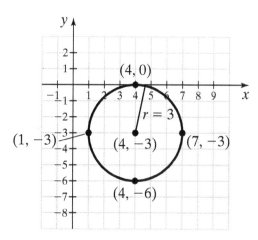

Figure 14 Graph of
$$x^2 + y^2 - 8x + 6y + 16 = 0$$

My animation summary

🔲 **Example 6** Convert the General Form of a Circle into Standard Form, Where $A \neq 1$ and $B \neq 1$

Write the equation $4x^2 + 4y^2 + 4x - 8y + 1 = 0$ in standard form; find the center, radius, and intercepts, and sketch the graph.

Solution Work through the animation to see that this equation is equivalent to the following equation in standard form:

$$\left(x + \frac{1}{2}\right)^2 + (y - 1)^2 = 1; \text{center} = \left(-\frac{1}{2}, 1\right), \text{and } r = 1$$

My animation summary

🔲 Work through the animation to verify that the one x-intercept is found at $\left(-\frac{1}{2}, 0\right)$, whereas the y-intercepts are found at $\left(0, 1 + \frac{\sqrt{3}}{2}\right)$ and $\left(0, 1 - \frac{\sqrt{3}}{2}\right)$. See the graph in Figure 15.

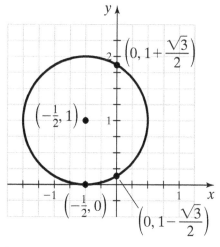

Figure 15 Graph of
$$4x^2 + 4y^2 + 4x - 8y + 1 = 0$$

You Try It   Work through this You Try It problem.

Work Exercises 18–28 in this textbook or in the MyMathLab Study Plan.

2.2 Exercises

Skill Check Exercises

For exercises SCE-1 through SCE-5, decide what number must be added to both sides of each equation to make the left side a perfect square trinomial, then factor the left side.

SCE-1. $x^2 + 10x = 3$ **SCE-2.** $x^2 - 5x = 1$ **SCE-3.** $y^2 - 8y = -7$

SCE-4. $y^2 - 9y = 2$ **SCE-5.** $x^2 + \dfrac{5}{3}x = -1$

For exercises SCE-6 through SCE-9, decide what numbers must be added to both sides of each equation to make the left side the sum of a perfect square trinomial in x and a perfect square trinomial in y.

SCE-6. $x^2 + 10x + y^2 - 8y = -4$ **SCE-7.** $x^2 - 5x + y^2 - 6y = 3$

SCE-8. $x^2 - 7x + y^2 + 11y = -2$ **SCE-9.** $x^2 + \dfrac{2}{3}x + y^2 + \dfrac{5}{7}y = -1$

In Exercises 1–10, write the standard form of the equation of each circle described.

1. Center $(0, 0)$, $r = 1$ 2. Center $(-2, 3)$, $r = 4$

3. Center $(1, -4)$, $r = 3$ 4. Center $\left(-\dfrac{1}{4}, -\dfrac{1}{3}\right)$, $r = 2$

5. Center $(0, 2)$, passes through the point $(4, -1)$

6. Center $(-4, 7)$, passes through the point $(2, 1)$

7. The endpoints of a diameter are $(0, 1)$ and $(6, -3)$

8. The endpoints of a diameter are $(2, -6)$ and $(6, 1)$

9. Center $(2, -3)$ and tangent to the x-axis

10. Center $(-4, 1)$ and tangent to the y-axis

In Exercises 11–17, find the center, radius, and intercepts of each circle and then sketch the graph.

SbS **11.** $x^2 + y^2 = 1$ SbS **12.** $x^2 + (y - 2)^2 = 4$

SbS **13.** $(x - 1)^2 + (y + 5)^2 = 16$ SbS **14.** $(x + 2)^2 + (y + 4)^2 = 36$

SbS **15.** $(x - 4)^2 + (y + 7)^2 = 12$ SbS **16.** $(x + 1)^2 + (y - 3)^2 = 20$

SbS **17.** $\left(x - \dfrac{1}{4}\right)^2 + \left(y + \dfrac{1}{2}\right)^2 = 4$

In Exercises 18–28, find the center, radius, and intercepts of each circle and then sketch the graph.

SbS **18.** $x^2 + y^2 + 2x - 4y + 1 = 0$

SbS **19.** $x^2 + y^2 - 10x + 6y + 18 = 0$

SbS **20.** $x^2 + y^2 - 4x - 8y + 19 = 0$

SbS **21.** $x^2 + y^2 + 2y - 8 = 0$

SbS **22.** $x^2 + y^2 - 6x - 12y - 5 = 0$

SbS **23.** $x^2 + y^2 - 3x - y - \dfrac{1}{2} = 0$

SbS **24.** $x^2 + y^2 + \dfrac{2}{3}x - \dfrac{1}{2}y - \dfrac{7}{18} = 0$

SbS **25.** $2x^2 + 2y^2 - 4x + 8y + 2 = 0$

SbS **26.** $16x^2 + 16y^2 - 16x + 8y - 11 = 0$

SbS **27.** $144x^2 + 144y^2 - 72x - 96y - 551 = 0$

SbS **28.** $36x^2 + 36y^2 + 12x + 72y - 35 = 0$

Brief Exercises

In Exercises 29–35, find the center and radius of each circle.

29. $x^2 + y^2 = 1$

30. $x^2 + (y - 2)^2 = 4$

31. $(x - 1)^2 + (y + 5)^2 = 16$

32. $(x + 2)^2 + (y + 4)^2 = 36$

33. $(x - 4)^2 + (y + 7)^2 = 12$

34. $(x + 1)^2 + (y - 3)^2 = 20$

35. $\left(x - \dfrac{1}{4}\right)^2 + \left(y + \dfrac{1}{2}\right)^2 = 4$

In Exercises 36–42, find intercepts of each circle.

36. $x^2 + y^2 = 1$

37. $x^2 + (y - 2)^2 = 4$

38. $(x - 1)^2 + (y + 5)^2 = 16$

39. $(x + 2)^2 + (y + 4)^2 = 36$

40. $(x - 4)^2 + (y + 7)^2 = 12$

41. $(x + 1)^2 + (y - 3)^2 = 20$

42. $\left(x - \dfrac{1}{4}\right)^2 + \left(y + \dfrac{1}{2}\right)^2 = 4$

In Exercises 43–49, sketch the graph of each circle.

43. $x^2 + y^2 = 1$

44. $x^2 + (y - 2)^2 = 4$

45. $(x - 1)^2 + (y + 5)^2 = 16$

46. $(x + 2)^2 + (y + 4)^2 = 36$

47. $(x - 4)^2 + (y + 7)^2 = 12$

48. $(x + 1)^2 + (y - 3)^2 = 20$

49. $\left(x - \dfrac{1}{4}\right)^2 + \left(y + \dfrac{1}{2}\right)^2 = 4$

In Exercises 50–60, find the center and radius of each circle.

50. $x^2 + y^2 + 2x - 4y + 1 = 0$

51. $x^2 + y^2 - 10x + 6y + 18 = 0$

52. $x^2 + y^2 - 4x - 8y + 19 = 0$

53. $x^2 + y^2 + 2y - 8 = 0$

54. $x^2 + y^2 - 6x - 12y - 5 = 0$

55. $x^2 + y^2 - 3x - y - \dfrac{1}{2} = 0$

56. $x^2 + y^2 + \dfrac{2}{3}x - \dfrac{1}{2}y - \dfrac{7}{18} = 0$

57. $2x^2 + 2y^2 - 4x + 8y + 2 = 0$

58. $16x^2 + 16y^2 - 16x + 8y - 11 = 0$

59. $144x^2 + 144y^2 - 72x - 96y - 551 = 0$

60. $36x^2 + 36y^2 + 12x + 72y - 35 = 0$

In Exercises 61–71, find intercepts of each circle.

61. $x^2 + y^2 + 2x - 4y + 1 = 0$

62. $x^2 + y^2 - 10x + 6y + 18 = 0$

63. $x^2 + y^2 - 4x - 8y + 19 = 0$

64. $x^2 + y^2 + 2y - 8 = 0$

65. $x^2 + y^2 - 6x - 12y - 5 = 0$

66. $x^2 + y^2 - 3x - y - \dfrac{1}{2} = 0$

67. $x^2 + y^2 + \dfrac{2}{3}x - \dfrac{1}{2}y - \dfrac{7}{18} = 0$

68. $2x^2 + 2y^2 - 4x + 8y + 2 = 0$

69. $16x^2 + 16y^2 - 16x + 8y - 11 = 0$

70. $144x^2 + 144y^2 - 72x - 96y - 551 = 0$

71. $36x^2 + 36y^2 + 12x + 72y - 35 = 0$

In Exercises 72–82, sketch the graph of each circle.

72. $x^2 + y^2 + 2x - 4y + 1 = 0$

73. $x^2 + y^2 - 10x + 6y + 18 = 0$

74. $x^2 + y^2 - 4x - 8y + 19 = 0$

75. $x^2 + y^2 + 2y - 8 = 0$

76. $x^2 + y^2 - 6x - 12y - 5 = 0$

77. $x^2 + y^2 - 3x - y - \dfrac{1}{2} = 0$

78. $x^2 + y^2 + \dfrac{2}{3}x - \dfrac{1}{2}y - \dfrac{7}{18} = 0$

79. $2x^2 + 2y^2 - 4x + 8y + 2 = 0$

80. $16x^2 + 16y^2 - 16x + 8y - 11 = 0$

81. $144x^2 + 144y^2 - 72x - 96y - 551 = 0$

82. $36x^2 + 36y^2 + 12x + 72y - 35 = 0$

2.3 Lines

THINGS TO KNOW

Before working through this section, be sure that you are familiar with the following concepts:

You Try It

1. Solving Linear Equations
    (Section 1.1)

You Try It

2. Plotting Ordered Pairs
    (Section 2.1)

OBJECTIVES

1   Determining the Slope of a Line

2   Sketching a Line Given a Point and the Slope

3   Finding the Equation of a Line Using the Point–Slope Form

4   Finding the Equation of a Line Using the Slope–Intercept Form

5   Writing the Equation of a Line in Standard Form

6   Finding the Slope and the y-Intercept of a Line in Standard Form

7   Sketching Lines by Plotting Intercepts

8   Finding the Equation of a Horizontal Line and a Vertical Line

---

OBJECTIVE 1   DETERMINING THE SLOPE OF A LINE

 *My video summary*

In this section, we study equations of lines that lie in the Cartesian plane. Before we learn about a line's equation, we must first establish a way to measure the steepness of a line. In mathematics, the steepness of a line can be measured by computing the line's **slope**. Every nonvertical line has slope. (Vertical lines are said to have no slope and are discussed in detail at the end of this section.) A line going up from left to right has **positive slope**, a line going down from left to right has **negative slope**, horizontal lines have **zero slope**, and vertical lines have **no slope** or **undefined slope**. See Figure 16.

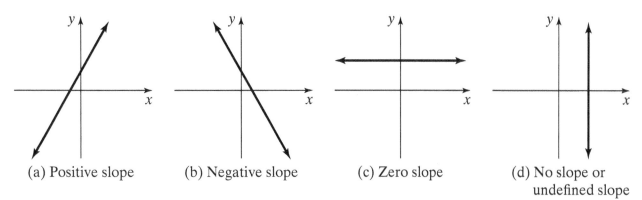

(a) Positive slope   (b) Negative slope   (c) Zero slope   (d) No slope or undefined slope

Figure 16

The slope can be computed by comparing the vertical change (the **rise**) to the horizontal change (the **run**). Given any two points on the line, the slope m, can be computed by taking the quotient of the rise over the run.

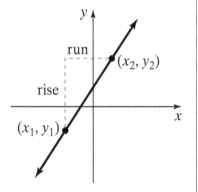

Definition Slope

If $x_1 \neq x_2$, the **slope** of a nonvertical line passing through distinct points (x_1, y_1) and (x_2, y_2) is given by

$$m = \frac{rise}{run} = \frac{\text{change in } y}{\text{change in } x} = \frac{y_2 - y_1}{x_2 - x_1}.$$

Note that any two points on the line can be used to calculate the slope. Therefore, the slope depends only on the line, and not on the choice of a pair of points that lie on it.

Example 1 Find the Slope of a Line

Find the slope of the line that passes through the indicated ordered pairs:

a. $(-2, 3)$ and $(2, -5)$ **b.** $(6, -4)$ and $(-5, 1)$

Solution

a. $m = \dfrac{y_2 - y_1}{x_2 - x_1} = \dfrac{-5 - 3}{2 - (-2)} = \dfrac{-8}{2 + 2} = -\dfrac{8}{4} = -2$

b. $m = \dfrac{y_2 - y_1}{x_2 - x_1} = \dfrac{1 - (-4)}{-5 - 6} = \dfrac{1 + 4}{-11} = \dfrac{5}{-11} = -\dfrac{5}{11}$

You Try It Work through this You Try It problem.

Work Exercises 1–4 in this textbook or in the MyMathLab Study Plan.

OBJECTIVE 2 SKETCHING A LINE GIVEN A POINT AND THE SLOPE

If we know a point on a line and the slope, we can quickly sketch the line. Watch this video to see how to sketch the line described in Example 2.

 *My video summary* ⊙ **Example 2** Sketch a Line Given a Point and the Slope

Sketch the line with slope $m = \dfrac{2}{3}$ that passes through the point $(-1, -4)$.

Also, find three more points located on the line.

Solution To sketch the line, plot the ordered pair $(-1, -4)$. We can now use the slope to plot another point. Starting at $(-1, -4)$, we rise two units and then run three units. In other words, we add 2 to the y-coordinate and 3 to the x-coordinate to get $(2, -2)$. We can continue to find more points on the line by adding 2 to each successive y-coordinate and 3 to each successive x-coordinate to get the points listed in Figure 17.

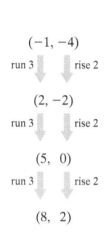

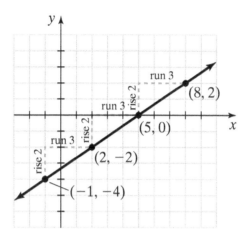

Figure 17 Line with slope passing through the point $(-1, -4)$

 You Try It   Work through this You Try It problem.

Work Exercises 5–8 in this textbook or in the MyMathLab Study Plan.

OBJECTIVE 3   FINDING THE EQUATION OF A LINE USING THE POINT–SLOPE FORM

Given the slope m of a line and a point on the line, (x_1, y_1), we can use what is known as the **point-slope form of the equation** to determine the equation of the line. The point–slope form of a line is given in the following box. Watch the video to see exactly how this equation is derived.

My video summary   ⊙ **Derivation of the Point–Slope Form of the Equation of a Line**

Point–Slope Form of the Equation of a Line

Given the slope m of a line and a point on the line (x_1, y_1), the point–slope form of the equation of a line is given by $y - y_1 = m(x - x_1)$.

Example 3 Find the Equation of a Line Given a Point and the Slope

Find an equation in point–slope form of the line with slope $m = \dfrac{2}{3}$ that passes through the point $(-1, -4)$.

Solution This is the same line that was discussed in Example 2. Because we are given the slope of the line and a point on that line, we can use the point–slope form to produce an equation for this specific line. Substitute $m = \dfrac{2}{3}$ and $(x_1, y_1) = (-1, -4)$ into the point–slope form $y - y_1 = m(x - x_1)$ to get the equation

$$y - (-4) = \frac{2}{3}(x - (-1)) \quad \text{or} \quad y + 4 = \frac{2}{3}(x + 1).$$

You Try It   Work through this You Try It problem.

Work Exercises 9–12 in this textbook or in the MyMathLab Study Plan.

The line used in Examples 2 and 3 is sketched in Figure 18. We see from Example 3 that the equation of this line in point–slope form is $y + 4 = \dfrac{2}{3}(x + 1)$. Is this the only equation of this line? Because the point $(8, 2)$ is also a point on this line, we can use the point–slope form with $m = \dfrac{2}{3}$ and $(x_1, y_1) = (8, 2)$ to obtain the equation $y - 2 = \dfrac{2}{3}(x - 8)$. We have now found two equations that describe the same line! Are these equations equivalent? Solving each equation for y, we see, in fact, that these equations are equivalent.

Equations $y + 4 = \dfrac{2}{3}(x + 1)$ and $y - 2 = \dfrac{2}{3}(x - 8)$ are equivalent

$y + 4 = \dfrac{2}{3}(x + 1)$	Write the original equations.	$y - 2 = \dfrac{2}{3}(x - 8)$
$y + 4 = \dfrac{2}{3}x + \dfrac{2}{3}$	Use the distributive property.	$y - 2 = \dfrac{2}{3}x - \dfrac{16}{3}$
$y = \dfrac{2}{3}x + \dfrac{2}{3} - 4$	Move constant term to the right.	$y = \dfrac{2}{3}x - \dfrac{16}{3} + 2$
$y = \dfrac{2}{3}x + \dfrac{2}{3} - \dfrac{12}{3}$	Get a common denominator.	$y = \dfrac{2}{3}x - \dfrac{16}{3} + \dfrac{6}{3}$
$y = \dfrac{2}{3}x - \dfrac{10}{3}$	⟵ The two equations are equivalent. ⟶	$y = \dfrac{2}{3}x - \dfrac{10}{3}$

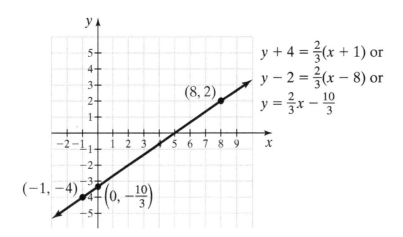

$y + 4 = \frac{2}{3}(x + 1)$ or

$y - 2 = \frac{2}{3}(x - 8)$ or

$y = \frac{2}{3}x - \frac{10}{3}$

Figure 18

It does not matter which point on the line we choose to determine the equation, as long as we use the correct slope.

OBJECTIVE 4 FINDING THE EQUATION OF A LINE USING THE SLOPE–INTERCEPT FORM

The equation $y = \dfrac{2}{3}x - \dfrac{10}{3}$ sketched in Figure 18 is said to be in *slope–intercept form* because $\dfrac{2}{3}$ describes the slope of the line, whereas the constant $-\dfrac{10}{3}$ is the *y*-intercept.

The slope–intercept form of a line is extremely important because every nonvertical line has exactly one slope–intercept equation.

> ### Slope–Intercept Form of the Equation of a Line
>
> Given the slope m of a line and the y-intercept b, the slope–intercept form of the equation of a line is given by $y = mx + b$.

To see how we can derive the slope intercept form from the point-slope form, watch this video.

My video summary   ⊙ **Derivation of the Slope-Intercept Form of the Equation of a Line**

Example 4 Find the Equation of a Line Given the Slope and y-Intercept

Find the equation of the line with slope $\dfrac{1}{4}$ and y-intercept 3, and write the equation in slope–intercept form.

Solution To find the equation of this line, we substitute $m = \dfrac{1}{4}$ and $b = 3$ into the slope–intercept form.

$$y = mx + b \qquad \text{Write the slope-intercept form.}$$

$$y = \frac{1}{4}x + 3 \qquad \text{Substitute } m = \frac{1}{4} \text{ and } b = 3.$$

We can sketch the graph of the equation $y = \dfrac{1}{4}x + 3$ by plotting the y-intercept at

the point $(0, 3)$ and then using the slope of $m = \dfrac{1}{4}$ to plot one more point. We then

draw a straight line through the two points to complete the sketch. See Figure 19.

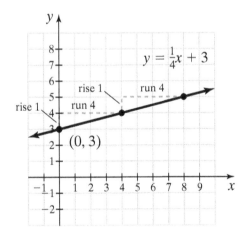

Figure 19 Graph of $y = \dfrac{1}{4}x + 3$

You Try It Work through this You Try It problem.

Work Exercises 13–16 in this textbook or in the MyMathLab Study Plan.

OBJECTIVE 5   WRITING THE EQUATION OF A LINE IN STANDARD FORM

The equation $y = \frac{1}{4}x + 3$ from Example 4 is in slope-intercept form. We can subtract $\frac{1}{4}x$ from both sides of this equation to obtain the equation $-\frac{1}{4}x + y = 3$. The equation $-\frac{1}{4}x + y = 3$ is said to be in **standard form.**

Standard Form Equation of a Line

The standard form of an equation of a line is given by $Ax + By = C$, where A, B, and C are real numbers such that A and B are both not zero.

Note that all coefficients of the equation $-\frac{1}{4}x + y = 3$ are rational numbers. When the coefficients are all rational numbers, we can eliminate any fractions by multiplying both sides of the equation by an appropriate constant. If we multiply the equation $-\frac{1}{4}x + y = 3$ by -4 we get the following.

$$-4\left(-\frac{1}{4}x + y\right) = 3(-4) \qquad \text{Multiply both sides of the equation by } -4.$$

$$x - 4y = -12 \qquad \text{Use the distributive property.}$$

The equations $y = \frac{1}{4}x + 3$, $-\frac{1}{4}x + y = 3$, and $x - 4y = -12$ all represent the same line. However, the equation $x - 4y = -12$ is often more desirable because of its simplicity with all integer coefficients and a positive leading coefficient.

Note   In this text (and in all exercises), whenever possible, the standard form of the line will be written using integer coefficients with $A \geq 0$.

My video summary   ⊙ **Example 5** Find the Equation of a Line Given Two Points on the Line

Find the equation of the line passing through the points $(-1, 3)$ and $(2, -4)$. Write the equation in point–slope form, slope–intercept form, and standard form.

Solution   First, we always find the slope.

$$m = \frac{-4 - 3}{2 - (-1)} = \frac{-7}{2 + 1} = -\frac{7}{3}.$$

Next, select either of the given points. Let's choose the first point $(-1, 3)$. Now that we have a point and the slope, we can use the point–slope form.

$$y - y_1 = m(x - x_1) \qquad \text{Write the point–slope equation.}$$

$$y - 3 = -\frac{7}{3}(x + 1) \qquad \text{Substitute } m = -\frac{7}{3} \text{ and } (x_1, y_1) = (-1, 3).$$

To find the slope-intercept form of the line, we solve the equation $y - 3 = -\dfrac{7}{3}(x + 1)$

for y to get $y = -\dfrac{7}{3}x + \dfrac{2}{3}$. To see how this is done, read these steps.

Finally, write the equation in standard form by eliminating the fractions of the slope-intercept form.

$$3 \cdot y = 3 \cdot \left(-\frac{7}{3}x + \frac{2}{3}\right) \qquad \text{Multiply by 3.}$$

$$3y = -7x + 2 \qquad \text{Use the distributive property.}$$

$$7x + 3y = 2 \qquad \text{Add } 7x \text{ to both sides.}$$

Three equations of this line are:

1. Point-slope form: $y - 3 = -\dfrac{7}{3}(x + 1)$ (or $y + 4 = -\dfrac{7}{3}(x - 2)$, or infinitely many others)

2. Slope-intercept form: $y = -\dfrac{7}{3}x + \dfrac{2}{3}$

3. Standard form: $7x + 3y = 2$

This line is sketched and labeled in Figure 20.

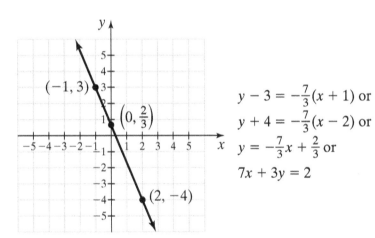

$y - 3 = -\frac{7}{3}(x + 1)$ or

$y + 4 = -\frac{7}{3}(x - 2)$ or

$y = -\frac{7}{3}x + \frac{2}{3}$ or

$7x + 3y = 2$

Figure 20 Line through the points $(-1, 3)$ and $(2, -4)$

You Try It   Work through this You Try It problem.

Work Exercises 17–20 in this textbook or in the MyMathLab Study Plan.

OBJECTIVE 6   FINDING THE SLOPE AND THE *y*-INTERCEPT OF A LINE IN STANDARD FORM

My video summary ⊙ Suppose we are given the standard form of a line $Ax + By = C$ with $B \neq 0$ and want to solve for y. To solve for y, we subtract Ax from both sides and divide by B.

$$Ax + By = C \qquad \text{Write the standard form of a line.}$$

$$By = -Ax + C \qquad \text{Subtract } Ax \text{ from both sides.}$$

$$\frac{By}{B} = -\frac{A}{B}x + \frac{C}{B} \qquad \text{Divide both sides by } B.$$

The standard form of a line $Ax + By = C$ with $B \neq 0$ is equivalent to the equation

$y = -\overset{\overbrace{m}}{\dfrac{A}{B}}x + \overset{\overbrace{b}}{\dfrac{C}{B}}$, which is the equation of a line in slope–intercept form. Thus, given

the standard form of a line $Ax + By = C$ with $B \neq 0$, the slope of the line is $m = -\dfrac{A}{B}$,

and the y-intercept is $b = \dfrac{C}{B}$. Watch the video to see how to find the slope and

y-intercept of a line given in standard form and to see how to work Example 6.

The Slope and y-Intercept of a Line Given in Standard Form

Given the standard form of a line $Ax + By = C$ with $B \neq 0$, the slope is given by

$m = -\dfrac{A}{B}$, and the y-intercept is $\dfrac{C}{B}$.

 *My video summary*

⊙ **Example 6** Find the Slope and y-Intercept of a Line in Standard Form

Find the slope and y-intercept and sketch the line $3x - 2y = 6$.

Solution Because $A = 3$ and $B = -2$, the slope is $m = -\dfrac{A}{B} = -\dfrac{3}{-2} = \dfrac{3}{2}$. The

y-intercept is $\dfrac{C}{B} = \dfrac{6}{-2} = -3$. To sketch this line, plot the y-intercept, and then rise 3

and run 2 to plot the point $(2, 0)$. Now connect the points with a straight line, as in Figure 21. Watch the video to see this solution worked out in its entirety.

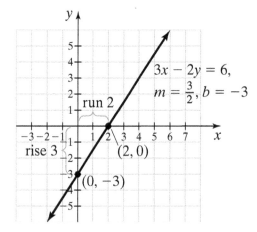

Figure 21 Line $3x - 2y = 6$

You Try It   Work through this You Try It problem.

Work Exercises 21–24 in this textbook or in the MyMathLab Study Plan.

OBJECTIVE 7 SKETCHING LINES BY PLOTTING INTERCEPTS

In Example 6, we were able to sketch the graph of the line $3x - 2y = 6$ by first plotting two points, and then drawing a straight line through them. These two points happened to be the x- and y-intercepts. The method of sketching a line by plotting intercepts is very useful and is illustrated again in Example 7.

 *My video summary*

⊗ **Example 7** Sketch a Line by Plotting Intercepts

Sketch the line $2x - 5y = 8$ by plotting intercepts.

Solution Find the x-intercept. Find the y-intercept.

The x-intercept is found by setting $y = 0$ and solving for x:

$$2x - 5(0) = 8$$
$$2x = 8$$
$$x = 4$$

The x-intercept is 4.

The y-intercept is found by setting $x = 0$ and solving for y:

$$2(0) - 5y = 8$$
$$-5y = 8$$
$$y = -\frac{8}{5}$$

The y-intercept is $-\frac{8}{5}$.

To sketch the line, plot the two intercepts by plotting the points $(4, 0)$ and $\left(0, -\frac{8}{5}\right)$ and then draw a straight line through the points, as in Figure 22. Watch

 *My video summary* ⊗ this video to see the complete solution to this example.

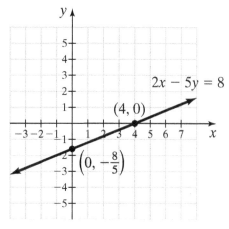

Figure 22 Line $2x - 5y = 8$

You Try It Work through this You Try It problem.

Work Exercises 25–28 in this textbook or in the MyMathLab Study Plan.

OBJECTIVE 8 FINDING THE EQUATIONS OF HORIZONTAL AND VERTICAL LINES

HORIZONTAL LINES

 *My video summary* ⊗ Suppose we want to determine the equation of the horizontal line that contains the point (a, b). To find this equation, we must first determine the slope. Because the line must also pass through the point $(0, b)$, we see that the slope of this line

is $m = \dfrac{b-b}{a-b} = \dfrac{0}{a} = 0$. Using the slope–intercept form of a line with $m = 0$ and y-intercept b, we see that the equation is $y = 0x + b$ or $y = b$. See Figure 23.

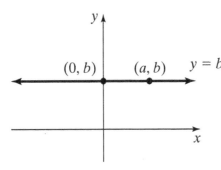

Figure 23 Equation of a horizontal line is $y = b$.

Therefore, we know that for any horizontal line that contains the point (a, b), the equation of that line is $y = b$, and the slope is $m = 0$.

VERTICAL LINES

 *My video summary*

> Vertical lines have **no slope** (also called an **undefined slope**). We can see this by looking at the vertical line that passes through the point (a, b). Because this line also passes through the x-intercept at the point $(a, 0)$, we see that the slope of this line is $m = \dfrac{b-0}{a-a} = \dfrac{b}{0}$, which is not a real number as division by zero is not defined. Because the x-coordinate of this vertical line is always equal to a regardless of the y-coordinate, we say that the equation of a vertical line is $x = a$. See Figure 24.

Therefore, we know that for any vertical line that contains the point (a, b), the equation of that line is $x = a$, and the slope is undefined.

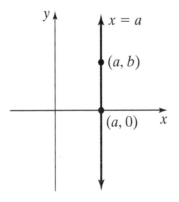

Figure 24 Equation of a vertical line is $x = a$.

Example 8 Find the Equations of Horizontal and Vertical Lines

a. Find the equation of the horizontal line passing through the point $(-1, 3)$.

b. Find the equation of the vertical line passing through the point $(-1, 3)$.

Solution

a. A horizontal line has the equation $y = b$, where b is the y-coordinate of any point on the line. Therefore, the equation is $y = 3$.

b. A vertical line has the equation $x = a$, where a is the x-coordinate of any point on the line. Therefore, the equation is $x = -1$.

Both lines are sketched in Figure 25.

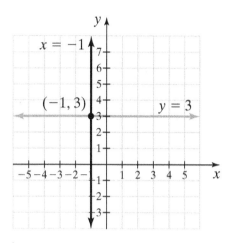

Figure 25 Graph of $y = 3$ and $x = -1$ ●

You Try It   Work through this You Try It problem.

Work Exercises 29–32 in this textbook or in the MyMathLab Study Plan.

Summary of Forms of Equations of Lines

1. $y - y_1 = m(x - x_1)$ — **Point–Slope Form**
   Slope is m, and (x_1, y_1) is a point on the line.

2. $y = mx + b$ — **Slope–Intercept Form**
   Slope is m, and y-intercept is at $(0, b)$.

3. $Ax + By = C$ — **Standard Form**
   A, B, and C are real numbers,
   with A and B both not zero and $A \geq 0$.

4. $y = b$ — **Horizontal Line**
   Slope is zero, and y-intercept is at $(0, b)$.

5. $x = a$ — **Vertical Line**
   Slope is undefined, and x-intercept is at $(a, 0)$.

2.3 Exercises

Skill Check Exercises

For exercises SCE-1 through SCE-2 solve each equation for the variable y.

SCE-1. $2x - y = 5$ **SCE-2.** $3x + 2y = 7$

For exercises SCE-3 through SCE-6, first use the distributive property then solve the equation for the variable y.

SCE-3. $y - 1 = 2(x + 3)$ **SCE-4.** $y + 3 = -5(x - 2)$

SCE-5. $y - 5 = \dfrac{2}{7}(x + 3)$ **SCE-6.** $y + 4 = -\dfrac{3}{5}(x - 1)$

In Exercises 1–4, find the slope of the line passing through the given points.

1. $(-5, 2)$ and $(2, -6)$

2. $(0, -7)$ and $(-4, -9)$

3.

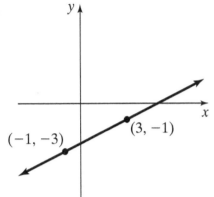

4.

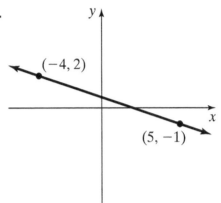

In Exercises 5–8, sketch the line with the given slope that passes through the indicated point.

5. Slope $= \dfrac{1}{2}$; line passes through the point $(-2, -2)$

6. Slope $= -\dfrac{4}{3}$; line passes through the point $(3, 8)$

7. Slope $= 0$; line passes through the point $(-5, -1)$

8. Slope is undefined; line passes through the point $(8, 0)$

In Exercises 9 –12, find the point-slope form of the line with the given slope that passes through the indicated point.

9. Slope $= \dfrac{1}{2}$; line passes through the point $(-2, -2)$

10. Slope $= -\dfrac{4}{3}$; line passes through the point $(3, 8)$

11. Slope $= 3$; line passes through the point $(-5, 3)$

12. Slope $= \dfrac{5}{11}$; line passes through the point $(-99, 88)$

In Exercises 13–16, find the slope-intercept form of the line with the given slope and *y*-intercept.

13. Slope $= 1$; *y*-intercept $= -2$

14. Slope $= -\dfrac{1}{6}$; *y*-intercept $= \dfrac{1}{2}$

15. Slope $= \dfrac{7}{9}$; *y*-intercept $= -\dfrac{8}{7}$

16. Slope $= 0$; *y*-intercept $= 5$

In Exercises 17–20, find the equation of the line passing through the indicated two points. Write the equation in point–slope form, slope–intercept form, and standard form.

17. $(1, 2)$ and $(-2, 5)$ **18.** $(-5, 7)$ and $(3, -5)$

19. $\left(-\dfrac{1}{2}, 1\right)$ and $\left(-3, \dfrac{2}{3}\right)$ **20.** $(-3, 4)$ and $(2, 4)$

In Exercises 21–24, given the equation of a line in standard form, determine the slope and y-intercept, and sketch the line.

21. $4x - y = 12$ **22.** $x - 3y = -9$ **23.** $5x + 7y = 12$ **24.** $8x - 11y = -23$

In Exercises 25–28, sketch the given line by plotting intercepts.

25. $4x - y = 12$ **26.** $x - 3y = -9$ **27.** $4x + 3y = 6$ **28.** $5x + 4y = -10$

29. Find the equation of the horizontal line passing through the point $(5, -2)$.

30. Find the equation of the vertical line passing through the point $(5, -2)$.

In Exercises 31 and 32, find the equation of the given line.

31.

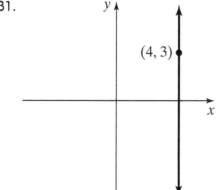

32.

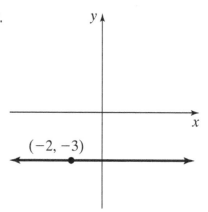

Brief Exercises

In Exercises 33–36, find the equation of the line passing through the indicated two points. Write the equation in point–slope form.

33. $(1, 2)$ and $(-2, 5)$ **34.** $(-5, 7)$ and $(3, -5)$

35. $\left(-\dfrac{1}{2}, 1\right)$ and $\left(-3, \dfrac{2}{3}\right)$ **36.** $(-3, 4)$ and $(2, 4)$

In Exercises 37–40, find the equation of the line passing through the indicated two points. Write the equation in slope–intercept form.

37. $(1, 2)$ and $(-2, 5)$ **38.** $(-5, 7)$ and $(3, -5)$

39. $\left(-\dfrac{1}{2}, 1\right)$ and $\left(-3, \dfrac{2}{3}\right)$ **40.** $(-3, 4)$ and $(2, 4)$

In Exercises 41–44, find the equation of the line passing through the indicated two points. Write the equation in standard form.

41. $(1, 2)$ and $(-2, 5)$ **42.** $(-5, 7)$ and $(3, -5)$ **43.** $\left(-\frac{1}{2}, 1\right)$ and $\left(-3, \frac{2}{3}\right)$ **44.** $(-3, 4)$ and $(2, 4)$

In Exercises 45–48, given the equation of a line in standard form, determine the slope and y-intercept.

45. $4x - y = 12$ **46.** $x - 3y = -9$ **47.** $5x + 7y = 12$ **48.** $8x - 11y = -23$

In Exercises 49–52, sketch the line whose equation is given in standard form.

49. $4x - y = 12$ **50.** $x - 3y = -9$ **51.** $5x + 7y = 12$ **52.** $8x - 11y = -23$

2.4 Parallel and Perpendicular Lines

THINGS TO KNOW

Before working through this section, be sure that you are familiar with the following concepts:

		VIDEO	ANIMATION	INTERACTIVE

You Try It   1. Determining the Slope of a Line (Section 2.3)

You Try It   2. Finding the Equation of a Line Using the Point–Slope Form (Section 2.3)

You Try It   3. Writing the Equation of a Line in Standard Form (Section 2.3)

You Try It   4. Finding the Slope and the y-Intercept of a Line in Standard Form (Section 2.3)

OBJECTIVES

1 Understanding the Definition of Parallel Lines

2 Understanding the Definition of Perpendicular Lines

3 Determining Whether Two Lines Are Parallel, Perpendicular, or Neither

4 Finding the Equations of Parallel and Perpendicular Lines

5 Solving a Geometric Application of Parallel and Perpendicular Lines

Given any two *distinct* lines in the Cartesian plane, the two lines will either intersect or not. In this section, we investigate the nature of two lines that do not intersect (**parallel lines**) and then discuss the special case of two lines that intersect at a right angle (**perpendicular lines**). These two cases are interesting because we need only know the slope of the two lines to determine whether the lines are parallel, perpendicular, or neither. Let's first consider two lines in the plane that do not intersect.

OBJECTIVE 1   UNDERSTANDING THE DEFINITION OF PARALLEL LINES

As stated, two lines are parallel if they do not intersect, or in other words, the lines do not share any common points. In Section 2.3, we learned about the slope of a line. It is imperative that you can compute the slope of a line given its equation. To review this concept, refer back to Section 2.3. Because parallel lines do not intersect, the ratio of the vertical change (rise) to the horizontal change (run) of each line must be equivalent (Figure 26). In other words, parallel lines have the same slope!

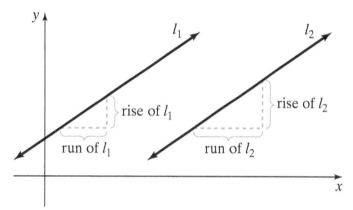

The ratios of the vertical rise to the horizontal run of any two parallel lines must be equal:

$$\frac{\text{rise of } l_1}{\text{run of } l_1} = \frac{\text{rise of } l_2}{\text{run of } l_2}$$

Figure 26

Theorem

Two distinct nonvertical lines in the Cartesian plane are parallel if and only if they have the same slope.

We can prove this theorem using a little geometry. We can show that if l_1 is parallel to l_2, then triangles $\triangle ABC$ and $\triangle DEF$ are similar. See Figure 27. Because the ratio of two corresponding sides of similar triangles are proportional, it follows that the slope of $l_1 = \dfrac{BC}{AB} = \dfrac{EF}{DE} =$ slope of l_2. Conversely, if the slope of l_1 is the same as the slope of l_2, then the two triangles are similar. Therefore, $\angle CAB \cong \angle FDE$ and thus the lines must be parallel.

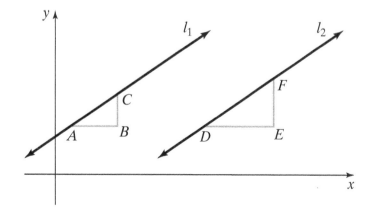

Figure 27 $\triangle ABC$ and $\triangle DEF$ are similar, so $\dfrac{BC}{AB} = \dfrac{EF}{DE}$. This implies that the slopes are equal.

Example 1 Show That Two Lines Are Parallel

Show that the lines $y = -\dfrac{2}{3}x - 1$ and $4x + 6y = 12$ are parallel.

Solution The line $y = -\dfrac{2}{3}x - 1$ is in slope–intercept form with slope $m = -\dfrac{2}{3}$.

The line $4x + 6y = 12$ is in standard form with slope $m = -\dfrac{A}{B} = -\dfrac{4}{6} = -\dfrac{2}{3}$.

Because the lines have the same slope, they must be parallel. See Figure 28.

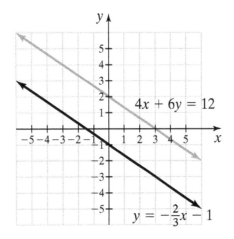

Figure 28 Parallel lines $y = -\dfrac{2}{3}x - 1$ and $4x + 6y = 12$

OBJECTIVE 2 UNDERSTANDING THE DEFINITION OF PERPENDICULAR LINES

If two *distinct* lines are not parallel, then they must intersect at a single point. If the two lines intersect at a right angle (90 degrees), then the lines are said to be **perpendicular**. The slopes of perpendicular lines have a special relationship that is stated in the following theorem.

Theorem

Two nonvertical lines in the Cartesian plane are perpendicular if and only if the product of their slopes is -1.

This theorem states that if line 1 with slope m_1 is perpendicular to line 2 with slope m_2, then $m_1 m_2 = -1$. Because the product of the slopes of nonvertical perpendicular lines is -1, it follows that

$$m_1 = -\frac{1}{m_2}.$$

This means that if $m_1 = \dfrac{a}{b}$, then $m_2 = -\dfrac{b}{a}$. In other words, the slopes of nonvertical perpendicular lines are **negative reciprocals** of each other, as in Figure 29.

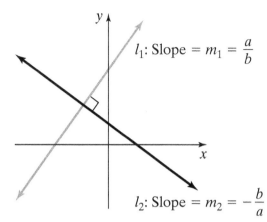

l_1: Slope $= m_1 = \dfrac{a}{b}$

l_2: Slope $= m_2 = -\dfrac{b}{a}$

Figure 29 Slopes of perpendicular lines are negative reciprocals of each other.

Example 2 Show That Two Lines Are Perpendicular

Show that the lines $3x - 6y = -12$ and $2x + y = 4$ are perpendicular.

Solution Both lines are in standard form. The first line, $3x - 6y = -12$, has slope $m_1 = -\dfrac{A}{B} = -\dfrac{3}{-6} = \dfrac{1}{2}$. The slope of the second line, $2x + y = 4$, has slope $m_2 = -\dfrac{A}{B} = -\dfrac{2}{1}$. Because the slopes of these lines are negative reciprocals of each other, this implies that the lines are perpendicular. See Figure 30.

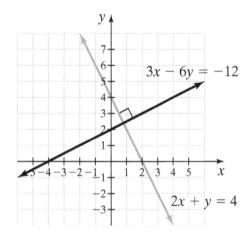

$3x - 6y = -12$

$2x + y = 4$

Figure 30 Perpendicular lines $3x - 6y = -12$ and $2x + y = 4$

Summary of Parallel and Perpendicular Lines

Given two nonvertical lines l_1 and l_2 such that the slope of line l_1 is $m_1 = \dfrac{a}{b}$,

1. l_2 is parallel to l_1 if and only if $m_2 = \dfrac{a}{b}$.

2. l_2 is perpendicular to l_1 if and only if $m_2 = -\dfrac{b}{a}$.

Note Any two distinct vertical lines are parallel to each other, whereas any horizontal line is perpendicular to any vertical line.

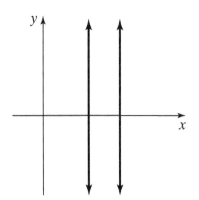

Figure 31 Any two distinct vertical lines are parallel.

Figure 32 Horizontal lines are perpendicular to vertical lines.

OBJECTIVE 3 DETERMINING WHETHER TWO LINES ARE PARALLEL, PERPENDICULAR, OR NEITHER

Given any two distinct lines, we are now able to quickly determine whether the lines are parallel to each other, perpendicular to each other, or neither by simply evaluating their slopes.

 *My video summary*

▶ **Example 3** Determine Whether Two Lines Are Parallel, Perpendicular, or Neither

For each of the following pairs of lines, determine whether the lines are parallel, perpendicular, or neither.

a. $3x - y = 4$
   $x + 3y = 7$

b. $y = \dfrac{1}{2}x + 3$
   $x + 2y = 1$

c. $x = -1$
   $x = 3$

Solution Watch the video to verify that

a. The lines are perpendicular.

b. The lines are neither parallel nor perpendicular.

c. The lines are parallel.

You Try It Work through this You Try It problem.

Work Exercises 1–6 in this textbook or in the MyMathLab Study Plan.

OBJECTIVE 4 FINDING THE EQUATIONS OF PARALLEL AND PERPENDICULAR LINES

 *My video summary*

▶ **Example 4** Find the Equation of a Parallel Line

Find the equation of the line parallel to the line $2x + 4y = 1$ that passes through the point $(3, -5)$. Write the answer in **point-slope form, slope–intercept form,** and **standard form.**

Solution Watch the video to see that the equations of the line parallel to $2x + 4y = 1$ that passes through the point $(3, -5)$ in all three forms are

$$y + 5 = -\frac{1}{2}(x - 3) \qquad \text{Point-slope form}$$

$$y = -\frac{1}{2}x - \frac{7}{2} \qquad \text{Slope-intercept form}$$

$$x + 2y = -7 \qquad \text{Standard form}$$

You Try It Work through this You Try It problem.

Work Exercises 7–11 in this textbook or in the MyMathLab Study Plan.

My video summary   ⊚ **Example 5** Find the Equation of a Perpendicular Line

Find the equation of the line perpendicular to the line $y = -5x + 2$ that passes through the point $(3, -1)$. Write the answer in slope-intercept form.

Solution The line $y = -5x + 2$ is in slope-intercept form with slope -5. We are looking for the line perpendicular to this line. Therefore, the slope of the new line must be the negative reciprocal of -5. Because we can rewrite -5 as $-\frac{5}{1}$, the negative reciprocal is $\frac{1}{5}$. Using a slope of $\frac{1}{5}$ and the point $(3, -1)$, we can obtain the point-slope equation $y + 1 = \frac{1}{5}(x - 3)$. Solving this equation for y, we obtain the slope-intercept form $y = \frac{1}{5}x - \frac{8}{5}$. You can see in Figure 33 how these two lines intersect at a right angle.

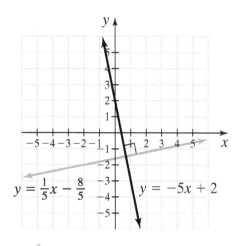

$y = \frac{1}{5}x - \frac{8}{5}$ $y = -5x + 2$ **Figure 33** Graph of perpendicular lines $y = -5x + 2$ and $y = \frac{1}{5}x - \frac{8}{5}$

You Try It Work through this You Try It problem.

Work Exercises 12–16 in this textbook or in the MyMathLab Study Plan.

OBJECTIVE 5 SOLVING A GEOMETRIC APPLICATION OF PARALLEL AND PERPENDICULAR LINES

In Section 2.1, we used the midpoint formula to determine whether four points in the plane form a parallelogram. (See Section 2.1, Example 5.) Another way to determine whether four points in the plane form a parallelogram is to determine the slopes of the sides opposite each other. If the slopes of the segments opposite one another are equal, then the four points must form a parallelogram. A special parallelogram in which all four sides have the same length is called a *rhombus*. In geometry, it can be shown that if the diagonals of a parallelogram are perpendicular, then the parallelogram is a rhombus. See Figures 34 and 35.

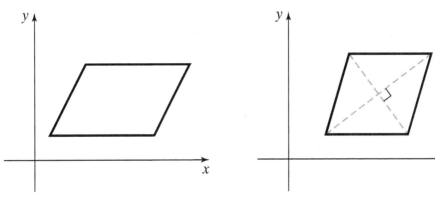

Figure 34 Parallelogram   **Figure 35** Rhombus

My video summary

⊘ Example 6 Determine Whether Four Points in the Plane Form a Parallelogram or a Rhombus

Do the points $A(0, 4)$, $B(3, 0)$, $C(9, 1)$, and $D(6, 5)$ form a parallelogram? If the points form a parallelogram, then is the parallelogram a rhombus?

Solution These four points do form a parallelogram because the slopes of the segments opposite one another are equal. Verify this for yourself, or watch the video to see the worked out solution. The parallelogram formed by these four points is not a rhombus because diagonal $\overline{AC}$ has slope $-\dfrac{1}{3}$, whereas diagonal $\overline{BD}$ has slope $\dfrac{5}{3}$ and $\left(-\dfrac{1}{3}\right)\left(\dfrac{5}{3}\right) = -\dfrac{5}{9} \neq -1$.

Therefore, the diagonals are *not* perpendicular. Watch the video to see the entire solution.   ●

You Try It Work through this You Try It problem.

Work Exercises 17–20 in this textbook or in the MyMathLab Study Plan.

2.4 Exercises

Skill Check Exercises

For exercises SCE-1 through SCE-2 solve each equation for the variable y.

SCE-1. $2x - y = 5$ **SCE-2.** $3x + 2y = 7$

For exercises SCE-3 through SCE-6, first use the distributive property then solve the equation for the variable y.

SCE-3. $y - 1 = 2(x + 3)$ **SCE-4.** $y + 3 = -5(x - 2)$

SCE-5. $y - 5 = \dfrac{2}{7}(x + 3)$ **SCE-6.** $y + 4 = -\dfrac{3}{5}(x - 1)$

In Exercises 1–6, decide whether the pair of lines are parallel, perpendicular, or neither.

1. $-4x - 2y = -9$
   $5x - 10y = 7$

2. $y = 3x - 2$
   $9x - 3y = 1$

3. $y = -2$
   $3y = 12$

4. $x = -1$
   $y = -1$

5. $4x - 3y = 7$
   $3x - 4y = -6$

6. $2x + 3y = 3$
   $y = -\dfrac{2}{3}x + 1$

In Exercises 7–9, find the equation of a line described as follows, and express your answer in point–slope form, slope–intercept form, and standard form.

SbS   7. Find the equation of the line parallel to the line $y = \dfrac{1}{4}x - 2$ that passes through the point $(-2, 3)$.

SbS   8. Find the equation of the line parallel to the line $3x - 5y = 1$ that passes through the point $(1, -4)$.

SbS   9. Find the equation of the line parallel to the line $y + 1 = 3(x - 5)$ that passes through the point $(-2, -5)$.

In Exercises 10 and 11, find the equation of a line described as follows, and express your answer in standard form.

10. Find the equation of the line parallel to the line $x = 7$ that passes through the point $(11, -3)$.

11. Find the equation of the line parallel to the line $y = 8$ that passes through the point $(11, -3)$.

In Exercises 12–14, find the equation of a line described as follows, and express your answer in point-slope form, slope-intercept form, and standard form.

SbS   12. Find the equation of the line perpendicular to the line $y = \dfrac{1}{4}x - 2$ that passes through the point $(-2, 3)$.

SbS   13. Find the equation of the line perpendicular to the line $3x - 5y = 1$ that passes through the point $(1, -4)$.

SbS **14.** Find the equation of the line perpendicular to the line $y + 1 = 3(x - 5)$ that passes through the point $(-2, -5)$.

In Exercises 15 and 16, find the equation of a line described as follows, and express your answer in standard form.

15. Find the equation of the line perpendicular to the line $x = 7$ that passes through the point $(11, -3)$.

16. Find the equation of the line perpendicular to the line $y = 8$ that passes through the point $(11, -3)$.

In Exercises 17–20, use slope to determine whether the quadrilateral with the given vertices forms a parallelogram. If it is a parallelogram, determine whether it is also a rhombus.

17. $A(-1, 2), B(3, 3), C(4, 7), D(0, 6)$ **18.** $A(2, 2), B(3, -1), C(5, 2), D(4, 5)$

19. $A(1, 1), B(5, 2), C(8, 5), D(3, 3)$ **20.** $A(1, 1), B(4, -3), C(7, 1), D(4, 5)$

Brief Exercises

21. Find the equation of the line parallel to the line $y = \dfrac{1}{4}x - 2$ that passes through the point $(-2, 3)$. Express your answer in point–slope form.

22. Find the equation of the line parallel to the line $3x - 5y = 1$ that passes through the point $(1, -4)$.
Express your answer in point–slope form.

23. Find the equation of the line parallel to the line $y + 1 = 3(x - 5)$ that passes through the point $(-2, -5)$. Express your answer in point–slope form.

24. Find the equation of the line parallel to the line $y = \dfrac{1}{4}x - 2$ that passes through the point $(-2, 3)$. Express your answer in slope–intercept form.

25. Find the equation of the line parallel to the line $3x - 5y = 1$ that passes through the point $(1, -4)$. Express your answer in slope–intercept form.

26. Find the equation of the line parallel to the line $y + 1 = 3(x - 5)$ that passes through the point $(-2, -5)$. Express your answer in slope–intercept form.

27. Find the equation of the line parallel to the line $y = \dfrac{1}{4}x - 2$ that passes through the point $(-2, 3)$.
Express your answer in standard form.

28. Find the equation of the line parallel to the line $3x - 5y = 1$ that passes through the point $(1, -4)$. Express your answer in standard form.

29. Find the equation of the line parallel to the line $y + 1 = 3(x - 5)$ that passes through the point $(-2, -5)$. Express your answer in standard form.

30. Find the equation of the line perpendicular to the line $y = \dfrac{1}{4}x - 2$ that passes through the point $(-2, 3)$. Express your answer in point–slope form.

31. Find the equation of the line perpendicular to the line $3x - 5y = 1$ that passes through the point $(1, -4)$. Express your answer in point-slope form.

32. Find the equation of the line perpendicular to the line $y + 1 = 3(x - 5)$ that passes through the point $(-2, -5)$. Express your answer in point-slope form.

33. Find the equation of the line perpendicular to the line $y = \dfrac{1}{4}x - 2$ that passes through the point $(-2, 3)$. Express your answer in slope–intercept form.

34. Find the equation of the line perpendicular to the line $3x - 5y = 1$ that passes through the point $(1, -4)$. Express your answer in slope–intercept form.

35. Find the equation of the line perpendicular to the line $y + 1 = 3(x - 5)$ that passes through the point $(-2, -5)$. Express your answer in slope–intercept form.

36. Find the equation of the line perpendicular to the line $y = \dfrac{1}{4}x - 2$ that passes through the point $(-2, 3)$. Express your answer in standard form.

37. Find the equation of the line perpendicular to the line $3x - 5y = 1$ that passes through the point $(1, -4)$. Express your answer in standard form.

38. Find the equation of the line perpendicular to the line $y + 1 = 3(x - 5)$ that passes through the point $(-2, -5)$. Express your answer in standard form.

Functions

CHAPTER THREE CONTENTS

3.1   Relations and Functions

THINGS TO KNOW

Before working through this section, be sure that you are familiar with the following concepts:

VIDEO   ANIMATION   INTERACTIVE

You Try It
1. Describing Intervals of Real Numbers (Section R.1)

You Try It
2. Solving Linear Inequalities (Section 1.7)   ⊙

You Try It
3. Solving Polynomial Inequalities (Section 1.9)   ⊙

You Try It
4. Solving Rational Inequalities (Section 1.9)   ⊙

You Try It
5. Graphing Equations by Plotting Points (Section 2.1)

OBJECTIVES

1 Understanding the Definitions of Relations and Functions

2 Determining Whether Equations Represent Functions

3 Using Function Notation; Evaluating Functions

4 Using the Vertical Line Test

5 Determining the Domain of a Function Given the Equation

Introduction to Section 3.1

Functions appear all around us and are a fundamental part of mathematics. Functions occur when one quantity depends on another. Here are just a few examples:

- The revenue generated by selling a certain product depends on the number of items sold.

- The volume of a cube depends on the length of an edge of the cube.

- For a person who works by the hour, the gross pay depends on the number of hours worked.

If you think about it, you can probably come up with countless situations where one thing, event, or occurrence depends on another. Before we officially define a function, we must first define a **relation**.

OBJECTIVE 1   UNDERSTANDING THE DEFINITIONS OF RELATIONS AND FUNCTIONS

 *My video summary*

> **Definition**   Relation
>
> ⊙ A **relation** is a correspondence between two sets A and B such that each element of set A corresponds to one or more elements of set B. Set A is called the **domain** of the relation, and set B is called the **range** of the relation.

Suppose that Adam, Patrice, and Scott are college roommates and their weights are 180 lb, 190 lb, and 180 lb, respectively. The pairing of the roommates' names with their weights is a relation. In this example, we are relating a set of names with a set of weights. The set of names {Adam, Patrice, Scott} is the domain of this relation, whereas the set of weights {180 lb, 190 lb} is the range. Notice that 180 lb is listed only once in the range. There is no need to list an element of either set more than once. See Figure 1.

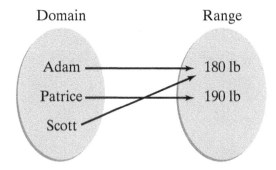

Figure 1

A relation can be represented as a set of ordered pairs with the elements of the domain listed first. The previous relation can be written as the set of ordered pairs {(Adam, 180 lb), (Patrice, 190 lb), (Scott, 180 lb)}.

Let's now switch the domain and range of this relation. The relation in Figure 2 can be written as the set of ordered pairs {(180 lb, Adam), (180 lb, Scott), (190 lb, Patrice)}.

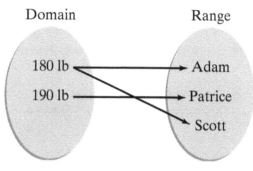

Figure 2

Suppose that a delivery man was to deliver a package to the roommate who weighs 180 lb. Because there are two people who weigh 180 lb, there is no way to know for sure who this package belongs to. This is because the element of 180 lb in the domain is associated with two distinct elements in the range. We say that this relation is not a **function**. For a relation to be a function, every element in the domain must correspond to exactly one element in the range.

Definition Function

A **function** is a relation such that for each element in the domain, there is *exactly one* corresponding element in the range.

The relation in Figure 1 is a function because for every name in the domain, there is exactly one corresponding weight in the range. The relation in Figure 2 is not a function because there is a domain value (180 lb) that corresponds to two range values (Adam and Scott).

My video summary   ⊙ **Example 1** Determine Whether a Relation Is a Function

Determine whether each relation is a function, and then find the domain and range.

a.

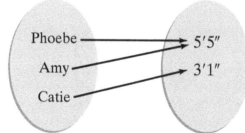

b.

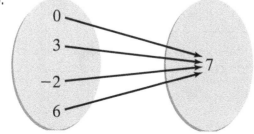

c. $\{(3, 7), (-3, 2), (-4, 5), (1, 4), (3, -4)\}$ d. $\{(-1, 2), (0, 2), (1, 5), (2, 3), (3, 2)\}$

Solution

a. This relation is a function. For each element in the domain, there is *exactly one* corresponding element in the range. The domain is {Phoebe, Amy, Catie}, and the range is {5′5″, 3′1″}.

b. This relation is a function. For each element in the domain, there is *exactly one* corresponding element in the range. The domain is {−2, 0, 3, 6}, and the range is {7}.

Note The elements of the domain and range are typically listed in ascending order when using set notation.

c. This relation is *not* a function because the element 3 in the domain corresponds to 7 and −4 in the range. The domain of this relation is {−4, −3, 1, 3}. The range is {−4, 2, 4, 5, 7}.

d. This relation is a function. For each element in the domain, there is *exactly one* corresponding element in the range. The domain is {−1, 0, 1, 2, 3}. The range is {2, 3, 5}.

My video summary ⊘ Watch this video to see a complete solution to this example.

If the domain and range of a relation are sets of real numbers, then the relation can be represented by plotting the ordered pairs in the Cartesian plane. The set of all *x*-coordinates is the domain of the relation, and the set of all *y*-coordinates is the range of the relation.

Example 2 Determine Whether a Relation Is a Function

Use the domain and range of the following relation to determine whether the relation is a function.

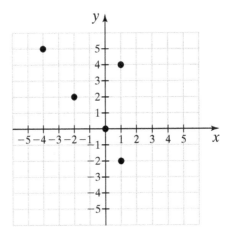

Solution The set of ordered pairs of this graph are {(−4, 5), (−2, 2), (0, 0), (1, −2), (1, 4)}. The domain is the set of all *x*-coordinates: {−4, −2, 0, 1}. The range is the set of all *y*-coordinates {−2, 0, 2, 4, 5}. This relation is *not* a function because the element 1 in the domain corresponds to two different elements (−2 and 4) in the range.

You Try It Work through the following You Try It problem.

Work Exercises 1–6 in this textbook or in the MyMathLab Study Plan.

OBJECTIVE 2   DETERMINING WHETHER EQUATIONS REPRESENT FUNCTIONS

In Example 2, we see that a *finite* number of ordered pairs in the Cartesian plane can represent a relation. Relations can also be described by *infinitely* many ordered pairs in the plane. Recall from Section 2.1 that the solution set of an equation in two variables is the set of all ordered pairs (x, y) for which the equation is true. For example, the ordered pairs $(1, 1)$, $(\sqrt{3}, 0)$, and $(1, -1)$ all satisfy the equation $x^2 + 2y^2 = 3$. (To verify, read this explanation.) Although this equation is a relation, it is not a function because $x = 1$ corresponds to both $y = 1$ and $y = -1$. To determine whether an equation represents a function, we must show that for any value in the domain, there is exactly one corresponding value in the range.

My interactive video summary

⊗ Example 3 Determine Whether an Equation Represents a Function

Determine whether the following equations represent y as a function of x:

a. $3x - 2y = -12$ **b.** $y = 3x^2 - x + 2$ **c.** $(x + 3)^2 + y^2 = 16$

Solution The equations in parts a and b both represent y as a function of x. Watch the interactive video to see parts a and b worked out in detail.

To determine whether the equation $(x + 3)^2 + y^2 = 16$ from part c represents y as a function of x, we can solve for y.

Solving the equation for y, we obtain $y = \pm\sqrt{16 - (x + 3)^2}$. To see how we solved for y, read these steps. Notice the symbol "$\pm$," which precedes the radical in the previous equation. This symbol indicates that there are two possible y-values for a given x-value. For example, if $x = -3$, then the corresponding y-values are

$$y = \pm\sqrt{16 - (-3 + 3)^2} = \pm\sqrt{16 - (0)^2} = \pm\sqrt{16} = \pm 4.$$

Therefore, the value of $x = -3$ in the domain corresponds to $y = 4$ and $y = -4$ in the range. Because there are two y-values that correspond to a single x-value, the original equation $(x + 3)^2 + y^2 = 16$ does not represent a function. Recall from Section 2.2 that the equation $(x + 3)^2 + y^2 = 16$ is the equation of a circle in standard form with center $(-3, 0)$ and $r = 4$. We can sketch this circle using the techniques discussed in Section 2.2, or we can use a graphing utility to graph the two equations $y = \sqrt{16 - (x + 3)^2}$ and $y = -\sqrt{16 - (x + 3)^2}$. For all x-values on the interval $(-7, 1)$, there are two corresponding values of y.

My interactive video summary

⊗ Watch the interactive video to see all three solutions worked out in detail.

Using Technology

$y_1 = \sqrt{16 - (x + 3)^2}$

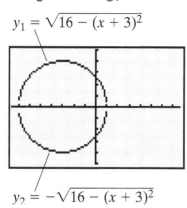

$y_2 = -\sqrt{16 - (x + 3)^2}$

The graph of $(x + 3)^2 + y^2 = 16$ was created using a graphing utility. The top half of the circle is $y = \sqrt{16 - (x + 3)^2}$, whereas the bottom half of the circle represents the equation $y = -\sqrt{16 - (x + 3)^2}$.

 **You Try It** Work through the following You Try It problem.

Work Exercises 7–12 in this textbook or in the MyMathLab Study Plan.

OBJECTIVE 3 USING FUNCTION NOTATION; EVALUATING FUNCTIONS

 My video summary

In Example 3, we see that the equation $3x - 2y = -12$ could be explicitly solved for y to obtain the function $y = \dfrac{3}{2}x + 6$. (See the interactive video from Example 3a.)

When an equation is explicitly solved for y, we say that "y is a function of x" or that the variable y depends on the variable x. Thus, x is the independent variable, and y is the dependent variable.

Instead of using the variable y, letters such as f, g, or h (and others) are commonly used for functions. For example, suppose we want to name a function f. Then, for any x-value in the domain, we call the y-value (or function value) $f(x)$. The symbol $f(x)$ is read as "the value of the function f at x" or simply "f of x." In our example, the function $y = \dfrac{3}{2}x + 6$ can be written as $f(x) = \dfrac{3}{2}x + 6$. The notation $f(x)$ is called **function notation**.

In our example, if $f(x) = \dfrac{3}{2}x + 6$, then the expression $f(-4)$ represents the range value (y-coordinate) when the domain value (x-coordinate) is -4. To this find value, replace x with -4 in the equation $f(x) = \dfrac{3}{2}x + 6$ to get $f(-4) = \dfrac{3}{2}(-4) + 6 = -6 + 6 = 0$. Thus, the ordered pair $(-4, 0)$ must lie on the graph of f. Table 1 shows this function evaluated at several values of x. The function $f(x) = \dfrac{3}{2}x + 6$ is a linear function whose graph is sketched in Figure 3.

⚠ **The symbol $f(x)$ does not mean f times x. The notation $f(x)$ refers to the value of the function at x.**

Table 1

Domain Value x	Range Value $f(x) = \dfrac{3}{2}x + 6$	Ordered Pair $(x, f(x))$
-4	$f(-4) = \dfrac{3}{2}(-4) + 6 = 0$	$(-4, 0)$
0	$f(0) = \dfrac{3}{2}(0) + 6 = 6$	$(0, 6)$
1	$f(1) = \dfrac{3}{2}(1) + 6 = \dfrac{15}{2}$	$\left(1, \dfrac{15}{2}\right)$
-3	$f(-3) = =\dfrac{3}{2}(-3) + 6 = \dfrac{3}{2}$	$\left(-3, \dfrac{3}{2}\right)$

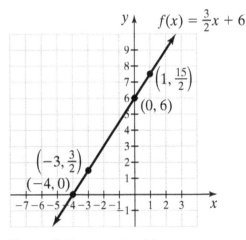

Figure 3 The graph of the function $f(x) = \dfrac{3}{2}x + 6$

My interactive video summary

⊘ **Example 4** Use Function Notation

Rewrite these equations using function notation where y is a function of x. Then answer the question following each equation.

a. $3x - y = 5$

What is the value of $f(4)$?

b. $x^2 - 2y + 1 = 0$

Does the point $(-2, 1)$ lie on the graph of this function?

c. $y + 7 = 0$

What is $f(x)$ when $x = 3$?

Solution You should watch the interactive video to verify the following:

a. This equation can be rewritten in function notation as $f(x) = 3x - 5$ and $f(4) = 7$.

b. This equation can be rewritten in function notation as $f(x) = \dfrac{1}{2}x^2 + \dfrac{1}{2}$. The point $(-2, 1)$ is *not* on the graph of this function because $f(-2) \neq 1$.

c. The equation $y + 7 = 0$ can be rewritten using function notation as $f(x) = -7$. When $x = 3$, the value of $f(x)$ is -7.

Example 5 Evaluate a Function

Given that $f(x) = x^2 + x - 1$, evaluate the following:

a. $f(0)$ b. $f(-1)$ c. $f(x + h)$ d. $\dfrac{f(x + h) - f(x)}{h}$

Solution

a. Substitute $x = 0$ into the function to get

$$f(0) = (0)^2 + (0) - 1 = -1$$

b. Substitute $x = -1$ into the function to get

$$f(-1) = (-1)^2 + (-1) - 1$$
$$= 1 - 1 - 1$$
$$= -1$$

 The expression $(-1)^2$ does not equal -1^2.

c. Substitute $x = x + h$ into the function to get

$$f(x + h) = (x + h)^2 + (x + h) - 1$$
$$= x^2 + 2xh + h^2 + x + h - 1$$

My video summary ⊗ **d.** The expression $\dfrac{f(x+h)-f(x)}{h}$ is called the **difference quotient** and

is very important in calculus. First, from part c, we see that
$f(x+h) = x^2 + 2xh + h^2 + x + h - 1$. So,

$$\frac{f(x+h)-f(x)}{h} = \frac{\overbrace{x^2 + 2xh + h^2 + x + h - 1}^{f(x+h)} - \overbrace{[x^2 + x - 1]}^{f(x)}}{h}$$

$$= \frac{x^2 + 2xh + h^2 + x + h - 1 - x^2 - x + 1}{h}$$

$$= \frac{2xh + h^2 + h}{h}$$

$$= \frac{h(2x + h + 1)}{h}$$

$$= 2x + h + 1$$

Therefore, $\dfrac{f(x+h)-f(x)}{h} = 2x + h + 1$. ●

Note that function notation is not restricted to the use of x for the independent variable or to the use of f for the name of the function, which is the dependent variable. Other letters commonly used in place of x are $h, t, r, a,$ and b. Letters such as A for area, V for volume, and d for distance are often used to represent the names of functions. In fact, any letter can be used to represent values of either the domain or the range.

You Try It   Work through the following You Try It problem.

Work Exercises 13–28 in this textbook or in the My MathLab **Study Plan.**

OBJECTIVE 4   USING THE VERTICAL LINE TEST

My video summary ⊗ We see in Figure 3 that the graph of the function $f(x) = \dfrac{3}{2}x + 6$ is created by

plotting several ordered pairs in the Cartesian plane. Although every function has a graph, it is *not* the case that every graph in the plane represents a function. A graph does not represent a function if two or more points on the graph with the same first coordinate have different second coordinates. The **vertical line test** can be used to quickly determine whether a graph represents a function. See Figure 4.

Vertical Line Test

A graph in the Cartesian plane is the graph of a function if and only if no vertical line intersects the graph more than once.

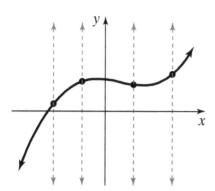

This graph is a function. (No vertical line intersects the graph more than once.)

This graph is not a function. (The graph does not pass the vertical line test.)

Figure 4

Example 6 Use the Vertical Line Test

Use the vertical line test to determine which of the following graphs represents the graph of a function.

a.
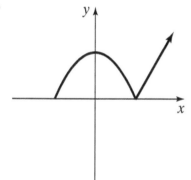

b.
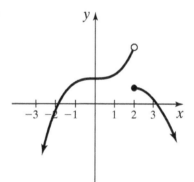

c.
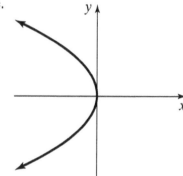

Solution

a. This graph represents a function because no vertical lines intersect the graph more than once.

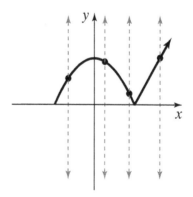

b. At first glance, it may appear that this graph does not represent a function. However, we see that at the point where $x = 2$, there is only one corresponding y-value. The "open dot" indicates that the function is not defined at that point. The vertical line $x = 2$ only intersects the graph once; therefore, the graph represents the graph of a function.

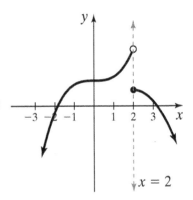

c. This graph does not represent a function because we can find several vertical lines that intersect the graph more than once.

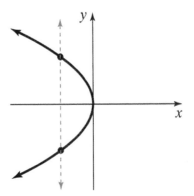

You Try It Work through the following You Try It problem.

Work Exercises 29–37 in this textbook or in the My MathLab **Study Plan.**

OBJECTIVE 5 DETERMINING THE DOMAIN OF A FUNCTION GIVEN THE EQUATION

My video summary ⊙ The domain of a function $y = f(x)$ is the set of all values of x for which the function is defined. In other words, a number $x = a$ is in the domain of a function f if $f(a)$ is a real number. For example, the domain of $f(x) = x^2$ is all real numbers because for any real number $x = a$, the value of $f(a) = a^2$ is also a real number.

It is very helpful to classify a function to determine its domain. For example, the function $f(x) = x^2$ belongs to a class of functions called **polynomial functions**. The domain of every polynomial function is all real numbers. Polynomial functions are discussed in great detail in Chapter 4.

Definition Polynomial Function

The function $f(x) = a_n x^n + a_{n-1}x^{n-1} + a_{n-2}x^{n-2} + \cdots + a_1 x + a_0$ is a **polynomial function** of degree n, where n is a nonnegative integer and $a_0, a_1, a_2, \ldots, a_n$ are real numbers. The domain of every polynomial function is $(-\infty, \infty)$.

Many functions can have restricted domains. For example, the quotient of two polynomial functions is called a **rational function**. The rational function $f(x) = \dfrac{x+5}{x-4}$ is defined everywhere except when $x = 4$ because the value $f(4) = \dfrac{9}{0}$ is undefined. Therefore, the domain of a rational function consists of all real numbers for which the denominator does not equal zero. We investigate rational functions further in Chapter 4.

Definition Rational Function

A **rational function** is a function of the form $f(x) = \dfrac{g(x)}{h(x)}$, where g and h are polynomial functions such that $h(x) \neq 0$.

The domain of a rational function is the set of all real numbers such that $h(x) \neq 0$.

Note If $h(x) = c$, where c is a real number, then we will consider the function $f(x) = g(x)/h(x) = g(x)/c$ to be a polynomial function.

Root functions can also have restricted domains. Consider the root function $f(x) = \sqrt{x + 1}$. The number $x = 3$ is in the domain because $f(3) = \sqrt{3 + 1} = \sqrt{4} = 2$ is a real number. However, -5 is not in the domain because $f(-5) = \sqrt{-5 + 1} = \sqrt{-4} = 2i$ is not a real number. Therefore, the domain of $f(x) = \sqrt{x + 1}$ consists of all values of x for which the radicand is greater than or equal to zero. The domain of f is the solution to the inequality $x + 1 \geq 0$.

$$x + 1 \geq 0$$

$$x \geq -1 \qquad \text{Subtract 1 from both sides.}$$

Therefore, the domain of f is $[-1, \infty)$. Root functions with roots that are odd numbers such as 3 or 5 *can have* negative radicands. Therefore, the domain of a root function of the form $f(x) = \sqrt[n]{g(x)}$, where n is an odd positive integer, consists of all real numbers for which $g(x)$ is defined.

Definition Root Function

The function $f(x) = \sqrt[n]{g(x)}$ is a **root function**, where n is an integer such that $n \geq 2$.

1. If n is *even*, the domain is the solution to the inequality $g(x) \geq 0$.

2. If n is *odd*, the domain is the set of all real numbers for which $g(x)$ is defined.

Following is a quick guide for finding the domain of three specific types of functions:

Class of Function	Form	Domain
Polynomial functions	$f(x) = a_n x^n + a_{n-1}x^{n-1} + \cdots + a_1 x + a_0$	Domain is $(-\infty, \infty)$.
Rational functions	$f(x) = \dfrac{g(x)}{h(x)}$, where g and h are polynomial functions such that $h(x) \neq 0$.	Domain is all real numbers such that $h(x) \neq 0$.
Root functions	$f(x) = \sqrt[n]{g(x)}$, where $g(x)$ is a function and n is an integer such that $n \geq 2$.	1. If n is even, the domain is the solution to the inequality $g(x) \geq 0$. 2. If n is odd, the domain is the set of all real numbers for which g is defined.

We study other classes of functions later in this text. For now, we limit our discussion to polynomial functions, rational functions, and root functions.

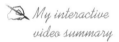
My interactive video summary

⊙ **Example 7 Find the Domain of a Function Given the Equation**

Classify each function as a polynomial function, rational function, or root function, and then find the domain. Write the domain in interval notation.

a. $f(x) = 2x^2 - 5x$

b. $f(x) = \dfrac{x}{x^2 - x - 6}$

c. $h(x) = \sqrt{x^2 - 2x - 8}$

d. $f(x) = \sqrt[3]{5x - 9}$

e. $k(x) = \sqrt[4]{\dfrac{x - 2}{x^2 + x}}$

Solution Work through the interactive video to verify the following:

a. The function $f(x) = 2x^2 - 5x$ is a polynomial function. The domain of f is all real numbers or $(-\infty, \infty)$.

b. The function $f(x) = \dfrac{x}{x^2 - x - 6}$ is a rational function. The domain of f is $\{x \mid x \neq -2, x \neq 3\}$. The domain in interval notation is $(-\infty, -2) \cup (-2, 3) \cup (3, \infty)$.

c. The function $h(x) = \sqrt{x^2 - 2x - 8}$ is a root function. The domain of h in interval notation is $(-\infty, -2] \cup [4, \infty)$.

d. The function $f(x) = \sqrt[3]{5x - 9}$ is a root function. The domain of f is all real numbers or $(-\infty, \infty)$.

e. The function $k(x) = \sqrt[4]{\dfrac{x - 2}{x^2 + x}}$ is a root function. The domain of k in interval notation is $(-1, 0) \cup [2, \infty)$. To see the solution to this entire example, work through this interactive video.

My interactive video summary

You Try It Work through the following You Try It problem.

Work Exercises 38–56 in this textbook or in the MyMathLab **Study Plan.**

3.1 Exercises

Skill Check Exercises

For exercises SCE-1 through SCE-6, solve the inequality. Write your answer using interval notation.

SCE-1. $2 - 3x \geq 0$

SCE-2. $x^2 + 4x - 21 \geq 0$

SCE-3. $(x - 1)(x + 4)(x - 3) \geq 0$

SCE-4. $\dfrac{x + 3}{x - 1} \geq 0$

SCE-5. $\dfrac{x^2 - 9}{x + 7} \geq 0$

SCE-6. $\dfrac{x - 2}{x^2 + 5x - 24} \geq 0$

For exercises SCE-7 and SCE-8, simplify the expression.

SCE-7. $6 - 9(x + h)^2 - (6 - 9x^2)$

SCE-8. $\sqrt{x + h} - 2(x + h) - (\sqrt{x} - 2x)$

In Exercises 1–6, find the domain and range of each relation, and then determine whether each relation represents a function.

1.

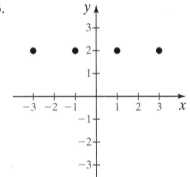

2.

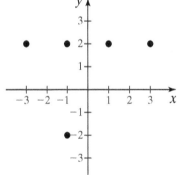

3. $\{(-1, 5), (0, 2), (1, 5), (2, 3), (1, 7)\}$

4. $\{(-1, 5), (0, 2), (1, 5), (2, 3), (3, 7)\}$

5.

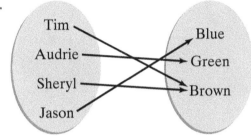

6.

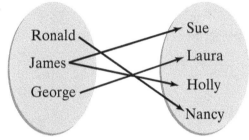

In Exercises 7–12, determine whether the equation represents y as a function of x.

7. $x^2 + y = 1$

8. $x^2 + y^2 = 1$

9. $x^3 + y^3 = 1$

10. $3xy = 6$

11. $3x - 4y = 10$

12. $6x + 1 = y^4$

In Exercises 13–21, evaluate the function at the indicated values.

13. $f(x) = 3x - 4$

a. $f(-3)$

b. $f\left(\dfrac{1}{2}\right)$

c. $f(x - 5)$

eText Screens 3.1-1–3.1-35

14. $f(x) = x^2 + 1$

   a. $f(2)$ **b.** $f(-2)$ **c.** $f(x - 1)$

15. $f(t) = \dfrac{t}{t + 1}$

   a. $f(-2)$ **b.** $f(7)$ **c.** $f(a)$

16. $V(r) = \dfrac{4}{3} \pi r^3$

   a. $V(3)$ **b.** $V(6)$ **c.** $V(r + 1)$

17. $f(x) = 2x - 5$

   a. $f(x + 1)$ **b.** $f(x + h)$

18. $f(x) = 8 - 7x$

   a. $f(x - 2)$ **b.** $f(x + h)$

19. $f(x) = x^2 - 2x$

   a. $f(x + 1)$ **b.** $f(x + h)$

20. $g(t) = t^3 - t$

   a. $g(t + 1)$ **b.** $g(t + h)$

21. $g(x) = \sqrt{x} - 2x$

   a. $g(4)$ **b.** $g(x + h)$

In Exercises 22–28, determine the difference quotient $\dfrac{f(x + h) - f(x)}{h}$.

SbS   **22.** $f(x) = 2x - 5$ SbS   **23.** $f(x) = 8 - 7x$ SbS   **24.** $f(x) = x^2 - 2x$

SbS   **25.** $f(x) = 6 - 9x^2$ SbS   **26.** $f(x) = 5x^2 - 4$ SbS   **27.** $f(x) = x^3 - x$

SbS   **28.** $f(x) = \sqrt{x} - 2x$

In Exercises 29–37, use the vertical line test to determine whether the graph represents a function.

29.

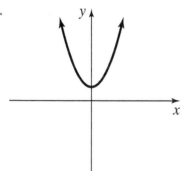

30.

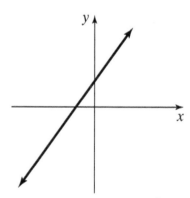

3-14   **Chapter 3**   Functions

31.

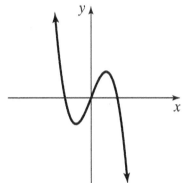

32.

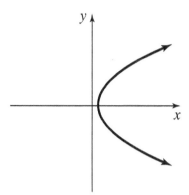

33.

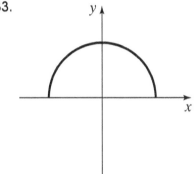

34.

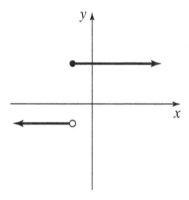

35.

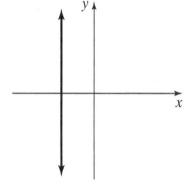

36.

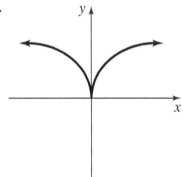

37.

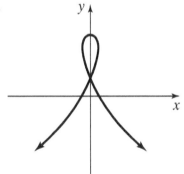

In Exercises 38–56, classify each function as a polynomial function, rational function, or root function, and then find the domain. Write the domain in interval notation.

38. $f(t) = 6t^4 + t^3 - 2t^2 + t - 5$

39. $h(t) = \dfrac{8 - t}{t^2}$

40. $f(x) = 8x^2 - x$

41. $g(x) = \sqrt{3 - x}$

42. $f(t) = \sqrt[3]{t - 5}$

43. $f(x) = \dfrac{2x - 1}{2x + 1}$

44. $f(x) = \sqrt[5]{\dfrac{x + 1}{x - 5}}$

45. $g(x) = 2x - 1$

46. $h(x) = \dfrac{x^2 + 4}{x^2 + x - 42}$

47. $f(x) = 4$

48. $h(t) = \sqrt[5]{t^4 - 1}$

49. $f(t) = \sqrt[4]{t^4 - 1}$

50. $g(t) = \dfrac{t + 5}{t^3 - 3t^2 - 4t}$

51. $f(x) = \sqrt{x^2 + 2x - 15}$

52. $h(x) = x^3 - 1$

53. $g(x) = \sqrt{\dfrac{x^2 - 9}{5}}$

54. $f(x) = \sqrt{\dfrac{5}{x^2 - 9}}$

55. $h(x) = \dfrac{1}{x^2 + 1}$

56. $f(x) = \sqrt[6]{\dfrac{2x - 6}{x^2 - x - 20}}$

Brief Exercises

In Exercises 57–63, determine the difference quotient $\dfrac{f(x + h) - f(x)}{h}$.

57. $f(x) = 2x - 5$

58. $f(x) = 8 - 7x$

59. $f(x) = x^2 - 2x$

60. $f(x) = 5x^2 - 4$

61. $f(x) = \sqrt{x} - 2x$

62. $f(x) = 6 - 9x^2$

63. $f(x) = x^3 - x$

3.2 Properties of a Function's Graph

THINGS TO KNOW

Before working through this section, be sure that you are familiar with the following concepts:

VIDEO   ANIMATION   INTERACTIVE

You Try It
1. Solving Higher-Order Polynomial Equations (Section 1.6)   ⊙

You Try It
2. Understanding the Definitions of Relations and Functions (Section 3.1)   ⊙

You Try It
3. Using the Vertical Line Test (Section 3.1)   ⊙

You Try It
4. Determining the Domain of a Function Given the Equation (Section 3.1)   ⊙

OBJECTIVES

1 Determining the Intercepts of a Function

2 Determining the Domain and Range of a Function from Its Graph

3 Determining Whether a Function Is Increasing, Decreasing, or Constant

4 Determining Relative Maximum and Relative Minimum Values of a Function

5 Determining Whether a Function Is Even, Odd, or Neither

6 Determining Information about a Function from a Graph

OBJECTIVE 1   DETERMINING THE INTERCEPTS OF A FUNCTION

 *My video summary*

▷ In Section 3.1, we discussed that every function has a graph in the Cartesian plane. In this section, we take a closer look at a function's graph by discussing some of its characteristics. A good place to start is to talk about a function's intercepts. An **intercept** of a function is a point on the graph of a function where the graph either crosses or touches a coordinate axis. There are two types of intercepts:

1. The y-intercept, which is the y-coordinate of the point where the graph crosses or touches the y-axis

2. The x-intercepts, which are the x-coordinates of the points where the graph crosses or touches the x-axis

THE *Y*-INTERCEPT

A function can have *at most* one y-intercept. The y-intercept exists if $x = 0$ is in the domain of the function. The y-intercept can be found by evaluating $f(0)$.

Example 1 Determine the y-Intercept of a Function

Find the y-intercept of the function $f(x) = -3x + 2$.

Solution Because $f(x) = -3x + 2$ is a polynomial function, we know that the domain of f is comprised of all real numbers. Thus, $x = 0$ is in the domain of f. The y-intercept is $f(0) = -3(0) + 2 = 2$. We say that the y-intercept of f is found at the point $(0, 2)$. The graph of f is the line sketched in Figure 5. You can see that the graph intersects the y-axis at $(0, 2)$.

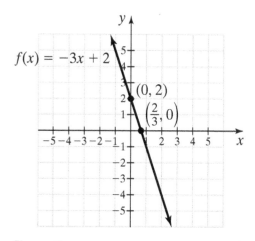

Figure 5

THE X-INTERCEPT(S)

A function may have several (even infinitely many) x-intercepts. The x-intercepts, also called real zeros, can be found by finding all *real solutions* to the equation $f(x) = 0$. Although a function function may have several zeros, only the real zeros are x-intercepts. The only real zero of $f(x) = -3x + 2$ is $x = \dfrac{2}{3}$ because the solution to the equation $-3x + 2 = 0$ is $x = \dfrac{2}{3}$. Notice from the graph in Figure 5 that the graph crosses the x-axis at the point $\left(\dfrac{2}{3}, 0\right)$.

Example 2 Find the Intercepts of a Function

Find all intercepts of the function $f(x) = x^3 - 2x^2 + x - 2$.

Solution Because $f(0) = -2$, the y-intercept is -2. To find the x-intercepts of f, we must find the real solutions to the equation $f(x) = 0$ or $x^3 - 2x^2 + x - 2 = 0$. To solve this equation, we can factor by grouping and solve to get the solution set $\{2, i, -i\}$. To see how we obtained this, read these steps. Because $x = 2$ is the only real solution, 2 is the only x-intercept.

Note   The solutions $x = i$ and $x = -i$ are zeros of f because $f(i) = 0$ and $f(-i) = 0$; however, these values do not represent x-intercepts because they are not real numbers.

Using Technology 🖥

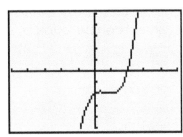

The graph of $f(x) = x^3 - 2x^2 + x - 2$ was created using a graphing utility. Notice that the graph crosses the x-axis at the point $(2, 0)$ and crosses the y-axis at $(0, -2)$. Therefore, the x-intercept is 2, and the y-intercept is -2.   ●

You Try It   Work through the following You Try It problem.

Work Exercises 1–6 in this textbook or in the MyMathLab Study Plan.

OBJECTIVE 2   DETERMINING THE DOMAIN AND RANGE OF A FUNCTION FROM ITS GRAPH

My video summary   ◉ In Section 3.1, we see that every function has a graph. Figure 6 illustrates how we can use a function's graph to determine the domain and range. The domain is the set of all input values (x-values), and the range is the set of all output values (y-values). In Figure 6, the domain is the interval $[a, b]$ while the range is the interval $[c, d]$.

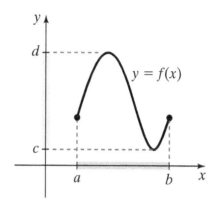

Figure 6 Domain and range of $y = f(x)$

Example 3 Find the Domain and Range of a Function from Its Graph

Use the graph of the following functions to determine the domain and range.

a.

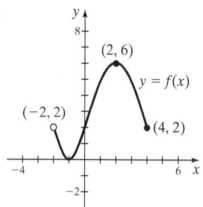

b.

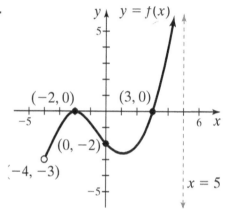

c.

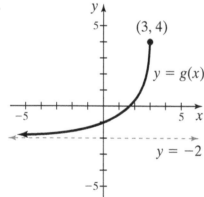

Solution

a. The open dot representing the ordered pair $(-2, 2)$ indicates that this point is not included on the graph, whereas the closed dot representing the ordered pair $(4, 2)$ indicates that this point is included on the graph. Therefore, the domain includes all x-values between -2, not including -2, and 4 inclusive. Thus, the domain in set-builder notation is $\{x \mid -2 < x \le 4\}$ or the interval $(-2, 4]$. The range includes all y-values between 0 and 6 inclusive because the points $(-1, 0)$ and $(2, 6)$ are both included on the graph. The range is $\{y \mid 0 \le y \le 6\}$ or the interval $[0, 6]$.

b. The vertical dashed line at $x = 5$ is called a **vertical asymptote** and indicates that the function will get arbitrarily close to the line $x = 5$ but will never intersect the line. Therefore, the domain of f in set-builder notation is $\{x \mid -4 < x < 5\}$ or the interval $(-4, 5)$. The range includes all y-values strictly greater than -3. Thus, the range is $\{y \mid y > -3\}$ or the interval $(-3, \infty)$.

c. To determine the domain of g, notice that the graph extends indefinitely to the left and includes all values of x less than or equal to 3. Thus, the domain is $\{x \mid x \leq 3\}$ or the interval $(-\infty, 3]$. The horizontal dashed line $y = -2$ is called a **horizontal asymptote** and indicates that the function will get arbitrarily close to the line $y = -2$. Thus, the range includes all y-values greater than -2 but less than or equal to 4. So, the range of g is $\{y \mid -2 < y \leq 4\}$ or the interval $(-2, 4]$. Watch this video to see the solution to this entre example.

My video summary

You Try It Work through the following You Try It problem.

Work Exercises 7–12 in this textbook or in the MyMathLab **Study Plan**.

OBJECTIVE 3 DETERMINING WHETHER A FUNCTION IS INCREASING, DECREASING, OR CONSTANT

My video summary A function f is said to be **increasing** on an open interval if the value of $f(x)$ gets larger as x gets larger on the interval. The graph of f rises from left to right on the interval in which f is increasing. Likewise, a function f is said to be **decreasing** on an open interval if the value of $f(x)$ gets smaller as x gets larger on the interval. The graph of f falls from left to right on the interval in which f is decreasing. A graph is **constant** on an open interval if the values of $f(x)$ do not change as x gets larger on the interval. In this case, the graph is a horizontal line on the interval. Figure 7 illustrates all three cases.

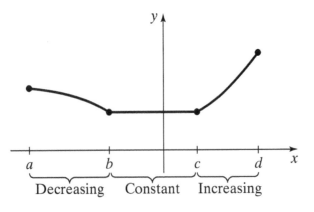

Figure 7 The function is increasing on the interval (c, d). The function is decreasing on the interval (a, b). The function is constant on the interval (b, c).

Example 4 Determine Whether a Function Is Increasing, Decreasing, or Constant from Its Graph

Given the graph of $y = f(x)$, determine whether the function is increasing, decreasing, or constant.

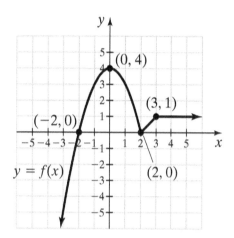

Solution The function is increasing on the interval $(-\infty, 0)$ and on the interval $(2, 3)$. The function is decreasing on the interval $(0, 2)$. The function is constant on the interval $(3, \infty)$. See the video to see the solution worked out in detail.

You Try It Work through the following You Try It problem.

Work Exercises 13–16 in this textbook or in the MyMathLab Study Plan.

OBJECTIVE 4 DETERMINING RELATIVE MAXIMUM AND RELATIVE MINIMUM VALUES OF A FUNCTION

My video summary

⊘ When a function changes from increasing to decreasing at a point $(c, f(c))$, then f is said to have a relative maximum at $x = c$. The relative maximum value is $f(c)$. See Figure 8(a). Similarly, when a function changes from decreasing to increasing at a point $(c, f(c))$, then f is said to have a relative minimum at $x = c$. The relative minimum value is $f(c)$. See Figure 8(b).

(a)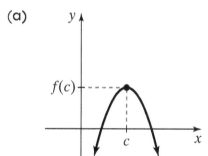

(b)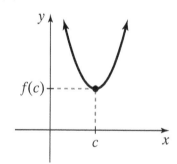

The relative maximum occurs at $x = c$, and the relative maximum value is $f(c)$.

The relative minimum occurs at $x = c$, and the relative minimum value is $f(c)$.

Figure 8

Note The word "relative" indicates that the function obtains a maximum or minimum value relative to some **open interval.** It is not necessarily the maximum (or minimum) value of the function on the entire domain.

Example 5 Determine Information from the Graph of a Function

Use the graph of $y = f(x)$ to answer each question.

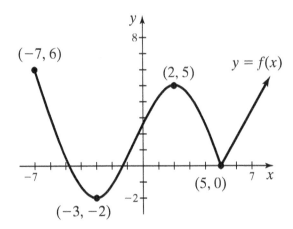

a. On what interval(s) is f increasing?

b. On what interval(s) is f decreasing?

c. For what value(s) of x does f have a relative minimum?

d. For what value(s) of x does f have a relative maximum?

e. What are the relative minima?

f. What are the relative maxima?

Solution

a. The function f is increasing on the interval $(-3, 2)$. The function is also increasing for values of x greater than 5; that is, f is also increasing on the interval $(5, \infty)$.

See the graph of $y = f(x)$.

b. The function f is decreasing on the intervals $(-7, -3)$ and $(2, 5)$.

c. The function f obtains a relative minimum at $x = -3$ and $x = 5$.

d. The function f obtains a relative maximum at $x = 2$.

e. The relative minima are $f(-3) = -2$ and $f(5) = 0$.

f. The relative maximum is $f(2) = 5$.

⚠ **In Example 5, the value $f(-7) = 6$ does not represent a relative maximum. This is because the function does not change from increasing to decreasing at the point $(-7, 6)$. Also, there is no open interval that includes $x = -7$ because the function is not defined for values of x less than -7.**

You Try It Work through the following You Try It problem.

Work Exercises 17–21 in this textbook or in the MyMathLab Study Plan.

OBJECTIVE 5   DETERMINING WHETHER A FUNCTION IS EVEN, ODD, OR NEITHER

My video summary   ⊘ The graphs of some functions are symmetric about the *y*-axis. For example, the graphs of the three functions sketched in Figure 9 are all symmetric about the *y*-axis.

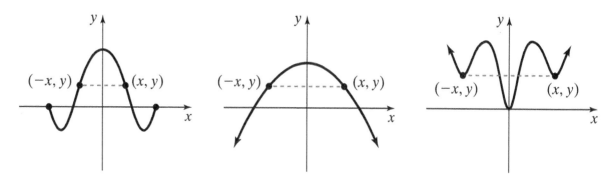

Figure 9 Functions whose graphs are symmetric about the *y*-axis are called even functions.

Imagine "folding" each graph of Figure 9 along the *y*-axis. If each graph was folded along the *y*-axis, the left and right halves of each graph would match. Functions whose graphs are symmetric about the *y*-axis are called *even functions*. For any point (x, y) on each graph, the point $(-x, y)$ also lies on the graph. Therefore, for any *x*-value in the domain, $f(x) = f(-x)$.

Definition   Even Functions

A function f is **even** if for every x in the domain, $f(x) = f(-x)$. The graph of an even function is symmetric about the *y*-axis. For each point (x, y) on the graph, the point $(-x, y)$ is also on the graph.

The graph of a function can also be symmetric about the origin. The functions sketched in Figure 10 illustrate this type of symmetry. Functions whose graphs are symmetric about the origin are called *odd functions*. If a function is an odd function, then for any point (x_1, y_1) on the graph of f in Quadrant I, there is a corresponding point $(-x_1, -y_1)$ on the graph of f in Quadrant III. Similarly, for any point (x_2, y_2) on the graph of f in Quadrant II, there is a corresponding point $(-x_2, -y_2)$ on the graph of f in Quadrant IV.

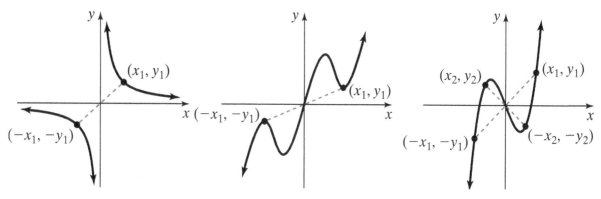

Figure 10 Functions whose graphs are symmetric about the origin are called odd functions.

For any point (x, y) on each graph in Figure 10, the point $(-x, -y)$ also lies on the graph. Therefore, for any x-value in the domain, $f(x) = -f(-x)$ or, equivalently, $-f(x) = f(-x)$.

Definition Odd Functions

A function f is **odd** if for every x in the domain, $-f(x) = f(-x)$. The graph of an odd function is symmetric about the origin. For each point (x, y) on the graph, the point $(-x, -y)$ is also on the graph.

Example 6 Determine Even and Odd Functions from the Graph

Determine whether each function is even, odd, or neither.

a.

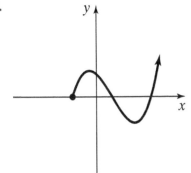

b.

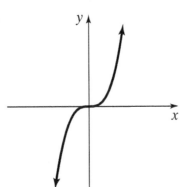

c.

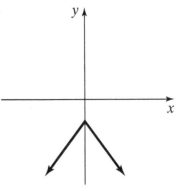

Solution

a. The graph is neither even nor odd because the graph has no symmetry.

b. The graph is symmetric about the origin; therefore, this is the graph of an odd function.

c. This graph is symmetric about the y-axis; thus, this represents the graph of an even function.

You Try It Work through the following You Try It problem.

Work Exercises 22–25 in this textbook or in the MyMathLab Study Plan.

My video summary ⊙ **Example 7** Determine Even and Odd Functions Algebraically

Determine whether each function is even, odd, or neither.

a. $f(x) = x^3 + x$ b. $g(x) = \dfrac{1}{x^2} + 7|x|$ c. $h(x) = 2x^5 - \dfrac{1}{x}$ d. $G(x) = x^2 + 4x$

Solution Watch the video to verify that the functions in parts a and c are odd functions. The function in part b is even, whereas the function in part d is neither even nor odd.

You Try It Work through the following You Try It problem.

Work Exercises 26–30 in this textbook or in the MyMathLab Study Plan.

OBJECTIVE 6   DETERMINING INFORMATION ABOUT A FUNCTION FROM A GRAPH

We are now ready to take the concepts learned in this section to answer questions about a function's graph. Carefully work through the animation of Example 8.

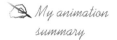
My animation summary

▶ **Example 8** Determine Information about a Function from a Graph

Use the graph of $y = f(x)$ to answer each question.

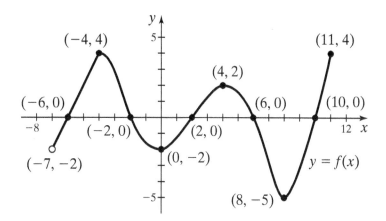

a. What is the y-intercept?

b. What are the real zeros of f?

c. Determine the domain and range of f.

d. Determine the interval(s) on which f is increasing, decreasing, and constant.

e. For what value(s) of x does f obtain a relative maximum? What are the relative maxima?

f. For what value(s) of x does f obtain a relative minimum? What are the relative minima?

g. Is f even, odd, or neither?

h. For what values of x is $f(x) = 4$?

i. For what values of x is $f(x) < 0$?

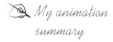
My animation summary

▶ **Solution** Carefully work through the animation to see the solution worked out in detail.   ●

You Try It   Work through this You Try It problem.

Work Exercises 31 and 32 in this textbook or in the MyMathLab **Study Plan.**

3.2 Exercises

Skill Check Exercises

In exercises SCE-1 through SCE-5, for each function, determine $f(-x)$.

SCE-1. $f(x) = -3x^5 + 4x^4 + 3x^3 - 7x^2 - 21$ **SCE-2.** $f(x) = x^4 - 7x^2 - 11$

SCE-3. $f(x) = -3x^7 + 8x^3 - x$

SCE-4. $f(x) = \dfrac{2x^2 - 4}{x}$

In Exercises 1–6, find the x-intercept(s) and the y-intercept of each function.

1. $f(x) = 2x - 1$

2. $g(x) = 3$

3. $f(t) = t^2 - t - 6$

4. $F(x) = 2x^2 - 7x - 3$

5. $h(x) = x^3 + 2x^2 - x - 2$

6. $f(x) = 2x^3 + 11x^2 - 6x$

In Exercises 7–12, use the graph to determine the domain and range of each function.

7.

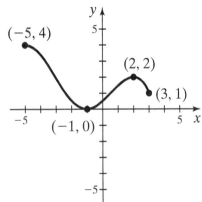

8.

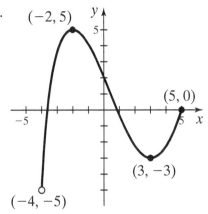

9.

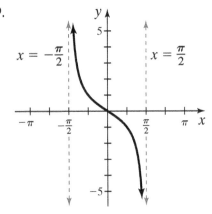

10.

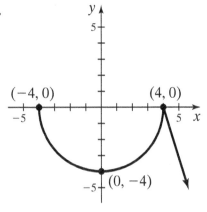

11.

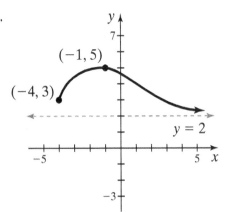

12.

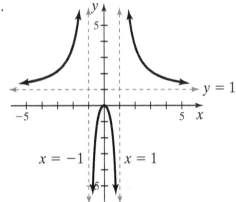

In Exercises 13–16, determine the interval(s) for which each function is (a) increasing, (b) decreasing, and (c) constant.

13.

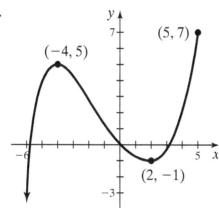

14.

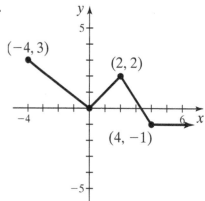

15.

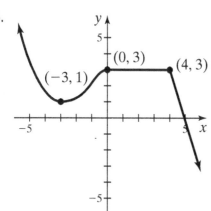

16.

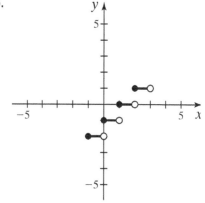

In Exercises 17–21, (a) For what value(s) of x does the function obtain a relative minimum? (b) find the relative minimum value, (c) For what value(s) of x does the function obtain a relative maximum? and (d) find the relative maximum value.

17.

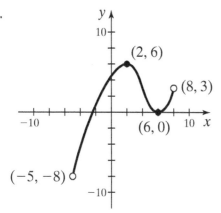

18.

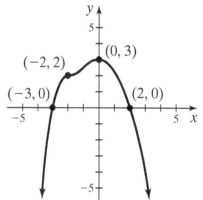

19.

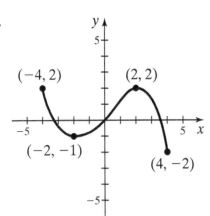

20.

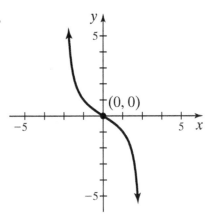

21.

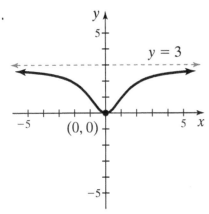

In Exercises 22–30 determine whether each function is even, odd, or neither.

22.

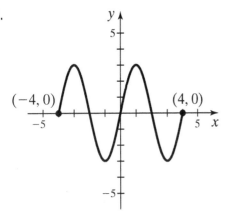

23.

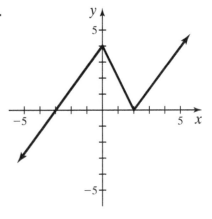

24.

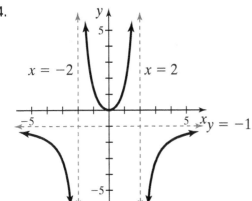

25.

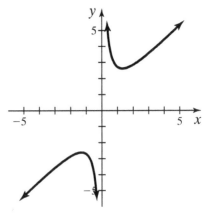

26. $f(x) = 2x^4 - x^2 + 1$ **27.** $g(x) = x^5 - x^2$ **28.** $h(x) = x^5 - x^3$

29. $F(x) = \dfrac{2x^2 - 4}{x}$ **30.** $G(x) = \sqrt{9 - x^2}$

31. Use the graph of $y = f(x)$ to answer each of the following questions:

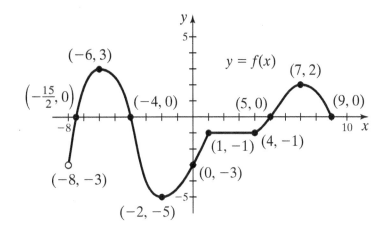

a. What is the domain of f?

b. What is the range of f?

c. On what interval(s) is f increasing, decreasing, or constant?

d. For what value(s) of x does f obtain a relative minimum? What are the relative minima?

e. For what value(s) of x does f obtain a relative maximum? What are the relative maxima?

f. What are the real zeros of f?

g. What is the y-intercept?

h. For what values of x is $f(x) \le 0$?

i. For how many values of x is $f(x) = -3$?

j. What is the value of $f(2)$?

32. Use the graph of $y = f(x)$ to answer each of the following questions:

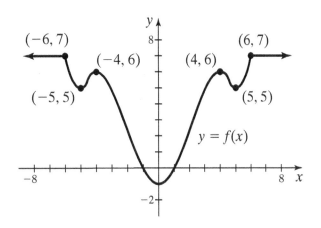

3.2 Properties of a Function's Graph **3-29**

a. What is the domain of f?

b. What is the range of f?

c. On what interval(s) is f increasing, decreasing, or constant?

d. For what value(s) of x does f obtain a relative minimum? What are the relative minima?

e. For what value(s) of x does f obtain a relative maximum? What are the relative maxima?

f. What are the x-intercept(s) of f?

g. Is f even, odd, or neither?

h. For what values of x is $f(x) > 0$?

i. True or false: $f(-3) > f(2)$

j. For what value(s) of x is $f(x) = 7$?

Brief Exercises

For Exercises 33–38, use the graph of $y = f(x)$ to answer each question.

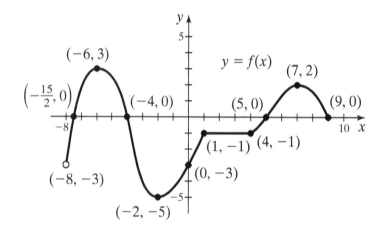

33. What is the domain and range of $y = f(x)$?

34. On what interval(s) is f increasing, decreasing, or constant?

35. a. For what value(s) of x does f obtain a relative minimum? What are the relative minima?

   b. For what value(s) of x does f obtain a relative maximum? What are the relative maxima?

36. What is the y-intercept of f?

37. For what values of x is $f(x) \leq 0$?

38. a. For how many values of x is $f(x) = 3$?

   b. What is the value of $f(2)$?

For Exercises 39–44, use the graph of $y = f(x)$ to answer each question.

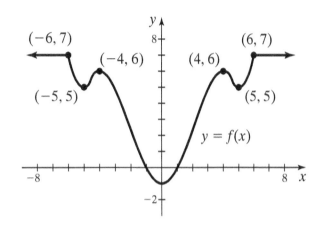

39. What is the domain and range of $y = f(x)$?

40. On what interval(s) is f increasing, decreasing, or constant?

41. a. For what value(s) of x does f obtain a relative minimum? What are the relative minima?

   b. For what value(s) of x does f obtain a relative maximum? What are the relative maxima?

42. What are the x-intercepts of f?

43. a. For what values of x is $f(x) > 0$?

   b. True or False: $f(-3) < f(-2)$?

   c. For what value(s) of x is $f(x) = 7$?

44. Is f even, odd or neither?

3.3 Graphs of Basic Functions; Piecewise Functions

THINGS TO KNOW

Before working through this section, be sure that you are familiar with the following concepts:

VIDEO   ANIMATION   INTERACTIVE

You Try It   1. Determining the Domain of a Function Given the Equation (Section 3.1)

You Try It   2. Determining the Domain and Range of a Function from Its Graph (Section 3.2)

OBJECTIVES

1 Sketching the Graphs of the Basic Functions

2 Analyzing Piecewise-Defined Functions

3 Solving Applications of Piecewise-Defined Functions

OBJECTIVE 1   SKETCHING THE GRAPHS OF THE BASIC FUNCTIONS

There are several functions whose graphs are worth memorizing. In this section, you learn about the graphs of nine basic functions. We begin by discussing the graphs of two specific linear functions. Recall that a linear function has the form $f(x) = mx + b$, where m is the slope of the line and b represents the y-coordinate of the y-intercept.

---

1. The constant function $f(x) = b$

The **constant function** is a linear function with $m = 0$. The graph of the constant function is a horizontal line.

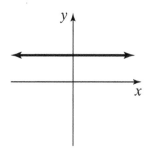

---

2. The identity function $f(x) = x$

The **identity function** defined by $f(x) = x$ is a linear function with $m = 1$ and $b = 0$. This function assigns to each number in the domain the exact same number in the range.

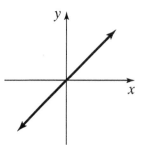

Of course, you should be able to sketch the graph of any linear function of the form $f(x) = mx + b$. To review how to sketch a line that is given in slope-intercept form, refer to Objective 4 in Section 2.3.

---

3. The square function $f(x) = x^2$

The **square function**, $f(x) = x^2$, assigns to each real number in the domain the square of that number in the range.

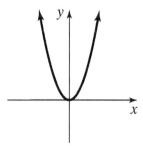

---

4. The cube function $f(x) = x^3$

The **cube function**, $f(x) = x^3$, assigns to each real number in the domain the cube of that number in the range.

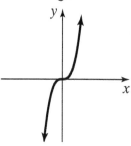

5. The absolute value function $f(x) = |x|$

The **absolute value function**, $f(x) = |x|$, assigns to each real number in the domain the absolute value of that number in the range.

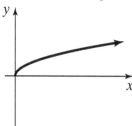

6. The square root function $f(x) = \sqrt{x}$

The **square root function**, $f(x) = \sqrt{x}$, is only defined for values of x that are greater than or equal to zero. It assigns to each real number in the domain the square root of that number in the range.

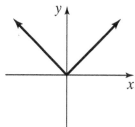

7. The cube root function $f(x) = \sqrt[3]{x}$

The **cube root function**, $f(x) = \sqrt[3]{x}$, is defined for all real numbers and assigns to each number in the domain the cube root of that number in the range.

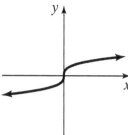

8. The reciprocal function $f(x) = \dfrac{1}{x}$

The **reciprocal function**, $f(x) = \dfrac{1}{x}$, is a rational function whose domain is

$\{x \,|\, x \neq 0\}$. It assigns to each number a in the domain its reciprocal, $\dfrac{1}{a}$, in the

range. The reciprocal function has two asymptotes. The y-axis (the line $x = 0$) is a vertical asymptote, and the x-axis (the line $y = 0$) is a horizontal asymptote.

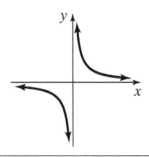

9. The greatest integer function $f(x) = [\![x]\!]$

The greatest integer function, $f(x) = [\![x]\!]$, (sometimes denoted $f(x) = \text{int}\,(x)$) assigns to each number in the domain the greatest integer less than or equal to that number in the range. For example, $[\![3.001]\!] = 3$, whereas $[\![2.999]\!] = 2$.

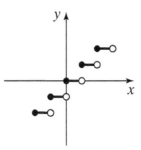

You Try It Work through the following You Try It problem.

Work Exercises 1–9 in this textbook or in the MyMathLab Study Plan.

OBJECTIVE 2 ANALYZING PIECEWISE-DEFINED FUNCTIONS

The absolute value function, $f(x) = |x|$, can also be defined by a rule that has two different pieces.

$$f(x) = |x| = \begin{cases} x \text{ if } x \geq 0 \\ -x \text{ if } x < 0 \end{cases}$$

You can see by the following graph that the left-hand piece is actually part of the line $y = -x$, whereas the right-hand piece is part of the line $y = x$.

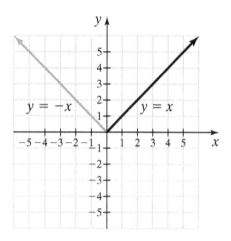

Functions defined by a rule that has more than one piece are called piecewise-defined functions.

My animation summary

Example 1 Sketch a Piecewise-Defined Function

Sketch the function $f(x) = \begin{cases} x^2 & \text{if } x < 1 \\ 1 - x & \text{if } x \geq 1 \end{cases}$.

Solution The graph of this piecewise function is sketched as follows. Watch the animation to see how to sketch this function.

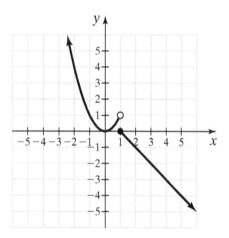

My video summary

Example 2 Analyze a Piecewise-Defined Function

Let $f(x) = \begin{cases} 1 & \text{if } x < -1 \\ \sqrt[3]{x} & \text{if } -1 \leq x < 0 \\ \dfrac{1}{x} & \text{if } x > 0 \end{cases}$.

a. Evaluate $f(-3), f(-1)$, and $f(2)$.

b. Sketch the graph of f.

c. Determine the domain of f.

d. Determine the range of f.

Solution Watch the video to view the solution.

You Try It Work through the following You Try It problem.

Work Exercises 10–15 in this textbook or in the MyMathLab Study Plan.

My video summary ⊚ **Example 3** Find a Rule That Defines a Piecewise Function

Find a rule that describes the following piecewise function:

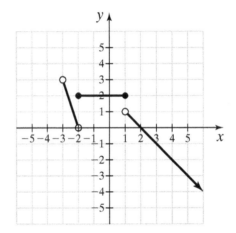

Solution Watch this video to see that we can define this piecewise function as

$$f(x) = \begin{cases} -3x - 6 & \text{if } -3 < x < -2 \\ 2 & \text{if } -2 \le x \le 1 \\ -x + 2 & \text{if } x > 1 \end{cases}$$

You Try It Work through the following You Try It problem.

Work Exercises 16–20 in this textbook or in the MyMathLab Study Plan.

OBJECTIVE 3 SOLVING APPLICATIONS OF PIECEWISE-DEFINED FUNCTIONS

Example 4 A Flat Tax Plan

Steve Forbes, a presidential candidate in 1996, proposed a flat tax to replace the existing U.S. income tax system. In his tax proposal, every adult would pay $0.00 in taxes on the first $13,000 earned. They would then pay a flat tax of 17% on everything over $13,000. Forbes's tax plan is actually a piecewise-defined function.

a. According to Forbes's plan, how much in taxes are owed for someone earning $50,000?

b. Find the piecewise function, $T(x)$, that describes the amount of taxes paid, T, as a function of the dollars earned, x, for Forbes's tax plan.

c. Sketch the piecewise function, $T(x)$.

Solution

a. On $50,000, a person would pay 0% on the first $13,000 and 17% on the remaining $37,000. Thus, the taxes owed would be $0.17(37,000) = \$6,290$.

b. For values of x less than or equal to 13,000, $T(x) = 0$. When x is greater than 13,000, the tax rate is 17% or $T(x) = 0.17(x - 13,000)$ for $x > 13,000$. Thus, the rule for computing the tax owed is

$$T(x) = \begin{cases} 0 & \text{if } x \leq 13,000 \\ 0.17(x - 13,000) & \text{if } x > 13,000 \end{cases}.$$

c. The graph of $T(x)$ is sketched as follows.

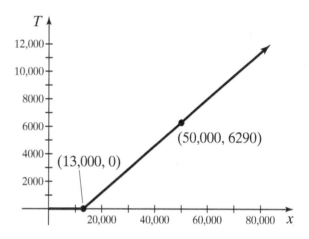

My video summary ⊙ **Example 5 Rent a Car**

Cheapo Rental Car Co. charges a flat rate of $85 to rent a car for up to 2 days. The company charges an additional $20 for each additional day (or part of a day).

a. How much does it cost to rent a car for $2\frac{1}{2}$ days? 3 days?

b. Write and sketch the piecewise function, $C(x)$, that describes the cost of renting a car as a function of the time, x, in days rented for values of x less than or equal to 5.

Solution The rule for this piecewise function is $C(x) = \begin{cases} 85 & \text{if} & x \leq 2 \\ 105 & \text{if} & 2 < x \leq 3 \\ 125 & \text{if} & 3 < x \leq 4 \\ 145 & \text{if} & 4 < x \leq 5 \end{cases}$

My video summary ⊙ and is sketched as follows. Watch the video for a detailed explanation.

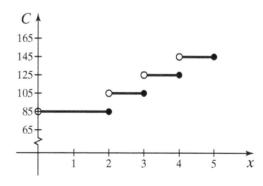

You Try It   Work through the following You Try It problem.

Work Exercises 21–24 in this textbook or in the MyMathLab Study Plan.

3.3 Exercises

In Exercises 1–9, sketch the graph of each function, and identify all properties that apply.

1. $f(x) = x^2$

2. $f(x) = \sqrt[3]{x}$

3. $f(x) = 2x - 1$

4. $f(x) = |x|$

5. $f(x) = -4$

6. $f(x) = x$

7. $f(x) = x^3$

8. $f(x) = \sqrt{x}$

9. $f(x) = \dfrac{1}{x}$

Properties

a. The function f is an even function.

b. The function f is an odd function.

c. The function f is increasing on the interval $(-\infty, \infty)$.

d. The function f is a linear function.

e. The domain of f is $(-\infty, \infty)$.

f. The range of f is $(-\infty, \infty)$.

In Exercises 10–15, (a) find $f(-2), f(0)$, and $f(2)$; (b) sketch the graph of each piecewise-defined function; (c) determine the domain of f; and (d) determine the range of f.

10. $f(x) = \begin{cases} 1 & \text{if } x < 2 \\ -3 & \text{if } x \geq 2 \end{cases}$

11. $f(x) = \begin{cases} 2x - 1 & \text{if } x \leq 1 \\ x + 2 & \text{if } x > 1 \end{cases}$

12. $f(x) = \begin{cases} x^2 & \text{if } x \leq 0 \\ \dfrac{1}{x} & \text{if } x > 0 \end{cases}$

13. $f(x) = \begin{cases} 3 & \text{if } x < -1 \\ |x| & \text{if } -1 \leq x < 2 \\ 1 & \text{if } x \geq 2 \end{cases}$

14. $f(x) = \begin{cases} x^2 & \text{if } x < 0 \\ 2 & \text{if } x = 0 \\ \sqrt{x} & \text{if } x > 0 \end{cases}$

15. $f(x) = \begin{cases} [\![x]\!] & \text{if } -3 \leq x \leq -1 \\ x + 2 & \text{if } -1 < x < 1 \\ x^3 & \text{if } x \geq 1 \end{cases}$

In Exercises 16–20, find the rule that describes each piecewise-defined function.

16.

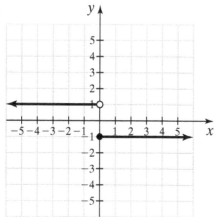

17.

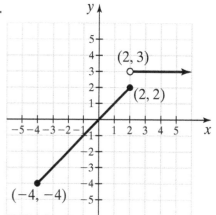

18.

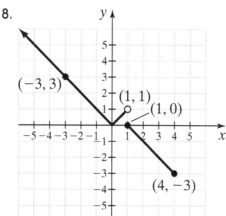

19.

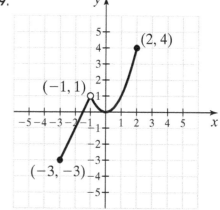

20.

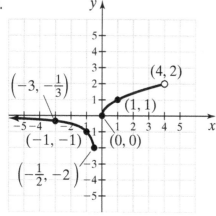

21. The currency in Norway is called the krone (plural: kroner) and is abbreviated NOK. Following is a table describing the 1999 class 1 individual tax rate on personal income in Norway.

Income	Tax Rate
NOK 0–26,300	0%
Over NOK 26,300	28%

a. How many kroner in taxes are owed for an individual earning NOK 30,000?

b. Find the piecewise function, $T(x)$, that describes the amount of taxes paid, T, as a function of kroner earned, x, for Norwegians paying class 1 personal income tax in Norway in 1999.

c. Sketch the piecewise function, $T(x)$.

22. An electric company charges $9.50 per month plus $0.07 per kilowatt-hour (kWh) for the first 350 kWh used per month and $0.05 per kWhr for anything over 350 kWh used.

a. How much is owed for using 280 kWh per month?

b. How much is owed for using 400 kWh per month?

c. Write a piecewise function, $C(x)$, where C is the cost charged per month as a function of the monthly kilowatt-hours used, x.

d. Sketch the piecewise function.

23. In 2007, the postage charged for a first-class mail parcel was $1.13 for the first ounce (or fraction of an ounce) and $0.17 for each additional ounce (or fraction of an ounce).

a. How much does it cost to send a 2.4-ounce first-class letter?

b. Write a piecewise function, $P(x)$, where P is the postage charged as a function of its weight in ounces, x, for values of x less than or equal to 5 ounces.

c. Sketch the piecewise function.

Source: United States Postal Service.

24. A cell phone company charges $25.00 per month for the first 500 minutes of use and $0.05 for each additional minute (or part of a minute).

a. How much is owed to the cell phone company if a person used 502.5 minutes?

b. Write a piecewise function, $C(x)$, where C is the cost charged per month as a function of the minutes, x, used for values of x less than or equal to 505 minutes.

c. Sketch the piecewise function.

Brief Exercises

In Exercises 25–30, (a) find $f(-2), f(0)$, and $f(2)$; (b) sketch the graph of each piecewise-defined function.

25. $f(x) = \begin{cases} 1 & \text{if } x < 2 \\ -3 & \text{if } x \geq 2 \end{cases}$

26. $f(x) = \begin{cases} 2x - 1 & \text{if } x \leq 1 \\ x + 2 & \text{if } x > 1 \end{cases}$

27. $f(x) = \begin{cases} x^2 & \text{if } x \leq \\ \dfrac{1}{x} & \text{if } x > 0 \end{cases}$

28. $f(x) = \begin{cases} 3 & \text{if } x < -1 \\ |x| & \text{if } -1 \leq x < 2 \\ 1 & \text{if } x \geq 2 \end{cases}$

29. $f(x) = \begin{cases} x^2 & \text{if } x < 0 \\ 2 & \text{if } x = 0 \\ \sqrt{x} & \text{if } x > 0 \end{cases}$

30. $f(x) = \begin{cases} [\![x]\!] & \text{if } -3 \leq x \leq -1 \\ x + 2 & \text{if } -1 < x < 1 \\ x^3 & \text{if } x \geq 1 \end{cases}$

3.4 Transformations of Functions

THINGS TO KNOW

Before working through this section, be sure that you are familiar with the following concepts:

You Try It

1. Determining the Domain of a Function Given the Equation (Section 3.1)

You Try It

2. Determining the Domain and Range of a Function from Its Graph (Section 3.2)

You Try It

3. Determining Whether a Function Is Even, Odd, or Neither (Section 3.2)

You Try It

4. Sketching the Graphs of the Basic Functions (Section 3.3)

OBJECTIVES

1 Using Vertical Shifts to Graph Functions

2 Using Horizontal Shifts to Graph Functions

3 Using Reflections to Graph Functions

4 Using Vertical Stretches and Compressions to Graph Functions

5 Using Horizontal Stretches and Compressions to Graph Functions

6 Using Combinations of Transformations to Graph Functions

Introduction to Section 3.4

In this section, we learn how to sketch the graphs of new functions using the graphs of known functions. Starting with the graph of a known function, we "transform" it into a new function by applying various transformations. Before we begin our discussion about transformations, it is critical that you know the graphs of the basic functions that are discussed in Section 3.3. Take a moment to review the basic functions.

REVIEW OF THE BASIC FUNCTIONS

Click on any function to review its graph.

The identity function	$f(x) = x$
The cube function	$f(x) = x^3$
The square root function	$f(x) = \sqrt{x}$
The reciprocal function	$f(x) = \dfrac{1}{x}$

The square function $f(x) = x^2$

The absolute value function $f(x) = |x|$

The cube root function $f(x) = \sqrt[3]{x}$

OBJECTIVE 1   USING VERTICAL SHIFTS TO GRAPH FUNCTIONS

My video summary   ⊙ **Example 1** Vertically Shift a Function

Sketch the graphs of $f(x) = |x|$ and $g(x) = |x| + 2$.

Solution Table 2 shows that for every value of x, the y-coordinate of the function g is always 2 greater than the y-coordinate of the function f. The two functions are sketched in Figure 11. The graph of $g(x) = |x| + 2$ is exactly the same as the graph of $f(x) = |x|$, except the graph of g is shifted *up* two units.

Table 2

x	$f(x) = |x|$	$g(x) = |x| + 2$
-3	3	5
-2	2	4
-1	1	3
0	0	2
1	1	3
2	2	4
3	3	5

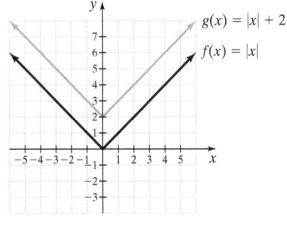

Figure 11

My video summary   ⊙ Watch this video to see the solution to this example.

We see from Example 1 that if c is a positive number, then $y = f(x) + c$ is the graph of f shifted *up* c units. It follows that for $c > 0$, the graph of $y = f(x) - c$ is the graph of f shifted *down* c units.

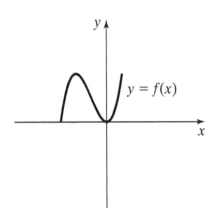

Vertical Shifts of Functions

If c is a positive real number,

The graph of $y = f(x) + c$ is obtained by shifting the graph of $y = f(x)$ vertically upward c units.

The graph of $y = f(x) - c$ is obtained by shifting the graph of $y = f(x)$ vertically downward c units.

Sketch $y = f(x) + c$.

☐ Click icon to animate

Sketch $y = f(x) - c$.

☐ Click icon to animate

You Try It Work through the following You Try It problem.

Work Exercises 1–8 in this textbook or in the MyMathLab Study Plan.

⊘ OBJECTIVE 2 USING HORIZONTAL SHIFTS TO GRAPH FUNCTIONS

To illustrate a horizontal shift, let $f(x) = x^2$ and $g(x) = (x + 2)^2$. Tables 3 and 4 show tables of values for f and g, respectively. The graphs of f and g are sketched in Figure 12. The graph of g is the graph of f shifted to the *left* two units.

Table 3

x	$f(x) = x^2$
-2	4
-1	1
0	0
1	1
2	4

Table 4

x	$g(x) = (x + 2)^2$
-4	4
-3	1
-2	0
-1	1
0	4

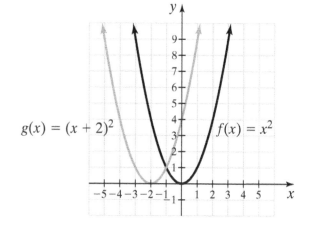

Figure 12

It follows that if c is a positive number, then $y = f(x + c)$ is the graph of f shifted to the *left* c units. For $c > 0$, the graph of $y = f(x - c)$ is the graph of f shifted to the *right* c units. At first glance, it appears that the rule for horizontal shifts is the opposite of what seems natural. Substituting $x + c$ for x causes the graph of $y = f(x)$ to be shifted to the left, whereas substituting $x - c$ for x causes the graph to shift to the right c units.

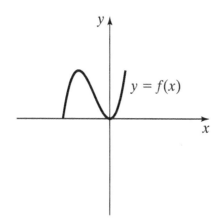

Sketch $y = f(x + c)$.

⬚ Click icon to animate

Sketch $y = f(x - c)$.

⬚ Click icon to animate

Horizontal Shifts of Functions

If c is a positive real number,

The graph of $y = f(x + c)$ is obtained by shifting the graph of $y = f(x)$ horizontally to the left c units.

The graph of $y = f(x - c)$ is obtained by shifting the graph of $y = f(x)$ horizontally to the right c units.

You Try It Work through the following You Try It problem.

Work Exercises 9–16 in this textbook or in the MyMathLab Study Plan.

My animation summary

📺 **Example 2** Combine Horizontal and Vertical Shifts

Use the graph of $y = x^3$ to sketch the graph of $g(x) = (x - 1)^3 + 2$.

Solution The graph of g is obtained by shifting the graph of the basic function $y = x^3$ first horizontally to the right one unit and then vertically upward two units. When doing a problem with multiple transformations, it is good practice to always perform the vertical transformation last. Click on the animate button to see how to sketch the graph of $g(x) = (x - 1)^3 + 2$.

📺 Click icon to animate

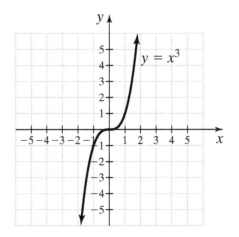

 You Try It Work through the following You Try It problem.

Work Exercises 17–24 in this textbook or in the MyMathLab Study Plan.

OBJECTIVE 3 USING REFLECTIONS TO GRAPH FUNCTIONS

My video summary

⊗ Given the graph of $y = f(x)$, what does the graph of $y = -f(x)$ look like?

Using a graphing utility with $y_1 = x^2$ and $y_2 = -x^2$, we can see that the graph of $y_2 = -x^2$ is the graph of $y_1 = x^2$ reflected about the **x-axis**.

Using Technology

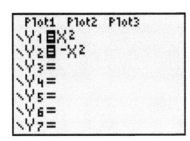

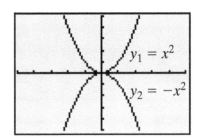

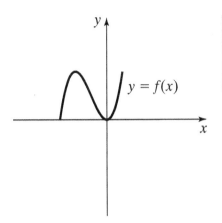

Reflections of Functions about the x-Axis

The graph of $y = -f(x)$ is obtained by reflecting the graph of $y = f(x)$ about the x-axis.

Sketch $y = -f(x)$.

▣ Click icon to animate

Functions can also be reflected about the y-axis. Given the graph of $y = f(x)$, the graph of $y = f(-x)$ will be the graph of $y = f(x)$ reflected about the y-axis. Using a graphing utility, we illustrate a y-axis reflection by letting $y_1 = \sqrt{x}$ and $y_2 = \sqrt{-x}$. You can see that the functions are mirror images of each other about the y-axis.

Using Technology

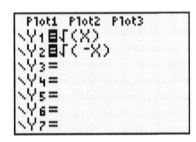

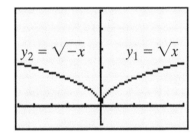

Reflections of Functions about the y-Axis

The graph of $y = f(-x)$ is obtained by reflecting the graph of $y = f(x)$ about the y-axis.

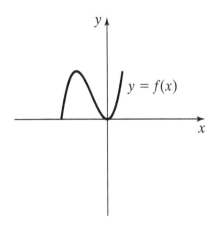

Sketch $y = f(-x)$.

▣ Click icon to animate

3.4 Transformations of Functions **3-45**

Example 3 Sketch Functions Using Reflections and Shifts

Use the graph of the basic function $y = \sqrt[3]{x}$ to sketch each graph.

a. $g(x) = -\sqrt[3]{x} - 2$

b. $h(x) = \sqrt[3]{1 - x}.$

Solution

a. Starting with the graph of $y = \sqrt[3]{x}$, we can obtain the graph of $g(x) = -\sqrt[3]{x} - 2$ by performing two transformations:

1. Reflect about the x-axis.

2. Vertically shift down two units.

See Figure 13.

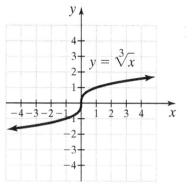

Start with the graph of the basic function $y = \sqrt[3]{x}$.

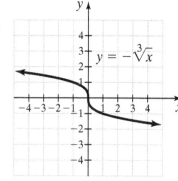

Sketch the graph of $y = -\sqrt[3]{x}.$

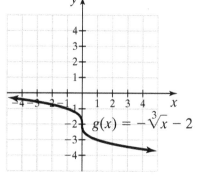

Sketch the graph of $g(x) = -\sqrt[3]{x} - 2.$

Figure 13

b. Starting with the graph of $y = \sqrt[3]{x}$, we can obtain the graph of $h(x) = \sqrt[3]{1 - x}$ by performing two transformations:

1. Horizontal shift left one unit.

2. Reflect about the y-axis.

See Figure 14.

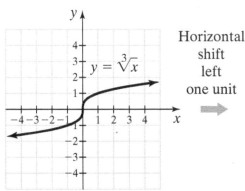

Horizontal shift left one unit

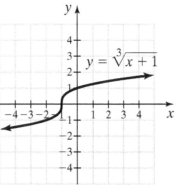

Reflect about the y-axis

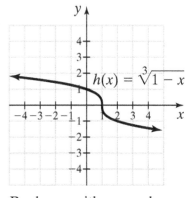

Start with the graph of the basic function $y = \sqrt[3]{x}$.

Sketch the graph of $y = \sqrt[3]{x + 1}$.

Replace x with $-x$, and sketch the graph of $h(x) = \sqrt[3]{-x + 1} = \sqrt[3]{1 - x}$.

Figure 14

You Try It Work through the following You Try It problem.

Work Exercises 25–38 in this textbook or in the MyMathLab Study Plan.

OBJECTIVE 4 USING VERTICAL STRETCHES AND COMPRESSIONS TO GRAPH FUNCTIONS

My video summary ⊙ **Example 4** Vertical Stretch and Compression

Use the graph of $f(x) = x^2$ to sketch the graph of $g(x) = 2x^2$.

Solution Notice in Table 5 that for each value of x, the y-coordinate of g is two times as large as the corresponding y-coordinate of f. We can see in Figure 15 that the graph of $f(x) = x^2$ is vertically stretched by a factor of 2 to obtain the graph of $g(x) = 2x^2$. In other words, for each point (a, b) on the graph of f, the graph of g contains the point $(a, 2b)$.

Table 5

x	$f(x) = x^2$	$g(x) = 2x^2$
-2	4	8
-1	1	2
0	0	0
1	1	2
2	4	8

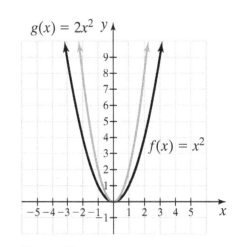

Figure 15

It follows from Example 4 that if $a > 1$, the graph of $y = af(x)$ is a **vertical stretch** of the graph of $y = f(x)$ and is obtained by multiplying each

y-coordinate on the graph of f by a factor of a. If $0 < a < 1$, then the graph of $y = af(x)$ is a **vertical compression** of the graph of $y = f(x)$. Table 6 and Figure 16 show the relationship between the graphs of the functions $f(x) = x^2$ and $h(x) = \dfrac{1}{2}x^2$.

Table 6

x	$f(x) = x^2$	$h(x) = \dfrac{1}{2}x^2$
-2	4	2
-1	1	$\dfrac{1}{2}$
0	0	0
0	1	$\dfrac{1}{2}$
2	4	2

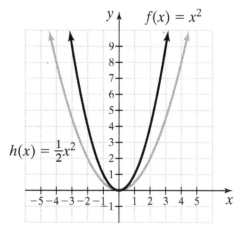

Figure 16

Vertical Stretches and Compressions of Functions

Suppose a is a positive real number:

The graph of $y = af(x)$ is obtained by multiplying each y-coordinate of $y = f(x)$ by a. If $a > 1$, the graph of $y = af(x)$ is a vertical stretch of the graph of $y = f(x)$. If $0 < a < 1$, the graph of $y = af(x)$ is a vertical compression of the graph of $y = f(x)$.

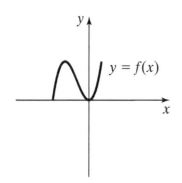

Sketch $y = af(x)$.
$(a > 1)$

🔲 Click icon to animate

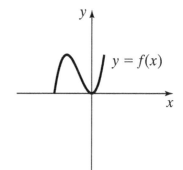

Sketch $y = af(x)$.
$(0 < a < 1)$

🔲 Click icon to animate

You Try It Work through the following You Try It problem.

Work Exercises 39–52 in this textbook or in the MyMathLab Study Plan.

OBJECTIVE 5 USING HORIZONTAL STRETCHES AND COMPRESSIONS TO GRAPH FUNCTIONS

My video summary ◉ The final transformation to discuss is a horizontal stretch or compression. A function, $y = f(x)$, will be horizontally stretched or compressed when x is multiplied by a positive number, $a \neq 1$, to obtain the new function, $y = f(ax)$.

Horizontal Stretches and Compressions of Functions

If a is a positive real number,

For $a > 1$, the graph of $y = f(ax)$ is obtained by dividing each x-coordinate of $y = f(x)$ by a. The resultant graph is a horizontal compression.

For $0 < a < 1$, the graph of $y = f(ax)$ is obtained by dividing each x-coordinate of $y = f(x)$ by a. The resultant graph is a horizontal stretch.

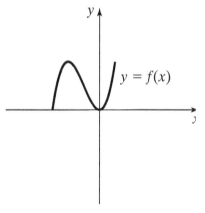

Sketch $y = f(ax)$. $(a > 1)$

▢ Click icon to animate

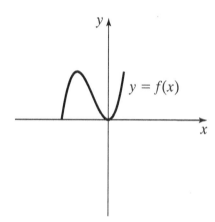

Sketch $y = f(ax)$. $(0 < a < 1)$

▢ Click icon to animate

Example 5 Horizontal Stretch and Compression

Use the graph of $f(x) = \sqrt{x}$ to sketch the graphs of $g(x) = \sqrt{4x}$ and $h(x) = \sqrt{\frac{1}{4}x}$.

Solution The graph of $f(x) = \sqrt{x}$ contains the ordered pairs $(0, 0)$, $(1, 1)$, $(4, 2)$. To sketch the graph of $g(x) = \sqrt{4x}$, we must divide each previous x-coordinate by 4. Therefore, the ordered pairs $(0, 0)$, $\left(\frac{1}{4}, 1\right)$, $(1, 2)$ must lie on the graph of g.

You can see that the graph of g is a horizontal compression of the graph of f. See Figure 17.

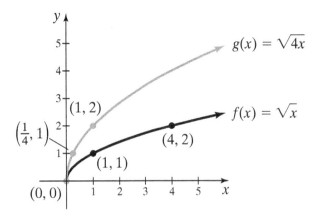

Figure 17 Graphs of $f(x) = \sqrt{x}$ and $g(x) = \sqrt{4x}$

Similarly, to sketch the graph of $h(x) = \sqrt{\dfrac{1}{4}x}$, we divide the x-coordinates of the ordered pairs of f by $\dfrac{1}{4}$ to get the ordered pairs $(0, 0), (4, 1), (16, 2)$. You can see that the graph of h is a horizontal stretch of the graph of f. See Figure 18.

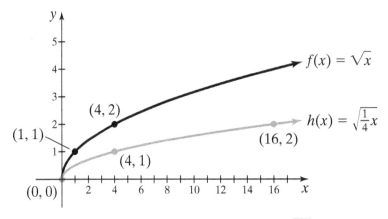

Figure 18 Graphs of $f(x) = \sqrt{x}$ and $g(x) = \sqrt{\dfrac{1}{4}x}$

You Try It   Work through the following You Try It problem.

Work Exercises 53–57 in this textbook or in the MyMathLab Study Plan.

OBJECTIVE 6   USING COMBINATIONS OF TRANSFORMATIONS TO GRAPH FUNCTIONS

You may encounter functions that combine many (if not all) of the transformations discussed in this section. When sketching a function that involves multiple transformations, it is important to follow a certain "order of operations." Following is the order in which each transformation is performed in this text:

1. Horizontal shifts

2. Horizontal stretches/compressions

3. Reflection about y-axis

4. Vertical stretches/compressions

5. Reflection about *x*-axis

6. Vertical shifts

Although different ordering is possible, the order above will always work.

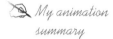

My animation summary

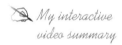 **Example 6** Combine Transformations

Use transformations to sketch the graph of $f(x) = -2(x + 3)^2 - 1$.

Solution Watch the animation to see how to sketch the function $f(x) = -2(x + 3)^2 - 1$ as seen in Figure 19.

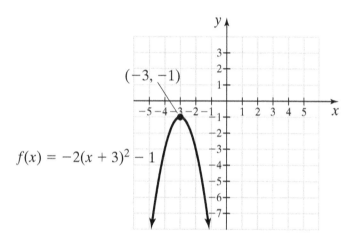

Figure 19 Graph of $f(x) = -2(x + 3)^2 - 1$

My interactive video summary

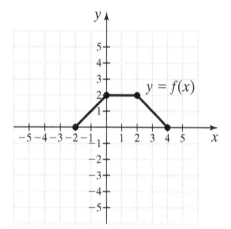 **Example 7** Combine Transformations

Use the graph of $y = f(x)$ to sketch each of the following functions.

a. $y = -f(2x)$ **b.** $y = 2f(x - 3) - 1$ **c.** $y = -\dfrac{1}{2}f(2 - x) + 3$

My interactive video summary

 **Solution** Watch the interactive video to see any one of the solutions worked out in detail.

a. The graph of $y = -f(2x)$ can be obtained from the graph of $y = f(x)$ using two transformations: (1) a horizontal compression and (2) a reflection about the *x*-axis. The resultant graph is sketched in Figure 20.

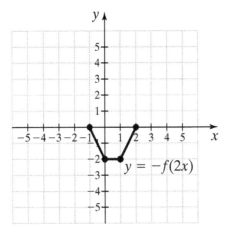

Figure 20 Graph of $y = -f(2x)$

b. The graph of $y = 2f(x - 3) - 1$ can be obtained from the graph of $y = f(x)$ using three transformations: (1) a horizontal shift to the right three units, (2) a vertical stretch by a factor of 2, and (3) a vertical shift down one unit. The resultant graph is sketched in Figure 21.

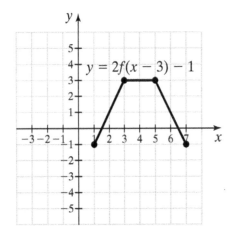

Figure 21 Graph of $y = 2f(x - 3) - 1$

c. The graph of $y = -\dfrac{1}{2}f(2 - x) + 3$ can be obtained from the graph of $y = f(x)$ using five transformations: (1) a horizontal shift to the left two units, (2) a reflection about the y-axis, (3) a vertical compression by a factor of $\dfrac{1}{2}$, (4) a reflection about the x-axis, and (5) a vertical shift up three units. The resultant graph is sketched in Figure 22.

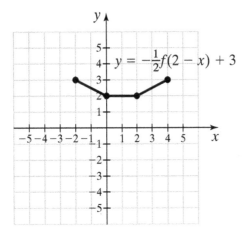

Figure 22 Graph of $y = -\dfrac{1}{2}f(2 - x) + 3$ ●

You Try It   Work through the following You Try It problem.

Work Exercises 58–68 in this textbook or in the MyMathLab Study Plan.

Summary of Transformation Techniques

Given a function $y = f(x)$ and a constant $c > 0$:

1. The graph of $y = f(x) + c$ is obtained by shifting the graph of $y = f(x)$ vertically upward c units.

2. The graph of $y = f(x) - c$ is obtained by shifting the graph of $y = f(x)$ vertically downward c units.

3. The graph of $y = f(x + c)$ is obtained by shifting the graph of $y = f(x)$ horizontally to the left c units.

4. The graph of $y = f(x - c)$ is obtained by shifting the graph of $y = f(x)$ horizontally to the right c units.

5. The graph of $y = -f(x)$ is obtained by reflecting the graph of $y = f(x)$ about the x-axis.

6. The graph of $y = f(-x)$ is obtained by reflecting the graph of $y = f(x)$ about the y-axis.

7. Suppose a is a positive real number. The graph of $y = af(x)$ is obtained by multiplying each y-coordinate of $y = f(x)$ by a.

   If $a > 1$, the graph of $y = af(x)$ is a vertical stretch of the graph of $y = f(x)$.

   If $0 < a < 1$, the graph of $y = af(x)$ is a vertical compression of the graph of $y = f(x)$.

8. Suppose a is a positive real number. The graph of $y = f(ax)$ is obtained by dividing each x-coordinate of $y = f(x)$ by a.

   If $a > 1$, the graph of $y = f(ax)$ is a horizontal compression of the graph of $y = f(x)$.

   If $0 < a < 1$, the graph of $y = f(ax)$ is a horizontal stretch of the graph of $y = f(x)$.

3.4 Exercises

In Exercises 1–6, use the graph of a known basic function and vertical shifting to sketch each function.

1. $f(x) = x^2 - 1$

2. $y = \sqrt{x} + 2$

3. $g(x) = \dfrac{1}{x} + 3$

4. $h(x) = \sqrt[3]{x} - 2$

5. $y = |x| - 3$

6. $g(x) = x^3 + 1$

7. Use the graph of $y = f(x)$ to sketch the graph of $y = f(x) - 1$. Label at least three points on the new graph.

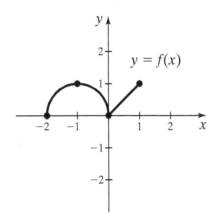

8. Use the graph of $y = f(x)$ to sketch the graph of $y = f(x) + 2$. Label at least three points on the new graph.

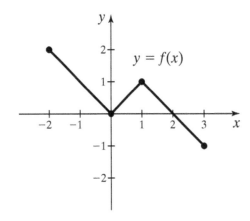

In Exercises 9–14, use the graph of a known basic function and horizontal shifts to sketch each function.

9. $f(x) = \sqrt[3]{x - 2}$

10. $g(x) = \dfrac{1}{x + 3}$

11. $y = \sqrt{x - 4}$

12. $h(x) = (x + 1)^3$

13. $k(x) = |x - 1|$

14. $y = (x - 3)^2$

15. Use the graph of $y = f(x)$ to sketch the graph of $y = f(x - 2)$. Label at least three points on the new graph.

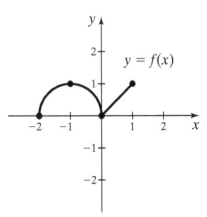

16. Use the graph of $y = f(x)$ to sketch the graph of $y = f(x + 2)$. Label at least three points on the new graph.

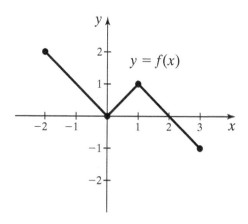

$y = f(x)$

In Exercises 17–22, use the graph of a known basic function and a combination of horizontal and vertical shifts to sketch each function.

17. $y = (x + 1)^2 - 2$

18. $f(x) = (x - 3)^2 + 1$

19. $y = \sqrt{x + 3} + 2$

20. $h(x) = \dfrac{1}{x - 2} + 3$

21. $f(x) = |x + 2| + 2$

22. $y = \sqrt[3]{x + 1} - 1$

23. Use the graph of $y = f(x)$ to sketch the graph of $y = f(x - 2) - 1$. Label at least three points on the new graph.

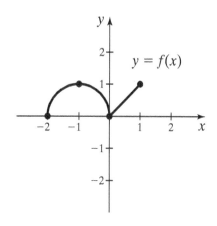

$y = f(x)$

24. Use the graph of $y = f(x)$ to sketch the graph of $y = f(x + 1) + 2$. Label at least three points on the new graph.

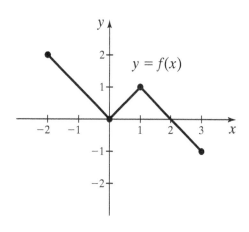

$y = f(x)$

In Exercises 25–36, use the graph of a known basic function and a combination of horizontal shifts, reflections, and vertical shifts to sketch each function.

25. $g(x) = -x^2 - 2$

26. $f(x) = -\dfrac{1}{x} + 1$

27. $h(x) = \sqrt{2 - x}$

28. $f(x) = |-1 - x|$

29. $h(x) = \sqrt[3]{-x - 2}$

30. $g(x) = -x^3 + 1$

31. $h(x) = -\sqrt[3]{x} + 2$

32. $g(x) = (3 - x)^2$

33. $h(x) = -\sqrt{x} - 1$

34. $f(x) = -|x| + 1$

35. $g(x) = (1 - x)^3$

36. $f(x) = \dfrac{1}{2 - x}$

37. Use the graph of $y = f(x)$ to sketch the graph of $y = -f(x)$. Label at least three points on the new graph.

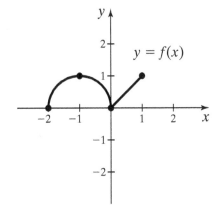

38. Use the graph of $y = f(x)$ to sketch the graph of $y = f(-x)$. Label at least three points on the new graph.

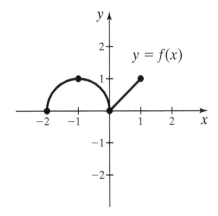

In Exercises 39–50, use the graph of a known basic function and a vertical stretch or vertical compression to sketch each function.

39. $f(x) = 3|x|$

40. $f(x) = \dfrac{1}{4}|x|$

41. $g(x) = 2\sqrt{x}$

42. $f(x) = \dfrac{1}{4}\sqrt{x}$

43. $f(x) = 3x^3$

44. $f(x) = \dfrac{1}{3}x^3$

45. $f(x) = 6\sqrt[3]{x}$

46. $f(x) = \dfrac{1}{2}\sqrt[3]{x}$

47. $g(x) = 3x^2$

48. $f(x) = \dfrac{1}{2}x^2$

49. $f(x) = \dfrac{4}{x}$

50. $y = \dfrac{1}{2x}$

51. Use the graph of $y = f(x)$ to sketch the graph of $y = 3f(x)$. Label at least three points on the new graph.

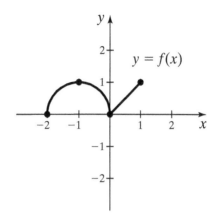

52. Use the graph of $y = f(x)$ to sketch the graph of $y = \dfrac{1}{2}f(x)$.

Label at least three points on the new graph.

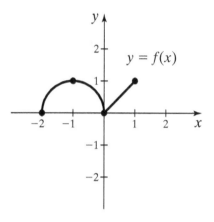

In Exercises 53–55, use the graph of a known basic function and a horizontal stretch or horizontal compression to sketch each function.

53. $y = \left| \dfrac{1}{4}x \right|$ **54.** $f(x) = \sqrt{2x}$ **55.** $g(x) = \sqrt[3]{3x}$

56. Use the graph of $y = f(x)$ to sketch the graph of $y = f(2x)$. Label at least three points on the new graph.

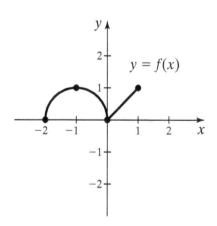

57. Use the graph of $y = f(x)$ to sketch the graph of $y = f\left(\dfrac{1}{2}x\right)$. Label at least three points on the new graph.

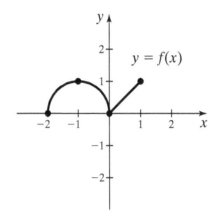

In Exercises 58–63, use the graph of a known basic function and a combination of transformations to sketch each function.

58. $f(x) = -(x - 2)^2 + 3$

59. $g(x) = \dfrac{1}{2}|x + 1| - 1$

60. $y = \dfrac{1}{x - 3} + 2$

61. $f(x) = 2\sqrt[3]{x} - 1$

62. $g(x) = -\dfrac{1}{2}(2 - x)^3 + 1$

63. $h(x) = 2\sqrt{4 - x} + 5$

In Exercises 64–68, use the graph of $y = f(x)$ to sketch each function. Label at least three points on each graph.

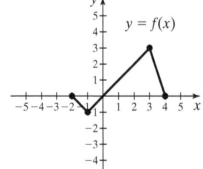

64. $y = -f(-x) - 1$

65. $y = \dfrac{1}{2}f(2 - x)$

66. $y = -2f(x + 1) + 2$

67. $y = 3 - 3f(x + 3)$

68. $y = -f(1 - x) - 2$

3.5 The Algebra of Functions; Composite Functions

THINGS TO KNOW

Before working through this section, be sure that you are familiar with the following concepts:

VIDEO   ANIMATION   INTERACTIVE

You Try It
1. Solving Compound Inequalities (Section 1.7)

You Try It
2. Solving Polynomial Inequalities (Section 1.9)

You Try It   3. Solving Rational Inequalities (Section 1.9) ⊘

You Try It   4. Determining the Domain of a Function
Given the Equation (Section 3.1)   ⊙

You Try It   5. Determining the Domain and Range   ⊘
of a Function from Its Graph (Section 3.2)

You Try It   6. Sketching the Graphs of the Basic
Functions (Section 3.3)

OBJECTIVES

1 Evaluating a Combined Function

2 Finding the Intersection of Intervals

3 Finding Combined Functions and Their Domains

4 Forming and Evaluating Composite Functions

5 Determining the Domain of Composite Functions

Introduction to Section 3.5

 *My video summary*   ⊘ In this section, we learn how to create a new function by combining two or more existing functions. First, we combine functions by adding, subtracting, multiplying, or dividing two existing functions. The addition, subtraction, multiplication, and division of functions is also known as the *algebra of functions*.

Algebra of Functions

Let f and g be functions, then for all x such that both $f(x)$ and $g(x)$ are defined, the sum, difference, product, and quotient of f and g exist and are defined as follows:

1. The sum of f and g: $(f + g)(x) = f(x) + g(x)$

2. The difference of f and g: $(f - g)(x) = f(x) - g(x)$

3. The product of f and g: $(fg)(x) = f(x)\, g(x)$

4. The quotient of f and g: $\left(\dfrac{f}{g}\right)(x) = \dfrac{f(x)}{g(x)}$ for all $g(x) \neq 0$

OBJECTIVE 1   EVALUATING A COMBINED FUNCTION

Example 1 Evaluate Combined Functions

Let $f(x) = \dfrac{12}{2x + 4}$ and $g(x) = \sqrt{x}$. Find each of the following.

a. $(f + g)(1)$ **b.** $(f - g)(1)$ **c.** $(fg)(4)$ **d.** $\left(\dfrac{f}{g}\right)(4)$

Solution

a. Because $f(1) = \dfrac{12}{2(1) + 4} = \dfrac{12}{6} = 2$ and $g(1) = \sqrt{1} = 1$, it follows from the sum of f and g that

$$(f + g)(1) = f(1) + g(1)$$
$$= 2 + 1$$
$$= 3$$

b. $(f - g)(1) = f(1) - g(1) = 2 - 1 = 1$

c. $f(4) = \dfrac{12}{2(4) + 4} = \dfrac{12}{12} = 1$ and $g(4) = \sqrt{4} = 2$, so $(fg)(4) = f(4)g(4) = (1)(2) = 2$

d. $\left(\dfrac{f}{g}\right)(4) = \dfrac{f(4)}{g(4)} = \dfrac{1}{2}$

 You Try It Work through the following You Try It problem.

Work Exercises 1–10 in this textbook or in the MyMathLab Study Plan.

My interactive
video summary

⊙ **Example 2** Evaluate Combined Functions Using a Graph

Use the graph to evaluate each expression or state that it is undefined.

a. $(f + g)(1)$ b. $(f - g)(0)$

c. $(fg)(4)$ d. $\left(\dfrac{f}{g}\right)(2)$

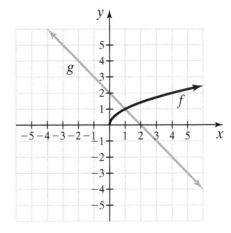

Solution Work through the interactive video to verify that

a. $(f + g)(1) = 2$ b. $(f - g)(0) = -2$

c. $(fg)(4) = -4$ d. $\left(\dfrac{f}{g}\right)(2)$ is undefined

 You Try It Work through the following You Try It problem.

Work Exercises 11–14 in this textbook or in the MyMathLab Study Plan.

OBJECTIVE 2 FINDING THE INTERSECTION OF INTERVALS

Before learning how to find the domain of $f + g, f - g, fg$, or $\dfrac{f}{g}$, you must be able to find the intersection of two or more intervals. The notation $A \cap B$ is used to represent the **intersection** of sets A and B. To find the intersection of two intervals, we must find the set of all numbers *common* to both intervals.

My video summary

⊘ **Example 3 Find the Intersection of Sets**

Find the intersection of the following sets and graph the set on a number line.

a. $[0, \infty) \cap (-\infty, 5]$

b. $((-\infty, -2) \cup (-2, \infty)) \cap [-4, \infty)$

Solution Watch the video to verify the solutions.

a. $[0, \infty) \cap (-\infty, 5] = [0, 5]$

b. $((-\infty, -2) \cup (-2, \infty)) \cap [-4, \infty) = [-4, -2) \cup (-2, \infty)$

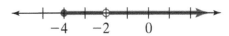

You Try It Work through the following You Try It problem.

Work Exercises 15–20 in this textbook or in the My Math Lab Study Plan.

⊘ OBJECTIVE 3 FINDING COMBINED FUNCTIONS AND THEIR DOMAINS

Consider the graph of f and g in Figure 23. Notice that the domain of f is the interval $[1, \infty)$ and the domain of g is the interval $[-1, 4)$. What is the domain of $f + g$? Because $(f + g)(x) = f(x) + g(x)$, for a number a to be in the domain of $f + g$ means that both $f(a)$ and $g(a)$ must be defined. For example, because $x = 0$ is not in the domain of f, $f(0)$ is undefined and $x = 0$ is *not* in the domain of $f + g$. Similarly, because $x = 4$ is not in the domain of g, $x = 4$ cannot be in the domain of $f + g$. Thus, for a number a to be in the domain of $f + g$, a must be in the domain of f, and a must be in the domain of g. In other words, the domain of $f + g$ must be the intersection of the two domains. The domain of $f + g$ in Figure 23 is $[1, 4)$.

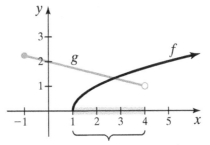

The intersection of the domain of f
and the domain of g is $[1, 4)$.   **Figure 23**

Domain of $f + g$, $f - g$, fg, and $\dfrac{f}{g}$

Suppose f is a function with domain A and g is a function with domain B, then

1. The domain of the sum, $f + g$, is the set of all x in $A \cap B$.

2. The domain of the difference, $f - g$, is the set of all x in $A \cap B$.

3. The domain of the product, fg, is the set of all x in $A \cap B$.

4. The domain of the quotient, $\dfrac{f}{g}$, is the set of all x in $A \cap B$ such that $g(x) \neq 0$.

 *My video summary*   ▶ **Example 4** Find $f + g$, $f - g$, fg, and $\dfrac{f}{g}$ and Their Domains

Let $f(x) = \dfrac{x + 2}{x - 3}$ and $g(x) = \sqrt{4 - x}$.

Find a. $f + g$, b. $f - g$, c. fg, d. $\dfrac{f}{g}$, and the domain of each.

Solution The domain of f is the interval $(-\infty, 3) \cup (3, \infty)$, and the domain of g is the interval $(-\infty, 4]$. The intersection of these two intervals is $(-\infty, 3) \cup (3, 4]$.

a. $(f + g)(x) = \dfrac{x + 2}{x - 3} + \sqrt{4 - x}$

The domain of $f + g$ is the set of all x in the intersection of the domain of f and the domain of g. Thus, the domain of $f + g$ is $(-\infty, 3) \cup (3, 4]$.

b. $(f - g)(x) = \dfrac{x + 2}{x - 3} - \sqrt{4 - x}$

The domain of $f - g$ is the set of all x in the intersection of the domain of f and the domain of g. Thus, the domain of $f - g$ is $(-\infty, 3) \cup (3, 4]$.

c. $(fg)(x) = \dfrac{(x + 2)\sqrt{4 - x}}{x - 3}$

The domain of fg is the set of all x in the intersection of the domain of f and the domain of g. Thus, the domain of fg is $(-\infty, 3) \cup (3, 4]$.

d. $\left(\dfrac{f}{g}\right)(x) = \dfrac{\dfrac{x + 2}{x - 3}}{\sqrt{4 - x}} = \dfrac{x + 2}{(x - 3)\sqrt{4 - x}}$

The domain of $\dfrac{f}{g}$ is the set of all x in the intersection of the domain of f and the domain of g such that $g(x) \neq 0$. Because $g(x) = 0$ when $x = 4$, we must exclude $x = 4$ from the domain of $\dfrac{f}{g}$. Thus, the domain of $\dfrac{f}{g}$ is $(-\infty, 3) \cup (3, 4)$.   ●

You Try It   Work through the following You Try It problem.

Work Exercises 21–32 in this textbook or in the MyMathLab Study Plan.

OBJECTIVE 4 FORMING AND EVALUATING COMPOSITE FUNCTIONS

 My video summary

⊗ Consider the function $f(x) = x^2$ and $g(x) = 2x + 1$. How could we find $f(g(x))$? To find $f(g(x))$, we substitute $g(x)$ for x in the function f to get

$$f(g(x)) = f(2x + 1) = (2x + 1)^2.$$

substitute $g(x)$ into f

The diagram in Figure 24 shows that given a number x, we first apply it to the function g to obtain $g(x)$. We then substitute $g(x)$ into f to get the result.

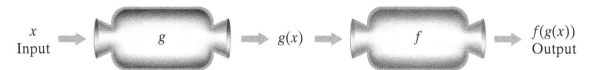

Figure 24 Composition of f and g

The function $f(g(x))$ is called a *composite function* because one function is "composed" of another function.

Definition Composite Function

Given functions f and g, the **composite function**, $f \circ g$, (also called the **composition of f and g**) is defined by

$$(f \circ g)(x) = f(g(x)),$$

provided $g(x)$ is in the domain of f.

⚠ **The composition of f and g does not equal the product of f and g: $(f \circ g)(x) \neq f(x)g(x)$. Also, the composition of f and g does not necessarily equal the composition of g and f, although this equality does exist for certain pairs of functions.**

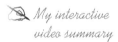 *My interactive video summary*

⊗ **Example 5** Form and Evaluate Composite Functions

Let $f(x) = 4x + 1, g(x) = \dfrac{x}{x - 2}$ and $h(x) = \sqrt{x + 3}$.

a. Find the function $f \circ g$.

b. Find the function $g \circ h$.

c. Find the function $h \circ f \circ g$.

d. Evaluate $(f \circ g)(4)$, or state that it is undefined.

e. Evaluate $(g \circ h)(1)$, or state that it is undefined.

f. Evaluate $(h \circ f \circ g)(6)$, or state that it is undefined.

My interactive video summary

⊗ **Solution** Work through the interactive video to verify the following:

a. $(f \circ g)(x) = 4\left(\dfrac{x}{x - 2}\right) + 1 = \dfrac{5x - 2}{x - 2}$

b. $(g \circ h)(x) = \dfrac{\sqrt{x+3}}{\sqrt{x+3}-2}$

c. $(h \circ f \circ g)(x) = \sqrt{\dfrac{5x-2}{x-2}+3} = \sqrt{\dfrac{8x-8}{x-2}}$

d. $(f \circ g)(4) = 9$

e. $(g \circ h)(1)$ is undefined

f. $(h \circ f \circ g)(6) = \sqrt{10}$

 You Try It Work through the following You Try It problem.

Work Exercises 33–56 in this textbook or in the MyMathLab Study Plan.

Example 6 Evaluate Composite Functions Using a Graph

Use the graph to evaluate each expression:

a. $(f \circ g)(4)$

b. $(g \circ f)(-3)$

c. $(f \circ f)(-1)$

d. $(g \circ g)(4)$

e. $(f \circ g \circ f)(1)$

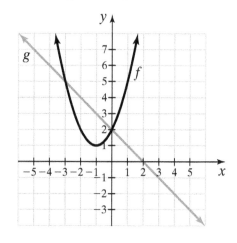

Solution Work through the interactive video to verify that

a. $(f \circ g)(4) = 2,$ **b.** $(g \circ f)(-3) = -3,$ **c.** $(f \circ f)(-1) = 5,$

d. $(g \circ g)(4) = 4,$ and   **e.** $(f \circ g \circ f)(1) = 5$

You Try It Work through the following You Try It problem.

Work Exercises 57 and 58 in this textbook or in the MyMathLab Study Plan.

OBJECTIVE 5 DETERMINING THE DOMAIN OF COMPOSITE FUNCTIONS

 *My video summary*

Suppose f and g are functions. For a number x to be in the domain of $f \circ g$, x must be in the domain of g *and* $g(x)$ must be in domain of f. To determine the domain of $f \circ g$, follow the two step process outlined below.

Determining the Domain of $f \circ g$

Step 1. Find the domain of g.

Step 2. Exclude from the domain of g all values of x for which $g(x)$ is not in the domain of f.

My interactive video summary

⊙ Example 7 Find the Domain of a Composite Function

Let $f(x) = \dfrac{-10}{x - 4}$, $g(x) = \sqrt{5 - x}$, and $h(x) = \dfrac{x - 3}{x + 7}$.

a. Find the domain of $f \circ g$.

b. Find the domain of $g \circ f$.

c. Find the domain of $f \circ h$.

d. Find the domain of $h \circ f$.

Solution

a. First, form the composite function $(f \circ g)(x) = \dfrac{-10}{\sqrt{5 - x} - 4}$. To find the domain of $f \circ g$, we follow these two steps:

Step 1. Find the domain of g.
The domain of $g(x) = \sqrt{5 - x}$ is $(-\infty, 5]$. The domain of $f \circ g$ cannot contain any values of x that are not in this interval; in other words, the domain of $f \circ g$ is a subset of $(-\infty, 5]$.

Step 2. Exclude from the domain of g all values of x for which $g(x)$ is not in the domain of f.

All real numbers except 4 are in the domain of f. This implies that $g(x)$ cannot equal 4 because $g(x)$ equal to 4 would make the denominator of f, $x - 4$, equal to 0. Thus, we must exclude all values of x such that $g(x) = 4$.

$$g(x) = 4$$
$$\sqrt{5 - x} = 4 \qquad \text{Substitute } \sqrt{5 - x} \text{ for } g(x).$$
$$5 - x = 16 \qquad \text{Square both sides.}$$
$$x = -11 \qquad \text{Solve for } x.$$

We must *exclude* $x = -11$ from the domain of $f \circ g$. Therefore, the domain of $f \circ g$ is all values of x less than 5 such that $x \neq -11$, or the interval $(-\infty, -11) \cup (-11, 5]$.

My interactive video summary

⊙ You should carefully work through this interactive video to see the entire solution to part (a) and to verify the following:

b. The domain of $g \circ f$ is the set $(-\infty, 2] \cup (4, \infty)$.

c. The domain of $f \circ h$ is the set $\left(-\infty, -\dfrac{31}{3}\right) \cup \left(-\dfrac{31}{3}, -7\right) \cup (-7, \infty)$.

d. The domain of $h \circ f$ is the set $(-\infty, 4) \cup \left(4, \dfrac{38}{7}\right) \cup \left(\dfrac{38}{7}, \infty\right)$.

Using Technology

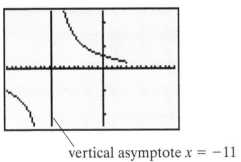

vertical asymptote $x = -11$

This screen shows the graph of $(f \circ g)(x) = \dfrac{-10}{\sqrt{5-x}-4}$. The vertical asymptote, $x = -11$, is shown in the graph. You can see from the graph that the domain is $(-\infty, -11) \cup (-11, 5]$.

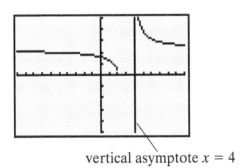

vertical asymptote $x = 4$

This screen represents the graph of $(g \circ f)(x) = \sqrt{5 + \dfrac{10}{x-4}}$. The vertical asymptote, $x = 4$, is shown in the graph. You can see that the domain is $(-\infty, 2] \cup (4, \infty)$.

You Try It Work through the following You Try It problem.

Work Exercises 59–66 in this textbook or in the MyMathLab Study Plan.

3.5 Exercises

Skill Check Exercises

For exercises SCE-1 through SCE-6, simplify each expression and write as a single fraction. Factor the numerator and denominator whenever possible.

SCE-1. $\dfrac{\dfrac{2x+1}{x-9}}{\dfrac{5x}{5x-2}}$

SCE-2. $\dfrac{\dfrac{x}{x^2-4}}{\dfrac{1}{x-5}}$

SCE-3. $\dfrac{\dfrac{x-7}{x^2-x-6}}{\dfrac{x+8}{x^2-49}}$

SCE-4. $\dfrac{8}{x+3} + 7$

SCE-5. $7\left(\dfrac{4}{x+6}\right) - 6$

SCE-6. $\dfrac{3}{\dfrac{3}{x+8} + 8}$

In Exercises 1–10, evaluate the following given that $f(x) = \sqrt{x+1}$, $g(x) = x^2 - 2$, and $h(x) = \dfrac{1}{2x+3}$.

1. $(f+g)(3)$

2. $(f-g)(8)$

3. $(g+h)(-2)$

4. $\left(\dfrac{f}{g}\right)(0)$

5. $(fh)(15)$

6. $(h-g)(4)$

7. $\left(\dfrac{g}{f}\right)(-1)$

8. $\left(\dfrac{f}{h}\right)(2)$

9. $(gf)(8)$

10. $(hh)\left(\dfrac{3}{2}\right)$

In Exercises 11–14, use the graph to evaluate the given expression, or state that it is undefined.

11.

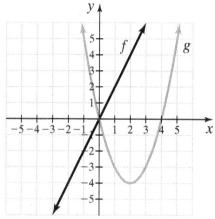

a. $(f + g)(0)$

b. $(f - g)(2)$

c. $(fg)(-1)$

d. $\left(\dfrac{f}{g}\right)(2)$

12.

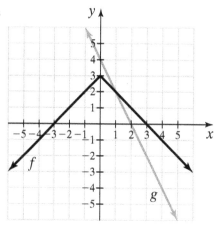

a. $(f + g)(4)$

b. $(g - f)(1)$

c. $(fg)(0)$

d. $\left(\dfrac{f}{g}\right)(2)$

13.

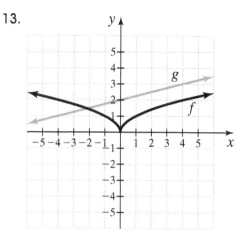

a. $(f + g)(4)$ b. $(f - g)(-4)$

c. $(gg)(4)$ d. $\left(\dfrac{f}{g}\right)(4)$

14.

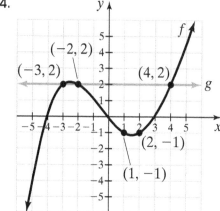

a. $(g + f)(4)$ b. $(g - f)(1)$

c. $(fg)(-3)$ d. $\left(\dfrac{f}{g}\right)(2)$

In Exercises 15–20, find the intersection of the given intervals and graph the set on a number line.

15. $(-\infty, \infty) \cap (-\infty, 3)$

16. $(-8, 5] \cap (-12, 3)$

17. $[0, \infty) \cap (-\infty, 4]$

18. $((-\infty, 1) \cup (1, \infty)) \cap (-\infty, \infty)$

19. $((-\infty, 0) \cup (0, \infty)) \cap [-3, \infty)$

20. $(-\infty, 10) \cap (-5, \infty) \cap (-\infty, 1]$

In Exercises 21–32, find **a.** $f + g$, **b.** $f - g$, **c.** fg, **d.** $\dfrac{f}{g}$, and the domain of each.

21. $f(x) = x^2 + 2, g(x) = x - 1$

22. $f(x) = x - 5, g(x) = x^2 - 4$

23. $f(x) = x + 5, g(x) = x^2 - 6x - 16$

24. $f(x) = \sqrt{x}, g(x) = x + 6$

25. $f(x) = \dfrac{1}{x}, g(x) = \sqrt{x - 1}$

26. $f(x) = \dfrac{2x + 1}{x + 3}, g(x) = \dfrac{x + 2}{3x + 5}$

27. $f(x) = \dfrac{x}{x^2 - 4}, g(x) = \dfrac{1}{x - 3}$

28. $f(x) = \dfrac{x - 4}{x^2 - x - 6}, g(x) = \dfrac{x + 5}{x^2 - 16}$

29. $f(x) = \sqrt{x + 2}, g(x) = \sqrt{2 - x}$

30. $f(x) = \sqrt{\dfrac{x + 1}{x + 2}}, g(x) = \sqrt{x - 3}$

31. $f(x) = \sqrt[3]{x + 1}, g(x) = \sqrt[3]{\dfrac{x - 1}{x + 2}}$

32. $f(x) = \sqrt{7 - x}, g(x) = \sqrt{x^2 - 2x - 15}$

In Exercises 33–44, let $f(x) = 3x + 1, g(x) = \dfrac{2}{x + 1}$, and $h(x) = \sqrt{x + 3}$.

33. Find the function $f \circ g$.

34. Find the function $g \circ f$.

35. Find the function $f \circ h$.

36. Find the function $g \circ h$.

37. Find the function $h \circ f$.

38. Find the function $h \circ g$.

39. Find the function $f \circ f$.

40. Find the function $g \circ g$.

41. Find the function $h \circ h$.

42. Find the function $f \circ g \circ h$.

43. Find the function $g \circ f \circ h$.

44. Find the function $h \circ f \circ g$.

In Exercises 45–56, evaluate the following composite functions given that $f(x) = 3x + 1, g(x) = \dfrac{2}{x + 1}$, and $h(x) = \sqrt{x + 3}$.

45. $(f \circ g)(0)$

46. $(f \circ h)(6)$

47. $(g \circ f)(1)$

48. $(g \circ h)(-2)$

49. $(h \circ f)(0)$

50. $(h \circ g)(3)$

51. $(f \circ f)(-1)$

52. $(g \circ g)(4)$

53. $(h \circ h)(1)$

54. $(f \circ g \circ h)(-2)$

55. $(h \circ f \circ g)(2)$

56. $(g \circ f \circ h)(6)$

In Exercises 57 and 58, use the graph to evaluate each expression.

57. a. $(f \circ g)(1)$

b. $(g \circ f)(-1)$

c. $(g \circ g)(0)$

d. $(f \circ f)(1)$

e. $(f \circ g \circ f)(0)$

f. $(g \circ f \circ g)(0)$

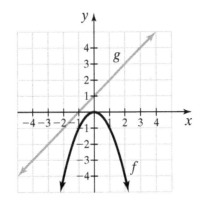

58. a. $(f \circ g)(1)$

   b. $(g \circ f)(-1)$

   c. $(g \circ g)(0)$

   d. $(f \circ f)(1)$

   e. $(f \circ g \circ f)(0)$

   f. $(g \circ f \circ g)(0)$

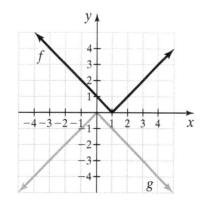

In Exercises 59–66, find the domain of $(f \circ g)(x)$ and $(g \circ f)(x)$.

59. $f(x) = x^2, g(x) = 2x - 1$

60. $f(x) = 3x - 5, g(x) = 2x^2 + 1$

61. $f(x) = x^2, g(x) = \sqrt{x}$

62. $f(x) = \dfrac{1}{x}, g(x) = x^2 - 4$

63. $f(x) = \dfrac{3}{x + 1}, g(x) = \dfrac{x}{x - 2}$

64. $f(x) = \dfrac{2x}{x - 3}, g(x) = \dfrac{x + 1}{x - 1}$

65. $f(x) = \sqrt{x}, g(x) = \dfrac{x + 3}{x - 2}$

66. $f(x) = \dfrac{1}{x - 2}, g(x) = \sqrt{4 - x}$

Brief Exercises

In Exercises 67–78, find **a.** $f + g$, **b.** $f - g$, **c.** fg, **d.** $\dfrac{f}{g}$,

67. $f(x) = x^2 + 2, g(x) = x - 1$

68. $f(x) = x - 5, g(x) = x^2 - 4$

69. $f(x) = x + 5, g(x) = x^2 - 6x - 16$

70. $f(x) = \sqrt{x}, g(x) = x + 6$

71. $f(x) = \dfrac{1}{x}, g(x) = \sqrt{x - 1}$

72. $f(x) = \dfrac{2x + 1}{x + 3}, g(x) = \dfrac{x + 2}{3x + 5}$

73. $f(x) = \dfrac{x}{x^2 - 4}, g(x) = \dfrac{1}{x - 3}$

74. $f(x) = \dfrac{x - 4}{x^2 - x - 6}, g(x) = \dfrac{x + 5}{x^2 - 16}$

75. $f(x) = \sqrt{x + 2}, g(x) = \sqrt{2 - x}$

76. $f(x) = \sqrt{\dfrac{x + 1}{x + 2}}, g(x) = \sqrt{x - 3}$

77. $f(x) = \sqrt[3]{x + 1}, g(x) = \sqrt[3]{\dfrac{x - 1}{x + 2}}$

78. $f(x) = \sqrt{7 - x}, g(x) = \sqrt{x^2 - 2x - 15}$

3.6 One-to-One Functions; Inverse Functions

THINGS TO KNOW

Before working through this section, be sure that you are familiar with the following concepts:

VIDEO   ANIMATION   INTERACTIVE

You Try It

1. Determining the Domain of a Function Given the Equation (Section 3.1)

You Try It

2. Determining the Domain and Range of a Function from Its Graph (Section 3.2)

You Try It

3. Analyzing Piecewise-Defined Functions (Section 3.3)

You Try It

4. Forming and Evaluating Composite Functions (Section 3.5)

OBJECTIVES

1 Understanding the Definition of a One-to-One Function

2 Determining If a Function Is One-to-One Using the Horizontal Line Test

3 Understanding and Verifying Inverse Functions

4 Sketching the Graphs of Inverse Functions

5 Finding the Inverse of a One-to-One Function

Introduction to Section 3.6

In this section, we study *inverse functions*. To illustrate an inverse function, let's consider the function $F(C) = \dfrac{9}{5}C + 32$. This function is used to convert a given temperature in degrees Celsius into its equivalent temperature in degrees Fahrenheit. For example,

$$F(25) = \frac{9}{5}(25) + 32 = 45 + 32 = 77$$

Thus, a temperature of 25°C corresponds to a temperature of 77°F. To convert a temperature of 77°F back into 25°C, we need a different function. The function that is used to accomplish this task is $C(F) = \dfrac{5}{9}(F - 32)$.

$$C(77) = \frac{5}{9}(77 - 32)$$

$$= \frac{5}{9}(45)$$

$$= 25$$

The function C is the *inverse* of the function F. In essence, these functions perform opposite actions (they "undo" each other). The first function converted 25°C into 77°F, whereas the second function converted 77°F back into 25°C.

The function F has an inverse function C. Later in this section, we examine a process for finding the inverse of a function, but keep the following in mind. We can find the inverse of many functions using this process, but that inverse will not always be a function. In this text, we are only interested in inverses that are functions, so we first develop a test to determine if a function has an inverse *function*. When the word *inverse* is used throughout the remainder of this section, we assume that we are referring to the inverse that is a function.

⊘ **OBJECTIVE 1** UNDERSTANDING THE DEFINITION OF A ONE-TO-ONE FUNCTION

First, we must define the concept of **one-to-one** functions.

Definition   One-to-One Function

A function f is **one-to-one** if for any values $a \neq b$ in the domain of f, $f(a) \neq f(b)$.

This definition suggests that a function is one-to-one if for any two *different* input values (domain values), the corresponding output values (range values) must be different. An alternate definition says that if two range values are the same, $f(u) = f(v)$, then the domain values must be the same; that is, $u = v$.

Alternate Definition of a One-to-One Function

A function f is **one-to-one** if for any two range values $f(u)$ and $f(v)$, $f(u) = f(v)$ implies that $u = v$.

The function sketched in Figure 25a is one-to-one because for any two distinct x-values in the domain ($a \neq b$), the function values or range values are not equal ($f(a) \neq f(b)$). In Figure 25b, we see that the function $y = g(x)$ is *not* one-to-one because we can easily find two different domain values that correspond to the same range value.

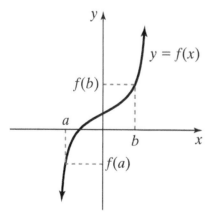

Figure 25a An example of a one-to-one function

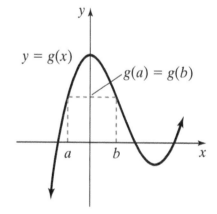

Figure 25b An example of a function that is not one-to-one

OBJECTIVE 2 DETERMINING IF A FUNCTION IS ONE-TO-ONE USING THE HORIZONTAL LINE TEST

My video summary ⊙ Notice in Figure 26 that the horizontal lines intersect the graph of $y = f(x)$ in at most one place, while we can find many horizontal lines that intersect the graph of $y = g(x)$ more than once. This gives us a visual example of how we can use horizontal lines to help us determine from the graph if a function is one-to-one. Using horizontal lines to determine if the graph of a function is one-to-one is known as the *horizontal line test*.

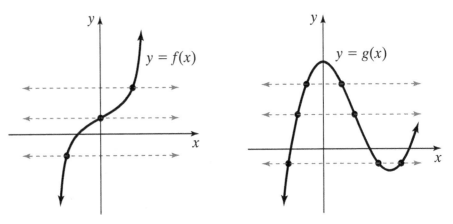

Figure 26 Drawing horizontal lines can help determine whether a graph represents a one-to-one function.

Horizontal Line Test

If every horizontal line intersects the graph of a function f at most once, then f is one-to-one.

My animation summary

▣ **Example 1 Determine If a Function Is One-to-One**

Determine if each function is one-to-one.

a.
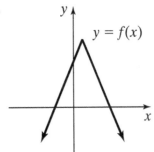

b.
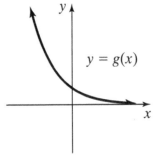

c. $f(x) = x^2 + 1, x \leq 0$

d. $f(x) = \begin{cases} 2x + 4 \text{ for } x \leq -1 \\ 2x - 6 \text{ for } x \geq 4 \end{cases}$

Solution The functions in parts b and c are one-to-one, whereas the functions in parts a and d are not one-to-one. Watch the animation to verify.

You Try It Work through the following You Try It problem.

Work Exercises 1–17 in this textbook or in the MyMathLab Study Plan.

OBJECTIVE 3   UNDERSTANDING AND VERIFYING INVERSE FUNCTIONS

 My video summary   ⊙ We are now ready to ask ourselves the question, "Why should we be concerned with one-to-one functions?"

Answer: Every one-to-one function has an inverse function!

> **Definition**   Inverse Function
>
> Let f be a one-to-one function with domain A and range B. Then f^{-1} is the **inverse function of f** with domain B and range A. Furthermore, if $f(a) = b$, then $f^{-1}(b) = a$.

According to the definition of an inverse function, the domain of f is exactly the same as the range of f^{-1}, and the range of f is the same as the domain of f^{-1}. Figure 27 illustrates that if the function f assigns a number a to b, then the inverse function will assign b back to a. In other words, if the point (a, b) is an ordered pair on the graph of f, then the point (b, a) must be on the graph of f^{-1}.

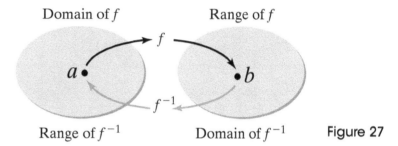

Figure 27

⚠ **Do not confuse f^{-1} with $\dfrac{1}{f(x)}$. The negative 1 in f^{-1} is *not* an exponent!**

 *My video summary*   ⊙ As with our opening example using an inverse function to convert a Fahrenheit temperature back into a Celsius temperature, inverse functions "undo" each other. For example, it can be shown that if $f(x) = x^3$, then the inverse of f is $f^{-1}(x) = \sqrt[3]{x}$. Note that

$$f(2) = (2)^3 = 8 \quad \text{and} \quad f^{-1}(8) = \sqrt[3]{8} = 2.$$

The function f takes the number 2 to 8, whereas f^{-1} takes 8 back to 2. Observe what happens if we look at the composition of f and f^{-1} and the composition of f^{-1} and f at specified values:

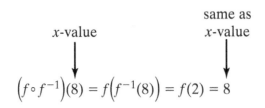

$$\left(f \circ f^{-1}\right)(8) = f\left(f^{-1}(8)\right) = f(2) = 8$$

same as
x-value   x-value

$$\left(f^{-1} \circ f\right)(2) = f^{-1}\left(f(2)\right) = f^{-1}(8) = 2$$

Because of the "undoing" nature of inverse functions, we get the following **composition cancellation equations:**

Composition Cancellation Equations

$f(f^{-1}(x)) = x$ for all x in the domain of f^{-1} and $f^{-1}(f(x)) = x$ for all x in the domain of f

---

These cancellation equations can be used to show if two functions are inverses of each other. We can see from our example that if $f(x) = x^3$ and $f^{-1}(x) = \sqrt[3]{x}$, then

$$f(f^{-1}(x)) = f(\sqrt[3]{x}) = (\sqrt[3]{x})^3 = x \quad \text{and} \quad f^{-1}(f(x)) = f^{-1}(x^3) = \sqrt[3]{x^3} = x.$$

My interactive video summary

⊘ **Example 2** Verify Inverse Functions

Show that $f(x) = \dfrac{x}{2x + 3}$ and $g(x) = \dfrac{3x}{1 - 2x}$ are inverse functions using the composition cancellation equations.

Solution To show that f and g are inverses of each other, we must show that $(f \circ g)(x) = x$ and $(g \circ f)(x) = x$. Work through the interactive video to verify that both composition cancellation equations are satisfied. ●

You Try It   Work through the following You Try It problem.

Work Exercises 18–24 in this textbook or in the MyMathLab Study Plan.

OBJECTIVE 4   SKETCHING THE GRAPHS OF INVERSE FUNCTIONS

If f is a one-to-one function, then we know that it must have an inverse function, f^{-1}. Given the graph of a one-to-one function f, we can obtain the graph of f^{-1} by simply interchanging the coordinates of each ordered pair that lies on the graph of f. In other words, for any point (a, b) on the graph of f, the point (b, a) must lie on the graph of f^{-1}. Notice in Figure 28 that the points (a, b) and (b, a)

are symmetric about the line $y = x$. Therefore, the graph of f^{-1} is a reflection of the graph of f about the line $y = x$. Figure 29 shows the graph of $f(x) = x^3$ and $f^{-1}(x) = \sqrt[3]{x}$. You can see that if the functions have any points in common, they must lie along the line $y = x$.

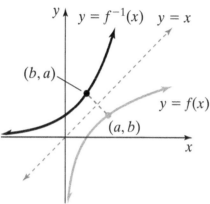

Figure 28 Graph of a one-to-one function and its inverse

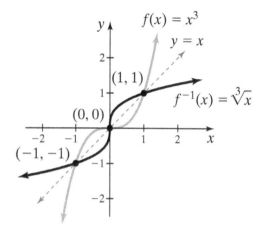

Figure 29 Graph of $f(x) = x^3$ and $f^{-1}(x) = \sqrt[3]{x}$

My animation summary

▣ Example 3 Sketch the Graph of a One-to-One Function and Its Inverse

Sketch the graph of $f(x) = x^2 + 1, x \le 0$, and its inverse. Also state the domain and range of f and f^{-1}.

Solution The graphs of f and f^{-1} are sketched in Figure 30. Notice how the graph of f^{-1} is a reflection of the graph of f about the line $y = x$. Also notice that the domain of f is the same as the range of f^{-1}. Likewise, the domain of f^{-1} is equivalent to the range of f.

View the animation to see exactly how to sketch f and f^{-1}.

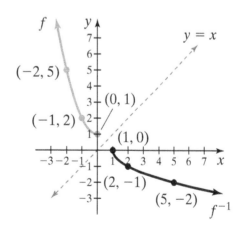

Domain of f: $(-\infty, 0]$ Domain of f^{-1}: $[1, \infty)$
Range of f: $[1, \infty)$ Range of f^{-1}: $(-\infty, 0]$

Figure 30 Graph of $f(x) = x^2 + 1, x \le 0$, and its inverse

Using Technology

$y_1 = (x^2 + 1)/(x \le 0)$

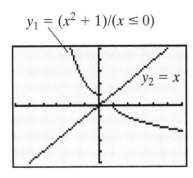

$y_2 = x$

Using a TI-83+, we can sketch the functions from Example 3 by letting $y_1 = (x^2 + 1)/(x \le 0)$. We can draw the inverse function by typing the command **DrawInv Y_1** in the calculator's main viewing window. The resultant graphs are shown here.

You Try It   Work through the following You Try It problem.

Work Exercises 25–31 in this textbook or in the MyMathLab Study Plan.

OBJECTIVE 5   FINDING THE INVERSE OF A ONE-TO-ONE FUNCTION

We are now ready to find the inverse of a one-to-one function algebraically. We know that if a point (x, y) is on the graph of a one-to-one function, then the point (y, x) is on the graph of its inverse function. We can use this information to develop a process for finding the inverse of a function algebraically simply by switching x and y in the original function to produce its inverse function.

We use as a motivating example the function $f(x) = x^2 + 1, x \le 0$, discussed in Example 3. To find the inverse of a one-to-one function, we follow the four-step process outlined here.

Step 1. Change $f(x)$ to y: $y = x^2 + 1$

Step 2. Interchange x and y: $x = y^2 + 1$

Step 3. Solve for y: $x - 1 = y^2$

$$\pm \sqrt{x - 1} = y$$

(Because the domain of f is $(-\infty, 0]$, the range of f^{-1} must be $(-\infty, 0]$. Therefore, we must use the negative square root or $y = -\sqrt{x - 1}$.)

Step 4. Change y to $f^{-1}(x)$: $f^{-1}(x) = -\sqrt{x - 1}$

Thus, the inverse of $f(x) = x^2 + 1, x \le 0$, is $f^{-1}(x) = -\sqrt{x - 1}$.

My animation summary

▶ Example 4 Find the Inverse of a Function

Find the inverse of the function $f(x) = \dfrac{2x}{1 - 5x}$, and state the domain and range of f and f^{-1}.

Solution Work through the animation, and follow the four-step process to verify that $f^{-1}(x) = \dfrac{x}{5x + 2}$. The domain of f is $\left(-\infty, \dfrac{1}{5}\right) \cup \left(\dfrac{1}{5}, \infty\right)$, whereas the domain of f^{-1} is $\left(-\infty, -\frac{2}{5}\right) \cup \left(-\frac{2}{5}, \infty\right)$. Because the range of f must be the domain of f^{-1} and the range of f^{-1} must be the domain of f, we get the following result:

Domain of f: $\left(-\infty, \frac{1}{5}\right) \cup \left(\frac{1}{5}, \infty\right)$ Domain of f^{-1}: $\left(-\infty, -\frac{2}{5}\right) \cup \left(-\frac{2}{5}, \infty\right)$

Range of f: $\left(-\infty, -\frac{2}{5}\right) \cup \left(-\frac{2}{5}, \infty\right)$ Range of f^{-1}: $\left(-\infty, \frac{1}{5}\right) \cup \left(\frac{1}{5}, \infty\right)$

You Try It   Work through the following You Try It problem.

Work Exercises 32–44 in this textbook or in the MyMathLab Study Plan.

Inverse Function Summary

1. The function f^{-1} exists if and only if the function f is one-to-one.

2. The domain of f is the same as the range of f^{-1}, and the range of f is the same as the domain of f^{-1}.

3. To verify that two one-to-one functions, f and g, are inverses of each other, we must use the composition cancellation equations to show that $f(g(x)) = g(f(x)) = x$.

4. The graph of f^{-1} is a reflection of the graph of f about the line $y = x$. That is, for any point (a, b) that lies on the graph of f, the point (b, a) must lie on the graph of f^{-1}.

5. To find the inverse of a one-to-one function, replace $f(x)$ with y, interchange the variables x and y, and solve for y. This is the function $f^{-1}(x)$.

3.6 Exercises

Skill Check Exercises

For exercises SCE-1 through SCE-2 simplify each expression and write as a single fraction.

SCE-1. $\dfrac{2\left(\dfrac{x+5}{x-1}\right) - 3}{\dfrac{x+5}{x-1}}$

SCE-2. $\dfrac{11\left(\dfrac{2x-4}{3+5x}\right) + 5}{4 - 7\left(\dfrac{2x-4}{3+5x}\right)}$

For exercises SCE-3 through SCE-8 solve each equation for the variable y.

SCE-3. $x = \dfrac{1}{4}y - 5$

SCE-4. $x = \dfrac{3y+9}{7}$

SCE-5. $x(7 - 5y) = 8y - 1$

SCE-6. $x = \dfrac{4y-5}{8y+4}$

SCE-7. $x = 6 - \sqrt[3]{y-3}$

SCE-8. $x = (y-4)^2 + 9$ for $y \le 4$

In Exercises 1–17, determine if each function is one-to-one.

1. $f(x) = 3x - 1$

2. $f(x) = 2x^2$

3. $f(x) = (x - 1)^2, x \geq 1$

4. $f(x) = (x - 1)^2, x \geq -1$

5. $f(x) = \dfrac{1}{x} - 2$

6. $f(x) = 4\sqrt{x}$

7. $f(x) = -2|x|$

8. $f(x) = 2$

9. $f(x) = (x + 1)^3 - 2$

10. $f(x) = \begin{cases} x + 4 & \text{if } x < -1 \\ 2x + 2 & \text{if } x \geq \dfrac{1}{2} \end{cases}$

11. $f(x) = \begin{cases} -3x & \text{if } x \leq 0 \\ 2 - x & \text{if } x \geq 2 \end{cases}$

12.

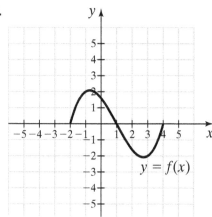

13.

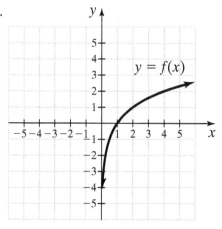

14.

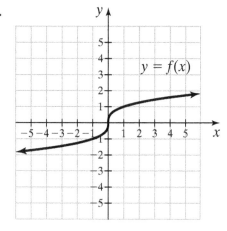

15.

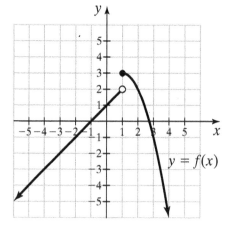

16.

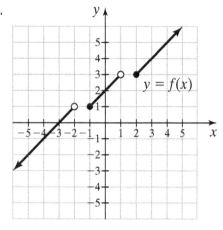

17.

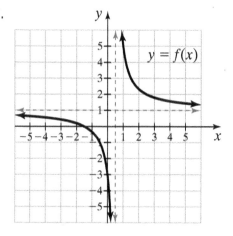

In Exercises 18–24, use the composition cancellation equations to verify that f and g are inverse functions.

18. $f(x) = \dfrac{3}{2}x - 4$ and $g(x) = \dfrac{2x + 8}{3}$

19. $f(x) = \dfrac{ax + b}{c}$ and $g(x) = \dfrac{cx - b}{a}$ for all real numbers a, b, and c such that $a \neq 0$, $c \neq 0$

20. $f(x) = (x - 1)^2, x \geq 1$, and $g(x) = \sqrt{x} + 1$

21. $f(x) = \dfrac{7}{x + 1}$ and $g(x) = \dfrac{7 - x}{x}$

22. $f(x) = \dfrac{x}{5 + 3x}$ and $g(x) = \dfrac{5x}{1 - 3x}$

23. $f(x) = 2\sqrt[3]{x - 1} + 3$ and $g(x) = \dfrac{(x - 3)^3}{8} + 1$

24. $f(x) = \dfrac{11x + 5}{4 - 7x}$ and $g(x) = \dfrac{4x - 5}{7x + 11}$

In Exercises 25–30, use the graph of f to sketch the graph of f^{-1}. Use the graphs to determine the domain and range of each function.

25.

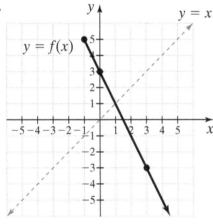

26.

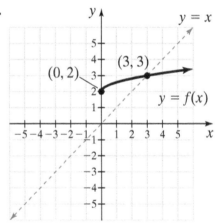

27.

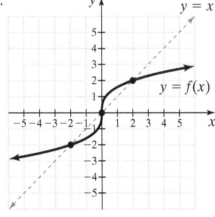

28.

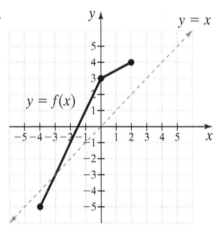

29.

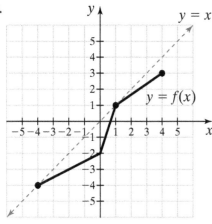

30.

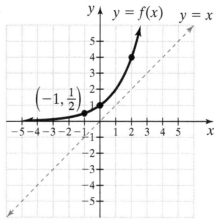

In Exercise 31, refer to the graph of the function $y = f(x)$.

31.

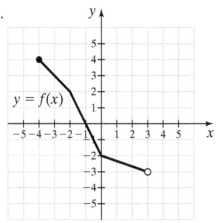

a. What is the domain of f^{-1}?

b. What is the range of f^{-1}?

c. What is the y-intercept of f^{-1}?

d. Evaluate $f^{-1}(0)$.

e. Evaluate $f^{-1}(2)$.

f. Evaluate $f^{-1}(-2)$.

g. Evaluate $f^{-1}(3)$.

h. Evaluate $f^{-1}(4)$.

In Exercises 32–44, write an equation for the inverse function, and then state the domain and range of f and f^{-1}. (*Hint:* In Exercises 43 and 44, complete the square to solve for y.)

32. $f(x) = \dfrac{1}{3}x - 5$

33. $f(x) = \dfrac{3x + 9}{7}$

34. $f(x) = \sqrt[3]{2x - 3}$

35. $f(x) = 1 - \sqrt[5]{x + 4}$

36. $f(x) = -x^2 - 2, x \geq 0$

37. $f(x) = (x + 3)^2 - 5, x \leq -3$

38. $f(x) = \dfrac{3}{x}$

39. $f(x) = \dfrac{1 - x}{2x}$

40. $f(x) = \dfrac{8x - 1}{7 - 5x}$

41. $f(x) = \dfrac{1 - 4x}{11 + 13x}$

42. $f(x) = \dfrac{4x - 5}{8x + 4}$

43. $f(x) = x^2 + 4x - 1, x \geq -2$

44. $f(x) = -2x^2 + 6x + 8, x \leq \dfrac{3}{2}$

Brief Exercises

In Exercises 45–50, use the graph of f to sketch the graph of f^{-1}.

45.

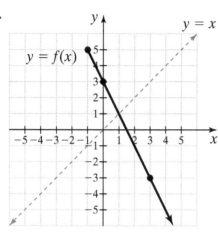

46.

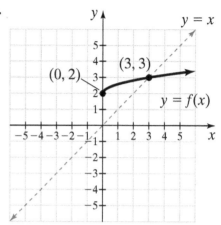

47.

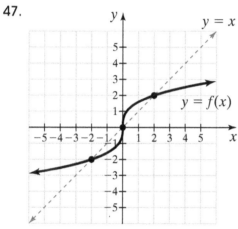

48.

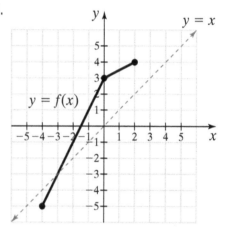

49.

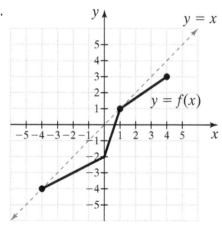

50.

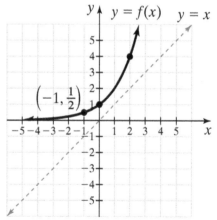

For Exercises 51–53, use the graph of the function $y = f(x)$ to answer each question.

51. a. What is the domain of f^{-1}?

   b. What is the range of f^{-1}?

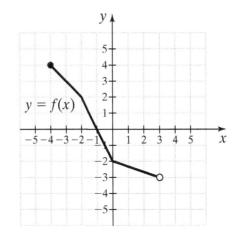

52. What is the y-intercept of f^{-1}?

53. Evaluate each of the following:

   a. $f^{-1}(0)$ **b.** $f^{-1}(2)$ **c.** $f^{-1}(-2)$ **d.** $f^{-1}(3)$ **e.** $f^{-1}(4)$

In Exercises 54–56, write an equation for the inverse of the given one-to-one function. (Hint: In Exercises 65 and 66, complete the square to solve for y.)

54. $f(x) = \dfrac{1}{3}x - 5$

55. $f(x) = \dfrac{3x + 9}{7}$

56. $f(x) = \sqrt[3]{2x - 3}$

57. $f(x) = 1 - \sqrt[5]{x + 4}$

58. $f(x) = -x^2 - 2, x \geq 0$

59. $f(x) = (x + 3)^2 - 5, x \leq -3$

60. $f(x) = \dfrac{3}{x}$

61. $f(x) = \dfrac{1 - x}{2x}$

62. $f(x) = \dfrac{8x - 1}{7 - 5x}$

63. $f(x) = \dfrac{1 - 4x}{11 + 13x}$

64. $f(x) = \dfrac{4x - 5}{8x + 4}$

65. $f(x) = x^2 + 4x - 1, x \geq -2$

66. $f(x) = -2x^2 + 6x + 8, x \leq \dfrac{3}{2}$

Polynomial and Rational Functions

CHAPTER FOUR CONTENTS

4.1 Quadratic Functions

THINGS TO KNOW

Before working through this section, be sure that you are familiar with the following concepts:

		VIDEO	ANIMATION	INTERACTIVE
You Try It	1. Solving Quadratic Equations by Factoring and the Zero Product Property (Section 1.4)	⊙		
You Try It	2. Solving Quadratic Equations by Completing the Square (Section 1.4)	⊙		
You Try It	3. Solving Quadratic Equations Using the Quadratic Formula (Section 1.4)			
You Try It	4. Determining the Domain and Range of a Function From Its Graph (Section 3.2)	⊙		
You Try It	5. Using Combinations of Transformations to Graph Functions (Section 3.4)			▭

OBJECTIVES

1 Understanding the Definition of a Quadratic Function and Its Graph

2 Graphing Quadratic Functions Written in Standard Form

3 Graphing Quadratic Functions by Completing the Square

4 Graphing Quadratic Functions Using the Vertex Formula

5 Determining the Equation of a Quadratic Function Given Its Graph

- -

OBJECTIVE 1   UNDERSTANDING THE DEFINITION OF A QUADRATIC FUNCTION
AND ITS GRAPH

My video summary   ⊙ In Section 1.4, we learned how to solve quadratic equations of the form
$ax^2 + bx + c = 0, a \neq 0$. In this section, we learn about *quadratic functions*.

> **Definition**   **Quadratic Function**
>
> A **quadratic function** is a function of the form $f(x) = ax^2 + bx + c$, where a, b,
> and c are real numbers with $a \neq 0$. Every quadratic function has a "u-shaped"
> graph called a *parabola*.

The function $f(x) = x^2$ is a quadratic function with $a = 1$, $b = 0$, and $c = 0$ whose
graph is shown in Figure 1a. The function $g(x) = -x^2$ is a quadratic function with
$a = -1$, $b = 0$, and $c = 0$ whose graph is shown in Figure 1b. Notice that both
graphs are parabolas.

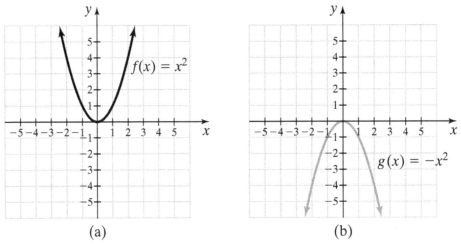

(a)   (b)

Figure 1 Graphs of $f(x) = x^2$ and $g(x) = -x^2$

A parabola either opens up or opens down, depending on the leading coeffi-
cient, a. If $a > 0$, as in Figure 1a, the parabola will "open up." If $a < 0$, as in Figure 1b,
the parabola will "open down."

Example 1 Determine Whether the Graph of a Quadratic Function Opens Up or Opens Down

Without graphing, determine whether the graph of the quadratic function
$f(x) = -3x^2 + 6x + 1$ opens up or opens down.

Solution Because the leading coefficient is $a = -3 < 0$, the graph of the quadratic function $f(x) = -3x^2 + 6x + 1$ must open down.   ●

Using Technology

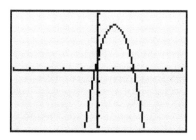

The graph of the quadratic function $f(x) = -3x^2 + 6x + 1$ is seen here using a graphing utility. You can see that the graph "opens down."

You Try It Work through the following You Try It problem.

Work Exercises 1–4 in this textbook or in the MyMathLab Study Plan.

CHARACTERISTICS OF A PARABOLA

Before we learn how to sketch the graphs of other quadratic functions, it is important to identify the five basic characteristics of a parabola. Work through the following animation, and click on each characteristic to get a detailed description.

My animation summary

1. Vertex

2. Axis of symmetry

3. y-Intercept

4. x-Intercept(s) or real zeros

5. Domain and range

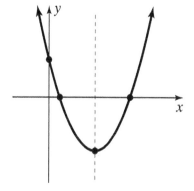

OBJECTIVE 2   GRAPHING QUADRATIC FUNCTIONS WRITTEN IN STANDARD FORM

In Section 3.4, Example 6, we sketched the graph of $f(x) = -2(x+3)^2 - 1$ using transformation techniques. Watch the animation to see how to sketch the function $f(x) = -2(x+3)^2 - 1$ using transformations. The graph of $f(x) = -2(x+3)^2 - 1$ is sketched in Figure 2. Note that the domain is $(-\infty, \infty)$ and the range is $(-\infty, -1]$.

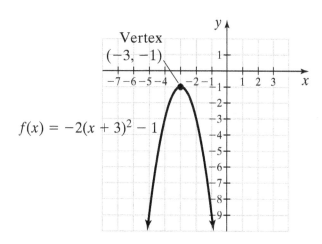

Figure 2 Graph of
$$f(x) = -2(x+3)^2 - 1$$

The function $f(x) = -2(x + 3)^2 - 1$ is an example of a quadratic function that is in **standard form** with vertex $(-3, -1)$.

Definition   Standard Form of a Quadratic Function

A quadratic function is in **standard form** if it is written as $f(x) = a(x - h)^2 + k$. The graph is a parabola with vertex (h, k). The parabola "opens up" if $a > 0$. The parabola "opens down" if $a < 0$.

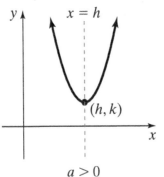

$a > 0$
Domain: $(-\infty, \infty)$
Range: $[k, \infty)$

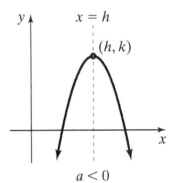

$a < 0$
Domain: $(-\infty, \infty)$
Range: $(-\infty, k]$

My video summary

⊙ Example 2 Graph a Quadratic Function That Is in Standard Form

Given that the quadratic function $f(x) = -(x - 2)^2 - 4$ is in standard form, address the following:

a. What are the coordinates of the vertex?

b. Does the graph "open up" or "open down"?

c. What is the equation of the axis of symmetry?

d. Find any x-intercepts.

e. Find the y-intercept.

f. Sketch the graph.

g. State the domain and range in interval notation.

Solution

a. The quadratic function $f(x) = -(x - 2)^2 - 4$ is in standard form. Therefore, we can find the vertex by determining the values of h and k.

$$f(x) = a(x - h)^2 + k \qquad \text{Quadratic function in standard form}$$

$$\boxed{a = -1} \quad \boxed{h = 2} \qquad \boxed{k = -4}$$

$$f(x) = -(x - 2)^2 - 4 \qquad \text{Given function}$$

Because $h = 2$ and $k = -4$, the vertex of this parabola is $(2, -4)$.

b. The leading coefficient is $a = -1 < 0$. Therefore, the parabola opens down.

c. The equation of the axis of symmetry is $x = 2$.

d. To find any x-intercepts, we must determine the real solution(s) to the equation $f(x) = 0$.

$$f(x) = -(x - 2)^2 - 4 \qquad \text{Write the original function.}$$
$$0 = -(x - 2)^2 - 4 \qquad \text{Replace } f(x) \text{ with 0 in } f(x) = -(x - 2)^2 - 4.$$
$$(x - 2)^2 = -4 \qquad \text{Add } (x - 2)^2 \text{ to both sides.}$$
$$x - 2 = \pm\sqrt{-4} \qquad \text{Use the square root property.}$$
$$x - 2 = \pm 2i \qquad \text{Recall that } \sqrt{-4} = \sqrt{4} \cdot \sqrt{-1} = 2i.$$
$$x = 2 \pm 2i \qquad \text{Add 2 to both sides.}$$

Because the equation $f(x) = 0$ has no real solutions, the parabola has no x-intercepts.

e. The y-intercept is $f(0) = -(0 - 2)^2 - 4 = -4 - 4 = -8$.

f. The vertex is $(2, -4)$, the parabola opens down, the equation of the axis of symmetry is $x = 2$, and the y-intercept is -8. Using this information, we can sketch the parabola as shown in Figure 3.

g. The domain of f is the interval $(-\infty, \infty)$. The range of f is the interval $(-\infty, -4]$.

 *My video summary* ◉ Watch the video to see each step worked out in detail.

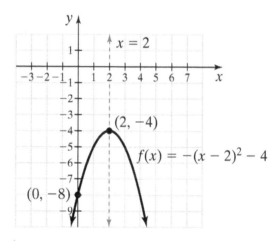

Figure 3 Graph of $f(x) = -(x - 2)^2 - 4$ ◉

You Try It Work through the following You Try It problem.

Work Exercises 5–10 in this textbook or in the MyMathLab **Study Plan.**

OBJECTIVE 3 GRAPHING QUADRATIC FUNCTIONS BY COMPLETING THE SQUARE

We can see from Example 2 that if a quadratic function is in standard form $f(x) = a(x - h)^2 + k$, it is fairly straightforward to determine its graph. Therefore, to graph a quadratic function of the form $f(x) = ax^2 + bx + c$, we convert it into standard form by completing the square as in Example 3.

Example 3 Write a Quadratic Function in Standard Form by Completing the Square and Sketch the Graph

Rewrite the quadratic function $f(x) = 2x^2 - 4x - 3$ in standard form, and then complete a) through g), as in Example 2.

Solution Our goal is to rewrite $f(x) = 2x^2 - 4x - 3$ in the form $f(x) = a(x - h)^2 + k$. To do this, we must complete the square.

$f(x) = 2x^2 - 4x - 3$ Write the original function.

$f(x) = (2x^2 - 4x) - 3$ Isolate the constant.

$f(x) = 2(x^2 - 2x) - 3$ Factor out the leading coefficient 2 from $2x^2 - 4x$.

$f(x) = 2(x^2 - 2x + 1) - 3 - 2$ Complete the square of $x^2 - 2x$. Note that we are adding $2 \cdot 1 = 2$, so we must subtract 2 as well!

$f(x) = 2(x - 1)^2 - 5$ Rewrite $(x^2 - 2x + 1)$ as a binomial squared.

 *My video summary*   ⊘ Thus, the quadratic function $f(x) = 2x^2 - 4x - 3$ written in standard form is $f(x) = 2(x - 1)^2 - 5$. Watch the video to verify the following:

a. The vertex is $(1, -5)$.

b. The parabola opens up.

c. The equation of the axis of symmetry is $x = 1$.

d. The two x-intercepts are $x = 1 + \dfrac{\sqrt{10}}{2} \approx 2.5811$ and $x = 1 - \dfrac{\sqrt{10}}{2} \approx -0.5811$.

e. The y-intercept is -3.

f. The graph of $f(x) = 2x^2 - 4x - 3$ is sketched in Figure 4.

g. The domain of f is the interval $(-\infty, \infty)$. The range of f is the interval $[-5, \infty)$.

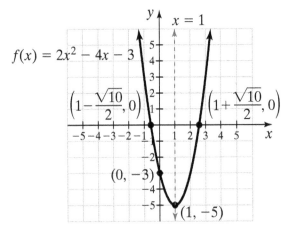

Figure 4 Graph of $f(x) = 2x^2 - 4x - 3$

You Try It Work through the following You Try It problem.

Work Exercises 11–20 in this textbook or in the MyMathLab Study Plan.

OBJECTIVE 4   GRAPHING QUADRATIC FUNCTIONS USING THE VERTEX FORMULA

 *My video summary*   ⊘ We can establish a "formula" for the vertex by completing the square on the quadratic function $f(x) = ax^2 + bx + c$. Work through the video to verify that the function $f(x) = ax^2 + bx + c$ is equivalent to $f(x) = a\left(x + \dfrac{b}{2a}\right)^2 + c - \dfrac{b^2}{4a}$.

We can now compare the function $f(x) = a\left(x + \dfrac{b}{2a}\right)^2 + c - \dfrac{b^2}{4a}$ to $f(x) = a(x - h)^2 + k$.

$$f(x) = a\left(x - \left(-\dfrac{b}{2a}\right)\right)^2 + \underbrace{c - \dfrac{b^2}{4a}}$$

$$\downarrow \qquad\qquad \downarrow$$

$$f(x) = a(x - \qquad h)^2 \;+\; k$$

We see that the coordinates of the vertex must be $\left(-\dfrac{b}{2a}, c - \dfrac{b^2}{4a}\right)$. Because the y-coordinate of the vertex can be found more easily by evaluating the function at the x-coordinate of the vertex than by memorizing the expression $c - \dfrac{b^2}{4a}$, we write the coordinates of the vertex as $\left(-\dfrac{b}{2a}, f\left(-\dfrac{b}{2a}\right)\right)$.

Formula for the Vertex of a Parabola

Given a quadratic function of the form $f(x) = ax^2 + bx + c, a \neq 0$, the vertex of the parabola is $\left(-\dfrac{b}{2a}, f\left(-\dfrac{b}{2a}\right)\right)$.

Example 4 Graph a Quadratic Function Using the Vertex Formula and Intercepts

Given the quadratic function $f(x) = -2x^2 - 4x + 5$, address the following:

a. Use the vertex formula to determine the vertex.

b. Does the graph "open up" or "open down"?

c. What is the equation of the axis of symmetry?

d. Find any x-intercepts.

e. Find the y-intercept.

f. Sketch the graph.

g. State the domain and range in interval notation.

Solution

a. The x-coordinate of the vertex is $-\dfrac{b}{2a} = -\dfrac{(-4)}{2(-2)} = -1$. The y-coordinate of the

vertex is $f\left(-\dfrac{b}{2a}\right) = f(-1) = -2(-1)^2 - 4(-1) + 5 = -2 + 4 + 5 = 7$. Therefore, the vertex is $(-1, 7)$.

b. Because $a = -2 < 0$, the parabola must open down.

c. The axis of symmetry is $x = -1$.

d. To find the x-intercepts, we solve the quadratic equation $-2x^2 - 4x + 5 = 0$.

Read these steps to verify that the x-intercepts are $x = -1 + \dfrac{\sqrt{14}}{2} \approx 0.8708$ and

$x = -1 - \dfrac{\sqrt{14}}{2} \approx -2.8708$.

e. The y-intercept is $f(0) = 5$.

f. Plotting the vertex and all intercepts, we get the graph sketched in Figure 5.

g. The domain of f is the interval $(-\infty, \infty)$. The range of f is the interval $(-\infty, 7]$.

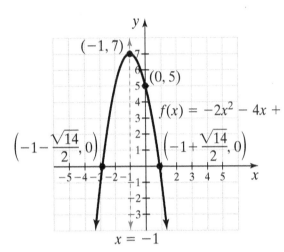

Figure 5 Graph of $f(x) = -2x^2 - 4x + 5$

Using Technology

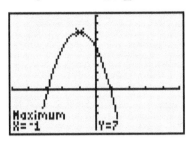

Using a graphing utility and the maximum function, we can verify that the vertex is $(-1, 7)$.

You Try It Work through the following You Try It problem.

Work Exercises 21–30 in this textbook or in the MyMathLab Study Plan.

OBJECTIVE 5 DETERMINING THE EQUATION OF A QUADRATIC FUNCTION GIVEN ITS GRAPH

Given the graph of a parabola, it may be necessary to find the equation of the quadratic function that it represents.

My video summary ⊚ **Example 5** Determine the Equation of a Quadratic Function Given Its Graph

Analyze the graph to address the following about the quadratic function it represents.

a. Is the leading coefficient positive or negative?

b. What is the value of h? What is the value of k?

c. What is the value of the leading coefficient a?

d. Write the equation of the function in standard form $f(x) = a(x - h)^2 + k$.

e. Write the equation of the function in the form $f(x) = ax^2 + bx + c$.

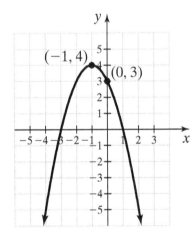

Solution

a. The leading coefficient is negative because the parabola opens down.

b. The value of h is the x-coordinate of the vertex, so $h = -1$. The value of k is the y-coordinate of the vertex, so $k = 4$.

c. Substitute the values of $h = -1$ and $k = 4$ into the standard form $f(x) = a(x - h)^2 + k$ to get $f(x) = a(x - (-1))^2 + 4$ or $f(x) = a(x + 1)^2 + 4$. To determine the value of a, we can use the fact that the graph contains the point $(0, 3)$, and therefore, $f(0) = 3$.

$f(x) = a(x + 1)^2 + 4$	Use the standard form with $h = -1$ and $k = 4$.
$f(0) = a(0 + 1)^2 + 4$	Evaluate $f(0)$.
$f(0) = 3$	The graph contains the point $(0, 3)$.
$a(0 + 1)^2 + 4 = 3$	Equate the two expressions for $f(0)$.
$a = -1$	Solve for a.

d. Substituting the values of $a = -1, h = -1$, and $k = 4$ into the equation $f(x) = a(x - h)^2 + k$, we get $f(x) = -(x + 1)^2 + 4$.

e. Writing the function in the form $f(x) = ax^2 + bx + c$, we get $f(x) = -x^2 - 2x + 3$. Watch the video to see each step worked out in detail.

You Try It   Work through the following You Try It problem.

Work Exercises 31–36 in this textbook or in the MyMathLab Study Plan.

4.1 Exercises

Skill Check Exercises

For exercises SCE-1 through SCE-3, decide what number must be added to each binomial to make a perfect square trinomial.

SCE-1. $x^2 + 10x$ **SCE-2.** $x^2 - 5x$ **SCE-3.** $x^2 + \dfrac{5}{3}x$

In Exercises 1–4, without graphing, determine whether the graph of each quadratic function opens up or opens down.

1. $f(x) = x^2 - 3$ **2.** $f(x) = -2x^2 + 4x + 1$

3. $f(x) = -4 + 5x^2 - 2x$ **4.** $f(x) = -\dfrac{2}{3}x^2 + x + 7$

In Exercises 5–10, given the quadratic function in standard form, address the following.

a. What are the coordinates of the vertex?

b. Does the graph "open up" or "open down"?

c. What is the equation of the axis of symmetry?

d. Find any x-intercepts.

e. Find the y-intercept.

f. Sketch the graph.

g. State the domain and range.

SbS **5.** $f(x) = (x - 2)^2 - 4$

SbS **6.** $f(x) = -(x + 1)^2 - 9$

SbS **7.** $f(x) = -2(x - 3)^2 + 2$

SbS **8.** $f(x) = \frac{1}{2}(x + 2)^2 + 2$

SbS **9.** $f(x) = -\frac{1}{4}(x - 4)^2 + 2$

SbS **10.** $f(x) = 3\left(x + \frac{1}{3}\right)^2 - 4$

In Exercises 11–20, first rewrite the quadratic function in standard form by completing the square, and then address the following.

a. What are the coordinates of the vertex?

b. Does the graph "open up" or "open down"?

c. What is the equation of the axis of symmetry?

d. Find any x-intercepts.

e. Find the y-intercept.

f. Sketch the graph.

g. State the domain and range in interval notation.

SbS **11.** $f(x) = x^2 - 8x$

SbS **12.** $f(x) = -x^2 - 4x + 12$

SbS **13.** $f(x) = 3x^2 + 6x - 4$

SbS **14.** $f(x) = 2x^2 - 5x - 3$

SbS **15.** $f(x) = -x^2 + 2x - 6$

SbS **16.** $f(x) = \frac{1}{2}x^2 + 6x + 1$

SbS **17.** $f(x) = -\frac{1}{3}x^2 - 2x + 5$

SbS **18.** $f(x) = -3x^2 + 7x + 5$

SbS **19.** $f(x) = x^2 + \frac{8}{3}x - 1$

SbS **20.** $f(x) = -\frac{1}{4}x^2 + 6x - 1$

In Exercises 21–30, use the quadratic function to address the following.

a. Use the vertex formula to determine the vertex.

b. Does the graph "open up" or "open down"?

c. What is the equation of the axis of symmetry?

d. Find any x-intercepts.

e. Find the y-intercept.

f. Sketch the graph.

g. State the domain and range in interval notation.

SbS **21.** $f(x) = x^2 - 8x$

SbS **22.** $f(x) = -x^2 - 4x + 8$

SbS **23.** $f(x) = 3x^2 + 6x - 4$

SbS **24.** $f(x) = 2x^2 - 5x - 3$

SbS **25.** $f(x) = -x^2 + 2x - 6$

SbS **26.** $f(x) = \frac{1}{2}x^2 + 6x + 1$

SbS **27.** $f(x) = -\frac{1}{3}x^2 - 9x + 5$

SbS **28.** $f(x) = -3x^2 + 7x + 5$

SbS **29.** $f(x) = x^2 + \frac{8}{3}x - 1$

SbS **30.** $f(x) = -\frac{1}{4}x^2 + 6x - 1$

In Exercises 31–36, analyze the graph to address the following about the quadratic function it represents.

a. Is the leading coefficient positive or negative?

b. What is the value of h? What is the value of k?

c. What is the value of the leading coefficient a?

d. Write the equation of the function in standard form $f(x) = a(x - h)^2 + k$.

e. Write the equation of the function in the form $f(x) = ax^2 + bx + c$.

SbS **31.**

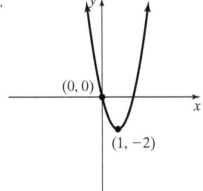

SbS **32.**

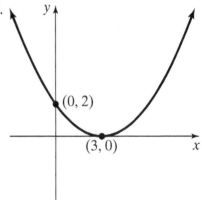

SbS **33.**

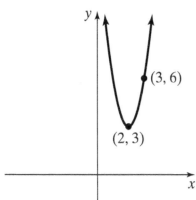

SbS **34.**

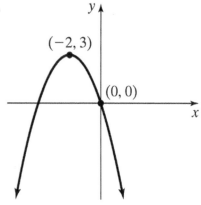

SbS **35.**

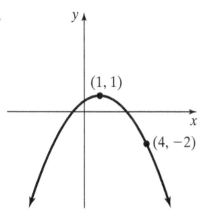

SbS **36.**

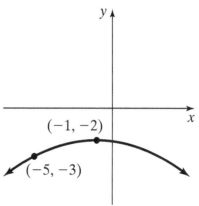

Brief Exercises

In Exercises 37–42, given the quadratic function in standard form, address the following.

    a. What are the coordinates of the vertex.

    b. Does the graph "open up" or "open down"?

    c. What is the equation of the axis of symmetry?

37. $f(x) = (x - 2)^2 - 4$

38. $f(x) = -(x + 1)^2 - 9$

39. $f(x) = -2(x - 3)^2 + 2$

40. $f(x) = \dfrac{1}{2}(x + 2)^2 + 2$

41. $f(x) = -\dfrac{1}{4}(x - 4)^2 + 2$

42. $f(x) = 3\left(x + \dfrac{1}{3}\right)^2 - 4$

In Exercises 43–48, given the quadratic function in standard form, address the following.

    a. Find any x-intercepts.

    b. Find the y-intercept.

43. $f(x) = (x - 2)^2 - 4$

44. $f(x) = -(x + 1)^2 - 9$

45. $f(x) = -2(x - 3)^2 + 2$

46. $f(x) = \dfrac{1}{2}(x + 2)^2 + 2$

47. $f(x) = -\dfrac{1}{4}(x - 4)^2 + 2$

48. $f(x) = 3\left(x + \dfrac{1}{3}\right)^2 - 4$

In Exercises 49–54, given the quadratic function in standard form, address the following.

    a. Sketch the graph.

    b. State the domain and range in interval notation.

49. $f(x) = (x - 2)^2 - 4$ **50.** $f(x) = -(x + 1)^2 - 9$ **51.** $f(x) = -2(x - 3)^2 + 2$

52. $f(x) = \frac{1}{2}(x + 2)^2 + 2$ **53.** $f(x) = -\frac{1}{4}(x - 4)^2 + 2$ **54.** $f(x) = 3\left(x + \frac{1}{3}\right)^2 - 4$

In Exercises 55–59, first rewrite the quadratic function in standard form by completing the square, and then address the following.

    a. What are the coordinates of the vertex.

    b. Does the graph "open up" or "open down"?

    c. What is the equation of the axis of symmetry?

55. $f(x) = 3x^2 + 6x - 4$ **56.** $f(x) = -x^2 + 2x - 6$ **57.** $f(x) = \frac{1}{2}x^2 + 6x + 1$

58. $f(x) = -3x^2 + 7x + 5$ **59.** $f(x) = -\frac{1}{4}x^2 + 6x - 1$

In Exercises 60–64, first rewrite the quadratic function in standard form by completing the square, and then address the following.

    a. Find any x-intercepts.

    b. Find the y-intercept.

60. $f(x) = 3x^2 + 6x - 4$ **61.** $f(x) = -x^2 + 2x - 6$ **62.** $f(x) = \frac{1}{2}x^2 + 6x + 1$

63. $f(x) = -3x^2 + 7x + 5$ **64.** $f(x) = -\frac{1}{4}x^2 + 6x - 1$

In Exercises 65–69, first rewrite the quadratic function in standard form by completing the square, and then address the following.

    a. Sketch the graph.

    b. State the domain and range in interval notation.

65. $f(x) = 3x^2 + 6x - 4$ **66.** $f(x) = -x^2 + 2x - 6$ **67.** $f(x) = \frac{1}{2}x^2 + 6x + 1$

68. $f(x) = -3x^2 + 7x + 5$ **69.** $f(x) = -\frac{1}{4}x^2 + 6x - 1$

In Exercises 70–74, use the quadratic function to address the following.

    a. Use the vertex formula to determine the coordinates of the vertex.

    b. Does the graph "open up" or "open down"?

    c. What is the equation of the axis of symmetry?

70. $f(x) = 3x^2 + 6x - 4$ **71.** $f(x) = -x^2 + 2x - 6$ **72.** $f(x) = \frac{1}{2}x^2 + 6x + 1$

73. $f(x) = -3x^2 + 7x + 5$ **74.** $f(x) = -\frac{1}{4}x^2 + 6x - 1$

In Exercise 75–79, use the quadratic function to address the following.

    a. Find any x-intercepts.

    b. Find any y-intercepts.

75. $f(x) = 3x^2 + 6x - 4$ **76.** $f(x) = -x^2 + 2x - 6$ **77.** $f(x) = \frac{1}{2}x^2 + 6x + 1$

78. $f(x) = -3x^2 + 7x + 5$ **79.** $f(x) = -\frac{1}{4}x^2 + 6x - 1$

In Exercises 80–84, use the quadratic function to address the following.

    a. Sketch the graph

    b. State the domain and range the interval notation.

80. $f(x) = 3x^2 + 6x - 4$ **81.** $f(x) = -x^2 + 2x - 6$ **82.** $f(x) = \frac{1}{2}x^2 + 6x + 1$

83. $f(x) = -3x^2 + 7x + 5$ **84.** $f(x) = -\frac{1}{4}x^2 + 6x - 1$

In Exercises 85–94, use the quadratic function to address the following.

    a. Does the function have a maximum or minimum value?

    b. What is this maximum or minimum value?

85. $f(x) = x^2 - 8x$ **86.** $f(x) = -x^2 - 4x + 8$ **87.** $f(x) = 3x^2 + 6x - 4$

88. $f(x) = 2x^2 - 5x - 3$ **89.** $f(x) = -x^2 + 2x - 6$ **90.** $f(x) = \frac{1}{2}x^2 + 6x + 1$

91. $f(x) = -\dfrac{1}{2}x^2 - 9x + 5$

92. $f(x) = -3x^2 + 7x + 5$

93. $f(x) = x^2 + \dfrac{8}{3}x - 1$

94. $f(x) = -\dfrac{1}{4}x^2 + 6x - 1$

In Exercises 95–100, given the graph of a quadratic function, write the equation of the function is standard form $f(x) = a(x - h)^2 + k$.

95.

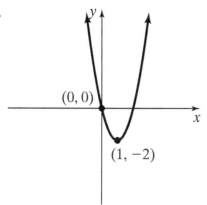

96.

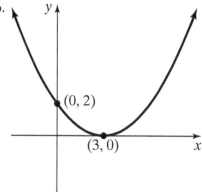

97.

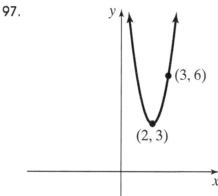

98.

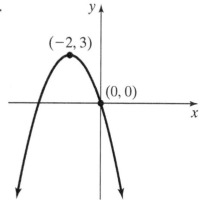

99.

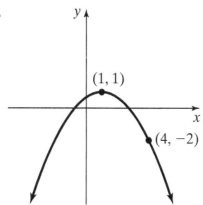

100.

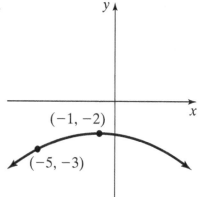

In Exercises 101–106, given the graph of a quadratic function, write the equation of the function in the form $f(x) = ax^2 + bx + c$.

101.

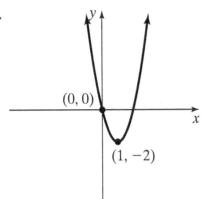

$(0, 0)$

$(1, -2)$

102.

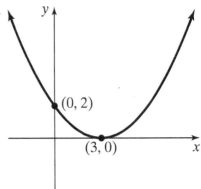

$(0, 2)$

$(3, 0)$

103.

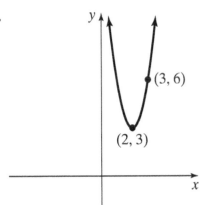

$(3, 6)$

$(2, 3)$

104.

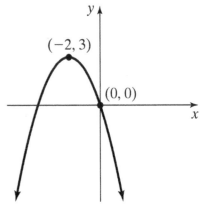

$(-2, 3)$

$(0, 0)$

105.

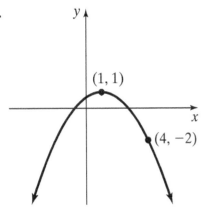

$(1, 1)$

$(4, -2)$

106.

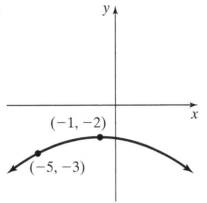

$(-1, -2)$

$(-5, -3)$

4.2 Applications and Modeling of Quadratic Functions

THINGS TO KNOW

Before working through this section, be sure that you are familiar with the following concepts:

VIDEO   ANIMATION   INTERACTIVE

You Try It

1. Graphing Quadratic Functions Using the Vertex Formula (Section 4.1)   ⊘

OBJECTIVES

1 Maximizing Projectile Motion Functions

2 Maximizing Functions in Economics

3 Maximizing Area Functions

....

Introduction to Section 4.2

My video summary ⊙ With application problems involving functions, we are often interested in finding the maximum or minimum value of the function. For example, a professional golfer may want to control the trajectory of a struck golf ball. Thus, he may be interested in determining the maximum height of the ball given certain parameters such as swing velocity and club angle. An economist may want to minimize a cost function or maximize a revenue or profit function. A builder with a fixed amount of fencing may want to maximize an area function. In a calculus course, you learn how to maximize or minimize a wide variety of functions. In this section, we concentrate only on quadratic functions. Quadratic functions are relatively easy to maximize or minimize because we know a formula for finding the coordinates of the vertex. Recall that if $f(x) = ax^2 + bx + c, a \neq 0$, we know that the coordinates of the vertex are $\left(-\dfrac{b}{2a}, f\left(-\dfrac{b}{2a} \right) \right)$. If $a > 0$, the parabola *opens up* and has a minimum value at the y-coordinate of the vertex. If $a < 0$, the parabola *opens down* and has a maximum value at the y-coordinate of the vertex. See Figure 6.

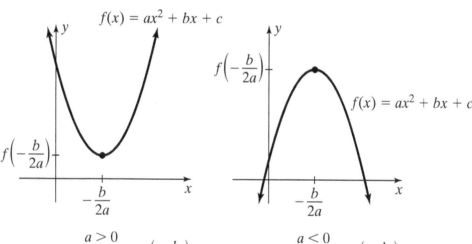

$a > 0$
Minimum value at $f\left(-\dfrac{b}{2a} \right)$

$a < 0$
Maximum value at $f\left(-\dfrac{b}{2a} \right)$

Figure 6

OBJECTIVE 1 MAXIMIZING PROJECTILE MOTION FUNCTIONS

We start our discussion by looking at the projectile motion model that was introduced in Section 1.5. An object launched, thrown, or shot vertically into the air with an initial velocity of v_0 meters/second from an initial height of h_0 meters above the ground can be modeled by the function $h(t) = -4.9t^2 + v_0 t + h_0$, where $h(t)$ is the height of the projectile t seconds after its departure.

Example 1 Launch a Toy Rocket

A toy rocket is launched with an initial velocity of 44.1 meters per second from a 1-meter-tall platform. The height h of the object at any time t seconds after launch is given by the function $h(t) = -4.9t^2 + 44.1t + 1$. How long after launch did it take the rocket to reach its maximum height? What is the maximum height obtained by the toy rocket?

Solution Because the function is quadratic with $a = -4.9 < 0$, we know that the graph is a parabola that opens down. Thus, the function obtains a maximum value at the vertex.

The t-coordinate of the vertex is $t = -\dfrac{b}{2a} = -\dfrac{44.1}{2(-4.9)} = \dfrac{-44.1}{-9.8} = 4.5$ seconds.

Therefore, the rocket reaches its maximum height 4.5 seconds after launch. The maximum height obtained by the rocket is $h(4.5) = -4.9(4.5)^2 + 44.1(4.5) + 1 = 100.225$ meters.

Using Technology

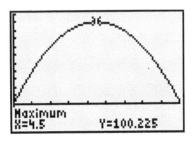

The graph of the quadratic function $h(t) = -4.9t^2 + 44.1t + 1$ is seen here using a graphing utility. Using the Maximum function, we can see that the maximum height is 100.225 meters when time is 4.5 seconds.

You Try It Work through the following You Try It problem.

Work Exercises 1–3 in this textbook or in the MyMathLab Study Plan.

Example 2 Horizontal and Vertical Projectile Motion

If an object is launched at an angle of 45 degrees from a 10-foot platform at 60 feet per second, it can be shown that the height of the object in feet is given by the quadratic function $h(x) = -\dfrac{32x^2}{(60)^2} + x + 10$, where x is the horizontal distance of the object from the platform. See Figure 7.

a. What is the height of the object when its horizontal distance from the platform is 20 feet? Round to two decimal places.

b. What is the horizontal distance from the platform when the object is at its maximum height?

c. What is the maximum height of the object?

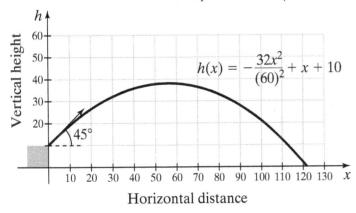

Horizontal distance

Figure 7

Solution

a. To find the height of the object when the horizontal distance from the platform is 20 feet, we must find the value of $h(20)$.

$$h(x) = -\frac{32x^2}{(60)^2} + x + 10 \qquad \text{Write the original function.}$$

$$h(20) = -\frac{32(20)^2}{(60)^2} + 20 + 10 \qquad \text{Substitute 20 for } x.$$

$$= -\frac{32(400)}{3,600} + 30 \qquad \text{Simplify.}$$

$$= -\frac{12,800}{3,600} + 30$$

$$\approx 26.44$$

When the horizontal distance from the platform is 20 feet, the height of the object is approximately 26.44 feet.

b. To find the horizontal distance of the object when it is at its maximum height, we must find the x-coordinate of the vertex. The x-coordinate of the vertex is found using the vertex formula $x = -\dfrac{b}{2a}$. Since $a = -\dfrac{32}{60^2}$ and $b = 1$, we get

$$x = -\frac{1}{2\left(\dfrac{-32}{60^2}\right)} = \frac{-1}{\dfrac{-64}{3,600}} = -1 \cdot \frac{3,600}{-64} = 56.25 \text{ feet}$$

Thus, the object obtains its maximum height at a horizontal distance of 56.25 feet from the base of the platform.

c. From part b, the object obtains its maximum height at a horizontal distance of 56.25 feet from the base of the platform. Therefore, the maximum height is

$$h(56.25) = -\frac{32(56.25)^2}{(60)^2} + 56.25 + 10 = 38.125 \text{ feet.}$$

You Try It   Work through the following You Try It problem.

Work Exercises 4–6 in this textbook or in the MyMathLab Study Plan.

OBJECTIVE 2   MAXIMIZING FUNCTIONS IN ECONOMICS

Revenue is defined as the dollar amount received by selling x items at a price of p dollars per item; that is, $R = xp$. For example, if a child sells 50 cups of lemonade at a price of \$.25 per cup, then the revenue generated is $R = \underbrace{(50)}_{x}\underbrace{(.25)}_{p} = \$12.50$.

A common model in economics states that as the quantity, x, increases, the price, p, tends to decrease. Likewise, if the quantity decreases, the price tends to increase.

 *My video summary*

⊙ Example 3 Maximize Revenue

Records can be kept on the price of shoes and the number of pairs sold in order to gather enough data to reasonably model shopping trends for a particular type of shoe. Demand functions of this type are often linear and can be developed using knowledge of the slope and equations of lines. Suppose that the marketing and research department of a shoe company determined that the price of a certain type of basketball shoe obeys the demand equation $p = -\frac{1}{50}x + 110$.

a. According to the demand equation, how much should the shoes sell for if 500 pairs of shoes are sold? 1,200 pairs of shoes?

b. What is the revenue if 500 pairs of shoes are sold? 1,200 pairs of shoes?

c. How many pairs of shoes should be sold in order to maximize revenue? What is the maximum revenue?

d. What price should be charged in order to maximize revenue?

Solution

a. The demand equation $p = -\frac{1}{50}x + 110$ is sketched in Figure 8. If 500 pairs of shoes are sold, the price should be $p(500) = -\frac{1}{50}(500) + 110 = -10 + 110 =$ $100. If 1,200 pairs of shoes are sold, the price should be $p(1,200) = -\frac{1}{50}(1,200) + 110 = -24 + 110 = 86.

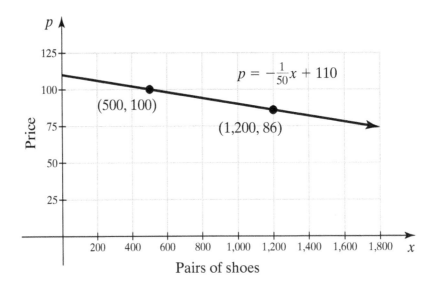

Figure 8

b. Because $R = xp$, we can substitute the demand equation, $p = -\frac{1}{50}x + 110$, for p to obtain the function $R(x) = \underbrace{x\left(-\frac{1}{50}x + 110\right)}_{p}$ or $R(x) = -\frac{1}{50}x^2 + 110x$. The graph of R is shown in Figure 9. The revenue generated by selling 500 pairs of shoes is $R(500) = \$50,000$, whereas the revenue generated by selling 1,200 pairs of shoes is $R(1,200) = \$103,200$.

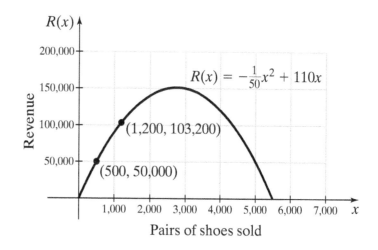

Figure 9

c. $R(x) = -\frac{1}{50}x^2 + 110x$ is a quadratic function with $a = -\frac{1}{50}$ and $b = 110$. Because a is less than 0, the function has a *maximum* value at the vertex. Therefore, the value of x that produces the maximum revenue is $x = -\frac{b}{2a} = -\frac{110}{2\left(-\frac{1}{50}\right)} =$

$\frac{-110}{-\frac{1}{25}} = 2{,}750$ pairs of shoes. The maximum revenue is $R(2{,}750) = \$151{,}250.$

See Figure 10.

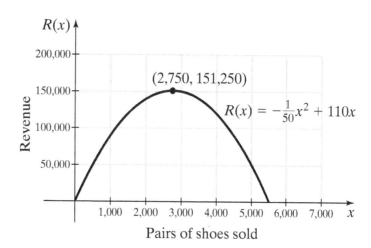

Figure 10

d. Using the demand equation, the price that should be charged to maximize revenue when selling 2,750 pairs of shoes is $p(2{,}750) = -\frac{1}{50}(2{,}750) + 110 = \$55.$

Watch this video to see entire solution to this example.

You Try It   Work through the following You Try It problem.

Work Exercises 7 and 8 in this textbook or in the MyMathLab Study Plan.

Example 4 Maximize Profit

To sell x waterproof CD alarm clocks, WaterTime, LLC, has determined that the price in dollars must be $p = 250 - 2x$, which is the demand equation. Each clock costs \$2 to produce, with fixed costs of \$4,000, producing the cost function of $C(x) = 2x + 4,000$.

a. Express the revenue R as a function of x.

b. Express the profit P as a function of x.

c. Find the value of x that maximizes profit. What is the maximum profit?

d. What is the price of the alarm clock that will maximize profit?

Solution

a. The equation for revenue is $R = xp$. Therefore, we can substitute the demand equation $p = 250 - 2x$ for p to obtain the function $R(x) = \underbrace{x(250 - 2x)}_{p}$ or

$$R(x) = -2x^2 + 250x.$$

b. Profit is equal to revenue minus cost.

$$P(x) = R(x) - C(x)$$
$$= -2x^2 + 250x - (2x + 4,000)$$
$$= -2x^2 + 248x - 4,000$$

c. The value of x that will produce the maximum profit is

$$x = -\frac{b}{2a} = -\frac{248}{2(-2)} = \frac{-248}{-4} = 62 \text{ alarm clocks}$$

The maximum profit is $P(62) = -2(62)^2 + 248(62) - 4,000 = \$3,688$.

d. Using the demand equation $p = 250 - 2x$ and the fact that profit will be maximized when 62 alarm clocks are produced, we obtain a price of $p = 250 - 2(62) = \$126$.

You Try It Work through the following You Try It problem.

Work Exercises 9–12 in this textbook or in the MyMathLab Study Plan.

My interactive video summary

⊙ Example 5 Maximize Revenue at a Country Club

A country club currently has 400 members who pay \$500 per month for membership dues. The country club's board members want to increase monthly revenue by *lowering* the monthly dues in hopes of attracting new members. A market research study has shown that for each \$1 decrease in the monthly membership price, two additional people will join the club. What price should the club charge to maximize revenue? What is the maximum revenue?

Solution We know that $R = xp$, where x is the number of members and p is the price of monthly membership. Currently, $R = (400)(\$500) = \$200,000$. If the price is reduced by \$1, two additional people will join the club, so the revenue generated would be $R = (402)(\$499) = \$200,598$. We can see that by reducing monthly dues to \$499, it can be anticipated that revenue will increase by \$598. We want to find the optimal price to charge in order to maximize revenue.

Because the number of members, x, depends on price, we can find an expression for x in terms of p:

$$x = \underbrace{400}_{\substack{\text{Number} \\ \text{of current} \\ \text{members}}} + \underbrace{2(500 - p)}_{\substack{\text{Additional two members} \\ \text{for each dollar decrease in} \\ \text{membership monthly dues}}}$$

We can now write the revenue as a function of p:

$$R = xp$$

$$R(p) = \underbrace{(400 + 2(500 - p))p}_{x} \quad \text{or}$$

$$R(p) = -2p^2 + 1,400p$$

The revenue is maximized when the price is $p = -\dfrac{b}{2a} = -\dfrac{1,400}{2(-2)} = \$350$.

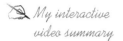 *My interactive video summary*

⊘ The maximum revenue is $R(350) = -2(350)^2 + 1,400(350) = \$245,000$. Watch the interactive video to see this problem worked out using an alternate method. ●

You Try It Work through the following You Try It problem.

Work Exercises **13–20** in this textbook or in the MyMathLab **Study Plan.**

OBJECTIVE 3 MAXIMIZING AREA FUNCTIONS

Suppose you are asked to build a rectangular fence that borders a river but you have only 3,000 feet of fencing available. No fencing will be required along the side of the river. How many ways can you build this fence? As it turns out, there are infinitely many ways to construct such a fence.

My video summary

⊘ Although there are infinitely many ways in which you could build this fence, there is only one way to build this fence in order to *maximize* the area! What should the length and width of the fence be in order to maximize the area? Watch this video to see how to determine the length and width of the fence that will maximize the area.

 *My video summary*   ⊙ **Example 6** Maximize Area

Mark has 100 feet of fencing available to build a rectangular pen for his hens and roosters. He wants to separate the hens and roosters by dividing the pen into two equal areas. What should the length of the center partition be in order to maximize the area? What is the maximum area?

Solution Watch the video to verify that the length of the partition should be $16\frac{2}{3}$ feet. The maximum area is $416\frac{2}{3}$ square feet.

You Try It   Work through the following You Try It problem.

Work Exercises 21–24 in this textbook or in the MyMathLab Study Plan.

4.2 Exercises

1. A baseball player swings and hits a pop fly straight up in the air to the catcher. The height of the baseball in meters t seconds after it is hit is given by the quadratic function $h(t) = -4.9t^2 + 34.3t + 1$. How long does it take for the baseball to reach its maximum height? What is the maximum height obtained by the baseball?

2. An object is launched vertically in the air at 36.75 meters per second from a 10-meter-tall platform. Using the projectile motion model, determine how long it will take for the object to reach its maximum height. What is the maximum height?

3. A toy rocket is shot vertically into the air from a 5-foot-tall launching pad with an initial velocity of 112 feet per second. Suppose the height of the rocket in feet t seconds after being launched can be modeled by the function $h(t) = -16t^2 + v_0 t + h_0$, where v_0 is the initial velocity of the rocket and h_0 is the initial height of the rocket. How long will it take for the rocket to reach its maximum height? What is the maximum height?

4. A football player punts a football at an inclination of 45 degrees to the horizontal at an initial velocity of 75 feet per second. The height of the football in feet is given by the quadratic function $h(x) = -\dfrac{32x^2}{(75)^2} + x + 4$, where x is the horizontal distance of the football from the point at which it was kicked. What was the horizontal distance of the football from the point at which it was kicked when the height of the ball is at a maximum? What is the maximum height of the football?

5. A golf ball is struck by a 60-degree golf club at an initial velocity of 88 feet per second. The height of the golf ball in feet is given by the quadratic function $h(x) = -\dfrac{16x^2}{(44)^2} + \dfrac{76.2}{44}x$, where x is the horizontal distance of the golf ball from the point of impact. What is the horizontal distance of the golf ball from the point of impact when the ball is at its maximum height? What is the maximum height obtained by the golf ball?

6. If a projectile is launched upward at an acute angle, then the height of the projectile at a horizontal distance of x units from the launch site is given by the quadratic function $h(x) = -\dfrac{1}{2}g\left(\dfrac{x}{w_0}\right)^2 + \left(\dfrac{v_0}{w_0}\right)x + h_0$, where g is acceleration due to gravity, w_0 is horizontal initial velocity, v_0 is vertical initial velocity, and h_0 is initial height of the object.

a. Find a formula describing the horizontal distance from the launch point where the projectile is at a maximum height in terms of g, w_0, and v_0.

b. Find a formula for the maximum height of the projectile in terms of g, v_0, and h_0.

c. Suppose $g = 32$, $w_0 = 64$, $v_0 = 64$, and $h_0 = 1$, use the formulas from parts a and b to find the horizontal distance from the launch point where the projectile is at a maximum height and find the maximum height.

7. The price p and the quantity x sold of a small flat screen television set obeys the demand equation $p = -.15x + 300$.

a. How much should be charged for the television set if there are 50 television sets in stock?

b. What quantity x will maximize revenue? What is the maximum revenue?

c. What price should be charged in order to maximize revenue?

8. The dollar price for a barrel of oil sold at a certain oil refinery tends to follow the demand equation $p = -\dfrac{1}{10}x + 72$, where x is the number of barrels of oil on hand (in millions).

a. How much should be charged for a barrel of oil if there are 4 million barrels on hand?

b. What quantity x will maximize revenue?

c. What price should be charged in order to maximize revenue?

In Exercises 9 and 10, use the fact that profit is defined as revenue minus cost or $P(x) = R(x) - C(x)$.

9. Rite-Cut riding lawnmowers obey the demand equation $p = -\dfrac{1}{20}x + 1{,}000$. The cost of producing x lawnmowers is given by the function $C(x) = 100x + 5{,}000$.

a. Express the revenue R as a function of x.

b. Express the profit P as a function of x.

c. Find the value of x that maximizes profit. What is the maximum profit?

d. What price should be charged in order to maximize profit?

10. The CarryItAll minivan, a popular vehicle among soccer moms, obeys the demand equation $p = -\dfrac{1}{40}x + 8{,}000$. The cost of producing x vans is given by the function $C(x) = 4{,}000x + 20{,}000$.

a. Express the revenue R as a function of x.

b. Express the profit P as a function of x.

c. Find the value of x that maximizes profit. What is the maximum profit?

d. What price should be charged in order to maximize profit?

11. Silver Scooter, Inc., finds that it costs $200 to produce each motorized scooter and that the fixed costs are $1,500. The price is given by $p = 600 - 5x$, where p is the price in dollars at which exactly x scooters will be sold. Find the quantity of scooters that Silver Scooter, Inc., should produce and the price it should charge to maximize profit. Find the maximum profit.

12. Amy, the owner of Amy's Pottery, can produce china pitchers at a cost of $5 each. She estimates her price function to be $p = 17 - 5x$, where p is the price at which exactly x pitchers will be sold per week. Find the number of pitchers that she should produce and the price that she should charge in order to maximize profit. Also find the maximum profit.

13. Lewiston Golf and Country Club currently has 300 members who pay $450 per month for membership dues. The country club's board members want to increase monthly revenue by *lowering* the monthly dues in hopes of attracting new members. A market research study has shown that for each $1 decrease in the monthly membership price, three additional people will join the club. What price should the club charge to maximize revenue? What is the maximum revenue?

14. An orange grove produces a profit of $90 per tree when there are 1,200 trees planted. Because of overcrowding, the profit per tree (for every tree in the grove) is reduced by 5 cents per tree for each additional tree planted. How many trees should be planted in order to maximize the total profit of the orange grove? What is the maximum profit?

15. All-Nite Fitness Health Club currently charges its 1,500 clients monthly membership dues of $45. The board of directors decides to increase the monthly membership dues. Market research states that each $1 increase in dues will result in the loss of five clients. How much should the club charge each month to optimize the revenue from monthly dues?

16. The marketing department of a local shoe company found that approximately 600 pairs of running shoes are sold monthly when the price of each pair is $90. It was also observed that for each $1 reduction in price, an additional 20 pairs of running shoes are sold monthly. What price should the shoe store charge for a pair of running shoes in order to maximize revenue?

17. Pearson Travel, Inc., charges $10 for a trip to a concert if 30 people travel in a group. But for each 1 person above the 30-person level, the charge will be reduced by $.20. How many people will maximize the total revenue for the company?

18. Sunshine Bicycle Rental Company at a California resort rents 100 bicycles per day at a flat rate of $10 per day for each bicycle. For each $1 increase in the rate, 5 fewer bikes are rented. At what rate should the bicycles be rented to produce the maximum daily revenue? What is the maximum daily revenue?

19. Cherrypickers, Inc., estimates from past records that if 30 trees are planted per acre, each tree will yield an average of 50 pounds of cherries per season. If for each additional tree planted per acre the average yield per tree is reduced by 1 pound, how many trees should be planted per acre to obtain the maximum yield per acre?

20. An orchard contains 30 apple trees, each of which yields about 400 apples over the growing season. The owner plans to add more trees, but each new tree will reduce the average yield per tree by about 10 apples over the season. How many trees should be added to maximize the total yield of apples? What would be the maximum yield for the apple orchard for the season?

21. A farmer has 1,800 feet of fencing available to enclose a rectangular area bordering a river. If no fencing is required along the river, find the dimensions of the fence that will maximize the area. What is the maximum area?

22. Jim wants to build a rectangular parking lot along a busy street but has only 2,500 feet of fencing available. If no fencing is required along the street, find the maximum area of the parking lot.

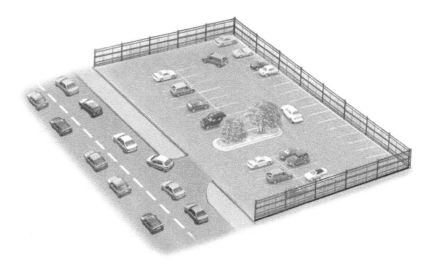

23. A rancher has 5,000 feet of fencing available to enclose a rectangular area bordering a river. He wants to separate his cows and horses by dividing the enclosure into two equal areas. If no fencing is required along the river, find the length of the center partition that will yield the maximum area.

24. A 32-inch wire is to be cut. One piece is to be bent into the shape of a square, whereas the other piece is to be bent into the shape of a rectangle whose length is twice the width. Find the width of the rectangle that will minimize the total area.

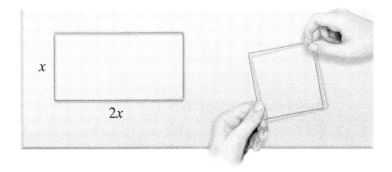

4.3 The Graphs of Polynomial Functions

THINGS TO KNOW

Before working through this section, be sure that you are familiar with the following concepts:

| | VIDEO | ANIMATION | INTERACTIVE |

You Try It   1. Determining the Intercepts of a Function (Section 3.2)

You Try It   2. Sketching the Graphs of the Basic Functions (Section 3.3)

You Try It   3. Using Transformations to Sketch Functions (Section 3.4)

OBJECTIVES

1 Understanding the Definition of a Polynomial Function

2 Sketching the Graphs of Power Functions

3 Determining the End Behavior of Polynomial Functions

4 Determining the Intercepts of a Polynomial Function

5 Determining the Real Zeros of Polynomial Functions and Their Multiplicities

6 Sketching the Graph of a Polynomial Function

7 Determining a Possible Equation of a Polynomial Function Given Its Graph

---

OBJECTIVE 1   UNDERSTANDING THE DEFINITION OF A POLYNOMIAL FUNCTION

   We have already studied linear functions of the form $f(x) = mx + b$, where $m \neq 0$, and quadratic functions of the form $f(x) = ax^2 + bx + c$, where $a \neq 0$. Both functions are examples of **polynomial functions**.

> **Definition**   Polynomial Function
>
> The function $f(x) = a_n x^n + a_{n-1} x^{n-1} + a_{n-2} x^{n-2} + \cdots + a_1 x + a_0$ is a polynomial function of degree n, where n is a nonnegative integer. The numbers $a_0, a_1, a_2, \ldots, a_n$ are called the **coefficients** of the polynomial function. The number a_n is called the **leading coefficient**, and a_0 is called the **constant coefficient**.

The important thing to remember from the definition is that the power (exponent) of each term must be an integer greater than or equal to zero. Although the coefficients can be any number (real or imaginary), we focus primarily on polynomial functions with real coefficients.

Example 1 Determine Whether a Function Is a Polynomial Function

Determine which functions are polynomial functions. If the function is a polynomial function, identify the degree, the leading coefficient, and the constant coefficient.

a. $f(x) = \sqrt{3}x^3 - 2x^2 - \dfrac{1}{6}$

b. $g(x) = 4x^5 - 3x^3 + x^2 + \dfrac{7}{x} - 3$

c. $h(x) = \dfrac{3x - x^2 + 7x^4}{9}$

Solution

a. The function $f(x) = \sqrt{3}x^3 - 2x^2 - \dfrac{1}{6}$ is a polynomial function because all powers of x are integers greater than or equal to 0. The highest power is 3, so this polynomial is of degree 3. The leading coefficient is $\sqrt{3}$, and the constant coefficient is $-\dfrac{1}{6}$.

b. The function $g(x) = 4x^5 - 3x^3 + x^2 + \dfrac{7}{x} - 3$ is *not* a polynomial function because the term $\dfrac{7}{x}$ can be rewritten as $7x^{-1}$ and -1 is *not* greater than or equal to 0.

c. The function $h(x) = \dfrac{3x - x^2 + 7x^4}{9}$ is a polynomial function of degree 4. Dividing each term by 9 and reordering the terms, we can write this polynomial function as $h(x) = \dfrac{7}{9}x^4 - \dfrac{1}{9}x^2 + \dfrac{1}{3}x$. The leading coefficient is $\dfrac{7}{9}$, and the constant coefficient is 0.

You Try It Work through the following You Try It problem.

Work Exercises 1–6 in this textbook or in the MyMathLab Study Plan.

OBJECTIVE 2 SKETCHING THE GRAPHS OF POWER FUNCTIONS

The most basic polynomial functions are monomial functions of the form $f(x) = ax^n$. Polynomial functions of this form are called **power functions**. Figure 11 shows the graphs of power functions for $a = 1$ and for $n = 1, 2, 3, 4,$ and 5.

You can see from Figure 11 that the graph of $f(x) = x^4$ closely resembles the graph of $f(x) = x^2$. The difference in the two graphs is that the graph of $f(x) = x^4$ flattens out near the origin and gets steep quicker than the graph of $f(x) = x^2$. Likewise, the graph of $f(x) = x^5$ resembles the graph of $f(x) = x^3$. In general if n is even, the graph of $f(x) = x^n$ will resemble the graph of $f(x) = x^2$. If n is odd, the graph of $f(x) = x^n$ will resemble the graph of $f(x) = x^3$.

Because we know the basic shape of $f(x) = x^n$, we can use the transformation techniques discussed in Section 3.4 to sketch more complicated polynomial functions. Read this summary for a quick review of the transformation techniques or refer to Section 3.4.

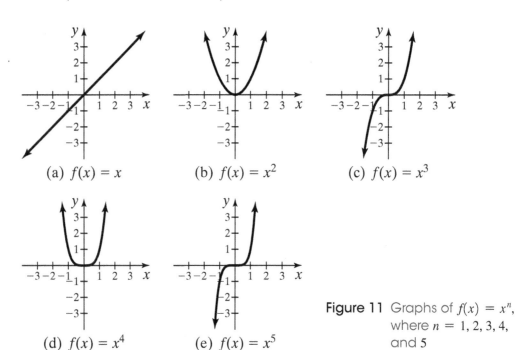

(a) $f(x) = x$ (b) $f(x) = x^2$ (c) $f(x) = x^3$

(d) $f(x) = x^4$ (e) $f(x) = x^5$

Figure 11 Graphs of $f(x) = x^n$, where $n = 1, 2, 3, 4$, and 5

My interactive video summary

⊙ **Example 2** Use Transformations to Sketch Polynomial Functions

Use transformations to sketch the following functions:

a. $f(x) = -x^6$ **b.** $f(x) = (x + 1)^5 + 2$ **c.** $f(x) = 2(x - 3)^4$

Solution

a. The graph of $f(x) = -x^6$ is obtained by reflecting the graph of $y = x^6$ about the x-axis. See Figure 12(a).

b. Beginning with the graph of $y = x^5$, we can obtain the graph of $f(x) = (x + 1)^5 + 2$ by horizontally shifting $y = x^5$ to the *left* one unit and then *up* two units. See Figure 12(b).

c. To sketch the graph of $f(x) = 2(x - 3)^4$, we start with the graph of $y = x^4$. We then horizontally shift the graph to the *right* three units and vertically stretch the graph by a factor of 2. See Figure 12(c).

Watch the interactive video for a detailed solution.

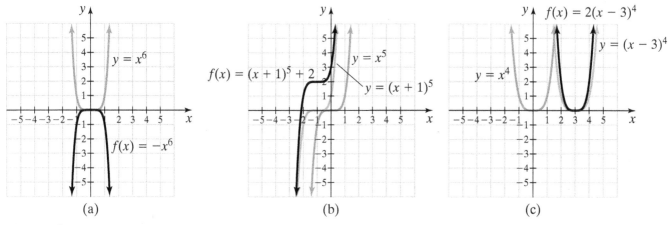

(a)   (b)   (c)

Figure 12

You Try It   Work through the following You Try It problem.

Work Exercises 7–15 in this textbook or in the MyMathLab Study Plan.

OBJECTIVE 3   DETERMINING THE END BEHAVIOR OF POLYNOMIAL FUNCTIONS

My video summary   To begin to sketch the graph of a more complicated polynomial function, it is necessary to understand the behavior of the graph as the x-coordinate gets large in the positive direction (approaches infinity) and as the x-coordinate gets small in the negative direction (approaches negative infinity). The nature of the graph of a polynomial function for large values of x in the positive and negative direction is known as the **end behavior**. The end behavior of the graph of a polynomial function depends on the leading term, that is, the term of highest degree. The graph of a polynomial function $f(x) = a_n x^n + a_{n-1} x^{n-1} + a_{n-2} x^{n-2} + \cdots + a_1 x + a_0$ has the same end behavior as the graph of the power function $f(x) = a_n x^n$.

To illustrate end behavior, consider the polynomial function $f(x) = x^5 - 2x^4 - 6x^3 + 8x^2 + 5x - 6$ sketched in Figure 13a and compare it to the graph of $f(x) = x^5$ sketched in Figure 13b. Notice that the "ends" of both graphs show similar behavior. The right-hand ends of both graphs increase without bound; that is, the y-values approach infinity as the x-values approach infinity. The left-hand ends of both graphs approach negative infinity as the x-values approach negative infinity.

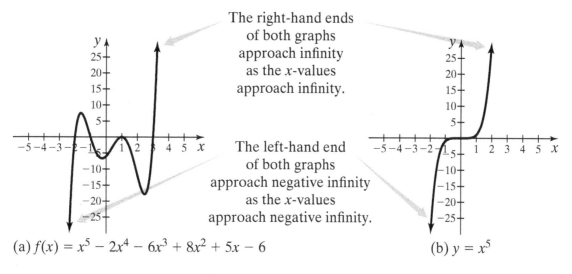

The right-hand ends of both graphs approach infinity as the x-values approach infinity.

The left-hand end of both graphs approach negative infinity as the x-values approach negative infinity.

(a) $f(x) = x^5 - 2x^4 - 6x^3 + 8x^2 + 5x - 6$ (b) $y = x^5$ Figure 13

To determine the end behavior of a polynomial function $f(x) = a_n x^n + a_{n-1} x^{n-1} + a_{n-2} x^{n-2} + \ldots + a_1 x + a_0$, we look at the leading term $a_n x^n$ and follow a two-step process.

Two-Step Process for Determining the End Behavior of a Polynomial Function $f(x) = a_n x^n + a_{n-1} x^{n-1} + a_{n-2} x^{n-2} + \cdots + a_1 x + a_0$

Step 1. Determine the sign of the leading coefficient a_n.

If $a_n > 0$, the right-hand behavior "finishes up."

If $a_n < 0$, the right-hand behavior "finishes down."

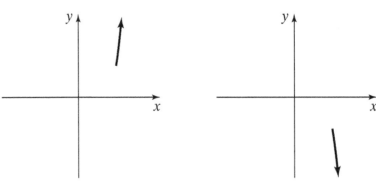

Step 2. Determine the degree.

If the degree n is odd, the graph has opposite left-hand and right-hand end behavior; that is, the graph "starts" and "finishes" in *opposite* directions.

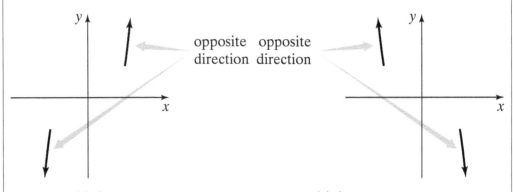

opposite opposite
direction direction

$a_n > 0$, odd degree $a_n < 0$, odd degree

If the degree n is even, the graph has the same left-hand and right-hand end behavior; that is, the graph "starts" and "finishes" in the *same* direction.

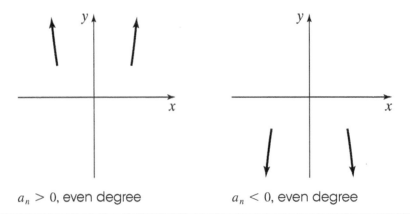

$a_n > 0$, even degree $a_n < 0$, even degree

My video summary

⊙ Example 3 Use End Behavior to Determine the Degree and Leading Coefficient

Use the end behavior of each graph to determine whether the degree is even or odd and whether the leading coefficient is positive or negative.

a.
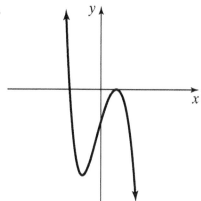

b.
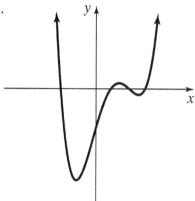

Solution

a. The leading coefficient is *negative* because the right-hand end behavior "ends down." The degree is *odd* because the graph has opposite left-hand and right-hand end behavior.

b. The leading coefficient is *positive* because the right-hand end behavior "ends up." The degree is *even* because the graph has the same left-hand and right-hand end behavior.

Watch the video to see a more detailed solution.

You Try It Work through the following You Try It problem.

Work Exercises 16–21 in this textbook or in the MyMathLab Study Plan.

OBJECTIVE 4 DETERMINING THE INTERCEPTS OF A POLYNOMIAL FUNCTION

Now that we know how to determine the end behavior of a polynomial function, it is time to find out what happens between the ends. We start by trying to locate the intercepts of the graph. Every polynomial function, $y = f(x)$, has a y-intercept that can be found by evaluating $f(0)$. Locating the x-intercepts is not that easy of a task. Recall that the number $x = c$ is called a **zero** of a function f if $f(c) = 0$. If c is a real number, then c is an x-intercept (see Section 3.2). Therefore, to find the x-intercepts of a polynomial function $y = f(x)$, we must find the real solutions of the equation $f(x) = 0$.

Example 4 Find the Intercepts of a Polynomial Function

Find the intercepts of the polynomial function $f(x) = x^3 - x^2 - 4x + 4$.

Solution The polynomial function $f(x) = x^3 - x^2 - 4x + 4$ has a y-intercept at $f(0) = (0)^3 - (0)^2 - 4(0) + 4 = 4$. Notice that the y-intercept is the same as the constant coefficient.

In general, every polynomial function of the form

$$f(x) = a_n x^n + a_{n-1}x^{n-1} + a_{n-2}x^{n-2} + \cdots + a_1 x + a_0$$

has a y-intercept at the constant coefficient a_0.

To find the x-intercepts of $f(x) = x^3 - x^2 - 4x + 4$, we must find the real solutions of the equation $x^3 - x^2 - 4x + 4 = 0$. This equation can be solved by factoring and using the zero product property.

$x^3 - x^2 - 4x + 4 = 0$	Write the equation $f(x) = 0$.
$x^2(x - 1) - 4(x - 1) = 0$	Factor by grouping.
$(x - 1)(x^2 - 4) = 0$	Factor out the common factor of $(x - 1)$.
$(x - 1)(x - 2)(x + 2) = 0$	Factor $(x^2 - 4)$ using difference of squares.

By the zero product property, $x - 1 = 0$, $x - 2 = 0$, or $x + 2 = 0$. Thus, the zeros of $f(x) = x^3 - x^2 - 4x + 4$ are $x = 1$, $x = 2$, and $x = -2$. Because the zeros are all real numbers, $x = 1$, $x = 2$, and $x = -2$ are the x-intercepts. By determining the end behavior and plotting the intercepts, we begin to get an idea of what the graph of $f(x) = x^3 - x^2 - 4x + 4$ looks like. See Figure 14.

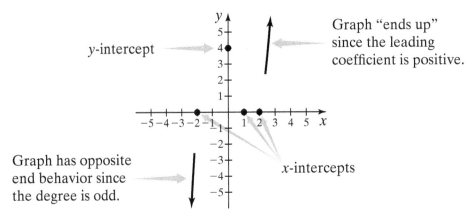

Figure 14 A portion of the graph of $f(x) = x^3 - x^2 - 4x + 4$

You Try It Work through the following You Try It problem.

Work Exercises 22–25 in this textbook or in the MyMathLab Study Plan.

OBJECTIVE 5 DETERMINING THE REAL ZEROS OF POLYNOMIAL FUNCTIONS AND THEIR MULTIPLICITIES

My video summary

In Example 4, we were able to factor the polynomial function $f(x) = x^3 - x^2 - 4x + 4$ into $f(x) = (x - 1)(x - 2)(x + 2)$. Once the polynomial function was in factored form, we were able to determine the three zeros $x = 1$, $x = 2$, and $x = -2$. In general, if $(x - c)$ is a factor of a polynomial function, then $x = c$ is a zero. The converse of the previous statement is also true. (If f is a polynomial function and $x = c$ is a zero, then $(x - c)$ is a factor.)

For example, consider the polynomial functions $f(x) = (x - 1)^2$ and $g(x) = (x - 1)^3$. Both functions have a real zero at $x = 1$ because $(x - 1)$ is a factor of both polynomials. For the function $f(x) = (x - 1)^2$, $x = 1$ is called a zero of multiplicity 2 because the factor $(x - 1)$ occurs two times in the factorization of f. Likewise, for $g(x) = (x - 1)^3$, $x = 1$ is a zero of multiplicity 3. Notice in Figure 15 that the graph of $f(x) = (x - 1)^2$ is *tangent* to (touches) the *x*-axis at $x = 1$, whereas the graph of $g(x) = (x - 1)^3$ *crosses* the *x*-axis at $x = 1$. The graph of a polynomial function will touch the *x*-axis at a real zero of even multiplicity and will cross the *x*-axis at a real zero of odd multiplicity.

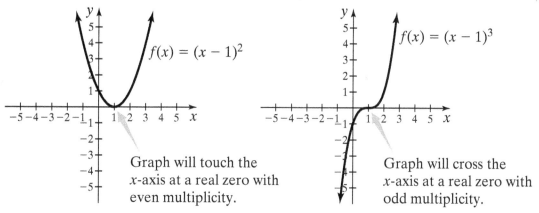

Graph will touch the *x*-axis at a real zero with even multiplicity.

Graph will cross the *x*-axis at a real zero with odd multiplicity.

Figure 15

Shape of the Graph of a Polynomial Function Near a Zero of Multiplicity k

Suppose c is a real zero of a polynomial function f of multiplicity k (where k is a positive integer); that is, $(x - c)^k$ is a factor of f. Then the shape of the graph of f near c is as follows:

If $k > 1$ is even, then the graph touches the *x*-axis at c.

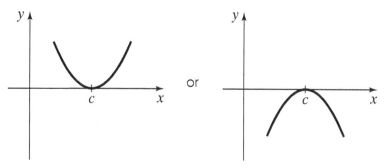

or

If $k \geq 1$ is odd, then the graph crosses the *x*-axis at c.

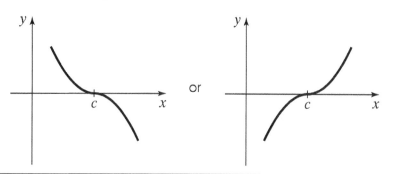

or

Example 5 Determine the Real Zeros (and Their Multiplicities) of a Polynomial Function

Find all real zeros of $f(x) = x(x^2 - 1)(x - 1)$. Determine the multiplicities of each zero, and decide whether the graph touches or crosses at each zero.

Solution Completely factor the polynomial function. We can use difference of squares to factor $x^2 - 1$.

$$f(x) = x(x^2 - 1)(x - 1) \qquad \text{Write the original function.}$$

$$f(x) = x(x + 1)(x - 1)(x - 1) \qquad \text{Factor using difference of squares.}$$

$$f(x) = x(x + 1)(x - 1)^2 \qquad \text{Rewrite } (x - 1)(x - 1) \text{ as } (x - 1)^2.$$

Because the first factor, x, is the same as $(x - 0)$, this implies that $x = 0$ is a zero of multiplicity 1. The next factor, $(x + 1)$, means that $x = -1$ is a zero of multiplicity 1.

The final factor, $(x - 1)^2$, means that $x = 1$ is a zero of multiplicity 2.

The graph of $f(x) = x(x + 1)(x - 1)^2$ must cross the x-axis at $x = 0$ and $x = -1$ and must touch the x-axis at $x = 1$. See the following calculator graph.

Using Technology

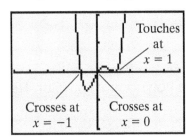

The graph of the function $f(x) = x(x + 1)(x - 1)^2$ is seen here using a graphing utility. Notice that the graph crosses the x-axis at $x = -1$ and $x = 0$. The graph touches the x-axis at $x = 1$. ⬤

You Try It Work through the following You Try It problem.

Work Exercises 26–33 in this textbook or in the MyMathLab Study Plan.

OBJECTIVE 6 SKETCHING THE GRAPH OF A POLYNOMIAL FUNCTION

If we are able to find the zeros of a polynomial function, we will be able to sketch an approximate graph. Earlier in this section, we were able to factor the polynomial function $f(x) = x^3 - x^2 - 4x + 4$ into $f(x) = (x - 1)(x - 2)(x + 2)$. Note that the three zeros $x = 1, x = 2$, and $x = -2$ have odd multiplicities. Therefore, the graph of f must cross the x-axis at each x-intercept. By determining the end behavior of this function and by plotting the x-intercepts and y-intercept, we can begin to sketch the graph as seen in Figure 16.

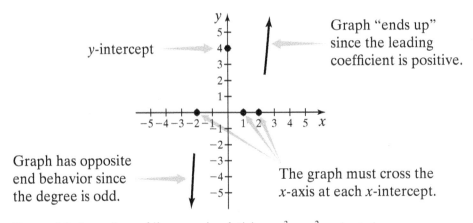

Figure 16 A portion of the graph of $f(x) = x^3 - x^2 - 4x + 4$

To complete the graph, we can create a table of values to plot additional points by choosing a test value between each of the zeros. Here, we choose $x = -1$ and $x = 1.5$. The approximate graph is sketched in Figure 17.

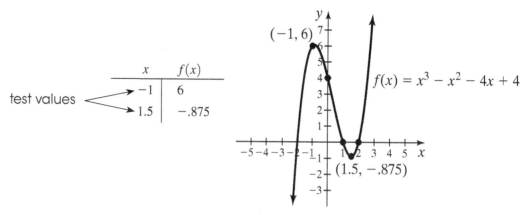

test values

x	$f(x)$
-1	6
1.5	$-.875$

Figure 17 The graph of $f(x) = x^3 - x^2 - 4x + 4$

We see from the graph in Figure 17 that there are two turning points. A polynomial function of degree n has at most $n - 1$ turning points. A turning point in which the graph changes from increasing to decreasing is also called a **relative maximum**. A turning point in which the graph changes from decreasing to increasing is called a **relative minimum**.

Without the use of calculus (or a graphing utility), there is no way to determine the precise coordinates of the relative minima and relative maxima. Using a graphing utility, we can use the Maximum and Minimum features to determine the coordinates of the relative maxima and minima as follows.

Using Technology 📷

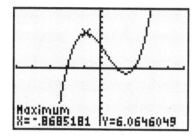

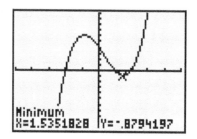

Most graphing calculators have a maximum and minimum feature that approximates the coordinates of a relative maximum or a relative minimum. The graphs above show the approximate coordinates of the relative maximum and relative minimum of $f(x) = x^3 - x^2 - 4x + 4$. Using calculus, it can be shown that the exact coordinates are $\left(\dfrac{1 - \sqrt{13}}{3}, \dfrac{70 + 26\sqrt{13}}{27} \right)$ and $\left(\dfrac{1 + \sqrt{13}}{3}, \dfrac{70 - 26\sqrt{13}}{27} \right)$, respectively.

Note Some texts use the terms *local maximum* and *local minimum* to describe a turning point.

We can summarize the sketching procedure with the following four-step process.

> **Four-Step Process for Sketching the Graph of a Polynomial Function**
>
> **Step 1.** Determine the end behavior.
>
> **Step 2.** Plot the y-intercept $f(0) = a_0$.
>
> **Step 3.** Completely factor f to find all real zeros and their multiplicities.
>
> **Step 4.** Choose a test value between each real zero and sketch the graph.

(Remember that without calculus, there is no way to determine the precise coordinates of the turning points.)

 *My interactive video summary*

⊙ **Example 6** Use the Four-Step Process to Sketch the Graph of a Polynomial Function

Use the four-step process to sketch the graphs of the following polynomial functions:

a. $f(x) = -2(x + 2)^2(x - 1)$

b. $f(x) = x^4 - 2x^3 - 3x^2$

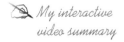 *My interactive video summary*

⊙ **Solution** The graphs of each polynomial are sketched in Figure 18. Work through the interactive video to see how to sketch these polynomial functions.

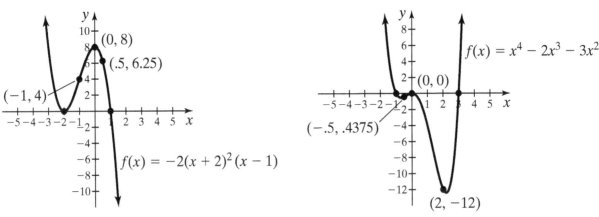

Figure 18 (a) Graph of $f(x) = -2(x + 2)^2(x - 1)$ (b) Graph of $f(x) = x^4 - 2x^3 - 3x^2$

You Try It Work through the following You Try It problem.

Work Exercises 34–44 in this textbook or in the MyMathLab Study Plan.

OBJECTIVE 7 DETERMINING A POSSIBLE EQUATION OF A POLYNOMIAL FUNCTION GIVEN ITS GRAPH

Now that we have identified the characteristics of a polynomial function and can create an approximate sketch of its graph, we should be able to analyze the graph of a polynomial function and describe features of its equation. The end behavior of the graph gives us information about the degree and leading coefficient, the y-intercept gives us the value of the constant coefficient, and the behavior of the graph at the x-intercepts gives us information about the multiplicity of the zeros. We can use this information to identify possible equations that the graph of a polynomial function might represent.

 *My interactive video summary*

⊙ **Example 7** Determine a Possible Equation of a Polynomial Function Given Its Graph

Analyze the graph to address the following about the polynomial function it represents.

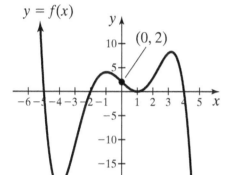

$y = f(x)$

a. Is the degree of the polynomial function even or odd?

b. Is the leading coefficient positive or negative?

c. What is the value of the constant coefficient?

d. Identify the real zeros, and state the multiplicity of each.

e. Select from this list a possible function that could be represented by this graph.

　i. $f(x) = -\dfrac{1}{20}(x + 5)(x + 2)(x - 1)^2(x - 4)$

　ii. $f(x) = -\dfrac{1}{800}(x + 5)^2(x + 2)^2(x - 1)(x - 4)^2$

　iii. $f(x) = \dfrac{1}{20}(x + 5)(x + 2)(x - 1)^2(x - 4)$

　iv. $f(x) = -\dfrac{1}{10}(x + 5)(x + 2)(x - 1)^2(x - 4)$

Solution

a. The degree of the polynomial function is odd because the graph "starts" and "finishes" in opposite directions.

b. The leading coefficient is negative because the right-hand behavior "finishes down."

c. The y-intercept is 2, so $f(0) = 2$.

d. The real zeros are -5, -2, 1, and 4. The zeros that have odd multiplicity are -5, -2, and 4 because the graph crosses the x-axis at these zeros. The only zero of even multiplicity is 1 because the graph touches the x-axis at $x = 1$.

My interactive video summary

⊙ e. The best choice is item i. Watch the interactive video for a more detailed explanation.

You Try It　Work through the following You Try It problem.

Work Exercises 45–47 in this textbook or in the My MathLab **Study Plan.**

4.3 Exercises

In Exercises 1–6, determine whether the function is a polynomial function. For each function that is a polynomial, identify the degree, the leading coefficient, and the constant coefficient.

1. $f(x) = \frac{1}{3}x^7 - 8x^3 + \sqrt{5}x - 8$

2. $g(x) = \frac{8x^3 - 5x + 3}{7x^2}$

3. $h(x) = 2x^{\frac{2}{3}} - 8x^2 + 5$

4. $f(x) = \frac{3x - 5x^3}{11}$

5. $G(x) = 7x^4 + 1 + 3x - 9x^5$

6. $f(x) = 12$

In Exercises 7–15, use the graph of a power function and transformations to sketch the graph of each polynomial function.

7. $f(x) = -x^5$

8. $f(x) = x^6 + 1$

9. $f(x) = -x^5 - 2$

10. $f(x) = \frac{1}{2}(x - 1)^4$

11. $f(x) = 2(x + 1)^5$

12. $f(x) = -(x - 2)^6$

13. $f(x) = (x - 2)^5 - 3$

14. $f(x) = -(x + 2)^4 + 1$

15. $f(x) = -(x - 1)^5 + 2$

In Exercises 16–21, use the end behavior of the graph of the polynomial function to

a. determine whether the degree is even or odd and

b. determine whether the leading coefficient is positive or negative.

16.

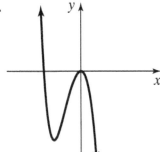

17.

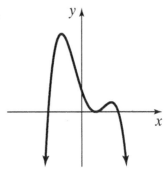

18.

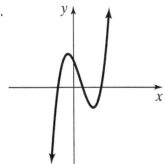

19.

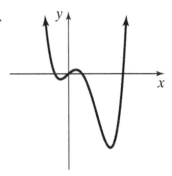

20.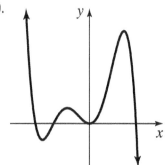

21.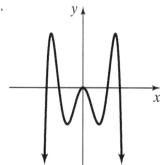

In Exercises 22–25, find the intercepts of each polynomial function.

22. $f(x) = x^2 - 3x - 10$

23. $f(x) = x^3 + 3x^2 - 16x - 48$

24. $f(x) = (x^2 + 1)\left(x - \dfrac{1}{2}\right)(x + 6)$

25. $f(x) = (x^2 - 9)(2 - x)$

In Exercises 26–33, determine the real zeros of each polynomial and their multiplicities. Then decide whether the graph touches or crosses the x-axis at each zero.

SbS 26. $f(x) = (x - 2)^2(x + 1)^3$

SbS 27. $f(x) = -(x + 1)(x + 3)^2$

SbS 28. $f(x) = x(x - 3)(x - 4)^4$

SbS 29. $f(x) = 3(x^2 - 2)^2$

SbS 30. $f(x) = -5(x - 2)(x^2 - 4)^3$

SbS 31. $f(x) = x^2(x^2 - x - 3)$

SbS 32. $f(x) = (x^2 - 3)(x + 1)^3$

SbS 33. $f(x) = -(x^2 - 7)^2(x + \sqrt{7})^3$

In Exercises 34–44, sketch each polynomial function using the four-step process.

SbS 34. $f(x) = x^2(x - 4)$

SbS 35. $f(x) = -x(x + 2)^2$

SbS 36. $f(x) = x^2(x + 2)(x - 2)$

SbS 37. $f(x) = -2(x - 1)^2(x + 1)$

SbS 38. $f(x) = -\dfrac{1}{2}x(x^2 - 4)^2$

SbS 39. $f(x) = x^4 + 2x^3 - 3x^2$

SbS 40. $f(x) = -2x^3 + 6x^2 + 8x$

SbS 41. $f(x) = x^3 + x^2 - 4x - 4$

SbS 42. $f(x) = -2x^3 - 4x^2 + 2x + 4$

SbS 43. $f(x) = x^4 + 2x^3 - 9x^2 - 18x$

SbS 44. $f(x) = -x^5 + 3x^4 + 9x^3 - 27x^2$

SbS 45. Analyze the graph to address the following about the polynomial function it represents.

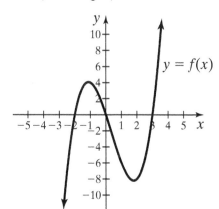

a. Is the degree of the polynomial function even or odd?

b. Is the leading coefficient positive or negative?

c. What is the value of the constant coefficient?

d. Identify the real zeros, and state the multiplicity of each.

e. Select from this list a possible function that could be represented by this graph.

i. $f(x) = -x(x + 2)(x - 3)$

ii. $f(x) = x(x + 2)(x - 3)$

iii. $f(x) = -x^3(x + 2)(x - 3)$

iv. $f(x) = (x + 2)(x - 3)$

SbS **46.** Analyze the graph to address the following about the polynomial function it represents.

a. Is the degree of the polynomial function even or odd?

b. Is the leading coefficient positive or negative?

c. What is the value of the constant coefficient?

d. Identify the real zeros, and state the multiplicity of each.

e. Select from this list a possible function that could be represented by this graph.

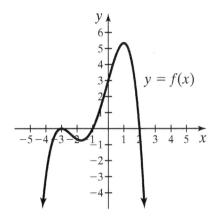

i. $f(x) = \dfrac{1}{4}(x + 3)(x + 1)^2(x - 2)^2$

ii. $f(x) = -\dfrac{1}{18}(x + 3)^3(x + 1)(x - 2)$

iii. $f(x) = -\dfrac{1}{6}(x + 3)^2(x - 1)(x + 2)$

iv. $f(x) = -\dfrac{1}{6}(x + 3)^2(x + 1)(x - 2)$

SbS **47.** Analyze the graph to address the following about the polynomial function it represents.

a. Is the degree of the polynomial function even or odd?

b. Is the leading coefficient positive or negative?

c. What is the value of the constant coefficient?

d. Identify the real zeros, and state the multiplicity of each.

e. Select from this list a possible function that could be represented by this graph.

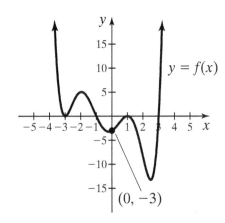

i. $f(x) = \dfrac{1}{6}(x + 3)^2(x + 1)(x - 1)^2(x - 3)$

ii. $f(x) = \dfrac{1}{9}(x + 3)^2(x + 1)(x - 1)^2(x - 3)$

iii. $f(x) = \dfrac{1}{3}(x + 3)^2(x + 1)(x - 1)^2(x - 3)$

iv. $f(x) = \dfrac{1}{12}(x + 3)^2(x + 1)(x - 1)^2(x - 3)$

Brief Exercises

In Exercises 48–58, sketch each polynomial function using the four-step process.

48. $f(x) = x^2(x - 4)$

49. $f(x) = -x(x + 2)^2$

50. $f(x) = x^2(x + 2)(x - 2)$

51. $f(x) = -2(x - 1)^2(x + 1)$

52. $f(x) = -\dfrac{1}{2}x(x^2 - 4)^2$

53. $f(x) = x^4 + 2x^3 - 3x^2$

54. $f(x) = -2x^3 + 6x^2 + 8x$

55. $f(x) = x^3 + x^2 - 4x - 4$

56. $f(x) = -2x^3 - 4x^2 + 2x + 4$

57. $f(x) = x^4 + 2x^3 - 9x^2 - 18x$

58. $f(x) = -x^5 + 3x^4 + 9x^3 - 27x^2$

In Exercises 59–61, select from a list a possible function that could be represented by the given graph.

59.

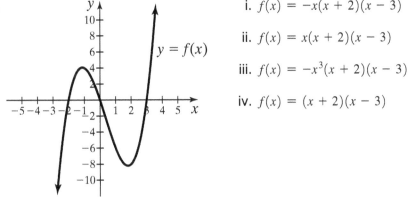

   i. $f(x) = -x(x + 2)(x - 3)$

   ii. $f(x) = x(x + 2)(x - 3)$

   iii. $f(x) = -x^3(x + 2)(x - 3)$

   iv. $f(x) = (x + 2)(x - 3)$

60.

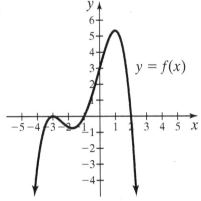

   i. $f(x) = \dfrac{1}{4}(x + 3)(x + 1)^2(x - 2)^2$

   ii. $f(x) = -\dfrac{1}{18}(x + 3)^3(x + 1)(x - 2)$

   iii. $f(x) = -\dfrac{1}{6}(x - 3)^2(x - 1)(x + 2)$

   iv. $f(x) = -\dfrac{1}{6}(x + 3)^2(x + 1)(x - 2)$

61.

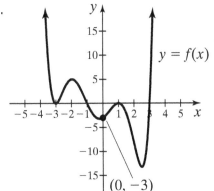

   i. $f(x) = \dfrac{1}{6}(x + 3)^2(x + 1)(x - 1)^2(x - 3)$

   ii. $f(x) = \dfrac{1}{9}(x + 3)^2(x + 1)(x - 1)^2(x - 3)$

   iii. $f(x) = \dfrac{1}{3}(x + 3)^2(x + 1)(x - 1)^2(x - 3)$

   iv. $f(x) = \dfrac{1}{12}(x + 3)^2(x + 1)(x - 1)^2(x - 3)$

4.4 Synthetic Division; The Remainder and Factor Theorems

THINGS TO KNOW

Before working through this section, be sure that you are familiar with the following concepts:

| | VIDEO | ANIMATION | INTERACTIVE |

 You Try It
1. Using Long Division to Divide Polynomials (Section R.4)

 You Try It
2. Simplifying Powers of i (Section 1.3)

You Try It
3. Determining the Intercepts of a Function (Section 3.2)

You Try It
4. Using Combinations of Transformations to Sketch Functions (Section 3.4)

You Try It
5. Sketching the Graph of a Polynomial Function (Section 4.3)

OBJECTIVES

1 Using the Division Algorithm
2 Using Synthetic Division
3 Using the Remainder Theorem
4 Using the Factor Theorem
5 Sketching the Graph of a Polynomial Function

OBJECTIVE 1   USING THE DIVISION ALGORITHM

Long division of polynomials was discussed in Section R.4. For instance, we saw in Example 7 of Section R.4 that if we divide $3x^4 + x^3 + 7x + 4$ by $x^2 - 1$, we get

$$\frac{3x^4 + x^3 + 7x + 4}{x^2 - 1} = 3x^2 + x + 3 + \frac{8x + 7}{x^2 - 1}.$$

Watch the video to see each step of the long division process. We see from this division that the quotient is $3x^2 + x + 3$ and the remainder is $8x + 7$. Letting $f(x) = 3x^4 + x^3 + 7x + 4$, $d(x) = x^2 - 1$, $q(x) = 3x^2 + x + 3$, and $r(x) = 8x + 7$, then

$$\frac{3x^4 + x^3 + 7x + 4}{x^2 - 1} = 3x^2 + x + 3 + \frac{8x + 7}{x^2 - 1} \text{ is equivalent to the equation}$$

$$\frac{f(x)}{d(x)} = q(x) + \frac{r(x)}{d(x)}.$$

If we multiply both sides of these equations by the divisor $d(x)$, we get

$$(x^2 - 1)\frac{3x^4 + x^3 + 7x + 4}{x^2 - 1} = (x^2 - 1)(3x^2 + x + 3) + \frac{8x + 7}{x^2 - 1}(x^2 - 1),$$

which is equivalent to

$$d(x)\frac{f(x)}{d(x)} \quad = \quad d(x)\,q(x) \quad + \frac{r(x)}{d(x)}d(x) \quad \text{or}$$

$$3x^4 + x^3 + 7x + 4 = (x^2 - 1)(3x^2 + x + 3) + (8x + 7), \text{ which is equivalent to}$$

$$f(x) = d(x)q(x) + r(x).$$

The equation $f(x) = d(x)q(x) + r(x)$ is used to check that the long division was done properly. The original polynomial should equal the product of the divisor and the quotient plus the remainder. This process is known as the **division algorithm**.

Division Algorithm

If $f(x)$ and $d(x)$ are polynomial functions with $d(x) \neq 0$ and the degree of $d(x)$ is less than or equal to the degree of $f(x)$, then unique polynomial functions $q(x)$ and $r(x)$ exist such that

$$f(x) = d(x)q(x) + r(x)$$

or equivalently

$$\frac{f(x)}{d(x)} = q(x) + \frac{r(x)}{d(x)},$$

where the remainder $r(x) = 0$ or is of degree less than the degree of $d(x)$.

Example 1 Use the Division Algorithm

Given the polynomials $f(x) = 4x^4 - 3x^3 + 5x - 6$ and $d(x) = x - 2$, find polynomial functions $q(x)$ and $r(x)$, and write $f(x)$ in the form of $f(x) = d(x)q(x) + r(x)$.

Solution We use long division inserting the term $0x^2$ into the dividend as a placeholder to make sure that all columns of the long division process line up properly.

$$
\begin{array}{r}
4x^3 + 5x^2 + 10x + 25 \\
x - 2 \overline{)4x^4 - 3x^3 + 0x^2 + 5x - 6} \\
\underline{4x^4 - 8x^3} \\
5x^3 + 0x^2 \\
\underline{5x^3 - 10x^2} \\
10x^2 + 5x \\
\underline{10x^2 - 20x} \\
25x - 6 \\
\underline{25x - 50} \\
44
\end{array}
$$

The process is complete because 44 is of lesser degree than the divisor, $x - 2$. Therefore, $q(x) = 4x^3 + 5x^2 + 10x + 25$ and $r(x) = 44$, so

$$4x^4 - 3x^3 + 5x - 6 = (x - 2)(4x^3 + 5x^2 + 10x + 25) + 44$$

Notice in Example 1 that the remainder, $r(x)$, was a constant. We know that the remainder must always be of degree less than the degree of the divisor. Therefore, if a polynomial $f(x)$ is divided by a first-degree polynomial, then the remainder must be a constant.

Corollary to the Division Algorithm

If a polynomial $f(x)$ of degree greater than or equal to 1 is divided by a polynomial $d(x)$ where $d(x)$ is of degree 1, then there exists a unique polynomial function $q(x)$ and a constant r such that

$$f(x) = d(x)q(x) + r.$$

OBJECTIVE 2 USING SYNTHETIC DIVISION

My video summary ⊙ If a polynomial $f(x)$ is divided by $d(x) = x - c$, then we can use a "shortcut" method" called **synthetic division** to find the quotient, $q(x)$, and the remainder, r. To illustrate synthetic division, we do a side-by-side comparison of long division and synthetic division using the polynomial function of Example 1.

Long Division

$$
\begin{array}{r}
4x^3 + 5x^2 + 10x + 25 \\
x - 2\overline{)4x^4 - 3x^3 + 0x^2 + 5x - 6} \\
\underline{4x^4 - 8x^3} \\
5x^3 + 0x^2 \\
\underline{5x^3 - 10x^2} \\
10x^2 + 5x \\
\underline{10x^2 - 20x} \\
25x - 6 \\
\underline{25x - 50} \\
44
\end{array}
$$

Synthetic Division

$$
\begin{array}{r|rrrrr}
2 & 4 & -3 & 0 & 5 & -6 \\
  &   & 8 & 10 & 20 & 50 \\
\hline
  & 4 & 5 & 10 & 25 & 44
\end{array}
$$

You can see how much more compact the synthetic division process is compared to long division. This is how it works.

Step 1. The constant coefficient, c, of the divisor $d(x) = x - c$ is written to the far left, whereas all coefficients of the polynomial $f(x)$ are written inside the symbol $\lfloor\underline{\quad}$. Once again, be sure to include the 0 for $0x^2$.

This is c in $x - c$. ⟶ $2\lfloor 4 \quad -3 \quad 0 \quad 5 \quad -6$

These are the coefficients of
$f(x) = 4x^4 - 3x^3 + 5x - 6.$

Step 2. Bring down the leading coefficient 4.

$$
\begin{array}{r|rrrrr}
2 & 4 & -3 & 0 & 5 & -6 \\
  & \downarrow & & & & \\
\hline
  & 4 & & & &
\end{array}
$$

Step 3. Multiply c times the leading coefficient that was just brought down. In this case, we multiply $2 \cdot 4$. The product (8 in this case) is written in the next column in the second row.

Multiply $2 \cdot 4 = 8$
$$
\begin{array}{r|rrrrr}
2 & 4 & -3 & 0 & 5 & -6 \\
  & \downarrow & 8 & & & \\
\hline
  & 4 & & & &
\end{array}
$$

Step 4. Add down the column and write the sum (5 in this case) in the bottom row.

$$\begin{array}{r|rrrr} 2 & 4 & -3 & 0 & 5 & -6 \\ & & 8 \\ \hline & 4 & 5 \end{array}$$

We now repeat this process by multiplying $c = 2$ times the value in the last row, always adding down the columns.

Multiply 2 · 5 = 10

$$\begin{array}{r|rrrrr} 2 & 4 & -3 & 0 & 5 & -6 \\ & & 8 & 10 \\ \hline & 4 & 5 \end{array}$$

Add 0 + 10 = 10

$$\begin{array}{r|rrrrr} 2 & 4 & -3 & 0 & 5 & -6 \\ & & 8 & 10 \\ \hline & 4 & 5 & 10 \end{array}$$

Multiply 2 · 10 = 20

$$\begin{array}{r|rrrrr} 2 & 4 & -3 & 0 & 5 & -6 \\ & & 8 & 10 & 20 \\ \hline & 4 & 5 & 10 \end{array}$$

Add 5 + 20 = 25

$$\begin{array}{r|rrrrr} 2 & 4 & -3 & 0 & 5 & -6 \\ & & 8 & 10 & 20 \\ \hline & 4 & 5 & 10 & 25 \end{array}$$

Multiply 2 · 25 = 50

$$\begin{array}{r|rrrrr} 2 & 4 & -3 & 0 & 5 & -6 \\ & & 8 & 10 & 20 & 50 \\ \hline & 4 & 5 & 10 & 25 \end{array}$$

Add −6 + 50 = 44

$$\begin{array}{r|rrrrr} 2 & 4 & -3 & 0 & 5 & -6 \\ & & 8 & 10 & 20 & 50 \\ \hline & 4 & 5 & 10 & 25 & 44 \end{array}$$

The last row now represents the quotient and the remainder. The last entry of the bottom row (44 in this case) is the remainder. The other numbers of the last row represent the coefficients of the quotient.

$$\begin{array}{r|rrrrr} 2 & 4 & -3 & 0 & 5 & -6 \\ & & 8 & 10 & 20 & 50 \\ \hline & 4 & 5 & 10 & 25 & 44 \end{array}$$

coefficients of $q(x) = 4x^3 + 5x^2 + 10x + 25$ ⟵ remainder $r = 44$

You can watch the **video** to see this example worked out. Whenever possible, synthetic division is used throughout the rest of this chapter.

⚠ **Synthetic division can only be used when the divisor $d(x)$ has the form $(x - c)$.**

 *My video summary*

▶ **Example 2** Use Synthetic Division to Find the Quotient and Remainder

Use synthetic division to divide $f(x)$ by $(x - c)$, and then write $f(x)$ in the form $f(x) = (x - c)q(x) + r$ for $f(x) = -2x^4 + 3x^3 + 7x^2 - x + 5$ divided by $x + 1$.

Solution The divisor $x + 1$ or $x - (-1)$ is of the form $x - c$ with $c = -1$ so we can use synthetic division. We start by writing $c = -1$ on the far left, putting the coefficients of f on the inside and bringing down the -2:

$$\begin{array}{r|rrrrr} -1 & -2 & 3 & 7 & -1 & 5 \\ & & \downarrow \\ \hline & -2 \end{array}$$

Using a series of multiplications and additions, we can complete the synthetic division process:

$$\begin{array}{r|rrrrr} -1 & -2 & 3 & 7 & -1 & 5 \\ & \downarrow & 2 & -5 & -2 & 3 \\ \hline & -2 & 5 & 2 & -3 & 8 \end{array}$$

Watch the video to see each step. We can see from the last entry of the bottom row that the remainder is 8. Thus,

$$q(x) = -2x^3 + 5x^2 + 2x - 3 \quad \text{and} \quad r = 8, \text{ so}$$

$$f(x) = -2x^4 + 3x^3 + 7x^2 - x + 5$$

$$= (x + 1)(-2x^3 + 5x^2 + 2x - 3) + 8$$

You Try It Work through the following You Try It problem.

Work Exercises 1–10 in this textbook or in the MyMathLab Study Plan.

OBJECTIVE 3 USING THE REMAINDER THEOREM

In Example 2, we see that the remainder is 8 when $f(x) = -2x^4 + 3x^3 + 7x^2 - x + 5$ is divided by $x + 1$. If we evaluate f at $x = -1$, we get

$$f(-1) = -2(-1)^4 + 3(-1)^3 + 7(-1)^2 - (-1) + 5$$

$$= -2(1) + 3(-1) + 7(1) + 1 + 5$$

$$= -2 - 3 + 7 + 1 + 5 = 8$$

It is no coincidence that $f(-1)$ is equal to the remainder. If we rewrite $f(x) = -2x^4 + 3x^3 + 7x^2 - x + 5$ in the form of $f(x) = (x + 1)(-2x^3 + 5x^2 + 2x - 3) + 8$ as in Example 2, we see that

$$f(-1) = \underbrace{(-1 + 1)(-2(-1)^3 + 5(-1)^2 + 2(-1) - 3) + 8}$$

$$= \qquad\qquad 0 \qquad\qquad\qquad + 8 = 8$$

Therefore, $f(-1) = 8$, and we can conclude that if a polynomial is divided by $x - c$, then the remainder is the value of $f(c)$. This is known as the **remainder theorem**.

Remainder Theorem

If a polynomial $f(x)$ is divided by $x - c$, then the remainder is $f(c)$.

To see why the remainder theorem is true, read this proof.

My video summary ⊙ **Example 3** Use the Remainder Theorem

Use the remainder theorem to find the remainder when $f(x)$ is divided by $x - c$.

a. $f(x) = 5x^4 - 8x^2 + 3x - 1; x - 2$

b. $f(x) = 3x^3 + 5x^2 - 5x - 6; x + 2$

Solution

a. We could use long division or synthetic division to find the remainder, but the remainder theorem says that the remainder must be $f(2)$.

$$f(2) = 5(2)^4 - 8(2)^2 + 3(2) - 1 = 80 - 32 + 6 - 1 = 53$$

Thus, the remainder is 53.

b. To find the remainder when $f(x) = 3x^3 + 5x^2 - 5x - 6$ is divided by $x + 2 = x - (-2)$, we must evaluate $f(-2)$.

$$f(-2) = 3(-2)^3 + 5(-2)^2 - 5(-2) - 6$$

$$= -24 + 20 + 10 - 6 = 0$$

Therefore, the remainder is 0.

You Try It   Work through the following You Try It problem.

Work Exercises 11–16 in this textbook or in the MyMathLab Study Plan.

OBJECTIVE 4   USING THE FACTOR THEOREM

In Example 3, we see that when $f(x) = 3x^3 + 5x^2 - 5x - 6$ is divided by $x + 2$, then the remainder is 0. The remainder of 0 implies that $x + 2$ divides $f(x)$ *evenly*, and therefore, $x + 2$ is a factor of f. In fact, by the remainder theorem, if $f(c) = 0$, then the remainder when a polynomial $f(x)$ is divided by $x - c$ must be zero. Conversely, if $x - c$ is a factor of $f(x)$, then $f(c)$ must equal zero. The last two statements can be rewritten as the following important theorem known as the factor theorem.

> **Factor Theorem**
>
> The polynomial $x - c$ is a factor of the polynomial $f(x)$ if and only if $f(c) = 0$.

My video summary   ⊙ **Example 4** Use the Factor Theorem to Determine whether $x - c$ Is a Factor of $f(x)$

Determine whether $x + 3$ is a factor of $f(x) = 2x^3 + 7x^2 + 2x - 3$.

Solution By the factor theorem, $x + 3$ is a factor of $f(x)$ if and only if $f(-3) = 0$. We can use synthetic division to determine the value of $f(-3)$.

$$
\begin{array}{r|rrrr}
-3 & 2 & 7 & 2 & -3 \\
   & \downarrow & -6 & -3 & 3 \\
\hline
   & 2 & 1 & -1 & 0 \longleftarrow f(-3) = 0
\end{array}
$$

The remainder $f(-3)$ is equal to zero, and therefore, $x + 3$ is a factor of $f(x)$. Furthermore, using the bottom row of the synthetic division, we can rewrite $f(x)$ as $f(x) = (x + 3)(2x^2 + x - 1)$.

You Try It   Work through the following You Try It problem.

Work Exercises 17–26 in this textbook or in the MyMathLab Study Plan.

OBJECTIVE 5   SKETCHING THE GRAPH OF A POLYNOMIAL FUNCTION

In Example 4, we show that $x + 3$ is a factor of $f(x) = 2x^3 + 7x^2 + 2x - 3$. Using synthetic division and the division algorithm, we were able to rewrite $f(x)$ as $f(x) = (x + 3)(2x^2 + x - 1)$. Because the quadratic function $2x^2 + x - 1$ can also be factored, we can now completely factor $f(x)$ as

$f(x) = 2x^3 + 7x^2 + 2x - 3$ Write the original polynomial.

$f(x) = (x + 3)(2x^2 + x - 1)$ Because $f(-3) = 0, x + 3$ is a factor (See Example 4).

$f(x) = (x + 3)(2x - 1)(x + 1)$ Factor $2x^2 + x - 1$ as $(2x - 1)(x + 1)$.

$f(x) = 2(x + 3)\left(x - \dfrac{1}{2}\right)(x + 1)$ Factor $(2x - 1)$ as $2\left(x - \dfrac{1}{2}\right)$.

The polynomial $f(x) = 2(x + 3)\left(x - \dfrac{1}{2}\right)(x + 1)$ is now said to be written in completely factored form. Because the polynomial is in completely factored form, we can follow the four-step process discussed in Section 4.3 to sketch the graph of $f(x)$. The graph of $f(x) = 2(x + 3)\left(x - \dfrac{1}{2}\right)(x + 1)$ is sketched in Figure 19. Watch this video to see how to use the four-step process to sketch this polynomial.

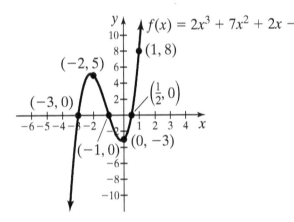

Figure 19 Graph of $f(x) = 2x^3 + 7x^2 + 2x - 3$

My video summary   ▶ **Example 5** Factor Completely and Sketch the Graph of a Polynomial Function

Given that $x = 2$ is a zero of $f(x) = x^3 - 6x + 4$, completely factor f and sketch its graph.

Solution We begin by using the synthetic division process to divide $f(x) = x^3 - 6x + 4$ by $x - 2$, remembering to use a 0 for the coefficient of the x^2-term.

$$
\begin{array}{r|rrrr}
2 & 1 & 0 & -6 & 4 \\
  & \downarrow & 2 & 4 & -4 \\
\hline
  & 1 & 2 & -2 & 0
\end{array}
$$

We see from the synthetic division process that we can rewrite $f(x)$ as $f(x) = (x - 2)(x^2 + 2x - 2)$. Because the quadratic $x^2 + 2x - 2$ does not factor with integer coefficients, we can use the quadratic formula or complete the square to find the zeros of $g(x) = x^2 + 2x - 2$. The zeros of $g(x) = x^2 + 2x - 2$ are $x = -1 + \sqrt{3} \approx .732$ and $x = -1 - \sqrt{3} \approx -2.732$. Because $x = -1 + \sqrt{3}$ is a zero of $g(x) = x^2 + 2x - 2$, this implies that $x - (-1 + \sqrt{3})$ is a factor of $x^2 + 2x - 2$. Similarly, $x - (-1 - \sqrt{3})$ is also a factor of $x^2 + 2x - 2$.

Thus, the complete factorization of f is $f(x) = (x - 2)\big(x - (-1 + \sqrt{3})\big)\big(x - (-1 - \sqrt{3})\big)$. Once again, we can now use the four-step process to sketch the graph seen in Figure 20. Watch the video to see this example worked out in detail.

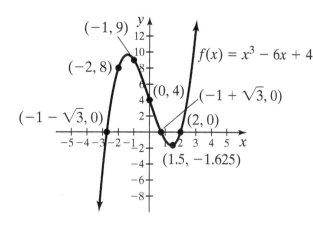

Figure 20 Graph of $f(x) = x^3 - 6x + 4$

You Try It   Work through the following You Try It problem.

Work Exercises 27–34 in this textbook or in the MyMathLab Study Plan.

4.4 Exercises

Skill Check Exercise

SCE-1. Find the quotient and remainder when $f(x) = 9x^4 - 2x^2 + 2x + 9$ is divided by $x^2 + 5$.

In Exercises 1–10, use synthetic division to divide $f(x)$ by $x - c$, and then write $f(x)$ in the form $f(x) = (x - c)q(x) + r$.

1. $f(x) = x^3 - 3x^2 + x - 5; x - 1$

2. $f(x) = x^3 + 5x^2 - 3x - 1; x + 1$

3. $f(x) = 6x^3 - 2x + 3; x - 2$

4. $f(x) = 2x^3 - x^2 + 4; x + 3$

5. $f(x) = x^4 + x^3 - x^2 - x - 4; x - 2$

6. $f(x) = -2x^4 + 3x^3 - x^2 + 7x + 2; x + 1$

7. $f(x) = -x^4 - x^3 - 5x; x + 4$

8. $f(x) = 3x^5 - x^2 - 2; x + 1$

9. $f(x) = x^3 - 1; x - 1$

10. $f(x) = x^5 + 1; x + 1$

In Exercises 11–16, use synthetic division and the remainder theorem to find the remainder when $f(x)$ is divided by $x - c$.

11. $f(x) = 2x^3 - 3x^2 - 5x - 7; x - 1$

12. $f(x) = x^4 + 3x^3 - 8x^2 - x + 2; x + 1$

13. $f(x) = x^5 - 2x^2 + x + 6; x - 3$

14. $f(x) = -4x^3 + 2x - 1; x + 3$

15. $f(x) = x^3 - 3x^2 + x - 5; x - i$

16. $f(x) = x^4 - x^3 - 2x^2 - x - 3; x + i$

In Exercises 17–26, use synthetic division and the factor theorem to determine whether $x - c$ is a factor of $f(x)$.

17. $f(x) = 2x^3 - 5x^2 + 7x - 4; x - 1$

18. $f(x) = 2x^3 - 5x^2 + 7x - 4; x + 1$

19. $f(x) = x^5 + x^4 - 3x^3 - 3x^2 - 4x - 4; x - 2$

20. $f(x) = x^5 + x^4 - 3x^3 - 3x^2 - 4x - 4; x + 2$

21. $f(x) = 2x^3 + 5x^2 - 8x - 5; x - \dfrac{1}{2}$

22. $f(x) = 2x^3 + 5x^2 - 8x - 5; x + \dfrac{1}{2}$

23. $f(x) = x^4 + x^3 + x^2 - 2x - 6; x - \sqrt{2}$

24. $f(x) = x^4 + x^3 + x^2 - 2x - 6; x + \sqrt{2}$

25. $f(x) = 3x^4 - x^3 - x^2 - x - 4; x - i$

26. $f(x) = x^3 + x^2 + 3x - 5; x - (1 + 2i)$

In Exercises 27–34, find the remaining zeros of $f(x)$ given that c is a zero. Then rewrite $f(x)$ in completely factored form and sketch its graph.

SbS 27. $f(x) = x^3 + 2x^2 - x - 2; c = 1$ is a zero

SbS 28. $f(x) = -x^3 + 5x^2 - 3x - 9; c = -1$ is a zero

SbS 29. $f(x) = 2x^3 + 7x^2 + 2x - 3; c = \dfrac{1}{2}$ is a zero

SbS 30. $f(x) = \dfrac{1}{2}x^3 - \dfrac{3}{2}x - 1; c = 2$ is a zero

SbS 31. $f(x) = -\dfrac{1}{3}x^3 - \dfrac{4}{3}x^2 - \dfrac{1}{3}x + 2; c = -3$ is a zero

SbS **32.** $f(x) = 3x^3 + 11x^2 + 8x - 4$; $c = -2$ is a zero of multiplicity 2

SbS **33.** $f(x) = 4x^4 + 5x^3 - 3x^2 - 5x - 1$; $c = -1$ is a zero of multiplicity 2

SbS **34.** $f(x) = 3x^5 - 5x^4 - 2x^3 + 6x^2 - x - 1$; $c = 1$ is a zero of multiplicity 3

Brief Exercises

In Exercises 35–42, find the remaining zeros of $f(x)$ given that c is a zero. Then rewrite $f(x)$ in completely factored form.

35. $f(x) = x^3 + 2x^2 - x - 2$; $c = 1$ is a zero

36. $f(x) = -x^3 + 5x^2 - 3x - 9$; $c = -1$ is a zero

37. $f(x) = 2x^3 + 7x^2 + 2x - 3$; $c = \dfrac{1}{2}$ is a zero

38. $f(x) = \dfrac{1}{2}x^3 - \dfrac{3}{2}x - 1$; $c = 2$ is a zero

39. $f(x) = -\dfrac{1}{3}x^3 - \dfrac{4}{3}x^2 - \dfrac{1}{3}x + 2$; $c = -3$ is a zero

40. $f(x) = 3x^3 + 11x^2 + 8x - 4$; $c = -2$ is a zero of multiplicity 2

41. $f(x) = 4x^4 + 5x^3 - 3x^2 - 5x - 1$; $c = -1$ is a zero of multiplicity 2

42. $f(x) = 3x^5 - 5x^4 - 2x^3 + 6x^2 - x - 1$; $c = 1$ is a zero of multiplicity 3

4.5 The Zeros of Polynomial Functions; The Fundamental Theorem of Algebra

THINGS TO KNOW

Before working through this section, be sure that you are familiar with the following concepts:

		VIDEO	ANIMATION	INTERACTIVE

You Try It
1. Sketching the Graph of a Polynomial Function (Section 4.3)

You Try It
2. Using the Remainder Theorem (Section 4.4)

You Try It
3. Using the Factor Theorem (Section 4.4)

You Try It
4. Sketching the Graph of a Polynomial Function (Section 4.4)

OBJECTIVES

1 Using the Rational Zeros Theorem

2 Using Descartes' Rule of Signs

3 Finding the Zeros of a Polynomial Function

4 Solving Polynomial Equations

5 Using the Complex Conjugate Pairs Theorem

6 Using the Intermediate Value Theorem

7 Sketching the Graphs of Polynomial Functions

Introduction to Section 4.5

The goal of this section is to be able to find the zeros of polynomial functions, or equivalently, we are interested in solving polynomial equations of the form $f(x) = 0$. To begin our discussion, we must take a look at a very important theorem that will help us create a list of potential zeros of a polynomial function.

OBJECTIVE 1   USING THE RATIONAL ZEROS THEOREM

In Example 5 of Section 4.4, we were able to find all real zeros of $f(x) = x^3 - 6x + 4$ given that $x = 2$ was a zero. You may want to work through this video to review this example. Because we knew that $x = 2$ was a zero, we were able to use the factor theorem and synthetic division to rewrite f as $f(x) = (x - 2)(x^2 + 2x - 2)$. We then solved the quadratic equation $x^2 + 2x - 2 = 0$ to find the other two zeros.

But what if we were not given the fact that $x = 2$ was a zero? How do we start looking for the zeros? Is there a systematic way to determine possible zeros of a polynomial function? The answer to this question is yes *if* we are given a polynomial with **integer** coefficients. If a polynomial has integer coefficients, then we are able to create a list of the **potential** rational zeros using the **rational zeros theorem**.

Rational Zeros Theorem

Let f be a polynomial function of the form $f(x) = a_n x^n + a_{n-1} x^{n-1} + a_{n-2} x^{n-2} + \ldots$
$+ a_1 x + a_0$ of degree $n \geq 1$, where each coefficient is an integer. If $\dfrac{p}{q}$ is

a rational zero of f (where $\dfrac{p}{q}$ is written in lowest terms), then p must be a

factor of the constant coefficient, a_0, and q must be a factor of the leading coefficient, a_n.

 The rational zeros theorem can only provide us with a list of possible rational zeros. It *does not* guarantee that the polynomial will have a zero from the list. Simply stated, if a polynomial with integer coefficients has a rational zero, then it must be on the list created using this theorem. Example 1 illustrates how this theorem works.

Example 1 Use the Rational Zeros Theorem

Use the rational zeros theorem to determine the potential rational zeros of each polynomial function.

$$f(x) = 4x^4 - 7x^3 + 9x^2 - x - 10$$

Solution Because f has integer coefficients, we can use the rational zeros theorem to create a list of **potential** rational zeros. The factors of the constant coefficient, -10, are $p = \pm 1, \pm 2, \pm 5, \pm 10$, whereas the factors of the leading coefficient, 4, are $q = \pm 1, \pm 2, \pm 4$. Listing all possibilities of $\dfrac{p}{q}$, we get

$$\frac{p}{q}: \pm 1, \ \pm 2, \ \pm 5, \ \pm 10, \ \pm \frac{1}{2}, \ \pm \frac{1}{4}, \ \pm \frac{5}{2}, \ \pm \frac{5}{4}.$$

Thus, if f has a rational zero, then it must appear on this list.

You Try It Work through the following You Try It problem.

Work Exercises 1–6 in this textbook or in the MyMathLab Study Plan.

OBJECTIVE 2   USING DESCARTES' RULE OF SIGNS

The French Mathematician René Descartes first described a rule for determining the number of positive or negative real zeros of a polynomial function. The rule, which is known as **Descartes' Rule of Signs**, is based on the number of variations in signs in a given polynomial function. A variation in sign occurs when successive nonzero terms of a polynomial function change from positive to negative or from negative to positive. Note that the term *nonzero* indicates that we do not consider terms with a zero coefficient.

Descartes' Rule of Signs

Let f be a polynomial function with real coefficients written in descending order.

1. The number of *positive* real zeros of f is equal to the number of variations in sign of $f(x)$ or is less than the number of variations in sign of $f(x)$ by a positive even integer.

   If $f(x)$ has one variation in sign, then f has exactly one positive real zero.

2. The number of *negative* real zeros of f is equal to the number of variations in sign of $f(-x)$ or is less than the number of variations in sign of $f(-x)$ by a positive even integer.

   If $f(-x)$ has one variation in sign, then f has exactly one negative real zero.

My video summary

▶ Example 2 Use Descartes' Rule of Signs

Determine the possible number of positive real zeros and negative real zeros of each polynomial function.

a. $f(x) = 2x^5 + 3x^4 + 8x^2 + 2x$ 　　　　**b.** $f(x) = 6x^4 + 13x^3 + 61x^2 + 8x - 10$

Solution

a. The polynomial function $f(x) = 2x^5 + 3x^4 + 8x^2 + 2x$ is written in descending order and has real coefficients, thus we can apply Descartes' Rule of Signs.

1. All coefficients of $f(x)$ are positive. Therefore, there are no variations in sign. Therefore, there are no positive real zeros.

2. To determine the number of possible negative real zeros, we must determine the number of variations in sign of $f(-x)$. We obtain $f(-x)$ by replacing x with $-x$ in the given polynomial function.

$$f(-x) = 2(-x)^5 + 3(-x)^4 + 8(-x)^2 + 2(-x) = -2x^5 + 3x^4 + 8x^2 - 2x$$

We now count the number of variations in sign of $f(-x)$.

$$f(-x) = \underbrace{-2x^5 + 3x^4}_{\substack{\text{variation} \\ \text{in sign}}} + \underbrace{8x^2 - 2x}_{\substack{\text{variation} \\ \text{in sign}}}$$

There are two variations in sign. Therefore there are 2 or $2 - 2 = 0$ possible negative real zeros.

b. The polynomial function $f(x) = 6x^4 + 13x^3 + 61x^2 + 8x - 10$ is written in descending order and has real coefficients, thus we can apply Descartes' Rule of Signs.

1. Count the number of variations in sign of $f(x)$.

$$f(x) = 6x^4 + 13x^3 + 61x^2 \underbrace{+ 8x - 10}_{\substack{\text{variation} \\ \text{in sign}}}$$

There is one variation in sign. Therefore, f has exactly one positive real zero.

2. We now count the number of variations in sign of $f(-x)$.

$$f(-x) = 6(-x)^4 + 13(-x)^3 + 61(-x)^2 + 8(-x) - 5$$

$$= \underbrace{6x^4 -}_{\substack{\text{variation} \\ \text{in sign}}} \underbrace{13x^3 +}_{\substack{\text{variation} \\ \text{in sign}}} \underbrace{61x^2 -}_{\substack{\text{variation} \\ \text{in sign}}} 8x - 5$$

There are three variations in sign. Therefore, there are 3 or $3 - 2 = 1$ possible negative real zeros.

You Try It Work through this You Try It problem.

Work exercises 7-10 in this textbook or in the MyMathLab Study Plan.

OBJECTIVE 3 FINDING THE ZEROS OF A POLYNOMIAL FUNCTION

Before we begin to find the zeros of polynomial functions, we should know how many zeros to expect. The fundamental theorem of algebra says that every polynomial of degree $n \geq 1$ has at least one complex zero.

Fundamental Theorem of Algebra

Every polynomial function of degree $n \geq 1$ has at least one complex zero.

Suppose $f(x)$ is a polynomial function of degree $n \geq 1$. By the fundamental theorem of algebra, f has at least one complex zero, call it c_1. By the factor theorem, $x - c_1$ is a factor of $f(x)$, and it follows that we can rewrite $f(x)$ as $f(x) = (x - c_1)q_1(x)$, where $q_1(x)$ is another polynomial. If $q_1(x)$ is of degree 1 or more, then we repeat the process again and rewrite $f(x)$ as $f(x) = (x - c_1)(x - c_2)q_2(x)$, where c_2 is a zero of $q_1(x)$. If $f(x)$ is a degree n polynomial, then we can repeat this process a total of n times to rewrite $f(x)$ as $f(x) = a(x - c_1)(x - c_2)(x - c_3) \cdots (x - c_n)$, where $c_1, c_2, c_3 \ldots c_n$ are zeros of $f(x)$ and a is the leading coefficient.

Therefore, every polynomial of degree n has n complex zeros and can be written in completely factored form. Note that some of the zeros could be the same and need to be counted each time, as indicated by the following theorem.

Number of Zeros Theorem

Every polynomial of degree n has n complex zeros provided each zero of multiplicity greater than 1 is counted accordingly.

 *My video summary*

⊙ Example 3 Find the Zeros of a Polynomial Function

Find all zeros of $f(x) = 6x^4 + 13x^3 + 61x^2 + 8x - 10$ and rewrite $f(x)$ in completely factored form.

Solution This is a degree 4 polynomial so there must be four complex zeros. The factors of the constant coefficient, -10, are $p = \pm 1, \pm 2, \pm 5, \pm 10$. The factors of the leading coefficient, 6, are $q = \pm 1, \pm 2, \pm 3, \pm 6$. By the rational zeros theorem, the list of possible rational zeros are

$$\frac{p}{q}: \pm 1, \pm 2, \pm 5, \pm 10, \pm \frac{1}{2}, \pm \frac{1}{3}, \pm \frac{1}{6}, \pm \frac{2}{3}, \pm \frac{5}{2}, \pm \frac{5}{3}, \pm \frac{5}{6}, \pm \frac{10}{3}.$$

Using Descartes' Rule of Signs we know that there must be exactly one positive real zero and either three or one negative real zeros. See Example 2b.

We go through the list of possible rational zeros looking for the one positive real zero. Remember, if $f\left(\dfrac{p}{q}\right) = 0$, then $\dfrac{p}{q}$ is a zero.

Eventually getting to $\dfrac{1}{3}$ and using synthetic division, we see that we obtain a zero.

$$
\begin{array}{r|rrrrr}
\frac{1}{3} & 6 & 13 & 61 & 8 & -10 \\
& \downarrow & 2 & 5 & 22 & 10 \\
\hline
& 6 & 15 & 66 & 30 & 0
\end{array}
$$

$\longleftarrow f(\frac{1}{3}) = 0$, so $\frac{1}{3}$ is a zero.

Thus, we can rewrite f as $f(x) = (x - \frac{1}{3})(6x^3 + 15x^2 + 66x + 30)$. We now try to find a zero of the polynomial $q_1(x) = 6x^3 + 15x^2 + 66x + 30$. There is no need to search for any more positive zeros since we have already found the one and only positive real zero of $f(x)$. Therefore, we have eliminated half of our list of possible rational zeros. The new list becomes

$$\frac{p}{q}: -1, -2, -5, -10, -\frac{1}{2}, -\frac{1}{3}, -\frac{1}{6}, -\frac{2}{3}, -\frac{5}{2}, -\frac{5}{3}, -\frac{5}{6}, -\frac{10}{3}.$$

Eventually getting to $-\dfrac{1}{2}$ and using synthetic division, we see that we obtain another zero.

$$
\begin{array}{r|rrrr}
-\dfrac{1}{2} & 6 & 15 & 66 & 30 \\
 & \downarrow & -3 & -6 & -30 \\
\hline
 & 6 & 12 & 60 & 0
\end{array}
$$
$\longleftarrow q_1\!\left(-\tfrac{1}{2}\right) = 0$, so $-\tfrac{1}{2}$ is a zero.

The remainder when $q_1(x) = 6x^3 + 15x^2 + 66x + 30$ is divided by $x + \dfrac{1}{2}$ is zero. This

implies that $-\dfrac{1}{2}$ is a zero. Using the bottom row of the synthetic division we see that

$f(x) = \left(x - \dfrac{1}{3}\right)(6x^3 + 15x^2 + 66x + 30) = \left(x - \dfrac{1}{3}\right)\left(x + \dfrac{1}{2}\right)(6x^2 + 12x + 60)$. Factoring

out a 6 from $6x^2 + 12x + 60$ we can rewrite $f(x)$ as $f(x) = 6\left(x - \dfrac{1}{3}\right)\left(x + \dfrac{1}{2}\right)(x^2 + 2x + 10)$.

We must now solve the quadratic equation $x^2 + 2x + 10 = 0$ to find the remaining two zeros. Solving either by completing the square or by using the quadratic

formula we get $x = -1 \pm 3i$. Thus the four zeros are $\dfrac{1}{3}, -\dfrac{1}{2}, -1 + 3i$ and $-1 - 3i$.

The completely factored form of this polynomial is

$f(x) = 6\left(x - \dfrac{1}{3}\right)\left(x + \dfrac{1}{2}\right)(x - (-1 + 3i))(x - (-1 - 3i))$.

My video summary

⊙ **Note** Once we factor a polynomial into the product of linear factors and a quadratic function of the form $f(x) = (x - c_1)(x - c_2)(x - c_3) \cdots (x - c_k)$ $(ax^2 + bx + c)$, we need simply to solve the quadratic equation $ax^2 + bx + c = 0$ to find the remaining two zeros. For example, if we want to find all zeros of $g(x) = 2x^3 - 3x^2 + 4x - 3$, we need only try to find *one* zero from the list created by the rational zeros theorem and then find the two zeros of the remaining quadratic function. See if you can find the zeros of $g(x) = 2x^3 - 3x^2 + 4x - 3$. Watch the video to see if you are correct!

You Try It Work through the following You Try It problem.

Work Exercises 11–19 in this textbook or in the MyMathLab Study Plan.

OBJECTIVE 4 SOLVING POLYNOMIAL EQUATIONS

Solving polynomial equations of the form $f(x) = 0$ is equivalent to finding the zeros of $f(x)$. For example, the solutions to the equation $6x^4 + 13x^3 + 61x^2 + 8x - 10 = 0$ are exactly the zeros of the polynomial function $f(x) = 6x^4 + 13x^3 + 61x^2 + 8x - 10$. Using the result from Example 2, we conclude that the real solutions of this equation

are $-\dfrac{1}{2}$ and $\dfrac{1}{3}$, whereas the two nonreal solutions are $-1 + 3i$ and $-1 - 3i$.

 **My video summary**

> **Example 4** Solve a Polynomial Equation

Solve $5x^5 - 9x^4 + 23x^3 - 35x^2 + 12x + 4 = 0$.

Solution Let $f(x) = 5x^5 - 9x^4 + 23x^3 - 35x^2 + 12x + 4$. The solutions to this equation must be the zeros of f. Using the rational zeros theorem, we see that the potential rational zeros are $\pm 1, \pm 2, \pm 4, \pm\frac{1}{5}, \pm\frac{2}{5}, \pm\frac{4}{5}$. Our goal is to attempt to find three zeros from this list and then solve the remaining quadratic to find the last two zeros. Watch the video to verify that the rational zeros of f are $x = 1$ (of multiplicity 2) and $-\frac{1}{5}$. Using synthetic division and factoring out the leading coefficient, we can rewrite f as $f(x) = 5(x - 1)^2\left(x + \frac{1}{5}\right)(x^2 + 4)$. The solutions to the quadratic equation $x^2 + 4 = 0$ are $\pm 2i$. Thus, the equation $5x^5 - 9x^4 + 23x^3 - 35x^2 + 12x + 4 = 0$ is equivalent to $5(x - 1)^2\left(x + \frac{1}{5}\right)(x - 2i)(x + 2i) = 0$.

Therefore, the solutions are $\left\{1, -\frac{1}{5}, 2i, -2i\right\}$.

 You Try It Work through the following You Try It problem.

Work Exercises 20–25 in this textbook or in the MyMathLab Study Plan.

OBJECTIVE 5   USING THE COMPLEX CONJUGATE PAIRS THEOREM

In Example 3, the two nonreal zeros of $f(x) = 6x^4 + 13x^3 + 61x^2 + 8x - 10$ are $-1 + 3i$ and $-1 - 3i$. In Example 4, $2i$ and $-2i$ are the two nonreal solutions of $5x^5 - 9x^4 + 23x^3 - 35x^2 + 12x + 4 = 0$. Notice that the nonreal zeros in both examples are complex conjugates of each other. This is no coincidence! In fact, the following theorem explains that complex zeros must occur in pairs.

Complex Conjugate Pairs Theorem

If $a + bi$ is a zero of a polynomial function with real coefficients, then the complex conjugate $a - bi$ is also a zero.

⚠ **To use the complex conjugate pairs theorem, it is imperative that the coefficients of the polynomial are real numbers. For example, the polynomial function $f(x) = x - i$ has a constant coefficient i. Note that $f(i) = i - i = 0$, and therefore, i is a zero of the polynomial function $f(x) = x - i$. However, the complex conjugate, $-i$, is not a zero because $f(-i) = -i - i = -2i$.**

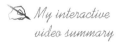 **My interactive video summary**

> **Example 5** Use the Complex Conjugate Pairs Theorem

Find the equation of a polynomial function f with real coefficients that satisfies the given conditions:

a. Fourth-degree polynomial function such that 1 is a zero of multiplicity 2 and $2 - i$ is also a zero.

b. Fifth-degree polynomial function sketched in Figure 21 given that $1 + 3i$ is a zero.

4.5 The Zeros of Polynomial Functions; The Fundamental Theorem of Algebra   **4-59**

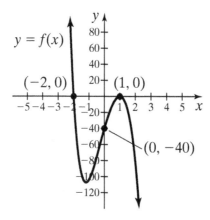

Figure 21

Solution

a. By the complex conjugate pairs theorem, because $2 - i$ is a zero, this implies that $2 + i$ is also a zero. The four zeros of f are $1, 1, 2 - i$, and $2 + i$. Using the factor theorem, we can write f as $f(x) = a(x - 1)(x - 1)(x - (2 - i))(x - (2 + i))$, where a is a constant. We can actually let a equal any number here because there are infinitely many polynomial functions that meet the stated conditions, and they all differ only by the leading coefficient. So, when no specific information is given to allow the exact leading coefficient to be determined, we choose a equal to 1 just for the sake of simplicity. Letting $a = 1$ and multiplying the factors together, we get $f(x) = x^4 - 6x^3 + 14x^2 - 14x + 5$. To see this multiplication worked out, read these steps or watch the interactive video.

b. Because we are looking for a fifth-degree polynomial, there must be five zeros (including multiplicities). From the graph, we see that $x = -2$ is a zero with odd multiplicity because the graph crosses the x-axis at $x = -2$. Similarly, $x = 1$ is a zero with even multiplicity because the graph is tangent to the x-axis at $x = 1$. By the complex conjugate pairs theorem, $1 + 3i$ and $1 - 3i$ are zeros. Thus, the completely factored form of f is $f(x) = a(x + 2)^m(x - 1)^n(x - (1 + 3i))(x - (1 - 3i))$, where m is odd and n is even.

Because f is a fifth-degree polynomial, m must equal 1 and n must equal 2. Therefore, $f(x) = a(x + 2)(x - 1)^2(x - (1 + 3i))(x - (1 - 3i))$. Multiplying the factors, we get $f(x) = a(x^5 - 2x^4 + 7x^3 + 8x^2 - 34x + 20)$. To see this multiplication worked out, read these steps or watch the interactive video. Unlike in part a, we cannot randomly assign a value for the leading coefficient because specific information has been given. To determine the value of a, we can use the fact that the graph contains the point $(0, -40)$ and, therefore, $f(0) = -40$.

$f(x) = a(x^5 - 2x^4 + 7x^3 + 8x^2 - 34x + 20)$	Write the polynomial function.
$f(0) = a(0^5 - 2(0)^4 + 7(0)^3 + 8(0)^2 - 34(0) + 20)$	Evaluate $f(0)$.
$f(0) = a(20)$	Simplify.
$f(0) = -40$	The graph contains the point $(0, -40)$.
$a(20) = -40$	Equate the two expressions for $f(0)$.
$a = -2$	Solve for a.

Thus,

$$f(x) = -2(x^5 - 2x^4 + 7x^3 + 8x^2 - 34x + 20) \text{ or } f(x) = -2x^5 + 4x^4 - 14x^3 - 16x^2 + 68x - 40.$$

You Try It   Work through the following You Try It problem.

Work Exercises 26–31 in this textbook or in the MyMathLab Study Plan.

OBJECTIVE 6   USING THE INTERMEDIATE VALUE THEOREM

A direct consequence of the complex conjugate pairs theorem is the fact that every *odd degree* polynomial with real coefficients has at least one real zero. This is true because complex zeros always occur in pairs; thus, the number of nonreal zeros of a polynomial with real coefficients is always *even*. Therefore, if f is a polynomial function of odd degree, then at least one of its zeros must be real. For example, the polynomial function $f(x) = x^3 + 2x - 1$ is of odd degree, so we know that there exists at least one real zero. By the rational zeros theorem, the only possible rational zeros are ± 1. But

$$f(1) = 1 + 2 - 1 = 2 \quad \text{and}$$
$$f(-1) = -1 - 2 - 1 = -4$$

Neither 1 nor -1 are zeros of f, so there are no real rational zeros. Therefore, the real zero(s) of f must be irrational. So how do we find the irrational zero(s)? The following theorem will help us approximate the irrational zeros to as many decimal places as we want.

Intermediate Value Theorem

Let f be a polynomial function with $a < b$. If $f(a)$ and $f(b)$ have opposite signs, then there exists at least one real zero strictly between a and b.

Although the proof of the intermediate value theorem involves calculus, you can see from Figure 22 that if $f(a)$ is negative and $f(b)$ is positive, then the graph of f *must* cross the x-axis somewhere between a and b.

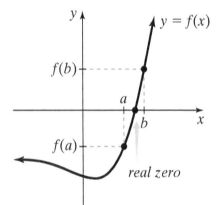

Figure 22 For a polynomial $y = f(x)$, if $f(a)$ is negative and $f(b)$ is positive, then there is a real zero somewhere between a and b.

My video summary   ◉ **Example 6** Use the Intermediate Value Theorem to Approximate a Real Zero

Find the real zero of $f(x) = x^3 + 2x - 1$ correct to two decimal places.

Solution Because f is of odd degree, there is at least one real zero. To approximate this zero, we must first find values of a and b such that $f(a)$ and $f(b)$ have opposite signs.

Note that for $a = 0$, $f(a) = f(0) = -1$. We must now find a number b such that $f(b)$ is positive. Using trial and error, we see that for $b = 1$, $f(b) = f(1) = 2$. Therefore, by the intermediate value theorem, there exists at least one real zero strictly between 0 and 1. We now evaluate the function at the decimal values .1, .2, .3, ... until we obtain another sign change:

$$f(.1) = -.7990$$

$$f(.2) = -.5920$$

$$f(.3) = -.3730$$

$$\left.\begin{array}{l} f(.4) = -.1360 \\ f(.5) = .1250 \end{array}\right\} \text{Sign change!}$$

Notice that $f(.4)$ and $f(.5)$ have opposite signs. Therefore, by the intermediate value theorem, there exists a zero strictly between .4 and .5. We repeat the process again, this time evaluating the function at .41, .42, .43, $\cdots$ until we obtain another sign change.

$$\left.\begin{array}{l} f(.45) = -.0089 \\ f(.46) = .01734 \end{array}\right\} \text{Sign change!}$$

My video summary ⊙ We conclude that the real zero lies between $x = .45$ and $x = .46$. Therefore, the real zero correct to two decimal places is $x = .45$. Using a graphing utility, we can see that the zero carried out to eight decimal places is $x = .45339765$. Watch this video to see the entire solution to this example. ●

Using Technology

$$f(x) = x^3 + 2x - 1$$

```
Zero
X=.45339765   Y=0
```

Use a graphing utility and the zero feature to approximate the real zero of $f(x) = x^3 + 2x - 1$.

You Try It Work through the following You Try It problem.

Work Exercises 32–37 in this textbook or in the MyMathLab Study Plan.

OBJECTIVE 7 SKETCHING THE GRAPHS OF POLYNOMIAL FUNCTIONS

Now that we have a procedure for finding the zeros of a polynomial function, we can modify the four-step process that was discussed in Section 4.3 and put it all together to sketch the graphs of polynomial functions of the form $f(x) = a_n x^n + a_{n-1}x^{n-1} + a_{n-2}x^{n-2} + \cdots + a_1 x + a_0$.

> **Steps for Sketching the Graphs of Polynomial Functions**
>
> **Step 1.** Determine the end behavior.
>
> **Step 2.** Plot the y-intercept $f(0) = a_0$.
>
> **Step 3.** Use the rational zeros theorem, the factor theorem and synthetic division, or the intermediate value theorem to find all zeros and completely factor f.
>
> **Step 4.** Choose a test value between each real zero and complete the graph.

My video summary ⊘ **Example 7** Sketch a Polynomial Function

Sketch the graph of $f(x) = 2x^5 - 5x^4 - 2x^3 + 7x^2 - 4x + 12$.

Solution Work through the video to see how to sketch this polynomial function. The polynomial is sketched in Figure 23.

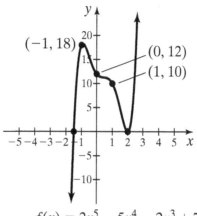

$$f(x) = 2x^5 - 5x^4 - 2x^3 + 7x^2 - 4x + 12$$

Figure 23 Graph of $f(x) = 2x^5 - 5x^4 - 2x^3 + 7x^2 - 4x + 12$

You Try It Work through the following You Try It problem.

Work Exercises 38–43 in this textbook or in the MyMathLab Study Plan.

4.5 Exercises

Skill Check Exercises

In exercises SCE-1 through SCE-3, for each function, determine $f(-x)$.

SCE-1. $f(x) = -3x^5 + 4x^4 + 3x^3 - 7x^2 - 21$ **SCE-2.** $f(x) = x^4 - 7x^2 - 11$

SCE-3. $f(x) = -3x^7 + 8x^3 - x$

In Exercises 1–6, use the rational zeros theorem to determine the potential rational zeros of each polynomial function. Do *not* find the zeros.

1. $f(x) = x^4 - 2$

2. $f(x) = x^3 - 3x + 16$

3. $f(x) = x^5 - 2x^4 + 8x^3 - x^2 + 2x - 12$

4. $f(x) = 3x^4 - 7x^2 - x + 8$

5. $f(x) = x^4 + 4 + 3x - 9x^5$

6. $f(x) = 5x^3 - 27x^6 + 10x^3 + 16 - 4x$

In Exercises 7-10, determine the possible number of positive real zeros and negative real zeros of each polynomial function using Descarates' rule of signs.

7. $f(x) = 2x^5 - 13x^4 + 16x^3 - 5x^2 + 8x - 10$

8. $f(x) = -8x^4 + 13x^3 - 11x^2 + 3x - 1$

9. $f(x) = 7x^4 - 11x^2 - 8x - 10$

10. $f(x) = -2x^6 - 11x^4 - 8x^2 - 10$

In Exercises 11–19, find all complex zeros of the given polynomial function, and write the polynomial in completely factored form.

11. $f(x) = 4x^3 - 5x^2 - 23x + 6$

12. $f(x) = 2x^3 + 7x^2 - 6x - 21$

13. $f(x) = -2x^3 - 7x^2 - 36x + 20$

14. $f(x) = -6x^4 + 19x^3 - 15x^2 - 3x + 5$

15. $f(x) = 3x^4 + 4x^3 + 73x^2 - 54x - 26$

16. $f(x) = -2x^5 - 11x^4 - 19x^3 - 8x^2 + 7x + 5$

17. $f(x) = 8x^5 + 36x^4 + 70x^3 - 5x^2 - 68x + 24$

18. $f(x) = 63x^5 + 33x^4 - 125x^3 - 67x^2 - 2x + 2$

19. $f(x) = 5x^6 - 18x^5 + 29x^4 - 32x^3 + 27x^2 - 14x + 3$

In Exercises 20–25, solve each polynomial equation in the complex numbers.

20. $x^3 - 1 = 0$

21. $x^4 - 1 = 0$

22. $x^3 - 4x^2 + x + 6 = 0$

23. $2x^4 + x^3 - 4x^2 + 11x - 10 = 0$

24. $3x^4 + 4x^3 + 73x^2 - 54x - 26 = 0$

25. $12x^4 + 56x^3 + 3x^2 - 13x - 2 = 0$

In Exercises 26–31, form a polynomial function $f(x)$ with real coefficients using the given information.

26. Form a third-degree polynomial function with real coefficients such that $5 + i$ and 2 are zeros.

27. Form a fourth-degree polynomial function with real coefficients such that $3 - 2i$ is a zero and 1 is a zero of multiplicity 2.

28. Form a fifth-degree polynomial function with real coefficients such that $3i$, $1 - 5i$, and -1 are zeros and $f(0) = -3$.

29. Form the third-degree polynomial function with real coefficients sketched here given that $2i$ is a zero.

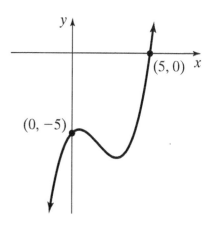

30. Form the fifth-degree polynomial function with real coefficients sketched here given that $1 + i$ is a zero.

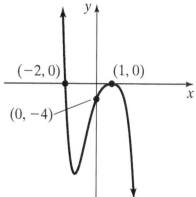

31. Form the fourth-degree polynomial function with real coefficients sketched here given that $-1 - 2i$ is a zero.

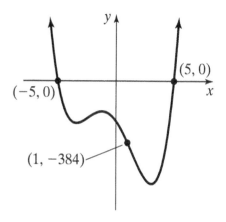

In Exercises 32–34, use the intermediate value theorem to show that the polynomial has a real zero on the given interval.

32. $f(x) = x^3 + 2x - 5; [1, 2]$

33. $f(x) = 3x^4 - x^3 + 8x^2 + x - 1; [0, 1]$

34. $f(x) = 2x^5 + 5x + 1; [-1, 0]$

In Exercises 35–37, use the intermediate value theorem to find the real zero correct to two decimal places.

35. $f(x) = x^3 + 2x - 5$

36. $f(x) = -x^3 + x^2 + x - 1$

37. $f(x) = 2x^5 + 5x + 1$

In Exercises 38–43, determine the end behavior, plot the y-intercept, find and plot all real zeros, and plot at least one test value between each intercept. Then connect the points with a smooth curve.

SbS **38.** $f(x) = x^3 + x^2 - x - 1$

SbS **39.** $f(x) = 4x^3 - 5x^2 - 23x + 6$

SbS **40.** $f(x) = -x^3 - 5x^2 - 3x + 9$

SbS **41.** $f(x) = -6x^4 + 19x^3 - 15x^2 - 3x + 5$

SbS **42.** $f(x) = 2x^5 + 11x^4 + 18x^3 + 8x^2 - 4x - 3$

SbS **43.** $f(x) = -6x^4 + x^3 + 53x^2 - 3x - 105$

Brief Exercises

In Exercises 44–49, sketch the graph of the given function.

44. $f(x) = x^3 + x^2 - x - 1$

45. $f(x) = 4x^3 - 5x^2 - 23x + 6$

46. $f(x) = -x^3 - 5x^2 - 3x + 9$

47. $f(x) = -6x^4 + 19x^3 - 15x^2 - 3x + 5$

48. $f(x) = 2x^5 + 11x^4 + 18x^3 + 8x^2 - 4x - 3$

49. $f(x) = -6x^4 + x^3 + 53x^2 - 3x - 105$

4.6 Rational Functions and Their Graphs

THINGS TO KNOW

Before working through this section, be sure that you are familiar with the following concepts:

VIDEO   ANIMATION   INTERACTIVE

You Try It

1. Determining the Domain of a Function Given the Equation (Section 3.1)

You Try It

2. Determining Whether a Function Is Even, Odd, or Neither (Section 3.2)

You Try It

3. Using Combinations of Transformations to Graph Functions (Section 3.4)

OBJECTIVES

1 Finding the Domain and Intercepts of Rational Functions

2 Identifying Vertical Asymptotes

3 Identifying Horizontal Asymptotes

4 Using Transformations to Sketch the Graphs of Rational Functions

5 Sketching Rational Functions Having Removable Discontinuities

6 Identifying Slant Asymptotes

7 Sketching Rational Functions

....

Introduction to Section 4.6

In this section, we investigate the properties and graphs of rational functions.

> **Definition** Rational Function
>
> A rational function is a function of the form $f(x) = \dfrac{g(x)}{h(x)}$, where g and h are polynomial functions such that $h(x) \neq 0$.

Note If $h(x) = c$, where c is a real number, then we will consider the function $f(x) = g(x)/h(x) = g(x)/c$ to be a polynomial function.

OBJECTIVE 1   FINDING THE DOMAIN AND INTERCEPTS OF RATIONAL FUNCTIONS

Because polynomial functions are defined for all values of x, it follows that rational functions are defined for all values of x *except* those for which the denominator $h(x)$ is equal to zero. If $f(x)$ has a y-intercept, it can be found by evaluating $f(0)$, provided that $f(0)$ is defined. If $f(x)$ has any x-intercepts, they can be found by solving the equation $g(x) = 0$ (provided that g and h do not share a common factor).

My video summary   ▶ **Example 1** Find the Domain and Intercepts of a Rational Function

Let $f(x) = \dfrac{x - 4}{x^2 + x - 6}$.

a. Determine the domain of f.

b. Determine the y-intercept (if any).

c. Determine any x-intercepts.

Solution Factor the denominator, and rewrite f as $f(x) = \dfrac{x - 4}{(x + 3)(x - 2)}$.

a. The domain of f is the set of all real numbers except those for which the denominator is equal to zero.

   The denominator, $(x + 3)(x - 2)$, is equal to zero when $x = -3$ or $x = 2$. Therefore, the domain is the set of all real numbers except -3 and 2 or $\{x \mid x \neq -3, x \neq 2\}$. In interval notation, the domain is $(-\infty, -3) \cup (-3, 2) \cup (2, \infty)$.

b. The y-intercept is $f(0) = \dfrac{-4}{-6} = \dfrac{2}{3}$.

c. To find any x-intercepts, we must solve the equation $x - 4 = 0$. The solution to this equation is $x = 4$. Thus, the only x-intercept is 4.

Watch the video to see the solution to Example 1 worked out in detail.

You Try It   Work through the following You Try It problem.

Work Exercises 1–6 in this textbook or in the MyMathLab Study Plan.

OBJECTIVE 2   IDENTIFYING VERTICAL ASYMPTOTES

To begin our discussion of the graphs of rational functions, we first look at a basic function that was introduced in Section 3.3, the reciprocal function, $f(x) = \dfrac{1}{x}$. The function $f(x) = \dfrac{1}{x}$ is defined everywhere except at $x = 0$; therefore, the domain of $f(x) = \dfrac{1}{x}$ is $(-\infty, 0) \cup (0, \infty)$. You can see from the graph of $f(x) = \dfrac{1}{x}$ sketched in Figure 24 that as the values of x get closer to 0 from the right side of 0, the values of $f(x)$ increase without bound. Mathematically, we say, "as x approaches zero from the right, $f(x)$ approaches infinity." In symbols, this is written as

$$f(x) \to \infty \quad \text{as} \quad x \to 0^+$$

The "+" symbol indicates that we are only looking at values of x that are located to the *right* of zero. Similarly, as x gets closer to 0 from the left side of 0, the values of $f(x)$ approach negative infinity. Symbolically, we can write the following:

$$f(x) \to -\infty \quad \text{as} \quad x \to 0^-$$

The "−" symbol indicates that we are only looking at values of x that are located to the *left* of zero.

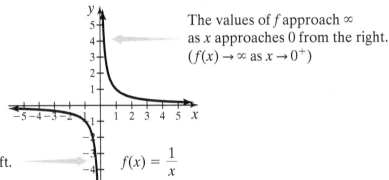

The values of f approach ∞ as x approaches 0 from the right. $(f(x) \to \infty$ as $x \to 0^+)$

The values of f approach $-\infty$ as x approaches 0 from the left. $(f(x) \to -\infty$ as $x \to 0^-)$

$f(x) = \dfrac{1}{x}$

Figure 24 Graph of $f(x) = \dfrac{1}{x}$

The graph of the function $f(x) = \dfrac{1}{x}$ gets closer and closer to the line $x = 0$ (the y-axis) as x gets closer and closer to 0 from either side of 0, but the graph never touches the line $x = 0$. The line $x = 0$ is said to be a **vertical asymptote** of the graph of $f(x) = \dfrac{1}{x}$. Many rational functions have vertical asymptotes. The vertical asymptotes occur when the graph of a function approaches positive or negative infinity as x approaches some finite number a as stated in the following definition.

Definition   Vertical Asymptote

The vertical line $x = a$ is a **vertical asymptote** of a function $y = f(x)$ if *at least* one of the following occurs:

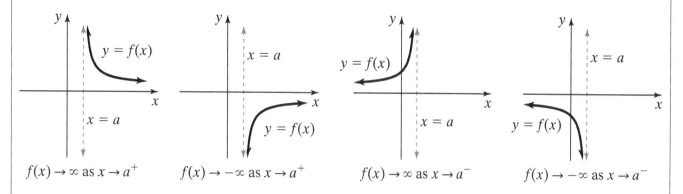

$f(x) \to \infty$ as $x \to a^+$ $f(x) \to -\infty$ as $x \to a^+$ $f(x) \to \infty$ as $x \to a^-$ $f(x) \to -\infty$ as $x \to a^-$

A rational function of the form $f(x) = \dfrac{g(x)}{h(x)}$, where $g(x)$ and $h(x)$ have no common factors, will have a vertical asymptote at $x = a$ if $h(a) = 0$.

⚠ **It is essential to cancel any common factors before locating the vertical asymptotes.**

The situation that arises when $g(x)$ and $h(x)$ share a common factor is discussed in objective 5 of this section.

My video summary   ▷ **Example 2** Find the Vertical Asymptotes of a Rational Function

Find the vertical asymptotes (if any) of the function $f(x) = \dfrac{x - 3}{x^2 + x - 6}$, and then sketch the graph near the vertical asymptotes.

Solution Factor the denominator of $f(x)$: $f(x) = \dfrac{x - 3}{(x + 3)(x - 2)}$.

Because the numerator and denominator have no common factors, f will have vertical asymptotes when the denominator, $(x + 3)(x - 2)$, is equal to zero or when $x = -3$ and $x = 2$. Therefore, the equations of the two vertical asymptotes are $x = -3$ and $x = 2$. To sketch the graph near the asymptotes, we must determine whether $f \to \infty$ or $f \to -\infty$ on either side of the vertical asymptote. To do this, we evaluate the sign of each factor of f by choosing a test value "close to" $x = -3$ and $x = 2$ from the left side and from the right side of $x = -3$ and $x = 2$.

Vertical Asymptote	$x = -3$		$x = 2$	
Approach from	Left	Right	Left	Right
Test Value	$x = -3.1$	$x = -2.9$	$x = 1.9$	$x = 2.1$
Substitute Test Value into $f(x) = \dfrac{x-3}{(x+3)(x-2)}$	$f(-3.1) = \dfrac{(-3.1-3)}{(-3.1+3)(-3.1-2)} = \dfrac{(-)}{(-)(-)} = -$	$f(-2.9) = \dfrac{(-2.9-3)}{(-2.9+3)(-2.9-2)} = \dfrac{(-)}{(+)(-)} = +$	$f(1.9) = \dfrac{(1.9-3)}{(1.9+3)(1.9-2)} = \dfrac{(-)}{(+)(-)} = +$	$f(2.1) = \dfrac{(2.1-3)}{(2.1+3)(2.1-2)} = \dfrac{(-)}{(+)(+)} = -$
Result	Sign is negative. $f \to -\infty$	Sign is positive. $f \to \infty$	Sign is positive. $f \to \infty$	Sign is negative. $f \to -\infty$

(!) **If there is an x-intercept near the vertical asymptote, it is essential to choose a test value that is between the x-intercept and the vertical asymptote.**

We see from the table that as the values of x approach -3 from the left side of -3, the values of f approach $-\infty$. Similarly, as the values of x approach -3 from the right side of -3, the values of f approach ∞. As x approaches 2 from the left side of 2, the values of f approach ∞. Finally, as the values of x approach 2 from the right side of 2, the values of f approach $-\infty$. Notice that we chose 2.1 as our test value for the case when x approaches 2 from the right. If we would have picked a test value greater than 3, our conclusions would have been incorrect. This is because $x = 3$ is an x-intercept and the graph to the right of 3 is above the x-axis. It is absolutely crucial to choose test values that are strictly between the x-intercept and the vertical asymptote.

My video summary (>) Watch the video to see this entire example worked out in detail. The graph of f near both vertical asymptotes is sketched in Figure 25.

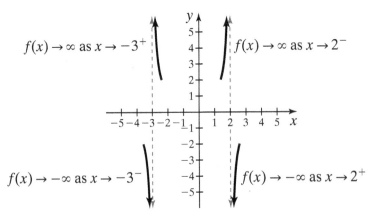

$f(x) \to \infty$ as $x \to -3^{+}$

$f(x) \to \infty$ as $x \to 2^{-}$

$f(x) \to -\infty$ as $x \to -3^{-}$

$f(x) \to -\infty$ as $x \to 2^{+}$

Figure 25 Graph of $f(x) = \dfrac{x-3}{x^2 + x - 6}$ near the vertical asymptotes

You Try It **Work through the following You Try It problem.**

My video summary ▶ **Example 3** Find the Vertical Asymptotes of a Rational Function

Find the vertical asymptotes (if any) of the function $f(x) = \dfrac{x+3}{x^2+x-6}$, and then sketch the graph near the vertical asymptotes.

Solution We start by factoring the denominator: $f(x) = \dfrac{x+3}{(x+3)(x-2)}$. Notice that the numerator and denominator share a common factor of $x+3$. Canceling the common factor, we get $f(x) = \dfrac{x+3}{(x+3)(x-2)} = \dfrac{1}{x-2}$ for $x \neq -3$. After cancellation, the only zero in the denominator is $x = 2$. Thus, the line $x = 2$ is the only vertical asymptote of the graph of f. To sketch the graph near the asymptote, we must determine the sign of the denominator as x approaches 2 from the left side and the right side of $x = 2$.

Vertical Asymptote	$x = 2$	
Approach from	**Left**	**Right**
Test Value	$x = 1.9$	$x = 2.1$
Substitute Test Value into $f(x) = \dfrac{1}{x-2}$ for $x \neq -3$.	$f(1.9) =$ $\dfrac{1}{(1.9-2)} =$ $\dfrac{1}{(-)} = -$	$f(2.1) =$ $\dfrac{1}{(2.1-2)} =$ $\dfrac{1}{(+)} = +$
Result	Sign is negative. $f \rightarrow -\infty$	Sign is positive. $f \rightarrow \infty$

My video summary ▶ We see that as the values of x approach 2 from the left side of 2, the values of f approach $-\infty$. Similarly, as the values of x approach 2 from the right side of 2, the values of f approach ∞. Watch the video to see this entire example worked out in detail.

A rough sketch of the graph of f near the vertical asymptote can be seen in Figure 26. We sketch the complete graph of f in Example 6.

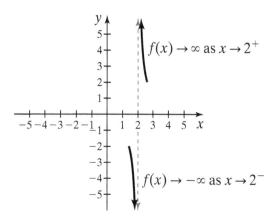

Figure 26 Graph of $f(x) = \dfrac{x+3}{x^2+x-6}$ near the vertical asymptote

You Try It   Work through the following You Try It problem.

Work Exercises 7–12 in this textbook or in the MyMathLab Study Plan.

OBJECTIVE 3   IDENTIFYING HORIZONTAL ASYMPTOTES

My video summary   ⊙ The graph of $f(x) = \dfrac{1}{x}$ is sketched again in Figure 27. The graph shows that as the values of x increase without bound (as $x \to \infty$), the values of f approach 0 ($f(x) \to 0$). Similarly, as the values of x decrease without bound (as $x \to -\infty$), the values of f approach 0 ($f(x) \to 0$). In this case, the line $y = 0$ is called a **horizontal asymptote** of the graph of $f(x) = \dfrac{1}{x}$.

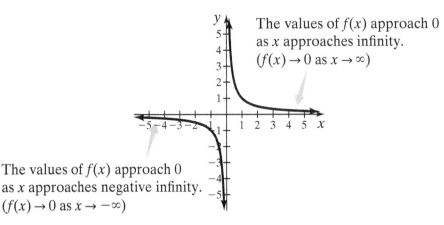

The values of $f(x)$ approach 0 as x approaches infinity. ($f(x) \to 0$ as $x \to \infty$)

The values of $f(x)$ approach 0 as x approaches negative infinity. ($f(x) \to 0$ as $x \to -\infty$)

Figure 27 The line $y = 0$ is a horizontal asymptote of the graph of $f(x) = \dfrac{1}{x}$.

Definition   Horizontal Asymptote

A horizontal line $y = H$ is a **horizontal asymptote** of a function f if the values of $f(x)$ approach some fixed number H as the values of x approach ∞ or $-\infty$.

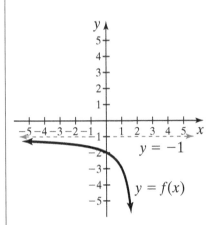

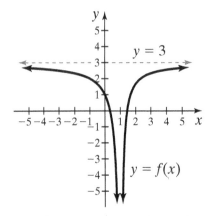

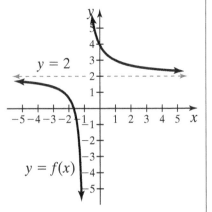

The line $y = -1$ is a horizontal asymptote because the values of $f(x)$ approach -1 as x approaches $-\infty$.

The line $y = 3$ is a horizontal asymptote because the values of $f(x)$ approach 3 as x approaches $\pm\infty$.

The line $y = 2$ is a horizontal asymptote because the values of $f(x)$ approach 2 as x approaches $\pm\infty$.

PROPERTIES OF HORIZONTAL ASYMPTOTES OF RATIONAL FUNCTIONS

- Although a rational function can have many vertical asymptotes, it can have, at most, one horizontal asymptote.

- The graph of a rational function will never intersect a vertical asymptote but may intersect a horizontal asymptote.

- A rational function $f(x) = \dfrac{g(x)}{h(x)}$ that is written in lowest terms (all common factors of the numerator and denominator have been canceled) will have a horizontal asymptote whenever the degree of $h(x)$ is greater than or equal to the degree of $g(x)$.

The third property listed here tells us that a rational function (written in lowest terms) will have a horizontal asymptote whenever the degree of the denominator is greater than or equal to the degree of the numerator. If the degree of the denominator is less than the degree of the numerator, then the rational function will not have a horizontal asymptote. We now summarize a technique for finding the horizontal asymptotes of a rational function.

FINDING HORIZONTAL ASYMPTOTES OF A RATIONAL FUNCTION

Let $f(x) = \dfrac{g(x)}{h(x)} = \dfrac{a_n x^n + a_{n-1} x^{n-1} + a_{n-2} x^{n-2} + \cdots + a_1 x + a_0}{b_m x^m + b_{m-1} x^{m-1} + b_{m-2} x^{m-2} + \cdots + b_1 x + b_0}$, $a_n \neq 0, b_m \neq 0$,

where f is written in lowest terms, n is the degree of g, and m is the degree of h.

- If $m > n$, then $y = 0$ is the horizontal asymptote.

- If $m = n$, then the horizontal asymptote is $y = \dfrac{a_n}{b_m}$, the ratio of the leading coefficients.

- If $m < n$, then there are no horizontal asymptotes.

 *My interactive video summary*

▶ **Example 4** Find the Horizontal Asymptotes of a Rational Function

Find the horizontal asymptote of the graph of each rational function or state that one does not exist.

a. $f(x) = \dfrac{x}{x^2 - 4}$ b. $f(x) = \dfrac{4x^2 - x + 1}{1 - 2x^2}$ c. $f(x) = \dfrac{2x^3 + 3x^2 - 2x - 2}{x - 1}$

Solution

a. $f(x) = \dfrac{x}{x^2 - 4}$

The degree of the denominator is *greater than* the degree of the numerator. Therefore, the graph has a horizontal asymptote whose equation is $y = 0$ (the x-axis). View the graph of f or work through the interactive video.

b. $f(x) = \dfrac{4x^2 - x + 1}{1 - 2x^2}$

The degree of the denominator is *equal* to the degree of the numerator. Thus, the graph has a horizontal asymptote. Note that the leading coefficient of the polynomial in the numerator is 4. The leading coefficient of the polynomial in the denominator is -2 because -2 is the coefficient of the term with the highest degree. Thus, the equation of the horizontal asymptote is $y = \dfrac{4}{-2}$ or $y = -2$. View the graph of f or work through the interactive video.

c. $f(x) = \dfrac{2x^3 + 3x^2 - 2x - 2}{x - 1}$

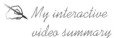

The degree of the denominator is *less than* the degree of the numerator. Therefore, the graph has no horizontal asymptotes. View the graph of f or work through the interactive video.

You Try It Work through the following You Try It problem.

Work Exercises 13–20 in this textbook or in the MyMathLab Study Plan.

OBJECTIVE 4   USING TRANSFORMATIONS TO SKETCH THE GRAPHS OF RATIONAL FUNCTIONS

We now introduce another basic rational function, $f(x) = \dfrac{1}{x^2}$. Like the reciprocal function, the domain of $f(x) = \dfrac{1}{x^2}$ is $(-\infty, 0) \cup (0, \infty)$. The denominator, x^2, is always greater than zero, and this implies that the range includes all values of y greater than zero. The graph has a y-axis vertical asymptote ($x = 0$) and an x-axis horizontal asymptote ($y = 0$). Following are the graphs of $f(x) = \dfrac{1}{x}$ and $f(x) = \dfrac{1}{x^2}$, along with some important properties of each. Knowing these two basic graphs will help us sketch more complicated rational functions.

Graphs of $f(x) = \dfrac{1}{x}$ and $f(x) = \dfrac{1}{x^2}$

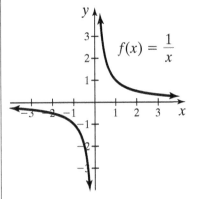

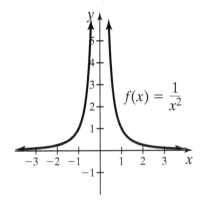

Properties of the graph of $f(x) = \dfrac{1}{x}$

Domain: $(-\infty, 0) \cup (0, \infty)$

Range: $(-\infty, 0) \cup (0, \infty)$

No intercepts

Vertical asymptote: $x = 0$

Horizontal asymptote: $y = 0$

Odd function $f(-x) = -f(x)$

The graph is symmetric about the origin.

Properties of the graph of $f(x) = \dfrac{1}{x^2}$

Domain: $(-\infty, 0) \cup (0, \infty)$

Range: $(0, \infty)$

No intercepts

Vertical asymptote: $x = 0$

Horizontal asymptote: $y = 0$

Even function $f(x) = f(-x)$

The graph is symmetric about the y-axis.

Sometimes we are able to use transformations to sketch the graphs of rational functions. To review this idea, read this summary of the transformation techniques that are discussed in Section 3.4.

My video summary ◉ **Example 5** Use Transformations to Sketch the Graph of a Rational Function

Use transformations to sketch the graph of $f(x) = \dfrac{-2}{(x+3)^2} + 1$.

Solution Because $x + 3$ is raised to the second power in the denominator, start with the graph of $y = \dfrac{1}{x^2}$. We can obtain the graph of $f(x) = \dfrac{-2}{(x+3)^2} + 1$ using the following sequence of transformations:

1. Horizontally shift the graph of $y = \dfrac{1}{x^2}$ to the left three units to obtain the graph of

$$y = \dfrac{1}{(x+3)^2}.$$

2. Vertically stretch the graph of $y = \dfrac{1}{(x+3)^2}$ by a factor of 2 to obtain the graph of

$$y = \dfrac{2}{(x+3)^2}.$$

3. Reflect the graph of $y = \dfrac{2}{(x+3)^2}$ about the x-axis to obtain the graph of

$$y = \dfrac{-2}{(x+3)^2}.$$

4. Vertically shift the graph of $y = \dfrac{-2}{(x+3)^2}$ up one unit to obtain the final graph

of $f(x) = \dfrac{-2}{(x+3)^2} + 1$.

You can see from the final graph in Figure 28 that the horizontal asymptote is now shifted up one unit to $y = 1$. The x-intercepts of $-3 - \sqrt{2}$ and $-3 + \sqrt{2}$ can be found by setting $f(x) = 0$ and solving for x. The y-intercept is $f(0) = \dfrac{7}{9}$. Watch the video to see each step worked out in detail.

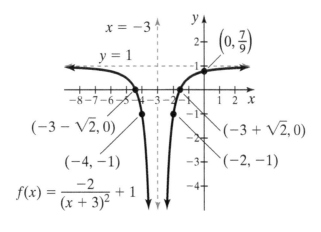

$x = -3$

$y = 1$

$\left(0, \dfrac{7}{9}\right)$

$(-3 - \sqrt{2}, 0)$

$(-3 + \sqrt{2}, 0)$

$(-4, -1)$

$(-2, -1)$

$f(x) = \dfrac{-2}{(x+3)^2} + 1$

Figure 28 Graph of $f(x) = \dfrac{-2}{(x+3)^2} + 1$

You Try It Work through the following You Try It problem.

Work Exercises 21–32 in this textbook or in the MyMathLab Study Plan.

OBJECTIVE 5 SKETCHING RATIONAL FUNCTIONS HAVING REMOVABLE DISCONTINUITIES

A rational function $f(x) = \dfrac{g(x)}{h(x)}$ may sometimes have a "hole" in its graph. In calculus, these holes are called **removable discontinuities**. For a rational function to have a removable discontinuity, $g(x)$ and $h(x)$ must share a common factor as in Example 6.

 *My video summary*

▶ **Example 6** Sketch a Rational Function Having a Removable Discontinuity

Sketch the graph of $f(x) = \dfrac{x^2 - 1}{x + 1}$, and find the coordinates of all removable discontinuities.

Solution The domain of f is $\{x \mid x \neq -1\}$. Factoring the numerator, we can rewrite f as $f(x) = \dfrac{(x+1)(x-1)}{x+1}$. Canceling the common factor of $x + 1$, we get $f(x) = x - 1$ for $x \neq -1$. Notice that the domain of this simplified function is still all real numbers *except* $x = -1$. Because there are no longer any factors containing the variable x in the denominator of the simplified function, there are no vertical asymptotes.

The graph of $f(x) = \dfrac{x^2 - 1}{x + 1}$ behaves exactly like the graph of the simplified function $f(x) = x - 1$ for $x \neq -1$. This graph will look like the graph of $y = x - 1$ except we must remove the point on the graph when $x = -1$. We can find this point by evaluating $y = x - 1$ at $x = -1$. We get $y = (-1) - 1 = -2$. Thus, the removable discontinuity is the point $(-1, -2)$. The graph no longer looks like a typical rational function because the entire denominator canceled leaving only $f(x) = x - 1$ for $x \neq -1$. In Figure 29, the graph is a line with a "hole" at the point $(-1, -2)$.

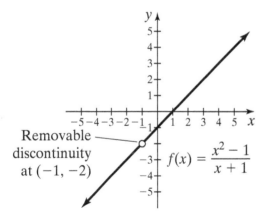

Figure 29 Graph of $f(x) = \dfrac{x^2 - 1}{x + 1}$

It is possible to have a rational function with a common factor in the numerator and denominator whose graph still looks like the graph of a rational function but has a "hole" in it, as in Example 7.

My video summary

> **Example 7** Sketch a Rational Function Having a Removable Discontinuity

Sketch the graph of $f(x) = \dfrac{x + 3}{x^2 + x - 6}$, and find the coordinates of all removable discontinuities.

Solution Notice that this is the function that was discussed in Example 3. Factoring the denominator, we get $f(x) = \dfrac{x + 3}{(x + 3)(x - 2)}$. The domain of $f(x)$ is all real numbers, except $x = 2$ and $x = -3$. Canceling the common factor of $x + 3$, we get $f(x) = \dfrac{1}{x - 2}$ for $x \neq -3$. Therefore, the graph of $f(x) = \dfrac{x + 3}{(x + 3)(x - 2)}$ will behave exactly like the graph of $f(x) = \dfrac{1}{x - 2}$ for $x \neq -3$. Notice that this graph will look like the graph of $y = \dfrac{1}{x - 2}$, except we must remember that $x = -3$ is *not* in the domain! There will be a removable discontinuity at the point where $x = -3$. We can find this point by evaluating $y = \dfrac{1}{x - 2}$ at $x = -3$. We get $y = \dfrac{1}{(-3) - 2} = -\dfrac{1}{5}$. Using transformations, we can shift the graph of $y = \dfrac{1}{x}$ to the right two units to obtain the graph of $y = \dfrac{1}{x - 2}$, remembering to remove the point $\left(-3, -\dfrac{1}{5}\right)$. See Figure 30.

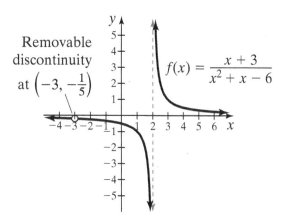

Removable discontinuity at $\left(-3, -\dfrac{1}{5}\right)$

$f(x) = \dfrac{x + 3}{x^2 + x - 6}$

Figure 30 Graph of $f(x) = \dfrac{x + 3}{x^2 + x - 6}$

⚠️ **Be careful when using graphing devices. Most graphing calculators will not show removable discontinuities!**

Using Technology

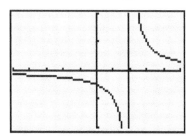

The graph of $y_1 = \dfrac{x + 3}{x^2 + x - 6}$ is shown using a graphing utility. The graph does not suggest that the function is undefined at $x = -3$. However, the table of values does illustrate this fact.

You Try It   Work through the following You Try It problem.

Work Exercises 33–38 in this textbook or in the MyMathLab Study Plan.

OBJECTIVE 6   IDENTIFYING SLANT ASYMPTOTES

Recall that the graph of a rational function $f(x) = \dfrac{g(x)}{h(x)}$, where f is written in lowest terms has a horizontal asymptote whenever the degree of $h(x)$ is greater than or equal to the degree of $g(x)$. But what happens if the degree of $h(x)$ is less than the degree of $g(x)$? If the degree of $h(x)$ is exactly 1 less than the degree of $g(x)$, then there will be a **slant asymptote**.

By the division algorithm, there exist polynomial functions $q(x)$ and $r(x)$ such that

$$\frac{g(x)}{h(x)} = q(x) + \frac{r(x)}{h(x)},$$

where the degree of $r(x)$ is less than the degree of $h(x)$. Because the degree of $h(x)$ is 1 less than the degree of $g(x)$, this means that the polynomial function $q(x)$ must be a linear function. Thus, we can rewrite $f(x)$ as

$$f(x) = mx + b + \frac{r(x)}{h(x)}.$$

Because the degree of $r(x)$ is less than the degree of $h(x)$, we know that as $x \to \pm\infty$, $\dfrac{r(x)}{h(x)} \to 0$. Therefore, as $x \to \pm\infty$, the graph of $f(x) = mx + b + \dfrac{r(x)}{h(x)}$ approaches the line $y = mx + b$. This line is the **slant asymptote**.

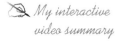 *My interactive video summary*

▶ **Example 8** Find the Slant Asymptote of a Rational Function

Find the slant asymptote of $f(x) = \dfrac{2x^2 + 3x - 2}{x - 1}$.

Solution The numerator and denominator have no common factors, and the degree of the denominator is exactly 1 less than the degree of the numerator. Thus, the function has a slant asymptote. Using synthetic division or long division, we can rewrite f as $f(x) = 2x + 5 + \dfrac{3}{x - 1}$ so the slant asymptote is the line $y = 2x + 5$.

The graph of $f(x) = \dfrac{2x^2 + 3x - 2}{x - 1}$ from Example 8 is sketched in Figure 31. You may work through this interactive video to see how to sketch this function. Notice in Figure 31 that the y-intercept is $f(0) = 2$. The x-intercepts are $x = \dfrac{1}{2}$ and $x = -2$, which are the solutions to the equation $2x^2 + 3x - 2 = 0$. The vertical asymptote is the line $x = 1$, and the slant asymptote is the line $y = 2x + 5$.

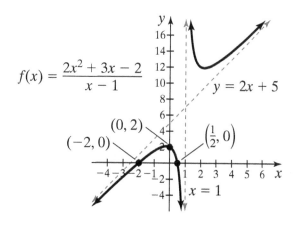

$$f(x) = \frac{2x^2 + 3x - 2}{x - 1}$$

$$y = 2x + 5$$

$(0, 2)$

$(-2, 0)$

$\left(\frac{1}{2}, 0\right)$

$x = 1$

Figure 31 Graph of

$$f(x) = \frac{2x^2 + 3x - 2}{x - 1}$$

You Try It   Work through the following You Try It problem.

Work Exercises 39–42 in this textbook or in the MyMathLab Study Plan.

OBJECTIVE 7   SKETCHING RATIONAL FUNCTIONS

Given a rational function of the form $f(x) = \dfrac{g(x)}{h(x)}$, we can sketch the graph of f by carefully following the nine-step procedure outlined here.

Steps for Graphing Rational Functions of the Form $f(x) = \dfrac{g(x)}{h(x)}$

Step 1. Find the domain.

Step 2. If $g(x)$ and $h(x)$ have common factors, cancel all common factors determining the x-coordinates of any removable discontinuities and rewrite f in lowest terms.

Step 3. Check for symmetry.

If $f(-x) = -f(x)$, then the graph of $f(x)$ is *odd* and thus symmetric about the origin.
If $f(x) = f(-x)$, then the graph of $f(x)$ is *even* and thus symmetric about the y-axis.

Step 4. Find the y-intercept by evaluating $f(0)$.

Step 5. Find the x-intercepts by finding the zeros of the numerator of f, being careful to use the new numerator if a common factor has been removed.

Step 6. Find the vertical asymptotes by finding the zeros of the denominator of f, being careful to use the new denominator if a common factor has been removed. Use test values to determine the behavior of the graph on each side of the vertical asymptotes.

Step 7. Determine whether the graph has any horizontal or slant asymptotes.

Step 8. Plot points, choosing values of x between each intercept and values of x on either side of all vertical asymptotes.

Step 9. Complete the sketch.

 *My video summary*   ▶ **Example 9** Sketch a Rational Function

Sketch the graph of $f(x) = \dfrac{x^3 + 2x^2 - 9x - 18}{x^3 + 6x^2 + 5x - 12}$.

Solution Watch the video to see each of the following steps worked out in detail:

1. We can use the rational zeros theorem and synthetic division to factor and rewrite f as $f(x) = \dfrac{(x + 3)(x - 3)(x + 2)}{(x + 3)(x - 1)(x + 4)}$. The domain is $\{x \mid x \neq -4, x \neq -3, x \neq 1\}$.

2. Because the numerator and denominator share a common factor of $x + 3$, there will be a removable discontinuity at $x = -3$. The function simplifies to $f(x) = \dfrac{(x - 3)(x + 2)}{(x - 1)(x + 4)}$ for $x \neq -3$ or $f(x) = \dfrac{x^2 - x - 6}{x^2 + 3x - 4}$ for $x \neq -3$. The removable discontinuity is the point $\left(-3, -\dfrac{3}{2}\right)$.

3. This function is neither odd nor even because $f(-x) \neq -f(x)$ and $f(x) \neq f(-x)$. Therefore, the graph is not symmetric about the origin or the y-axis.

4. The y-intercept is $f(0) = \dfrac{3}{2}$.

5. The x-intercepts occur when $(x - 3)(x + 2) = 0$. The x-intercepts are $x = -2$ and $x = 3$.

6. The equations of the two vertical asymptotes are $x = -4$ and $x = 1$. Using test values, we can determine the behavior of the graph near the vertical asymptotes.

   As $x \to -4^-$, $f(x) \to \infty$. The graph approaches positive infinity as x approaches -4 from the left.
   As $x \to -4^+$, $f(x) \to -\infty$. The graph approaches negative infinity as x approaches -4 from the right.
   As $x \to 1^-$, $f(x) \to \infty$. The graph approaches positive infinity as x approaches 1 from the left.
   As $x \to 1^+$, $f(x) \to -\infty$. The graph approaches negative infinity as x approaches 1 from the right.

7. Because the degree of the denominator is equal to the degree of the numerator, the graph of f has a horizontal asymptote at $y = \dfrac{1}{1} = 1$.

8. We can evaluate the function at several values of x. Note that $f(-5) = 4$, $f\left(\dfrac{1}{2}\right) = \dfrac{25}{9}$, and $f(2) = -\dfrac{2}{3}$. Hence, the graph must pass through the points $(-5, 4)$, $\left(\dfrac{1}{2}, \dfrac{25}{9}\right)$, and $\left(2, -\dfrac{2}{3}\right)$.

My video summary ▶ 9. The completed graph is sketched in Figure 32. Watch the video to see each step worked out in detail.

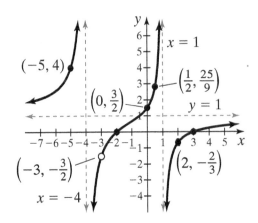

Figure 32 Graph of $f(x) = \dfrac{x^3 + 2x^2 - 9x - 18}{x^3 + 6x^2 + 5x - 12}$

The graph seen in Figure 32 does not have any symmetry. Some rational functions, however, do have symmetry so it is important not to skip step 3. For example, the rational function $f(x) = \dfrac{x}{x^2 - 4}$ is symmetric about the origin because $f(-x) = -f(x)$. See if you can follow the nine-step graphing process to sketch this graph. To see how to sketch the function $f(x) = \dfrac{x}{x^2 - 4}$, watch this video.

You Try It Work through the following You Try It problem.

Work Exercises 43–54 in this textbook or in the MyMathLab Study Plan.

4.6 Exercises

In Exercises 1–6, for each rational function, answer parts a to c.

a. Determine the domain.

b. Determine the y-intercept (if any).

c. Determine any x-intercepts.

1. $f(x) = \dfrac{x}{x^2 - 1}$

2. $f(x) = \dfrac{x}{x^2 + 1}$

3. $f(x) = \dfrac{x^2 - 1}{x}$

4. $f(x) = \dfrac{3x - 1}{x + 2}$

5. $f(x) = \dfrac{x^2 + x - 12}{x^2 + x}$

6. $f(x) = \dfrac{x^3 - 2x^2 - 5x + 6}{x^2 - 4x - 5}$

In Exercises 7–12, find all vertical asymptotes, and create a rough sketch of the graph near each asymptote.

7. $f(x) = \dfrac{3}{x + 1}$

8. $f(x) = \dfrac{x}{x - 2}$

9. $f(x) = \dfrac{x - 1}{x^2 - 1}$

10. $f(x) = \dfrac{x + 3}{x^2 - 6x + 8}$

11. $f(x) = \dfrac{x^2 - 2x - 15}{x^2 + x - 2}$

12. $f(x) = \dfrac{x^2 - 9}{x^3 - 2x^2 - 5x + 6}$

In Exercises 13–20, find the equation of all horizontal asymptotes (if any) of each rational function.

13. $f(x) = \dfrac{3}{x + 1}$

14. $f(x) = \dfrac{x^2 - 1}{x + 1}$

15. $f(x) = \dfrac{3x - 1}{5x + 2}$

16. $f(x) = \dfrac{2x^3 - x^2 + 1}{x^2 + 1}$

17. $f(x) = \dfrac{3x^2 - 4x^3}{7x^3 + x^2 - x + 1}$

18. $f(x) = \dfrac{9x^2}{4 - x^2}$

19. $f(x) = \dfrac{1}{x^3 + x^2 - 1}$

20. $f(x) = \dfrac{3x^2 - 2x - 9}{11x^2 + x - 2}$

In Exercises 21–32, use transformations of $y = \dfrac{1}{x}$ or $y = \dfrac{1}{x^2}$ to sketch each rational function. Label all intercepts, and find the equation of all asymptotes.

SbS **21.** $f(x) = \dfrac{1}{x} - 2$

SbS **22.** $f(x) = \dfrac{1}{x^2} + 2$

SbS **23.** $f(x) = \dfrac{1}{x - 1}$

SbS **24.** $f(x) = \dfrac{1}{(x - 1)^2}$

SbS **25.** $f(x) = \dfrac{1}{x + 1} - 3$

SbS **26.** $f(x) = \dfrac{1}{(x + 3)^2} - 2$

SbS **27.** $f(x) = \dfrac{-1}{x - 2} - 3$

SbS **28.** $f(x) = \dfrac{-1}{(x + 1)^2} + 2$

SbS **29.** $f(x) = \dfrac{2}{x + 1}$

SbS **30.** $f(x) = \dfrac{2}{(x - 3)^2}$

SbS **31.** $f(x) = \dfrac{-2}{x + 1} - 1$

SbS **32.** $f(x) = \dfrac{-2}{(x - 1)^2} + 3$

In Exercises 33–38, identify the coordinates of all removable discontinuities, and sketch the graph of each rational function. Label all intercepts, and find the equations of all asymptotes.

SbS **33.** $f(x) = \dfrac{x^2 - 9}{x - 3}$

SbS **34.** $f(x) = \dfrac{x^2 + 3x - 4}{x + 4}$

SbS **35.** $f(x) = \dfrac{x - 1}{x^2 - x}$

SbS **36.** $f(x) = \dfrac{x + 2}{x^2 - x - 6}$

SbS **37.** $f(x) = \dfrac{x + 2}{(x + 2)(x - 1)^2}$

SbS **38.** $f(x) = \dfrac{x - 1}{x^3 + 3x^2 - 4}$

In Exercises 39–42, find the slant asymptote of the graph of each rational function.

39. $f(x) = \dfrac{2x^2 - 1}{x}$

40. $f(x) = \dfrac{x^2 - 2x + 5}{x + 1}$

41. $f(x) = \dfrac{x^3 - 1}{x^2 + x}$

42. $f(x) = \dfrac{2x^4 - 1}{x^3 - x}$

In Exercises 43–54, follow the nine-step graphing strategy to sketch the graph of each rational function. Be sure to label all intercepts and find the equations of all asymptotes. Also label any removable discontinuities.

SbS **43.** $f(x) = \dfrac{2x + 4}{x - 1}$

SbS **44.** $f(x) = \dfrac{2x + 4}{x^2 - x - 2}$

SbS **45.** $f(x) = \dfrac{x^2 - 1}{x^2 + 1}$

SbS **46.** $f(x) = \dfrac{2x^2 - 5x - 3}{x + 2}$

SbS **47.** $f(x) = \dfrac{1}{x^2 - 9}$

SbS **48.** $f(x) = \dfrac{x}{x^2 - 9}$

SbS **49.** $f(x) = \dfrac{x^2 + 2x - 3}{x + 1}$

SbS **50.** $f(x) = \dfrac{2x^2 + 4x - 6}{x^2 - 4x + 3}$

SbS **51.** $f(x) = \dfrac{x^3 - 3x^2 - x + 3}{x^2 - x - 6}$

SbS **52.** $f(x) = \dfrac{x^3 - 7x + 6}{x^3 - 5x^2 + 2x + 8}$

SbS **53.** $f(x) = \dfrac{2x^3 - 3x^2 - 3x + 2}{x^3 - 3x^2 - 6x + 8}$

SbS **54.** $f(x) = \dfrac{2x^3 - 3x^2 - 3x + 2}{x^2 - 9}$

Brief Exercises

In Exercises 55–60, determine the domain of each rational function.

55. $f(x) = \dfrac{x}{x^2 - 1}$

56. $f(x) = \dfrac{x}{x^2 + 1}$

57. $f(x) = \dfrac{x^2 - 1}{x}$

58. $f(x) = \dfrac{3x - 1}{x + 2}$

59. $f(x) = \dfrac{x^2 + x - 12}{x^2 + x}$

60. $f(x) = \dfrac{x^3 - 2x^2 - 5x + 6}{x^2 - 4x - 5}$

In Exercises 61–66, determine the vertical asymptotes of each rational function.

61. $f(x) = \dfrac{3}{x + 1}$

62. $f(x) = \dfrac{x}{x - 2}$

63. $f(x) = \dfrac{x - 1}{x^2 - 1}$

64. $f(x) = \dfrac{x + 3}{x^2 - 6x + 8}$

65. $f(x) = \dfrac{x^2 - 2x - 15}{x^2 + x - 2}$

66. $f(x) = \dfrac{x^2 - 9}{x^3 - 2x^2 - 5x + 6}$

In Exercise 67–96, sketch the graph of each rational function.

67. $f(x) = \dfrac{1}{x} - 2$

68. $f(x) = \dfrac{1}{x^2} + 2$

69. $f(x) = \dfrac{1}{x - 1}$

70. $f(x) = \dfrac{1}{(x - 1)^2}$

71. $f(x) = \dfrac{1}{x + 1} - 3$

72. $f(x) = \dfrac{1}{(x + 3)^2} - 2$

73. $f(x) = \dfrac{-1}{x - 2} - 3$

74. $f(x) = \dfrac{-1}{(x + 1)^2} + 2$

75. $f(x) = \dfrac{2}{x + 1}$

76. $f(x) = \dfrac{2}{(x - 3)^2}$

77. $f(x) = \dfrac{-2}{x+1} - 1$

78. $f(x) = \dfrac{-2}{(x-1)^2} + 3$

79. $f(x) = \dfrac{x^2 - 9}{x - 3}$

80. $f(x) = \dfrac{x^2 + 3x - 4}{x + 4}$

81. $f(x) = \dfrac{x - 1}{x^2 - x}$

82. $f(x) = \dfrac{x + 2}{x^2 - x - 6}$

83. $f(x) = \dfrac{x + 2}{(x + 2)(x - 1)^2}$

84. $f(x) = \dfrac{x - 1}{x^3 + 3x^2 - 4}$

85. $f(x) = \dfrac{2x + 4}{x - 1}$

86. $f(x) = \dfrac{2x + 4}{x^2 - x - 2}$

87. $f(x) = \dfrac{x^2 - 1}{x^2 + 1}$

88. $f(x) = \dfrac{2x^2 - 5x - 3}{x + 2}$

89. $f(x) = \dfrac{1}{x^2 - 9}$

90. $f(x) = \dfrac{x}{x^2 - 9}$

91. $f(x) = \dfrac{x^2 + 2x - 3}{x + 1}$

92. $f(x) = \dfrac{2x^2 + 4x - 6}{x^2 - 4x + 3}$

93. $f(x) = \dfrac{x^3 - 3x^2 - x + 3}{x^2 - x - 6}$

94. $f(x) = \dfrac{x^3 - 7x + 6}{x^3 - 5x^2 + 2x + 8}$

95. $f(x) = \dfrac{2x^3 - 3x^2 - 3x + 2}{x^3 - 3x^2 - 6x + 8}$

96. $f(x) = \dfrac{2x^3 - 3x^2 - 3x + 2}{x^2 - 9}$

4.7 Variation

THINGS TO KNOW

Before working through this section, be sure that you are familiar with the following concepts:

VIDEO ANIMATION INTERACTIVE

You Try It

1. Converting Verbal Statements into Mathematical Statements (Section 1.2)

OBJECTIVES

1 Solve Application Problems Involving Direct Variation

2 Solve Application Problems Involving Inverse Variation

3 Solve Application Problems Involving Combined Variation

....

OBJECTIVE 1 SOLVE APPLICATION PROBLEMS INVOLVING DIRECT VARIATION

In application problems, we are often concerned with exploring how one quantity varies with respect to other quantities. Variation equations allow us to show how one quantity changes with respect to one or more additional quantities.

In this section we will discuss direct variation, inverse variation, and combined variation. We start by introducing **direct variation**.

Direct Variation

For an equation of the form

$$y = kx,$$

we say that y **varies directly** with x, or y is **proportional to** x. The constant k is called the **constant of variation** or the **proportionality constant**.

For direct variation, the ratio of the two quantities is constant (the constant of variation). For example, consider $y = kx$ and $y = kx^3$.

$$\underbrace{y = kx}_{\substack{y \text{ varies} \\ \text{directly} \\ \text{with } x}} \rightarrow \underbrace{\frac{y}{x}}_{\substack{\text{Ratio of the} \\ \text{quantities} \\ y \text{ and } x}} = \underbrace{k}_{\substack{\text{Constant of} \\ \text{variation}}} \qquad \underbrace{y = kx^3}_{\substack{y \text{ varies} \\ \text{directly with} \\ \text{the cube} \\ \text{of } x}} \rightarrow \underbrace{\frac{y}{x^3}}_{\substack{\text{Ratio of the} \\ \text{quantities} \\ y \text{ and } x^3}} = \underbrace{k}_{\substack{\text{Constant of} \\ \text{variation}}}$$

Problems involving variation can generally be solved using the following guidelines.

Solving Variation Problems

Step 1. Translate the problem into an equation that models the situation.

Step 2. Substitute given values for the variables into the equation and solve for the constant of variation, k.

Step 3. Substitute the value for k into the equation to form the general model.

Step 4. Use the general model to answer the question posed in the problem.

Example 1 Kinetic Energy

The kinetic energy of an object in motion varies directly with the square of its speed. If a van traveling at a speed of 30 meters per second has 945,000 joules of kinetic energy, how much kinetic energy does it have if it is traveling at a speed of 20 meters per second?

Solution Follow the guidelines for solving variation problems.

Step 1. We are told that the kinetic energy of an object in motion varies directly with the square of its speed. If we let $K =$ kinetic energy and $s =$ speed, we can translate the problem statement into the model

$$K = ks^2,$$

where k is the constant of variation.

Step 2. To determine the value of k, we use the fact that the kinetic energy is 945,000 joules when the velocity is 30 meters per second.

$$945,000 = k(30)^2 \qquad \text{Substitute 945,000 for } K \text{ and 30 for } s.$$

$$945,000 = 900k \qquad \text{Simplify } 30^2.$$

$$1050 = k \qquad \text{Divide both sides by 900.}$$

Step 3. The constant of variation is 1050, so the general model is $K = 1050s^2$.

Step 4. We want to determine the kinetic energy of the van if its speed is 20 meters per second. Substituting 20 for s, we find

$$K = 1050(20)^2 = 1050(400) = 420,000.$$

The van will have 420,000 joules of kinetic energy if it is traveling at a speed of 20 meters per second.

You Try It Work through this You Try It problem.

Work Exercises 1–2 in this textbook or in the MyMathLab Study Plan.

My video summary ⊘ **Example 2 Measuring Leanness**

The Ponderal Index measure of leanness states that body mass varies directly with the cube of height. If a "normal" person who is 1.2 m tall has a body mass of 21.6 kg, then what is the body mass of a "normal" person that is 1.8 m tall?

Solution Follow the guidelines for solving variation problems.

Step 1. We are told that weight varies directly with the cube of height. If we let $w = $ weight and $h = $ height, we can translate the problem statement into the model

$$w = kh^3,$$

where k is the constant of variation.

Step 2. To determine the value of k, we use the fact that a normal person who is 1.2 meters tall has a mass of 21.6 kg.

$21.6 = k(1.2)^3$	Substitute 1.2 for h and 21.6 for w.
$21.6 = 1.728k$	Simplify $(1.2)^3$
$12.5 = k$	Divide both sides by 1.728.

Step 3. The constant of variation is 12.5, so the general model is $w = 12.5h^3$.

Use the general model to determine the body mass of a normal person who is 1.8 m tall. View the solution, or watch this video to see the entire detailed solution.

You Try It Work through this You Try It problem.

Work Exercises 3–4 in this textbook or in the MyMathLab Study Plan.

OBJECTIVE 2 SOLVE APPLICATION PROBLEMS INVOLVING INVERSE VARIATION

We now discuss **inverse variation**. Inverse variation means that one variable is a constant multiple of the reciprocal of another variable.

Inverse Variation

For equations of the form

$$y = \frac{k}{x} \quad \text{or} \quad y = k \cdot \frac{1}{x},$$

we say that y **varies inversely** with x, or y is **inversely proportional to** x. The constant k is called the **constant of variation**.

For inverse variation, the product of the two quantities is constant (the constant of variation). For example, consider $y = \dfrac{k}{x}$ and $y = \dfrac{k}{x^2}$.

$$\underbrace{y = \dfrac{k}{x}}_{\substack{y \text{ varies} \\ \text{inversely} \\ \text{with } x}} \rightarrow \underbrace{xy}_{\substack{\text{Product} \\ \text{of the} \\ \text{quantities} \\ y \text{ and } x}} = \underbrace{k}_{\substack{\text{Constant of} \\ \text{variation}}} \qquad \underbrace{y = \dfrac{k}{x^2}}_{\substack{y \text{ varies inversely} \\ \text{with the square} \\ \text{of } x}} \rightarrow \underbrace{x^2 y}_{\substack{\text{Product of} \\ \text{the quantities} \\ y \text{ and } x^2}} = \underbrace{k}_{\substack{\text{Constant of} \\ \text{variation}}}$$

Example 3 Density of an Object

For a given mass, the density of an object is inversely proportional to its volume. If 50 cubic centimeters of an object with a density of $28\dfrac{\text{g}}{\text{cm}^3}$ is compressed to 40 cubic centimeters, what would be its new density?

Solution Follow the guidelines for solving variation problems.

Step 1. We are told that the density of an object with a given mass varies inversely with its volume. If we let $D =$ density and $V =$ volume, we can translate the problem statement into the model

$$D = \dfrac{k}{V},$$

where k is the constant of variation.

Step 2. To determine the value of k, we use the fact that the density is 28 g/cm^3 when the volume is 50 cm^3.

$$28 = \dfrac{k}{50} \qquad \text{Substitute 28 for } D \text{ and 50 for } V.$$

$$(28)(50) = k \qquad \text{Multiply both sides by 50.}$$

$$1400 = k \qquad \text{Simplify.}$$

Step 3. The constant of variation is 1400, so the general model is $D = \dfrac{1400}{V}$.

Step 4. We want to determine the density of the object if the volume is compressed to 40 cm^3. Substituting 40 for V, we find

$$D = \dfrac{1400}{40} = 35 \text{ g/cm}^3.$$

The density of the compressed object would be 35 g/cm^3.

You Try It Work through this You Try It problem.

Work Exercises 5–6 in this textbook or in the MyMathLab Study Plan.

My video summary ⊙ **Example 4 Shutter Speed**

The shutter speed, S, of a camera varies inversely as the square of the aperture setting, f. If the shutter speed is 125 for an aperture of 5.6, what is the shutter speed if the aperture is 1.4?

Solution Follow the guidelines for solving variation problems.

Step 1. We are told that shutter speed varies inversely with the square of the aperture setting. Letting $S =$ shutter speed and $f =$ aperture setting, we can translate the problem statement into the model

$$S = \frac{k}{f^2},$$

where k is the constant of variation.

Step 2. To determine the value of k, we use the fact that an aperture setting of 5.6 corresponds to a shutter speed of 125.

$$125 = \frac{k}{(5.6)^2} \qquad \text{Substitute 5.6 for } f \text{ and 125 for } S.$$

$$125 = \frac{k}{31.36} \qquad \text{Simplify } (5.6)^2.$$

$$3920 = k \qquad \text{Multiply both sides by 31.36.}$$

Step 3. The constant of variation is 3920, so the general model is $S = \dfrac{3920}{f^2}$.

Use the general model to determine the shutter speed for an aperture setting of 1.4. View the solution, or watch this video to see the entire detailed solution. ●

You Try It Work through this You Try It problem.

Work Exercises 7–8 in this textbook or in the My MathLab **Study Plan.**

OBJECTIVE 3   SOLVE APPLICATION PROBLEMS INVOLVING COMBINED VARIATION

When a variable is related to more than one other variable, we call this **combined variation**. Some examples of combined variation are

$$y = k\frac{x}{z} \quad y = kxz \quad y = k\frac{w^2x}{z}.$$

In the first example we say that y varies directly as x and inversely as z. In the second example we say that y varies directly as x and z. In the third example we say that y varies directly as x and the square of w, and inversely as z. In general, variables directly related to the dependent variable occur in the numerator while variables inversely related occur in the denominator.

$$y = k\frac{x \leftarrow \text{ directly related}}{z \leftarrow \text{ inversely related}} \qquad y = kxz \leftarrow \text{directly related} \qquad y = k\frac{w^2x \leftarrow \text{ directly related}}{z \leftarrow \text{ inversely related}}$$

We solve these problems the same way we solved other variation problems. First we find the constant of variation, then use it to help answer the question.

When a variable is directly proportional to the product of two or more other variables, such as $y = kxz$, this is often called **joint variation**. In this case we would say that y varies jointly as x and z.

Example 5 Volume of a Conical Tank

The number of gallons of a liquid that can be stored in a conical tank is directly proportional to the area of the base of the tank and its height (joint variation). A tank with a base area of 1200 square feet and a height of 15 feet holds 45,000 gallons of liquid. How tall must the tank be to hold 75,000 gallons of liquid if its base area is 1500 square feet?

Solution Follow the guidelines for solving variation problems.

Step 1. We are told that the volume is directly related to the area of the base of the tank and its height. If we let V = volume, B = base area, and h = height, we can translate the problem statement into the model

$$V = kBh \leftarrow \text{joint variation } (B \text{ and } h \text{ both direct}),$$

where k is the constant of variation.

Step 2. To determine the value of k, we use the fact that a tank with a base area of 1200 square feet and a height of 15 feet holds 45,000 gallons.

$45{,}000 = k(1200)(15)$	Substitute 1200 for B, 15 for h, and 45,000 for V.
$45{,}000 = 18{,}000k$	Simplify.
$2.5 = k$	Divide both sides by 18,000.

Step 3. The constant of variation is 2.5, so the general model is $V = 2.5Bh$.

Step 4. We want to determine the height of a tank with a base area of 1500 square feet that holds 75,000 gallons. Substituting 75,000 for V and 1500 for B, we then solve for h.

$75{,}000 = 2.5(1500)h$	Substitute 1500 for B and 75,000 for V.
$75{,}000 = 3750h$	Simplify.
$20 = h$	Divide both sides by 3750.

The tank would need to be 20 feet tall.

You Try It Work through this You Try It problem.

Work Exercises 9–10 in this textbook or in the MyMathLab Study Plan.

 *My video summary*

⊙ Example 6 Electrical Resistance

The resistance of a wire varies directly with the length of the wire and inversely with the square of its radius. A wire with a length of 500 cm and a radius of 0.5 cm has a resistance of 15 ohms. Determine the resistance of an 800 cm piece of similar wire with a radius of 0.8 cm.

Solution Follow the guidelines for solving variation problems.

Step 1. We are told that resistance varies directly with the length of the wire and inversely with the square of its radius. Letting R = resistance, L = length, and r = radius, we can translate the problem statement into the model

$$R = k\frac{L}{r^2} \begin{matrix} \leftarrow \text{length (direct)} \\ \leftarrow \text{square of the radius (inverse)} \end{matrix}$$

where k is the constant of variation.

Step 2. To determine the value of k, we use the fact that a wire of length 500 cm and radius 0.5 cm has a resistance of 15 ohms.

$$15 = k\frac{(500)}{(0.5)^2} \qquad \text{Substitute 500 for } L, 0.5 \text{ for } r, \text{ and 15 for } R.$$

$$15 = 2000k \qquad \text{Simplify.}$$

$$0.0075 = k \qquad \text{Divide both sides by 2000.}$$

Use the constant of variation to write a general model for this situation. Then use the model to determine the resistance of an 800 cm piece of similar wire with a radius of 0.8 cm. View the solution, or watch this video to see the entire detailed solution.

 You Try It Work through this You Try It problem.

Work Exercises 11–12 in this textbook or in the MyMathLab **Study Plan.**

 *My video summary* ⊙ **Example 7** Burning Calories

For a fixed speed, the number of calories burned while jogging varies jointly with the mass of the jogger (in kg) and the time spent jogging (in minutes). If a 100-kg man jogs for 40 minutes and burns 490 calories, how many calories will a 130-kg man burn if he jogs for 60 minutes at the same speed?

Solution Letting C = calories burned, W = weight in kg, and T = time jogging in minutes, we have the model $C = kWT$, where k is the constant of variation. Solve the problem on your own. View the solution, or watch this video to see the entire detailed solution.

You Try It Work through this You Try It problem.

Work Exercises 13–14 in this textbook or in the MyMathLab **Study Plan.**

4.7 Exercises

In Exercises 1–14, solve the variation problems.

1. The water pressure on a scuba diver is directly proportional to the depth of the diver. If the pressure on a diver is 13.5 psi when she is 30 feet below the surface, how far below the surface will she be when the pressure is 18 psi?

2. The length of a simple pendulum varies directly with the square of its period. If a pendulum of length 2.25 meters has a period of 3 seconds, how long is a pendulum with a period of 8 seconds?

3. For a fixed water flow rate, the amount of water that can be pumped through a pipe varies directly as the square of the diameter of the pipe. In one hour, a pipe with an 8-inch diameter can pump 400 gallons of water. Assuming the same water flow rate, how much water could be pumped through a pipe that is 12 inches in diameter?

4. The distance an object falls varies directly with the square of the time it spends falling. If a ball falls 19.6 meters after falling for 2 seconds, how far will it fall after 9 seconds?

5. The value of a car is inversely proportional to its age. If a car is worth \$8100 when it is 4 years old, how old will it be when it is worth \$3600?

6. For a given voltage, the resistance of a circuit is inversely related to its current. If a circuit has a resistance of 5 ohms and a current of 12 amps, what is the resistance if the current is 20 amps?

7. The weight of an object within Earth's atmosphere varies inversely with the square of the distance of the object from Earth's center. If a low Earth orbit satellite weighs 100 kg on Earth's surface (6400 km), how much will it weigh in its orbit 800 km above Earth?

8. The intensity of a light varies inversely as the square of the distance from the light source. If the intensity from a light source 3 feet away is 8 lumens, what is the intensity at a distance of 2 feet?

9. For a car loan using simple interest at a given rate, the amount of interest charged varies jointly with the loan amount and the time of the loan (in years). If a \$15,000 car loan earns \$2175 in simple interest over 5 years, how much interest will a \$32,000 car loan earn over 6 years?

10. The horsepower of a water pump varies jointly with the weight of the water moved (in lb) and the length of the discharge head (in feet). A 3.1-horsepower pump can move 341 lb of water against a discharge head of 300 feet. What is the horsepower of a pump that moves 429 lb of water against a discharge head of 400 feet?

11. The pressure of a gas in a container varies directly with its temperature and inversely with its volume. At a temperature of 90 Kelvin and a volume of 10 cubic meters, the pressure is 32.4 kilograms per square meter. What is the pressure if the temperature is increased to 100 Kelvin and the volume is decreased to 8 cubic meters?

12. In milling operations, the spindle speed S (in revolutions per minute) is directly related to the cutting speed C (in feet per minute) and inversely related to the tool diameter D (in inches). A milling cut taken with a 2-inch high-speed drill and a cutting speed of 70 feet per minute has a spindle speed of 133.7 revolutions per minute. What is the spindle speed for a cut taken with a 4-inch high-speed drill and a cutting speed of 50 feet per minute?

13. The load that can be supported by a rectangular beam varies jointly as the width of the beam and the square of its height, and inversely as the length of the beam. A beam 12 feet long, with a width of 6 inches and a height of 4 inches can support a maximum load of 900 pounds. If a similar board has a width of 8 inches and a height of 5 inches, how long must it be to support 1200 pounds?

14. The power produced by a windmill varies directly with the square of its diameter and the cube of the wind speed. A fan with a diameter of 2.5 meters produces 270 watts of power if the wind speed is 3 m/s. How much power would a similar fan produce if it had a diameter of 4 meters and the wind speed was 4 m/s?

Exponential and Logarithmic Functions and Equations

CHAPTER FIVE CONTENTS

5.1 Exponential Functions

THINGS TO KNOW

Before working through this section, be sure that you are familiar with the following concepts:

| | VIDEO | ANIMATION | INTERACTIVE |

You Try It

1. Using Combinations of Transformations to Graph Functions (Section 3.4)

You Try It

2. Determining if a Function Is One-to-One Using the Horizontal Line Test (Section 3.6)

OBJECTIVES

1 Understanding the Characteristics of Exponential Functions

2 Sketching the Graphs of Exponential Functions Using Transformations

3 Solving Exponential Equations by Relating the Bases

4 Solving Applications of Exponential Functions

OBJECTIVE 1   UNDERSTANDING THE CHARACTERISTICS OF EXPONENTIAL FUNCTIONS

Many natural phenomena and real-life applications can be modeled using exponential functions. Before we define the exponential function, it is important to remember how to manipulate exponential expressions because this skill is necessary when solving certain equations involving exponents. In Section R.3, expressions of the form b^r were evaluated for rational numbers r. For example,

$$3^2 = 9, \quad 4^{-2} = \frac{1}{4^2} = \frac{1}{16}, \quad \text{and} \quad 27^{-2/3} = \frac{1}{27^{2/3}} = \frac{1}{(\sqrt[3]{27})^2} = \frac{1}{(3)^2} = \frac{1}{9}$$

My video summary ⊙ In this section, we extend the meaning of b^r to include all **real** values of r by defining the exponential function $f(x) = b^x$. Watch this video for an explanation of the exponential function.

Definition Exponential Function

An **exponential function** is a function of the form $f(x) = b^x$, where x is any real number and $b > 0$ such that $b \neq 1$.

The constant, b, is called the base of the exponential function.

Notice in the definition that the base, b, must be positive and must not equal 1. If $b = 1$, then the function $f(x) = 1^x$ is equal to 1 for all x and is hence equivalent to the constant function $f(x) = 1$. If b were negative, then $f(x) = b^x$ would not be defined for all real values of x.

For example, if $b = -4$, then $f\left(\frac{1}{2}\right) = (-4)^{1/2} = \sqrt{-4} = 2i$, which is *not* a positive real number.

To illustrate the basic shapes of exponential functions we create a table of values and sketch the graph of $y = b^x$ for $b = 2, 3, \frac{1}{2}$ and $\frac{1}{3}$. See Table 1 and Figure 1.

Table 1

x	$y = 2^x$	$y = 3^x$	$y = \left(\dfrac{1}{2}\right)^x$	$y = \left(\dfrac{1}{3}\right)^x$
-2	$2^{-2} = \dfrac{1}{2^2} = \dfrac{1}{4}$	$3^{-2} = \dfrac{1}{3^2} = \dfrac{1}{9}$	$\left(\dfrac{1}{2}\right)^{-2} = 2^2 = 4$	$\left(\dfrac{1}{3}\right)^{-2} = 3^2 = 9$
-1	$2^{-1} = \dfrac{1}{2^1} = \dfrac{1}{2}$	$3^{-1} = \dfrac{1}{3^1} = \dfrac{1}{3}$	$\left(\dfrac{1}{2}\right)^{-1} = 2^1 = 2$	$\left(\dfrac{1}{3}\right)^{-1} = 3^1 = 3$
0	$2^0 = 1$	$3^0 = 1$	$\left(\dfrac{1}{2}\right)^0 = 1$	$\left(\dfrac{1}{3}\right)^0 = 1$
1	$2^1 = 2$	$3^1 = 3$	$\left(\dfrac{1}{2}\right)^1 = \dfrac{1}{2}$	$\left(\dfrac{1}{3}\right)^1 = \dfrac{1}{3}$
2	$2^2 = 4$	$3^2 = 9$	$\left(\dfrac{1}{2}\right)^2 = \dfrac{1}{4}$	$\left(\dfrac{1}{3}\right)^2 = \dfrac{1}{9}$

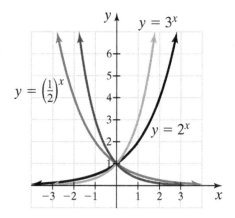

Figure 1 Graphs of $y = 2^x$, $y = 3^x$, $y = \left(\dfrac{1}{2}\right)^x$ and $y = \left(\dfrac{1}{3}\right)^x$

My video summary ⊙ Any positive number b, where $b \neq 1$, can be used as the base of an exponential function. However, there is one number that appears as the base in exponential applications more than any other number. This number is called the **natural base** and is symbolized using the letter e. The number e is an irrational number that is defined as the value of the expression $\left(1 + \dfrac{1}{n}\right)^n$ as n approaches infinity. Table 2 shows the values of the expression $\left(1 + \dfrac{1}{n}\right)^n$ for increasingly large values of n.

You can see from Table 2 that as the values of n get large, the value e (rounded to six decimal places) is 2.718282. The function $f(x) = e^x$ is called the **natural exponential function**.

Table 2

n	$\left(1 + \dfrac{1}{n}\right)^{n}$
1	2
2	2.25
10	2.5937424601
100	2.7048138294
1,000	2.7169239322
10,000	2.7181459268
100,000	2.7182682372
1,000,000	2.7182804693
10,000,000	2.7182816925
100,000,000	2.7182818149

Because $2 < e < 3$, it follows that the graph of $f(x) = e^x$ lies between the graphs of $y = 2^x$ and $y = 3^x$, as seen in Figure 2.

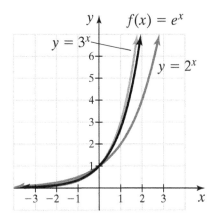

Figure 2 The graphs of $y = 2^x$, $y = 3^x$, and the graph of the natural exponential function $f(x) = e^x$.

You can see from the graphs sketched in Figure 1 and Figure 2 that all of the exponential functions intersect the y-axis at the point $(0, 1)$. This is true because $b^0 = 1$ for all nonzero values of b. For values of $b > 1$, the graph of $y = b^x$ increases rapidly as the values of x approach positive infinity ($b^x \to \infty$ as $x \to \infty$). In fact, the larger the base, the faster the graph will grow. Also, for $b > 1$, the graph of $y = b^x$ decreases quickly, approaching 0 as the values of x approach negative infinity ($b^x \to 0$ as $x \to -\infty$). Thus, the x-axis (the line $y = 0$) is a horizontal asymptote.

However, for $0 < b < 1$, the graph decreases quickly, approaching the horizontal asymptote $y = 0$ as the values of x approach positive infinity ($b^x \to 0$ as $x \to \infty$), whereas the graph increases rapidly as the values of x approach negative infinity ($b^x \to \infty$ as $x \to -\infty$). The preceding statements, along with some other characteristics of the graphs of exponential functions, are outlined on the following page.

Characteristics of Exponential Functions

For $b > 0$, $b \neq 1$, the exponential function with base b is defined by $f(x) = b^x$.

The domain of $f(x) = b^x$ is $(-\infty, \infty)$, and the range is $(0, \infty)$. The graph of $f(x) = b^x$ has one of the following two shapes:

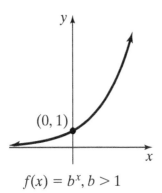

$$f(x) = b^x, b > 1 \qquad\qquad f(x) = b^x, 0 < b < 1$$

The graph intersects the y-axis at $(0, 1)$.
$b^x \to \infty$ as $x \to \infty$
$b^x \to 0$ as $x \to -\infty$
The line $y = 0$ is a horizontal asymptote.
The function is one to one.

The graph intersects the y-axis at $(0, 1)$.
$b^x \to 0$ as $x \to \infty$
$b^x \to \infty$ as $x \to -\infty$
The line $y = 0$ is a horizontal asymptote.
The function is one to one.

It is important that you are able to use your calculator to evaluate exponential expressions. Most calculators have an $\boxed{e^x}$ key. Find this special key on your calculator and evaluate the expressions in the following example.

Example 1 Evaluate Exponential Expressions

Evaluate each exponential expressions correctly to six decimal places.

a. $\left(\dfrac{3}{7}\right)^{0.6}$ **b.** $9\left(\dfrac{5}{8}\right)^{-0.375}$ **c.** e^2 **d.** $e^{-0.534}$ **e.** $1{,}000e^{0.013}$

Solution

a. $\left(\dfrac{3}{7}\right)^{0.6} \approx 0.601470$ **b.** $9\left(\dfrac{5}{8}\right)^{-0.375} \approx 10.734640$

For parts c–e, we use the $\boxed{e^x}$ key on a calculator to get

c. $e^2 \approx 7.389056$ **d.** $e^{-0.534} \approx 0.586255$ **e.** $1{,}000e^{0.013} \approx 1{,}013.084867$

You Try It Work through this You Try It problem.

Work Exercises 1–7 in this textbook or in the MyMathLab Study Plan.

My interactive video summary

⊙ **Example 2** Sketch the Graph of an Exponential Function

Sketch the graph of each exponential function.

a. $f(x) = \left(\dfrac{2}{3}\right)^x$ **b.** $f(x) = -2e^x$

Solution

a. Because the base of the exponential function is $\frac{2}{3}$, which is between 0 and 1, the graph must approach the x-axis as the value of x approaches positive infinity. The graph intersects the y-axis at $(0, 1)$. We can find a few more points by choosing some negative and positive values of x:

$$f(-2) = \left(\frac{2}{3}\right)^{-2} = \left(\frac{3}{2}\right)^{2} = \frac{9}{4}$$

$$f(-1) = \left(\frac{2}{3}\right)^{-1} = \left(\frac{3}{2}\right)^{1} = \frac{3}{2}$$

$$f(1) = \left(\frac{2}{3}\right)^{1} = \frac{2}{3}$$

$$f(2) = \left(\frac{2}{3}\right)^{2} = \frac{4}{9}$$

My interactive video summary

⊙ We can complete the graph by connecting the points with a smooth curve. Watch this video to see how to sketch the complete function as seen in Figure 3.

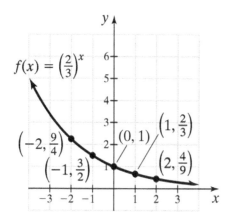

Figure 3 The graph of $f(x) = \left(\frac{2}{3}\right)^{x}$.

b. We know that the value of e^x is greater than zero for all values of x. Thus, the values of $-2e^x$ must be less than zero for all values of x. This means that the graph of the function $f(x) = -2e^x$ must be below the x-axis for all values of x.

The y-intercept is $f(0) = -2e^0 = -2(1) = -2$. Therefore, the graph intersects the y-axis at $(0, -2)$. We can find a few more points by choosing some arbitrary values of x.

$$f(-2) = -2e^{-2} = -2\left(\frac{1}{e^2}\right) \approx -0.27067$$

$$f(-1) = -2e^{-1} = -2\left(\frac{1}{e}\right) \approx -0.73576$$

$$f(1) = -2e^{1} = -2(e) \approx -5.43656$$

My interactive video summary

⊙ We can complete the graph by connecting the points with a smooth curve. Watch this interactive video to see how to sketch the function seen in Figure 4. Note that you could also use transformation techniques to sketch the graph of $f(x) = -2e^x$. We will use transformation techniques to sketch the graphs of exponential functions in Objective 2.

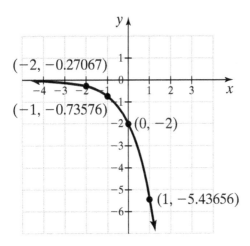

Figure 4 The graph of $f(x) = -2e^x$.

You Try It Work through this You Try It problem.

Work Exercises 8–14 in this textbook or in the MyMathLab Study Plan.

Example 3 Determine an Exponential Function Given the Graph

Find the exponential function $f(x) = b^x$ whose graph is given as follows.

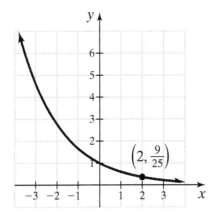

Solution From the point $\left(2, \dfrac{9}{25}\right)$, we see that $f(2) = \dfrac{9}{25}$. Thus,

$$f(x) = b^x \qquad \text{Write the exponential function } f(x) = b^x.$$

$$f(2) = b^2 \qquad \text{Evaluate } f(2).$$

$$f(2) = \frac{9}{25} \qquad \text{The graph contains the point } \left(2, \frac{9}{25}\right).$$

$$b^2 = \frac{9}{25} \qquad \text{Equate the two expressions for } f(2).$$

Therefore, we are looking for a constant b such that $b^2 = \dfrac{9}{25}$. Using the square root property, we get

$$\sqrt{b^2} = \pm\sqrt{\frac{9}{25}}$$

$$b = \pm\frac{3}{5}$$

By definition of an exponential function, $b > 0$; thus, $b = \dfrac{3}{5}$.

Therefore, this is the graph of $f(x) = \left(\dfrac{3}{5}\right)^x$.

You Try It   Work through this You Try It problem.

Work Exercises 15–21 in this textbook or in the MyMathLab Study Plan.

OBJECTIVE 2   SKETCHING THE GRAPHS OF EXPONENTIAL FUNCTIONS USING TRANSFORMATIONS

Often we can use the transformation techniques that are discussed in Section 3.4 to sketch the graph of an exponential function. For example, the graph of $f(x) = 3^x - 1$ can be obtained by vertically shifting the graph of $y = 3^x$ down one unit. You can see in Figure 5 that the y-intercept of $f(x) = 3^x - 1$ is $(0, 0)$ and the horizontal asymptote is the line $y = -1$.

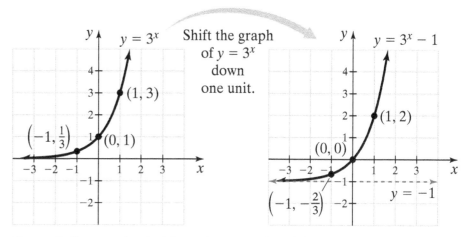

Figure 5 The graph of $f(x) = 3^x - 1$ can be obtained by vertically shifting the graph of $y = 3^x$ down one unit.

My video summary   ⊘ **Example 4** Use Transformations to Sketch an Exponential Function

Use transformations to sketch the graph of $f(x) = -2^{x+1} + 3$.

Solution

Starting with the graph of $y = 2^x$, we can obtain the graph of $f(x) = -2^{x+1} + 3$ through a series of three transformations:

1. Horizontally shift the graph of $y = 2^x$ to the left one unit, producing the graph of $y = 2^{x+1}$.

2. Reflect the graph of $y = 2^{x+1}$ about the x-axis, producing the graph of $y = -2^{x+1}$.

3. Vertically shift the graph of $y = -2^{x+1}$ up three units, producing the graph of $f(x) = -2^{x+1} + 3$.

The graph of $f(x) = -2^{x+1} + 3$ is shown in Figure 6. Watch the video to see every step.

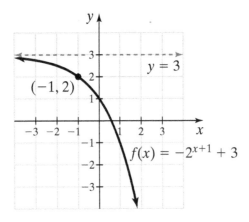

Figure 6 Graph of
$$f(x) = -2^{x+1} + 3$$

Notice that the graph of $f(x) = -2^{x+1} + 3$ in Figure 6 has a y-intercept. We can find the y-intercept by evaluating $f(0) = -2^{0+1} + 3 = -2 + 3 = 1$. Also notice that the graph has an x-intercept. Can you find it? Recall that to find an x-intercept, we need to set $f(x) = 0$ and solve for x.

$f(x) = -2^{x+1} + 3$	Write the original function.
$0 = -2^{x+1} + 3$	Substitute 0 for $f(x)$.
$-3 = -2^{x+1}$	Subtract 3 from both sides.
$3 = 2^{x+1}$	Multiply both sides by -1.

We now have a problem! We are going to need a way to solve for a variable that appears in an exponent. Stay tuned for Section 5.4.

 *My video summary*

⊙ **Example 5** Use Transformations to Sketch Natural Exponential Functions

Use transformations to sketch the graph of $f(x) = -e^x + 2$. Determine the domain, range, and y-intercept, and find the equation of any asymptotes.

Solution We can sketch the graph of $f(x) = -e^x + 2$ through a series of the following two transformations.

Start with the graph of $y = e^x$.

1. Reflect the graph of $y = e^x$ about the x-axis, producing the graph of $y = -e^x$.

2. Vertically shift the graph of $y = -e^x$ up two units, producing the graph of $f(x) = -e^x + 2$.

The graph of $f(x) = -e^x + 2$ is shown in Figure 7. Watch the video to see each step. The domain of f is the interval $(-\infty, \infty)$. The range of f is the interval $(-\infty, 2)$. The y-intercept is 1, and the equation of the horizontal asymptote is $y = 2$.

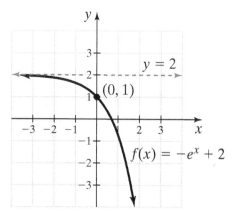

Figure 7 Graph of $f(x) = -e^x + 2$

You Try It   Work through this You Try It problem.

Work Exercises 22–34 in this textbook or in the MyMathLab Study Plan.

OBJECTIVE 3   SOLVING EXPONENTIAL EQUATIONS BY RELATING THE BASES

One important property of all exponential functions is that they are one to one functions. You may want to review Section 3.6, which discusses one to one functions in detail. The function $f(x) = b^x$ is one to one because the graph of f passes the horizontal line test. In Section 3.6, the alternate definition of one to one is stated as follows:

A function f is one to one if for any two range values $f(u)$ and $f(v)$, $f(u) = f(v)$ implies that $u = v$.

Using this definition and letting $f(x) = b^x$, we can say that if $b^u = b^v$, then $u = v$. In other words, if the bases of an exponential equation of the form $b^u = b^v$ are the same, then the exponents must be the same. Solving exponential equations with this property is known as the **method of relating the bases** for solving exponential equations.

Method of Relating the Bases

If an exponential equation can be written in the form $b^u = b^v$, then $u = v$.

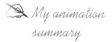
My animation summary

▶ **Example 6** Use the Method of Relating the Bases to Solve Exponential Equations

Solve the following equations:

a. $8 = \dfrac{1}{16^x}$ **b.** $\dfrac{1}{27^x} = \left(\sqrt[4]{3}\right)^{x-2}$

Solution

a. Work through the animation to see how to obtain a solution of $x = -\dfrac{3}{4}$.

b. Work through the animation to see how to obtain a solution of $x = \dfrac{2}{13}$.

My interactive video summary

⊙ **Example 7** Use the Method of Relating the Bases to Solve Natural Exponential Equations

Use the method of relating the bases to solve each exponential equation:

a. $e^{3x-1} = \dfrac{1}{\sqrt{e}}$ b. $\dfrac{e^{x^2}}{e^{10}} = (e^x)^3$

Solution Work through the interactive video to verify that the solutions are as follows:

a. $x = \dfrac{1}{6}$ b. $x = -2$ or $x = 5$ ●

You Try It Work through this You Try It problem.

Work Exercises 35–52 in this textbook or in the MyMathLab Study Plan.

OBJECTIVE 4 SOLVING APPLICATIONS OF EXPONENTIAL FUNCTIONS

Exponential functions are used to describe many real-life situations and natural phenomena. We now look at some examples.

Example 8 Learn to Hit a 3-Wood on a Golf Driving Range

Most golfers find that their golf skills improve dramatically at first and then level off rather quickly. For example, suppose that the distance (in yards) that a typical beginning golfer can hit a 3-wood after t weeks of practice on the driving range is given by the exponential function $d(t) = 225 - 100e^{-0.7t}$. This function has been developed after many years of gathering data on beginning golfers.

How far can a typical beginning golfer initially hit a 3-wood? How far can a typical beginning golfer hit a 3-wood after 1 week of practice on the driving range? After 5 weeks? After 9 weeks? Round to the nearest hundredth yard.

Solution Initially, when $t = 0$, $d(0) = 225 - 100e^0 = 225 - 100 = 125$ yards. Therefore, a typical beginning golfer can hit a 3-wood 125 yards.

After 1 week of practice on the driving range, a typical beginning golfer can hit a 3-wood $d(1) = 225 - 100e^{-0.7(1)} \approx 175.34$ yards.

After 5 weeks of practice, $d(5) = 225 - 100e^{-0.7(5)} \approx 221.98$ yards.

After 9 weeks of practice, $d(9) = 225 - 100e^{-0.7(9)} \approx 224.82$ yards.   ●

Using a graphing utility, we can sketch the graph of $d(t) = 225 - 100e^{-0.7t}$. You can see from the graph in Figure 8 that the distance increases rather quickly and then tapers off toward a horizontal asymptote of 225 yards.

Using Technology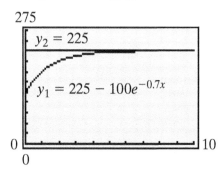

Figure 8 A TI-83 Plus was used to sketch the function $d(t) = 225 - 100e^{-0.7t}$ and the horizontal asymptote $y = 225$.

You Try It Work through this You Try It problem.

Work Exercises 53–56 in this textbook or in the MyMathLab Study Plan.

COMPOUND INTEREST

A real-life application of exponential functions is the concept of **compound interest**, that is, interest that is paid on both principal and interest. First, we take a look at how **simple interest** is accrued. If an investment of P dollars is invested at r percent annually (written as a decimal) using simple interest, then the interest earned after 1 year is Pr dollars. Adding this interest to the original investment yields a total amount, A, of

$$A = \underbrace{P}_{\substack{\text{Original} \\ \text{investment}}} + \underbrace{Pr}_{\substack{\text{Interest} \\ \text{earned}}} = P(1 + r).$$

If this amount is reinvested at the same interest rate, then the total amount after 2 years becomes

$$A = \underbrace{P(1 + r)}_{\substack{\text{Total investment} \\ \text{after 1 year}}} + \underbrace{P(1 + r)r}_{\substack{\text{Interest} \\ \text{earned}}} = P(1 + r)(1 + r) = P(1 + r)^2.$$

Reinvesting this amount for a third year gives an amount of $P(1 + r)^3$. Continuing this process for k years, we can see that the amount becomes $A = P(1 + r)^k$. This is an exponential function with base $1 + r$.

We can now modify this formula to obtain another formula that will model interest that is compounded periodically throughout the year(s). When interest is compounded periodically, then k no longer represents the number of years but rather the number of pay periods. If interest is paid n times per year for t years, then $k = nt$ pay periods. Thus, in the formula $A = P(1 + r)^k$, we substitute nt for k and get $A = P(1 + r)^{nt}$.

In the earlier simple interest model, the variable r was used to represent annual interest. In the periodically compound interest model being developed here with n pay periods per year, the interest rate per pay period is no longer r but rather $\dfrac{r}{n}$.

Thus, in the formula $A = P(1 + r)^{nt}$, we replace r with $\dfrac{r}{n}$ and get the periodic compound interest formula $A = P\left(1 + \dfrac{r}{n}\right)^{nt}$.

Periodic Compound Interest Formula

Periodic compound interest can be calculated using the formula

$$A = P\left(1 + \frac{r}{n}\right)^{nt},$$

where

A = Total amount after t years
P = Principal (original investment)
r = Interest rate per year
n = Number of times interest is compounded per year
t = Number of years

Example 9 Calculate Compound Interest

Which investment results in the greatest total amount after 25 years?

Investment A: $12,000 invested at 3% compounded monthly
Investment B: $10,000 invested at 3.9% compounded quarterly

Solution Investment A: $P = 12{,}000, r = .03, n = 12, t = 25$:

$$A = 12{,}000\left(1 + \frac{.03}{12}\right)^{12(25)} \approx \$25{,}380.23$$

Investment B: $P = 10{,}000, r = .039, n = 4, t = 25$:

$$A = 10{,}000\left(1 + \frac{.039}{4}\right)^{4(25)} \approx \$26{,}386.77$$

Investment B will yield the most money after 25 years.

You Try It Work through this You Try It problem.

Work Exercises **57–59** in this textbook or in the MyMathLab Study Plan.

CONTINUOUS COMPOUND INTEREST

Some banks use **continuous compounding**; that is, they compound the interest every fraction of a second every day! If we start with the formula for periodic compound interest, $A = P\left(1 + \frac{r}{n}\right)^{nt}$, and let n (the number of times the interest is compounded each year) approach infinity, we can derive the formula $A = Pe^{rt}$, which is the formula for continuous compound interest. Work through this animation to see exactly how this formula is derived.

Continuous Compound Interest Formula

Continuous compound interest can be calculated using the formula

$$A = Pe^{rt},$$

where

A = Total amount after t years
P = Principal
r = Interest rate per year
t = Number of years

Example 10 Calculate Continuous Compound Interest

How much money would be in an account after 5 years if an original investment of $6,000 was compounded continuously at 4.5%? Compare this amount to the same investment that was compounded daily. Round to the nearest cent.

Solution First, the amount after 5 years compounded continuously is $A = Pe^{rt} = 6{,}000e^{.045(5)} \approx \$7{,}513.94$. The same investment compounded daily yields an amount of $A = P\left(1 + \dfrac{r}{n}\right)^{nt} = 6{,}000\left(1 + \dfrac{.045}{365}\right)^{365(5)} \approx \$7{,}513.83$.

Continuous compound interest yields $.11 more interest after 5 years!

You Try It   Work through this You Try It problem.

Work Exercises 60–62 in this textbook or in the MyMathLab Study Plan.

PRESENT VALUE FOR PERIODIC COMPOUND INTEREST

Sometimes investors want to know how much money to invest now to reach a certain investment goal in the future. This amount of money, P, is known as the **present value** of A dollars. To find a formula for present value, start with the formula for periodic compound interest and solve the formula for P:

$$A = P\left(1 + \frac{r}{n}\right)^{nt}$$ Use the periodic compound interest formula.

$$\frac{A}{\left(1 + \dfrac{r}{n}\right)^{nt}} = \frac{\cancel{P\left(1 + \dfrac{r}{n}\right)^{nt}}}{\cancel{\left(1 + \dfrac{r}{n}\right)^{nt}}}$$ Divide both sides by $\left(1 + \dfrac{r}{n}\right)^{nt}$.

$$A\left(1 + \frac{r}{n}\right)^{-nt} = P$$ Rewrite $\dfrac{1}{\left(1 + \dfrac{r}{n}\right)^{nt}}$ as $\left(1 + \dfrac{r}{n}\right)^{-nt}$.

The formula $P = A\left(1 + \dfrac{r}{n}\right)^{-nt}$ is known as the **present value formula for periodic compound interest**.

Present Value Formula for Periodic Compound Interest

Present value can be calculated using the formula

$$P = A\left(1 + \frac{r}{n}\right)^{-nt},$$

where

P = Principal (original investment)
A = Total amount after t years
r = Interest rate per year
n = Number of times interest is compounded per year
t = Number of years

PRESENT VALUE FOR CONTINUOUS COMPOUND INTEREST

To find a formula for present value on money that is compounded continuously, we start with the formula for continuous compound interest and solve for P.

$$A = Pe^{rt}$$ Write the continuous compound interest formula.

$$\frac{A}{e^{rt}} = P\frac{e^{rt}}{e^{rt}}$$ Divide both sides by e^{rt}.

$$Ae^{-rt} = P$$ Rewrite $\frac{1}{e^{rt}}$ as e^{-rt}.

Present Value Formula for Continuous Compound Interest

The present value of A dollars after t years of continuous compound interest, with interest rate r, is given by the formula

$$P = Ae^{-rt}.$$

Example 11 Determine Present Value

a. Find the present value of \$8,000 if interest is paid at a rate of 5.6% compounded quarterly for 7 years. Round to the nearest cent.

b. Find the present value of \$18,000 if interest is paid at a rate of 8% compounded continuously for 20 years. Round to the nearest cent.

Solution

a. Using the present value formula for periodic compound interest
$$P = A\left(1 + \frac{r}{n}\right)^{-nt}$$ with $A = \$8,000, r = .056, n = 4$, and $t = 7$, we get

$$P = A\left(1 + \frac{r}{n}\right)^{-nt}$$ Write the present value formula for periodic compound interest.

$$P = 8,000\left(1 + \frac{.056}{4}\right)^{-(4)(7)} \approx 5,420.35$$ Substitute and use a calculator to simplify.

Therefore, the present value of \$8,000 in 7 years at 5.6% compounded quarterly is \$5,420.35.

b. Using the present value formula for continuous compound interest $P = Ae^{-rt}$ with $A = \$18,000, r = .08$, and $t = 20$, we get

$$P = Ae^{-rt}$$ Write the present value formula for continuous compound interest.

$$P = (18,000)e^{-(.08)(20)} \approx \$3,634.14.$$ Substitute and use a calculator to simplify.

You Try It Work through this You Try It problem.

Work Exercises 63–67 in this textbook or in the MyMathLab Study Plan.

EXPONENTIAL GROWTH MODEL

You have probably heard that some populations grow exponentially. Most populations grow at a rate proportional to the size of the population. In other words, the larger the population, the faster the population grows. With this in mind, it can be shown in a more advanced math course that the mathematical model that can describe population growth is given by the function $P(t) = P_0 e^{kt}$.

Exponential Growth

A model that describes the population, P, after a certain time, t, is

$$P(t) = P_0 e^{kt},$$

where $P_0 = P(0)$ is the initial population and $k > 0$ is a constant called the **relative growth rate**. (*Note:* k may be given as a percent.)

The graph of the exponential growth model is shown in Figure 9. Notice that the graph has a y-intercept of P_0.

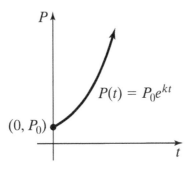

Figure 9 Graph of the exponential growth model $P(t) = P_0 e^{kt}$

My video summary ⊘ **Example 12** Population Growth

The population of a small town follows the exponential growth model $P(t) = 900e^{.015t}$, where t is the number of years after 1900.

Answer the following questions, rounding each answer to the nearest whole number:

a. What was the population of this town in 1900?

b. What was the population of this town in 1950?

c. Use this model to predict the population of this town in 2012.

Solution

a. The initial population was $P(0) = 900e^{.015(0)} = 900$.

b. Because 1950 is 50 years after 1900, we must evaluate $P(50)$.
   $P(50) = 900e^{.015(50)} \approx 1,905$.

c. In the year 2012, we can predict that the population will be
   $P(112) = 900e^{.015(112)} \approx 4,829$.

 *My video summary*   ⊙ **Example 13** Determine the Initial Population

Twenty years ago, the Idaho Fish and Game Department introduced a new breed of wolf into a certain Idaho forest. The current wolf population in this forest is now estimated at 825, with a relative growth rate of 12%.

Use the exponential growth model $P(t) = P_0 e^{kt}$ to answer the following questions. Round each answer to the nearest whole number.

a. How many wolves did the Idaho Fish and Game Department initially introduce into this forest?

b. How many wolves can be expected after another 20 years?

Solution

a. The relative growth rate is .12, so we use the exponential growth model $P(t) = P_0 e^{.12t}$. Because $P(20) = 825$, we get

$$P(20) = P_0 e^{.12(20)} \qquad \text{Substitute 20 for } t.$$

$$825 = P_0 e^{.12(20)} \qquad \text{Substitute 825 for } P(20).$$

$$P_0 = \frac{825}{e^{.12(20)}} \approx 75 \qquad \text{Solve for } P_0.$$

Therefore, the Idaho Fish and Game Department initially introduced 75 wolves into the forest.

b. Because $P_0 = 75$, we can use the exponential growth model $P(t) = 75 e^{.12t}$. In another 20 years, the value of t will be 40. Thus, we must evaluate $P(40)$.

$$P(40) = 75 e^{.12(40)} \approx 9{,}113$$

Therefore, we can expect the wolf population to be approximately 9,113 in another 20 years. Watch this video to see the entire solution to this example.   ●

You Try It   Work through this You Try It problem.

Work Exercises 68–71 in this textbook or in the MyMathLab Study Plan.

5.1 Exercises

Skill Check Exercises

For exercises SCE-1 through SCE-5, rewrite each expression in the form 2^u, 3^u, or 5^u where u is a constant or an algebraic expression.

SCE-1. 27 SCE-2. $\dfrac{1}{2^3}$ SCE-3. $\sqrt[4]{3}$ SCE-4. $\dfrac{1}{5^{3x}}$ SCE-5. $\left(\sqrt[3]{3}\right)^x$

SCE-6. $\left(\dfrac{1}{2}\right)^{x-1}$ SCE-7. $\dfrac{25}{\sqrt[4]{5}}$ SCE-8. $\dfrac{81}{\sqrt[5]{3^x}}$ SCE-9. $\dfrac{3^{x^2}}{9^x}$ SCE-10. $\dfrac{4^x}{2^{-x^2}}$

In Exercises 1–7, use a calculator to approximate each exponential expression to six decimal places.

1. $\left(\dfrac{2}{11}\right)^{0.7}$ **2.** $4\left(\dfrac{3}{13}\right)^{-0.254}$ **3.** e^3 **4.** $e^{-0.2}$ **5.** $e^{1/3}$ **6.** $100e^{-.123}$ **7.** $\sqrt{2}e^{\pi}$

In Exercises 8–14, sketch the graph of each exponential function. Label the y-intercept and at least two other points on the graph using both positive and negative values of x.

8. $f(x) = 4^x$ **9.** $f(x) = \left(\dfrac{1}{4}\right)^x$ **10.** $f(x) = \left(\dfrac{3}{2}\right)^x$ **11.** $f(x) = (.4)^x$

12. $f(x) = (2.7)^x$ **13.** $f(x) = -3e^x$ **14.** $f(x) = 2\left(\dfrac{1}{e}\right)^x$

In Exercises 15–21, determine the correct exponential function of the form $f(x) = b^x$ whose graph is given.

15.

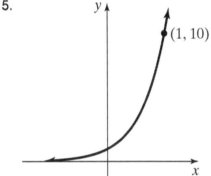

16.

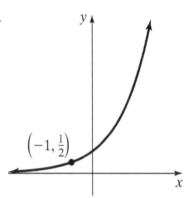

17.

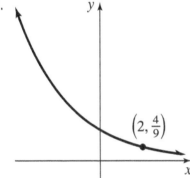

18.

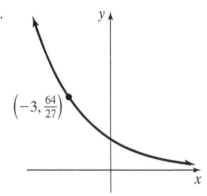

19.

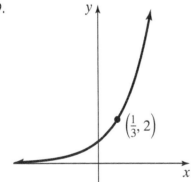

20.

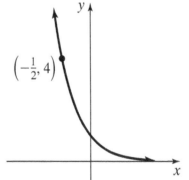

21.

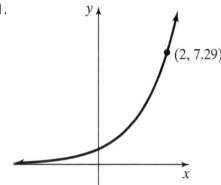

$(2, 7.29)$

In Exercises 22–34, use the graph of $y = 2^x$, $y = e^x$, or $y = 3^x$ and transformations to sketch each exponential function. Determine the domain and range. Also, determine the y-intercept, and find the equation of the horizontal asymptote.

SbS **22.** $f(x) = 2^{x-1}$

SbS **23.** $f(x) = 3^x - 1$

SbS **24.** $f(x) = -3^{x+2}$

SbS **25.** $f(x) = -2^{x+1} - 1$

SbS **26.** $f(x) = \left(\dfrac{1}{2}\right)^{x+1}$

SbS **27.** $f(x) = \left(\dfrac{1}{3}\right)^x - 3$

SbS **28.** $f(x) = 2^{-x} + 1$

SbS **29.** $f(x) = 3^{1-x} - 2$

SbS **30.** $f(x) = e^{x-1}$

SbS **31.** $f(x) = e^x - 1$

SbS **32.** $f(x) = -e^{x+2}$

SbS **33.** $f(x) = -e^{x+1} - 1$

SbS **34.** $f(x) = e^{-x} - 2$

In Exercises 35–52, solve each exponential equation using the method of "relating the bases" by first rewriting the equation in the form $b^u = b^v$.

35. $2^x = 16$

36. $3^{x-1} = \dfrac{1}{9}$

37. $\sqrt{5} = 25^x$

38. $\left(\sqrt[3]{3}\right)^x = 9$

39. $\dfrac{1}{\sqrt[5]{8}} = 2^x$

40. $\dfrac{9}{\sqrt[4]{3}} = \left(\dfrac{1}{27}\right)^x$

41. $(49)^x = \left(\dfrac{1}{7}\right)^{x-1}$

42. $\dfrac{125}{\sqrt[3]{5^x}} = \left(\dfrac{1}{25^x}\right)$

43. $\dfrac{3^{x^2}}{9^x} = 27$

44. $2^{x^3} = \dfrac{4^x}{2^{-x^2}}$

45. $e^x = \dfrac{1}{e^2}$

46. $e^{5x+2} = \sqrt[3]{e}$

47. $\dfrac{1}{e^x} = \dfrac{\sqrt{e}}{e^{1-x}}$

48. $(e^{x^2})^2 = e^8$

49. $e^{x^2} = (e^x) \cdot e^{12}$

50. $e^{x^2} = \dfrac{e^3}{(e^x)^5}$

51. $\dfrac{e^{x^3}}{e^x} = \dfrac{e^{2x^2}}{e^2}$

52. $e^{x^3} \cdot e^6 = \dfrac{(e^{2x^2})^2}{e^x}$

53. Typically, weekly sales will drop off rather quickly after the end of an advertising campaign. This drop in sales is known as *sales decay*. Suppose that the gross sales S, in hundreds of dollars, of a certain product is given by the exponential function $S(t) = 3{,}000(1.5^{-0.3t})$, where t is the number of weeks after the end of the advertising campaign.

Answer the following questions, rounding each answer to the nearest whole number:

a. What was the level of sales immediately after the end of the advertising campaign when $t = 0$?

b. What was the level of sales 1 week after the end of the advertising campaign?

c. What was the level of sales 5 weeks after the end of the advertising campaign?

54. Most people who start a serious weight-lifting regimen initially notice a rapid increase in the maximum amount of weight that they can bench press. After a few weeks, this increase starts to level off. The following function models the maximum weight, w, that a particular person can bench press in pounds at the end of t weeks of working out.

$$w(t) = 250 - 120e^{-0.3t}$$

Answer the following questions, rounding each answer to the nearest whole number:

a. What is the maximum weight that this person can bench press initially?

b. What is the maximum weight that this person can bench press after 3 weeks of weight lifting?

c. What is the maximum weight that this person can bench press after 7 weeks of weight lifting?

55. *Escherichia coli* bacteria reproduce by simple cell division, which is known as binary fission. Under ideal conditions, a population of *E. coli* bacteria can double every 20 minutes. This behavior can be modeled by the exponential function $N(t) = N_0(2^{.05t})$, where t is in minutes and N_0 is the initial number of *E. coli* bacteria.

Answer the following questions, rounding each answer to the nearest bacteria:

a. If the initial number of *E. coli* bacteria is five, how many bacteria will be present in 3 hours?

b. If the initial number of *E. coli* bacteria is eight, how many bacteria will be present in 3 hours?

c. If the initial number of *E. coli* bacteria is eight, how many bacteria will be present in 10 hours?

56. A wildlife-management research team noticed that a certain forest had no rabbits, so they decided to introduce a rabbit population into the forest for the first time. The rabbit population will be controlled by wolves and other predators. This rabbit population can be modeled by the function $R(t) = \dfrac{960}{.6 + 23.4e^{-.045t}}$, where t is the number of weeks after the research team first introduced the rabbits into the forest.

Answer the following questions, rounding each answer to the nearest whole number:

a. How many rabbits did the wildlife management research team bring into the forest?

b. How many rabbits can be expected after 10 weeks?

c. How many rabbits can be expected after the first year?

d. What is the expected rabbit population after 4 years? 5 years?
    What can the expected rabbit population approach as time goes on?

Use the periodic compound interest formula to solve Exercises 57–59.

57. Suppose that $9,000 is invested at 3.5% compounded quarterly. Find the total amount of this investment after 10 years. Round to the nearest cent.

58. Suppose that you have $5,000 to invest. Which investment yields the greater return over a 10-year period: 7.35% compounded daily or 7.4% compounded quarterly?

59. Which investment yields the greatest total amount?

Investment A: $4,000 invested for 5 years compounded semiannually at 8%
Investment B: $5,000 invested for 4 years compounded quarterly at 4.5%

Use the continuous compound interest formula to solve Exercises 60–62.

60. An original investment of $6,000 earns 6.25% interest compounded continuously. What will the investment be worth in 2 years? 20 years? Round to the nearest cent.

61. How much more will an investment of $10,000 earning 5.5% compounded continuously for 9 years earn compared to the same investment at 5.5% compounded quarterly for 9 years? Round to the nearest cent.

62. Suppose your great-great grandfather invested $500 earning 6.5% interest compounded continuously 100 years ago. How much would his investment be worth today? Round to the nearest cent.

Use the present value formula for periodic compound interest to solve Exercises 63–65.

63. Find the present value of $10,000 if interest is paid at a rate of 4.5% compounded semiannually for 12 years. Round to the nearest cent.

64. Find the present value of $1,000,000 if interest is paid at a rate of 9.5% compounded monthly for 8 years. Round to the nearest cent.

65. How much money would you have to invest at 10% compounded semiannually so that the total investment has a value of $2,205 after 1 year? Round to the nearest cent.

Use the present value formula for continuous compound interest to solve Exercises 66–67.

66. Find the present value of $16,000 if interest is paid at a rate of 4.5% compounded continuously for 10 years. Round to the nearest cent.

67. Which has the lower present value:

a. $20,000 if interest is paid at a rate of 5.18% compounded continuously for 2 years or

b. $25,000 if interest is paid at a rate of 3.8% compounded continuously for 30 months?

68. The population of a rural city follows the exponential growth model $P(t) = 2,000e^{.035t}$, where t is the number of years after 1995.

a. What was the population of this city in 1995?

b. What is the relative growth rate as a percent?

c. Use this model to approximate the population in 2030, rounding to the nearest whole number.

69. The relative growth rate of a certain bacteria colony is 25%.

Suppose there are 10 bacteria initially.

Use the exponential growth model $P(t) = P_0e^{kt}$ to answer the following.

a. Find a function that describes the population of the bacteria after t hours.

b. How many bacteria should we expect after 1 day? Round to the nearest whole number.

70. In 2006, the population of a certain American city was 18,221. If the relative growth rate has been 6% since 1986, what was the population of this city in 1986? Round to the nearest whole number. Use the exponential growth model $P(t) = P_0e^{kt}$.

71. In 1970, a wildlife resource management team introduced a certain rabbit species into a forest for the first time. In 2004, the rabbit population had grown to 7,183. The relative growth rate for this rabbit species is 20%.

    Use the exponential growth model $P(t) = P_0 e^{kt}$ to answer the following questions. Round each answer to the nearest whole number.

    a. How many rabbits did the wildlife resource management team introduce into the forest in 1970?

    b. How many rabbits can be expected in the year 2025?

Brief Exercises

In Exercises 72–84, use the graph of $y = 2^x$, $y = e^x$, or $y = 3^x$ and transformations to sketch each exponential function. Determine the domain and range. Also, determine the y-intercept, and find the equation of the horizontal asymptote.

72. $f(x) = 2^{x-1}$

73. $f(x) = 3^x - 1$

74. $f(x) = -3^{x+2}$

75. $f(x) = -2^{x+1} - 1$

76. $f(x) = \left(\dfrac{1}{2}\right)^{x+1}$

77. $f(x) = \left(\dfrac{1}{3}\right)^x - 3$

78. $f(x) = 2^{-x} + 1$

79. $f(x) = 3^{1-x} - 2$

80. $f(x) = e^{x-1}$

81. $f(x) = e^x - 1$

82. $f(x) = -e^{x+2}$

83. $f(x) = -e^{x+1} - 1$

84. $f(x) = e^{-x} - 2$

85. The population of a rural city follows the exponential growth model $P(t) = 2,000e^{.035t}$, where t is the number of years after 1995.

    Use this model to approximate the population in 2030, rounding to the nearest whole number.

86. The relative growth rate of a certain bacteria colony is 25%. Suppose there are 10 bacteria initially. Use the exponential growth model $P(t) = P_0 e^{kt}$ to determine the number of bacteria that should be expected after 1 day.

87. In 1970, a wildlife resource management team introduced a certain rabbit species into a forest for the first time. In 2004, the rabbit population had grown to 7,183. The relative growth rate for this rabbit species is 20%. Use the exponential growth model $P(t) = P_0 e^{kt}$ to determine the number of rabbits that can be expected in the year 2025. Round the answer to the nearest whole number.

5.2 Logarithmic Functions

THINGS TO KNOW

Before working through this section, be sure that you are familiar with the following concepts:

VIDEO   ANIMATION   INTERACTIVE

You Try It

1. Solving Polynomial Inequalities (Section 1.9)

You Try It

2. Solving Rational Inequalities (Section 1.9)

You Try It   3. Determining If a Function Is One to One Using the Horizontal Line Test (Section 3.6)

You Try It   4. Sketching the Graph of an Inverse Function (Section 3.6)

You Try It   5. Using the Composition Cancellation Properties of Inverse Functions (Section 3.6)

You Try It   6. Finding the Equation of an Inverse Function (Section 3.6)

You Try It   7. Sketching the Graphs of Exponential Functions (Section 5.1)

You Try It   8. Sketching the Graphs of Exponential Functions Using Transformations (Section 5.1)

You Try It   9. Solving Exponential Equations by Relating the Bases (Section 5.1)

OBJECTIVES

1 Understanding the Definition of a Logarithmic Function

2 Evaluating Logarithmic Expressions

3 Understanding the Properties of Logarithms

4 Using the Common and Natural Logarithms

5 Understanding the Characteristics of Logarithmic Functions

6 Sketching the Graphs of Logarithmic Functions Using Transformations

7 Finding the Domain of Logarithmic Functions

OBJECTIVE 1   UNDERSTANDING THE DEFINITION OF A LOGARITHMIC FUNCTION

My video summary   Every exponential function of the form $f(x) = b^x$, where $b > 0$ and $b \neq 1$, is one to one and thus has an inverse function. (You may want to refer to Section 3.6 to review one to one functions and inverse functions.) Remember, given the graph of a one to one function f, the graph of the inverse function is a reflection about the line $y = x$. That is, for any point (a, b) that lies on the graph of f, the point (b, a) must lie on the graph of f^{-1}. In other words, the graph of f^{-1} can be obtained by simply interchanging the x and y coordinates of the ordered pairs of $f(x) = b^x$. Watch this video to see how to sketch the graph of $f(x) = b^x$ and its inverse.

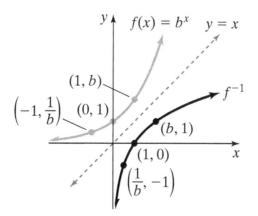

Figure 10 Graph of $f(x) = b^x, b > 1$, and its inverse function

The graphs of f and f^{-1} are sketched in Figure 10, but what is the equation of the inverse of $f(x) = b^x$? To find the equation of f^{-1}, we follow the four-step process for finding inverse functions that is discussed in Section 3.6.

Step 1. Change $f(x)$ to y: $\quad y = b^x$

Step 2. Interchange x and y: $\quad x = b^y$

Step 3. Solve for y: $\quad\quad\quad$??

Before we can solve for y we must introduce the following definition:

Definition Logarithmic Function

For $x > 0, b > 0$, and $b \neq 1$, the **logarithmic function** with base b is defined by

$$y = \log_b x \text{ if and only if } x = b^y$$

The equation $y = \log_b x$ is said to be in **logarithmic form,** whereas the equation $x = b^y$ is in **exponential form.** We can now continue to find the inverse of $f(x) = b^x$ by completing steps 3 and 4.

Step 3. Solve for y: $\quad\quad x = b^y$ can be written as $y = \log_b x$

Step 4. Change y to $f^{-1}(x)$: $\quad f^{-1}(x) = \log_b x$

In general, if $f(x) = b^x$ for $b > 0$ and $b \neq 1$, then the inverse function is $f^{-1}(x) = \log_b x$. For example, the inverse of $f(x) = 2^x$ is $f^{-1}(x) = \log_2 x$, which is read as "the log base 2 of x." We revisit the graphs of logarithmic functions later on in this section, but first it is very important to understand the definition of the logarithmic function and practice how to go back and forth writing exponential equations as logarithmic equations and vice versa.

My video summary ⓥ **Example 1 Change from Exponential Form to Logarithmic Form**

Write each exponential equation as an equation involving a logarithm.

a. $2^3 = 8$ $\quad\quad\quad\quad$ **b.** $5^{-2} = \dfrac{1}{25}$ $\quad\quad\quad\quad$ **c.** $1.1^M = z$

Solution We use the fact that the equation $x = b^y$ is equivalent to the equation $y = \log_b x$.

a. $2^3 = 8$ is equivalent to $\log_2 8 = 3$.

b. $5^{-2} = \dfrac{1}{25}$ is equivalent to $\log_5 \dfrac{1}{25} = -2$.

c. $1.1^M = z$ is equivalent to $\log_{1.1} z = M$.

Note that the exponent of the original (exponential) equation ends up by itself on the right side of the second (logarithmic) equation. Therefore, a logarithmic expression can be thought of as describing the exponent of a certain exponential equation.

 Watch the video to see this example worked out in more detail.

You Try It Work through this You Try It problem.

Work Exercises 1–6 in this textbook or in the MyMathLab Study Plan.

  ⊙ **Example 2** Change from Logarithmic Form to Exponential Form

Write each logarithmic equation as an equation involving an exponent.

a. $\log_3 81 = 4$ b. $\log_4 16 = y$ c. $\log_{3/5} x = 2$

Solution We use the fact that the equation $y = \log_b x$ is equivalent to the equation $x = b^y$.

a. $\log_3 81 = 4$ is equivalent to $3^4 = 81$.

b. $\log_4 16 = y$ is equivalent to $4^y = 16$.

c. $\log_{3/5} x = 2$ is equivalent to $\left(\dfrac{3}{5}\right)^2 = x$.

Watch the video to see this example worked out in more detail.

You Try It Work through this You Try It problem.

Work Exercises 7–11 in this textbook or in the MyMathLab Study Plan.

OBJECTIVE 2   EVALUATING LOGARITHMIC EXPRESSIONS

Because a logarithmic expression represents the exponent of an exponential equation, it is possible to evaluate many logarithms by inspection or by creating the corresponding exponential equation. Remember that the expression $\log_b x$ is the exponent to which b must be raised to in order to get x. For example, suppose we are to evaluate the expression $\log_4 64$. To evaluate this expression, we must ask ourselves, "4 raised to what power is 64?" Because $4^3 = 64$, we conclude that $\log_4 64 = 3$. For some logarithmic expressions, it is often convenient to create an exponential equation and use the method of relating the bases for solving exponential equations. For more complex logarithmic expressions, additional techniques are required. These techniques are discussed in Sections 5.4 and 5.5.

My interactive video summary

⊙ **Example 3 Evaluate Logarithmic Expressions**

Evaluate each logarithm:

a. $\log_5 25$ b. $\log_3 \dfrac{1}{27}$ c. $\log_{\sqrt{2}} \dfrac{1}{4}$

Solution

a. To evaluate $\log_5 25$, we must ask, "5 raised to what exponent is 25?" Because $5^2 = 25$, $\log_5 25 = 2$.

b. The expression $\log_3 \dfrac{1}{27}$ requires more analysis. In this case, we ask, "3 raised to what exponent is $\dfrac{1}{27}$?" Suppose we let y equal this exponent. Then $3^y = \dfrac{1}{27}$. To solve for y, we can use the method of relating the bases for solving exponential equations.

$$3^y = \frac{1}{27} \qquad \text{Write the exponential equation.}$$

$$3^y = \frac{1}{3^3} \qquad \text{Rewrite 27 as } 3^3.$$

$$3^y = 3^{-3} \qquad \text{Use } \frac{1}{b^n} = b^{-n}.$$

$$y = -3 \qquad \text{Use the method of relating the bases. (If } b^u = b^v, \text{ then } u = v.)$$

Thus, $\log_3 \dfrac{1}{27} = -3$.

My interactive video summary

⊙ c. Watch the interative video to verify that $\log_{\sqrt{2}} \dfrac{1}{4} = -4$ and to see each solution worked out in detail.

You Try It Work through this You Try It problem.

Work Exercises 12–18 in this textbook or in the MyMathLab Study Plan.

OBJECTIVE 3   UNDERSTANDING THE PROPERTIES OF LOGARITHMS

Because $b^1 = b$ for any real number b, we can use the definition of the logarithmic function ($y = \log_b x$ if and only if $x = b^y$) to rewrite this expression as $\log_b b = 1$. Similarly, because $b^0 = 1$ for any real number b, we can rewrite this expression as $\log_b 1 = 0$. These two general properties are summarized as follows.

General Properties of Logarithms

For $b > 0$ and $b \neq 1$,

1. $\log_b b = 1$

2. $\log_b 1 = 0$.

In Section 3.6, we see that a function f and its inverse function f^{-1} satisfy the following two composition cancellation equations:

$$f(f^{-1}(x)) = x \text{ for all } x \text{ in the domain of } f^{-1} \text{ and}$$

$$f^{-1}(f(x)) = x \text{ for all } x \text{ in the domain of } f$$

If $f(x) = b^x$, then $f^{-1}(x) = \log_b x$. Applying the two composition cancellation equations, we get

$$f(f^{-1}(x)) = b^{\log_b x} = x \text{ and}$$

$$f^{-1}(f(x)) = \log_b b^x = x.$$

Cancellation Properties of Exponentials and Logarithms

For $b > 0$ and $b \neq 1$,

1. $b^{\log_b x} = x$

2. $\log_b b^x = x.$

Example 4 Use the Properties of Logarithms

Use the properties of logarithms to evaluate each expression.

a. $\log_3 3^4$ **b.** $\log_{12} 12$ **c.** $7^{\log_7 13}$ **d.** $\log_8 1$

Solution

a. By the second cancellation property, $\log_3 3^4 = 4$.

b. Because $\log_b b = 1$ for all $b > 0$ and $b \neq 1$, $\log_{12} 12 = 1$.

c. By the first cancellation property, $7^{\log_7 13} = 13$.

d. Because $\log_b 1 = 0$ for all $b > 0$ and $b \neq 1$, $\log_8 1 = 0$.

You Try It Work through this You Try It problem.

Work Exercises 19–26 in this textbook or in the My Math**Lab Study Plan.**

OBJECTIVE 4 USING THE COMMON AND NATURAL LOGARITHMS

There are two bases that are used more frequently than any other base. They are base 10 and base e. (Refer to Section 5.1 to review the natural base e). Because our counting system is based on the number 10, the base 10 logarithm is known as the **common logarithm**. Instead of using the notation $\log_{10} x$ to denote the common logarithm, it is usually abbreviated without the subscript 10 as simply $\log x$. The base e logarithm is called the **natural logarithm** and is abbreviated as $\ln x$ instead of $\log_e x$. Most scientific calculators are equipped with a $\boxed{\log}$ key and a $\boxed{\ln}$ key. We can apply the definition of the logarithmic function for the base 10 and for the base e logarithm as follows.

Definition Common Logarithmic Function

For $x > 0$, the **common logarithmic function** is defined by

$$y = \log x \text{ if and only if } x = 10^y.$$

Definition   Natural Logarithmic Function

For $x > 0$, the **natural logarithmic function** is defined by

$$y = \ln x \text{ if and only if } x = e^{y}.$$

 *My video summary*   ⊘ **Example 5** Change from Exponential Form to Logarithmic Form

Write each exponential equation as an equation involving a common logarithm or natural logarithm.

a. $e^{0} = 1$ **b.** $10^{-2} = \dfrac{1}{100}$ **c.** $e^{K} = w$

Solution

a. $e^{0} = 1$ is equivalent to $\ln 1 = 0$.

b. $10^{-2} = \dfrac{1}{100}$ is equivalent to $\log\left(\dfrac{1}{100}\right) = -2$.

c. $e^{K} = w$ is equivalent to $\ln w = K$.

Watch the video to see this example worked out in more detail.

You Try It   Work through this You Try It problem.

Work Exercises 27–31 in this textbook or in the MyMathLab **Study Plan.**

 *My video summary*   ⊘ **Example 6** Change from Logarithmic Form to Exponential Form

Write each logarithmic equation as an equation involving an exponent.

a. $\log 10 = 1$ **b.** $\ln 20 = Z$ **c.** $\log (x - 1) = T$

Solution

a. $\log 10 = 1$ is equivalent to $10^{1} = 10$.

b. $\ln 20 = Z$ is equivalent to $e^{Z} = 20$.

c. $\log (x - 1) = T$ is equivalent to $10^{T} = x - 1$.

Watch the video to see this example worked out in more detail.

You Try It   Work through this You Try It problem.

Work Exercises 32–35 in this textbook or in the MyMathLab **Study Plan.**

 *My video summary*   ⊘ **Example 7** Evaluate Common and Natural Logarithmic Expressions

Evaluate each expression without the use of a calculator.

a. $\log 100$ **b.** $\ln \sqrt{e}$ **c.** $e^{\ln 51}$ **d.** $\log 1$

Solution

a. $\log 100 = 2$ because $10^{2} = 100$ by the definition of the logarithmic function or $\log 100 = \log 10^{2} = 2$ by cancellation property (2).

b. $\ln \sqrt{e} = \ln e^{1/2} = \dfrac{1}{2}$ by cancellation property (2).

c. $e^{\ln 51} = 51$ by cancellation property (1).

d. $\log 1 = 0$ by general property (1).

Watch the video to see this example worked out in more detail.

You Try It   Work through this You Try It problem.

Work Exercises 36–43 in this textbook or in the MyMathLab Study Plan.

OBJECTIVE 5   UNDERSTANDING THE CHARACTERISTICS OF LOGARITHMIC FUNCTIONS

To sketch the graph of a logarithmic function of the form $f(x) = \log_b x$, where $b > 0$ and $b \neq 1$, follow these three steps:

Steps for Sketching Logarithmic Functions of the Form $f(x) = \log_b x$

Step 1. Start with the graph of the exponential function $y = b^x$, labeling several ordered pairs.

Step 2. Because $f(x) = \log_b x$ is the inverse of $y = b^x$, we can find several points on the graph of $f(x) = \log_b x$ by reversing the coordinates of the ordered pairs of $y = b^x$.

Step 3. Plot the ordered pairs from step 2, and complete the graph of $f(x) = \log_b x$ by connecting the ordered pairs with a smooth curve. The graph of $f(x) = \log_b x$ is a reflection of the graph of $y = b^x$ about the line $y = x$.

My video summary   ⊘ **Example 8** Sketch the Graph of a Logarithmic Function

Sketch the graph of $f(x) = \log_3 x$.

Solution

Step 1. The graph of $y = 3^x$ passes through the points $\left(-1, \dfrac{1}{3}\right)$, $(0, 1)$, and $(1, 3)$. See Figure 11.

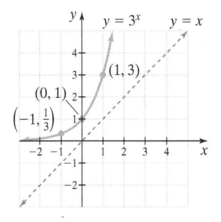

Figure 11 Graph of $y = 3^x$

Step 2. We reverse the three ordered pairs from step 1 to obtain the following three points: $\left(\frac{1}{3}, -1\right)$, $(1, 0)$, and $(3, 1)$.

Step 3. Plot the points $\left(\frac{1}{3}, -1\right)$, $(1, 0)$, and $(3, 1)$, and connect them with a smooth curve to obtain the graph of $f(x) = \log_3 x$. See Figure 12.

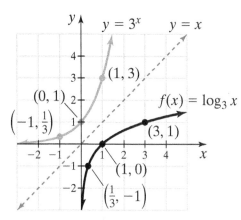

Figure 12 Graph of $y = 3^x$ and $f(x) = \log_3 x$

Notice in Figure 12 that the y-axis is a vertical asymptote of the graph of $f(x) = \log_3 x$. Every logarithmic function of the form $y = \log_b x$, where $b > 0$ and $b \neq 1$ has a vertical asymptote at the y-axis. The graphs and the characteristics of logarithmic functions are outlined as follows. Watch this video to see each step of this graphing process.

Characteristics of Logarithmic Functions

For $b > 0$, $b \neq 1$, the logarithmic function with base b is defined by $y = \log_b x$. The domain of $f(x) = \log_b x$ is $(0, \infty)$, and the range is $(-\infty, \infty)$. The graph of $f(x) = \log_b x$ has one of the following two shapes.

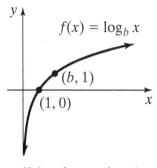

$f(x) = \log_b x, b > 1$

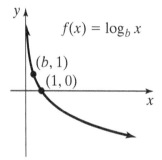

$f(x) = \log_b x, 0 < b < 1$

- The graph intersects the x-axis at $(1, 0)$.

- The graph contains the point $(b, 1)$.

- The graph is increasing on the interval $(0, \infty)$.

- The y-axis $(x = 0)$ is a vertical asymptote.

- The function is one to one.

- The graph intersects the x-axis at $(1, 0)$.

- The graph contains the point $(b, 1)$.

- The graph is decreasing on the interval $(0, \infty)$.

- The y-axis $(x = 0)$ is a vertical asymptote.

- The function is one to one.

You Try It   Work through this You Try It problem.

Work Exercises 44 and 45 in this textbook or in the MyMathLab Study Plan.

OBJECTIVE 6   SKETCHING THE GRAPHS OF LOGARITHMIC FUNCTIONS USING TRANSFORMATIONS

Often we can use the transformation techniques that are discussed in Section 3.4 to sketch the graph of logarithmic functions.

My video summary   ▷ **Example 9** Use Transformations to Sketch the Graph of a Logarithmic Function

Sketch the graph of $f(x) = -\ln(x + 2) - 1$.

Solution Recall that the function $y = \ln x$ has a base of e, where $2 < e < 3$. This means that the graph of $y = \ln x$ is increasing on the interval $(0, \infty)$ and contains the points $(1, 0)$ and $(e, 1)$. Starting with the graph of $y = \ln x$, we can obtain the graph of $f(x) = -\ln(x + 2) - 1$ through the following series of transformations:

1. Shift the graph of $y = \ln x$ horizontally to the left two units to obtain the graph of $y = \ln(x + 2)$.

2. Reflect the graph of $y = \ln(x + 2)$ about the x-axis to obtain the graph of $y = -\ln(x + 2)$.

3. Shift the graph of $y = -\ln(x + 2)$ vertically down one unit to obtain the final graph of $f(x) = -\ln(x + 2) - 1$.

My video summary   ▷ The graph of $f(x) = -\ln(x + 2) - 1$ is sketched in Figure 13. You can see from the graph that the domain of $f(x) = -\ln(x + 2) - 1$ is $(-2, \infty)$. The vertical asymptote is $x = -2$, and the x-intercept is $\left(\dfrac{1}{e} - 2, 0\right)$. Watch the video to see each step worked out in detail.

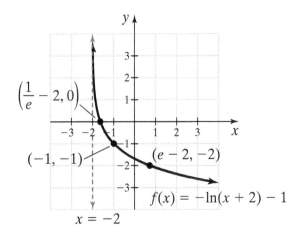

Figure 13 Graph of $f(x) = -\ln(x + 2) - 1$

You Try It   Work through this You Try It problem.

Work Exercises 46–55 in this textbook or in the MyMathLab Study Plan.

OBJECTIVE 7 FINDING THE DOMAIN OF LOGARITHMIC FUNCTIONS

In Example 9, we sketch the function $f(x) = -\ln(x + 2) - 1$ and observe that the domain was $(-2, \infty)$. We do not have to sketch the graph of a logarithmic function to determine the domain. The domain of a logarithmic function consists of all values of x for which the **argument** of the logarithm is greater than zero. In other words, if $f(x) = \log_b[g(x)]$, then the domain of f can be found by solving the inequality $g(x) > 0$. For example, given the function $f(x) = -\ln(x + 2) - 1$ from Example 9, we can determine the domain by solving the linear inequality $x + 2 > 0$. Solving this inequality for x, we obtain $x > -2$. Thus, the domain of $f(x) = -\ln(x + 2) - 1$ is $(-2, \infty)$.

Example 10 is a bit more challenging because the argument of the logarithm is a rational expression.

My interactive video summary

▶ **Example 10** Find the Domain of a Logarithmic Function with a Rational Argument

Find the domain of $f(x) = \log_5\left(\dfrac{2x - 1}{x + 3}\right)$.

Solution To find the domain of f, we must find all values of x for which the argument $\dfrac{2x - 1}{x + 3}$ is greater than zero. That is, you must solve the rational inequality $\left(\dfrac{2x - 1}{x + 3}\right) > 0$. See Section 1.9 if you need help remembering how to solve this inequality. By the techniques discussed in Section 1.9, we find that the solution to $\left(\dfrac{2x - 1}{x + 3}\right) > 0$ is $x < -3$ or $x > \dfrac{1}{2}$. Therefore, the domain of $f(x) = \log_5\left(\dfrac{2x - 1}{x + 3}\right)$ in set notation is $\left\{x \mid x < -3 \text{ or } x > \dfrac{1}{2}\right\}$. In interval notation, the domain is $(-\infty, -3) \cup \left(\dfrac{1}{2}, \infty\right)$. Watch the interactive video to see this problem worked out in detail.

You Try It Work through this You Try It problem.

Work Exercises 56–62 in this textbook or in the MyMathLab Study Plan.

5.2 Exercises

In Exercises 1–6, write each exponential equation as an equation involving a logarithm.

1. $3^2 = 9$

2. $16^{1/2} = 4$

3. $2^{-3} = \dfrac{1}{8}$

4. $\sqrt{2}^\pi = W$

5. $\left(\dfrac{1}{3}\right)^t = 27$

6. $7^{5k} = L$

In Exercises 7–11, write each logarithmic equation as an exponential equation.

7. $\log_5 1 = 0$

8. $\log_7 343 = 3$

9. $\log_{\sqrt{2}} 8 = 6$

10. $\log_4 K = L$

11. $\log_a (x - 1) = 3$

In Exercises 12–18, evaluate each logarithm without the use of a calculator.

12. $\log_2 8$

13. $\log_6 \sqrt{6}$

14. $\log_3 \dfrac{1}{9}$

15. $\log_{\sqrt{5}} 25$

16. $\log_4 \left(\dfrac{1}{\sqrt[5]{64}} \right)$

17. $\log_{1/7} \sqrt[3]{7}$

18. $\log_{0.1} 100$

In Exercises 19–26, use the properties of logarithms to evaluate each expression without the use of a calculator.

19. $2^{\log_2 11}$

20. $\log_4 4$

21. $\log_9 1$

22. $\log_7 7^{-3}$

23. $\log_a a, \, a > 1$

24. $5^{\log_5 M}, \, M > 0$

25. $\log_y 1, \, y > 0$

26. $\log_x x^{20}, \, x > 1$

In Exercises 27–31, write each exponential equation as an equation involving a common logarithm or a natural logarithm.

27. $10^3 = 1,000$

28. $e^{-1} = \dfrac{1}{e}$

29. $e^k = 2$

30. $10^e = M$

31. $e^{10} = Z$

In Exercises 32–35, write each logarithmic equation as an exponential equation.

32. $\ln 1 = 0$

33. $\log (1,000,000) = 6$

34. $\log K = L$

35. $\ln Z = 4$

In Exercises 36–43, evaluate each expression without the use of a calculator, and then verify your answer using a calculator.

36. $\log 10,000$

37. $\log \left(\dfrac{1}{1,000} \right)$

38. $\ln 1$

39. $\ln \sqrt[3]{e^2}$

40. $10^{\log e}$

41. $e^{\ln 49}$

42. $\log 10^6$

43. $\ln e + \ln e^3$

In Exercises 44–55, sketch each logarithmic function. Label at least two points on the graph, and determine the domain and the equation of any vertical asymptotes.

S b S **44.** $h(x) = \log_4 x$

S b S **45.** $g(x) = \log_{\frac{1}{3}} x$

S b S **46.** $f(x) = \log_2 (x) - 1$

S b S **47.** $f(x) = \log_5 (x - 1)$

S b S **48.** $f(x) = \ln(x + 1)$

S b S **49.** $f(x) = -\ln (x)$

S b S **50.** $f(x) = -\ln (x - 2)$

S b S **51.** $f(x) = -2 + \ln x$

S b S **52.** $f(x) = 3 - \ln x$

S b S **53.** $y = \log_{1/2} (x + 1) + 2$

S b S **54.** $y = \log_3 (1 - x)$

S b S **55.** $h(x) = -\dfrac{1}{2}\log_3 (x + 3) + 1$

In Exercises 56–62, find the domain of each logarithmic function.

56. $f(x) = \log (-x)$

57. $f(x) = \log_{1/4} (2x + 6)$

58. $f(x) = \ln (1 - 3x)$

59. $f(x) = \log_2 (x^2 - 9)$

60. $f(x) = \log_7 (x^2 - x - 20)$

61. $f(x) = \ln \left(\dfrac{x + 5}{x - 8} \right)$

62. $f(x) = \log \left(\dfrac{x^2 - x - 6}{x + 10} \right)$

Brief Exercises

In Exercises 63–74, sketch each logarithmic function. Label at least two points on the graph.

63. $h(x) = \log_4 x$

64. $g(x) = \log_{\frac{1}{3}} x$

65. $f(x) = \log_2 (x) - 1$

66. $f(x) = \log_5 (x - 1)$

67. $f(x) = \ln (x + 1)$

68. $f(x) = -\ln (x)$

69. $f(x) = -\ln (x - 2)$

70. $f(x) = -2 + \ln x$

71. $f(x) = 3 - \ln x$

72. $y = \log_{1/2} (x + 1) + 2$

73. $y = \log_3 (1 - x)$

74. $h(x) = -\dfrac{1}{2}\log_3 (x + 3) + 1$

5.3 Properties of Logarithms

THINGS TO KNOW

Before working through this section, be sure you are familiar with the following concepts:

VIDEO   ANIMATION   INTERACTIVE

You Try It

1. Solving Exponential Equations by Relating the Bases (Section 5.1)

You Try It

2. Change from Exponential Form to Logarithmic Form (Section 5.2)

You Try It

3. Change from Logarithmic Form to Exponential Form (Section 5.2)

You Try It

4. Evaluating Logarithmic Expressions (Section 5.2)

You Try It

5. Using the Common and Natural Logarithms (Section 5.2)

You Try It

6. Finding the Domain of Logarithmic Functions (Section 5.2)

OBJECTIVES

1 Using the Product Rule, Quotient Rule, and Power Rule for Logarithms

2 Expanding and Condensing Logarithmic Expressions

3 Solving Logarithmic Equations Using the Logarithm Property of Equality

4 Using the Change of Base Formula

OBJECTIVE 1   USING THE PRODUCT RULE, QUOTIENT RULE, AND POWER RULE FOR LOGARITHMS

In this section, we learn how to manipulate logarithmic expressions using properties of logarithms. Understanding how to use these properties will help us solve exponential and logarithmic equations that are encountered in the next section. Recall from Section 5.3 the general properties and cancellation properties of logarithms. We now look at three additional properties of logarithms.

> **Properties of Logarithms**
>
> If $b > 0, b \neq 1, u$ and v represent positive numbers and r is any real number, then
>
> $$\log_b uv = \log_b u + \log_b v$$ product rule for logarithms
>
> $$\log_b \frac{u}{v} = \log_b u - \log_b v$$ quotient rule for logarithms
>
> $$\log_b u^r = r \log_b u$$ power rule for logarithms

To prove the product rule and quotient rule for logarithms, we use properties of exponents and the method of relating the bases to solve exponential equations. The power rule for logarithms is a direct result of the product rule. Click on a video proof link above to see a proof of one or more of these properties.

My video summary ⊙ **Example 1** Use the Product Rule

Use the product rule for logarithms to expand each expression. Assume $x > 0$.

a. $\ln(5x)$ **b.** $\log_2(8x)$

Solution

a. $\ln(5x) = \ln 5 + \ln x$ — Use the product rule for logarithms.

b. $\log_2(8x) = \log_2 8 + \log_2 x$ — Use the product rule for logarithms

$\qquad\qquad = 3 + \log_2 x$ — Use the **definition of the logarithmic function** to rewrite $\log_2 8$ as 3 because $2^3 = 8$.

⚠ $\log_b(u + v)$ **is not equivalent to** $\log_b u + \log_b v$.

 You Try It Work through this You Try It problem.

My video summary ⊙ **Example 2** Use the Quotient Rule

Use the quotient rule for logarithms to expand each expression. Assume $x > 0$.

a. $\log_5\left(\dfrac{12}{x}\right)$ **b.** $\ln\left(\dfrac{x}{e^5}\right)$

Solution

a. $\log_5\left(\dfrac{12}{x}\right) = \log_5 12 - \log_5 x$ — Use the quotient rule for logarithms.

b. $\ln\left(\dfrac{x}{e^5}\right) = \ln x - \ln e^5$ — Use the quotient rule for logarithms

$\qquad\qquad = \ln x - 5$ — Use cancellation property (2) to rewrite $\ln e^5$ as 5.

⚠ $\log_b(u - v)$ **is not equivalent to** $\log_b u - \log_b v$, **and** $\dfrac{\log_b u}{\log_b v}$ **is not equivalent to** $\log_b u - \log_b v$.

 You Try It Work through this You Try It problem.

My video summary ⊙ **Example 3** Use the Power Rule

Use the power rule for logarithms to rewrite each expression. Assume $x > 0$.

a. $\log 6^3$ **b.** $\log_{1/2} \sqrt[4]{x}$

Solution

a. $\log 6^3 = 3 \log 6$ Use the power rule for logarithms.

b. $\log_{1/2} \sqrt[4]{x} = \log_{1/2} x^{1/4}$ Rewrite the fourth root of x using a rational exponent.

$\quad = \dfrac{1}{4} \log_{1/2} x$ Use the power rule for logarithms.

The process of using the power rule to simplify a logarithmic expression is often casually referred to as "bringing down the exponent."

 $(\log_b u)^r$ **is** *not* **equivalent to** $r \log_b u.$

You Try It   Work through this You Try It problem.

Work Exercises 1–10 in this textbook or in the MyMathLab Study Plan.

OBJECTIVE 2   EXPANDING AND CONDENSING LOGARITHMIC EXPRESSIONS

Sometimes it is necessary to combine several properties of logarithms to expand a logarithmic expression into the sum and/or difference of logarithms or to condense several logarithms into a single logarithm.

My interactive video summary

⊙ **Example 4 Expand a Logarithmic Expression**

Use properties of logarithms to expand each logarithmic expression as much as possible.

a. $\log_7 \left(49 x^3 \sqrt[5]{y^2} \right)$ **b.** $\ln \left(\dfrac{(x^2 - 4)}{9 e^{x^3}} \right)$

Solution

a. $\log_7 \left(49 x^3 \sqrt[5]{y^2} \right)$ Write the original expression.

$\quad = \log_7 49 + \log_7 x^3 \sqrt[5]{y^2}$ Use the product rule.

$\quad = \log_7 49 + \log_7 x^3 + \log_7 \sqrt[5]{y^2}$ Use the product rule again.

$\quad = \log_7 49 + \log_7 x^3 + \log_7 y^{2/5}$ Rewrite $\sqrt[5]{y^2}$ using a rational exponent.

$\quad = 2 + 3 \log_7 x + \dfrac{2}{5} \log_7 y$ Rewrite $\log_7 49$ as 2 and use the power rule.

b. $\ln \left(\dfrac{(x^2 - 4)}{9 e^{x^3}} \right) = \ln \left(\dfrac{(x - 2)(x + 2)}{9 e^{x^3}} \right)$ Factor the expression in the numerator.

$\quad = \ln (x - 2)(x + 2) - \ln 9 e^{x^3}$ Use the quotient rule.

$\quad = \ln (x - 2) + \ln (x + 2) - \left[\ln 9 + \ln e^{x^3} \right]$ Use the product rule twice.

$\quad = \ln (x - 2) + \ln (x + 2) - \left[\ln 9 + x^3 \right]$ Use cancellation property (2) to rewrite $\ln e^{x^3}$ as x^3.

$\quad = \ln (x - 2) + \ln (x + 2) - \ln 9 - x^3$ Simplify.

Watch the interactive video to see this example worked out in detail.

You Try It Work through this You Try It problem.

Work Exercises 11–20 in this textbook or in the MyMathLab Study Plan.

My interactive video summary

▶ **Example 5** Condense a Logarithmic Expression

Use properties of logarithms to rewrite each expression as a single logarithm.

a. $\dfrac{1}{2}\log(x-1) - 3\log z + \log 5$

b. $\dfrac{1}{3}(\log_3 x - 2\log_3 y) + \log_3 10$

Solution

a. $\dfrac{1}{2}\log(x-1) - 3\log z + \log 5$ Write the original expression.

  $= \log(x-1)^{1/2} - \log z^3 + \log 5$ Use the power rule twice.

  $= \log\dfrac{(x-1)^{1/2}}{z^3} + \log 5$ Use the quotient rule.

  $= \log\dfrac{5(x-1)^{1/2}}{z^3}$ or $\log\dfrac{5\sqrt{x-1}}{z^3}$ Use the product rule.

b. $\dfrac{1}{3}(\log_3 x - 2\log_3 y) + \log_3 10$ Write the original expression.

  $= \dfrac{1}{3}(\log_3 x - \log_3 y^2) + \log_3 10$ Use the power rule.

  $= \dfrac{1}{3}\log_3 \dfrac{x}{y^2} + \log_3 10$ Use the quotient rule.

  $= \log_3\left(\dfrac{x}{y^2}\right)^{1/3} + \log_3 10$ Use the power rule.

  $= \log_3\left[10\left(\dfrac{x}{y^2}\right)^{1/3}\right]$ or $\log_3\left[10\sqrt[3]{\dfrac{x}{y^2}}\right]$ Use the product rule.

Watch the interactive video to see this example worked out in detail.

You Try It Work through this You Try It problem.

Work Exercises 21–33 in this textbook or in the MyMathLab Study Plan.

OBJECTIVE 3 SOLVING LOGARITHMIC EQUATIONS USING THE LOGARITHM PROPERTY OF EQUALITY

Remember that all logarithmic functions of the form $f(x) = \log_b x$ for $b > 0$ and $b \neq 1$ are one to one. In Section 3.6, the alternate definition of **one to one** stated that

A function f is one to one if, for any two range values $f(u)$ and $f(v)$, $f(u) = f(v)$ implies that $u = v$.

Using this definition and letting $f(x) = \log_b x$, we can say that if $\log_b u = \log_b v$, then $u = v$. In other words, if the bases of a logarithmic equation of the form $\log_b u = \log_b v$ are equal, then the arguments must be equal. This is known as the **logarithm property of equality**.

Logarithm Property of Equality

If a logarithmic equation can be written in the form $\log_b u = \log_b v$, then $u = v$. Furthermore, if $u = v$, then $\log_b u = \log_b v$.

The second statement of the logarithm property of equality says that if we start with the equation $u = v$, then we can rewrite the equation as $\log_b u = \log_b v$. This process is often casually referred to as "taking the log of both sides."

My interactive video summary

⊙ Example 6 Use the Logarithm Property of Equality to Solve Logarithmic Equations

Solve the following equations:

a. $\log_7 (x - 1) = \log_7 12$ b. $2 \ln x = \ln 16$

Solution

a. Because the base of each logarithm is 7, we can use the logarithm property of equality to eliminate the logarithms.

$\log_7 (x - 1) = \log_7 12$	Write the original equation.
$(x - 1) = 12$	If $\log_b u = \log_b v$, then $u = v$.
$x = 13$	Solve for x.

b.
$2 \ln x = \ln 16$	Write the original expression.
$\ln x^2 = \ln 16$	Use the power rule.
$x^2 = 16$	If $\log_b u = \log_b v$, then $u = v$.
$x = \pm 4$	Use the square root property.

The domain of $\ln x$ is $x > 0$; this implies that $x = -4$ is an extraneous solution, and hence, we must discard it. Therefore, this equation has only one solution, $x = 4$.

Watch this interactive video to see the entire solution to this example.   ●

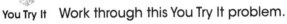

You Try It Work through this You Try It problem.

Work Exercises 34–39 in this textbook or in the My Math**Lab Study Plan.**

OBJECTIVE 4 USING THE CHANGE OF BASE FORMULA

Most scientific calculators are equipped with a $\boxed{\log}$ key and a $\boxed{\ln}$ key to evaluate common logarithms and natural logarithms. But how do we use a calculator to evaluate logarithmic expressions having a base other than 10 or e? The answer is to use the following **change of base formula**.

Change of Base Formula

For any positive base $b \neq 1$ and for any positive real number u, then

$$\log_b u = \frac{\log_a u}{\log_a b},$$

where a is any positive number such that $a \neq 1$.

My video summary ⊘ To see the proof of the change of base formula, watch this video.

The change of base formula allows us to change the base of a logarithmic expression into a ratio of two logarithms using any base we choose. For example, suppose we are given the logarithmic expression $\log_3 10$. We can use the change of base formula to write this logarithm as a quotient of logarithms involving any base we choose:

$$\log_3 10 = \frac{\log_7 10}{\log_7 3} \quad \text{or} \quad \log_3 10 = \frac{\log_2 10}{\log_2 3} \quad \text{or} \quad \log_3 10 = \frac{\log 10}{\log 3} \quad \text{or} \quad \log_3 10 = \frac{\ln 10}{\ln 3}$$

In each of the previous four cases, we introduced a new base (7, 2, 10, and e, respectively). However, if we want to use a calculator to get a numerical approximation of $\log_3 10$, then it really only makes sense to change $\log_3 10$ into an expression involving base 10 or base e because these are the only two bases most calculators can handle.

Note $\log_3 10 = \dfrac{\log 10}{\log 3} \approx 2.0959$ or $\log_3 10 = \dfrac{\ln 10}{\ln 3} \approx 2.0959$

Example 7 Use the Change of Base Formula

Approximate the following expressions. Round each to four decimal places.

a. $\log_9 200$ b. $\log_{\sqrt{3}} \pi$

Solution

a. $\log_9 200 = \dfrac{\log 200}{\log 9} \approx 2.4114$ or $\log_9 200 = \dfrac{\ln 200}{\ln 9} \approx 2.4114$

b. $\log_{\sqrt{3}} \pi = \dfrac{\log \pi}{\log \sqrt{3}} \approx 2.0840$ or $\log_{\sqrt{3}} \pi = \dfrac{\ln \pi}{\ln \sqrt{3}} \approx 2.0840$

You Try It Work through this You Try It problem.

Work Exercises 40–43 in this textbook or in the MyMathLab Study Plan.

My video summary ⊘ **Example 8** Use the Change of Base Formula and Properties of Logarithms

Use the change of base formula and the properties of logarithms to rewrite as a single logarithm involving base 2.

$$\log_4 x + 3 \log_2 y$$

Solution To use properties of logarithms, the base of each logarithmic expression must be the same. We use the change of base formula to rewrite $\log_4 x$ as a logarithmic expression involving base 2:

$$\log_4 x + 3 \log_2 y = \frac{\log_2 x}{\log_2 4} + 3 \log_2 y \qquad \text{Use the change of base formula}$$
$$\log_4 x = \frac{\log_2 x}{\log_2 4}.$$

$$= \frac{\log_2 x}{2} + 3 \log_2 y \qquad \text{Rewrite } \log_2 4 \text{ as 2 because } 2^2 = 4.$$

$$= \frac{1}{2} \log_2 x + 3 \log_2 y \qquad \text{Rewrite } \frac{\log_2 x}{2} \text{ as } \frac{1}{2} \log_2 x.$$

$$= \log_2 x^{1/2} + \log_2 y^3 \qquad \text{Use the power rule.}$$

$$= \log_2 x^{1/2} y^3 \quad \text{or} \quad \log_2 \sqrt{x} y^3 \qquad \text{Use the product rule.}$$

Therefore, the expression $\log_4 x + 3 \log_2 y$ is equivalent to $\log_2 \sqrt{x} y^3$. Note that we could have chosen to rewrite the original expression as a single logarithm involving base 4. Watch the video to see that the expression $\log_4 x + 3 \log_2 y$ is also equivalent to $\log_4 xy^6$.

You Try It Work through this You Try It problem.

Work Exercises 44–47 in this textbook or in the MyMathLab Study Plan.

 *My video summary*

⊙ **Example 9** Use the Change of Base Formula to Solve Logarithmic Equations

Use the change of base formula and the properties of logarithms to solve the following equation:

$$2 \log_3 x = \log_9 16$$

Solution Watch the video to see how the change of base formula and the power rule for logarithms can be used to solve this equation.

You Try It Work through this You Try It problem.

Work Exercises 48–51 in this textbook or in the MyMathLab Study Plan.

5.3 Exercises

In Exercises 1–10, use the product rule, quotient rule, or power rule to expand each logarithmic expression. Wherever possible, evaluate logarithmic expressions.

1. $\log_4 (xy)$
2. $\log \left(\frac{9}{t} \right)$
3. $\log_5 y^3$
4. $\log_3 (27w)$
5. $\ln 5e^2$

6. $\log_9 \sqrt[4]{k}$
7. $\log 100P$
8. $\ln \left(\frac{e^5}{r} \right)$
9. $\log_{\sqrt{2}} 8x$
10. $\log_2 \left(\frac{M}{32} \right)$

In Exercises 11–20, use the properties of logarithms to expand each logarithmic expression. Wherever possible, evaluate logarithmic expressions.

11. $\log_7 x^2 y^3$

12. $\ln \dfrac{a^2 b^3}{c^4}$

13. $\log \dfrac{\sqrt{x}}{10y^3}$

14. $\log_3 9(x^2 - 25)$

15. $\log_2 \sqrt{4xy}$

16. $\log_5 \dfrac{\sqrt{5x^5}}{\sqrt[3]{25y^4}}$

17. $\ln \dfrac{\sqrt[5]{ez}}{\sqrt{x-1}}$

18. $\log_3 \sqrt[4]{\dfrac{x^2 y^5}{9}}$

19. $\ln \left[\dfrac{x+1}{(x^2-1)^3} \right]^{2/3}$

20. $\log \dfrac{(10x)^3 \sqrt{x-4}}{(x^2-16)^5}$

In Exercises 21–33, use properties of logarithms to rewrite each expression as a single logarithm. Wherever possible, evaluate logarithmic expressions.

21. $\log_b A + \log_b C$

22. $\log_4 M - \log_4 N$

23. $2 \log_8 x + \dfrac{1}{3} \log_8 y$

24. $\log 20 + \log 5$

25. $\log_2 80 - \log_2 5$

26. $\ln \sqrt{x} - \dfrac{1}{3} \ln x + \ln \sqrt[4]{x}$

27. $\log_5 (x - 2) + \log_5 (x + 2)$

28. $\log_3 (x + 1) - \log_3 (x + 4) - \log_3 \sqrt{6x}$

29. $\ln(x - 1) - \ln \sqrt{x + 5} - \ln 4x$

30. $\ln(x - 3)^2 + \ln 7x^2$

31. $\log_9 (x^2 - 5x + 6) - \log_9 (x^2 - 4) + \log_9 (x + 2)$

32. $\log (x - 3) + 2 \log (x + 3) - \log (x^3 + x^2 - 9x - 9)$

33. $\dfrac{1}{2} \left[\ln (x - 1)^2 - \ln (2x^2 - x - 1)^4 \right] + 2 \ln (2x + 1)$

In Exercises 34–39, use the properties of logarithms and the logarithm property of equality to solve each logarithmic equation.

34. $\log_3 (2x + 1) = \log_3 11$

35. $\log_{11} \sqrt{x} = \log_{11} 6$

36. $2 \log (x + 5) = \log 12 + \log 3$

37. $\ln 5 + \ln x = \ln 7 + \ln (3x - 2)$

38. $\log_7 (x + 6) - \log_7 (x + 2) = \log_7 x$

39. $\log_2 (x + 3) + \log_2 (x - 4) = \log_2 (x + 12)$

In Exercises 40–43, use the change of base formula and a calculator to approximate each logarithmic expression. Round your answers to four decimal places.

40. $\log_4 51$

41. $\log_7 0.8$

42. $\log_{1/5} 72$

43. $\log_{\sqrt{7}} 100$

In Exercises 44–47, use the change of base formula and the properties of logarithms to rewrite each expression as a single logarithm in the indicated base.

44. $\log_3 x + 4\log_9 w$, base 3

45. $\log_5 x + \log_{1/5} x^3$, base 5

46. $\log_{16} x^4 + \log_8 y^3 + \log_4 w^2$, base 2

47. $\log_{e^2} x^5 + \log_{e^3} x^6 + \log_{e^4} x^{12}$, base e

In Exercises 48–51, solve each logarithmic equation.

48. $\log_2 x = \log_4 25$

49. $\log_{1/3} x = \log_3 20$

50. $\log_5 x = \log_{\sqrt{5}} 6$

51. $2\ln x = \log_{e^3} 125$

5.4 Exponential and Logarithmic Equations

THINGS TO KNOW

Before working through this section, be sure that you are familiar with the following concepts:

| | VIDEO | ANIMATION | INTERACTIVE |

 You Try It
1. Solving Exponential Equations by Relating the Bases (Section 5.1)

 You Try It
2. Changing from Exponential to Logarithmic Form (Section 5.2)

 You Try It
3. Changing from Logarithmic to Exponential Form (Section 5.2)

 You Try It
4. Using the Cancellation Properties of Exponentials and Logarithms (Section 5.2)

 You Try It
5. Expanding and Condensing Logarithmic Expressions (Section 5.3)

 You Try It
6. Solving Logarithmic Equations Using the Logarithm Property of Equality (Section 5.3)

OBJECTIVES

1 Solving Exponential Equations

2 Solving Logarithmic Equations

In this section, we learn how to solve exponential and logarithmic equations. The techniques and strategies learned in this section help us solve applied problems that are discussed in Section 5.5. We start by developing a strategy to solve exponential equations.

OBJECTIVE 1   SOLVING EXPONENTIAL EQUATIONS

We have already solved exponential equations using the method of relating the bases. For example, we can solve the equation $4^{x+3} = \frac{1}{2}$ by converting the base on both sides of the equation to base 2. To see how to solve this equation, read these steps.

But suppose we are given an exponential equation in which the bases cannot be related, such as $2^{x+1} = 3$. Remember in Section 5.1, Example 4, we wanted to find the one x-intercept of the graph of $f(x) = -2^{x+1} + 3$ (Figure 14).

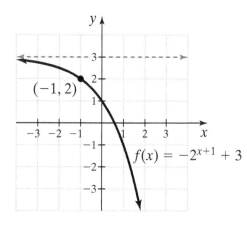

Figure 14 Graph of $f(x) = -2^{x+1} + 3$

To find the x-intercept of $f(x) = -2^{x+1} + 3$, we need to set $f(x) = 0$ and solve for x.

$$f(x) = -2^{x+1} + 3$$
$$0 = -2^{x+1} + 3$$
$$2^{x+1} = 3$$

In Section 5.1, we could not solve this equation for x because we had not yet defined the logarithm. We can now use some properties of logarithms to solve this equation. Recall the following logarithmic properties.

If $u = v$, then $\log_b u = \log_b v$.   **logarithm property of equality**

$\log_b u^r = r \log_b u$ **power rule for logarithms**

We can use these two properties to solve the equation $2^{x+1} = 3$ and thus determine the x-intercept of $f(x) = -2^{x+1} + 3$. We solve the equation $2^{x+1} = 3$ in Example 1.

 *My video summary*   ⊘ **Example 1 Solve an Exponential Equation**

Solve $2^{x+1} = 3$.

Solution

$2^{x+1} = 3$	Write the original equation.
$\ln 2^{x+1} = \ln 3$	Use the logarithm property of equality.
$(x + 1) \ln 2 = \ln 3$	Use the power rule for logarithms.
$x \ln 2 + \ln 2 = \ln 3$	Use the distributive property.
$x \ln 2 = \ln 3 - \ln 2$	Subtract ln 2 from both sides.
$x = \dfrac{\ln 3 - \ln 2}{\ln 2}$	Divide both sides by ln 2.

The solution to Example 1 verifies that the x-intercept of $f(x) = -2^{x+1} + 3$ is $x = \dfrac{\ln 3 - \ln 2}{\ln 2} \approx .5850$. See Figure 15.

When we cannot easily relate the bases of an exponential equation, as in Example 1, we use logarithms and their properties to solve them. The methods used to solve exponential equations are outlined as follows.

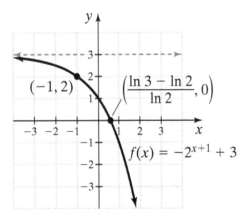

Figure 15 Graph of $f(x) = -2^{x+1} + 3$

Solving Exponential Equations

- If the equation can be written in the form $b^u = b^v$, then solve the equation $u = v$.

- If the equation cannot easily be written in the form $b^u = b^v$,

  1. Use the logarithm property of equality to "take the log of both sides" (typically using base 10 or base e).

  2. Use the power rule of logarithms to "bring down" any exponents.

  3. Solve for the given variable.

My interactive video summary

⊙ Example 2 Solve Exponential Equations

Solve each equation. For part b, round to four decimal places.

a. $3^{x-1} = \left(\dfrac{1}{27}\right)^{2x+1}$

b. $7^{x+3} = 4^{2-x}$

Solution

a. Watch the interactive video, or read these steps to see that the solution is $x = -\dfrac{2}{7}$.

b. We cannot easily use the method of relating the bases because we cannot easily write both 7 and 4 using a common base. Therefore, we use logarithms to solve.

$7^{x+3} = 4^{2-x}$	Write the original equation.
$\ln 7^{x+3} = \ln 4^{2-x}$	Use the logarithm property of equality.
$(x + 3) \ln 7 = (2 - x) \ln 4$	Use the power rule for logarithms.
$x \ln 7 + 3 \ln 7 = 2 \ln 4 - x \ln 4$	Use the distributive property.

$x \ln 7 + x \ln 4 = 2 \ln 4 - 3 \ln 7$ Add $x \ln 4$ to both sides, and subtract $3 \ln 7$ from both sides.

$x (\ln 7 + \ln 4) = 2 \ln 4 - 3 \ln 7$ Factor out an x from the left-hand side.

$$x = \frac{2 \ln 4 - 3 \ln 7}{\ln 7 + \ln 4}$$ Divide both sides by $\ln 7 + \ln 4$.

$$= \frac{\ln 16 - \ln 343}{\ln 28}$$ Use the power rule for logarithms in the numerator, and use the product rule for logarithms in the denominator.

$$= \frac{\ln\left(\dfrac{16}{343}\right)}{\ln 28}$$ Use the quotient rule for logarithms to rewrite $\ln 16 - \ln 343$ as $\ln\left(\dfrac{16}{343}\right)$.

$$\approx -.9199$$ Use a calculator to round to four decimal places.

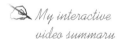 *My interactive video summary*

⊙ Watch this interactive video to see the entire solution to this example.

You Try It   Work through this You Try It problem.

Work Exercises 1–10 in this textbook or in the My MathLab **Study Plan.**

 *My interactive video summary*

⊙ **Example 3** Solve Exponential Equations Involving the Natural Exponential Function

Solve each equation. Round to four decimal places.

a. $25e^{x-5} = 17$ **b.** $e^{2x-1} \cdot e^{x+4} = 11$

Solution

a. First Isolate the exponential term on the left by dividing both sides of the equation by 25.

$$25e^{x-5} = 17$$ Write the original equation.

$$e^{x-5} = \frac{17}{25}$$ Divide both sides by 25.

Now use the natural logarithm and the logarithm property of equality to solve for x.

$$\ln e^{x-5} = \ln\frac{17}{25}$$ Use the logarithm property of equality.

$$x - 5 = \ln\frac{17}{25}$$ Use cancellation property (2) to rewrite $\ln e^{x-5}$ as $x - 5$.

$$x = \ln\frac{17}{25} + 5$$ Add 5 to both sides.

$$\approx 4.6143$$ Use a calculator to round to four decimal places.

b. $e^{2x-1} \cdot e^{x+4} = 11$ Write the original equation.

$e^{(2x-1)+(x+4)} = 11$ Use $b^m \cdot b^n = b^{m+n}$.

$e^{3x+3} = 11$ Combine like terms in the exponent.

$\ln e^{3x+3} = \ln 11$ Use the logarithm property of equality.

$3x + 3 = \ln 11$ Use cancellation property (2) to rewrite $\ln e^{3x+3}$ as $3x + 3$.

$3x = \ln 11 - 3$ Subtract 3 from both sides.

$x = \dfrac{\ln 11 - 3}{3}$ Divide both sides by 3.

$\approx -.2007$ Use a calculator to round to four decimal places.

Watch the interactive video to see the solutions to this example worked out in detail.

You Try It   Work through this You Try It problem.

Work Exercises 11–15 in this textbook or in the My Math**Lab Study Plan.**

OBJECTIVE 2   SOLVING LOGARITHMIC EQUATIONS

We now turn our attention to solving logarithmic equations. In Section 5.3, we learned how to solve certain logarithmic equations by using the logarithm property of equality. That is, if we can write a logarithmic equation in the form $\log_b u = \log_b v$, then $u = v$. Before we look at an example, let's review three of the properties of logarithms.

> **Properties of Logarithms**
>
> If $b > 0, b \neq 1$, u and v represent positive numbers and r is any real number, then
>
> $$\log_b uv = \log_b u + \log_b v \qquad \textbf{product rule for logarithms}$$
>
> $$\log_b \frac{u}{v} = \log_b u - \log_b v \qquad \textbf{quotient rule for logarithms}$$
>
> $$\log_b u^r = r \log_b u \qquad \textbf{power rule for logarithms}$$

My video summary   ⊙ **Example 4** Solve a Logarithmic Equation Using the Logarithm Property of Equality

Solve $2 \log_5 (x - 1) = \log_5 64$.

Solution   We can use the power rule for logarithms and the logarithmic property of equality to solve.

$2 \log_5 (x - 1) = \log_5 64$ Write the original equation.

$\log_5 (x - 1)^2 = \log_5 64$ Use the power rule.

$(x - 1)^2 = 64$ Use the logarithm property of equality.

$$x - 1 = \pm 8 \qquad \text{Use the square root property.}$$

$$x = 1 \pm 8 \qquad \text{Solve for } x.$$

$$x = 9 \quad \text{or} \quad x = -7 \qquad \text{Simplify.}$$

Recall that the domain of a logarithmic function must contain only positive numbers; thus, $x - 1$ must be positive. Therefore, the solution of $x = -7$ must be discarded. The only solution is $x = 9$. You may want to review how to determine the domain of a logarithmic function, which is discussed in Section 5.2.

 When solving logarithmic equations, it is important to always verify the solutions. Logarithmic equations often lead to extraneous solutions, as in Example 4.

When a logarithmic equation cannot be written in the form $\log_b u = \log_b v$, as in Example 4, we adhere to the steps outlined as follows:

Solving Logarithmic Equations

1. Determine the domain of the variable.

2. Use properties of logarithms to combine all logarithms, and write as a single logarithm, if needed.

3. Eliminate the logarithm by rewriting the equation in exponential form. To review how to change from logarithmic form to exponential form, view Example 2 from Section 5.2.

4. Solve for the given variable.

5. Check for any extraneous solutions. Verify that each solution is in the domain of the variable.

You Try It   Work through this You Try It problem.

Work Exercises 16–19 in this textbook or in the MyMathLab Study Plan.

 *My video summary* ◉ **Example 5** Solve a Logarithmic Equation

Solve $\log_4 (2x - 1) = 2$.

Solution The domain of the variable in this equation is the solution to the inequality $2x - 1 > 0$ or $x > \dfrac{1}{2}$. Thus, our solution must be greater than $\dfrac{1}{2}$. Because the equation involves a single logarithm, we can proceed to the third step.

$$\log_4 (2x - 1) = 2 \qquad \text{Write the original equation.}$$

$$4^2 = 2x - 1 \qquad \text{Rewrite in exponential form.}$$

$$16 = 2x - 1 \qquad \text{Simplify.}$$

$$17 = 2x \qquad \text{Add 1 to both sides.}$$

$$x = \frac{17}{2} \qquad \text{Divide by 2.}$$

Because the solution satisfies the inequality, there are no extraneous solutions. We can verify the solution by substituting $x = \dfrac{17}{2}$ into the original equation.

Check:

$$\log_4 (2x - 1) = 2 \qquad \text{Write the original equation.}$$

$$\log_4 \left(2\left(\frac{17}{2}\right) - 1\right) \overset{?}{=} 2 \qquad \text{Substitute } x = \frac{17}{2}.$$

$$\log_4 (17 - 1) \overset{?}{=} 2 \qquad \text{Simplify.}$$

$$\log_4 (16) = 2 \qquad \text{This is a true statement because } 4^2 = 16.$$

You Try It Work through this You Try It problem.

Work Exercises 20–24 in this textbook or in the MyMathLab Study Plan.

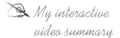
My interactive video summary

⊙ **Example 6** Solve a Logarithmic Equation

Solve $\log_2 (x + 10) + \log_2 (x + 6) = 5$.

Solution

The domain of the variable in this equation is the solution to the compound inequality $x + 10 > 0$ and $x + 6 > 0$. The solution to this compound inequality is $x > -6$. (You may want to review compound inequalities from Section 1.7.)

$$\log_2 (x + 10) + \log_2 (x + 6) = 5 \qquad \text{Write the original equation.}$$

$$\log_2 (x + 10)(x + 6) = 5 \qquad \text{Use the product rule.}$$

$$(x + 10)(x + 6) = 2^5 \qquad \text{Rewrite in exponential form.}$$

$$x^2 + 16x + 60 = 32 \qquad \text{Simplify.}$$

$$x^2 + 16x + 28 = 0 \qquad \text{Subtract 32 from both sides.}$$

$$(x + 14)(x + 2) = 0 \qquad \text{Factor.}$$

$$x = -14 \quad \text{or} \quad x = -2 \qquad \text{Use the zero product property to solve.}$$

Because the domain of the variable is $x > -6$, we must *exclude* the solution $x = -14$. Therefore, the only solution to this logarithmic equation is $x = -2$. Work through the interactive video to see this solution worked out in detail.

You Try It Work through this You Try It problem.

Work Exercises 25–33 in this textbook or in the MyMathLab Study Plan.

Example 7 Solve a Logarithmic Equation

Solve $\ln (x - 4) - \ln (x - 5) = 2$. Round to four decimal places.

Solution The domain of the variable is the solution to the compound inequality $x - 4 > 0$ and $x - 5 > 0$. The solution to this compound inequality is $x > 5$. (You may want to review compound inequalities from Section 1.7.)

$$\ln (x - 4) - \ln (x - 5) = 2 \qquad \text{Write the original equation.}$$

$$\ln \left(\frac{x - 4}{x - 5}\right) = 2 \qquad \text{Use the quotient rule.}$$

$$e^2 = \frac{x-4}{x-5}$$ Rewrite in exponential form.

$$e^2(x-5) = x-4$$ Multiply both sides by $x-5$.

$$e^2x - 5e^2 = x-4$$ Use the distributive property.

$$e^2x - x = 5e^2 - 4$$ Add $5e^2$ to both sides and subract x from both sides.

$$x(e^2 - 1) = 5e^2 - 4$$ Factor out an x from the left-hand side.

$$x = \frac{5e^2 - 4}{e^2 - 1} \approx 5.1565$$ Solve for x. Use a calculator to round to four decimal places.

We approximate the exact answer $x = \dfrac{5e^2 - 4}{e^2 - 1}$ in order to verify that the solution is in the domain of the variable. In this example, we see that 5.1565 is clearly greater than 5. In some cases, we may need to use the exact answer to verify a solution to a logarithmic equation. To see this verification for Example 7, read these steps.

You Try It Work through this You Try It problem.

Work Exercises 34 and 35 in this textbook or in the My MathLab **Study Plan.**

5.4 Exercises

Skill Check Exercises

For exercises SCE-1 through SCE-8, evaluate the expression using a calculator. Round your answer to 4 decimal places.

SCE-1. $\ln\left(\dfrac{7}{2}\right)$

SCE-2. $\dfrac{5e^2 - 4}{e^2 - 1}$

SCE-3. $\dfrac{\ln 3 + \ln 5}{\ln 5}$

SCE-4. $\dfrac{\ln \pi - 3\ln 4}{\ln \pi - 2\ln 4}$

SCE-5. $\dfrac{\ln\left(\dfrac{17}{3}\right)}{.00235}$

SCE-6. $\dfrac{\ln\left(\dfrac{77}{131}\right)}{\ln\left(\dfrac{120}{131}\right)}$

SCE-7. $\dfrac{\ln 2}{12\ln\left(1 + \dfrac{.06}{12}\right)}$

SCE-8. $\dfrac{-\ln 65}{\ln\left(\dfrac{\ln\left(\dfrac{37}{117}\right)}{20}\right)}$

In Exercises 1–15, solve each exponential equation. For irrational solutions, round to four decimal places.

1. $3^x = 5$

2. $2^{x/3} = 19$

3. $4^{x^2 - 2x} = 64$

4. $3^{x+7} = -20$

5. $(1.52)^{-3x/7} = 11$

6. $\left(\dfrac{1}{5}\right)^{x-1} = 25^x$

7. $8^{4x-7} = 11^{5+x}$

8. $3(9)^{x-1} = (81)^{2x+1}$

9. $(3.14)^x = \pi^{1-2x}$

10. $7(2 - 10^{4x-2}) = 8$

11. $e^x = 2$

12. $150e^{x-4} = 5$

13. $e^{x-3} \cdot e^{3x+7} = 24$

14. $2(e^{x-1})^2 \cdot e^{3-x} = 80$

15. $8e^{-x/3} \cdot e^x = 1$

In Exercises 16–33, solve each logarithmic equation.

16. $\log_4 (x + 1) = \log_4 (6x - 5)$

17. $\log_3 (x^2 - 21) = \log_3 4x$

18. $2 \log_5 (3 - x) - \log_5 2 = \log_5 18$

19. $2 \ln x - \ln (2x - 3) = \ln 2x - \ln (x - 1)$

20. $\log_2 (4x - 7) = 3$

21. $\log (1 - 5x) = 2$

22. $\log_3 (2x - 5) = -2$

23. $\log_x 3 = -1$

24. $\log_{x/2} 16 = 2$

25. $\log_2 (x - 2) + \log_2 (x + 2) = 5$

26. $\log_7 (x + 9) + \log_7 (x + 15) = 1$

27. $\log_3 (3x + 1) - \log_3 (x - 2) = 2$

28. $\log_6 (x - 8) = 2 - \log_6 (x + 8)$

29. $\log_4 (x + 21) - 2 = -\log_4 (x + 6)$

30. $2 - \log_5 (5x + 3) + \log_5 (x - 1) = 0$

31. $\log_4 (x - 7) + \log_4 x = \dfrac{3}{2}$

32. $\ln 3 + \ln \left(x^2 + \dfrac{2x}{3} \right) = 0$

33. $\log_2 (x - 4) + \log_2 (x + 6) = 2 + \log_2 x$

In Exercises 34 and 35, solve each logarithmic equation. Round to four decimal places.

34. $\ln x - \ln (x + 6) = 1$

35. $\ln (x + 3) - \ln (x - 2) = 4$

5.5 Applications of Exponential and Logarithmic Functions

THINGS TO KNOW

Before working through this section, be sure you are familiar with the following concepts:

VIDEO   ANIMATION   INTERACTIVE

You Try It

1. Solving Applications of Exponential Functions (Compound Interest) (Section 5.1)

You Try It

2. Solving Applications of Exponential Functions (Exponential Growth) (Section 5.1)

You Try It

3. Understanding the Properties of Logarithms (Section 5.2)

OBJECTIVES

1 Solving Compound Interest Applications
2 Exponential Growth and Decay
3 Solving Logistic Growth Applications
4 Using Newton's Law of Cooling

Introduction to Section 5.5

We have seen that exponential functions appear in a wide variety of settings, including biology, chemistry, physics, and business. In this section, we revisit some applications that are discussed previously in this chapter and then introduce several new applications. The difference between the applications presented earlier and the applications presented in this section is that we are now equipped with the tools necessary to solve for variables that appear as exponents. We start with applications involving **compound interest**. You may want to review periodic compound interest and continuous compound interest from Section 5.1 before proceeding.

OBJECTIVE 1 SOLVING COMPOUND INTEREST APPLICATIONS

In Section 5.1, the formulas for compound interest and continuous compound interest are defined as follows.

Compound Interest Formulas

Periodic Compound Interest Formula

$$A = P\left(1 + \frac{r}{n}\right)^{nt}$$

Continuous Compound Interest Formula

$$A = Pe^{rt},$$

where

A = Total amount after t years
P = Principal (original investment)
r = Interest rate per year
n = Number of times interest is compounded per year
t = Number of years

 *My video summary*

⊚ Example 1 Find the Doubling Time

How long will it take (in years and months) for an investment to double if it earns 7.5% compounded monthly?

Solution

We use the periodic compound interest formula $A = P\left(1 + \dfrac{r}{n}\right)^{nt}$ with $r = .075$ and $n = 12$ and solve for t. Notice that the principal is not given. As it turns out, any value of P will suffice. If the principal is P, then the amount needed to double the investment is $A = 2P$. We now have all of the information necessary to solve for t:

$$A = P\left(1 + \frac{r}{n}\right)^{nt}$$ Use the periodic compound interest formula.

$$2P = P\left(1 + \frac{.075}{12}\right)^{12t}$$ Substitute the appropriate values.

$$2 = \left(1 + \frac{.075}{12}\right)^{12t}$$ Divide both sides by P.

$$2 = (1.00625)^{12t}$$ Simplify within the parentheses.

$$\ln 2 = \ln (1.00625)^{12t}$$ Use the logarithm property of equality.

$$\ln 2 = 12t \ln (1.00625)$$ Use the power rule: $\log_b u^r = r \log_b u$.

$$t = \frac{\ln 2}{12 \ln (1.00625)}$$ Divide both sides by $12 \ln (1.00625)$.

$$t \approx 9.27 \text{ years}$$ Round to two decimal places.

Note that $.27$ years $= .27$ years $\times \dfrac{12 \text{ months}}{1 \text{ year}} = 3.24$ months. Because the interest is compounded at the end of each month, the investment will not double until 9 years and 4 months.

 *My video summary*

▶ Example 2 Continuous Compound Interest

Suppose an investment of $5,000 compounded continuously grew to an amount of $5,130.50 in 6 months. Find the interest rate, and then determine how long it will take for the investment to grow to $6,000. Round the interest rate to the nearest hundredth of a percent and the time to the nearest hundredth of a year.

Solution Because the investment is compounded continuously, we use the formula $A = Pe^{rt}$.

We are given that $P = 5,000$, so $A = 5,000e^{rt}$. In 6 months, or when $t = 0.5$ years, we know that $A = 5,130.50$. Substituting these values into the compound interest formula will enable us to solve for r:

$$5,130.50 = 5,000e^{r(0.5)}$$ Substitute the appropriate values.

$$\frac{5,130.50}{5,000} = e^{0.5r}$$ Divide by 5,000.

$$\ln \left(\frac{5,130.50}{5,000}\right) = \ln e^{0.5r}$$ Use the logarithm property of equality.

$$\ln \left(\frac{5,130.50}{5,000}\right) = 0.5r$$ Use cancellation property (2) to rewrite $\ln e^{0.5r}$ as $0.5r$.

$$r = \frac{\ln \left(\dfrac{5,130.50}{5,000}\right)}{0.5} \approx .051530\cdot$$ Divide by 0.5.

My video summary ⊙ Therefore, the interest rate is 5.15%. To find the time that it takes for the investment to grow to $6,000, we use the formula $A = Pe^{rt}$, with $A = 6,000$, $P = 5,000$, and $r = 0.0515$, and solve for t. Watch the video to verify that it will take approximately 3.54 years.

You Try It   Work through this You Try It problem.

Work Exercises 1–5 in this textbook or in the MyMathLab Study Plan.

OBJECTIVE 2   EXPONENTIAL GROWTH AND DECAY

In Section 5.1, the exponential growth model is introduced. This model is used when a population grows at a rate proportional to the size of its current population. This model is often called the **uninhibited growth** model. We review this exponential growth model and sketch the graph in Figure 16.

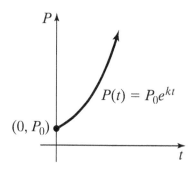

Figure 16 Graph of $P(t) = P_0e^{kt}$ for $k > 0$

Exponential Growth

A model that describes the exponential uninhibited growth of a population, P, after a certain time, t, is

$$P(t) = P_0e^{kt},$$

where $P_0 = P(0)$ is the initial population and $k > 0$ is a constant called the **relative growth rate**. (*Note*: k is sometimes given as a percent.)

My video summary ⊙ **Example 3** Population Growth

The population of a small town grows at a rate proportional to its current size. In 1900, the population was 900. In 1920, the population had grown to 1,600. What was the population of this town in 1950? Round to the nearest whole number.

Solution

Using the model $P(t) = P_0e^{kt}$, we must first determine the constants P_0 and k. The initial population was 900 in 1900 so $P_0 = 900$. Therefore, $P(t) = 900e^{kt}$. To find k, we use the fact that in 1920, or when $t = 20$, the population was 1,600; thus,

$$P(20) = 900e^{k(20)} = 1,600 \qquad \text{Substitute } P(20) = 1,600.$$

$$900e^{20k} = 1,600$$

$$e^{20k} = \frac{16}{9} \qquad \text{Divide by 900 and simplify.}$$

$$\ln e^{20k} = \ln \frac{16}{9}$$ Use the logarithm property of equality.

$$20k = \ln \frac{16}{9}$$ Use cancellation property (2) to rewrite $\ln e^{20k}$ as $20k$.

$$k = \frac{\ln\left(\dfrac{16}{9}\right)}{20}$$ Divide by 20.

The function that models the population of this town at any time t is given by $P(t) = 900e^{\frac{\ln(16/9)}{20}t}$. To determine the population in 1950, or when $t = 50$, we evaluate $P(50)$:

$$P(50) = 900e^{\frac{\ln(16/9)}{20}(50)} \approx 3{,}793$$

Some populations exhibit *negative exponential growth*. In other words, the population, quantity, or amount *decreases* over time. Such models are called **exponential decay** models. The only difference between an exponential growth model and an exponential decay model is that the constant, k, is less than zero. See Figure 17.

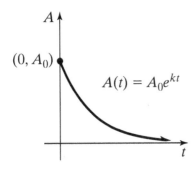

Figure 17 Graph of $A(t) = A_0 e^{kt}$ for $k < 0$

Exponential Decay

A model that describes the exponential decay of a population, quantity, or amount A, after a certain time, t, is

$$A(t) = A_0 e^{kt},$$

where $A_0 = A(0)$ is the initial quantity and $k < 0$ is a constant called the **relative decay constant**. (*Note:* k is sometimes given as a percent.)

HALF-LIFE

Every radioactive element has a half-life, which is the required time for a given quantity of that element to decay to half of its original mass. For example, the half-life of cesium-137 is 30 years. Thus, it takes 30 years for any amount of cesium-137 to decay to $\frac{1}{2}$ of its original mass. It takes an additional 30 years to decay to $\frac{1}{4}$ of its original mass and so on. See Figure 18, and view the animation that illustrates the half-life of cesium-137.

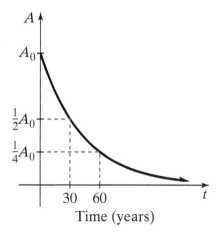

Figure 18 Half-life of cesium-137

My video summary ⊙ **Example 4** Radioactive Decay

Suppose that a meteorite is found containing 4% of its original krypton-99. If the half-life of krypton-99 is 80 years, how old is the meteorite? Round to the nearest year.

Solution We use the formula $A(t) = A_0 e^{kt}$, where A_0 is the original amount of krypton-99. We first must find the constant k. To find k, we use the fact that the half-life of krypton-99 is 80 years. Therefore, $A(80) = \frac{1}{2}A_0$. Because $A(80) = A_0 e^{k(80)}$, we can set $\frac{1}{2}A_0 = A_0 e^{k(80)}$ and solve for k.

$\frac{1}{2}A_0 = A_0 e^{k(80)}$ Half of the original amount will be present in 80 years.

$\frac{1}{2} = e^{80k}$ Divide both sides by A_0.

$\ln \frac{1}{2} = \ln e^{80k}$ Use the logarithm property of equality.

$\ln \frac{1}{2} = 80k$ Use cancellation property (2) to rewrite $\ln e^{80k}$ as $80k$.

$\dfrac{\ln \frac{1}{2}}{80} = k$ Divide both sides by 80.

$\dfrac{-\ln 2}{80} = k$ $\ln \frac{1}{2} = \ln 1 - \ln 2 = 0 - \ln 2 = -\ln 2$

Now that we know $k = \dfrac{-\ln 2}{80}$, our function becomes $A(t) = A_0 e^{\frac{-\ln 2}{80}t}$. To find out the age of the meteorite, we set $A(t) = .04A_0$ because the meteorite now contains 4% of the original amount of krypton-99.

$.04A_0 = A_0 e^{\frac{-\ln 2}{80}t}$ Substitute $.04A_0$ for $A(t)$.

$.04 = e^{\frac{-\ln 2}{80}t}$ Divide both sides by A_0.

$\ln .04 = \ln e^{\frac{-\ln 2}{80}t}$ Use the logarithm property of equality.

5.5 Applications of Exponential and Logarithmic Functions   **5-55**

$$\ln .04 = \frac{-\ln 2}{80} t$$

Use cancellation property (2) to rewrite $\ln e^{\frac{-\ln 2}{80} t}$ as $\frac{-\ln 2}{80} t$.

$$\frac{\ln .04}{\left(\frac{-\ln 2}{80}\right)} = t \approx 372 \text{ years}$$

Divide both sides by $\frac{-\ln 2}{80}$.

The meteorite is about 372 years old.

Watch this video to see the solution to this example worked out in detail.

You Try It Work through this You Try It problem.

Work Exercises 6–12 in this textbook or in the MyMathLab Study Plan.

OBJECTIVE 3   SOLVING LOGISTIC GROWTH APPLICATIONS

The uninhibited exponential growth model $P(t) = P_0 e^{kt}$ for $k > 0$ is used when there are no outside limiting factors such as predators or disease that affect the population growth. When such outside factors exist, scientists often use a **logistic model** to describe the population growth. One such logistic model is described and sketched in Figure 19.

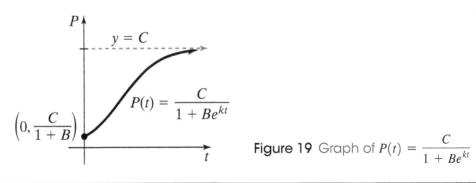

Figure 19 Graph of $P(t) = \dfrac{C}{1 + Be^{kt}}$

Logistic Growth

A model that describes the logistic growth of a population P at any time t is given by the function

$$P(t) = \frac{C}{1 + Be^{kt}},$$

where B, C, and k are constants with $C > 0$ and $k < 0$.

The number C is called the **carrying capacity**. In the logistic model, the population will approach the value of the carrying capacity over time but never exceed it. You can see in the graph sketched in Figure 19 that the graph of the logistic growth model approaches the horizontal asymptote $y = C$.

⊘ **Example 5** Logistic Growth

Ten goldfish were introduced into a small pond. Because of limited food, space, and oxygen, the carrying capacity of the pond is 400 goldfish. The goldfish population at any time t, in days, is modeled by the logistic growth function $F(t) = \dfrac{C}{1 + Be^{kt}}$.

If 30 goldfish are in the pond after 20 days,

a. Find B.

b. Find k.

c. When will the pond contain 250 goldfish? Round to the nearest whole number.

Solution

a. The carrying capacity is 400; thus, $C = 400$. Also, initially (at $t = 0$) there were 10 goldfish, so $F(0) = 10$. Therefore,

$$10 = \frac{400}{1 + Be^{k(0)}} \qquad \text{Substitute } C = 400, t = 0, \text{ and } f(0) = 10.$$

$$10 = \frac{400}{1 + B} \qquad \text{Evaluate } e^0 = 1.$$

$$10 + 10B = 400 \qquad \text{Multiply both sides by } 1 + B.$$

$$B = 39 \qquad \text{Solve for } B.$$

b. Use the function $F(t) = \dfrac{400}{1 + 39e^{kt}}$ and the fact that $F(20) = 30$ (there are 30 goldfish after 20 days) to solve for k.

$$30 = \frac{400}{1 + 39e^{k(20)}} \qquad \text{Substitute } F(20) = 30.$$

$$30(1 + 39e^{20k}) = 400 \qquad \text{Multiply both sides by } 1 + 39e^{20k}.$$

$$30 + 1{,}170e^{20k} = 400 \qquad \text{Use the distributive property.}$$

$$1{,}170e^{20k} = 370 \qquad \text{Subtract 30 from both sides.}$$

$$e^{20k} = \frac{370}{1{,}170} \qquad \text{Divide both sides by 1,170.}$$

$$e^{20k} = \frac{37}{117} \qquad \text{Simplify.}$$

$$\ln e^{20k} = \ln \frac{37}{117} \qquad \text{Use the logarithm property of equality.}$$

$$20k = \ln \frac{370}{117} \qquad \text{Use cancellation property (2) to rewrite } \ln e^{20k} \text{ as } 20k.$$

$$k = \frac{\ln \dfrac{37}{117}}{20} \qquad \text{Divide both sides by 20.}$$

c. Use the function $F(t) = \dfrac{400}{1 + 39e^{\frac{\ln \frac{37}{117}}{20}t}}$, and then find t when $F(t) = 250$.

By repeating the exact same process as in part b, we find that it will take approximately 73 days until there are 250 goldfish in the pond. Watch the interactive video to verify the solution.

You Try It   Work through this You Try It problem.

Work Exercises 13–15 in this textbook or in the MyMathLab Study Plan.

OBJECTIVE 4   USING NEWTON'S LAW OF COOLING

Newton's law of cooling states that the temperature of an object changes at a rate proportional to the difference between its temperature and that of its surroundings. It can be shown in a more advanced course that the function describing Newton's law of cooling is given by the following.

Newton's Law of Cooling

The temperature T of an object at any time t is given by

$$T(t) = S + (T_0 - S)e^{kt}$$

where T_0 is the original temperature of the object, S is the constant temperature of the surroundings, and k is the cooling constant.

View the animation to see how this function behaves.

My video summary   ⊙ **Example 6** Newton's Law of Cooling

Suppose that the temperature of a cup of hot tea obeys Newton's law of cooling. If the tea has a temperature of $200°F$ when it is initially poured and 1 minute later has cooled to $189°F$ in a room that maintains a constant temperature of $69°F$, determine when the tea reaches a temperature of $146°F$. Round to the nearest minute.

Solution

We start using the formula for Newton's law of cooling with $T_0 = 200$ and $S = 69$.

$$T(t) = S + (T_0 - S)e^{kt}$$ Use Newton's law of cooling formula.

$$T(t) = 69 + (200 - 69)e^{kt}$$ Substitute $T_0 = 200$ and $S = 69$.

$$T(t) = 69 + 131e^{kt}$$ Simplify.

We now proceed to find k. The object cools to $189°F$ in 1 minute; thus, $T(1) = 189$. Therefore,

$$189 = 69 + 131e^{k(1)}$$ Substitute $t = 1$ and $T(1) = 189$.

$$120 = 131e^k$$ Subtract 69 from both sides.

$$\frac{120}{131} = e^k$$ Divide both sides by 131.

$$\ln \frac{120}{131} = \ln e^k$$ Use the logarithm property of equality.

$$\ln \frac{120}{131} = k$$ Use cancellation property (2) to rewrite $\ln e^k$ as k.

Now that we know the cooling constant $k = \ln\dfrac{120}{131}$, we can use the function $T(t) = 69 + 131e^{\ln\frac{120}{131}t}$ and determine the value of t when $T(t) = 146$.

$146 = 69 + 131e^{\ln\frac{120}{131}t}$ Set $T(t) = 146$.

$77 = 131e^{\ln\frac{120}{131}t}$ Subtract 69 from both sides.

$\dfrac{77}{131} = e^{\ln\frac{120}{131}t}$ Divide both sides by 131.

$\ln\dfrac{77}{131} = \ln e^{\ln\frac{120}{131}t}$ Use the logarithm property of equality.

$\ln\dfrac{77}{131} = \ln\dfrac{120}{131}t$ Use cancellation property (2) to rewrite $\ln e^{\ln\frac{120}{131}t}$ as $\ln\dfrac{120}{131}t$.

$\dfrac{\ln\dfrac{77}{131}}{\ln\dfrac{120}{131}} = t$ Divide both sides by $\ln\dfrac{120}{131}$.

$t \approx 6$ minutes   Use a calculator to approximate the time rounded to the nearest minute.

So, it takes approximately 6 minutes for the tea to cool to 146°F.

Using Technology

The graph of $T(t) = 69 + 131e^{\ln\frac{120}{131}t}$, which describes the temperature of the tea t minutes after being poured, was created using a graphing utility. Note that the line $y = 69$, which represents the temperature of the surroundings, is a horizontal asymptote.

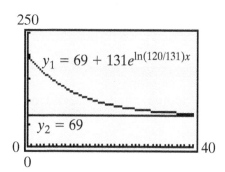

$y_1 = 69 + 131e^{\ln(120/131)x}$

$y_2 = 69$

 You Try It Work through this You Try It problem.

Work Exercises 16–18 in this textbook or in the MyMathLab Study Plan.

5.5 Exercises

1. Jimmy invests $15,000 in an account that pays 6.25% compounded quarterly. How long (in years and months) will it take for his investment to reach $20,000?

2. How long (in years and months) will it take for an investment to double at 9% compounded monthly?

3. How long will it take for an investment to triple if it is compounded continuously at 8%? Round to 2 decimal places.

4. What is the interest rate necessary for an investment to quadruple after 8 years of continuous compound interest? Round to the nearest hundredth of a percent.

5. Marsha and Jan both invested money on March 1, 2005. Marsha invested $5,000 at Bank A, where the interest was compounded quarterly. Jan invested $3,000 at Bank B, where the interest was compounded continuously. On March 1, 2007, Marsha had a balance of $5,468.12, whereas Jan had a balance of $3,289.09. What was the interest rate at each bank? Round to the nearest tenth of a percent.

6. The population of Adamsville grew from 9,000 to 15,000 in 6 years. Assuming uninhibited exponential growth, what is the expected population in an additional 4 years? Round to the nearest whole number.

7. During a research experiment, it was found that the number of bacteria in a culture grew at a rate proportional to its size. At 8:00 AM, there were 2,000 bacteria present in the culture. At noon, the number of bacteria grew to 2,400. How many bacteria will there be at midnight? Round to the nearest whole number.

8. A skull cleaning factory cleans animal skulls such as deer, buffalo, and other types of animal skulls using flesh-eating beetles to clean the skulls. The factory owner started with only 10 adult beetles. After 40 days, the beetle population grew to 30 adult beetles. How long did it take before the beetle population reached 10,000 beetles? Round to the nearest whole number.

9. The population of a Midwest industrial town decreased from 210,000 to 205,000 in just 3 years. Assuming that this trend continues, what will the population be after an additional 3 years? Round to the nearest whole number.

10. A certain radioactive isotope is leaked into a small stream. Three hundred days after the leak, 2% of the original amount of the substance remained. Determine the half-life of this radioactive isotope. Round to the nearest whole number.

11. Radioactive iodine-131 is a by-product of certain nuclear reactors. On April 26, 1986, one of the nuclear reactors in Chernobyl, Ukraine, a republic of the former Soviet Union, experienced a massive release of radioactive iodine. Fortunately, iodine-131 has a very short half-life of 8 days. Estimate the percentage of the original amount of iodine-131 released by the Chernobyl explosion, 5 days after the explosion. Round to 2 decimal places.

12. Superman is rendered powerless when exposed to 50 or more grams of kryptonite. A 500-year-old rock that originally contained 300 grams of kryptonite was recently stolen from a rock museum by Superman's enemies. The half-life of kryptonite is known to be 200 years.

   a. How many grams of kryptonite are still contained in the stolen rock? Round to two decimal places.

   b. For how many years can this rock be used by Superman's enemies to render him powerless? Round to the nearest whole number.

13. The logistic growth model $H(t) = \dfrac{6,000}{1 + 2e^{-.65t}}$ represents the number of families that own a home in a certain small (but growing) Idaho city t years after 1980.

   a. What is the maximum number of families that will own a home in this city?

   b. How many families owned a home in 1980?

   c. In what year did 5,920 families own a home?

SbS **14.** The number of students that hear a rumor on a small college campus t days after the rumor starts is modeled by the logistic function $R(t) = \dfrac{3{,}000}{1 + Be^{kt}}$. Determine the following if 8 students initially heard the rumor and 100 students heard the rumor after 1 day.

   a. What is the carrying capacity for the number of students who will hear the rumor?

   b. Find B.

   c. Find k.

   d. How long will it take 2,900 students to hear the rumor?

SbS **15.** In 1999, 1,500 runners entered the inaugural Run-for-Your-Life marathon in Joppetown, USA. In 2005, 21,500 runners entered the race. Because of the limited number of hotels, restaurants, and portable toilets in the area, the carrying capacity for the number of racers is 61,500. The number of racers at any time, t, in years, can be modeled by the logistic function $P(t) = \dfrac{C}{1 + Be^{kt}}$.

   a. What is the value of C?

   b. Find B.

   c. Find k.

   d. In what year should at least 49,500 runners be expected to run in the race? Round to the nearest year.

16. Estabon poured himself a hot beverage that had a temperature of 198°F and then set it on the kitchen table to cool. The temperature of the kitchen was a constant 75°F. If the drink cooled to 180°F in 2 minutes, how long will it take for the drink to cool to 100°F?

17. Police arrive at a murder scene at 1:00 AM and immediately record the body's temperature, which was 92°F. At 2:30 AM, after thoroughly inspecting and fingerprinting the area, they again took the temperature of the body, which had dropped to 85°F. The temperature of the crime scene has remained at a constant 60°F. Determine when the person was murdered. (Assume that the victim was healthy at the time of death. That is, assume that the temperature of the body at the time of death was 98.6°F.)

18. Jodi poured herself a cold soda that had an initial temperature of 40°F and immediately went outside to sunbathe where the temperature was a steady 99°F. After 5 minutes, the temperature of the soda was 47°F. Jodi had to run back into the house to answer the phone. What is the expected temperature of the soda after an additional 10 minutes?

Brief Exercises

19. The logistic growth model $H(t) = \dfrac{6{,}000}{1 + 2e^{-.65t}}$ represents the number of families that own a home in a certain small (but growing) Idaho city t years after 1980. In what year did 5,920 families own a home?

20. The number of students that hear a rumor on a small college campus t days after the rumor starts is modeled by the logistic function $R(t) = \dfrac{3{,}000}{1 + Be^{kt}}$. If 8 students initially heard the rumor and 100 students heard the after 1 day, then how long will it take 2,900 students to hear the rumor?

21. In 1999, 1,500 runners entered the inaugural Run-for-Your-Life marathon in Joppetown, USA. In 2005, 21,500 runners entered the race. Because of the limited number of hotels, restaurants, and portable toilets in the area, the carrying capacity for the number of racers is 61,500. The number of racers at any time, t, in years, can be modeled by the logistic function $P(t) = \dfrac{C}{1 + Be^{kt}}$. In what year should at least 49,500 runners be expected to run in the race? Round to the nearest year.

CHAPTER SIX

Conic Sections

CHAPTER SIX CONTENTS

Introduction to Conic Sections

INTRODUCTION TO CONIC SECTIONS

In this chapter, we focus on the geometric study of conic sections. Conic sections (or conics) are formed when a plane intersects a pair of right circular cones. The surface of the cones comprises the set of all line segments that intersect the outer edges of the circular bases of the cones and pass through a fixed point. The fixed point is called the **vertex** of the cone, and the line segments are called the **elements**. See Figure 1.

When a plane intersects a right circular cone, a conic section is formed. The four conic sections that can be formed are circles, ellipses, parabolas, and hyperbolas. Click on one of the following four animations to see how each conic section is formed.

Figure 1 Pair of right circular cones

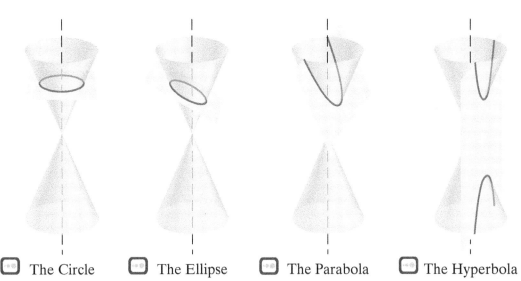

The Circle   The Ellipse   The Parabola   The Hyperbola

Because we studied circles in Section 2.2, they are not covered again in this chapter. It may, however, be useful to review circles before going on. Watch the video to review how to write the equation of a circle in standard form by completing the square. We also need to be able to complete the square to write each of the other conic sections in standard form. These conic sections are introduced in the following order:

Section 6.1 The Parabola

Section 6.2 The Ellipse

Section 6.3 The Hyperbola

DEGENERATE CONIC SECTIONS

It is worth noting that when the circle, ellipse, parabola, or hyperbola is formed, the intersecting plane does not pass through the vertex of the cones. When a plane does intersect the vertex of the cones, a degenerate conic section is formed. The three degenerate conic sections are a point, a line, and a pair of intersecting lines. See Figure 2. We do not concern ourselves with degenerate conic sections in this chapter.

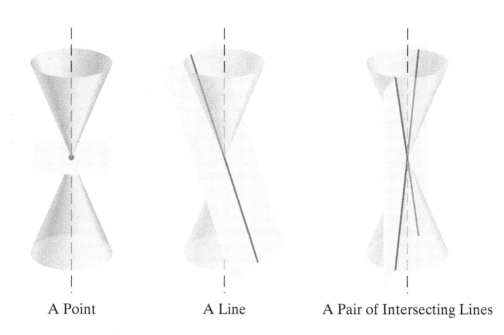

A Point   A Line   A Pair of Intersecting Lines

Figure 2 Degenerate conic sections

6.1 The Parabola

THINGS TO KNOW

Before working through this section, be sure that you are familiar with the following concepts:

VIDEO   ANIMATION   INTERACTIVE

You Try It

1. Finding the Distance between Two Points Using the Distance Formula (Section 2.1)

You Try It

2. Converting the General Form of a Circle into Standard Form (Section 2.2)

You Try It

3. Finding the Equations of Horizontal and Vertical Lines (Section 2.3)

OBJECTIVES

1 Determining the Equation of a Parabola with a Vertical Axis of Symmetry

2 Determining the Equation of a Parabola with a Horizontal Axis of Symmetry

3 Determining the Equation of a Parabola, Given Information about the Graph

4 Completing the Square to Find the Equation of a Parabola in Standard Form

5 Solving Applied Problems Involving Parabolas

OBJECTIVE 1 DETERMINING THE EQUATION OF A PARABOLA WITH A VERTICAL AXIS OF SYMMETRY

In Section 4.1, we studied quadratic functions of the form $f(x) = ax^2 + bx + c$, $a \neq 0$. We learned that every quadratic function has a u-shaped graph called a *parabola*. You may want to review the different characteristics of a parabola. Work through the following animation, and click on each characteristic to get a detailed description.

 *My animation summary*

CHARACTERISTICS OF A PARABOLA

1. Vertex

2. Axis of symmetry

3. y-Intercept

4. x-Intercept(s) or real zeros

5. Domain and range

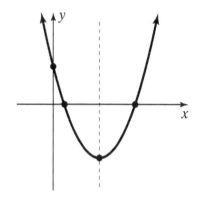

 *My animation summary*

In Section 4.1, we studied quadratic functions and parabolas from an algebraic point of view. We now look at parabolas from a geometric perspective. We see in the introduction to this chapter that when a plane is parallel to an element of the cone, the plane will intersect the cone in a parabola. (Click on the animation.)

The set of points that define the parabola formed by the intersection described previously is stated in the following geometric definition of the parabola.

Geometric Definition of the Parabola

A **parabola** is the set of all points in a plane equidistant from a fixed point F and a fixed line D. The fixed point is called the **focus**, and the fixed line is called the **directrix**.

The Parabola

My video summary   ▶ Watch the video to see how to sketch the parabola seen in Figure 3 using the geometric definition of the parabola.

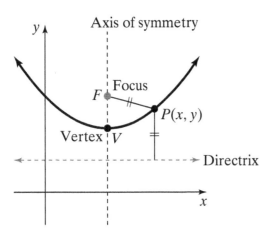

Figure 3 The distance from any point P on the parabola to the focus is the same as the distance from point P to the directrix.

In Figure 3, we can see that for any point $P(x, y)$ that lies on the graph of the parabola, the distance from point P to the focus is exactly the same as the distance from point P to the directrix. Similarly, because the vertex, V, lies on the graph of the parabola, the distance from V to the focus must also be the same as the distance from V to the directrix. Therefore, if the distance from V to F is p units, then the distance from V to the directrix is also p units. If the coordinates of the vertex in Figure 3 are (h, k), then the coordinates of the focus must be $(h, k + p)$ and the equation of the directrix is $y = k - p$. We can use this information and the fact that the distance from $P(x, y)$ to the focus is equal to the distance from $P(x, y)$ to the directrix to derive the equation of a parabola. The equation of the parabola with a vertical axis of symmetry is derived in Appendix A.

Equation of a Parabola in Standard Form with a Vertical Axis of Symmetry

The equation of a parabola with a vertical axis of symmetry is $(x - h)^2 = 4p(y - k)$,

where
  the vertex is $V(h, k)$,
  $|p|$ = distance from the vertex to the focus = distance from the vertex to the directrix,
  the focus is $F(h, k + p)$, and
  the equation of the directrix is $y = k - p$.

The parabola opens *upward* if $p > 0$ or *downward* if $p < 0$.

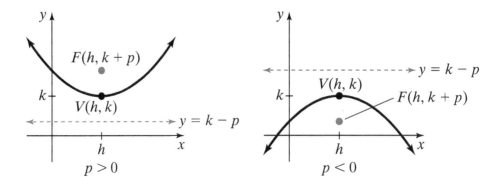

Example 1 Find the Vertex, Focus, and Directrix of a Parabola and Sketch Its Graph

Find the vertex, focus, and directrix of the parabola $x^2 = 8y$ and sketch its graph.

Solution Notice that we can rewrite the equation $x^2 = 8y$ as $(x - 0)^2 = 8(y - 0)$. We can now compare the equation $(x - 0)^2 = 8(y - 0)$ to the standard form equation $(x - h)^2 = 4p(y - k)$ to see that $h = 0$ and $k = 0$; hence, the vertex is at the origin, $(0, 0)$. To find the focus and directrix, we need to find p.

$$4p = 8$$

$$p = 2 \quad \text{Divide both sides by 4.}$$

Because the value of p is positive, the parabola opens upward and the focus is located two units vertically *above* the vertex, whereas the directrix is the horizontal line located two units vertically *below* the vertex. The focus has coordinates $(0, 2)$, and the equation of the directrix is $y = -2$. The graph is shown in Figure 4.

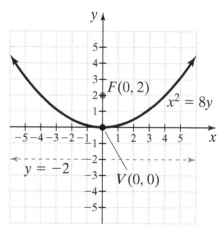

Figure 4

My video summary ⊙ **Example 2** Find the Vertex, Focus, and Directrix of a Parabola and Sketch Its Graph

Find the vertex, focus, and directrix of the parabola $-(x + 1)^2 = 4(y - 3)$ and sketch its graph.

Solution Watch the video to verify that the vertex has coordinates $(-1, 3)$, the focus has coordinates $(-1, 2)$, and the equation of the directrix is $y = 4$. The graph of this parabola is sketched in Figure 5.

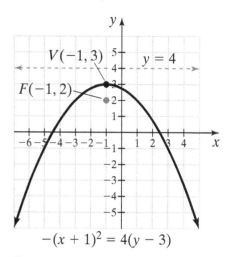

Figure 5

You Try It Work Exercises 1–5 in this textbook or in the MyMathLab Study Plan.

OBJECTIVE 2   DETERMINING THE EQUATION OF A PARABOLA WITH A HORIZONTAL AXIS OF SYMMETRY

In Examples 1 and 2, the graphs of both parabolas had vertical axes of symmetry. The graph of a parabola could also have a horizontal axis of symmetry and open "sideways." We derive the standard form of the parabola with a horizontal axis of symmetry in much the same way as we did with the parabola with a vertical axis of symmetry. This equation is derived in Appendix A.

Equation of a Parabola in Standard Form with a Horizontal Axis of Symmetry

The equation of a parabola with a horizontal axis of symmetry is $(y - k)^2 = 4p(x - h)$,

where
  the vertex is $V(h, k)$,
  $|p|$ = distance from the vertex to the focus = distance from the vertex to the directrix,
  the focus is $F(h + p, k)$, and
  the equation of the directrix is $x = h - p$.

The parabola opens *right* if $p > 0$ or *left* if $p < 0$.

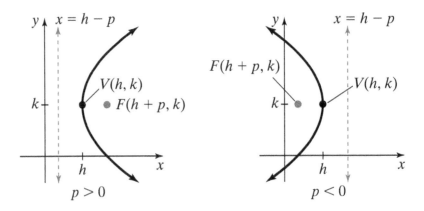

$p > 0$ $p < 0$

 *My video summary* ⊘ **Example 3** Find the Vertex, Focus, and Directrix of a Parabola and Sketch Its Graph

Find the vertex, focus, and directrix of the parabola $(y - 3)^2 = 8(x + 2)$ and sketch its graph.

Solution Watch the video to verify that the vertex has coordinates $(-2, 3)$, the focus has coordinates $(0, 3)$, and the equation of the directrix is $x = -4$. The graph of this parabola is sketched in Figure 6.

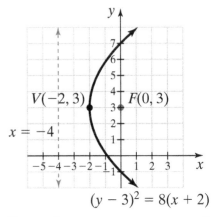

$$(y - 3)^2 = 8(x + 2)$$

Figure 6

You Try It Work Exercises 6–10 in this textbook or in the MyMathLab Study Plan.

OBJECTIVE 3 DETERMINING THE EQUATION OF A PARABOLA, GIVEN INFORMATION ABOUT THE GRAPH

It is often necessary to determine the equation of a parabola, given certain information. It is always useful to first determine whether the parabola has a vertical axis of symmetry or a horizontal axis of symmetry. Try to work through Examples 4 and 5. Then watch the corresponding video solutions to determine whether you are correct.

My video summary ⊘ **Example 4** Find the Equation of a Parabola

Find the standard form of the equation of the parabola with focus $\left(-3, \dfrac{5}{2}\right)$ and directrix $y = \dfrac{11}{2}$.

Solution Watch the video to see that the equation of this parabola is $(x + 3)^2 = -6(y - 4)$. The graph is shown in Figure 7.

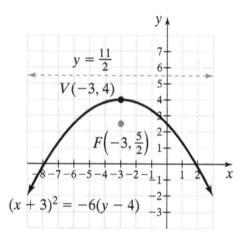

Figure 7

My video summary ⊘ **Example 5** Find the Vertex, Focus, and Directrix of a Parabola and Sketch Its Graph

Find the standard form of the equation of the parabola with focus $(4, -2)$ and vertex $\left(\dfrac{13}{2}, -2\right)$.

Solution Watch the video to see that the equation of this parabola is $(y + 2)^2 = -10\left(x - \dfrac{13}{2}\right)$. The graph is shown in Figure 8.

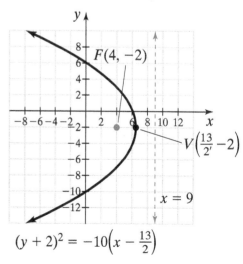

$$(y + 2)^2 = -10\left(x - \dfrac{13}{2}\right)$$

Figure 8

Using Technology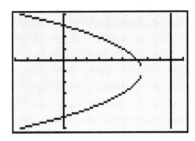
We can use a graphing utility to graph the
parabola from Example 5 by solving the
equation for y.

$$(y + 2)^2 = -10\left(x - \frac{13}{2}\right)$$

$$y + 2 = \pm\sqrt{-10\left(x - \frac{13}{2}\right)}$$

$$y = -2 \pm \sqrt{-10\left(x - \frac{13}{2}\right)}$$

Figure 9

Using $y_1 = -2 + \sqrt{-10\left(x - \frac{13}{2}\right)}$, and $y_2 = -2 - \sqrt{-10\left(x - \frac{13}{2}\right)}$, we obtain the
graph seen in Figure 9.

Note   The directrix was created using the calculator's DRAW feature.

You Try It   Work Exercises 11–19 in this textbook or in the MyMathLab **Study Plan.**

OBJECTIVE 4   COMPLETING THE SQUARE TO FIND THE EQUATION OF A PARABOLA IN STANDARD FORM

If the equation of a parabola is in the standard form of $(x - h)^2 = 4p(y - k)$ or
$(y - k)^2 = 4p(x - h)$, it is not too difficult to determine the vertex, focus, and directrix and sketch its graph. However, the equation might not be given in standard
form. If this is the case, we complete the square on the variable that is squared
to rewrite the equation in standard form as in Example 6.

My video summary   ⊙ **Example 6** Rewrite the Equation of a Parabola in Standard
Form by Completing the Square

Find the vertex, focus, and directrix and sketch the graph of the parabola
$x^2 - 8x + 12y = -52$.

Solution   Because x is squared, we will complete the square on the variable x.

$x^2 - 8x + 12y = -52$	Write the original equation.
$x^2 - 8x \quad\quad = -12y - 52$	Subtract $12y$ from both sides.
$x^2 - 8x + 16 = -12y - 52 + 16$	Complete the square by adding 16 to both sides.
$(x - 4)^2 = -12y - 36$	Factor and simplify.
$(x - 4)^2 = -12(y + 3)$	Factor.

My video summary   ⊙ The equation is now in standard form with vertex $(4, -3)$ and $4p = -12$ so
$p = -3$. The parabola must open down because the variable x is squared and
$p < 0$. The focus is located three units below the vertex, whereas the directrix is
three units above the vertex. Thus, the focus has coordinates $(4, -6)$, and the
equation of the directrix is $y = 0$ or the x-axis. You can watch the video to see
this solution worked out in detail. The graph is shown in Figure 10.

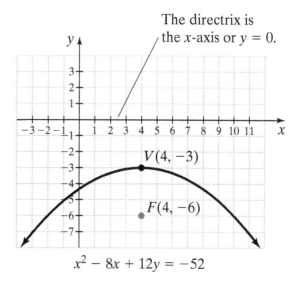

The directrix is the x-axis or $y = 0$.

$V(4, -3)$

$F(4, -6)$

$x^2 - 8x + 12y = -52$

Figure 10

You Try It Work Exercises 20–23 in this textbook or in the MyMathLab Study Plan.

OBJECTIVE 5 SOLVING APPLIED PROBLEMS INVOLVING PARABOLAS

The Romans were one of the first civilizations to use the engineering properties of parabolic structures in their creation of arch bridges. The cables of many suspension bridges, such as the Golden Gate Bridge in San Francisco, span from tower to tower in the shape of a parabola.

Parabolic surfaces are used in the manufacture of many satellite dishes, search lights, car headlights, telescopes, lamps, heaters, and other objects. This is because parabolic surfaces have the property that incoming rays of light or radio waves traveling parallel to the axis of symmetry of a parabolic reflector or receiver will reflect off the parabolic surface and travel directly toward the antenna that is placed at the focus. See Figure 11. When a light source such as the headlight of a car is placed at the focus of a parabolic reflector, the light reflects off the surface outward, producing a narrow beam of light and thus maximizing the output of illumination.

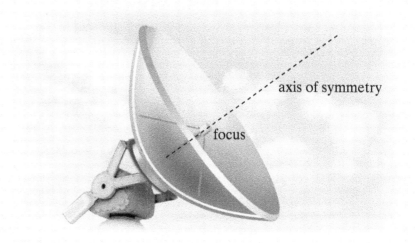

axis of symmetry

focus

Figure 11 Incoming rays reflect off the parabolic surface toward the antenna placed at the focus.

Example 7 Find the Focus of a Parabolic Microphone

Parabolic microphones can be seen on the sidelines of professional sporting events so that television networks can capture audio sounds from the players on the field. If the surface of a parabolic microphone is 27 centimeters deep and has a diameter of 72 centimeters at the top, where should the microphone be placed relative to the vertex of the parabola?

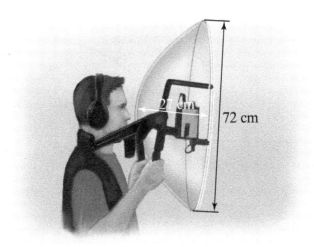

27 cm

72 cm

Solution We can draw a parabola with the vertex at the origin representing the center cross section of the parabolic microphone. The equation of this parabola in standard form is $x^2 = 4py$. Substitute the point $(36, 27)$ into the equation to get

$$x^2 = 4py$$
$$(36)^2 = 4p(27)$$
$$1{,}296 = 108p$$
$$p = 12.$$

The microphone must be placed 12 centimeters from the vertex.

You Try It   Work Exercises 24–28 in this textbook or in the MyMathLab Study Plan.

6.1 Exercises

In Exercises 1–10, determine the vertex, focus, and directrix of the parabola and sketch its graph.

SbS   **1.** $x^2 = 16y$

SbS   **2.** $x^2 = -8y$

SbS   **3.** $(x - 1)^2 = -12(y - 4)$

SbS   **4.** $(x + 3)^2 = 6(y - 1)$

SbS   **5.** $(x + 2)^2 = 5(y + 6)$

SbS   **6.** $y^2 = 4x$

SbS   **7.** $y^2 = -8x$

SbS   **8.** $(y - 5)^2 = -4(x - 2)$

SbS   **9.** $(y + 3)^2 = 20(x - 4)$

SbS   **10.** $(y + 4)^2 = 9(x + 3)$

In Exercises 11–19, find the equation in standard form of the parabola described.

11. The focus has coordinates $(2, 0)$, and the equation of the directrix is $x = -2$.

12. The focus has coordinates $\left(0, -\dfrac{1}{2}\right)$, and the equation of the directrix is $y = \dfrac{1}{2}$.

13. The focus has coordinates $(3, -5)$, and the equation of the directrix is $y = -1$.

14. The focus has coordinates $(2, 4)$, and the equation of the directrix is $x = -4$.

15. The vertex has coordinates $\left(-\dfrac{11}{4}, -2\right)$, and the focus has coordinates $(-3, -2)$.

16. The vertex has coordinates $\left(4, -\dfrac{1}{4}\right)$, and the focus has coordinates $\left(4, \dfrac{1}{4}\right)$.

17. The vertex has coordinates $(-3, 4)$, and the equation of the directrix is $x = -7$.

18. Find the equations of the two parabolas in standard form that have a horizontal axis of symmetry and a focus at the point $(0, 4)$, and that pass through the origin.

19. Find the equations of the two parabolas in standard form that have a vertical axis of symmetry and a focus at the point $\left(2, \dfrac{3}{2}\right)$, and that pass through the origin.

In Exercises 20–23, determine the vertex, focus, and directrix of the parabola and sketch its graph.

SbS **20.** $x^2 + 10x = 5y - 10$

SbS **21.** $y^2 - 12y + 6x + 30 = 0$

SbS **22.** $x^2 - 2x = -y + 1$

SbS **23.** $y^2 + x + 6y = -10$

24. A parabolic eavesdropping device is used by a CIA agent to record terrorist conversations. The parabolic surface measures 120 centimeters in diameter at the top and is 90 centimeters deep at its center. How far from the vertex should the microphone be located?

25. A parabolic space heater is 18 inches in diameter and 8 inches deep. How far from the vertex should the heat source be located to maximize the heating output?

26. A large NASA parabolic satellite dish is 22 feet across and has a receiver located at the focus 4 feet from its base. The satellite dish should be how deep?

27. A parabolic arch bridge spans 160 feet at the base and is 40 feet above the water at the center. Find the equation of the parabola if the vertex is placed at the point $(0, 40)$. Can a sailboat that is 35 feet tall fit under the bridge 30 feet from the center?

28. The cable between two 40-meter towers of a suspension bridge is in the shape of a parabola that just touches the bridge halfway between the towers. The two towers are 100 meters apart. Vertical cables are spaced every 10 meters along the bridge. What are the lengths of the vertical cables located 30 meters from the center of the bridge?

Brief Exercises

In Exercises 29–42, sketch the graph of each parabola.

29. $x^2 = 16y$

30. $x^2 = -8y$

31. $(x - 1)^2 = -12(y - 4)$

32. $(x + 3)^2 = 6(y - 1)$

33. $(x + 2)^2 = 5(y + 6)$

34. $y^2 = 4x$

35. $y^2 = -8x$

36. $(y - 5)^2 = -4(x - 2)$

37. $(y + 3)^2 = 20(x - 4)$

38. $(y + 4)^2 = 9(x + 3)$

39. $x^2 + 10x = 5y - 10$

40. $y^2 - 12y + 6x + 30 = 0$

41. $x^2 - 2x = -y + 1$

42. $y^2 + x + 6y = -10$

6.2 The Ellipse

THINGS TO KNOW

Before working through this section, be sure that you are familiar with the following concepts:

VIDEO   ANIMATION   INTERACTIVE

You Try It

1. Finding the Distance between Two Points Using the Distance Formula (Section 2.1)

You Try It

2. Completing the Square to Find the Equation of a Parabola in Standard Form (Section 6.1)

OBJECTIVES

1 Sketching the Graph of an Ellipse

2 Determining the Equation of an Ellipse, Given Information about the Graph

3 Completing the Square to Find the Equation of an Ellipse in Standard Form

4 Solving Applied Problems Involving Ellipses

 *My animation summary*

Introduction to Section 6.2

When a plane intersects a right circular cone at an angle between 0 and 90 degrees to the axis of the cone, the conic section formed is an ellipse. (Click on the animation.)

The set of points in the plane that defines an ellipse formed by the intersection of a plane and a cone described previously is stated in the following geometric definition.

> **Definition** Geometric Definition of the Ellipse
>
> An **ellipse** is the set of all points in a plane, the sum of whose distances from two fixed points is a positive constant. The two fixed points, F_1 and F_2, are called the *foci*.

The previous geometric definition implies that for any two points P and Q that lie on the graph of the ellipse, the sum of the distance between P and F_1 plus the distance between P and F_2 is equal to the sum of the distance between Q and F_1 plus the distance between Q and F_2. In symbols, we write $d(P, F_1) + d(P, F_2) = d(Q, F_1) + d(Q, F_2)$. See Figure 12.

The Ellipse

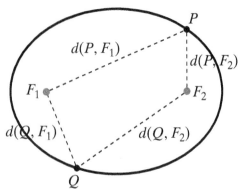

Figure 12 For any points P and Q on an ellipse,
$$d(P, F_1) + d(P, F_2) = d(Q, F_1) + d(Q, F_2).$$

OBJECTIVE 1   SKETCHING THE GRAPH OF AN ELLIPSE

My video summary

◉ An ellipse has two **axes of symmetry**. The longer axis, which is the line segment that connects the two vertices, is called the *major axis*. The foci are always located along the major axis. The shorter axis is called the *minor axis*. It is the line segment perpendicular to the major axis that passes through the center having endpoints that lie on the ellipse. Watch the **video** to see how to sketch the two ellipses seen in Figure 13. In Figure 13a, the ellipse has a horizontal major axis. The ellipse in Figure 13b has a vertical major axis.

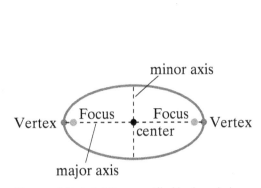

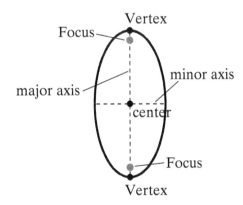

Figure 13 (a) Ellipse with Horizontal Major Axis

(b) Ellipse with Vertical Major Axis

Consider the ellipse with a **horizontal major axis** centered at (h, k) as shown in Figure 14, where $c > 0$ is the distance between the center and a focus, $a > 0$ is the distance between the center and one of the vertices, and $b > 0$ is the distance from the center to an endpoint of the minor axis.

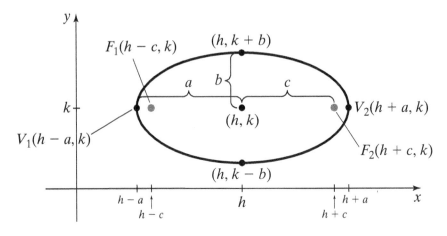

Figure 14

Before we can derive the equation of this ellipse, we must establish the following two facts:

Fact 1: The sum of the distances from any point on the ellipse to the two foci is $2a$.

Fact 2: $b^2 = a^2 - c^2$, or equivalently, $c^2 = a^2 - b^2$.

Click on **Fact 1** or **Fact 2** to see how these two facts are established in Appendix A.

Once we have established these two facts, we can derive the equation of the ellipse. The equation of an ellipse is derived in Appendix A. We now state the two standard equations of an ellipse.

Equation of an Ellipse in Standard Form with Center (h, k)

Horizontal Major Axis

$$\frac{(x - h)^2}{a^2} + \frac{(y - k)^2}{b^2} = 1$$

- $a > b > 0$

- Foci:
  $F_1(h - c, k)$ and $F_2(h + c, k)$

- Vertices:
  $V_1(h - a, k)$ and $V_2(h + a, k)$

- Endpoints of minor axis:
  $(h, k - b)$ and $(h, k + b)$

- $c^2 = a^2 - b^2$

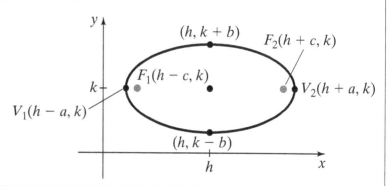

Equation of an Ellipse in Standard Form with Center (h, k)

Vertical Major Axis

$$\frac{(x - h)^2}{b^2} + \frac{(y - k)^2}{a^2} = 1$$

- $a > b > 0$

- Foci: $F_1(h, k - c)$ and $F_2(h, k + c)$

- Vertices: $V_1(h, k - a)$ and $V_2(h, k + a)$

- Endpoints of minor axis:
  $(h - b, k)$ and $(h + b, k)$

- $c^2 = a^2 - b^2$

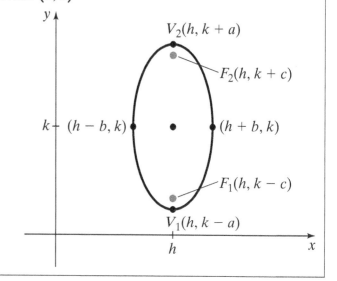

Notice that the two equations of the ellipse are identical except for the placement of a^2. If the equation of an ellipse is in standard form, we can quickly determine whether the ellipse has a horizontal major axis or a vertical major axis by looking at the denominator. If the larger denominator, a^2, appears under the expression $(x - h)^2$, then the ellipse has a horizontal major axis. If the larger denominator appears under the expression $(y - k)^2$, then the ellipse has a vertical major axis. If the denominators are equal ($a^2 = b^2$), then the ellipse is a circle!

If $h = 0$ and $k = 0$, then the ellipse is centered at the origin. Ellipses centered at the origin have the following equations.

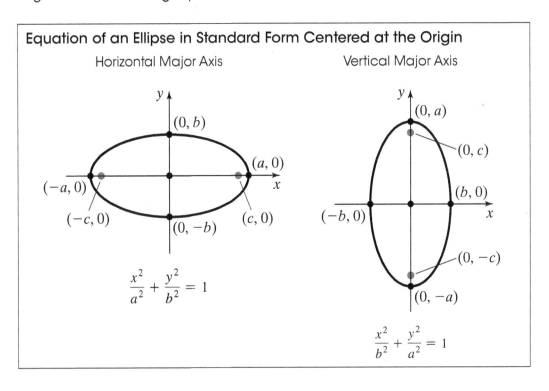

Equation of an Ellipse in Standard Form Centered at the Origin

Horizontal Major Axis

Vertical Major Axis

$$\frac{x^2}{a^2} + \frac{y^2}{b^2} = 1$$

$$\frac{x^2}{b^2} + \frac{y^2}{a^2} = 1$$

My video summary ⊙ **Example 1** Sketch the Graph of an Ellipse Centered at the Origin

Sketch the graph of the ellipse $\dfrac{x^2}{25} + \dfrac{y^2}{4} = 1$, and label the center, foci, and vertices.

Solution Watch the video to see how to sketch the ellipse shown in Figure 15.

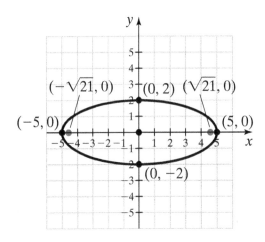

Figure 15

You Try It Work Exercises 1–4 in this textbook or in the MyMathLab Study Plan.

My video summary ⊘ **Example 2** Sketch the Graph of an Ellipse

Sketch the graph of the ellipse $\dfrac{(x + 2)^2}{20} + \dfrac{(y - 3)^2}{36} = 1$, and label the center, foci, and vertices.

Solution Note that the larger denominator appears under the y-term. This indicates that the ellipse has a vertical major axis. Watch the video to verify that the center of the ellipse is $(-2, 3)$. The foci have coordinates $(-2, -1)$ and $(-2, 7)$, the vertices have coordinates $(-2, -3)$ and $(-2, 9)$, and the coordinates of the endpoints of the minor axis are $(-2 - 2\sqrt{5}, 3)$ and $(-2 + 2\sqrt{5}, 3)$. The graph is shown in Figure 16.

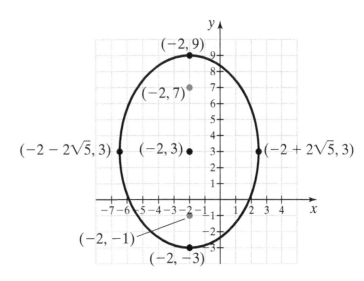

Figure 16

You Try It   Work Exercises 5–8 in this textbook or in the MyMathLab Study Plan.

OBJECTIVE 2   DETERMINING THE EQUATION OF AN ELLIPSE, GIVEN INFORMATION ABOUT THE GRAPH

It is often necessary to determine the equation of an ellipse given certain information. It is always useful to first determine whether the ellipse has a horizontal major axis or a vertical major axis. Try to work through Examples 3 and 4. Then watch the solutions to the corresponding videos to determine whether you are correct.

My video summary ⊘ **Example 3** Determine the Equation of an Ellipse

Find the standard form of the equation of the ellipse with foci at $(-6, 1)$ and $(-2, 1)$ such that the length of the major axis is eight units.

Solution Watch the video to verify that this is an ellipse centered at $(-4, 1)$ with a horizontal major axis such that $a = 4$ and $b = \sqrt{12}$. The equation in standard form is

$$\frac{(x + 4)^2}{4^2} + \frac{(y - 1)^2}{(\sqrt{12})^2} = 1 \quad \text{or} \quad \frac{(x + 4)^2}{16} + \frac{(y - 1)^2}{12} = 1.$$

The graph of this ellipse is shown in Figure 17.

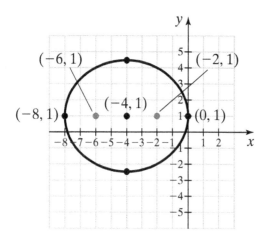

Figure 17 $\dfrac{(x + 4)^2}{16} + \dfrac{(y - 1)^2}{12} = 1$

Using Technology

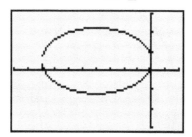

Figure 18 We can use a graphing utility to graph the ellipse from Example 3 by solving the equation for y:

$$y = 1 \pm \sqrt{12\left(1 - \dfrac{(x + 4)^2}{16}\right)}$$

(View these **steps** to see how to solve for y.) Using

$$y_1 = 1 + \sqrt{12\left(1 - \dfrac{(x + 4)^2}{16}\right)} \quad \text{and} \quad y_2 = 1 - \sqrt{12\left(1 - \dfrac{(x + 4)^2}{16}\right)},$$

we obtain the graph seen in Figure 18.

You Try It Work Exercises 9–18 in this textbook or in the My Math Lab Study Plan.

My video summary ⊙ **Example 4** Determine the Equation of an Ellipse

Determine the equation of the ellipse with foci located at $(0, 6)$ and $(0, -6)$ that passes through the point $(-5, 6)$.

Solution To find the equation, we can use Fact 1 and Fact 2 of ellipses. Watch the video to see how we can use these two facts to determine that the equation of this ellipse is $\dfrac{x^2}{45} + \dfrac{y^2}{81} = 1$.

You Try It Work Exercises 19 and 20 in this textbook or in the My Math Lab Study Plan.

OBJECTIVE 3 COMPLETING THE SQUARE TO FIND THE EQUATION OF AN ELLIPSE IN STANDARD FORM

If the equation of an ellipse is in the standard form of

$$\frac{(x-h)^2}{a^2} + \frac{(y-k)^2}{b^2} = 1 \text{ or } \frac{(x-h)^2}{b^2} + \frac{(y-k)^2}{a^2} = 1,$$

it is not too difficult to determine the center and foci and sketch its graph. However, the equation might not be given in standard form. If this is the case, we complete the square on both variables to rewrite the equation in standard form as in Example 5.

 *My video summary*

▷ Example 5 Rewrite the Equation of an Ellipse in Standard Form by Completing the Square

Find the center and foci and sketch the ellipse $36x^2 + 20y^2 + 144x - 120y - 396 = 0$.

Solution Rearrange the terms, leaving some room to complete the square and move any constants to the right-hand side:

$36x^2 + 144x \quad + 20y^2 - 120y \quad = 396$	Rearrange the terms.

Then factor and complete the square.

$36(x^2 + 4x \quad) + 20(y^2 - 6y \quad) = 396$	Factor out 36 and 20.
$36(x^2 + 4x + 4) + 20(y^2 - 6y + 9) = 396 + 144 + 180$	Complete the square on x and y. Remember to add $36 \cdot 4 = 144$ and $20 \cdot 9 = 180$ to the right side.
$36(x + 2)^2 + 20(y - 3)^2 = 720$	Factor the left side, and simplify the right side.
$\dfrac{36(x + 2)^2}{720} + \dfrac{20(y - 3)^2}{720} = \dfrac{720}{720}$	Divide both sides by 720.
$\dfrac{(x + 2)^2}{20} + \dfrac{(y - 3)^2}{36} = 1$	Simplify.

 *My video summary*

▷ The equation is now in standard form. Watch the video to see the solution to this example worked out in detail. Notice that this is the exact same ellipse that we sketched in Example 2.

 *My video summary*

▷ To see how to sketch this ellipse, refer to Example 2 or watch the last part of this video.

You Try It Work Exercises 21–24 in this textbook or in the MyMathLab **Study Plan**.

OBJECTIVE 4 SOLVING APPLIED PROBLEMS INVOLVING ELLIPSES

Ellipses have many applications. The planets of our solar system travel around the Sun in an elliptical orbit with the Sun at one focus. Some comets, such as Halley's comet, also travel in elliptical orbits. See Figure 19. (Some comets are only seen in our solar system once because they travel in hyperbolic orbits.) We saw in Section 6.1 that the parabola had a special reflecting property where incoming rays of light or radio waves traveling parallel to the axis of symmetry of a parabolic reflector or receiver will reflect off the parabolic surface and travel directly toward the antenna that is placed at the focus.

(eText Screens 6.2-1–6.2-30)

Figure 19 Planets travel around the Sun in an elliptical orbit.

Ellipses have a similar reflecting property. When light or sound waves originate from one focus of an ellipse, the waves will reflect off the surface of the ellipse and travel directly toward the other focus. See Figure 20. This reflecting property is used in a medical procedure called *sound wave lithotripsy* in which the patient is placed in an elliptical tank with the kidney stone placed at one focus. An ultrasound wave emitter is positioned at the other focus. The sound waves reflect off the walls of the tank directly to the kidney stone, thus obliterating the stone into fragments that are easily passed naturally through the patient's body. See Example 6.

Figure 20 Sound waves or light rays emitted from one focus of an ellipse reflect off the surface directly to the other focus.

Example 6 Position a Patient During Kidney Stone Treatment

A patient is placed in an elliptical tank that is 200 centimeters long and 80 centimeters wide to undergo sound wave lithotripsy treatment for kidney stones. Determine where the sound emitter and the stone should be positioned relative to the center of the ellipse.

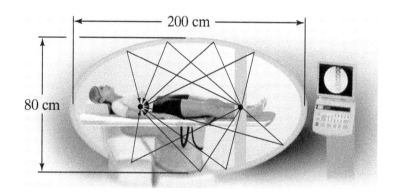

Solution The kidney stone and the sound emitter must be placed at the foci of the ellipse. We know that the major axis is 200 centimeters. Therefore, the vertices must be 100 centimeters from the center. Similarly, the endpoints of the minor axis are located 40 centimeters from the center.

To find c, we use the fact that $c^2 = a^2 - b^2$.

$$c^2 = 100^2 - 40^2$$
$$c^2 = 8{,}400$$
$$c = \sqrt{8{,}400} \approx 91.65$$

The stone and the sound emitter should be positioned approximately 91.65 centimeters from the center of the tank on the major axis.

You Try It Work Exercises 25–31 in this textbook or in the MyMathLab Study Plan.

6.2 Exercises

In Exercises 1–8, determine the center, foci, and vertices of the ellipse and sketch its graph.

SbS **1.** $\dfrac{x^2}{16} + \dfrac{y^2}{4} = 1$

SbS **2.** $\dfrac{x^2}{12} + \dfrac{y^2}{25} = 1$

SbS **3.** $4x^2 + 9y^2 = 36$

SbS **4.** $20x^2 + 5y^2 = 100$

SbS **5.** $\dfrac{(x-2)^2}{36} + \dfrac{(y-4)^2}{25} = 1$

SbS **6.** $\dfrac{(x-1)^2}{9} + \dfrac{(y+4)^2}{49} = 1$

SbS **7.** $9(x-5)^2 + 25(y+1)^2 = 225$

SbS **8.** $(x+7)^2 + 9(y-2)^2 = 9$

In Exercises 9–18, determine the standard equation of each ellipse using the given graph or the stated information.

SbS **9.**

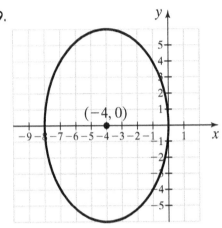

SbS **10.**

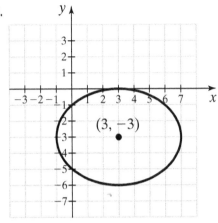

11. Foci at $(-5, 0)$ and $(5, 0)$; length of the major axis is fourteen units.

12. Foci at $(8, -1)$ and $(-2, -1)$; length of the major axis is twelve units.

13. Vertices at $(-6, 10)$ and $(-6, 0)$; length of the minor axis is eight units.

14. Center at $(4, 5)$; vertical minor axis with length sixteen units; $c = 6$.

15. Center at $(-1, 4)$; vertex at $(-1, 8)$; focus at $(-1, 7)$.

16. Vertices at $(2, -7)$ and $(2, 5)$; focus at $(2, 3)$.

17. Center at $(4, 1)$; focus at $(4, 8)$; ellipse passes through the point $(6, 1)$.

18. Center at $(1, 3)$; focus at $(1, 6)$; ellipse passes through the point $(2, 3)$.

In Exercises 19 and 20, determine the standard equation of each ellipse by first using **Fact 1** to find a and then by using **Fact 2** to find b.

19. The foci have coordinates $(-2, 0)$ and $(2, 0)$. The ellipse contains the point $(2, 3)$.

20. The foci have coordinates $(0, -8)$ and $(0, 8)$. The ellipse contains the point $(-2, 6)$.

In Exercises 21–24, complete the square to write each equation in the form

$$\frac{(x - h)^2}{a^2} + \frac{(y - k)^2}{b^2} = 1 \quad \text{or} \quad \frac{(x - h)^2}{b^2} + \frac{(y - k)^2}{a^2} = 1.$$

Determine the center, foci, and vertices of the ellipse and sketch its graph.

SbS **21.** $16x^2 + 20y^2 + 64x - 40y - 236 = 0$

SbS **22.** $9y^2 + 16x^2 + 224x + 54y + 721 = 0$

SbS **23.** $x^2 + 4x + 16y^2 - 32y + 4 = 0$

SbS **24.** $50x^2 + y^2 + 20y = 0$

25. An elliptical arch railroad tunnel 16 feet high at the center and 30 feet wide is cut through the side of a mountain. Find the equation of the ellipse if a vertex is represented by the point (0, 16).

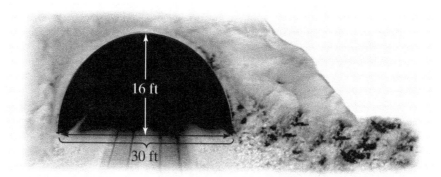

26. A 78-inch by 36-inch door contains a decorative glass elliptical pattern. One vertex of the ellipse is located 9 inches from the top of the door. The other vertex is located 27 inches from the bottom of the door. The endpoints of the minor axis are located 9 inches from each side. Find the equation of the elliptical pattern. (Assume that the center of the elliptical pattern is at the origin.)

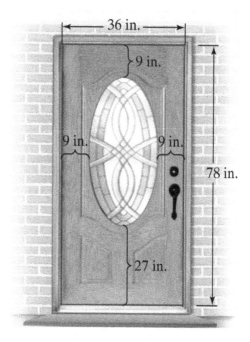

27. A rectangular playing field lies in the interior of an elliptical track that is 50 yards wide and 140 yards long. What is the width of the rectangular playing field if the width is located 10 yards from either vertex?

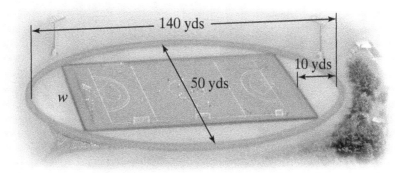

28. A patient is placed in an elliptical tank that is 180 centimeters long and 100 centimeters wide to undergo sound wave lithotripsy treatment for kidney stones. Determine where the sound emitter and the stone should be positioned relative to the center of the tank. (Round to the nearest hundredth of a centimeter.)

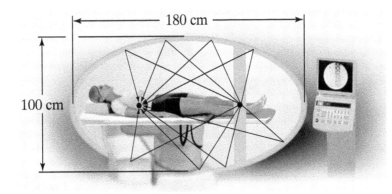

29. A window is constructed with the top half of an ellipse on top of a square. The square portion of the window has a 36-inch base. If the window is 50 inches tall at its highest point, find the height, h, of the window 14 inches from the center of the base. (Round the height to the nearest hundredth of an inch.)

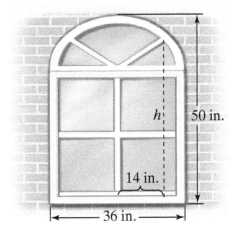

30. A government spy satellite is in an elliptical orbit around the Earth with the center of the Earth at a focus. If the satellite is 150 miles from the surface of the Earth at one vertex of the orbit and 500 miles from the surface of the Earth at the other vertex, find the equation of the elliptical orbit. (Assume that the Earth is a sphere with a diameter of 8,000 miles.)

31. An elliptical arch bridge spans 160 feet. The elliptical arch has a maximum height of 40 feet. What is the height of the arch at a distance of 10 feet from the center?

Brief Exercises

In Exercises 32–43, sketch the graph of each ellipse.

32. $\dfrac{x^2}{16} + \dfrac{y^2}{4} = 1$

33. $\dfrac{x^2}{12} + \dfrac{y^2}{25} = 1$

34. $4x^2 + 9y^2 = 36$

35. $20x^2 + 5y^2 = 100$

36. $\dfrac{(x-2)^2}{36} + \dfrac{(y-4)^2}{25} = 1$

37. $\dfrac{(x-1)^2}{9} + \dfrac{(y+4)^2}{49} = 1$

38. $9(x-5)^2 + 25(y+1)^2 = 225$

39. $(x+7)^2 + 9(y-2)^2 = 9$

40. $16x^2 + 20y^2 + 64x - 40y - 236 = 0$

41. $9y^2 + 16x^2 + 224x + 54y + 721 = 0$

42. $x^2 + 4x + 16y^2 - 32y + 4 = 0$

43. $50x^2 + y^2 + 20y = 0$

6.3 The Hyperbola

THINGS TO KNOW

Before working through this section, be sure that you are familiar with the following concepts:

VIDEO   ANIMATION   INTERACTIVE

You Try It

1. Finding the Distance between Two Points Using the Distance Formula (Section 2.1)

You Try It

2. Finding the Equation of a Line Using the Point–Slope Form (Section 2.3)

You Try It

3. Completing the Square to Find the Equation of an Ellipse in Standard Form (Section 6.2)

OBJECTIVES

1 Sketching the Graph of a Hyperbola

2 Determining the Equation of a Hyperbola in Standard Form

3 Completing the Square to Find the Equation of a Hyperbola in Standard Form

4 Solving Applied Problems Involving Hyperbolas

Introduction to Section 6.3

My animation summary

When a plane intersects two right circular cones at the same time, the conic section formed is a hyperbola. (Click on the animation.)

The set of points in the plane that defines a hyperbola formed by the intersection of a plane and the cones described previously is stated in the following geometric definition.

Definition   Geometric Definition of the Hyperbola

A **hyperbola** is the set of all points in a plane, the difference of whose distances from two fixed points is a positive constant. The two fixed points, F_1 and F_2, are called the foci.

The Hyperbola

Notice that the previous geometric definition is very similar to the geometric definition of the ellipse. Recall that for a point to lie on the graph of an ellipse, the **sum** of the distances from the point to the two foci is constant. For a point to lie on the graph of a hyperbola, the **difference** between the distances from the point to the two foci is constant. Because subtraction is not commutative, we consider the absolute value of the difference in the distances between a point on the hyperbola and the foci to ensure that the constant is positive. Thus, for any two points P and Q that lie on the graph of a hyperbola, $|d(P, F_1) - d(P, F_2)| = d(Q, F_1) - d(Q, F_2)|$. See Figure 21.

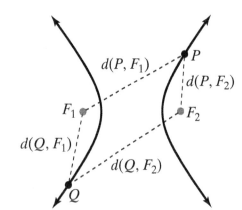

Figure 21 For any points P and Q that lie on the graph of a hyperbola, $|d(P, F_1) - d(P, F_2)| = |d(Q, F_1) - d(Q, F_2)|$.

OBJECTIVE 1   SKETCHING THE GRAPH OF A HYPERBOLA

My video summary

⊚ The graph of a hyperbola has two branches. These branches look somewhat like parabolas, but they are certainly not because the branches do not satisfy the geometric definition of the parabola. Every hyperbola has a center, two vertices, and two foci. The vertices are located at the endpoints of an invisible line segment called the **transverse axis**. The transverse axis is either parallel to the x-axis (horizontal transverse axis) or parallel to the y-axis (vertical transverse axis). The center of a hyperbola is located midway between the two vertices (or two foci). The hyperbola has another invisible line segment called the **conjugate axis** that passes through the center and lies perpendicular to the transverse axis. Each branch of the hyperbola approaches (but never intersects) a pair of lines called *asymptotes*. A **reference rectangle** is typically used as a guide to help sketch the asymptotes. The reference rectangle is a rectangle whose midpoints of each side are the vertices of the hyperbola or the endpoints of the conjugate axis.

My video summary ⊘ The asymptotes pass diagonally through opposite corners of the reference rectangle. Watch the video to see how to sketch the graphs of the two hyperbolas shown in Figure 22.

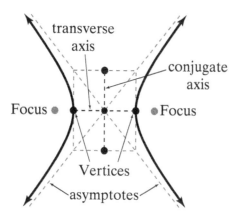

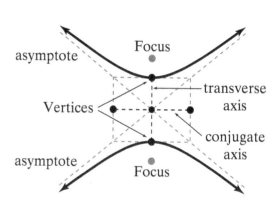

Figure 22 (a) Hyperbola with a Horizontal Transverse Axis

(b) Hyperbola with a Vertical Transverse Axis

To derive the equation of a hyperbola, consider a hyperbola with a horizontal transverse axis centered at (h, k). If the distance between the center and either vertex is $a > 0$, then the coordinates of the vertices are $V_1(h - a, k)$ and $V_2(h + a, k)$, and the length of the transverse axis is equal to $2a$. If $c > 0$ is the distance between the center and either foci, then the foci have coordinates $F_1(h - c, k)$ and $F_2(h + c, k)$. See Figure 23.

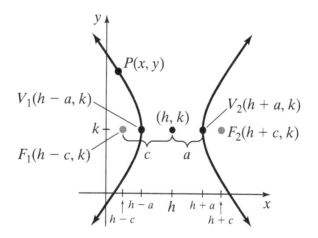

Figure 23

By the geometric definition of a hyperbola, we know that for any point $P(x, y)$ that lies on the hyperbola, the difference of the distances from P to the two foci is a constant. It can be shown that this constant is equal to $2a$, the length of the transverse axis. We now state this fact and denote it as Fact 1 for Hyperbolas.

FACT 1 FOR HYPERBOLAS

Given the foci of a hyperbola F_1 and F_2 and any point P that lies on the graph of the hyperbola, the difference of the distances between P and the foci is equal to $2a$. In other words, $|d(P, F_1) - d(P, F_2)| = 2a$. The constant $2a$ represents the length of the transverse axis.

We prove Fact 1 for Hyperbolas in Appendix A.

Once we know that the constant stated in the geometric definition is equal to $2a$, we can derive the equation of a hyperbola. The equation of a hyperbola is derived in Appendix A. We now state the two standard equations of hyperbolas.

Equation of a Hyperbola in Standard Form with Center (h, k)

Horizontal Transverse Axis

$$\frac{(x - h)^2}{a^2} - \frac{(y - k)^2}{b^2} = 1$$

- Foci: $F_1(h - c, k)$ and $F_2(h + c, k)$
- Vertices: $V_1(h - a, k)$ and $V_2(h + a, k)$
- Endpoints of conjugate axis:
  $(h, k - b)$ and $(h, k + b)$
- $b^2 = c^2 - a^2$
- Asymptotes: $y - k = \pm\dfrac{b}{a}(x - h)$

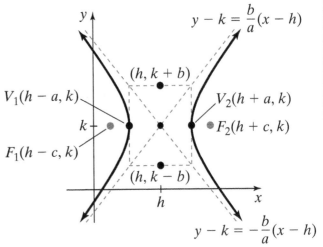

Equation of a Hyperbola in Standard Form with Center (h, k)

Vertical Transverse Axis

$$\frac{(y - k)^2}{a^2} - \frac{(x - h)^2}{b^2} = 1$$

- Foci: $F_1(h, k - c)$ and $F_2(h, k + c)$
- Vertices: $V_1(h, k - a)$ and $V_2(h, k + a)$
- Endpoints of conjugate axis:
  $(h - b, k)$ and $(h + b, k)$
- $b^2 = c^2 - a^2$
- Asymptotes: $y - k = \pm\dfrac{a}{b}(x - h)$

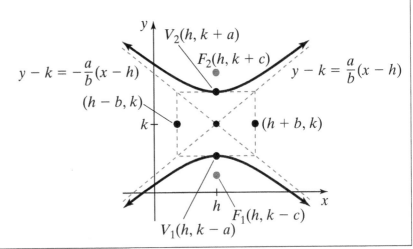

The two hyperbola equations are nearly identical except for two differences. The term involving $(x - h)^2$ is positive and the term involving $(y - k)^2$ is negative in the equation of a hyperbola with the horizontal transverse axis. The signs are reversed in the equation with a vertical transverse axis. Also note that a^2 appears in the denominator of the positive squared term in each equation. We can derive the equations of the asymptotes by using the **point-slope form** of the equation of a line.

Note If $h = 0$ and $k = 0$, then the hyperbola is centered at the origin. Hyperbolas centered at the origin have the following equations.

Standard Equations of a Hyperbola with the Center at the Origin

$$\frac{x^2}{a^2} - \frac{y^2}{b^2} = 1 \qquad\qquad \frac{y^2}{a^2} - \frac{x^2}{b^2} = 1$$

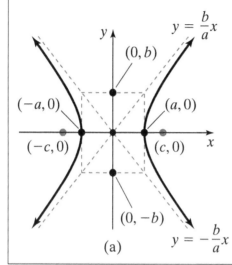

(a)

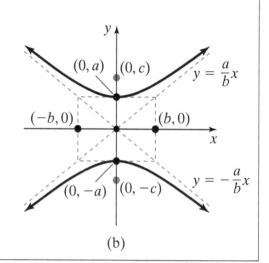

(b)

My interactive video summary

◉ Example 1 Sketch the Graph of a Hyperbola in Standard Form

Sketch the following hyperbolas. Determine the center, transverse axis, vertices, and foci and find the equations of the asymptotes.

a. $\dfrac{(y-4)^2}{36} - \dfrac{(x+5)^2}{9} = 1$ **b.** $25x^2 - 16y^2 = 400$

Solution Watch the interative video to see how to sketch each hyperbola shown in Figure 24(a) and Figure 24(b).

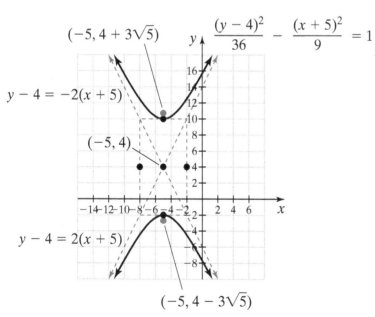

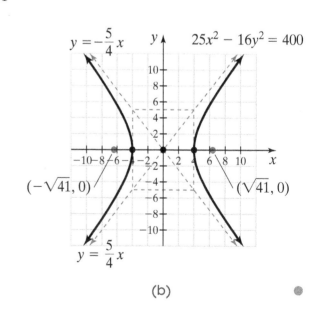

Figure 24   (a)   (b)

You Try It   Work Exercises 1–6 in this textbook or in the MyMathLab Study Plan.

6.3 The Hyperbola   6-29

OBJECTIVE 2   DETERMINING THE EQUATION OF A HYPERBOLA IN STANDARD FORM

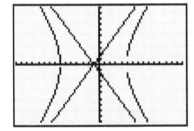 *My video summary*   ◉ **Example 2** Find the Equation of a Hyperbola

Find the equation of the hyperbola with the center at $(-1, 0)$, a focus at $(-11, 0)$, and a vertex at $(5, 0)$.

Solution   A focus and a vertex lie along the x-axis. This indicates that the hyperbola has a horizontal transverse axis. Because the center is at $(-1, 0)$, we know that the equation of the hyperbola is $\dfrac{(x + 1)^2}{a^2} - \dfrac{y^2}{b^2} = 1$. Watch the video to verify that the equation in standard form is $\dfrac{(x + 1)^2}{36} - \dfrac{y^2}{64} = 1$. The equations of the asymptotes are $y = -\dfrac{4}{3}(x + 1)$ and $y = \dfrac{4}{3}(x + 1)$. The graph of this hyperbola is shown in Figure 25.

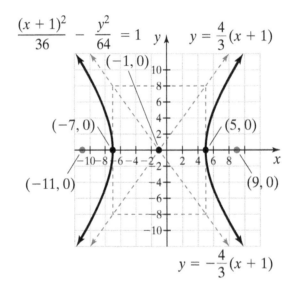

Figure 25

Using Technology

Figure 26 We can use a graphing utility to graph the hyperbola from Example 2 by solving the equation for y:

$$y = \pm\frac{4}{3}\sqrt{(x + 1)^2 - 36}$$

(View these steps to see how to solve for y.) Using

$$y_1 = \frac{4}{3}\sqrt{(x + 1)^2 - 36}, \; y_2 = -\frac{4}{3}\sqrt{(x + 1)^2 - 36}, \; y_3 = -\frac{4}{3}(x + 1),$$

and $y_4 = -\dfrac{4}{3}(x + 1)$, we obtain the graph seen in Figure 26.   ●

You Try It   Work Exercises 7–14 in this textbook or in the MyMathLab Study Plan.

OBJECTIVE 3 COMPLETING THE SQUARE TO FIND THE EQUATION OF A HYPERBOLA IN STANDARD FORM

If the equation of a hyperbola is in the standard form of

$$\frac{(x-h)^2}{a^2} - \frac{(y-k)^2}{b^2} = 1 \quad \text{or} \quad \frac{(y-k)^2}{a^2} - \frac{(x-h)^2}{b^2} = 1,$$

then it is not too difficult to determine the center, vertices, foci, and asymptotes and sketch its graph. However, the equation might not be given in standard form. If this is the case, we may need to complete the square on both variables as in Example 3.

My video summary ⊘ **Example 3** Rewrite the Equation of a Hyperbola in Standard Form by Completing the Square

Find the center, vertices, foci, and equations of asymptotes and sketch the hyperbola $12x^2 - 4y^2 - 72x - 16y + 140 = 0$.

Solution Rearrange the terms leaving some room to complete the square, and move any constants to the right-hand side:

$$12x^2 - 72x \quad - 4y^2 - 16y \quad = -140 \qquad \text{Rearrange the terms.}$$

Then factor and complete the square.

$$12(x^2 - 6x \quad) - 4(y^2 + 4y \quad) = -140 \qquad \text{Factor out 12 and } -4.$$

$$12(x^2 - 6x + 9) - 4(y^2 + 4y + 4) = -140 + 108 - 16 \qquad \begin{array}{l}\text{Complete the square on } x \text{ and } y. \text{ Remember to} \\ \text{add } 12\cdot 9 = 108 \text{ and } -4\cdot 4 = -16 \text{ to the right side.}\end{array}$$

$$12(x-3)^2 - 4(y+2)^2 = -48 \qquad \text{Factor the left side, and simplify the right side.}$$

$$\frac{12(x-3)^2}{-48} - \frac{4(y+2)^2}{-48} = \frac{-48}{-48} \qquad \text{Divide both sides by } -48.$$

$$-\frac{(x-3)^2}{4} + \frac{(y+2)^2}{12} = 1 \qquad \text{Simplify.}$$

$$\frac{(y+2)^2}{12} - \frac{(x-3)^2}{4} = 1 \qquad \text{Rewrite the equation.}$$

The equation is now in standard form. Watch the video to see this example worked out in detail. You should verify that this is the equation of a hyperbola with a vertical transverse axis with center $(3, -2)$. The vertices have coordinates $(3, -2 - 2\sqrt{3})$ and $(3, -2 + 2\sqrt{3})$. The foci have coordinates $(3, -6)$ and $(3, 2)$. The equations of the asymptotes are $y + 2 = -\sqrt{3}(x - 3)$ and $y + 2 = \sqrt{3}(x - 3)$. ●

You Try It Work Exercises 15–20 in this textbook or in the MyMathLab Study Plan.

OBJECTIVE 4 SOLVING APPLIED PROBLEMS INVOLVING HYPERBOLAS

Hyperbolas have many applications. We saw in Section 6.2 that the planets in our solar system and some comets, such as Halley's comet, travel through our solar system in elliptical orbits. However, some comets are only seen once in our solar system because they travel through the solar system on the path of a hyperbola with the Sun at a focus. On October 14, 1947, Chuck Yeager became the first person to break the sound barrier. As an airplane moves faster than the speed of sound, a cone-shaped shock wave is produced. The cone intersects the ground in the shape of a hyperbola. When two rocks are simultaneously tossed into a calm pool of water, ripples move outward in the form of concentric circles. These circles intersect in points that form a hyperbola. Hyperbolas can be used to locate ships by sending radio signals simultaneously from radio transmitters placed at some fixed distance apart. A device measures the difference in the time it takes the radio signals to reach the ship. The equation of a hyperbola can then be determined to describe the current path of the ship. If three transmitters are used, two hyperbolic equations can be determined. The precise location of the ship can be determined by finding the intersection of the two hyperbolas. This system of locating ships is known as long-range navigation or LORAN. See Example 4.

Example 4 Use a Hyperbola to Locate a Ship

One transmitting station is located 100 miles due east from another transmitting station. Each station simultaneously sends out a radio signal. The signal from the west tower is received by a ship $\dfrac{1{,}600}{3}$ microseconds after the signal from the east tower. If the radio signal travels at 0.18 miles per microsecond, find the equation of the hyperbola on which the ship is presently located.

Solution Start by plotting the two foci of the hyperbola at points $F_1(-50, 0)$ and $F_2(50, 0)$. These two points represent the position of the two transmitting towers. Note that $c = 50$. Because the hyperbola is centered at the origin with a horizontal transverse axis, the equation must be of the form $\dfrac{x^2}{a^2} - \dfrac{y^2}{b^2} = 1$. The difference in the distances from the two transmitters to the ship is

$\left(\dfrac{1{,}600}{3}\text{ microseconds}\right) \cdot \left(0.18\,\dfrac{\text{miles}}{\text{microsecond}}\right) = 96$ miles. This distance represents the constant stated in the Fact 1 for Hyperbolas. Therefore, $2a = 96$, so $a = 48$ or $a^2 = 2{,}304$. To find b^2, we use the fact that $b^2 = c^2 - a^2$.

$$b^2 = c^2 - a^2$$
$$b^2 = 50^2 - 48^2$$
$$b^2 = 196$$

We now substitute the values of $a^2 = 2{,}304$ and $b^2 = 196$ into the previous equation to obtain the equation $\dfrac{x^2}{2{,}304} - \dfrac{y^2}{196} = 1$.

You Try It Work Exercises 21–25 or in this textbook in the MyMathLab Study Plan.

6.3 Exercises

In Exercises 1–6, determine the center, transverse axis, vertices, foci, and the equations of the asymptotes and sketch the hyperbola.

SbS **1.** $\dfrac{x^2}{16} - \dfrac{y^2}{9} = 1$ SbS **2.** $\dfrac{y^2}{9} - \dfrac{x^2}{16} = 1$ SbS **3.** $\dfrac{(y-4)^2}{25} - \dfrac{(x-2)^2}{36} = 1$

SbS **4.** $\dfrac{(x+1)^2}{9} - \dfrac{(y+3)^2}{49} = 1$ SbS **5.** $20x^2 - 5y^2 = 100$ SbS **6.** $20(x-1)^2 - 16(y-3)^2 = -320$

In Exercises 7–12, determine the standard equation of the hyperbola with the given characteristics and sketch the graph.

7. The center is at $(0,0)$, a focus is at $(5,0)$, and a vertex is at $(3,0)$.

8. The center is at $(0,0)$, a focus is at $(0,10)$, and a vertex is at $(0,-6)$.

9. The center is at $(4,-4)$, a focus is at $(6,-4)$, and a vertex is at $(5,-4)$.

10. The center is at $(-6,-1)$, a focus is at $(-6,-9)$, and a vertex is at $(-6,-5)$.

11. The foci are at $(9,3)$ and $(9,9)$; the vertex is at $(9,8)$.

12. The vertices are at $(-2,-3)$ and $(10,-3)$; an asymptote has equation $y + 3 = \dfrac{7}{6}(x-4)$.

In Exercises 13 and 14, determine the equation of the hyperbola with the given characteristics and sketch the graph. *Hint:* $|d(P, F_1) - d(P, F_2)| = 2a$.

13. The hyperbola has foci with coordinates $F_1(0, -6)$ and $F_2(0, 6)$ and passes through the point $P(8, 10)$.

14. The hyperbola has foci with coordinates $F_1(-6, 1)$ and $F_2(10, 1)$ and passes through the point $P(10, 13)$.

In Exercises 15–20, complete the square to write each equation in the form
$$\frac{(x-h)^2}{a^2} - \frac{(y-k)^2}{b^2} = 1 \quad \text{or} \quad \frac{(y-k)^2}{a^2} - \frac{(x-h)^2}{b^2} = 1.$$
Determine the center, vertices, foci, endpoints of the conjugate axis, and the equations of the asymptotes of the hyperbola, and sketch its graph.

SbS **15.** $x^2 - y^2 - 4x + 2y - 1 = 0$

SbS **16.** $x^2 - y^2 + 8x - 6y + 9 = 0$

SbS **17.** $y^2 - 9x^2 - 12y - 36x - 9 = 0$

SbS **18.** $x^2 - 16y^2 + 10x + 64y - 55 = 0$

SbS **19.** $25y^2 - 144x^2 + 1{,}728x + 400y + 16 = 0$

SbS **20.** $49y^2 - 576x^2 - 98y - 1{,}152x - 28{,}751 = 0$

21. A light on a wall produces a shadow in the shape of a hyperbola. If the distance between the two vertices of the hyperbola is 14 inches and the distance between the two foci is 16 inches, find the equation of the hyperbola.

22. This figure shows the hyperbolic orbit of a comet with the center of the Sun positioned at a focus, 100 million miles from the origin. (The units are in millions of miles.) The comet will be 50 million miles from the center of the Sun at its nearest point during the orbit. Find the equation of the hyperbola describing the comet's orbit.

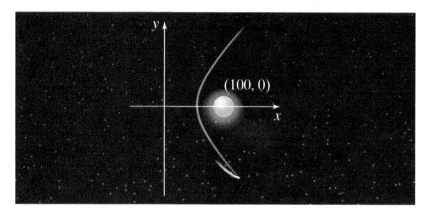

23. A nuclear power plant has a large cooling tower with sides curved in the shape of a hyperbola. The radius of the base of the tower is 60 meters. The radius of the top of the tower is 50 meters. The sides of the tower are 60 meters apart at the closest point located 90 meters above the ground.

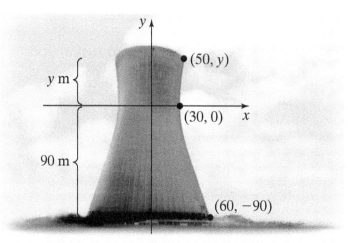

   a. Find the equation of the hyperbola that describes the sides of the cooling tower. (Assume that the center is at the origin.)

   b. Determine the height of the tower. (Round your answer to the nearest meter.)

24. One transmitting station is located 80 miles north of another transmitting station. Each station simultaneously sends out a radio signal. The signal from the north station is received 200 microseconds after the signal from the south station. If the radio signal travels at 0.18 miles per microsecond, find the equation of the hyperbola on which the ship is presently located.

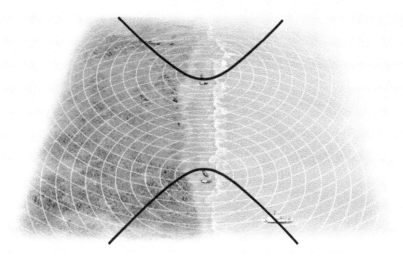

Brief Exercises

In Exercises 25–36, sketch the graph of each hyperbola.

25. $\dfrac{x^2}{16} - \dfrac{y^2}{9} = 1$

26. $\dfrac{y^2}{9} - \dfrac{x^2}{16} = 1$

27. $\dfrac{(y-4)^2}{25} - \dfrac{(x-2)^2}{36} = 1$

28. $\dfrac{(x+1)^2}{9} - \dfrac{(y+3)^2}{49} = 1$

29. $20x^2 - 5y^2 = 100$

30. $20(x-1)^2 - 16(y-3)^2 = -320$

31. $x^2 - y^2 - 4x + 2y - 1 = 0$

32. $x^2 - y^2 + 8x - 6y + 9 = 0$

33. $y^2 - 9x^2 - 12y - 36x - 9 = 0$

34. $x^2 - 16y^2 + 10x + 64y - 55 = 0$

35. $25y^2 - 144x^2 + 1{,}728x + 400y + 16 = 0$

36. $49y^2 - 576x^2 - 98y - 1{,}152x - 28{,}751 = 0$

CHAPTER SEVEN

Systems of Equations and Inequalities

CHAPTER SEVEN CONTENTS

7.1 Solving Systems of Linear Equations in Two Variables

THINGS TO KNOW

Before working through this section, be sure that you are familiar with the following concepts:

VIDEO   ANIMATION   INTERACTIVE

You Try It 1. Applications of Linear Equations (Section 1.2)

You Try It 2. Writing the Equation of a Line in Standard Form (Section 2.3)

You Try It 3. Sketching Lines by Plotting Intercepts (Section 2.3)

You Try It 4. Understanding the Definition of Parallel Lines (Section 2.4)

OBJECTIVES

1 Verifying Solutions to a System of Linear Equations in Two Variables

2 Solving a System of Linear Equations Using the Substitution Method

3 Solving a System of Linear Equations Using the Elimination Method

4 Solving Applied Problems Using a System of Linear Equations

....

OBJECTIVE 1 VERIFYING SOLUTIONS TO A SYSTEM OF LINEAR EQUATIONS
IN TWO VARIABLES

 *My video summary*

⊙ In Section 1.1, a linear equation in one variable was defined as an equation that can be written in the form $ax + b = 0$, where a and b are real numbers with $a \neq 0$. This definition can be extended to more variables as follows.

Definition   Linear Equation in n Variables

A linear equation in n variables is an equation that can be written in the form $a_1x_1 + a_2x_2 + \cdots + a_nx_n = b$ for variables $x_1, x_2, \ldots x_n$ and real numbers $a_1, a_2, \ldots a_n, b$, where at least one of $a_1, a_2, \ldots a_n$ is nonzero.

For example, $2x - 5y = -7$ is a linear equation in two variables, whereas $4x_1 + 9x_2 - 17x_3 = 11$ is a linear equation in three variables. The important thing to remember is that all variables of a linear equation have an exponent of 1. For much of this chapter, we learn how to solve **systems of linear equations**. We start our discussion of linear systems by considering linear systems in two variables.

Definition   System of Linear Equations in Two Variables

A system of linear equations in two variables is the collection of two linear equations in two variables considered simultaneously. The solution to a system of equations in two variables is the set of all ordered pairs for which *both* equations are true.

Consider the following three systems of linear equations in two variables.

$$3x - 2y = -9 \qquad -3x_1 + 2x_2 = 11 \qquad \sqrt{2}a + \quad b = \pi$$
$$x + \ y = \ \ 2 \qquad 8x_1 - 9x_2 = -2 \qquad -4a + \sqrt{3}b = 1$$

In each system, different variables were used. It does not matter what we call the two variables as long as both equations are linear. According to the definition of a system of linear equations in two variables, the solution is the set of all ordered pairs for which both equations are true. We verify that the ordered pair $(-1, 3)$ is a solution to the system $\begin{array}{l} 3x - 2y = -9 \\ x + \ y = \ \ 2 \end{array}$ in Example 1.

Example 1 Verify the Solution to a System of Linear Equations in Two Variables

Show that the ordered pair $(-1, 3)$ is a solution to the system $\begin{aligned} 3x - 2y &= -9 \\ x + y &= 2 \end{aligned}$.

Solution We can verify that the ordered pair $(-1, 3)$ is a solution to this system by substituting -1 for x and 3 for y in both equations.

First Equation   Second Equation

$$3x \quad - \quad 2y = -9 \qquad\qquad\qquad\qquad\qquad\qquad x + y = 2$$

$$3(-1) - 2(3) \overset{?}{=} -9 \qquad \text{Substitute } x = -1 \text{ and } y = 3. \qquad (-1) + (3) \overset{?}{=} 2$$

$$-9 = -9 \checkmark \qquad \text{Both equations are true.} \qquad\qquad 2 = 2 \checkmark$$

Both equations yielding true statements indicate that the ordered pair $(-1, 3)$ is a solution to this system.

You Try It Work through this You Try It problem.

Work Exercises 1–4 in this textbook or in the My Math Lab **Study Plan.**

If a linear system has at least one solution, like the system shown in Example 1, it is said to be **consistent**, and the solution is the set of all ordered pairs that satisfy both equations. If the system does not have a solution, it is said to be **inconsistent**.

The solution to a system of two linear equations involving two variables can be viewed geometrically. Because the graph of each equation of the system is a line, we can sketch the two lines and geometrically view the solution. There are only three possibilities for the graph of a linear system of equations involving two variables:

1. The lines can have different slopes and thus intersect at a single point (Figure 1a). Systems that have at least one solution are said to be **consistent**. This consistent system has exactly one solution so it is said to be **consistent** and **independent**.

2. The two equations can be different representations of the same line and have the same graph (Figure 1b). In this case, there are infinitely many solutions. Because this system has at least one solution, it is said to be **consistent**. This consistent system has infinitely many solutions so it is called **consistent** and **dependent**.

3. The two lines can be parallel (have the same slope) and thus have no intersecting point (Figure 1c). This system is said to be **inconsistent** and has no solution.

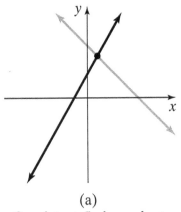
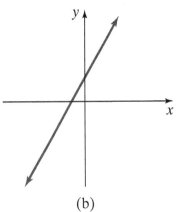
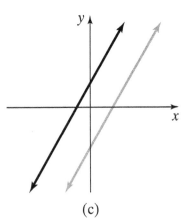

(a)	(b)	(c)
Consistent, Independent	Consistent, Dependent	Inconsistent,
One solution	**Infinitely many solutions**	**No solution**
Two lines are different, having one common point. The lines have different slopes.	Two lines are the same, having infinitely many common points. The lines have the same slope and same y-intercepts.	Two lines are different, having no common points. The lines have the same slope but different y-intercepts.

Figure 1

We now learn two algebraic methods used for solving linear systems involving two equations and two variables. These methods are known as the *substitution method* and the *elimination method*.

OBJECTIVE 2   SOLVING A SYSTEM OF LINEAR EQUATIONS USING THE SUBSTITUTION METHOD

 *My video summary*

⊘ The substitution method involves solving one of the equations for one variable in terms of the other, and then substituting that expression into the other equation. The substitution method can be summarized in the following four steps.

Solving a System of Equations by the Method of Substitution

Step 1. Choose an equation and solve for one variable in terms of the other variable.

Step 2. Substitute the expression from step 1 into the other equation.

Step 3. Solve the equation in one variable.

Step 4. Substitute the value found in step 3 into one of the original equations to find the value of the other variable.

The substitution method is used to solve the system in Example 2.

Example 2 Solve a System of Equations Using the Substitution Method

Solve the following system using the method of substitution: $\begin{aligned} 2x - 3y &= -5 \\ x + y &= 5 \end{aligned}$.

Solution

Step 1. If possible, choose an equation in which at least one coefficient is 1. In this example, both coefficients of the second equation are 1. We can choose to solve the second equation for either variable. In this case, we solve the second equation for y:

$$x + y = 5 \qquad \text{Write the second equation.}$$

$$y = 5 - x \qquad \text{Subtract } x \text{ from both sides.}$$

Step 2. Substitute $5 - x$ for y in the first equation:

$$2x - 3(5 - x) = -5$$

Step 3. Solve for x:

$$2x - 3(5 - x) = -5 \qquad \text{Rewrite the equation from step 2.}$$

$$2x - 15 + 3x = -5 \qquad \text{Use the distributive property.}$$

$$5x - 15 = -5 \qquad \text{Combine like terms.}$$

$$5x = 10 \qquad \text{Add 15 to both sides.}$$

$$x = 2 \qquad \text{Divide both sides by 5.}$$

Step 4. Because $x = 2$, we can solve for y by substituting $x = 2$ into one of the original equations. Here we use the simplified equation found in step 1.

$$y = 5 - (2) = 3$$

The solution of this system is the ordered pair $(2, 3)$. You can see in Figure 2 that this is the intersection of the two lines of this system.

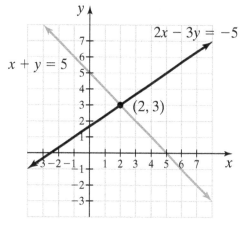

Figure 2 The solution to the system
$$2x - 3y = -5$$
$$x + y = 5$$
is the ordered pair $(2, 3)$.

You Try It Work through this You Try It problem.

Work Exercises 5–9 in this textbook or in the MyMathLab Study Plan.

OBJECTIVE 3 SOLVING A SYSTEM OF LINEAR EQUATIONS USING THE ELIMINATION METHOD

My video summary ⊙ Another method used to solve a system of two equations is called the elimination method. The elimination method involves adding the two equations together in an attempt to eliminate one of the variables. To accomplish this task, the coefficients of one of the variables must differ only in sign. This can be done by multiplying one or both of the equations by a suitable constant. The elimination method can be summarized in the following five steps.

Solving a System of Equations by the Method of Elimination

Step 1. Choose a variable to eliminate.

Step 2. Multiply one or both equations by an appropriate nonzero constant so that the sum of the coefficients of one of the variables is zero.

Step 3. Add the two equations together to obtain an equation in one variable.

Step 4. Solve the equation in one variable.

Step 5. Substitute the value obtained in step 4 into one of the original equations to solve for the other variable.

The elimination method is used to solve the system in Example 3.

Example 3 Solve a System of Equations Using the Elimination Method

Solve the following system using the method of elimination: $\begin{array}{r} -2x + 5y = 29 \\ 3x + 2y = 4 \end{array}$.

Solution

Step 1. Here we choose to eliminate the variable x.

Step 2. Multiply both sides of the first equation by 3, and multiply both sides of second equation by 2, thus making the coefficient of the x-terms in the two equations equal to -6 and 6, respectively.

$$-2x + 5y = 29 \xrightarrow{\text{Multiply by 3.}} 3(-2x + 5y) = 3(29) \longrightarrow -6x + 15y = 87$$

$$3x + 2y = 4 \xrightarrow{\text{Multiply by 2.}} 2(3x + 2y) = 2(4) \longrightarrow 6x + 4y = 8$$

Step 3. We now add the two new equations together:

$$-6x + 15y = 87$$
$$\underline{6x + 4y = 8}$$
$$19y = 95$$

Step 4. Solve for y:

$$19y = 95$$

$$y = 5 \qquad \text{Divide both sides by 19.}$$

Step 5. To solve for x, we substitute $y = 5$ into one of the original equations. Here, we choose the first equation:

$$-2x + 5y = 29 \qquad \text{Write the original first equation.}$$
$$-2x + 5(5) = 29 \qquad \text{Substitute 5 for } y \text{ in the first equation.}$$
$$-2x + 25 = 29 \qquad \text{Simplify.}$$
$$-2x = 4 \qquad \text{Subtract 25 from both sides.}$$
$$x = -2 \qquad \text{Divide both sides by } -2.$$

Therefore, the solution is the ordered pair $(-2, 5)$.

Using Technology

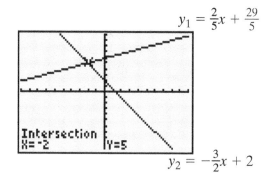

$$y_1 = \frac{2}{5}x + \frac{29}{5}$$

$$y_2 = -\frac{3}{2}x + 2$$

We can solve the system in Example 3 using a graphing utility by first solving both equations for y:

$$-2x + 5y = 29 \longrightarrow y_1 = \frac{2}{5}x + \frac{29}{5}$$

$$3x + 2y = 4 \longrightarrow y_2 = -\frac{3}{2}x + 2$$

Graph y_1 and y_2, and then use the INTERSECT feature to find that the point of intersection is $(-2, 5)$.

You Try It Work through this You Try It problem.

Work Exercises 10–14 in this textbook or in the MyMathLab Study Plan.

The systems in Examples 2 and 3 had one solution. We now look at an example of a system with no solution, and then solve a system that has infinitely many solutions.

Example 4 Solve a System with No Solution

Solve the system $\begin{array}{l} x - 2y = 11 \\ -2x + 4y = \ \ 8 \end{array}$.

Solution We can solve this system using either the substitution method or the elimination method. In this case, we use the elimination method.

Step 1. We choose to eliminate the variable x because it has a coefficient of 1 in the first equation.

Step 2. Multiply both sides of the first equation by 2, thus making the coefficient of the x-terms in the two equations equal to 2 and -2, respectively.

$$x - 2y = 11 \xrightarrow{\text{Multiply by 2}} 2x - 4y = 22$$
$$-2x + 4y = \ \ 8 \xrightarrow{\text{Leave equation alone}} -2x + 4y = \ \ 8$$

Step 3. We now add the two new equations together:

$$\begin{array}{r} 2x - 4y = 22 \\ \underline{-2x + 4y = \ \ 8} \\ 0 = 30 \end{array}$$

Notice how both variables were eliminated, leaving us with a **contradiction**. The number 0 is never equal to 30. This means that there is no solution to this system. This is an example of an **inconsistent** system. You can see in Figure 3 that the graphs of the two linear equations from this system are parallel. Thus, the two lines do not intersect, and the system has no solution.

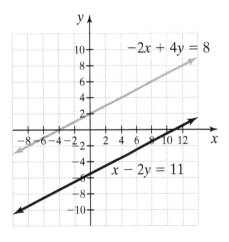

Figure 3 The linear system $\begin{array}{l} x - 2y = 11 \\ -2x + 4y = 8 \end{array}$ has no solution.

Example 5 Solve a System Having Infinitely Many Solutions

Solve the system $\begin{array}{l} -3x + 6y = 9 \\ x - 2y = -3 \end{array}$.

Solution We can solve this system using either the substitution method or the elimination method. Here, we choose the substitution method.

Step 1. We start by solving the second equation for x:

$$x - 2y = -3 \qquad \text{Write the second equation.}$$

$$x = 2y - 3 \qquad \text{Add } 2y \text{ to both sides.}$$

Step 2. Substitute $2y - 3$ for x in the first equation:

$$-3(2y - 3) + 6y = 9$$

Step 3. Solve for y:

$$-3(2y - 3) + 6y = 9 \qquad \text{Rewrite the equation from step 2.}$$

$$-6y + 9 + 6y = 9 \qquad \text{Use the distributive property.}$$

$$9 = 9 \qquad \text{Combine like terms.}$$

The equation simplifies to the identity $9 = 9$, which is true for *any* value of y. Therefore, this system has infinitely many solutions. Note that we could have multiplied the second equation in the system by -3 to obtain the first equation. Because of this, both equations represent the same line and thus have the same graph. See Figure 4. The solution is the set of all ordered pairs that lie on the graph of either equation.

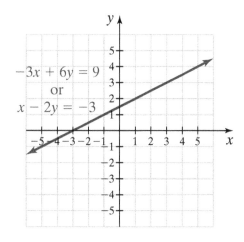

Figure 4 The system $\begin{aligned} -3x + 6y &= 9 \\ x - 2y &= -3 \end{aligned}$ has infinitely many solutions.

We can write the solution to this system in **ordered pair notation**. Recall in step 1 that we wrote x in terms of y as $x = 2y - 3$; thus, we can write the solution in ordered pair notation as $(2y - 3, y)$, where y is any real number. For example,

If $y = -2$, then $x = 2(-2) - 3 = -7$, so the ordered pair $(-7, -2)$ is a solution.

If $y = 0$, then $x = 2(0) - 3 = -3$, so the ordered pair $(-3, 0)$ is another solution.

Similarly, we could have solved either equation for y to obtain $y = \dfrac{1}{2}x + \dfrac{3}{2}$.

So, an alternative way to describe the solution to this system in ordered pair notation is $\left(x, \dfrac{1}{2}x + \dfrac{3}{2} \right)$.

You Try It Work through this You Try It problem.

Work Exercises 15–20 in this textbook or in the MyMathLab Study Plan.

OBJECTIVE 4 SOLVING APPLIED PROBLEMS USING A SYSTEM OF LINEAR EQUATIONS

Applied problems can involve two or more unknown quantities. Sometimes we are able to use a single equation involving one variable to solve such problems as seen in Section 1.2. However, it is often easier to use two variables and create a system of two equations. The following steps are based on the four-step strategy for problem solving that was first introduced in Section 1.2.

Five-Step Strategy for Problem Solving Using Systems of Equations

Step 1. Read the problem several times until you have an understanding of what is being asked. If possible, create diagrams, charts, or tables to assist you. } Understand the problem.

Step 2. Choose variables that describe each unknown quantity.

Step 3. Write a system of equations using the given information and the variables. } Devise a plan.

(continued)

Step 4.	Carefully solve the system of equations using the method of elimination or substitution.	} Carry out the plan.
Step 5.	Make sure that you have answered the question, and check all answers to ensure they make sense.	} Look back.

We start with an example that was seen in Section 1.2. This time, we use two variables to solve the problem.

Example 6 Determine the Number of Touchdowns Thrown

Roger Staubach and Terry Bradshaw were both quarterbacks in the National Football League. In 1973, Staubach threw three touchdown passes more than twice the number of touchdown passes thrown by Bradshaw. If the total number of touchdown passes between Staubach and Bradshaw was 33, how many touchdown passes did each player throw?

Solution

Step 1. After carefully reading the problem, we see that we are trying to figure out how many touchdown passes were thrown by each player.

Step 2. Let

$$S = \text{number of touchdowns thrown by Staubach}$$

$$B = \text{number of touchdowns thrown by Bradshaw}$$

Step 3. The combined number of touchdown passes is 33, so the first equation is

$$\underbrace{\substack{Staubach's \\ touchdown\ passes}}_{S} + \underbrace{\substack{Bradshaw's \\ touchdown\ passes}}_{B} = \underbrace{\substack{Total\ touchdown \\ passes}}_{33}$$

To find the second equation, we use the fact that Staubach threw three more than twice the number of touchdowns thrown by Bradshaw. Therefore, the second equation is $S = 2B + 3$. This equation can be rewritten as $S - 2B = 3$. Therefore, the system of equations is

$$S + B = 33$$
$$S - 2B = 3$$

Step 4. Using the method of elimination and multiplying the first equation by 2, we obtain the system

$$2S + 2B = 66$$
$$S - 2B = 3$$

We can now add the two equations together to eliminate the variable B.

$$2S + 2B = 66$$
$$\underline{S - 2B = 3}$$
$$3S = 69$$

Dividing by 3 gives $S = 23$. To find B, replace S with 23 in the original first equation:

$$S + B = 33 \qquad \text{Write the original first equation.}$$

$$23 + B = 33 \qquad \text{Replace } S \text{ with 23.}$$

$$B = 10 \qquad \text{Subtract 23 from both sides.}$$

Therefore, Staubach threw 23 touchdown passes, and Bradshaw threw 10 touchdown passes.

Step 5. To check, we see that the total number of touchdown passes is $23 + 10 = 33$ and the number of touchdown passes thrown by Staubach is exactly 3 more than twice the number of touchdown passes thrown by Bradshaw. •

Example 7 Find the Number of Tickets Sold at a Jazz Festival

During one night at the jazz festival, 2,100 tickets were sold. Adult tickets sold for $12, and child tickets sold for $7. If the receipts totaled $22,100, how many of each type of ticket were sold?

Solution

Step 1. We are asked to determine the number of adult tickets and the number of child tickets sold.

Step 2. Let

$$A = \text{number of adult tickets}$$

$$C = \text{number of child tickets}$$

Step 3. The number of adult tickets plus the number of child tickets totals 2,100, so we obtain our first equation

$$\underbrace{A}_{\substack{\text{Number of adult} \\ \text{tickets}}} + \underbrace{C}_{\substack{\text{Number of child} \\ \text{tickets}}} = \underbrace{2{,}100}_{\substack{\text{Total number of} \\ \text{tickets}}}$$

To find the second equation, we use the fact that the value of an adult ticket is $12 and the value of a child ticket is $7; therefore,

$$\underbrace{12A}_{\substack{\text{Total value of} \\ \text{adult tickets}}} + \underbrace{7C}_{\substack{\text{Total value of} \\ \text{child tickets}}} = \underbrace{22{,}100}_{\substack{\text{Total value of} \\ \text{all tickets}}}$$

Thus, the system of equations is $\begin{array}{r} A + C = 2{,}100 \\ 12A + 7C = 22{,}100 \end{array}$.

Step 4. Using the method of elimination and multiplying the first equation by -7, we obtain the system

$$-7A - 7C = -14{,}700$$

$$12A + 7C = 22{,}100$$

Adding the equations together eliminates the variable C:

$$-7A - 7C = -14,700$$
$$12A + 7C = 22,100$$
$$5A = 7,400$$

Dividing by 5 gives $A = 1,480$. To find C, replace A with 1,480 in the original first equation:

$A + C = 2,100$	Write the original first equation.
$1,480 + C = 2,100$	Replace A with 1,480.
$C = 620$	Subtract 1,480 from both sides.

Therefore, there were 1,480 adult tickets sold and 620 child tickets sold.

Step 5. To check, we see that the total number of tickets is $1,480 + 620 = 2,100$ and the total value of all tickets sold is $(\$12)(1,480) + (\$7)(620) = \$22,100$.

Following are two more examples. Although the solutions to these examples are not worked out for you in the text, you may view them in detail by clicking on the appropriate video link.

 *My video summary*

▷ **Example 8** Mix Beans in a Food Plant

Twin City Foods, Inc., created a 10-lb bean mixture that sells for $5.75 by mixing lima beans and green beans. If lima beans sell for $.70 per pound and green beans sell for $.50 per pound, how many pounds of each bean went into the mixture?

Solution Watch the video to see that the mixture contains $3\frac{3}{4}$ pounds of lima beans and $6\frac{1}{4}$ pounds of green beans.

 *My video summary*

▷ **Example 9** Find the Speed of an Airplane

A small airplane flies from Seattle, Washington, to Portland, Oregon—a distance of 150 miles. Because the pilot encountered a strong headwind, the trip took 1 hour and 15 minutes. On the return flight, the wind is still blowing at the same speed. If the return trip took 45 minutes, what was the average speed of the airplane in still air? What was the speed of the wind?

Solution Watch the video to verify that the speed of the plane is 160 mph and the speed of the wind is 40 mph.

You Try It Work through this You Try It problem.

Work Exercises 21–30 in this textbook or in the MyMathLab Study Plan.

7.1 Exercises

In Exercises 1–4, two ordered pairs are given. Determine whether each ordered pair is a solution to the given system.

1. $\begin{aligned} -4x - 2y &= -2 \\ 5x + 7y &= 16 \end{aligned}$ Ordered pairs: $(-1, 3), (-3, 7)$

2. $\begin{aligned} \frac{1}{2}x_1 - \frac{1}{3}x_2 &= -\frac{7}{24} \\ -\frac{1}{4}x_1 + \frac{1}{2}x_2 &= \frac{5}{16} \end{aligned}$ Ordered pairs: $\left(-\frac{1}{2}, \frac{1}{4}\right), \left(-\frac{1}{4}, \frac{1}{2}\right)$

3. $\begin{aligned} .3a + .2b &= -.12 \\ -.7a - .1b &= -.05 \end{aligned}$ Ordered pairs: $(.2, -.9), (-.3, .4)$

4. $\begin{aligned} -2x + y &= -1 \\ \frac{2}{3}x - \frac{1}{3}y &= \frac{1}{3} \end{aligned}$ Ordered pairs: $(-2, 5), (3, 5)$

In Exercises 5–9, use the method of substitution to solve each system of linear equations.

5. $\begin{aligned} x &= 4 - y \\ 3x - 2y &= -3 \end{aligned}$

6. $\begin{aligned} x + 4y &= 5 \\ 2x - y &= -8 \end{aligned}$

7. $\begin{aligned} 2x - y &= 3 \\ -8x + 3y &= -8 \end{aligned}$

8. $\begin{aligned} 5x + 4y &= -6 \\ 8x - 4y &= -72 \end{aligned}$

9. $\begin{aligned} -\frac{2}{3}x + \frac{1}{2}y &= -\frac{1}{3} \\ -\frac{1}{2}x + \frac{3}{4}y &= -\frac{1}{2} \end{aligned}$

In Exercises 10–14, use the method of *elimination* to solve each system of linear equations

10. $\begin{aligned} x + 14y &= 20 \\ -x + 14y &= 15 \end{aligned}$

11. $\begin{aligned} 3x - 2y &= 20 \\ 2x + 6y &= -38 \end{aligned}$

12. $\begin{aligned} 8x - 6y &= -6 \\ 5x - 7y &= -33 \end{aligned}$

13. $\begin{aligned} .1x - .4y &= -2.4 \\ .7x - y &= -9.6 \end{aligned}$

14. $\begin{aligned} \frac{7}{16}x - \frac{3}{4}y &= -\frac{5}{2} \\ \frac{3}{4}x + \frac{5}{2}y &= 26 \end{aligned}$

In Exercises 15–20, use the substitution method or the elimination method to solve each system. If the system has infinitely many solutions, express the ordered pairs in terms of x or y as in Example 4.

15. $\begin{aligned} x + y &= 3 \\ -2x - 2y &= 1 \end{aligned}$

16. $\begin{aligned} 3x - 12y &= 6 \\ -2x + 8y &= -4 \end{aligned}$

17. $\begin{aligned} 8x - y &= -13 \\ y &= -8x \end{aligned}$

18. $\begin{aligned} -\frac{1}{2}x + \frac{2}{3}y &= -\frac{1}{3} \\ \frac{3}{2}x - 2y &= 1 \end{aligned}$

19. $\begin{aligned} 8x + 4y &= 5 \\ 2x + y &= -4 \end{aligned}$

20. $\begin{aligned} y &= \frac{5}{7}x - 9 \\ -15x + 21y &= -189 \end{aligned}$

7.1 Solving Systems of Linear Equations in Two Variables   7-13

In Exercises 21–30, use a system of equations to solve each problem.

21. Together, teammates Pedro and Ricky got 2,673 base hits last season. Pedro had 281 more hits than Ricky. How many hits did each player have?

22. Benjamin & Associates, a real estate developer, recently built 200 condominiums in McCall, Idaho. The condos were either two-bedroom units or three-bedroom units. If the total number of rooms in the entire complex is 507, how many two-bedroom units are there? How many three-bedroom units are there?

23. On a certain hot summer's day, 492 people used the public swimming pool. The daily prices are $1.50 for children and $2.50 for adults. The receipts for admission totaled $929. How many children and how many adults swam at the public pool that day?

24. Scott invested a total of $5,900 at two separate banks. One bank pays simple interest of 12% per year, whereas the other pays simple interest at a rate of 7% per year. If Scott earned $568 in interest during a single year, how much did he have on deposit in each bank?

25. Thursday is ladies' night at the Slurp and Burp Bar and Grill. All adult beverages are $2.50 for men and $1.50 for women. A total of 956 adult beverages were sold last Thursday night. If the Slurp and Burp sold a total of $1,996 in adult beverages last Thursday night, how many adult beverages were sold to women?

26. Farmer Brown planted corn and wheat on his 1,200 acres of land. The cost of planting and harvesting corn (which includes seed, planting, fertilizer, machinery, labor, and other costs) is $285 per acre. The cost of planting and harvesting wheat is $130 per acre. If Farmer Brown's total cost was $245,900, how many acres of corn did he plant?

27. An airplane encountered a headwind during a flight between Joppetown and Jawsburgh which took 3 hours and 36 minutes. The return flight took 3 hours. If the distance from Joppetown to Jawsburgh is 1800 miles, find the airspeed of the plane (the speed of the plane in still air) and the speed of the wind, assuming both remain constant.

28. Gabby's piggy bank contains nickels and quarters worth $9.10. If she has 46 coins in all, how many of each does she have?

29. Ned, the owner of Ned's Nut Shop, sells peanuts for $3.25 per pound and cashews for $6.00 per pound. Ned wants to create a 58.5-lb barrel of mixed nuts by mixing the peanuts and cashews together and sell it for $4.00 per pound. How many pounds of each should Ned use?

30. Angela, the head of the office party planning committee, sent two coworkers, Kevin and Dwight, to the party store to purchase party hats and party whistles. Kevin purchased five packs of party hats and four packs of party whistles and paid a total of $23.01. Dwight purchased four packs of party hats and five packs of party whistles, and paid a total of $20.91. When they returned to the office, Angela informed them that they were supposed to buy a total of five packs of party hats and five packs of party whistles and sent them back to the store to straighten things out. How much should they be refunded?

7.2 Solving Systems of Linear Equations in Three Variables Using the Elimination Method

THINGS TO KNOW

Before working through this section, be sure that you are familiar with the following concepts:

| | | VIDEO | ANIMATION | INTERACTIVE |

You Try It
1. Solving a System of Linear Equations Using the Substitution Method (Section 7.1)

You Try It
2. Solving a System of Linear Equations Using the Elimination Method (Section 7.1)

OBJECTIVES

1 Verifying the Solution of a System of Linear Equations in Three Variables

2 Solving a System of Linear Equations Using the Elimination Method

3 Solving Consistent, Dependent Systems of Linear Equations in Three Variables

4 Solving Inconsistent Systems of Linear Equations in Three Variables

5 Solving Applied Problems Using a System of Linear Equations Involving Three Variables

---

OBJECTIVE 1   VERIFYING THE SOLUTION OF A SYSTEM OF LINEAR EQUATIONS IN THREE VARIABLES

In Section 7.1 we solved systems of linear equations in two variables. In this section (and in Section 7.3) we will learn techniques that will enable us to solve systems of linear equations in three variables. An equation such as $2x + 3y + 4z = 12$ is called a **linear equation in three variables** because there are three variables and each variable term is linear.

> **Definition**   Linear Equation in Three Variables
>
> A **linear equation in three variables** is an equation that can be written in the form $Ax + By + Cz = D$, where $A, B, C,$ and D are real numbers, and $A, B,$ and C are not all equal to 0.

When two or more linear equations in three variables are considered simultaneously, then this collection of linear equations forms a system of linear equations in three variables.

> **Definition** System of Linear Equations in Three Variables
>
> A **system of linear equations in three variables** is a collection of linear equations in three variables considered simultaneously. A solution to a system of linear equations in three variables is an **ordered triple** that satisfies all equations in the system.

Recall from Section 7.1 that if a system of two linear equations has one unique solution, then the solution can be geometrically represented by the intersection of two lines. The two lines intersect at exactly one point (or at a unique ordered pair.) Similarly, if a system of three linear equations in three variables has a unique solution, then the solution can be geometrically represented by the intersection of three planes. The three planes intersect at exactly one point (or at a unique ordered triple.) See Figure 5.

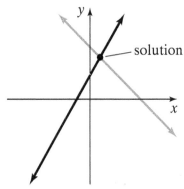

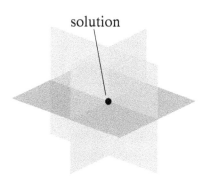

The figure above represents a system of two linear equations in two variables that has a unique solution. The solution can be represented by the point of intersection of two lines.

The figure above represents a system of three linear equations in three variables that has a unique solution. The solution can be represented by the point of intersection of three planes.

Figure 5

The following is an example of a system of three linear equations in three variables:

$$2x + 3y + 4z = 12$$
$$x - 2y + 3z = 0$$
$$-x + y - 2z = -1$$

The solution to this linear system is the set of all *ordered triples* (x, y, z) that satisfy all three equations. In Example 1, we show that the ordered triple $(1, 2, 1)$ is a solution to this linear system.

 *My video summary*

⊙ **Example 1** Verify the Solution of a System of Linear Equations

Verify that the ordered triple $(1, 2, 1)$ is a solution to the following system of linear equations:

$$2x + 3y + 4z = 12$$
$$x - 2y + 3z = 0$$
$$-x + y - 2z = -1$$

Solution To verify that the ordered triple $(1, 2, 1)$ is a solution, we must substitute $x = 1$, $y = 2$, and $z = 1$ in each of the three equations and determine whether a true statement occurs.

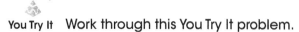

$$2(1) + 3(2) + 4(1) \stackrel{?}{=} 12 \longrightarrow 2 + 6 + 4 \stackrel{?}{=} 12 \longrightarrow 12 = 12 \ \checkmark \text{ True}$$

$$(1) - 2(2) + 3(1) \stackrel{?}{=} 0 \longrightarrow 1 - 4 + 3 \stackrel{?}{=} 0 \longrightarrow 0 = 0 \ \checkmark \text{ True}$$

$$-(1) + (2) - 2(1) \stackrel{?}{=} -1 \longrightarrow -1 + 2 - 2 \stackrel{?}{=} -1 \longrightarrow -1 = -1 \ \checkmark \text{ True}$$

All three statements are true. This implies that the ordered triple $(1, 2, 1)$ is a solution to this system.

You Try It Work through this You Try It problem.

Work Exercises 1 and 2 in this textbook or in the MyMathLab Study Plan.

OBJECTIVE 2 SOLVING A SYSTEM OF LINEAR EQUATIONS USING THE ELIMINATION METHOD

A system of three linear equations in three variables can be geometrically represented by three planes in space. In Example 1, the system has one unique solution, the ordered triple $(1, 2, 1)$. If we were to graph the three planes represented by the three linear equations in Example 1, we would see that the three planes intersect at exactly one point. Systems of equations that have at least one solution are called **consistent systems**. A system that has exactly one solution is called a **consistent, independent** system. It is possible that a system of three linear equations in three variables has infinitely many solutions. These systems are called **consistent, dependent** systems. It is also possible for a system of three linear equations in three variables to have no solution at all. These systems are called **inconsistent systems**. Figure 6 illustrates some possibilities of consistent systems and inconsistent systems.

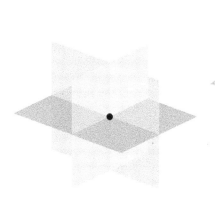

Consistent, Independent
One solution

Consistent, Dependent
Infinitely many solutions
(Planes intersect at a line.)

Consistent, Dependent
Infinitely many solutions
(Three equations describe the same plane.)

Figure 6 Continues

Inconsistent
No solution
(Three planes are parallel.)

Inconsistent
No solution
(Two planes are parallel.)

Inconsistent
No solution
(Planes intersect two at a time.)

Figure 6

To solve systems of linear equations in three variables by graphing would require us to graph planes in three dimensions, which is not a practical task. Instead, we will focus in this section on solving systems of linear equations in three variables by applying the elimination method. This method is similar to the elimination method used when solving systems of linear equations in two variables. See Section 7.1 to review this method. (In Section 7.3, an alternative method will be introduced.)

The goal of the elimination method when solving a system of three linear equations in three variables is to reduce the system of three equations in three variables down to a system of two equations in two variables. At that point, we can reduce the two-equation system to a single equation in one variable and easily solve that equation as we did in Section 7.1. Then, using back substitution, we can find the values of the other two variables.

Before looking at an example, we present some guidelines for solving systems of linear equations in three variables by elimination.

Guidelines for Solving a System of Linear Equations in Three Variables by Elimination

Step 1. Write each equation In standard form. Write each equation in the form $Ax + By + Cz = D$ lining up the variable terms. Number the equations to keep track of them.

Step 2. Eliminate a variable from one pair of equations. Use the elimination method to eliminate a variable from any two of the original three equations, leaving one equation in two variables,

Step 3. Eliminate the same variable again. Use a different pair of the original equations and eliminate the same variable again, leaving one equation in two variables.

Step 4. Solve the system of linear equations in two variables. Use the resulting equations from steps 2 and 3 to create and solve the corresponding system of linear equations in two variables by substitution or elimination.

Step 5. **Use back substitution to find the value of the third variable.**
Substitute the results from step 4 into any of the original equations to find the value of the remaining variable.

Step 6. **Check the solution.** Check the proposed solution in each equation of the system and write the solution set.

My video summary ⊚ **Example 2** Solving Systems of Linear Equations in Three Variables

Solve the following system:

$$2x + 3y + 4z = 12$$
$$x - 2y + 3z = 0$$
$$-x + y - 2z = -1$$

Solution We follow our guidelines for solving systems of linear equations in three variables. Follow the steps below, or watch this video for a detailed solution.

Step 1. The equations are already in standard form and all the variables are lined up. We rewrite the system and number each equation.

$$2x + 3y + 4z = 12 \qquad (1)$$
$$x - 2y + 3z = 0 \qquad (2)$$
$$-x + y - 2z = -1 \qquad (3)$$

Step 2. We can eliminate any of the variables. For convenience, we will eliminate the variable x from equations **(1)** and **(2)**. We can do this by multiplying equation **(2)** by -2 and adding the equations together.

$$2x + 3y + 4z = 12 \qquad (1)$$
$$\underline{-2x + 4y - 6z = 0} \qquad \text{Multiply (2) by } -2.$$
$$7y - 2z = 12 \qquad \text{Add to produce new equation (4).}$$

How do you know which variable to eliminate? Click here to find out.

Step 3. We need to eliminate the same variable, x, from a different pair of equations. We will use equations **(2)** and **(3)** for the second pair. Since the coefficients of x in these equations have the same absolute value, but opposite signs, we can eliminate the variable by simply adding the equations.

$$x - 2y + 3z = 0 \qquad (2)$$
$$\underline{-x + y - 2z = -1} \qquad (3)$$
$$-y + z = -1 \qquad \text{Add to produce new equation (5).}$$

Step 4. Combining equations **(4)** and **(5)**, we form a system of linear equations in two variables.

$$7y - 2z = 12 \qquad (4)$$
$$-y + z = -1 \qquad (5)$$

To solve this system, we use the elimination method by multiplying equation **(5)** by 7 and adding the result to equation **(4)** to eliminate the variable y.

$$7y - 2z = 12 \qquad \text{(4)}$$

$$\underline{-7y + 7z = -7} \qquad \text{Multiply (5) by 7.}$$

$$5z = 5 \qquad \text{Add.}$$

$$z = 1 \qquad \text{Divide by 5.}$$

Since $z = 1$, we back-substitute this into **(5)** to solve for y.

$$-y + z = -1 \qquad \text{(5)}$$

$$-y + (1) = -1 \qquad \text{Substitute 1 for } z.$$

$$-y + 1 = -1 \qquad \text{Simplify.}$$

$$-y = -2 \qquad \text{Subtract 1.}$$

$$y = 2 \qquad \text{Divide by } -1.$$

When back-substituting, we can use either of the equations, but often one equation is preferred over another. Can you see why we chose to use equation **(5)** instead of equation **(4)**? Click here for an explanation.

Step 5. Substitute 2 for y and 1 for z in any of the original equations, **(1)**, **(2)**, or **(3)**, and solve for x. We will back-substitute using equation **(2)**.

$$x - 2y + 3z = 0 \qquad \text{(2)}$$

$$x - 2(2) + 3(1) = 0 \qquad \text{Substitute 2 for } y \text{ and 1 for } z.$$

$$x - 4 + 3 = 0 \qquad \text{Multiply.}$$

$$x - 1 = 0 \qquad \text{Simplify.}$$

$$x = 1 \qquad \text{Add 1.}$$

The solution to the system is the ordered triple $(1, 2, 1)$.

Step 6. Click here to check your answer.

You Try It Work through the following You Try It problem.

Work Exercises 3–5 in this textbook or in the MyMathLab Study Plan.

My video summary ⊙ **Example 3** Solving Systems of Linear Equations in Three Variables Involving Fractions

Solve the following system:

$$\frac{1}{2}x + \quad y + \frac{2}{3}z = 2$$

$$\frac{3}{4}x + \frac{5}{2}y - 2z = -7$$

$$x + 4y + 2z = 4$$

Solution If an equation in the system contains fractions, then it is often helpful to clear the fractions first. After doing this, follow the guidelines for solving a system of linear equations in three variables. Click here to check your answer, or watch this video for a detailed solution.

You Try It   Work through the following You Try It problem.

Work Exercises 6–8 this textbook or in the My MathLab Study Plan.

My video summary   ⊘ **Example 4** Solving Systems of Linear Equations in Three Variables with Missing Terms

Solve the following system:

$$2x + y = 13$$
$$3x - 2y + z = 8$$
$$x + 2y - 3z = 5$$

Solution   We follow our guidelines for solving a system of linear equations in three variables.

Step 1. First, we rewrite and number each equation, lining up the variables. Notice that the first equation has a variable term missing, so we put a gap in its place.

$$2x + y \qquad = 13 \qquad \textbf{(1)}$$
$$3x - 2y + z = 8 \qquad \textbf{(2)}$$
$$x + 2y - 3z = 5 \qquad \textbf{(3)}$$

Step 2. We can eliminate any of the variables, but we notice that one equation already has z eliminated. By selecting z as the variable to eliminate, we can move directly to **step 3.**

Step 3. Looking at equations **(2)** and **(3)**, we might be tempted to add these equations to eliminate y. However, in step 2, we selected z as the variable to eliminate, so we need to eliminate z again in this step.

Continue working through the solution. Click here to check your answer, or watch this video for a detailed solution.

You Try It   Work through the following You Try It problem.

Work Exercises 9-11 this textbook or in the My MathLab Study Plan.

OBJECTIVE 3   SOLVING CONSISTENT, DEPENDENT SYSTEMS OF LINEAR EQUATIONS IN THREE VARIABLES

Recall from Section 7.1 that if a system of equations has at least one solution, then the system is consistent. If the system has infinitely many solutions, then the system is consistent and **dependent**. For the two-variable case, a consistent, dependent system can be geometrically represented by a pair of coinciding lines. For the three-variable case, consistent, dependent systems can be geometrically represented by three planes that intersect at a line or by three equations that all describe the same plane. See Figure 7.

Consistent, Dependent
Infinitely many solutions
(Planes intersect at a line.)

Consistent, Dependent
Infinitely many solutions
(Three equations describe the same plane.)

Figure 7 A geometric representation of consistent, dependent systems in three variables.

When solving systems of linear equations in two variables, encountering an identity automatically meant that the system was dependent and had infinitely many ordered pair solutions. (See Example 5 of Section 7.1.) However, unlike the two-variable case, identifying dependent systems in three variables takes a bit more work because we must consider more than one pairing of the equations in the system. Obtaining an identity with one pairing of equations is not sufficient to say that the system is dependent as we will see later in Step 2 of Example 6. However, during the solution process, if we encounter an identity after successfully obtaining a two-variable system of equations, then the system is dependent and the system has infinitely many solutions.

 *My video summary*   ⊙ **Example 5** Solving Systems of Dependent Linear Equations in Three Variables

Solve the following system.

$$x + 2y + 3z = 10 \quad (1)$$
$$x + y + z = 7 \quad (2)$$
$$3x + 2y + z = 18 \quad (3)$$

Solution Eliminating the variable x using equations **(1)** and **(2)** results in the equation $-y - 2z = -3$. Eliminating the variable x again using equations **(1)** and **(3)** results in the equation $-4y - 8z = -12$. Combining these two equations gives the two-variable system:

$$-y - 2z = -3 \quad (4)$$
$$-4y - 8z = -12 \quad (5)$$

We can multiply equation **(4)** by -4 to get the new system:

$$4y + 8z = 12 \quad (6)$$
$$-4y - 8z = -12 \quad (5)$$

We now add equations **(6)** and **(5)**.

$$4y + 8z = 12 \quad \text{(6)}$$
$$\underline{-4y - 8z = -12} \quad \text{(5)}$$
$$0 = 0$$

We see that adding equations **(6)** and **(5)** results in the identity $0 = 0$. Therefore, this two-variable system has infinitely many solutions and the system is said to be dependent. Solve either equation **(6)** or **(5)** for either variable. Solving for the variable y we get $y = 3 - 2z$.

Since $y = 3 - 2z$, we back-substitute this into (1) to solve for x.

$$x + 2(3 - 2z) + 3z = 10 \qquad \text{Substitute } 3 - 2z \text{ for } y \text{ in equation (1).}$$
$$x + 6 - 4z + 3z = 10 \qquad \text{Multiply.}$$
$$x + 6 - z = 10 \qquad \text{Simplify.}$$
$$x = 4 + z \qquad \text{Solve for } x.$$

Therefore, $x = 4 + z$ and $y = 3 - 2z$. Notice that the value of x and y depend on the choice of the variable z. The variable z is free to be any real number. Consequently, we say that z is a *free variable*.

Recall from Example 5 of Section 7.1 that we can write the solutions to a dependent system of two equations in two variables in ordered pair notation. Similarly, the solution to a dependent system of three equations in three variables can be written in **ordered triple notation**. The solution to this dependent system can be written in ordered triple notation as $(4 + z, 3 - 2z, z)$ where z is free to be any real number.

Thus, this system has infinitely many solutions. For example, letting $z = 0, z = -1$, and $z = 3$, we get the following ordered triples as solutions to this system:

$$z = 0: \quad (4, 3, 0)$$
$$z = -1: \quad (3, 5, -1)$$
$$z = 3; \quad (7, -3, 3)$$

Try substituting these ordered triples into the original system to verify that all three are solutions.

Watch this video to see every step of the solution process to this system of three linear equations in three variables.

You Try It   Work through the following You Try It problem.

Work Exercises 12–15 this textbook or in the MyMathLab Study Plan.

OBJECTIVE 4   SOLVING INCONSISTENT SYSTEMS OF LINEAR EQUATIONS IN THREE VARIABLES

Recall from **Section 7.1** that a system of linear equations in two variables could have no solution. Linear systems having no solutions are called **inconsistent systems**. For the two-variable case, this occurred if the two linear equations represented parallel lines because parallel lines have no points in common. For the three-variable case, a system is **inconsistent** if all three planes have no points in common. This happens if all three planes are parallel or if two of the planes are parallel. A third possibility of an inconsistent system is when the planes intersect two at a time. Figure 8 geometrically illustrates three types of inconsistent systems in three variables.

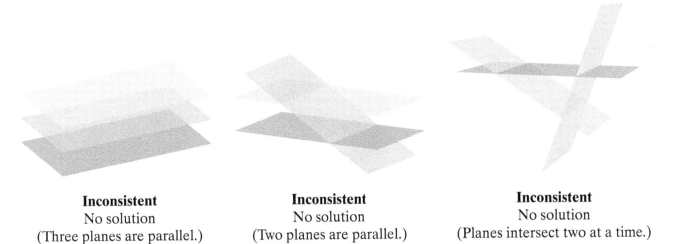

Inconsistent	**Inconsistent**	**Inconsistent**
No solution	No solution	No solution
(Three planes are parallel.)	(Two planes are parallel.)	(Planes intersect two at a time.)

Figure 8 A geometric representation of inconsistent systems in three variables.

Recall from **Example 4** of Section 7.1 that when solving an inconsistent system of two equations we will eventually obtain a **contradiction**. This is also true in the three-variable case. If we encounter a contradiction during the solution process, then the system is inconsistent and there is no solution. We illustrate this in Example 6.

Example 6 Solving Systems of Linear Equations in Three Variables with No Solution

Solve the following system:

$$x - y + 2z = 5$$
$$3x - 3y + 6z = 15$$
$$-2x + 2y - 4z = 7$$

Solution We follow our guidelines for solving a system of linear equations in three variables.

Step 1. The equations are already in standard form with all the variables lined up. We rewrite the system and number each equation.

$$x - y + 2z = 5 \qquad (1)$$
$$3x - 3y + 6z = 15 \qquad (2)$$
$$-2x + 2y - 4z = 7 \qquad (3)$$

Step 2. For convenience, we will eliminate the variable x from equations **(1)** and **(2)**. To do this, we multiply equation **(1)** by -3 and add the equations together.

$$-3x + 3y - 6z = -15 \qquad \text{Multiply (1) by } -3.$$
$$\underline{3x - 3y + 6z = 15} \qquad (2)$$
$$0 = 0 \qquad \text{Add}$$

The last line, $0 = 0$, is an **identity**. In the two-variable case, we would stop and say that the system had an infinite number of solutions. However, in systems of three variables, this is not necessarily the case.

Step 3. Let's continue our process and eliminate the variable x from the pairing of equations **(1)** and **(3)**. To do this, we multiply equation **(1)** by 2 and add the equations.

$$\begin{array}{ll} 2x - 2y + 4z = 10 & \text{Multiply (1) by 2.} \\ \underline{-2x + 2y - 4z = 7} & \text{(3)} \\ \qquad\qquad 0 = 17 & \text{Add.} \end{array}$$

The last line, $0 = 17$, is a **contradiction**, so the system is **inconsistent** and has no solution. Click here to find out why this occurs.

You Try It Work through the following You Try It problem.

Work Exercises 12–17 this textbook or in the MyMathLab **Study Plan.**

OBJECTIVE 5 SOLVING APPLIED PROBLEMS USING A SYSTEM OF LINEAR EQUATIONS IN THREE VARIABLES

Systems of linear equations can be used to solve a wide variety of applications. Following are two examples. We will solve each example using the elimination method.

 *My video summary*

▶ **Example 7** Real-Time Strategy Game

While playing a real-time strategy game, Joel created military units to defend his town: warriors, skirmishers, and archers. Warriors require 20 units of food and 50 units of gold. Skirmishers require 25 units of food and 35 units of wood. Archers require 32 units of wood and 32 units of gold. If Joel used 506 units of gold, 606 units of wood, and 350 units of food to create the units, how many of each type of military unit did he create?

Solution We want to find the number of each type of military unit created. There are three types of units: warriors, skirmishers, and archers. Each unit requires a certain amount of gold, wood, and food. We know how much of each resource is needed for each unit, and we know the total amount of each resource that is used. We summarize this information in the following table.

	Each Warrior	Each Skirmisher	Each Archer	Total
Units of Gold	50	0	32	506
Units of Wood	0	35	32	606
Units of Food	20	25	0	350

Let W, S, and A represent the number of warriors, skirmishers, and archers respectively. We can obtain the three equations based on the total units of gold, wood, and food used.

GOLD

We base our first equation on the total amount of gold used. Each warrior requires 50 units of gold, each skirmisher requires 0 units of gold, and each archer requires 32 units of gold. Joel used a total of 506 units of gold. Thus, we obtain the first equation

Units of gold for warriors		Units of gold for skirmishers		Units of gold for archers		Total units of gold
$50W$	$+$	$0S$	$+$	$32A$	$=$	$506.$

WOOD

The second equation is based on the total amount of wood used. Each warrior requires 0 units of wood, each skirmisher requires 35 units of wood, and each archer requires 32 units of wood. Joel used a total of 606 units of wood. Using this information, we obtain the second equation

Units of wood for warriors		Units of wood for skirmishers		Units of wood for archers		Total units of wood
$0W$	$+$	$35S$	$+$	$32A$	$=$	$606.$

FOOD

The third equation is based on the total amount of food used. Each warrior required 20 units of food, each skirmisher requires 25 units of food, and each archer requires 0 units of food. Joel used a total of 350 units of food. We now write the third equation

Units of food for warriors		Units of food for skirmishers		Units of food for archers		Total units of food
$20W$	$+$	$25S$	$+$	$0A$	$=$	$350.$

Using these three equations, we form the following system.

$$50W + 0S + 32A = 506$$

$$0W + 35S + 32A = 606$$

$$20W + 25S + 0A = 350$$

Now solve this system using the elimination method. Once you have solved this system, check your answers, or watch this video for a detailed solution.

 *My video summary*

◎ Example 8 Buy Clothes Online

Wendy ordered 30 T-shirts online for her three children. The small T-shirts cost $4 each, the medium T-shirts cost $5 each, and the large T-shirts were $6 each. She spent $40 more purchasing the large T-shirts than the small T-shirts, Wendy's total bill was $154, How many T-shirts of each size did she buy?

Solution Let S, M, and L represent the number of small, medium, and large T-shirts, respectively. Because a total of 30 T-shirts were purchased, we obtain the first equation

$$S + M + L = 30$$

Because small T-shirts cost $4, we know that the total amount spent on small T-shirts is $4S$. Similarly, the total spent on medium T-shirts is $5M$, and the total spent on large T-shirts is $6L$. Therefore, because the total value of the T-shirts was $154, we obtain the second equation

$$4S + 5M + 6L = 154$$

Finally, Wendy spent $40 more buying large T-shirts than small T-shirts, so $6L = 4S + 40$, which is equivalent to the equation

$$4S - 6L = -40$$

Thus, we obtain the system

$$\begin{aligned} S + M + L &= 30 \\ 4S + 5M + 6L &= 154, \\ 4S \phantom{{}+ 5M} - 6L &= -40 \end{aligned}$$

Watch this video to see how to use the elimination method to determine that Wendy purchased 8 small T-shirts, 10 medium T-shirts, and 12 large T-shirts.

You Try It Work through the following You Try It problem.

Work Exercises 18–23 in this textbook or in the MyMathLab Study Plan.

7.2 Exercises

In Exercises 1 and 2, two ordered triples are given. Determine whether each ordered triple is a solution of the given system.

1. $(-1, 1, -2), (1, -1, 2)$

$$\begin{aligned} x + y + z &= -2 \\ -x - 3y - 2z &= 2 \\ 2x - 2y + 5z &= -14 \end{aligned}$$

2. $(2, -1, 4), (-2, 1, -4)$

$$\begin{aligned} \tfrac{1}{2}x + 3y - z &= 6 \\ -x + y + \tfrac{1}{4}z &= 2 \\ x - 4y + z &= -10 \end{aligned}$$

In Exercises 3–11, solve each system of linear equations using the elimination method.

3.
$$\begin{aligned} x + y + z &= 4 \\ 2x - y - 2z &= -10 \\ -x - y + 3z &= 8 \end{aligned}$$

4.
$$\begin{aligned} x - 2y + z &= 6 \\ 2x + y - 3z &= -3 \\ x - 3y + 3z &= 10 \end{aligned}$$

5.
$$\begin{aligned} x - 2y + 2z &= 2 \\ 3x + 2y - 2z &= -1 \\ x - y - 2z &= 0 \end{aligned}$$

6.
$$\begin{aligned} x - \tfrac{1}{2}y + \tfrac{1}{2}z &= -3 \\ x + y - z &= 0 \\ -3x - 3y + 4z &= 1 \end{aligned}$$

7.
$$\begin{aligned} \tfrac{1}{3}x - \tfrac{2}{3}y + z &= 0 \\ \tfrac{1}{2}x - \tfrac{3}{4}y + z &= -\tfrac{1}{2} \\ -2x - y + z &= 7 \end{aligned}$$

8.
$$\begin{aligned} x + y + 10z &= 3 \\ \tfrac{1}{2}x - y + z &= -\tfrac{5}{6} \\ -2x + 3y - 5z &= \tfrac{7}{3} \end{aligned}$$

9.
$$\begin{aligned} -4x + 5y + 9z &= -9 \\ x - 2y + z &= 0 \\ 2y - 8z &= 8 \end{aligned}$$

10.
$$\begin{aligned} 2x + 2y + z &= 9 \\ x + z &= 4 \\ 4y - 3z &= 17 \end{aligned}$$

11.
$$\begin{aligned} x - y &= 7 \\ y - z &= 2 \\ x + z &= 1 \end{aligned}$$

In Exercises 12–17, solve each system of linear equations. If the system has infinitely many solutions, describe the solution with the equation of a plane or an ordered triple in terms of one variable.

12. $2x + 6y - 4z = 8$
$-x - 3y + 2z = -4$
$x + 3y - 2z = 4$

13. $2x - y + z = -6$
$x - \dfrac{1}{2}y + \dfrac{1}{2}z = -3$
$4x - 2y + 2z = -12$

14. $x + 2y - z = 11$
$x + 3y - 2z = 14$
$3x + 7y - 4z = 36$

15. $x - y + z = 5$
$2x + 3y - 3z = -5$
$3x + 2y - 2z = 0$

16. $x - 4y + 2z = 7$
$\dfrac{1}{2}x - 2y + z = 1$
$-3x + y - 4z = 9$

17. $4x - y + z = 8$
$x + y + 3z = 2$
$3x - 2y - 2z = 5$

In Exercises 18–23, Write a system of linear equations in three variables and then use the elimination method to solve the system.

18. While playing a real-time strategy game. Arvin created military units for a battle: long swordsmen, spearmen, and cross-bowmen. Long swordsmen require 60 units of food and 20 units of gold. Spearmen require 35 units of food and 25 units of wood, Crossbowmen require 25 units of wood and 45 units of gold. If Arvin used 1975 units of gold. 1375 units of wood, and 1900 units of food to create the units, how many of each type of military unit did he create?

19. The concession stand at a school basketball tournament sells hot dogs, hamburgers, and chicken sandwiches. During one game, the stand sold 16 hot dogs, 14 hamburgers, and 8 chicken sandwiches for a total of $89.00. During a second game, the stand sold 10 hot dogs. 13 hamburgers. and 5 chicken sandwiches for a total of $66.25. During a third game, the stand sold 4 hot dogs, 7 hamburgers, and 7 chicken sandwiches for a total of $49.75. Determine the price of each product.

20. Ben ordered 35 pizzas for an office party. He ordered three types: cheese, supreme, and pepperoni. Cheese pizza costs $9 each, pepperoni pizza costs $12 each, and supreme pizza costs $15 each. He spent exactly twice as much on the pepperoni pizzas as he did on the cheese pizzas. If Ben spent $420, how many pizzas of each type did he buy?

21. On opening night of the play *The Music Man,* 1010 tickets were sold for a total of $10,300. Adult tickets cost $12 each, children's tickets cost $10 each, and senior citizen tickets cost $7 each. If the total number of adult and children tickets sold exceeded twice the number of senior citizen tickets sold by 170 tickets, then how many tickets of each type were sold?

22. Tyler Hansbrough was the leading scorer of the 2009 NCAA Basketball champions, the North Carolina Tar Heels. Hansbrough scored a total of 722 points during the 2009 season. He made 26 more one-point free throws than two-point field goals, and his number of two-point field goals was two less than 25 times his number of three-point field goals. How many free throws, two-point field goals, and three-point field goals did Tyler Hansbrough make during the 2009 season? (*Source;* espn.com)

23. The number of new Facebook users, y (in millions), between September 2008 and March 2009 can be modeled by the equation $y = ax^2 + bx + c$, where x represents the age of the user. Using the ordered-pair solutions (15, 1), (35, 7), and (55, 3), create a system of linear equations in three variables for $a, b,$ and c. Do this by substituting each ordered-pair solution into the model, creating an equation in three variables. Solve the resulting system to find the coefficients of the model. Then use the model to predict the number of new Facebook users who were 25 years old. (*Source:* www.facebook.com)

7.3 Solving Systems of Linear Equations in Three Variables Using Gaussian Elimination and Gauss-Jordan Elimination

THINGS TO KNOW

Before working through this section, be sure that you are familiar with the following concepts:

| | VIDEO | ANIMATION | INTERACTIVE |

You Try It

1. Solving a System of Linear Equations Using the Substitution Method (Section 7.1)

You Try It

2. Solving a System of Linear Equations Using the Elimination Method (Section 7.1)

You Try It

3. Verifying the Solution of a System of Linear Equations in Three Variables (Section 7.2)

OBJECTIVES

1 Solving a System of Linear Equations Using Gaussian Elimination

2 Using an Augmented Matrix to Solve a System of Linear Equations

3 Solving Consistent, Dependent Systems of Linear Equations in Three Variables

4 Solving Inconsistent Systems of Linear Equations in Three Variables

5 Determining Whether a System Has No Solution or Infinitely Many Solutions

6 Solving Linear Systems Having Fewer Equations Than Variables

7 Solving Applied Problems Using a System of Linear Equations Involving Three Variables

---

OBJECTIVE 1 SOLVING A SYSTEM OF LINEAR EQUATIONS USING GAUSSIAN ELIMINATION

In Section 7.2, we used the method of elimination to solve systems of three equations involving three variables. In this section, will learn two more similar techniques that can be used to solve such systems. Before we learn these techniques, first consider the following system of equations.

$$
\begin{aligned}
x - 2y + 3z &= 0 \quad &(1) \\
- y + z &= -1 \quad &(2) \\
5z &= 5 \quad &(3)
\end{aligned}
$$

Notice that equation **(3)** has only one variable, equation **(2)** contains two variables, and equation **(1)** has three variables. A system of this form is said to be in **triangular form**. Note that we can easily solve equation **(3)** by dividing both sides of the equation by 5 to get the solution $z = 1$. Using back substitution, we can solve for y and x.

Using the value of $z = 1$ in equation **(2)**, we can solve for y:

$$-y + z = -1 \qquad \text{Write equation (2).}$$
$$-y + (1) = -1 \qquad \text{Substitute 1 for } z.$$
$$-y = -2 \qquad \text{Subtract 1 from both sides.}$$
$$y = 2 \qquad \text{Divide both sides by } -1.$$

Finally, using $z = 1$ and $y = 2$ in equation **(1)** gives

$$x - 2y + 3z = 0 \qquad \text{Write equation (1).}$$
$$x - 2(2) + 3(1) = 0 \qquad \text{Substitute 2 for } y \text{ and 1 for } z.$$
$$x - 1 = 0 \qquad \text{Simplify.}$$
$$x = 1 \qquad \text{Add 1 to both sides.}$$

Therefore, the solution to this system of equations is the ordered triple $(1, 2, 1)$.

The process of writing a system of three linear equations in three variables into an equivalent system that is in triangular form and then using back substitution to solve for each variable is called **Gaussian elimination,** named after the famous German mathematician Carl Friedrich Gauss.

When using Gaussian elimination, there are three algebraic operations that we can use to reduce a system of linear equations into an equivalent system in triangular form. These algebraic operations are called the **elementary row operations** and are listed below.

ELEMENTARY ROW OPERATIONS

The following algebraic operations will result in an equivalent system of linear equations:

1. Interchange any two equations.

2. Multiply any equation by a nonzero constant.

3. Add a multiple of one equation to another equation.

Because Gaussian elimination can be a lengthy process, we use shorthand notation to document the elementary row operation performed at each step. This documentation is important because it will help you "retrace" your steps in case you make a mistake or want to study your work later. We use R_i to describe the i th equation of the system and R_j to describe the j th equation of the system. The shorthand notation is outlined as follows.

NOTATION USED TO DESCRIBE ELEMENTARY ROW OPERATIONS

Notation	Meaning
$R_i \Leftrightarrow R_j$	Interchange Rows i and j.
$kR_i \rightarrow \text{New } R_i$	k times Row i becomes New Row i.
$kR_i + R_j \rightarrow \text{New } R_j$	k times Row i plus Row j becomes New Row j.

Example 1 illustrates how we can use Gaussian elimination to solve a system of three linear equations. You should carefully work through Example 1 and take notes while you watch the video solution.

My video summary ⊗ **Example 1** Use Gaussian Elimination to Solve a System of Three Linear Equations

For the following system, use elementary row operations to find an equivalent system in triangular form and then use back substitution to solve the system:

$$2x + 3y + 4z = 12$$

$$x - 2y + 3z = 0$$

$$-x + y - 2z = -1$$

My video summary ⊗ **Solution** Our goal is to use a series of elementary row operations to eliminate the variable x in one of the equations and eliminate the variables x and y in another equation. We typically start by trying to make the first coefficient of the first equation equal to 1. We can accomplish this by interchanging the first two equations. The other four elementary row operations are outlined as follows. You can also watch the video to see each step worked out in detail.

$$
\begin{array}{l}
2x + 3y + 4x = 12 \\
x - 2y + 3z = 0 \\
-x + y - 2z = -1
\end{array}
\xrightarrow{R_1 \Leftrightarrow R_2}
\begin{array}{l}
x - 2y + 3z = 0 \\
2x + 3y + 4z = 12 \\
-x + y - 2z = -1
\end{array}
\xrightarrow{-2R_1 + R_2 \to \text{New } R_2}
\begin{array}{l}
x - 2y + 3z = 0 \\
7y - 2z = 12 \\
-x + y - 2z = -1
\end{array}
$$

$$
\xrightarrow{R_1 + R_3 \to \text{New } R_3}
\begin{array}{l}
x - 2y + 3z = 0 \\
7y - 2z = 12 \\
-y + z = -1
\end{array}
\xrightarrow{R_2 \Leftrightarrow R_3}
\begin{array}{l}
x - 2y + 3z = 0 \\
-y + z = -1 \\
7y - 2z = 12
\end{array}
\xrightarrow{7R_2 + R_3 \to \text{New } R_3}
\begin{array}{l}
x - 2y + 3z = 0 \\
-y + z = -1 \\
5z = 5
\end{array}
$$

Now that the linear system is written in triangular form, we can solve the third equation for z and then use back substitution to solve for the other two variables. You should verify that the solution to this system is the ordered triple $(1, 2, 1)$. ●

You Try It Work through this You Try It problem.

Work Exercises 1-4 in this textbook or in the MyMathLab Study Plan.

OBJECTIVE 2 USING AN AUGMENTED MATRIX TO SOLVE A SYSTEM OF LINEAR EQUATIONS

In Example 1, the variables x, y, and z were used. We actually could have used any three variables to describe this system. As it turns out, we do not need variables at all to solve a system of linear equations using Gaussian elimination! We need only the coefficients of each variable to perform the elementary row operations. We can simplify the Gaussian elimination process by simplifying the notation. Instead of writing three equations at each step, we only write the coefficients at each step. We accomplish this by writing the coefficients in a rectangular array called an **augmented matrix**. For example, the system from Example 1 can be written using an augmented matrix as follows:

System of Equations from Example 1   **Corresponding Augmented Matrix**

$$2x + 3y + 4z = 12$$

$$x - 2y + 3z = 0$$

$$-x + y - 2z = -1$$

$$
\left[\begin{array}{ccc|c}
2 & 3 & 4 & 12 \\
1 & -2 & 3 & 0 \\
-1 & 1 & -2 & -1
\end{array}\right]
$$

Notice that each row of the corresponding augmented matrix represents the coefficients of each equation. The vertical bar is used to separate the coefficients of each variable from the constant coefficient that appears on the right side of each equation. We can perform the exact same elementary row operations as in Example 1 by writing an augmented matrix at each step. We now illustrate this process. For clarity, the corresponding system of equations is listed below each augmented matrix.

$$\begin{bmatrix} 2 & 3 & 4 & | & 12 \\ 1 & -2 & 3 & | & 0 \\ -1 & 1 & -2 & | & -1 \end{bmatrix} \xrightarrow{R_1 \Leftrightarrow R_2} \begin{bmatrix} 1 & -2 & 3 & | & 0 \\ 2 & 3 & 4 & | & 12 \\ -1 & 1 & -2 & | & -1 \end{bmatrix} \xrightarrow{-2R_1+R_2 \to \text{New } R_2} \begin{bmatrix} 1 & -2 & 3 & | & 0 \\ 0 & 7 & -2 & | & 12 \\ -1 & 1 & -2 & | & -1 \end{bmatrix} \xrightarrow{R_1+R_3 \to \text{New } R_3}$$

$$\begin{aligned} 2x + 3y + 4z &= 12 \\ x - 2y + 3z &= 0 \\ -x + y - 2z &= -1 \end{aligned} \qquad \begin{aligned} x - 2y + 3z &= 0 \\ 2x + 3y + 4z &= 12 \\ -x + y - 2z &= -1 \end{aligned} \qquad \begin{aligned} x - 2y + 3z &= 0 \\ 7y - 2z &= 12 \\ -x + y - 2z &= -1 \end{aligned}$$

$$\begin{bmatrix} 1 & -2 & 3 & | & 0 \\ 0 & 7 & -2 & | & 12 \\ 0 & -1 & 1 & | & -1 \end{bmatrix} \xrightarrow{R_2 \Leftrightarrow R_3} \begin{bmatrix} 1 & -2 & 3 & | & 0 \\ 0 & -1 & 1 & | & -1 \\ 0 & 7 & -2 & | & 12 \end{bmatrix} \xrightarrow{7R_2+R_3 \to \text{New } R_3} \begin{bmatrix} 1 & -2 & 3 & | & 0 \\ 0 & -1 & 1 & | & -1 \\ 0 & 0 & 5 & | & 5 \end{bmatrix}$$ This matrix is now in triangular form. (All 0's below the diagonal.)

$$\begin{aligned} x - 2y + 3z &= 0 \\ 7y - 2z &= 12 \\ -y + z &= -1 \end{aligned} \qquad \begin{aligned} x - 2y + 3z &= 0 \\ -y + z &= -1 \\ 7y - 2z &= 12 \end{aligned} \qquad \begin{aligned} x - 2y + 3z &= 0 \\ -y + z &= -1 \\ 5z &= 5 \end{aligned}$$

Once we have a matrix written in triangular form, we can use the last row to solve for z and then use back substitution as before to solve for the remaining variables.

Just as in Example 1, the solution to this system is the ordered triple $(1, 2, 1)$.

 *My video summary*

⊘ Example 2 Solve a Linear System Using an Augmented Matrix (Triangular Form)

Create an augmented matrix and solve the following linear system using Gaussian elimination by writing an equivalent system in triangular form:

$$\begin{aligned} x + 2y - z &= 3 \\ x - 3y - 2z &= 11 \\ -x - 2y + 2z &= -6 \end{aligned}$$

Solution

The augmented matrix that corresponds to the linear system is

$$\begin{bmatrix} 1 & 2 & -1 & | & 3 \\ 1 & -3 & -2 & | & 11 \\ -1 & -2 & 2 & | & -6 \end{bmatrix}$$

We now proceed to use the following elementary row operations:

$$\begin{bmatrix} 1 & 2 & -1 & | & 3 \\ 1 & -3 & -2 & | & 11 \\ -1 & -2 & 2 & | & -6 \end{bmatrix} \xrightarrow{(-1)R_1+R_2 \to \text{New } R_2} \begin{bmatrix} 1 & 2 & -1 & | & 3 \\ 0 & -5 & -1 & | & 8 \\ -1 & -2 & 2 & | & -6 \end{bmatrix} \xrightarrow{R_1+R_3 \to \text{New } R_3} \begin{bmatrix} 1 & 2 & -1 & | & 3 \\ 0 & -5 & -1 & | & 8 \\ 0 & 0 & 1 & | & -3 \end{bmatrix}$$

The augmented matrix is now in triangular form. Looking at the last row, we see that $z = -3$. The second row corresponds to the equation $-5y - z = 8$. Therefore,

$$-5y - z = 8 \qquad \text{Use the augmented matrix to write equation (2).}$$
$$-5y - (-3) = 8 \qquad \text{Substitute } -3 \text{ for } z.$$
$$-5y + 3 = 8 \qquad \text{Simplify.}$$
$$-5y = 5 \qquad \text{Subtract 3 from both sides.}$$
$$y = -1 \qquad \text{Divide both sides by } -5.$$

Because $z = -3$ and $y = -1$, we can now use the first row to solve for x. You should verify that $x = 2$. Therefore, the solution to this linear system is the ordered triple $(2, -1\ -3)$.

You Try It Work through this You Try It problem.

Work Exercises 5–10 in this textbook or in the MyMathLab Study Plan.

 *My video summary* ⊙ **TRIANGULAR FORM, ROW-ECHELON FORM, AND REDUCED ROW-ECHELON FORM**

When solving a system of three linear equations with three unknowns, only triangular form is needed to find the solution. However, it is often useful to further reduce the augmented matrix into one of two other simpler forms known as row-echelon form or reduced row-echelon form.

We see in Example 3 that the linear system can be reduced into the following triangular form:

Triangular Form

$$\begin{bmatrix} 1 & 2 & -1 & \bigm| & 3 \\ 0 & -5 & -1 & \bigm| & 8 \\ 0 & 0 & 1 & \bigm| & -3 \end{bmatrix}$$

Triangular form only requires zeros below the diagonal.

To reduce this matrix into row-echelon form requires that all zeros remain below the diagonal and that all coefficients along the diagonal are 1. If we multiply the second row of the previous matrix by $-\dfrac{1}{5}$, we get the following equivalent matrix written in row-echelon form:

Row-Echelon Form

Row-echelon form requires zeros below the diagonal . . .
$$\begin{bmatrix} 1 & 2 & -1 & \bigm| & 3 \\ 0 & 1 & \frac{1}{5} & \bigm| & -\frac{8}{5} \\ 0 & 0 & 1 & \bigm| & -3 \end{bmatrix}$$
. . . and 1's down the diagonal.

Reduced row-echelon form requires zeros below *and* above the diagonal and 1's down the diagonal. We can reduce the previous matrix into reduced row-echelon form by performing the following three elementary row operations.

$$\begin{bmatrix} 1 & 2 & -1 & | & 3 \\ 0 & 1 & \frac{1}{5} & | & -\frac{8}{5} \\ 0 & 0 & 1 & | & -3 \end{bmatrix} \xrightarrow{(-2)R_2 + R_1 \to \text{New } R_1} \begin{bmatrix} 1 & 0 & -\frac{7}{5} & | & \frac{31}{5} \\ 0 & 1 & \frac{1}{5} & | & -\frac{8}{5} \\ 0 & 0 & 1 & | & -3 \end{bmatrix} \xrightarrow{\left(-\frac{1}{5}\right)R_3 + R_2 \to \text{New } R_2}$$

$$\begin{bmatrix} 1 & 0 & -\frac{7}{5} & | & \frac{31}{5} \\ 0 & 1 & 0 & | & -1 \\ 0 & 0 & 1 & | & -3 \end{bmatrix} \xrightarrow{\left(\frac{7}{5}\right)R_3 + R_1 \to \text{New } R_1} \begin{bmatrix} 1 & 0 & 0 & | & 2 \\ 0 & 1 & 0 & | & -1 \\ 0 & 0 & 1 & | & -3 \end{bmatrix} \begin{matrix} \Rightarrow x = 2 \\ \Rightarrow y = -1 \\ \Rightarrow z = -3 \end{matrix}$$

As you can see from the final matrix, we have now obtained all zeros above and below the diagonal. This final matrix is in reduced row-echelon form. The advantage to reducing the augmented matrix into reduced row-echelon form is that you can immediately read the solution by looking down the last column. In the previous reduced row-echelon matrix, you can clearly see that the solution is $x = 2$, $y = -1$, and $z = -3$. The process of reducing a system into reduced row-echelon form is called **Gauss-Jordan elimination**. We use Gauss-Jordan elimination from this point forward.

 *My video summary*

⊙ **Example 3** Solve a Linear System Using Gauss-Jordan Elimination

Solve the following system using Gauss-Jordan elimination:

$$x_1 + x_2 + x_3 = -1$$
$$x_1 + 2x_2 + 4x_3 = 3$$
$$x_1 + 3x_2 + 9x_3 = 3$$

Solution The augmented matrix that corresponds to the linear system is

$$\begin{bmatrix} 1 & 1 & 1 & | & -1 \\ 1 & 2 & 4 & | & 3 \\ 1 & 3 & 9 & | & 3 \end{bmatrix}$$

We need to reduce this matrix into reduced row-echelon form. We must have 1s down the diagonal and zeros everywhere else to the left of the vertical bar. We can accomplish this by performing the following seven elementary row operations. Watch the video for a detailed step-by-step explanation.

$$\begin{bmatrix} 1 & 1 & 1 & | & -1 \\ 1 & 2 & 4 & | & 3 \\ 1 & 3 & 9 & | & 3 \end{bmatrix} \xrightarrow{(-1)R_1 + R_2 \to \text{New } R_2} \begin{bmatrix} 1 & 1 & 1 & | & -1 \\ 0 & 1 & 3 & | & 4 \\ 1 & 3 & 9 & | & 3 \end{bmatrix} \xrightarrow{(-1)R_1 + R_3 \to \text{New } R_3}$$

$$\begin{bmatrix} 1 & 1 & 1 & | & -1 \\ 0 & 1 & 3 & | & 4 \\ 0 & 2 & 8 & | & 4 \end{bmatrix} \xrightarrow{(-1)R_2 + R_1 \to \text{New } R_1} \begin{bmatrix} 1 & 0 & -2 & | & -5 \\ 0 & 1 & 3 & | & 4 \\ 0 & 2 & 8 & | & 4 \end{bmatrix} \xrightarrow{(-2)R_2 + R_3 \to \text{New } R_3}$$

$$\begin{bmatrix} 1 & 0 & -2 & | & -5 \\ 0 & 1 & 3 & | & 4 \\ 0 & 0 & 2 & | & -4 \end{bmatrix} \xrightarrow{\left(\frac{1}{2}\right)R_3 \to \text{New } R_3} \begin{bmatrix} 1 & 0 & -2 & | & -5 \\ 0 & 1 & 3 & | & 4 \\ 0 & 0 & 1 & | & -2 \end{bmatrix} \xrightarrow{(-3)R_3 + R_2 \to \text{New } R_2}$$

$$\begin{bmatrix} 1 & 0 & -2 & | & -5 \\ 0 & 1 & 0 & | & 10 \\ 0 & 0 & 1 & | & -2 \end{bmatrix} \xrightarrow{2R_3 + R_1 \to \text{New } R_1} \begin{bmatrix} 1 & 0 & 0 & | & -9 \\ 0 & 1 & 0 & | & 10 \\ 0 & 0 & 1 & | & -2 \end{bmatrix} \begin{aligned} &\Rightarrow x_1 = -9 \\ &\Rightarrow x_2 = 10 \\ &\Rightarrow x_3 = -2 \end{aligned}$$

You can see by looking at the last matrix that $x_1 = -9$, $x_2 = 10$, and $x_3 = -2$. Thus, the solution is the ordered triple $(-9, 10, -2)$.

You Try It   Work through this You Try It problem.

Work Exercises 11–19 in this textbook or in the MyMathLab Study Plan.

OBJECTIVE 3   SOLVING CONSISTENT, DEPENDENT SYSTEMS OF LINEAR EQUATIONS IN THREE VARIABLES

So far in this section, every system of equations that we have encountered had one unique solution. Geometrically, this happens when the three planes described by each equation intersect at exactly one point. Systems of equations that have at least one solution are called **consistent systems**. When a system has *exactly one solution* then the system is called a **consistent, independent system**. It is possible for a system of three linear equations in three variables to have infinitely many solutions. If the system has infinitely many solutions, then the system is called a **consistent, dependent** system. For the two-variable case, a consistent, dependent system can be geometrically represented by a pair of coinciding lines. For the three-variable case, consistent, dependent systems can be geometrically represented by three planes that intersect at a line or by three equations that all describe the same plane. See Figure 9.

Consistent, Dependent
Infinitely many solutions
(Planes intersect at a line.)

Consistent, Dependent
Infinitely many solutions
(Three equations describe the same plane.)

Figure 9 A geometric representation of consistent, dependent systems in three variables.

When using Gauss-Jordan elimination, we can determine that a system of three linear equations is a dependent system if one of the rows of the augmented matrix reduces to a row consisting of all zeros. We illustrate this in the following example.

 *My video summary*

⊙ **Example 4** Solve System of Equations with Infinitely Many Solutions

Use Gauss-Jordan elimination to solve the system:

$$x + 2y + 3z = 10$$

$$x + y + z = 7$$

$$3x + 2y + z = 18$$

Solution Write the system using an augmented matrix, and use a series of elementary row operations to attempt to reduce this matrix into reduced row-echelon form. Watch the video to see each step worked out in detail.

$$\begin{bmatrix} 1 & 2 & 3 & | & 10 \\ 1 & 1 & 1 & | & 7 \\ 3 & 2 & 1 & | & 18 \end{bmatrix} \xrightarrow{(-1)R_1 + R_2 \rightarrow New\ R_2} \begin{bmatrix} 1 & 2 & 3 & | & 10 \\ 0 & -1 & -2 & | & -3 \\ 3 & 2 & 1 & | & 18 \end{bmatrix} \xrightarrow{(-3)R_1 + R_3 \rightarrow New\ R_3}$$

$$\xrightarrow{(-2)R_2 + R_1 \rightarrow New\ R_1}$$

$$\begin{bmatrix} 1 & 2 & 3 & | & 10 \\ 0 & -1 & -2 & | & -3 \\ 0 & -4 & -8 & | & -12 \end{bmatrix} \xrightarrow{(-1)R_2 \rightarrow New\ R_2} \begin{bmatrix} 1 & 2 & 3 & | & 10 \\ 0 & 1 & 2 & | & 3 \\ 0 & -4 & -8 & | & -12 \end{bmatrix}$$

$$\begin{bmatrix} 1 & 0 & -1 & | & 4 \\ 0 & 1 & 2 & | & 3 \\ 0 & -4 & -8 & | & -12 \end{bmatrix} \xrightarrow{4R_2 + R_3 \rightarrow New\ R_3} \begin{bmatrix} 1 & 0 & -1 & | & 4 \\ 0 & 1 & 2 & | & 3 \\ 0 & 0 & 0 & | & 0 \end{bmatrix}$$

Notice that the last row of the final augmented matrix consists entirely of zeros. This corresponds to the identity $0 = 0$, which is true for every value of x, y, and z. The first two rows correspond to the equations

$$x \quad - z = 4$$

$$y + 2z = 3$$

We can solve the first equation for x and the second equation for y to obtain the equations

$$x = 4 + z$$

$$y = 3 - 2z$$

Notice that the variables x and y *depend* on the choice of z. Thus, the system is said to be **dependent**. In this case, the variable z is free to be any real number of our choosing. Consequently, we say that z is a **free variable**. The solutions of this system are all ordered triples of the form $(4 + z, 3 - 2z, z)$. Thus, this system has infinitely many solutions. For example, letting $z = 0$, $z = -1$, and $z = 3$, we get the following ordered triples as solutions to this system:

$$z = 0: \quad (4, 3, 0)$$

$$z = -1: \quad (3, 5, -1)$$

$$z = 3: \quad (7, -3, 3)$$

Try substituting these ordered triples into the original system to verify that all three are solutions. ●

You Try It Work through this You Try It problem.

Work Exercises 20–23 in this textbook or in the My Math Lab **Study Plan.**

OBJECTIVE 4 SOLVING INCONSISTENT SYSTEMS OF LINEAR EQUATIONS IN THREE VARIABLES

Recall from Section 7.1 that a system of linear equations in two variables could have no solution. Linear systems having no solutions are called **inconsistent systems.** For the two-variable case, this occurred if the two linear equations represented parallel lines because parallel lines have no points in common. For the three-variable case, a system is **inconsistent** if all three planes have no points in common. This happens if all three planes are parallel or if two of the planes are parallel. A third possibility of an inconsistent system occurs when the planes intersect two at a time. Figure 10 geometrically illustrates three types of inconsistent systems in three variables.

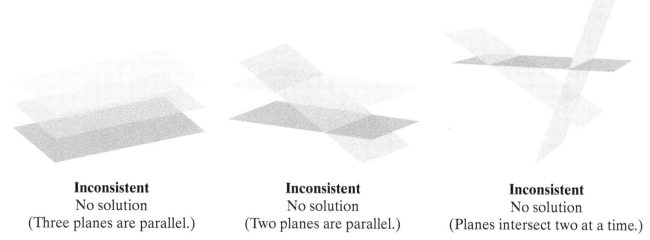

Inconsistent
No solution
(Three planes are parallel.)

Inconsistent
No solution
(Two planes are parallel.)

Inconsistent
No solution
(Planes intersect two at a time.)

Figure 10 A geometric representation of inconsistent systems in three variables.

Recall from **Example 4** of Section 7.1 that when solving an inconsistent system of two equations we will eventually obtain a contradiction. This is also true in the three-variable case. When using Gauss-Jordan elimination, we can determine that a system of three linear equations is an inconsistent system if the three entries on the left-hand side of the augmented matrix reduces to all zeros while the entry on the right-hand side of the augmented matrix is a non-zero constant. We illustrate this in Example 5.

 *My video summary*

⊘ **Example 5** Solve a System of Equations with No Solution

Use Gauss-Jordan elimination to solve the system:

$$x - y + 2z = 4$$
$$-x + 3y + z = -6$$
$$x + y + 5z = 3$$

Solution Write the system using an augmented matrix and use a series of elementary row operations to attempt to reduce this matrix into reduced row-echelon form. Watch the video to see each step worked out in detail.

$$\begin{bmatrix} 1 & -1 & 2 & | & 4 \\ -1 & 3 & 1 & | & -6 \\ 1 & 1 & 5 & | & 3 \end{bmatrix} \xrightarrow{R_1 + R_2 \to \text{New } R_2} \begin{bmatrix} 1 & -1 & 2 & | & 4 \\ 0 & 2 & 3 & | & -2 \\ 1 & 1 & 5 & | & 3 \end{bmatrix} \xrightarrow{(-1)R_1 + R_3 \to \text{New } R_3}$$

$$\begin{bmatrix} 1 & -1 & 2 & | & 4 \\ 0 & 2 & 3 & | & -2 \\ 0 & 2 & 3 & | & -1 \end{bmatrix} \xrightarrow{(-1)R_2 + R_3 \to \text{New } R_3} \begin{bmatrix} 1 & -1 & 2 & | & 4 \\ 0 & 2 & 3 & | & -2 \\ 0 & 0 & 0 & | & 1 \end{bmatrix}$$

Looking at the final augmented matrix, we see that the last row represents the equation $0x + 0y + 0z = 1$ or, simply, $0 = 1$. Obviously, this is a contradiction (zero can never equal one). Therefore, we say that the system is inconsistent and has no solution.

 You Try It Work through this You Try It problem.

Work Exercises 24–26 in this textbook or in the MyMathLab Study Plan.

OBJECTIVE 5 DETERMINING WHETHER A SYSTEM HAS NO SOLUTION OR INFINITELY MANY SOLUTIONS

We have seen an example of a consistent, dependent system (a system having infinitely many solutions) and an example of an inconsistent system (a system having no solution). Recall that we can identify a consistent, dependent system when we encounter an identity during the Gauss-Jordan solution process whereas we can identify an inconsistent system when we encounter a contradiction.

My interactive video summary

⊙ **Example 6** Determine Whether a System Has No Solution or Infinitely Many Solutions

Each augmented matrix in row reduced form is equivalent to the augmented matrix of a system of linear equations in variables $x, y,$ and z. Determine whether the system is a consistent, dependent system or an inconsistent system. If it is a consistent, dependent system, the describe the solution as in Example 4.

$$\textbf{a.} \begin{bmatrix} 1 & 0 & -2 & 5 \\ 0 & 1 & 3 & -2 \\ 0 & 0 & 0 & 0 \end{bmatrix} \qquad \textbf{b.} \begin{bmatrix} 1 & 0 & 0 & -4 \\ 0 & 1 & 0 & 6 \\ 0 & 0 & 0 & 10 \end{bmatrix} \qquad \textbf{c.} \begin{bmatrix} 1 & 0 & 0 & 3 \\ 0 & 1 & -2 & 4 \\ 0 & 0 & 0 & 0 \end{bmatrix}$$

Solution Watch the interactive video to verify that the augmented matrices in parts a and c represent consistent, dependent systems, whereas the augmented matrix in part b represents an inconsistent system. The infinite solutions to the system in part a can be described by all ordered triples of the form $(5 + 2z, -2 - 3z, z)$. The infinite solutions to the system in part c can be described by all ordered triples of the form $(3, 4 + 2z, z)$ or $\left(3, y, \frac{1}{2}y - 2\right)$.

 You Try It Work through this You Try It problem.

Work exercises 27–32 in this textbook or in the MyMathLab Study Plan.

OBJECTIVE 6 SOLVING LINEAR SYSTEMS HAVING FEWER EQUATIONS THAN VARIABLES

In every linear system that we have encountered thus far, the number of equations in the system was exactly the same as the number of variables. However, this does not always have to be the case. For example, suppose we encounter a system of linear equations in three variables that has only two equations.

We can geometrically illustrate such a system by sketching two planes. There are only three possible scenarios for such a system:

1. The two planes can be distinct parallel planes.

2. The two planes can intersect at a straight line.

3. The two planes can coincide; that is, the two equations describe the same plane. (See Figure 11).

Two planes are parallel.
No solution

Two planes intersect at a line.
Infinitely many solutions

Two planes coincide.
Infinitely many solutions

Figure 11

Example 7 Solve a System Having Fewer Equations Than Variables

Solve the linear system using Gauss-Jordan elimination:

$$x + y + z = 1$$
$$2x - 2y + 6z = 10$$

Solution We write the system using an augmented matrix and obtain the following sequence of equivalent matrices.

$$\begin{bmatrix} 1 & 1 & 1 & | & 1 \\ 2 & -2 & 6 & | & 10 \end{bmatrix} \xrightarrow{(-2)R_1 + R_2 \to \text{New } R_2} \begin{bmatrix} 1 & 1 & 1 & | & 1 \\ 0 & -4 & 4 & | & 8 \end{bmatrix} \xrightarrow{\left(-\frac{1}{4}\right)R_2 \to \text{New } R_2}$$

$$\begin{bmatrix} 1 & 1 & 1 & | & 1 \\ 0 & 1 & -1 & | & -2 \end{bmatrix} \xrightarrow{(-1)R_2 + R_1 \to \text{New } R_1} \begin{bmatrix} 1 & 0 & 2 & | & 3 \\ 0 & 1 & -1 & | & -2 \end{bmatrix}$$

The last augmented matrix is in row reduced form. We see that the original system is equivalent to the system

$$x + 2z = 3$$
$$y - z = -2$$

We can solve for x and y in terms of z. Thus, z is a free variable, and we find that

$$x = 3 - 2z$$
$$y = z - 2$$
$$z = Free$$

The solution can be written in the form $(3 - 2z, z - 2, z)$.

You Try It Work through this You Try It problem.

Work exercises 33–36 in this textbook or in the MyMathLab **Study Plan.**

OBJECTIVE 7   SOLVING APPLIED PROBLEMS USING A SYSTEM OF LINEAR EQUATIONS
INVOLVING THREE VARIABLES

Systems of linear equations in three variables and Gauss-Jordan elimination can
be used to solve a wide variety of applications. Following are two examples.

My video summary   ⊘ **Example 8** Buy Clothes Online

Wendy ordered 30 T-shirts online for her three children. The small T-shirts cost
$4 each, the medium T-shirts cost $5 each, and the large T-shirts were $6 each.
She spent $40 more purchasing the large T-shirts than the small T-shirts, Wendy's
total bill was $154. How many T-shirts of each size did she buy?

Solution Let S, M, and L represent the number of small, medium, and large
T-shirts, respectively. Because a total of 30 T-shirts were purchased, we obtain the
first equation

$$S + M + L = 30$$

Because small T-shirts cost $4, we know that the total amount spent on small T-shirts
is $4S$. Similarly, the total spent on medium T-shirts is $5M$, and the total spent on large
T-shirts is $6L$. Therefore, because the total value of the T-shirts was $154, we obtain
the second equation

$$4S + 5M + 6L = 154$$

Finally, Wendy spent $40 more buying large T-shirts than small T-shirts, so $6L = 4S + 40$,
which is equivalent to the equation

$$4S - 6L = -40$$

Thus, we obtain the system
$$\begin{aligned} S + M + L &= 30 \\ 4S + 5M + 6L &= 154. \\ 4S \quad\;\; - 6L &= -40 \end{aligned}$$

We now rewrite this system using the corresponding augmented matrix. Note
that we must place a 0 in the middle column of the last row of the augmented
matrix to represent the coefficient of the variable M of the third equation.

$$\begin{bmatrix} 1 & 1 & 1 & | & 30 \\ 4 & 5 & 6 & | & 154 \\ 4 & 0 & -6 & | & -40 \end{bmatrix}$$

My video summary   ⊘ Watch the video to see how to use Gauss-Jordan elimination to rewrite the
matrix in the following reduced row-echelon form:

$$\begin{bmatrix} 1 & 0 & 0 & | & 8 \\ 0 & 1 & 0 & | & 10 \\ 0 & 0 & 1 & | & 12 \end{bmatrix}$$

Looking down the last column of this augmented matrix, we see that Wendy
purchased 8 small T-shirts, 10 medium T-shirts, and 12 large T-shirts.

You Try It   Work through this You Try It problem.

Work exercises 37–41 in this textbook or in the MyMathLab **Study Plan.**

It takes two points to determine the equation of a line. Similarly, it takes three non-collinear points to determine the equation of the quadratic function whose graph passes through the points. Example 6 shows how to find the equation of a quadratic function given three points that lie on its graph. You may want to review quadratic functions from Section 4.1.

Example 9 Use Three Points to Determine a Quadratic Function

Determine the quadratic function whose graph passes through the three points $(1, -9)$, $(-1, -5)$, and $(-3, 7)$.

Solution A quadratic function has the form $f(x) = ax^2 + bx + c$. The points $(1, -9)$, $(-1, -5)$, and $(-3, 7)$ lie on the graph of f so

$$f(1) = a(1)^2 + b(1) + c \quad = -9 \longrightarrow a + b + c = -9$$

$$f(-1) = a(-1)^2 + b(-1) + c = -5 \longrightarrow a - b + c = -5$$

$$f(3) = a(-3)^2 + b(-3) + c = \quad 7 \longrightarrow 9a - 3b + c = \quad 7$$

Therefore, we obtain the system and corresponding augmented matrix:

System of Equations   Corresponding Augmented Matrix

$$\begin{aligned} a + b + c &= -9 \\ a - b + c &= -5 \\ 9a - 3b + c &= 7 \end{aligned} \qquad \begin{bmatrix} 1 & 1 & 1 & -9 \\ 1 & -1 & 1 & -5 \\ 9 & -3 & 1 & 7 \end{bmatrix}$$

Using Gauss-Jordan elimination, we can rewrite this matrix in the following reduced row-echelon form. (View these steps to see the entire solution.)

$$\begin{bmatrix} 1 & 0 & 0 & 1 \\ 0 & 1 & 0 & -2 \\ 0 & 0 & 1 & -8 \end{bmatrix}$$

We can see from the last column that $a = 1$, $b = -2$, and $c = -8$. Therefore, the quadratic function that passes through the points $(1, -9)$, $(-1, -5)$, and $(-3, 7)$ is $f(x) = x^2 - 2x - 8$.

You can see in Figure 12 that the graph of $f(x) = x^2 - 2x - 8$ passes through the three given points.

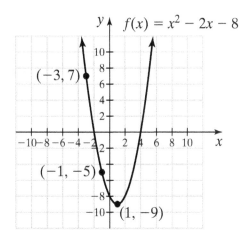

Figure 12
Graph of the quadratic function $f(x) = x^2 - 2x - 8$

You Try It   Work through this You Try It problem.

Work exercises 42 and 43 in this textbook or in the MyMathLab Study Plan.

7.3 Exercises

In Exercises, 1–4 use Gaussian elimination to solve each linear system by finding an equivalent system in triangular form.

1. $\begin{aligned} x + y + z &= 4 \\ 2x - y - 2z &= -10 \\ -x - y + 3z &= 8 \end{aligned}$

2. $\begin{aligned} x - 2y + z &= 6 \\ 2x + y - 3z &= -3 \\ x - 3y + 3z &= 10 \end{aligned}$

3. $\begin{aligned} -4x_1 + 5x_2 + 9x_3 &= -9 \\ x_1 - 2x_2 + x_3 &= 0 \\ 2x_2 - 8x_3 &= 8 \end{aligned}$

4. $\begin{aligned} 2x_1 + 2x_2 + x_3 &= 9 \\ x_1 + x_3 &= 4 \\ 4x_2 - 3x_3 &= 17 \end{aligned}$

In Exercises 5–10, use Gaussian elimination and matrices to solve each system of linear equations. Write your final augmented matrix in triangular form and then solve for each variable using back substitution.

5. $\begin{aligned} x + y + z &= 0 \\ x - 2y + 3z &= -7 \\ -x - y + 4z &= 6 \end{aligned}$

6. $\begin{aligned} 2x - y + z &= -6 \\ x + y - z &= 0 \\ -3x - 3y + 4z &= 1 \end{aligned}$

7. $\begin{aligned} x_1 + 2x_2 + 3x_3 &= 11 \\ 3x_1 + 8x_2 + 5x_3 &= 27 \\ -x_1 + x_2 + 2x_3 &= 2 \end{aligned}$

8. $\begin{aligned} 2x_1 + 3x_3 &= 13 \\ -x_1 + 3x_2 &= -8 \\ 2x_1 - x_2 + 4x_3 &= 21 \end{aligned}$

9. $\begin{aligned} x_1 + x_2 + 10x_3 &= 3 \\ 3x_1 - 6x_2 + 6x_3 &= -5 \\ -2x_1 + 3x_2 + 5x_3 &= \frac{7}{5} \end{aligned}$

10. $\begin{aligned} x_1 - 2x_2 + 2x_3 &= 2 \\ 3x_1 + 2x_2 - 2x_3 &= -1 \\ x_1 - x_2 + 2x_3 &= 0 \end{aligned}$

In Exercises 11–26 use Gauss-Jordan elimination to solve each linear system and determine whether the system has a unique solution, no solution, or infinitely many solutions. If the system has infinitely many solutions, describe the solution as an ordered triple involving a free variable.

11. $\begin{aligned} x - y &= 7 \\ y - z &= 2 \\ x + z &= 1 \end{aligned}$

12. $\begin{aligned} x - 2y + 3z &= 0 \\ 4x - 6y + 8z &= -4 \\ -2x - y + z &= 7 \end{aligned}$

13. $\begin{aligned} -x_1 + 2x_2 - x_3 &= -15 \\ 3x_1 + x_2 + 4x_3 &= 34 \\ 2x_1 - 4x_2 + 3x_3 &= 40 \end{aligned}$

14. $\begin{aligned} -2x - y + 3z &= 0 \\ 4x + 3y + z &= -1 \\ x + y - 2z &= -\frac{1}{2} \end{aligned}$

15. $\begin{aligned} x_1 - x_3 &= 1 \\ -x_1 + 2x_2 - 3x_3 &= 12 \\ 2x_1 - 4x_2 &= -6 \end{aligned}$

16. $\begin{aligned} 2x_1 + x_2 + 6x_3 &= 1 \\ -x_1 - 3x_3 &= -2 \\ 5x_1 + 4x_2 &= -7 \end{aligned}$

17. $\begin{aligned} 2x - y + z &= -6 \\ x + y - z &= 0 \\ -3x - 3y + 4z &= 1 \end{aligned}$

18. $\begin{aligned} -x + 2y - z &= -15 \\ 3x + y + 4z &= 34 \\ 2x - 4y + 3z &= 40 \end{aligned}$

19. $\begin{aligned} x - 2y + 3z &= 0 \\ 4x - 6y + 8z &= -4 \\ -2x - y + z &= 7 \end{aligned}$

20. $\begin{aligned} x + y + z &= 3 \\ 2x + y - z &= 4 \\ 3x + y - 3z &= 5 \end{aligned}$

21. $\begin{aligned} x + y - z &= 2 \\ x + 2y - 3z &= 6 \\ 2x - 3y + 8z &= -16 \end{aligned}$

22. $\begin{aligned} x + y + z &= 2 \\ 4x + 2y + 3z &= 9 \\ 3x - y + z &= 8 \end{aligned}$

23. $\begin{aligned} x + 4y + 2z &= 5 \\ 3x + 7y + z &= 0 \\ 2x + 5y + z &= 1 \end{aligned}$

$$\begin{aligned} x - 3y + 5z &= 6 \\ \textbf{24. } x - 4y + 13z &= 16 \\ x - 2y - 3z &= -5 \end{aligned} \qquad \begin{aligned} x + y - z &= 3 \\ \textbf{25. } x + 2y + 2z &= 8 \\ 2x + y - 5z &= 2 \end{aligned} \qquad \begin{aligned} x + y + z &= 2 \\ \textbf{26. } 3x + y - 2z &= 8 \\ 2x - y - 8z &= 5 \end{aligned}$$

In Exercises 27–32, each augmented matrix in row reduced form is equivalent to the augmented matrix of a system of linear equations in variables x, y, and z. Determine whether the system is dependent or inconsistent. If the system is dependent, determine which variable is free and describe the solution as an ordered triple in terms of the free variable. See Example 6.

27. $\left[\begin{array}{ccc|c} 1 & 0 & 0 & 6 \\ 0 & 1 & 0 & 5 \\ 0 & 0 & 0 & -2 \end{array}\right]$ **28.** $\left[\begin{array}{ccc|c} 1 & 0 & -2 & -5 \\ 0 & 1 & 1 & 4 \\ 0 & 0 & 0 & 0 \end{array}\right]$ **29.** $\left[\begin{array}{ccc|c} 1 & 3 & 0 & 5 \\ 0 & 1 & 1 & 8 \\ 0 & 0 & 0 & 0 \end{array}\right]$

30. $\left[\begin{array}{ccc|c} 1 & 1 & 2 & 8 \\ 0 & 0 & 0 & 2 \\ 0 & 0 & 1 & 5 \end{array}\right]$ **31.** $\left[\begin{array}{ccc|c} 1 & 2 & 0 & 1 \\ 0 & 0 & 0 & 0 \\ 0 & 0 & 1 & -4 \end{array}\right]$ **32.** $\left[\begin{array}{ccc|c} 1 & 0 & 2 & -6 \\ 0 & 1 & -3 & 8 \\ 0 & 0 & 0 & 0 \end{array}\right]$

In Exercises 33–36, use Gauss-Jordan elimination to solve each linear system and determine whether the system has a unique solution, no solution, or infinitely many solutions. If the system has infinitely many solutions, describe the solution as an ordered triple involving a free variable.

33. $\begin{aligned} x + 2y - z &= 11 \\ x + 3y - 2z &= 14 \end{aligned}$ **34.** $\begin{aligned} x - 2y + z &= -2 \\ -3x + 6y + 3z &= -6 \end{aligned}$ **35.** $\begin{aligned} x + y + 5z &= 11 \\ 3x + y - z &= 7 \end{aligned}$

36. $\begin{aligned} x - y + z &= 5 \\ 3x + 2y - 2z &= 0 \end{aligned}$

In Exercises 37–43, write a system of linear equations in three variables and then use matrices to solve the system using Gaussian elimination or Gauss-Jordan elimination.

37. Ben was in charge of ordering 35 pizzas for the office party. He ordered three types of pizzas: cheese, supreme, and pepperoni. The cheese pizzas cost $9 each, the pepperoni pizzas cost $12 each, and the supreme pizzas cost $15 each. He spent exactly twice as much on the pepperoni pizzas as he did on the cheese pizzas. If Ben spent a total of $420 on pizza, how many pizzas of each type did he buy?

38. Nine hundred forty nine tickets were sold on opening night of the play "The Music Man" at the Lewiston Civic Theater. The total receipts for the performance were $10,967. Adult tickets cost $14 apiece, children tickets cost $11 apiece, and senior citizen tickets cost $8 apiece. If the combined number of adult and children tickets exceeded twice the number of senior citizen tickets by 151, determine how many tickets of each type were sold.

39. Gary and Larry's Tire Shop sells three types of tires: high performance, ultra performance, and extreme performance. High-performance tires sell for $50/tire. Ultra performance tires sell for $60/tire. Extreme performance tires sell for $80/tire. Last week, the revenue generated by the sale of ultra performance tires exceeded the revenue from the sale of extreme performance tires by $160. If the tire shop sold a total of 238 tires last week for a total revenue of $14,440, how many of each type of tire were sold?

40. Amanda invested a total of $2,200 into three separate accounts that pay 6%, 8%, and 9% annual interest. Amanda has three times as much invested in the account that pays 9% as she does in the account that pays 6%. If the total interest for the year is $178, how much did Amanda invest at each rate?

41. Adam and Murph entered the inaugural Joppetown triathlon, which consisted of running, biking, and swimming. Adam averaged 8 mph during the running segment, 2 mph during the swimming segment, and 15 mph during the biking segment. Murph averaged 6 mph running, 3 mph swimming, and 20 mph biking. Adam's total time was 5 hours and 25 minutes. Murph's total time was 4 hours and 35 minutes. If the total of all three segments is 40 miles, how long was each segment?

42. Determine the quadratic function whose graph passes through the three points $(0, 0)$, $(-4, 36)$, and $(4, 60)$.

43. Determine the quadratic function whose graph passes through the three points $(-1, 2)$, $(3, 10)$, and $(-2, 15)$.

7.4 Partial Fraction Decomposition

THINGS TO KNOW

Before working through this section, be sure that you are familiar with the following concepts:

| | VIDEO | ANIMATION | INTERACTIVE |

You Try It
1. Solving a System of Linear Equations Using the Substitution Method (Section 7.1) ⊘

You Try It
2. Solving a System of Linear Equations Using the Elimination Method (Section 7.1) ⊘

You Try It
3. Solving a System of Linear Equations Using Gaussian Elimination (Section 7.2) ⊘

OBJECTIVES

1 Decomposing Rational Expressions of the Form $\dfrac{P(x)}{Q(x)}$, where $Q(x)$ has Only Distinct Linear Factors

2 Decomposing Rational Expressions of the Form $\dfrac{P(x)}{Q(x)}$, where $Q(x)$ has a Repeated Linear Factor

3 Decomposing Rational Expressions of the Form $\dfrac{P(x)}{Q(x)}$, where $Q(x)$ has a Distinct Prime Quadratic Factor

4 Decomposing Rational Expressions of the Form $\dfrac{P(x)}{Q(x)}$, where $Q(x)$ has a Repeated Prime Quadratic Factor

Introduction to Section 7.4

Recall that a rational expression in one variable is of the form $\frac{P(x)}{Q(x)}$, where $P(x)$ and $Q(x)$ are polynomials such that $Q(x) \neq 0$. When combining two or more rational expressions into a single rational expression, we must first determine the lowest common denominator (LCD). For example, suppose that we wish to subtract the rational expression $\frac{7}{x-1}$ from the rational expression $\frac{1}{x+3}$. The LCD is the product $(x-1)(x+3)$. The result is shown below.

$$\frac{1}{x+3} - \frac{7}{x-1} = \frac{1(x-1) - 7(x+3)}{(x+3)(x-1)} = \frac{x-1-7x-21}{(x+3)(x-1)} = \frac{-6x-22}{x^2+2x-3}$$

In this section, we will attempt to reverse this procedure. That is, we will start with a single rational expression and write it as the sum or difference of two or more simpler rational expressions. This process is called **partial fraction decomposition**. Using the example from above, we say that the partial fraction decomposition of the expression $\frac{-6x-22}{x^2+2x-3}$ is $\frac{1}{x+3} - \frac{7}{x-1}$. The two fractions $\frac{1}{x+3}$ and $\frac{-7}{x-1}$ are called **partial fractions**. Partial fraction decomposition is a procedure that is especially useful and often necessary in calculus.

The partial fraction decomposition of a rational expression of the form $\frac{P(x)}{Q(x)}$ can be performed when the following two criteria are satisfied:

1. The polynomials $P(x)$ and $Q(x)$ share no common factors.

2. The degree of $P(x)$ is less than the degree of $Q(x)$.

In all examples and exercises in this section, the two criteria above are always satisfied. We start by introducing rational expressions that involve distinct linear factors within the denominator.

OBJECTIVE 1 DECOMPOSING RATIONAL EXPRESSIONS OF THE FORM $\frac{P(x)}{Q(x)}$, WHERE $Q(x)$ HAS ONLY DISTINCT LINEAR FACTORS

Recall that a linear factor is an expression of the form $ax + b$ where $a \neq 0$. If the denominator of $\frac{P(x)}{Q(x)}$ is the product of n distinct linear factors, then the partial fraction decomposition of $\frac{P(x)}{Q(x)}$ will be of the form

$$\frac{P(x)}{Q(x)} = \frac{A_1}{a_1 x + b_1} + \frac{A_2}{a_2 x + b_2} + \cdots + \frac{A_n}{a_n x + b_n}$$

where $A_1, A_2, \ldots, A_n$ are constants to be determined.

We start with an example of a rational expression that contains two distinct linear factors in the denominator.

Example 1 Decomposing a Rational Expression in which the Denominator has Only Distinct Linear Factors

Determine the partial fraction decomposition of $\dfrac{x + 10}{2x^2 + 5x - 3}$.

Solution Start by factoring the denominator: $2x^2 + 5x - 3 = (2x - 1)(x + 3)$. We see that the denominator is the product of two distinct linear factors. Therefore, we introduce the two partial fractions $\dfrac{A}{2x - 1}$ and $\dfrac{B}{x + 3}$ to create the following equation:

$$\frac{x + 10}{(2x - 1)(x + 3)} = \frac{A}{2x - 1} + \frac{B}{x + 3}$$

Note that for convenience we use A and B to represent the unknown constants instead of A_1 and A_2.

To eliminate the denominators, multiply both sides of the rational equation by $(2x - 1)(x + 3)$.

$(2x - 1)(x + 3)\dfrac{x + 10}{(2x - 1)(x + 3)} = \left(\dfrac{A}{2x - 1} + \dfrac{B}{x + 3}\right)(2x - 1)(x + 3)$ Multiply both sides by $(2x - 1)(x + 3)$.

$\cancel{(2x - 1)}\cancel{(x + 3)}\dfrac{x + 10}{\cancel{(2x - 1)}\cancel{(x + 3)}} = \dfrac{A}{\cancel{(2x - 1)}}\cancel{(2x - 1)}(x + 3) + \dfrac{B}{\cancel{x + 3}}(2x - 1)\cancel{(x + 3)}$ Use the distributive property and cancel common factors.

$x + 10 = A(x + 3) + B(2x - 1)$ Write the resulting new equation.

We can now write the polynomial on the right-hand side in descending order by collecting like terms.

$x + 10 = Ax + 3A + 2Bx - B$ Use the distributive property.

$x + 10 = Ax + 2Bx + 3A - B$ Rearrange the terms.

$x + 10 = (A + 2B)x + (3A - B)$ Factor out an x. The polynomial on the right-hand side is now written in descending order.

Note that the polynomial on the left-hand side must be equivalent to the polynomial on the right-hand side. Thus, the corresponding coefficients must be equivalent. Therefore, the coefficient of the x-term on the left-hand side, 1, is equivalent to the coefficient of the x-term on the right-hand side, $A + 2B$.

$$1x + 10 = (A + 2B)x + (3A - B)$$
$$A + 2B = 1$$

Similarly, the coefficient of the constant term on the left-hand side, 10, is equivalent to the coefficient of the constant term on the right-hand side, $3A - B$.

$$x + 10 = (A + 2B)x + (3A - B)$$
$$3A - B = 10$$

The result of equating the coefficients of the left-hand side with the corresponding coefficients on the right-hand side leads to the following system of equations:

$$A + 2B = 1$$

$$3A - B = 10$$

We can solve the system of equations using the method of substitution or the method of elimination to obtain $A = 3$ and $B = -1$. Therefore,

$$\frac{x + 10}{(2x - 1)(x + 3)} = \frac{3}{2x - 1} - \frac{1}{x + 3}.$$

Note It is often possible to determine some or all of the constants of the partial fraction decomposition using a short-cut method that involves choosing appropriate values of x and substituting those values into the equation found after eliminating the denominators. We choose values of x that will make one of the linear factors equal to zero. This may result in the elimination of one or more of the variables that represent the undetermined constants. Using the equation $x + 10 = A(x + 3) + B(2x - 1)$ found in Example 1, we see that the linear factor $(x + 3)$ is equal to zero when $x = -3$. Thus, we can substitute $x = -3$ into the equation to eliminate the variable A to get the following:

$x + 10 = A(x + 3) + B(2x - 1)$	Start with the equation found after eliminating the denominators.
$-3 + 10 = A(-3 + 3) + B(2(-3) - 1)$	Substitute $x = -3$ to eliminate terms involving the variable A.
$7 = -7B$	Simplify.
$-1 = B$	Solve for B.

Similarly, we can substitute $x = \dfrac{1}{2}$ to get the following:

$x + 10 = A(x + 3) + B(2x - 1)$	Start with the equation found after eliminating the denominators.
$\dfrac{1}{2} + 10 = A\left(\dfrac{1}{2} + 3\right) + B\left(2\left(\dfrac{1}{2}\right) - 1\right)$	Substitute $x = \dfrac{1}{2}$ to eliminate terms involving the variable B.
$\dfrac{21}{2} = \dfrac{7}{2}A$	Simplify.
$3 = A$	Solve for A.

The values of $A = 3$ and $B = -1$ were precisely the values found in the solution to Example 1.

We now establish a seven-step procedure for determining the partial fraction decomposition of a rational expression of the form $\dfrac{P(x)}{Q(x)}$.

Steps for Determining the Partial Fraction Decomposition of $\dfrac{P(x)}{Q(x)}$

Step 1. Factor $Q(x)$.

Step 2. Set up an equation with $\dfrac{P(x)}{Q(x)}$ on the left-hand side and the correct partial fractions with the undetermined constants on the right-hand side.

Step 3. Multiply both sides of the equation created in step 2 by $Q(x)$ to eliminate all denominators.

Step 4*. If possible, choose appropriate values of x to readily solve for one or more undetermined constants. This may conclude the partial fraction decomposition procedure. If not, go on to step 5.

Step 5. Write the polynomial on the right-hand side of the equation found in step 3 in descending order.

Step 6. Equate the coefficients of $P(x)$ with the corresponding coefficients of the polynomial on the right-hand side of the equation, thus creating a system of equations.

Step 7. Solve the system of equations.

*Note that although step 4 usually simplifies the partial fraction decomposition process, it is not necessary.

 *My video summary*

⊘ **Example 2** Decomposing a Rational Expression in which the Denominator has Only Distinct Linear Factors

Determine the partial fraction decomposition of $\dfrac{5x^2 - x + 1}{x^3 + 3x^2 - 4x}$.

Solution

Step 1. Factor the denominator: $x^3 + 3x^2 - 4x = x(x - 1)(x + 4)$. We see that the denominator is the product of three distinct linear factors.

Step 2. Introduce the three partial fractions $\dfrac{A}{x}, \dfrac{B}{x - 1}$, and $\dfrac{C}{x + 4}$ to create the following equation:

$$\frac{5x^2 - x + 1}{x^3 + 3x^2 - 4x} = \frac{A}{x} + \frac{B}{x - 1} + \frac{C}{x + 4}$$

Step 3. Multiply both sides of the equation from Step 2 by $x(x - 1)(x + 4)$ to get

$$5x^2 - x + 1 = A(x - 1)(x + 4) + Bx(x + 4) + Cx(x - 1)$$

Step 4. We can choose $x = 0, x = 1$, and $x = -4$ to solve for each constant.

$$x = 0: 5(0)^2 - (0) + 1 = A(0 - 1)(0 + 4) + B(0)(0 + 4) + C(0)(0 - 1)$$

$$1 = -4A$$

$$-\frac{1}{4} = A$$

$$x = 1: 5(1)^2 - (1) + 1 = A(1 - 1)(1 + 4) + B(1)(1 + 4) + C(1)(1 - 1)$$

$$5 = 5B$$

$$1 = B$$

$$x = -4: 5(-4)^2 - (-4) + 1 = A(-4 - 1)(-4 + 4) + B(-4)(-4 + 4) + C(-4)(-4 - 1)$$

$$85 = 20C$$

$$\frac{17}{4} = C$$

Therefore, $A = -\frac{1}{4}$, $B = 1$, and $C = \frac{17}{4}$. All constants have been determined so there is no need to go on to step 5. Thus, the partial fraction decomposition is

$$\frac{5x^2 - x + 1}{x^3 + 3x^2 - 4x} = \frac{-\frac{1}{4}}{x} + \frac{1}{x - 1} + \frac{\frac{17}{4}}{x + 4}.$$

My video summary ⊘ Watch this video to see every step of this solution.

You Try It Work through this You Try It problem.

Work Exercises 1–14 in this textbook or in the MyMathLab Study Plan.

OBJECTIVE 2 DECOMPOSING RATIONAL EXPRESSIONS OF THE FORM $\frac{P(x)}{Q(x)}$, WHERE $Q(x)$ HAS A REPEATED LINEAR FACTOR

A repeated linear factor is of the form $(ax + b)^n$, where n is an integer such that $n \geq 2$. If the denominator of $\frac{P(x)}{Q(x)}$ has a repeated linear factor, then for every repeated linear factor of the form $(ax + b)^n$, we introduce n partial fractions of the form

$$\frac{A_1}{ax + b} + \frac{A_2}{(ax + b)^2} + \cdots + \frac{A_n}{(ax + b)^n}$$

Example 3 Setting up the Partial Fraction Decomposition of $\frac{P(x)}{Q(x)}$, where $Q(x)$ has Repeated Linear Factors

Set up the partial fraction decomposition for $\dfrac{x}{x^2(3x - 1)^3(x - 5)}$.

Do not solve for the constants.

Solution The denominator has two repeated linear factors, x^2 and $(3x - 1)^3$, and one distinct linear factor, $(x - 5)$. The correct partial fraction decomposition is of the form

$$\frac{x}{x^2(3x - 1)^3(x - 5)} = \frac{A}{x} + \frac{B}{x^2} + \frac{C}{3x - 1} + \frac{D}{(3x - 1)^2} + \frac{E}{(3x - 1)^3} + \frac{F}{x - 5},$$

where $A, B, C, D, E,$ and F are constants to be determined.

Notice that for convenience we use $A, B, C, \ldots$ instead of $A_1, A_2, A_3, \ldots$.

Example 4 illustrates how to follow the steps for determining the partial fraction decomposition of $\dfrac{P(x)}{Q(x)}$, where $Q(x)$ has a repeated linear factor.

My video summary ⊙ **Example 4** Decomposing a Rational Expression in which the Denominator Has a Repeated Linear Factor

Determine the partial fraction decomposition of $\dfrac{x-1}{x(x-2)^2}$.

Solution

Step 1. The denominator is already factored.

Step 2. Introduce the three partial fractions $\dfrac{A}{x}, \dfrac{B}{x-2},$ and $\dfrac{C}{(x-2)^2}$ to create the following equation:

$$\frac{x-1}{x(x-2)^2} = \frac{A}{x} + \frac{B}{x-2} + \frac{C}{(x-2)^2}$$

Step 3. Multiply both sides of the rational equation from step 2 by $x(x-2)^2$ to obtain the equation

$$x - 1 = A(x-2)^2 + Bx(x-2) + Cx$$

Step 4. We can solve for C by choosing $x = 2$.

$$x = 2: \quad 2 - 1 = A(2-2)^2 + B(2)(2-2) + C(2)$$

$$1 = 2C$$

$$\frac{1}{2} = C$$

Step 5. Write the polynomial on the right-hand side of the equation found in step 3 in descending order by collecting like terms. View these steps to see that we can write the equation found in step 3 as

$$x - 1 = (A + B)x^2 + (-4A - 2B + C)x + 4A$$

Step 6. The coefficient of the x^2-term on the left-hand side, 0, is equivalent to the coefficient of the x^2-term on the right-hand side, $A + B$.

$$0x^2 + x - 1 = (A + B)x^2 + (-4A - 2B + C)x + 4A$$

$$A + B = 0$$

The coefficient of the x-term on the left-hand side, 1, is equivalent to the coefficient of the x-term on the right-hand side, $-4A - 2B + C$.

$$1x - 1 = (A + B)x^2 + (-4A - 2B + C)x + 4A$$

$$-4A - 2B + C = 1$$

The coefficient of the constant term on the left-hand side, -1, is equivalent to the coefficient of the constant term on the right-hand side, $4A$.

$$x - 1 = (A + B)x^2 + (-4A - 2B + C)x + 4A$$

The result of equating the coefficients of the left-hand side with the corresponding coefficients on the right-hand side leads to the following system of equations:

$$A + B \quad\quad = 0$$

$$-4A - 2B + C = 1$$

$$4A \quad\quad\quad = -1$$

Step 7. We know from step 4 that $C = \dfrac{1}{2}$. Solving the third equation for A gives $A = -\dfrac{1}{4}$. Substituting $A = -\dfrac{1}{4}$ into the first equation, we get $B = \dfrac{1}{4}$. Therefore,

$$\frac{x - 1}{x(x - 2)^2} = \frac{-\dfrac{1}{4}}{x} + \frac{\dfrac{1}{4}}{x - 2} + \frac{\dfrac{1}{2}}{(x - 2)^2}.$$

Watch this video to see every step of this partial fraction decomposition process.

You Try It Work through this You Try It problem.

Work Exercises 15–26 in this textbook or in the MyMathLab Study Plan.

OBJECTIVE 3 DECOMPOSING RATIONAL EXPRESSIONS OF THE FORM $\dfrac{P(x)}{Q(x)}$, WHERE $Q(x)$ HAS A DISTINCT PRIME QUADRATIC FACTOR

A quadratic factor has the form $ax^2 + bx + c$, where $a \neq 0$. If the quadratic factor will not factor into the product of two linear factors using integer coefficients, then the quadratic factor is prime. If the denominator of $\dfrac{P(x)}{Q(x)}$ has a prime quadratic factor, then for every prime quadratic factor of the form $ax^2 + bx + c$, we introduce a partial fraction of the form

$$\frac{A_1 x + B_1}{ax^2 + bx + c}.$$

Note For every linear factor of $Q(x)$, we introduced a partial fraction containing a constant over the linear factor, $\dfrac{A_1}{ax + b}$. For every prime quadratic factor of $Q(x)$, we now introduce a partial fraction containing a linear factor over the prime quadratic factor, $\dfrac{A_1 x + B_1}{ax^2 + bx + c}$.

Example 5 Setting up the Partial Fraction Decomposition of $\dfrac{P(x)}{Q(x)}$, where $Q(x)$ has Distinct Prime Quadratic Factors

Set up the partial fraction decomposition for $\dfrac{5x^3 - 7x^2 - 8x + 1}{(x^2 + 1)(2x^2 + x + 7)(5x - 3)^2}$. Do not solve for the constants.

Solution The denominator has two prime quadratic factors, $x^2 + 1$ and $2x^2 + x + 7$, and one repeated linear factor of the form $(5x - 3)^2$. The correct partial fraction decomposition is of the form $\dfrac{5x^3 - 7x^2 - 8x + 1}{(x^2 + 1)(2x^2 + x + 7)(3x - 1)^2} =$ $\dfrac{Ax + B}{x^2 + 1} + \dfrac{Cx + D}{2x^2 + x + 7} + \dfrac{E}{5x - 3} + \dfrac{F}{(5x - 3)^2}$ where $A, B, C, D, E,$ and F are constants to be determined.

Notice that for convenience we use $A, B, C, D, \ldots$ instead of $A_1, B_1, A_2, B_2, \ldots$. ●

My video summary ⊙ **Example 6** Decomposing a Rational Expression in which the Denominator has a Prime Quadratic Factor

Determine the partial fraction decomposition of $\dfrac{8x^2 + 7}{x^3 - 1}$.

Solution

Step 1. The denominator is the difference of two cubes and factors as $x^3 - 1 = (x - 1)(x^2 + x + 1)$. The quadratic factor, $x^2 + x + 1$, does not factor with integer coefficients and is thus a prime quadratic factor.

Step 2. Introduce the two partial fractions $\dfrac{A}{x - 1}$ and $\dfrac{Bx + C}{x^2 + x + 1}$ to create the following equation:

$$\frac{8x^2 + 7}{x^3 - 1} = \frac{A}{x - 1} + \frac{Bx + C}{x^2 + x + 1}$$

Step 3. Multiply both sides of the rational equation from step 2 by $(x - 1)(x^2 + x + 1)$ to obtain the equation

$$8x^2 + 7 = A(x^2 + x + 1) + (Bx + C)(x - 1)$$

Step 4. We can solve for A by choosing $x = 1$.

$$x = 1: 8(1)^2 + 7 = A((1)^2 + 1 + 1) + (B(1) + C)(1 - 1)$$

$$15 = 3A$$

$$5 = A$$

Step 5. Write the polynomial on the right-hand side of the equation found in step 3 in descending order by collecting like terms. View these steps to see that we can write the equation found in step 3 as

$$8x^2 + 7 = (A + B)x^2 + (A - B + C)x + (A - C)$$

Step 6. Equating the coefficients of $8x^2 + 7$ with the corresponding coefficients of $(A + B)x^2 + (A - B + C)x + (A - C)$ leads to the following system of equations:

$$A + B = 8$$

$$A - B + C = 0$$

$$A - C = 7$$

Step 7. Using the fact that $A = 5$, we can use the first equation to get $B = 3$. We can then use the second or third equation to get $C = -2$.

Thus, the partial fraction decomposition is $\dfrac{8x^2 + 7}{x^3 - 1} = \dfrac{5}{x - 1} + \dfrac{3x - 2}{x^2 + x + 1}$.

You Try It Work through this You Try It problem.

Work Exercises 27–35 in this textbook or in the MyMathLab Study Plan.

OBJECTIVE 4 DECOMPOSING RATIONAL EXPRESSIONS OF THE FORM $\dfrac{P(x)}{Q(x)}$, WHERE $Q(x)$ HAS A REPEATED PRIME QUADRATIC FACTOR

A repeated prime quadratic factor has the form $(ax^2 + bx + c)^n$, where $a \neq 0$ and where n is an integer such that $n \geq 2$. If the denominator of $\dfrac{P(x)}{Q(x)}$ has a repeated prime quadratic factor, then for every repeated prime quadratic factor of the form $(ax^2 + bx + c)^n$, we introduce n partial fractions of the form

$$\frac{A_1 x + B_1}{ax^2 + bx + c} + \frac{A_2 x + B_2}{(ax^2 + bx + c)^2} + \cdots + \frac{A_n x + B_n}{(ax^2 + bx + c)^n}$$

Example 7 Setting up the Partial Fraction Decomposition of $\dfrac{P(x)}{Q(x)}$, where $Q(x)$ has Repeated Prime Quadratic Factors

Set up the partial fraction decomposition for $\dfrac{x^4 + x^3 + x^2 + x + 1}{(x^2 + 4)^3(3x^2 + 10x + 1)^2}$.

Do not solve for the constants.

Solution The denominator has two repeated prime quadratic factors of the form $(x^2 + 4)^3$ and $(3x^2 + 10x + 1)^2$. The correct partial fraction decomposition is of the form

$$\frac{x^4 + x^3 + x^2 + x + 1}{(x^2 + 4)^3(3x^2 + 10x + 1)^2} = \frac{Ax + B}{x^2 + 4} + \frac{Cx + D}{(x^2 + 4)^2} + \frac{Ex + F}{(x^2 + 4)^3} + \frac{Gx + H}{3x^2 + 10x + 1} + \frac{Ix + J}{(3x^2 + 10x + 1)^2},$$

where $A, B, C, D, E, F, F, H, I$, and J are constants to be determined. Notice that for convenience we use $A, B, C, D, \ldots$ instead of $A_1, B_1, A_2, B_2, \ldots$.

My video summary ⊙ **Example 8** Decomposing a Rational Expression in which the Denominator has a Repeated Prime Quadratic Factor

Determine the partial fraction decomposition of $\dfrac{3x^4 - 5x^3 + 7x^2 + x - 2}{(x - 1)(x^2 + 1)^2}$.

Solution

Step 1. The denominator is already factored.

Step 2. Introduce the three partial fractions

$$\frac{A}{x - 1}, \frac{Bx + C}{x^2 + 1}, \text{and } \frac{Dx + E}{(x^2 + 1)^2}$$

to create the following equation:

$$\frac{3x^4 - 5x^3 + 7x^2 + x - 2}{(x - 1)(x^2 + 1)^2} = \frac{A}{x - 1} + \frac{Bx + C}{x^2 + 1} + \frac{Dx + E}{(x^2 + 1)^2}$$

Step 3. Multiply both sides of the rational equation from step 2 by $(x - 1)(x^2 + 1)^2$ to obtain the equation

$$3x^4 - 5x^3 + 7x^2 + x - 2 = A(x^2 + 1)^2 + (Bx + C)(x - 1)(x^2 + 1) + (Dx + E)(x - 1).$$

Step 4. Choosing $x = 1$ will let us solve for A.

$$x = 1: 3(1)^4 - 5(1)^3 + 7(1)^2 + (1) - 2 = A((1)^2 + 1)^2 + (B(1) + C)(1 - 1)((1)^2 + 1) + (D(1) + E)(1 - 1)$$

$$4 = 4A$$

$$1 = A$$

Step 5. Write the polynomial on the right-hand side of the equation from step 3 in descending order by collecting like terms. View these steps to see that we can write the equation found in step 3 as

$$3x^4 - 5x^3 + 7x^2 + x - 2 = (A + B)x^4 + (-B + C)x^3 + (2A + B - C + D)x^2$$
$$+ (-B + C - D - E)x + (A - C - E).$$

Step 6. Equating the coefficients of $3x^4 - 5x^3 + 7x^2 + x - 2$ with the corresponding coefficients of $(A + B)x^4 + (-B + C)x^3 + (2A + B - C + D)x^2 + (-B + C - D + E)x + (A - C - E)$ leads to the following system of equations:

$$A + B = 3$$

$$-B + C = -5$$

$$2A + B - C + D = 7$$

$$-B + C - D + E = 1$$

$$A - C - E = -2$$

Step 7. Because $A = 1$, we can use the first equation to get $B = 2$. Substituting $B = 2$ into the second equation yields $C = -3$. Substituting the known quantities into the third equation gives $D = 0$. It follows from the fourth or fifth equation that $E = 6$.

 *My video summary*

 Therefore, the partial fraction decomposition is $\dfrac{3x^4 - 5x^3 + 7x^2 + x - 2}{(x - 1)(x^2 + 1)^2} =$

$\dfrac{1}{x - 1} + \dfrac{2x - 3}{x^2 + 1} + \dfrac{6}{(x^2 + 1)^2}$. Watch this video to see each step of this solution.

You Try It Work through this You Try It problem.

Work Exercises 36–41 in this textbook or in the MyMathLab Study Plan.

7.4 Exercises

In Exercises 1–41, determine the partial fraction decomposition of each rational expression.

1. $\dfrac{3}{x(x-3)}$

2. $\dfrac{4x+30}{(x+9)(x+3)}$

3. $\dfrac{x}{(x-1)(x-2)}$

4. $\dfrac{x}{(x-2)(x+4)}$

5. $\dfrac{x-7}{(x-3)(x-5)}$

6. $\dfrac{6}{2x^2-11x+5}$

7. $\dfrac{x}{x^2+2x-35}$

8. $\dfrac{x}{x^2-3x+2}$

9. $\dfrac{8x^2-22x+20}{x(x+4)(x+5)}$

10. $\dfrac{x}{12x^2-17x-7}$

11. $\dfrac{7x^2-x-12}{x(x^2-1)}$

12. $\dfrac{7x^2-x-16}{x^3-x}$

13. $\dfrac{x^2}{(x^2-25)(x+1)}$

14. $\dfrac{2x+4}{x^3-x^2-42x}$

15. $\dfrac{-7x+11}{(x-5)^2}$

16. $\dfrac{x^2-6x+4}{(x-2)^3}$

17. $\dfrac{7x+197}{(x-4)^2(x+5)}$

18. $\dfrac{x^2}{(x-1)^2(x+4)}$

19. $\dfrac{x+3}{x^3-2x^2+x}$

20. $\dfrac{x^2-x+30}{x^3+6x^2}$

21. $\dfrac{4x^2}{(x-2)^2(x+2)^2}$

22. $\dfrac{x-4}{(x+3)(x+5)^2}$

23. $\dfrac{7x^3-2}{x^2(x+1)^3}$

24. $\dfrac{2}{(x-1)(x^2-1)}$

25. $\dfrac{6x+5}{x^4-25x^2}$

26. $\dfrac{2x^2+3x+6}{x^3-4x^2-16x+64}$

27. $\dfrac{6}{x(x^2+6)}$

28. $\dfrac{18x+3}{(x-1)(x^2+x+1)}$

29. $\dfrac{8x^2-38x+22}{(x-6)(x^2+5)}$

30. $\dfrac{-4}{x^3-27}$

31. $\dfrac{x^2-111}{x^4-x^2-72}$

32. $\dfrac{x+7}{x^2(x^2+9)}$

33. $\dfrac{-39x+52}{(x+3)^2(x^2+4)}$

34. $\dfrac{6x^2+11x+31}{x^3+3x^2+4x+12}$

35. $\dfrac{x^2+3x+9}{(x+3)(x^2-12x+6)}$

36. $\dfrac{x^3+x^2+5}{(x^2+4)^2}$

37. $\dfrac{5x^3+14x^2+22x-37}{(x^2+4x+8)^2}$

38. $\dfrac{4x^3+2x^2+15x+4}{(x^2+3)^3}$

39. $\dfrac{3x-1}{x(5x^2+3)^2}$

40. $\dfrac{x^4-2x^3+25x^2+5x+70}{(x+2)(x^2+4)^2}$

41. $\dfrac{3x^4+4x^2-2x+1}{x^6+2x^4+x^2}$

7.5 Systems of Nonlinear Equations

THINGS TO KNOW

Before working through this section, be sure that you are familiar with the following concepts:

		VIDEO	ANIMATION	INTERACTIVE
You Try It	1. Sketching the Graph of a Circle (Section 2.2)	⊗		
You Try It	2. Sketching Lines by Plotting Intercepts (Section 2.3)	⊗		
You Try It	3. Graphing Quadratic Functions (Section 4.1)	⊗		
You Try It	4. Use Transformations to Sketch the Graph of a Rational Function (Section 4.6)	⊗		
You Try It	5. Sketching the Graph of a Logarithmic Function (Section 5.3)	⊗		
You Try It	6. Solving Logarithmic Equations (Section 5.5)			⊗
You Try It	7. Sketching the Graph of a Parabola (Section 6.1)	⊗		
You Try It	8. Sketching the Graph of an Ellipse (Section 6.2)	⊗		
You Try It	9. Sketching the Graph of a Hyperbola (Section 6.3)			⊗
You Try It	10. Solving a System of Linear Equation Using the Method of Substitution (Section 7.1)	⊗		
You Try It	11. Solving a System of Linear Equations Using the Method of Elimination (Section 7.1)	⊗		

OBJECTIVES

1 Determining the Number of Solutions to a System of Nonlinear Equations

2 Solving a System of Nonlinear Equations Using the Substitution Method

3 Solving a System of Nonlinear Equations Using Substitution, Elimination, or Graphing

4 Solving Applied Problems Using a System of Nonlinear Equations

Introduction to Section 7.5

In Section 7.1 we solved systems of linear equations in two variables. The solution to a system of linear equations in two variables is represented by the set of all ordered pairs that satisfy both equations. Graphically, the solution can be represented by the point (or points) of intersection of the two lines.

In this section, we turn our attention to **nonlinear systems**. Systems of nonlinear equations in two variables contain at least one equation that is nonlinear. That is, at least one of the equations is *not* of the form $Ax + By = C$. Just like with linear systems, the real solutions to a nonlinear system can be graphically represented by all points of intersection between the graphs of both equations. In this section it is essential that you can sketch the graphs of nonlinear equations such as circles, parabolas, ellipses, and hyperbolas. Being able to sketch the equations of a nonlinear system will help you understand the nature of the solutions. Before we solve a system of nonlinear equations, take a moment to practice sketching the graphs of these familiar nonlinear equations. Click on a You Try It icon below to practice sketching the graphs of circles, parabolas, ellipses, and hyperbolas.

You Try It Practice sketching the graph of a circle that is in standard form.

You Try It Practice sketching the graph of a circle that is in general form.

You Try It Practice sketching the graph of a parabola that opens up or down (a quadratic function).

You Try It Practice sketching the graph of a parabola that opens left or right.

You Try It Practice sketching the graph of an ellipse that is in standard form.

You Try It Practice sketching the graph of an ellipse that is not in standard form.

You Try It Practice sketching the graph of a hyperbola that is in standard form.

You Try It Practice sketching the graph of a hyperbola that is not in standard form.

You Try It Practice sketching the graph of a rational function.

You Try It Practice sketching the graph of a logarithmic function.

OBJECTIVE 1   DETERMINING THE NUMBER OF SOLUTIONS TO A SYSTEM OF NONLINEAR EQUATIONS

As stated previously, the real solutions to a system of nonlinear equations can be represented by the points of intersection between the graphs of each equation. In Example 1, three different systems of nonlinear equations are shown. Without solving the systems, we can determine the number of solutions of each system by sketching the graph of each equation.

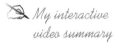

⊙ Example 1 Determining the Number of Solutions to a Nonlinear System

For each system of nonlinear equations, sketch the graph of each equation of the system and then determine the number of real solutions to each system. Do not solve the system.

a.
$$x^2 + y^2 = 25$$
$$x - y = 1$$

b.
$$x - y^2 = 4$$
$$x - y = 6$$

c.
$$x^2 + y^2 = 9$$
$$x^2 - y = 3$$

Solution

a. The graph of the equation $x^2 + y^2 = 25$ is a circle centered at the origin with a radius of 5 units. The graph of $x - y = 1$ is a line with a slope of 1 and a y-intercept of -1. The two graphs are shown in Figure 13.

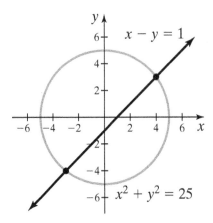

Figure 13

Because the graphs in Figure 13 intersect at two points, the system of nonlinear equations has two real solutions. Watch this interactive video to see how to sketch each equation.

b. The graph of the equation $x - y^2 = 4$ is a parabola with a horizontal axis of symmetry. The parabola has a vertex at the point $(4, 0)$ and opens to the right.

The graph of $x - y = 6$ is a line with a slope of 1 having a y-intercept of -6 and an x-intercept of 6. The two graphs are shown in Figure 14.

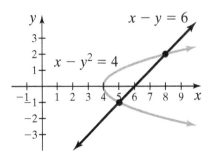

Figure 14

⊙ Because the graphs in Figure 14 intersect at two points, the system of nonlinear equations has two real solutions. Watch this interactive video to see how to sketch each equation.

c. The graph of the equation $x^2 + y^2 = 9$ is a circle centered at the origin with a radius of 3 units. The equation $x^2 - y = 3$ is equivalent to the equation $y = x^2 - 3$, which is a quadratic function whose graph is a parabola opening up having a vertex at the point $(0, -3)$. The two graphs are shown in Figure 15.

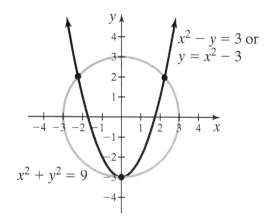

Figure 15

My interactive video summary

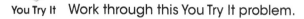 Because the graphs in Figure 15 intersect at three points, the system of nonlinear equations has three real solutions. Watch this interactive video to see how to sketch each equation.

You Try It Work through this You Try It problem.

Work Exercises 1–8 in this textbook or in the MyMathLab Study Plan.

OBJECTIVE 2 SOLVING A SYSTEM OF NONLINEAR EQUATIONS USING THE SUBSTITUTION METHOD

The substitution method for solving nonlinear systems of equations involves solving one of the equations for one variable in terms of the other and then substituting that expression into the other equation. This method is very similar to the substitution method used for solving a system of linear equations that was first introduced in Section 7.1. The substitution method for solving nonlinear systems of equations can be summarized in the following five steps.

Solving a System of Nonlinear Equations by the Substitution Method

Step 1. Choose an equation and solve for one variable (or expression) in terms of the other variable.

Step 2. Substitute the expression from step 1 into the other equation.

Step 3. Solve the equation in one variable.

Step 4. Substitute the value(s) found in step 3 into one of the original equations to find the value(s) of the other variable.

Step 5. Check each solution by substituting all proposed solutions into the other equation of the system.

Note Many nonlinear systems can be solved using more than one method. However, systems in which one equation involves all squared variable terms while the other equation is linear must be algebraically solved using the substitution method. The system shown in Example 2 is an example of a nonlinear system that must be solved algebraically using the method of substitution.

My video summary ⊙ **Example 2** Solve a System of Nonlinear Equations Using the Substitution Method

Determine the real solutions to the following system using the substitution method.

$$x^2 + y^2 = 25$$

$$x - y = 1$$

Solution

Step 1. If one of the equations of a nonlinear system is linear, then always solve the linear equation for one of the variables. Here we solve the linear equation for x.

$$x - y = 1 \qquad \text{Write the second equation.}$$

$$x = y + 1 \qquad \text{Add } y \text{ to both sides to solve for } x.$$

Step 2. Substitute $y + 1$ for x in the first equation.

$$(y + 1)^2 + y^2 = 25$$

Step 3. Solve for y:

$$
\begin{array}{ll}
(y + 1)^2 + y^2 = 25 & \text{Rewrite the equation from step 2.} \\
y^2 + 2y + 1 + y^2 = 25 & \text{Square: } (y + 1)^2 = y^2 + 2y + 1. \\
2y^2 + 2y - 24 = 0 & \text{Combine like terms.} \\
y^2 + y - 12 = 0 & \text{Divide both sides by 2.} \\
(y + 4)(y - 3) = 0 & \text{Factor.} \\
y + 4 = 0 \quad \text{or} \quad y - 3 = 0 & \text{Use the zero product property.} \\
y = -4 \quad \text{or} \qquad y = 3 & \text{Solve for } y.
\end{array}
$$

Step 4. We can solve for x by substituting $y = -4$ and $y = 3$ into either one of the original equations. Here, we choose the first equation $x^2 + y^2 = 25$.

$$
\begin{array}{ll}
y = -4: \quad x^2 + (-4)^2 = 25 & \text{Substitute } y = -4 \text{ into the first equation.} \\
x^2 + 16 = 25 & \text{Square.} \\
x^2 = 9 & \text{Subtract 16 from both sides.} \\
x = \pm 3 & \text{Take the square root of both sides.}
\end{array}
$$

Thus, for $y = -4$, the possibilities for x are $x = 3$ and $x = -3$. This implies that two possible solutions to the nonlinear system are the ordered pairs $(3, -4)$ and $(-3, -4)$.

In a similar fashion, we can find the value(s) of x that correspond to $y = 3$. View these steps to verify that the values of x that correspond to $y = 3$ are $x = \pm 4$. Therefore, two more possible solutions to the nonlinear system are the ordered pairs $(4, 3)$ and $(-4, 3)$.

Step 5. The four proposed solutions are the ordered pairs $(3, -4)$, $(-3, -4)$, $(4, 3)$, and $(-4, 3)$. We know that these four solutions check in the first equation, $x^2 + y^2 = 25$. Now we must check each proposed solution by substituting the appropriate values of x and y into the other equation, $x - y = 1$.

$(3, -4)$: $\qquad 3 - (-4) \overset{?}{=} 1 \longrightarrow 3 + 4 \overset{?}{=} 1 \longrightarrow 7 \overset{?}{=} 1 \times \longrightarrow (3, -4)$ is **not** a solution.

$(-3, -4)$: $\qquad -3 - (-4) \overset{?}{=} 1 \longrightarrow -3 + 4 \overset{?}{=} 1 \longrightarrow 1 \overset{?}{=} 1 \checkmark \longrightarrow (-3, -4)$ is a solution.

$(4, 3)$: $\qquad 4 - (3) \overset{?}{=} 1 \longrightarrow 4 - 3 \overset{?}{=} 1 \longrightarrow 1 \overset{?}{=} 1 \checkmark \longrightarrow (4, 3)$ is a solution.

$(-4, 3)$: $\qquad -4 - (3) \overset{?}{=} 1 \longrightarrow -4 - 3 \overset{?}{=} 1 \longrightarrow -7 \overset{?}{=} 1 \times \longrightarrow (-4, 3)$ is **not** a solution.

 My video summary

▷ We see that the only two solutions to the nonlinear system are $(-3, -4)$ and $(4, 3)$. Figure 16 shows the graphs of the two equations of the system. Note that the only two points of intersection are precisely the two points $(-3, -4)$ and $(4, 3)$. Watch this video to see every step of the solution to the given nonlinear system.

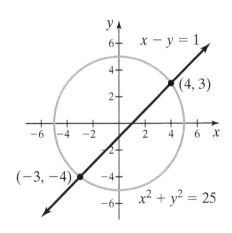

Figure 16 The solutions to the nonlinear system $\begin{array}{l} x^2 + y^2 = 25 \\ x \ - y \ = 1 \end{array}$ are the ordered pairs $(-3, -4)$ and $(4, 3)$.

⚠ **It is absolutely critical to always check the solutions to a system of nonlinear equations by substituting each proposed solution into both equations of the system. In the previous example, there were four proposed solutions to the nonlinear system, but only two of those solutions checked in both equations.**

 My video summary

▷ **Example 3** Solve a System of Nonlinear Equations Using the Substitution Method

Determine the real solutions to the following system using the substitution method.

$$5x^2 - y^2 = 25$$
$$2x \ + y \ = 0$$

Solution

Step 1. The second equation is linear. Thus, solve the second equation for one of the variables. Here we solve the linear equation for y.

$$2x + y = 0 \qquad \text{Write the second equation.}$$

$$y = -2x \qquad \text{Subtract } 2x \text{ from both sides to solve for } y.$$

Step 2. Substitute $-2x$ for y in the first equation.

$$5x^2 - (-2x)^2 = 25$$

Step 3. Solve for x:

$5x^2 - (-2x)^2 = 25$	Rewrite the equation from step 2.
$5x^2 - 4x^2 = 25$	Square.
$x^2 = 25$	Combine like terms.
$x = \pm 5$	Take the square root of both sides.

Step 4. We can solve for x by substituting $x = 5$ or $x = -5$ into one of the original equations. Here, we choose the second equation of the form $y = -2x$.

$$x = 5: \; y = -2\,(5) = -10 \qquad x = -5: \; y = -2\,(-5) = 10$$

We see that the possible solutions are $(5, -10)$ and $(-5, 10)$.

Step 5. You should verify that $(5, -10)$ and $(-5, 10)$ are both solutions to the non-linear system. View this verification process. Figure 17 shows the graphical solution to this system. Note that the graph of the first equation is a hyperbola and the graph of the second equation is a line.

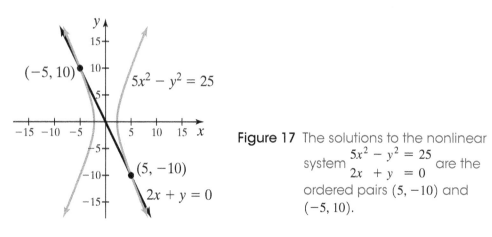

Figure 17 The solutions to the nonlinear system $\begin{aligned}5x^2 - y^2 &= 25 \\ 2x + y &= 0\end{aligned}$ are the ordered pairs $(5, -10)$ and $(-5, 10)$.

My video summary ◉ **Example 4 Solve a System of Nonlinear Equations Using the Substitution Method**

Determine the real solutions to the following system using the substitution method.

$$x^2 + 2y^2 = 18$$

$$xy = 4$$

Solution Try solving this nonlinear system on your own using the five-step process for solving a nonlinear system by the method of substitution.

When you think that you have accurately solved this system, view the solution or watch this video to see every step of the solution process.

You Try It Work through this You Try It problem.

Work Exercises 9–20 in this textbook or in the MyMathLab Study Plan.

OBJECTIVE 3   SOLVING A SYSTEM OF NONLINEAR EQUATIONS USING SUBSTITUTION, ELIMINATION, OR GRAPHING

We saw in Section 7.1 that systems of linear equations can be solved by adding a multiple of one equation to the other equation in attempt to eliminate one of the variables. This elimination method also works for certain nonlinear systems as well. Elimination is especially useful when both equation contain an x^2-term or a y^2-term. The elimination method can be summarized in the following six steps:

Solving a System of Nonlinear Equations by the Elimination Method

Step 1. Choose a variable to eliminate.

Step 2. Multiply one or both equations by an appropriate nonzero constant so that the sum of the coefficients of one of the terms of both equations is zero.

Step 3. Add the two equations together to obtain an equation in one variable.

Step 4. Solve the equation in one variable.

Step 5. Substitute the value(s) obtained from step 4 into one of the original equations to solve for the other variable.

Step 6. Check each solution by substituting all proposed solutions into the other equation of the system.

My interactive video summary

▶ **Example 5** Solve a System of Nonlinear Equations Using the Elimination Method

Determine the real solutions to the following system.

$$x^2 + y^2 = 9$$
$$x^2 - y \ \ = 3$$

Solution

We can solve this system using either the substitution method or the elimination method. Here we choose the elimination method.

Step 1. Note that both equations contain an x^2-term. Thus, we can eliminate the variable x.

Step 2. Multiply both sides of the first equation by -1.

$$x^2 + y^2 = 9 \quad \xrightarrow{\text{Multiply by } -1.} \quad -x^2 - y^2 = -9$$

$$x^2 - y \ \ = 3 \quad \xrightarrow{\text{Leave equation alone.}} \quad x^2 - y \ \ = \ \ 3$$

Step 3. Now add the two new equations together:

$$\begin{array}{r} -x^2 - y^2 = -9 \\ x^2 - y \ \ = \ \ 3 \\ \hline -y^2 - y \ \ = -6 \end{array}$$

Step 4. View these steps to verify that the solution to the quadratic equation $-y^2 - y = -6$ are $y = -3$ or $y = 2$.

Step 5. Solve for x by substituting $y = -3$ and $y = 2$ into one of the original equations. Here, we choose the second equation.

$y = -3$: $x^2 - (-3) = 3$ Substitute the appropriate values into the second equation. $y = 2$: $x^2 - (2) = 3$

$x^2 + 3 = 3$ Simplify. $x^2 - 2 = 3$

$x^2 = 0$ Isolate x^2. $x^2 = 5$

$x = 0$ Take the square root of both sides. $x = \pm\sqrt{5}$

We see that the three possible solutions are $(0, -3)$, $(\sqrt{5}, 2)$, and $(-\sqrt{5}, 2)$.

Step 6. You should take a moment to check that each of the three possible solutions also satisfies the first equation by substituting each ordered pair into the equation $x^2 + y^2 = 9$. Thus the solutions are the ordered pairs $(0, -3)$, $(\sqrt{5}, 2)$, and $(-\sqrt{5}, 2)$.

My interactive video summary

⊙ We could have solved this system using the substitution method. Watch this interactive video to see how to solve this system using the substitution method or the elimination method. Figure 18 shows the graphical solution to this nonlinear system.

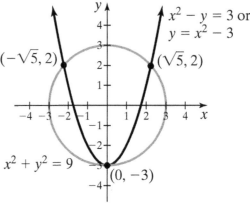

Figure 18 The solutions to the nonlinear system $\begin{aligned} x^2 + y^2 &= 9 \\ x^2 - y &= 3 \end{aligned}$ are the ordered pairs $(0, -3)$, $(\sqrt{5}, 2)$, and $(-\sqrt{5}, 2)$. ●

You Try It Work through this You Try It problem.

Work Exercises 21–29 in this textbook or in the MyMathLab Study Plan.

My video summary

⊙ **Example 6** Solve a System of Nonlinear Equations

Determine the real solutions to the following system.

$$y = \log(3x + 1) - 5$$

$$y = \log(x - 2) - 4$$

Solution We can solve this system using either the substitution method or the elimination method. Here we choose the substitution method.

Step 1. The variable y is isolated in both equations.

Step 2. Substitute $\log(3x + 1) - 5$ for y in the second equation.

$$\log(3x + 1) - 5 = \log(x - 2) - 4$$

Step 3. Solve for x:

$\log(3x + 1) - 5 = \log(x - 2) - 4$	Rewrite the equation from step 2.
$\log(3x + 1) - \log(x - 2) = 1$	Subtract $\log(x - 2)$ from both sides and add 5 to both sides.
$\log\left(\dfrac{3x + 1}{x - 2}\right) = 1$	Use the quotient rule for logarithms.
$\dfrac{3x + 1}{x - 2} = 10^1$	Rewrite in exponential form.
$3x + 1 = 10(x - 2)$	Multiply both sides by $x - 2$.
$3x + 1 = 10x - 20$	Use the distributive property.
$21 = 7x$	Combine like terms.
$3 = x$	Divide both sides by 7.

Step 4. We can solve for y by substituting $x = 3$ into one of the original equations. Here, we choose the second equation of the form $y = \log(x - 2) - 4$.

$x = 3$: $y = \log(3 - 2) - 4$	Substitute $x = 3$ into the second equation.
$y = \log(1) - 4$	Simplify.
$y = -4$	Substitute 0 for $\log(1)$.

The only proposed solution is the ordered pair $(3, -4)$.

Step 5. Check the proposed solution by substituting the values of $x = 3$ and $y = -4$ into the first equation, $y = \log(3x + 1) - 5$.

$(3, -4)$: $-4 \overset{?}{=} \log(3(3) + 1) - 5 \longrightarrow -4 \overset{?}{=} \log(10) - 5 \longrightarrow -4 \overset{?}{=} 1 - 5 \longrightarrow -4 \overset{?}{=} -4\ \checkmark$

The ordered pair $(3, -4)$ satisfies both equations and thus the solution is verified. Figure 19 shows the graphical solution to this nonlinear system.

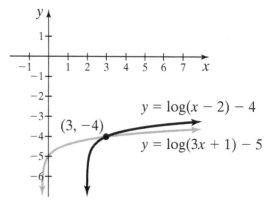

Figure 19 The solution to the nonlinear system $\begin{aligned} y &= \log(3x + 1) - 5 \\ y &= \log(x - 2) - 4 \end{aligned}$ is the ordered pair $(3, -4)$.

You Try It   Work through this You Try It problem.

Work Exercises 30–31 in this textbook or in the MyMathLab Study Plan.

Example 7 Solve a System of Nonlinear Equations

Determine the real solutions to the following system.

$$y = \ln x$$
$$y = 1 - x$$

Solution First try using the substitution method by substituting the expression $1 - x$ for y in the first equation. This gives the following equation in one variable:

$$1 - x = \ln x$$

This equation cannot be solved using algebraic methods discussed in this eText. Solving by elimination yields a similar equation that cannot be solved using algebraic methods discussed in this eText.

Therefore, we can try to sketch the graphs of each equation and see if we can determine any points of intersection.

The graph of the natural logarithmic function, $y = \ln x$, and the graph of the linear function $y = 1 - x$, are shown below in Figure 20.

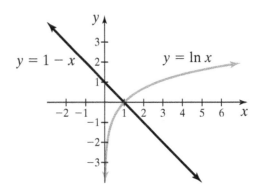

Figure 20 The graph of the nonlinear system
$$y = \ln x$$
$$y = 1 - x$$

It appears from Figure 20 that the graphs of $y = \ln x$ and $y = 1 - x$ intersect at the point $(1, 0)$. Check this proposed solution by substituting the values of $x = 1$ and $y = 0$ into both equations.

$$(1, 0): \quad \begin{array}{ll} 0 \overset{?}{=} \ln(1) & \quad 0 \overset{?}{=} 0 \checkmark \\ \longrightarrow & \\ 0 \overset{?}{=} 1 - 1 & \quad 0 \overset{?}{=} 0 \checkmark \end{array}$$

We see that ordered pair $(1, 0)$ satisfies both equations and thus the solution is verified.

Using Technology

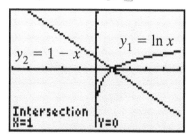

We can solve the nonlinear system from Example 7 using a graphing utility. Graph $y_1 = \ln x$ and $y_2 = 1 - x$, and then use the INTERSECT feature to find that the point of intersection is $(1, 0)$.

You Try It   Work through this You Try It problem.

Work Exercise 32 in this textbook or in the MyMathLab Study Plan.

OBJECTIVE 4   SOLVING APPLIED PROBLEMS USING A SYSTEM OF NONLINEAR EQUATIONS

Many applications involving unknown quantities and many geometric applications can be solved using systems of nonlinear equations.

Example 8 Determining Two Unknown Numbers Using a System of Nonlinear Equations

Find two positive numbers such that the sum of the squares of the two numbers is 25 and the difference between the two numbers is 1.

Solution Let x and y represent the two numbers.

The sum of the squares of the two numbers is 25. Therefore, $x^2 + y^2 = 25$.
The difference between the two numbers is 1. Thus, $x - y = 1$.
We must therefore solve the following nonlinear system of equations:

$$x^2 + y^2 = 25$$

$$x - y = 1$$

My video summary   This is exactly the same system seen in Example 2. Refer to the solution to Example 2 or watch the video solution to Example 2 to see that the solutions to this system are the ordered pairs $(-3, -4)$ and $(4, 3)$. Therefore, the only positive numbers that have the given relationship are 3 and 4.

You Try It   Work through this You Try It problem.

Work Exercises 33–37 in this textbook or in the MyMathLab Study Plan.

Example 9 Solve an Applied Problem Using a System of Nonlinear Equations

An open box (with no lid) has a rectangular base. The height of the box is equal in length to the shortest side of the base. What are the dimensions of the box if the volume is 176 cubic inches and the surface area is 164 square inches?

Solution First, draw a box with a rectangular base and label the lengths of the sides of the base x and y, where x represents the shorter of the two sides. Then, the height of the box is x inches. See figure below.

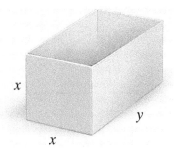

The surface area of the box is the equal to the sum of the areas of the five sides of the box, which is given to be 164 square inches. Therefore, $x^2 + x^2 + xy + xy + xy = 164$. This equation simplifies to $2x^2 + 3xy = 164$.

The volume of the box is 176 cubic inches, which is equal to the length times the width times the height. Thus, $x^2y = 176$. This gives the following system of nonlinear equations:

$$2x^2 + 3xy = 164$$

$$x^2y = 176$$

We can solve this system using the five-step substitution method.

Step 1. Solve the second equation for y.

$$x^2y = 176 \qquad \text{Write the second equation.}$$

$$y = \frac{176}{x^2} \qquad \text{Divide both sides by } x^2.$$

Step 2. Substitute $y = \dfrac{176}{x^2}$ for y in the first equation.

$$2x^2 + 3x\left(\frac{176}{x^2}\right) = 164$$

Step 3. Solve for x:

$$2x^2 + 3x\left(\frac{176}{x^2}\right) = 164 \qquad \text{Rewrite the equation from step 2.}$$

$$2x^2 + \frac{528}{x} = 164 \qquad \text{Multiply and cancel common factors.}$$

$$2x^3 + 528 = 164x \qquad \text{Multiply both sides by } x$$

$$2x^3 - 164x + 528 = 0 \qquad \text{Subtract } 164x \text{ from both sides.}$$

$$x^3 - 82x + 264 = 0 \qquad \text{Divide both sides by 2.}$$

To solve the equation $x^3 - 82x + 264 = 0$, we can use the rational zeros theorem and synthetic division to verify that $x - 4$ is a factor of $x^3 - 82x + 264$.

$$
\begin{array}{r|rrrr}
4 & 1 & 0 & -82 & 264 \\
  &   & 4 & 16 & -264 \\
\hline
  & 1 & 4 & -66 & 0
\end{array}
$$

The remainder is 0 when $x^3 - 82x + 264$ is divided by $x - 4$. Therefore, $x - 4$ is a factor of $x^3 - 82x + 264$.

Therefore, the polynomial $x^3 - 82x + 264$ factors as $(x - 4)(x^2 + 4x - 66)$. We can now solve the equation $x^3 - 82x + 264 = 0$.

$(x - 4)(x^2 + 4x - 66) = 0$ Write the left-hand side as the product of a linear factor and a quadratic factor.

$x - 4 = 0$ or $x^2 + 4x - 66 = 0$ Use the zero product property.

The solution to the equation $x - 4 = 0$ is $x = 4$. Solving the quadratic equation $x^2 + 4x - 66 = 0$ by completing the square gives $x = -2 \pm \sqrt{70}$. (View the steps to the completing the square process.) Because x represents the length of a side of the box, we can disregard the negative value $x = -2 - \sqrt{70}$. Therefore, the two possible values of x are $x = 4$ or $x = -2 + \sqrt{70}$.

Step 4. We can solve for x by substituting $x = 4$ and $x = -2 + \sqrt{70} \approx 6.37$ into one of the original equations. Here, we choose the second equation of the form $y = \dfrac{176}{x^2}$.

$x = 4$: $y = \dfrac{176}{(4)^2} = \dfrac{176}{16} = 11$ $x = -2 + \sqrt{70} \approx 6.37$: $y = \dfrac{176}{(-2 + \sqrt{70})^2} \approx 4.34$

As stated in the original problem, the value of x had to be less than the value of y. Therefore, the only possible solution is $x = 4$ and $y = 11$.

Step 5. You should verify that the values of $x = 4$ and $y = 11$ also satisfy the equation $2x^2 + 3xy = 164$. Therefore, the dimensions of the box are 4 inches by 11 inches by 4 inches.

You Try It   Work through this You Try It problem.

Work Exercises 38–42 in this textbook or in the MyMathLab Study Plan.

7.5 Exercises

For Exercises 1–8 a system of nonlinear equations is given. Sketch the graph of each equation of the system and then determine the number of real solutions to each system. Do not solve the system.

1. $\begin{aligned} x^2 - y &= -4 \\ 2x - y &= -5 \end{aligned}$ **2.** $\begin{aligned} x^2 + y^2 &= 16 \\ x + y &= 2 \end{aligned}$ **3.** $\begin{aligned} x - y^2 &= 0 \\ x - 2y &= -1 \end{aligned}$

4. $\begin{aligned} -x^2 + y^2 &= 16 \\ 8x - y &= 8 \end{aligned}$ **5.** $\begin{aligned} 2x^2 - y &= 3 \\ x^2 + 4y^2 &= 16 \end{aligned}$ **6.** $\begin{aligned} x^2 + y^2 &= 25 \\ 4x^2 - 16y^2 &= 100 \end{aligned}$

7. $\begin{aligned} x + y^2 &= -2 \\ x^2 + y^2 &= 4 \end{aligned}$ **8.** $\begin{aligned} 4x^2 + y^2 &= 36 \\ xy &= -2 \end{aligned}$

For Exercises 9–20, determine the real solutions to each system of nonlinear equations using the substitution method.

9. $\begin{aligned} x + y &= 5 \\ y &= x^2 - 15 \end{aligned}$ **10.** $\begin{aligned} x^2 + y^2 &= 100 \\ x - y &= 2 \end{aligned}$ **11.** $\begin{aligned} y &= x^2 - 2x + 7 \\ y &= x^2 - 8x + 1 \end{aligned}$

12. $\begin{aligned} y^2 &= x^2 - 64 \\ 3y &= x - 8 \end{aligned}$ **13.** $\begin{aligned} x + y &= 2 \\ y &= x^2 - 7x + 10 \end{aligned}$ **14.** $\begin{aligned} y &= x + 2 \\ 4x^2 + y^2 &= 4 \end{aligned}$

15. $\begin{aligned} x + y &= 2 \\ (x + 3)^2 + (y + 2)^2 &= 25 \end{aligned}$

16. $\begin{aligned} xy &= 30 \\ 3x - y &= -9 \end{aligned}$

17. $\begin{aligned} xy &= 5 \\ x^2 + y^2 &= 26 \end{aligned}$

18. $\begin{aligned} 3x^2 + y^2 &= 12 \\ xy &= 3 \end{aligned}$

19. $\begin{aligned} x - y &= 1 \\ x^2 - xy - y^2 &= -19 \end{aligned}$

20. $\begin{aligned} 7x^2 - 3xy + 9y^2 &= 91 \\ x + 3y &= 10 \end{aligned}$

For Exercises 21–32, determine the real solutions to each system of nonlinear equations.

21. $\begin{aligned} x^2 + y^2 &= 80 \\ x^2 - y^2 &= 48 \end{aligned}$

22. $\begin{aligned} 4x^2 - 5y^2 &= -16 \\ 5x^2 + 2y^2 &= 13 \end{aligned}$

23. $\begin{aligned} x^2 + y^2 &= 9 \\ x^2 + (y - 1)^2 &= 4 \end{aligned}$

24. $\begin{aligned} x^2 + y^2 &= 16 \\ y^2 - 3x &= 16 \end{aligned}$

25. $\begin{aligned} x^2 - 4y^2 + 60 &= 0 \\ 6x^2 + y^2 &= 40 \end{aligned}$

26. $\begin{aligned} 9x^2 - 5y^2 &= -9 \\ 10x^2 + 2y^2 &= 7 \end{aligned}$

27. $\begin{aligned} x^2 + 4xy &= 6 \\ 2x^2 - xy &= 3 \end{aligned}$

28. $\begin{aligned} 2x^2 + y^2 &= 1 \\ x^2 - 5y^2 &= -15 \end{aligned}$

29. $\begin{aligned} x^3 + y &= 0 \\ 2x^2 + y &= 0 \end{aligned}$

30. $\begin{aligned} y &= \log_2(3x + 7) - 6 \\ y &= \log_2(x + 5) - 5 \end{aligned}$

31. $\begin{aligned} y &= \log_4(x + 22) + 3 \\ y &= 5 - \log_4(x + 7) \end{aligned}$

32. $\begin{aligned} y &= \log_2 x \\ y &= 6 - x \end{aligned}$

33. Find two numbers such that the sum of the two numbers is 22 and the product of the two numbers is 112.

34. Find two positive numbers such that the sum of the squares of the two numbers is 353 and the difference between the two numbers is 9.

35. Find two positive numbers such that the ratio of the two numbers is 6 to 5 and the product of the two numbers is 270.

36. The sum of the squares of the digits of a positive two-digit number is 97. The difference between the two digits is 5. Find this two-digit number.

37. The product of the three digits of a positive three-digit number is 56. The units digit is 2 less than the tens digit. If the difference between the hundreds digit and the tens digit is 3, then find this three-digit number.

38. The area of a rectangle is 78 square feet and the perimeter is 38 feet. Find the dimensions of the rectangle.

39. The area of a right triangle is 52 square meters. The length of the hypotenuse is $\sqrt{233}$ meters. Find the lengths of the legs of the right triangle.

40. The diagonal of a rectangular pen is 50 feet. The width of the pen is 7 feet less than the length. Find the length of the pen. (Round the length to one decimal place.)

41. An open box (with no lid) has a square base and four sides of equal height. What are the dimensions of the box if the volume is 972 cubic inches and the surface area is 513 square inches?

42. An open box (with no lid) has a rectangular base. The height of the box is equal in length to the shortest side of the base. What are the dimensions of the box if the volume is 468 cubic centimeters and the surface area is 306 square centimeters?

7.6 Systems of Inequalities

THINGS TO KNOW

Before working through this section, be sure that you are familiar with the following concepts:

| | VIDEO | ANIMATION | INTERACTIVE |

You Try It 1. Sketching the Graph of a Circle (Section 2.2) — VIDEO

You Try It 2. Sketching Lines by Plotting Intercepts (Section 2.3) — VIDEO

You Try It 3. Graphing Quadratic Functions (Section 4.1) — VIDEO

You Try It 4. Sketching the Graph of a Parabola (Section 6.1) — VIDEO

You Try It 5. Sketching the Graph of an Ellipse (Section 6.2) — VIDEO

You Try It 6. Sketching the Graph of a Hyperbola (Section 6.3) — INTERACTIVE

You Try It 7. Solving a System of Nonlinear Equations Using the Method of Substitution (Section 7.5) — VIDEO

You Try It 8. Solving a System of Nonlinear Equations Using the Method of Elimination (Section 7.5) — VIDEO

OBJECTIVES

1 Determine If an Ordered Pair Is a Solution to an Inequality in Two Variables

2 Graphing a Linear Inequality in Two Variables

3 Graphing a Nonlinear Inequality in Two Variables

4 Determining If an Ordered Pair Is a Solution to a System of Inequalities in Two Variables

5 Graphing a System of Linear Inequalities in Two Variables

6 Graphing a System of Nonlinear Inequalities in Two Variables

....

OBJECTIVE 1   DETERMINE IF AN ORDERED PAIR IS A SOLUTION TO AN INEQUALITY IN TWO VARIABLES

In Section 1.7 we solved linear inequalities in one variable. In Section 1.9 we solved polynomial inequalities and rational inequalities. Polynomial and rational inequalities were examples of nonlinear inequalities in one variable. The solution set for inequalities in one variable is the set of all values of the variable that make the inequality true. We graph the solution set of inequalities in one variable on a number line.

The solution set to an inequality in two variables is the set of all ordered pairs that satisfy the inequality. Thus, given an inequality in the variables x and y, we can determine if the ordered pair (a, b) is a solution to the inequality by substituting the value of a in for x and b in for y. If a true statement results after this substitution, then (a, b) is a solution to the inequality.

 *My video summary* ⊙ **Example 1** Determining If an Ordered Pair Is a Solution to an Inequality in Two Variables

Determine if the given ordered pair is a solution to the inequality $-2x + y^2 \leq 4$.

a. $(2, -1)$ **b.** $(-2, 0)$ **c.** $\left(-\dfrac{1}{2}, \dfrac{7}{2}\right)$

Solution

a. $-2x + y^2 \leq 4$ Write the original inequality.

      $-2(2) + (-1)^2 \overset{?}{\leq} 4$ Substitute 2 for x and -1 for y.

            $-3 \overset{?}{\leq} 4$ ✓   Simplify and evaluate. A true statement results.

The final statement is true, so $(2, -1)$ is a solution to the inequality. Work through parts (b) and (c) by yourself, then check your answers, or watch this video solution.

You Try It Work through this You Try It problem.

Work Exercises 1–4 in this textbook or in the MyMathLab Study Plan.

OBJECTIVE 2   GRAPHING A LINEAR INEQUALITY IN TWO VARIABLES

We now take our first look at graphing the solution set to an inequality in two variables. We start by graphing the solution to a **linear inequality in two variables.** Recall that a linear equation in two variables is of the form $Ax + By = C$, where A and B are not both equal to zero. Replacing the equal sign of this equation with an inequality symbol results in the following definition.

Definition   Linear Inequality in Two Variables

A **linear inequality in two variables** is an inequality that can be written in the form $Ax + By < C$, where A, B, and C are real numbers, and A and B are not both equal to zero.

Note: The inequality symbol "$<$" can be replaced with $>$, $\leq$, or $\geq$.

Before sketching the solution to a linear inequality in two variables, we must first focus on the related linear equation. For example, to find the solutions to $x + y > 3$ or $x + y < 3$, we look at the related linear equation $x + y = 3$. For an ordered pair to satisfy this equation, the sum of its coordinates must be 3. Note that the ordered pairs $(-2, 5)$, $(1, 2)$, and $(3, 0)$ are all solutions to the equation $x + y = 3$ because the sum of the coordinates of each ordered pair is 3. Plotting and connecting these points gives the line shown in Figure 21.

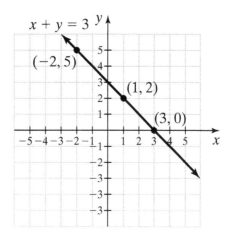

Figure 21

The line $x + y = 3$ divides the coordinate plane into two **half-planes**, an *upper half-plane* and a *lower half-plane*. In Figure 22, the upper half-plane is shaded blue and the lower half-plane is shaded red.

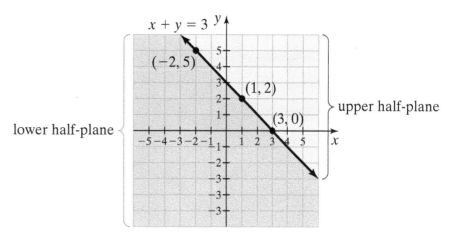

Figure 22 The line $x + y = 3$ divides the coordinate plane into two half-planes.

Let's now examine some points that lie in each half-plane to see which ones satisfy the inequalities $x + y > 3$ or $x + y < 3$.

First, choose any point that lies in the upper half-plane, say $(4, 2)$. The sum of its coordinates is $4 + 2 = 6$, which is larger than 3. Choose any other point in this region, such as $(0, 4)$. The sum of its coordinates is $0 + 4 = 4$, which is also greater than 3. For *any point* we choose in the upper half-plane, the sum of its coordinates will be greater than 3. This means that any ordered pair from the upper half-plane is a solution to the inequality $x + y > 3$. Thus, the area shaded blue in Figure 23 represents the set of all ordered pairs that are solutions to the inequality $x + y > 3$.

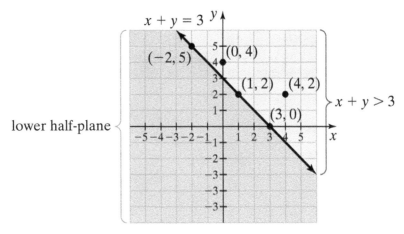

Figure 23 The blue shaded upper half-plane represents the solution to the inequality $x + y > 3$.

Repeat this process for the lower half-plane to see that the area shaded red in Figure 24 represents the set of all ordered-pair solutions to the inequality $x + y < 3$. For example, test the points $(0, 0)$, $(-1, 1)$, and $(-2, -2)$ from the lower half-plane. View this verification process to see that all three ordered pairs satisfy the inequality $x + y < 3$. Note that we will always use the point $(0, 0)$ as a test point unless it lies along the boundary line.

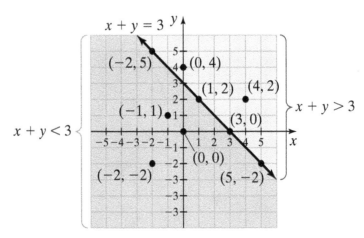

Figure 24 The red shaded lower half-plane represents the solution to the inequality $x + y < 3$.

The equation $x + y = 3$ acts as a **boundary line** that separates the solutions of the two inequalities $x + y > 3$ and $x + y < 3$. Based on this information, we can define a set of **steps for graphing linear inequalities in two variables**.

Steps for Graphing Linear Inequalities in Two Variables

Step 1. Find the boundary line for the inequality by replacing the inequality symbol with an equal sign and graphing the resulting equation. If the inequality is strict, graph the boundary line using a dashed line. If the inequality is non-strict, graph the boundary line using a solid line.

Step 2. Choose a **test point** that does not belong to the boundary line and determine if it is a solution to the inequality.

Step 3. If the test point is a solution to the inequality, then shade the half-plane that contains the test point. If the test point is not a solution to the inequality, then shade the half-plane that does not contain the test point. The shaded area represents the set of all ordered-pair solutions to the inequality.

Note When graphing a linear inequality in two variables, using a dashed line is similar to using an open circle when graphing a linear inequality in one variable on a number line. Likewise, using a solid line is similar to using a solid circle. Do you see why?

My interactive video summary

⊘ **Example 2** Graphing a Linear Inequality in Two Variables

Graph each inequality.

a. $x - 2y \geq 4$ **b.** $3y < 2x$ **c.** $x < -2$

Solution

a. We follow the three-step process for Graphing Linear Inequalities in Two Variables.

Step 1. The boundary line is $x - 2y = 4$. We graph the line as a solid line because the inequality is non-strict. See Figure 25a.

Step 2. We choose the test point $(0, 0)$ and check to see if it satisfies the inequality.

$$x - 2y \overset{?}{\geq} 4 \qquad \text{Write the original inequality.}$$

$$0 - 2(0) \overset{?}{\geq} 4 \qquad \text{Substitute 0 for } x \text{ and 0 for } y.$$

$$0 \overset{?}{\geq} 4 \times \qquad \text{Simplify and evaluate. A false statement results.}$$

The point $(0, 0)$ is not a solution to the inequality.

Step 3. Because the test point is not a solution to the inequality, we shade the half-plane that does not contain $(0, 0)$. See Figure 25b. The shaded region, including the boundary line, represents all ordered-pair solutions to the inequality $x - 2y \geq 4$.

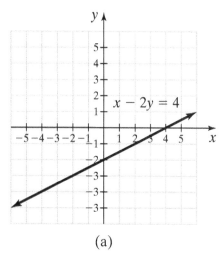
(a)

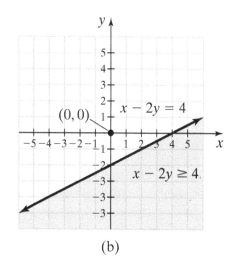
(b)

Figure 25

My interactive video summary

 b. For the inequality $3y < 2x$, the boundary line is $3y = 2x$ (or $y = \dfrac{2}{3}x$.)

Because the inequality is strict, we graph the boundary line using a dashed line. Choose a test point and complete the graph. Note here that we cannot use the point $(0, 0)$ as a test point because it lies along the boundary line. When you have completed the work, view the graph of the solution or watch this interactive video to see each step of the solution process.

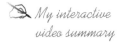
My interactive video summary

 c. For the inequality $x < -2$, the boundary line is the vertical line $x = -2$. Because the inequality is strict, we graph the boundary line using a dashed line. Choose a test point and complete the graph. When you have completed the work, view the graph of the solution or watch this interactive video to see each step of the solution process.

You Try It Work through this You Try It problem.

Work Exercises 5–14 in this textbook or in the MyMathLab Study Plan.

OBJECTIVE 3 GRAPHING A NONLINEAR INEQUALITY IN TWO VARIABLES

The procedure used for graphing a nonlinear inequality in two variables is nearly the same as the procedure used for sketching the graph of a linear inequality in two variables.

Steps for Graphing Nonlinear Inequalities in Two Variables

Step 1. Replace the inequality symbol with an equal sign and graph the resulting equation. If the inequality is strict, sketch the graph using dashes. If the inequality is non-strict, sketch the graph using a solid curve. This graph divides the coordinate plane into two or more regions.

Step 2. Choose one test point that does not belong to the graph of the equation from step 1 and determine if it is a solution to the inequality.

Step 3. If the test point is a solution to the inequality, shade the region that contains the test point. If the test point is not a solution to the inequality, shade the region that does not include the test point.

Note In order to be certain that all of the solution set has been identified, it may be necessary to choose more than one test point when the graph of a nonlinear inequality in two variables divides the coordinate plane into multiple regions.

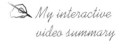
My interactive video summary

⊘ **Example 3** Graphing a Linear Inequality in Two Variables

Graph each inequality.

a. $x^2 + y^2 \geq 9$ 　　　　 b. $9y^2 - 4x^2 \leq 36$ 　　　　 c. $x - y^2 > 1$

Solution

a. We follow the steps for Graphing Nonlinear Inequalities in Two Variables.

Step 1. The inequality $x^2 + y^2 \geq 9$ is non-strict, so we graph the equation $x^2 + y^2 = 9$ (which is a circle centered at the origin having a radius of 3 units) using a continuous solid curve. See Figure 26a.

Step 2. We choose the test point $(0, 0)$, which lies inside the circle.

$$x^2 + y^2 \overset{?}{\geq} 9 \qquad \text{Write the original inequality.}$$

$$0^2 + 0^2 \overset{?}{\geq} 9 \qquad \text{Substitute 0 for } x \text{ and 0 for } y.$$

$$0 \overset{?}{\geq} 9 \times \qquad \text{Simplify and evaluate. A false statement results.}$$

The point $(0, 0)$ that lies inside the circle is not a solution to the inequality.

Step 3. Because the test point $(0, 0)$ is not a solution to the inequality, we shade the region that lies outside of the circle. See Figure 26b. The shaded region, including the graph of the circle, represents all ordered-pair solutions to the inequality $x^2 + y^2 \geq 9$.

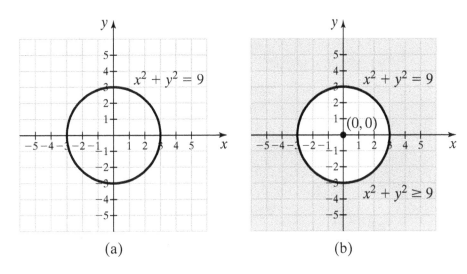

(a)   (b)   **Figure 26**

My interactive video summary

⊙ **b.** For the inequality $9y^2 - 4x^2 \leq 36$, the inequality is non-strict, so we graph the equation $9y^2 - 4x^2 = 36$ (which is a hyperbola centered at the origin having a vertical transverse axis) using a solid curve. Choose a test point and complete the graph. When you have completed the work, view the graph of the solution or watch this interactive video to see each step of the solution process.

My interactive video summary

⊙ **c.** For the inequality $x - y^2 > 1$, the inequality is strict, so we graph the equation $x - y^2 = 1$ (which is a parabola that opens to the right having a vertex at the point $(1, 0)$) using dashes. Choose a test point and complete the graph. When you have completed the work, view the graph of the solution or watch this interactive video to see each step of the solution process.

You Try It Work through this You Try It problem.

Work Exercises 15–24 in this textbook or in the MyMathLab Study Plan.

OBJECTIVE 4   DETERMINING IF AN ORDERED PAIR IS A SOLUTION
TO A SYSTEM OF INEQUALITIES IN TWO VARIABLES

We now consider systems of inequalities in two variables.

Definition   System of Inequalities in Two Variables

A **system of inequalities in two variables** is the collection of two or more inequalities in two variables considered together. If all inequalities are linear, then the system is a **system of linear inequalities in two variables**. If at least one inequality is nonlinear, then the system is a **system of nonlinear inequalities in two variables**.

Example 4 illustrates a system of linear inequalities in two variables and a system of nonlinear inequalities in two variables.

An ordered pair is a solution to a system of inequalities in two variables if the ordered pair satisfies all inequalities in the system.

My video summary   ⟩ **Example 4** Determining If an Ordered Pair Is a Solution to a System of Inequalities in Two Variables

Determine which ordered pairs are solutions to the given system.

a. $\begin{aligned} 2x - 3y &\le 9 \\ 2x - y &\ge -1 \end{aligned}$ i. $(1, -2)$ ii. $(-1, 2)$ iii. $(3, -1)$

b. $\begin{aligned} x^2 - y^2 &\le 25 \\ x - y &> 1 \end{aligned}$ i. $(-1, -3)$ ii. $(0, 5)$ iii. $(-3, 4)$

Solution

a. Watch this video to verify that the ordered pairs $(1, -2)$ and $(3, -1)$ are solutions to the system but $(-1, 2)$ is not a solution.

b. Watch this video to verify $(-1, -3)$ is a solution whereas the ordered pairs $(0, 5)$ and $(-3, 4)$ are not solutions.

You Try It   Work through this You Try It problem.

Work Exercises 25–30 in this textbook or in the MyMathLab Study Plan.

OBJECTIVE 5   GRAPHING A SYSTEM OF LINEAR INEQUALITIES IN TWO VARIABLES

My animation summary   ⊡ The graph of a system of inequalities in two variables is the intersection of the graphs of each inequality in the system. We start by sketching **the graph of a system of linear inequalities in two variables**. Carefully watch this animation to see how to graph the following system of linear inequalities.

$$2x - 3y \leq 9$$
$$2x - y \geq -1$$

We obtain the graph of a system of linear inequalities by graphing each inequality in the system and finding the region they have in common, if any. Thus, we follow the two-step process outlined below.

Steps for Graphing Systems of Linear Inequalities in Two Variables

Step 1. Use the Steps for Graphing Linear Inequalities in Two Variables to graph each inequality.

Step 2. Determine where the shaded regions overlap, if any. This overlapped (or shared) region represents the set of all solutions to the system of inequalities.

My interactive video summary

⊙ **Example 5** Graphing a Linear Inequality in Two Variables

Graph each system of linear inequalities in two variables.

a. $\begin{aligned} x + y &> 2 \\ 2x - y &\leq 6 \end{aligned}$

b. $\begin{aligned} x - 3y &> 6 \\ 2x - 6y &< -9 \end{aligned}$

c. $\begin{aligned} 4x &> y \\ x - 3y &< 9 \\ x + y &< 4 \end{aligned}$

Solution

a. To graph the system $\begin{aligned} x + y &> 2 \\ 2x - y &\leq 6 \end{aligned}$, we follow the Steps for Graphing Systems of Linear Inequalities in Two Variables.

   Step 1. **Graph $x + y > 2$.** The inequality $x + y > 2$ is strict, so we graph the boundary line $x + y = 2$ using a dashed line. The test point $(0, 0)$ does not satisfy the inequality $x + y > 2$ because $0 + 0 > 2$ is false, so we shade the half-plane that does not contain $(0, 0)$. See Figure 27a.

   Graph $2x - y \leq 6$. The inequality $2x - y \leq 6$ is non-strict, so we graph the boundary line $2x - y = 6$ using a solid line. The test point $(0, 0)$ satisfies the inequality $2x - y \leq 6$ because $2(0) - 0 \leq 6$ is true, so we shade the half-plane that contains $(0, 0)$. See Figure 27b.

   Step 2. We determine the region shared by both inequalities in the system. This is the purple-shaded region in Figure 27c. Any point in this region is a solution to this system of linear inequalities in two variables.

My interactive video summary

⊙ Watch this interactive video for a detailed solution.

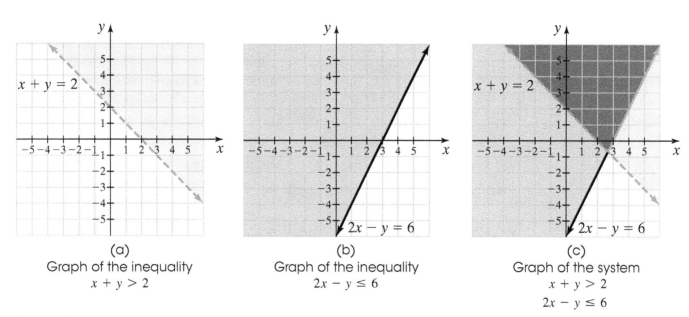

(a)
Graph of the inequality
$x + y > 2$

(b)
Graph of the inequality
$2x - y \leq 6$

(c)
Graph of the system
$x + y > 2$
$2x - y \leq 6$

Figure 27

> **Only the ordered pairs in the purple-shaded region in Figure 27c, including the portion of the solid line $2x - y = 6$, are solutions to the system. Note that none of the points on the boundary line $x + y = 2$ are solutions to the system. Furthermore, the two boundary lines intersect at the point $\left(\dfrac{8}{3}, -\dfrac{2}{3}\right)$.**
>
> **View these steps to see how to determine this point of intersection. Because the inequality $x + y > 2$ is strict, this point is not part of the solution.**

  b. To graph the system $\begin{array}{l} x - 3y > 6 \\ 2x - 6y < -9 \end{array}$, we once again follow the Steps for Graphing Systems of Linear Inequalities in Two Variables.

  Step 1. **Graph $x - 3y > 6$.** The inequality $x - 3y > 6$ is strict, so we graph the boundary line $x - 3y = 6$ using a dashed line. The test point $(0, 0)$ does not satisfy the inequality $x - 3y > 6$, so we shade the half-plane that does not contain $(0, 0)$. See Figure 28a.

  Graph $2x - 6y < -9$. The inequality $2x - 6y < -9$ is strict, so we graph the boundary line $2x - 6y = -9$ using a dashed line. The test point $(0, 0)$ does not satisfy the inequality $2x - 6y < -9$, so we shade the half-plane that does not contain $(0, 0)$. See Figure 28b.

  Step 2. The boundary lines $x - 3y = 6$ and $2x - 6y = -9$ are parallel, and the shaded regions are on opposite sides of these parallel lines. Therefore, no region is shaded by both inequalities in the system. See Figure 28c. There are no ordered pairs that satisfy both inequalities, so there is no solution to the system.

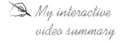 *My interactive video summary*   ⊙ Watch this interactive video for a detailed solution

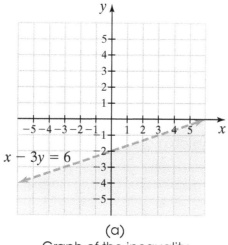

(a)
Graph of the inequality
$x - 3y > 6$

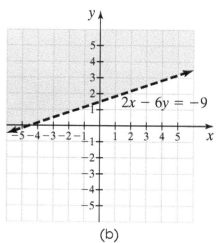

(b)
Graph of the inequality
$2x - 6y < -9$

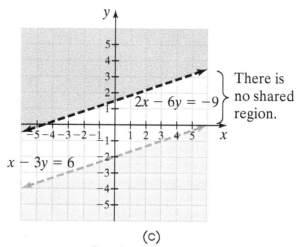

(c)
Graph of the system
$x - 3y > 6$
$2x - 6y < -9$

Figure 28

My interactive video summary

$$4x > y$$

c. To graph the system $x - 3y < 9$, we once again follow the Steps for Graphing
$$x + y < 4$$
Systems of Linear Inequalities in Two Variables. Each inequality in the system is strict, so all of the boundary lines are dashed lines. Graph all three inequalities on the same coordinate plane and find the region shared by all three inequalities, if any. Once you are finished, you can view the solution or watch this interactive video to see a detailed solution. ●

You Try It   Work through this You Try It problem.

Work Exercises 31–45 in this textbook or in the MyMathLab Study Plan.

OBJECTIVE 6   GRAPHING A SYSTEM OF NONLINEAR INEQUALITIES IN TWO VARIABLES

The procedure used for graphing a system of nonlinear inequalities in two variables is nearly the same as the procedure used for graphing a system of linear inequalities. First, graph the solution to each system by shading the appropriate regions. Then, determine the region where the shaded regions overlap.

Steps for Graphing Systems of Nonlinear Inequalities in Two Variables

Step 1. Use the Steps for Graphing Nonlinear Inequalities in Two Variables to graph each inequality.

Step 2. Determine where the shaded regions overlap, if any. This overlapped (or shared) region represents the set of all solutions to the system of inequalities.

My interactive video summary

⊚ **Example 6** Graphing a System of Nonlinear Inequalities in Two Variables

Graph each system of nonlinear inequalities in two variables.

a. $\begin{aligned} x^2 + y^2 &\leq 25 \\ x - y &> 1 \end{aligned}$
b. $\begin{aligned} x - y^2 &\geq 4 \\ x - y &\leq 6 \end{aligned}$
c. $\begin{aligned} 4x^2 + 9y^2 &\leq 36 \\ x^2 + y &\leq -2 \end{aligned}$

Solution

a. To graph the system $\begin{aligned} x^2 + y^2 &\leq 25 \\ x - y &> 1 \end{aligned}$, we follow the steps for Graphing Systems of Nonlinear Inequalities in Two Variables.

Step 1. **Graph $x^2 + y^2 \leq 25$.** The inequality $x^2 + y^2 \leq 25$ is non-strict, so we sketch the graph of $x^2 + y^2 = 25$ (which is a circle centered at the origin having a radius of 5 units) using a solid curve. The test point $(0, 0)$ satisfies the inequality $x^2 + y^2 \leq 25$ because $0 + 0 \leq 25$ is true, so we shade the region inside the circle containing $(0, 0)$. See Figure 29a.

Graph $x - y > 1$. The inequality $x - y > 1$ is strict, so we graph the boundary line $x - y = 1$ using a dashed line. The test point $(0, 0)$ does not satisfy the inequality $x - y > 1$ because $0 - 0 > 1$ is false, so we shade the half-plane that does not contain $(0, 0)$. See Figure 29b.

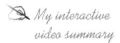
My interactive video summary

⊚ **Step 2.** We determine the region shared by both inequalities in the system. This is the purple-shaded region inside the circle and below the line. See Figure 29c. Any point in this region is a solution to this system of nonlinear inequalities in two variables. Watch this interactive video for a detailed solution. Note that the two graphs intersect at the points $(-3, -4)$ and $(4, 3)$. We can determine the coordinates of these points using the method of substitution. See Example 2 from Section 7.5 or watch this video to see how to find these points. These points of intersection are not solutions to the system of inequalities because no points lying on the dashed boundary line $x - y = 1$ can be solutions. Thus, we represent these points using an open circle.

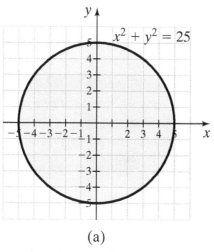

(a)

Graph of the inequality
$x^2 + y^2 \leq 25$

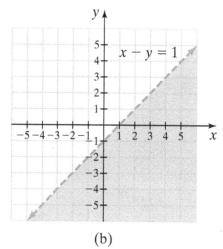

(b)

Graph of the inequality
$x - y > 1$

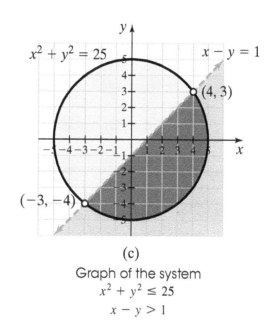

(c)

Graph of the system
$x^2 + y^2 \leq 25$
$x - y > 1$

Figure 29

b. To graph the system $\begin{aligned} x - y^2 &\geq 4 \\ x - y &\leq 6 \end{aligned}$, we follow the steps for Graphing Systems of Nonlinear Inequalities in Two Variables.

Step 1. Graph $x - y^2 \geq 4$. The inequality $x - y^2 \geq 4$ is non-strict, so we sketch the graph of $x - y^2 = 4$ (which is parabola that opens to the right having a vertex at the point $(4, 0)$) using a solid curve. The test point $(0, 0)$ does not satisfy the inequality $x - y^2 \geq 4$ because $0 - 0 \geq 4$ is false, so we shade the region inside the parabola. See Figure 30a.

Graph $x - y \leq 6$. The inequality $x - y \leq 6$ is non-strict, so we graph the boundary line $x - y = 6$ using a solid line. The test point $(0, 0)$ satisfies the inequality $x - y \leq 6$ because $0 - 0 \leq 6$ is true, so we shade the half-plane that contains $(0, 0)$. See Figure 30b.

My interactive video summary

Step 2. We determine the region shared by both inequalities in the system. This is the purple-shaded region seen in **Figure 30c** inside the parabola and to the left of the line. Any point in this region is a solution to this system of nonlinear inequalities in two variables. Watch this **interactive video** for a detailed solution.

Note that the two graphs intersect at the points $(5, -1)$ and $(8, 2)$. These points are part of the solution to the system of inequalities because both inequalities are non-strict. Therefore, we represent these points using a solid circle.

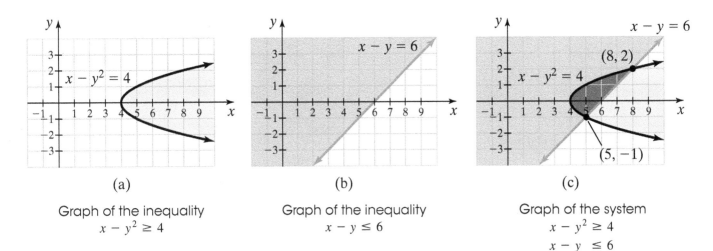

(a)

Graph of the inequality
$x - y^2 \geq 4$

(b)

Graph of the inequality
$x - y \leq 6$

(c)

Graph of the system
$x - y^2 \geq 4$
$x - y \leq 6$

Figure 30

My interactive video summary

c. To graph the system $\begin{array}{l} 4x^2 + 9y^2 \leq 36 \\ x^2 + y \leq -2 \end{array}$, we once again follow the Steps for Graphing Systems of Nonlinear Inequalities in Two Variables. Both inequalities are non-strict, so we sketch the graphs of the equations $4x^2 + 9y^2 = 36$ and $x^2 + y = -2$ using solid curves. Note that the graph of the equation $4x^2 + 9y^2 = 36$ is an ellipse centered at the origin having a horizontal major axis. The graph of the equation $x^2 + y = -2$ is a parabola opening down having a vertex at the point $(0, -2)$. Try graphing this system of nonlinear inequalities on your own. When you have completed the work, view the graph of this system or watch this interactive video to see that the two inequalities have exactly one point in common, the point $(0, -2)$. Therefore, the ordered pair $(0, -2)$ is the only solution to this system of nonlinear inequalities.

You Try It   Work through this You Try It problem.

Work Exercises 46–58 in this textbook or in the MyMathLab Study Plan.

7.6 Exercises

In Exercises 1–4, determine if each ordered pair is a solution to the given inequality.

1. $2x + 5y > 10$
   a. $(-5, 4)$
   b. $(2, 3)$
   c. $\left(\dfrac{5}{2}, \dfrac{3}{5}\right)$

2. $6x - 5y \leq 30$
   a. $(0, 0)$
   b. $(2, -4)$
   c. $(2.5, -1.5)$

3. $x^2 - 2y^2 \geq 8$
   a. $(-4, -4)$
   b. $(3, -5)$
   c. $(6.5, 4.1)$

4. $x^2 - xy < -3$
   a. $(-1, -4)$
   b. $(2, 5)$
   c. $\left(-\dfrac{1}{2}, -\dfrac{9}{2}\right)$

In Exercises 5–14, graph each inequality.

5. $2x - y \geq -3$

6. $3x + 2y > 6$

7. $-x + 3y < -9$

8. $x + y \leq 0$

9. $y \leq \dfrac{5}{2}x - 1$

10. $4y < -3x$

11. $\dfrac{1}{2}x + \dfrac{2}{3}y > \dfrac{5}{6}$

12. $0.6x - 1.8y \leq 2.4$

13. $x \geq -1$

14. $y < 3$

In Exercises 15–24, graph each inequality.

15. $x^2 + y^2 \geq 49$

16. $x^2 + y^2 < 25$

17. $x^2 - y > 3$

18. $y - x^2 \leq 2$

19. $x^2 + 9y^2 < 81$

20. $4x^2 + y^2 \geq 4$

21. $x - y^2 \leq 2$

22. $y^2 > 4 - x$

23. $x^2 - 4y^2 \geq 16$

24. $y^2 - 9x^2 < 9$

In Exercises 25–30, determine if each ordered pair is a solution to the given system of inequalities in two variables.

25. $\begin{array}{l} x - 2y < 6 \\ 3x + y > 2 \end{array}$
   a. $(1, 0)$
   b. $(4, -1)$
   c. $(3, -2)$

26. $\begin{array}{l} 3x - 2y \geq 6 \\ 4x + 3y \leq 9 \end{array}$
   a. $(5, -3)$
   b. $(0, 0)$
   c. $(2, -2)$

27. $\begin{array}{l} x - y > -8 \\ 5x - 2y \leq 10 \\ 4x + 3y > 12 \end{array}$
   a. $(2, 4)$
   b. $(1, 1)$
   c. $(-1, 6)$

28. $\begin{array}{l} x - y^2 > 4 \\ x - y \leq 6 \end{array}$
   a. $(4, 0)$
   b. $(5, -1)$
   c. $(6, 1)$

29. $\begin{array}{l} 4x^2 + 9y^2 < 36 \\ x^2 + y \leq 2 \end{array}$
   a. $(-2, 0)$
   b. $(0, 2)$
   c. $(-1, 1)$

30. $\begin{array}{l} x - y^2 > 1 \\ x + y^2 > -3 \\ y - 2x \leq 3 \end{array}$
   a. $\left(\dfrac{3}{2}, 0\right)$
   b. $(2, 1)$
   c. $\left(\dfrac{5}{4}, \dfrac{1}{4}\right)$

In Exercises 31–45, graph each system of linear inequalities in two variables.

31. $x + y \geq 4$
$2x - y \leq 5$

32. $y < -2x$
$y > -3x$

33. $2x - y < 6$
$y \geq \dfrac{3}{2}x + 3$

34. $x - 3y \leq 6$
$x < 3$

35. $\dfrac{1}{3}x - \dfrac{1}{2}y \leq 1$
$y \geq -2$

36. $x > -4$
$y \geq 1$

37. $3x - 5y \geq 15$
$-3x + 5y \geq 15$

38. $x > 3$
$x < -2$

39. $6x - 8y < 4$
$-3x + 4y \leq 12$

40. $y \geq 2$
$y \leq 5$

41. $2x - y > 7$
$y < 2x + 5$

42. $2x - 3y \geq 12$
$-x + 1.5y < 6$

43. $3x + 2y \leq 4$
$3x - 2y \geq -16$
$x - 2y \leq 4$

44. $2x - y > -3$
$5x + y > 0$
$x > 1$

45. $2x + y \leq 4$
$x + 3y \leq 6$
$x \geq 0$
$y \geq 0$

In Exercises 46–58, graph each system of linear inequalities in two variables.

46. $x^2 + y^2 \leq 36$
$x + y < 4$

47. $x^2 + y^2 > 4$
$x^2 + y^2 \leq 25$

48. $x + y \leq 2$
$y \geq x^2 - 3$

49. $x - y^2 \geq 0$
$x - 3y > -2$

50. $x^2 + y^2 \leq 64$
$x^2 - y \leq 8$

51. $x^2 + y^2 \leq 9$
$y - x^2 > 0$

52. $y - x^2 > -7$
$y - x \leq -4$

53. $x^2 + y^2 \geq 4$
$x - y \leq 2$

54. $y \geq x^2 - 2x$
$y < -2x^2 + 4x$

55. $x + y^2 \leq 2$
$x^2 + 4x + y \geq -4$

56. $x^2 + y^2 \leq 4$
$2x - y \geq 4$

57. $9x^2 + 4y^2 < 36$
$x - y^2 \geq 0$

58. $x^2 + 4y^2 \leq 16$
$x^2 - y^2 > 4$

CHAPTER EIGHT
Matrices

CHAPTER EIGHT CONTENTS

8.1 Matrix Operations

THINGS TO KNOW

Before working through this section, be sure that you are familiar with the following concept:

VIDEO   ANIMATION   INTERACTIVE

You Try It

1. Solving a System of Three Equations in Three Variables Using Gauss-Jordan Elimination (Section 7.3)

OBJECTIVES

1 Understanding the Definition of a Matrix

2 Adding and Subtracting Matrices

3 Scalar Multiplication of Matrices

4 Multiplication of Matrices

5 Applications of Matrix Multiplication

OBJECTIVE 1   UNDERSTANDING THE DEFINITION OF A MATRIX

 *My video summary*

⊙ In Section 7.3, we introduced matrices to display the coefficients of a system of linear equations in order to simplify the Gaussian elimination (or Gauss-Jordan elimination) process. For example, the linear system

$$\begin{aligned} 2x + 3y + 4z &= 12 \\ x - 2y + 3z &= 0 \\ -x + y - 2z &= -1 \end{aligned}$$

can be represented by the augmented matrix

$$\begin{bmatrix} 2 & 3 & 4 & | & 12 \\ 1 & -2 & 3 & | & 0 \\ -1 & 1 & -2 & | & -1 \end{bmatrix}$$

The vertical bar was used to separate the left- and right-hand sides of the equations. We only used this vertical bar for convenience. We could have written this same matrix without the vertical bar as

$$\begin{bmatrix} 2 & 3 & 4 & 12 \\ 1 & -2 & 3 & 0 \\ -1 & 1 & -2 & -1 \end{bmatrix}$$

This matrix has three rows and four columns. Therefore, we say that the size of this matrix is 3 by 4 (or 3×4). In general, a matrix is any rectangular array of numbers of any size.

Definition Matrix

A **matrix** is a rectangular array of numbers arranged in a fixed number, m, of rows and a fixed number, n, of columns. The size of a matrix is $m \times n$. The numbers in the array are called entries. Each entry is labeled a_{ij}, where i is the ith row of the entry and j is the jth column of the entry.

$$A = \begin{bmatrix} a_{11} & a_{12} & \cdots & a_{1n} \\ a_{21} & a_{22} & \cdots & a_{2n} \\ \vdots & \vdots & \vdots & \vdots \\ a_{m1} & a_{m2} & \cdots & a_{mn} \end{bmatrix}$$

Two matrices A and B are said to be equal ($A = B$) provided both matrices are of the same size and the corresponding entries are the same.

We typically use capital letters to name a matrix. Following are some examples of matrices of various sizes:

$$A = \begin{bmatrix} 3 & -1 & 8 \\ 2 & -2 & 9 \\ 5 & 7 & -6 \end{bmatrix}, \quad B = [6 \ -5], \quad C = \begin{bmatrix} 2 & -1 \\ 3 & 1 \end{bmatrix}, \quad D = \begin{bmatrix} 3 & -1 \\ 9 & 6 \\ 2 & 7 \end{bmatrix},$$

      3×3 1×2 2×2 3×2

Square matrix   Row matrix   Square matrix   Matrix

$$E = \begin{bmatrix} 3 \\ 7 \\ 1 \\ -3 \end{bmatrix}, \quad F = \begin{bmatrix} 1 & 0 & 0 & 0 \\ 0 & 1 & 0 & 0 \\ 0 & 0 & 1 & 0 \\ 0 & 0 & 0 & 1 \end{bmatrix}$$

     4×1 4×4

Column matrix   Identity matrix

Matrices A, C, and F are square matrices because the number of rows equals the number of columns. (The size of matrix A is 3×3, the size of matrix C is 2×2, and the size of matrix F is 4×4.) Matrix B is called a **row matrix** because it only has one row. Similarly, matrix E is called a **column matrix** because it only has one column. Matrix F is an example of an **identity matrix**. An identity matrix is a square matrix with 1's down the diagonal (from left to right) and zeros everywhere else. (You will see why we call matrix F an identity matrix when we define matrix multiplication.) Each number in a matrix is called an entry.

We typically use lowercase letters and subscripts to denote each entry. For example, in matrix A, a_{23} refers to the entry of matrix A that is located in the second row and the third column. Therefore, $a_{23} = 9$. Most graphing calculators come equipped with matrix editors. Figure 1 shows matrix A with the entry $a_{23} = 9$ highlighted.

Using Technology

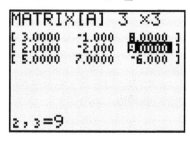

Figure 1 Using a graphing calculator to create a matrix

You Try It   Work through this You Try It problem.

Work Exercises 1–4 in this textbook or in the MyMathLab Study Plan.

OBJECTIVE 2   ADDING AND SUBTRACTING MATRICES

Just like with real numbers, we can add and subtract two matrices. To add or subtract matrices, they must be the *same size*. To add matrices, we simply add the corresponding entries. Similarly, to subtract matrices, we subtract the corresponding entries. Example 1 illustrates the addition and subtraction of two 3×4 matrices.

 *My video summary*

⊙ Example 1 Add and Subtract Matrices

Let $A = \begin{bmatrix} 2 & -1 & 0 & 3 \\ -1 & 4 & -2 & 5 \\ 5 & -3 & 7 & 9 \end{bmatrix}$ and $B = \begin{bmatrix} -3 & -2 & 6 & 1 \\ 0 & 4 & -5 & 1 \\ -1 & 4 & 2 & 7 \end{bmatrix}$.

a. Find matrix $A + B$.

b. Find matrix $B + A$.

c. Find matrix $A - B$.

Solution

a. Adding each corresponding entry of matrix A and matrix B, we obtain the matrix

$$A + B = \begin{bmatrix} (2 + (-3)) & (-1 + (-2)) & (0 + 6) & (3 + 1) \\ (-1 + 0) & (4 + 4) & (-2 + (-5)) & (5 + 1) \\ (5 + -1) & (-3 + 4) & (7 + 2) & (9 + 7) \end{bmatrix} = \begin{bmatrix} -1 & -3 & 6 & 4 \\ -1 & 8 & -7 & 6 \\ 4 & 1 & 9 & 16 \end{bmatrix}$$

b.

$$B + A = \begin{bmatrix} (-3 + 2) & (-2 + (-1)) & (6 + 0) & (1 + 3) \\ (0 + (-1)) & (4 + 4) & (-5 + (-2)) & (1 + 5) \\ (-1 + 5) & (4 + (-3)) & (2 + 7) & (7 + 9) \end{bmatrix} = \begin{bmatrix} -1 & -3 & 6 & 4 \\ -1 & 8 & -7 & 6 \\ 4 & 1 & 9 & 16 \end{bmatrix}$$

c. Subtracting each corresponding entry of matrix B from matrix A, we obtain the matrix

$$A - B = \begin{bmatrix} (2 - (-3)) & (-1 - (-2)) & (0 - 6) & (3 - 1) \\ (-1 - 0) & (4 - 4) & (-2 - (-5)) & (5 - 1) \\ (5 - (-1)) & (-3 - 4) & (7 - 2) & (9 - 7) \end{bmatrix} = \begin{bmatrix} 5 & 1 & -6 & 2 \\ -1 & 0 & 3 & 4 \\ 6 & -7 & 5 & 2 \end{bmatrix}$$

Notice in Example 1 that $A + B = B + A$. This is no coincidence. Matrices, like real numbers, are said to be **commutative** under the operation of addition. That is, the order in which we add two matrices does not matter. However, matrix subtraction is *not* commutative. Matrices are also **associative** under addition, which means $(A + B) + C = A + (B + C)$ provided that $A, B,$ and C are the same size.

You Try It   Work through this You Try It problem.

Work Exercises 5–7 in this textbook or in the My Math Lab **Study Plan.**

OBJECTIVE 3   SCALAR MULTIPLICATION OF MATRICES

We can multiply any matrix by a real number. We call this real number a **scalar** and this operation is called scalar multiplication. Scalar multiplication is performed by multiplying every entry in the matrix by that scalar. For example, suppose matrix A is given by

$$A = \begin{bmatrix} a_{11} & a_{12} & \cdots & a_{1n} \\ a_{21} & a_{22} & \cdots & a_{2n} \\ \vdots & \vdots & \vdots & \vdots \\ a_{m1} & a_{m2} & \cdots & a_{mn} \end{bmatrix}$$

then if c is any real number, we define matrix cA as

$$cA = \begin{bmatrix} ca_{11} & ca_{12} & \cdots & ca_{1n} \\ ca_{21} & ca_{22} & \cdots & ca_{2n} \\ \vdots & \vdots & \vdots & \vdots \\ ca_{m1} & ca_{m2} & \cdots & ca_{mn} \end{bmatrix}$$

My video summary   ⊚ **Example 2** Scalar Multiplication and Addition of Matrices

If $A = \begin{bmatrix} -4 & 0 \\ 3 & 1 \end{bmatrix}$ and $B = \begin{bmatrix} 6 & 3 \\ -3 & -9 \end{bmatrix}$, find the following matrices: $-2A$, $\frac{1}{3}B$, and $-2A + \frac{1}{3}B$.

Solution Work through the video to verify that

$$-2A = \begin{bmatrix} 8 & 0 \\ -6 & -2 \end{bmatrix}, \quad \frac{1}{3}B = \begin{bmatrix} 2 & 1 \\ -1 & -3 \end{bmatrix}, \quad \text{and } -2A + \frac{1}{3}B = \begin{bmatrix} 10 & 1 \\ -7 & -5 \end{bmatrix}$$

Figure 2 shows how a graphing utility can be used to perform scalar multiplication and matrix addition.

Using Technology

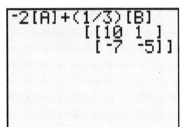

Figure 2 Using a graphing utility to compute matrix $-2A + \frac{1}{3}B$

You Try It   Work through this You Try It problem.

Work Exercises 8–10 in this textbook or in the MyMathLab Study Plan.

OBJECTIVE 4   MULTIPLICATION OF MATRICES

My video summary   Unlike with real numbers, we cannot always multiply two matrices. The product of two matrices is defined only if the number of columns of the first matrix is equal to the number of rows of the second matrix. If the number of columns of the first matrix does not equal the number of rows of the second matrix, then we say that the product matrix is not defined.

MATRIX MULTIPLICATION

If A is an $m \times n$ matrix and B is an $n \times q$ matrix, then there exists an $m \times q$ product matrix AB. The element in the ith row and jth column of matrix AB is equal to the sum of the products of the corresponding entries from row i of matrix A and column j of matrix B.

We now illustrate matrix multiplication by multiplying the 2×3 and 3×2 matrices. Watch the video to see every step of the multiplication process.

$$
\underset{2 \times 3}{\overset{A}{\begin{bmatrix} 1 & 1 & 2 \\ 2 & 4 & -3 \end{bmatrix}}}
\underset{3 \times 2}{\overset{B}{\begin{bmatrix} 0 & -5 \\ 4 & -1 \\ 1 & 3 \end{bmatrix}}}
= \begin{bmatrix} (1)(0)+(1)(4)+(2)(1) & (1)(-5)+(1)(-1)+(2)(3) \\ (2)(0)+(4)(4)+(-3)(1) & (2)(-5)+(4)(-1)+(-3)(3) \end{bmatrix}
= \underset{2 \times 2}{\overset{AB}{\begin{bmatrix} 6 & 0 \\ 13 & -23 \end{bmatrix}}}.
$$

Same

Size of AB is 2×2

Notice that matrix AB has the same number of rows as matrix A and the same number of columns as matrix B. This will always be the case. In other words, if we multiply an $m \times n$ matrix by an $n \times q$ matrix, then the size of the resultant matrix will be $m \times q$.

My video summary   **Example 3** Matrix Multiplication

Given matrices $A = \begin{bmatrix} 1 & 1 & 2 \\ 2 & 4 & -3 \\ 3 & 6 & -5 \end{bmatrix}$, $B = \begin{bmatrix} 0 & -1 & 1 \\ -2 & 1 & 3 \\ 4 & 5 & -3 \end{bmatrix}$, and $C = \begin{bmatrix} 1 & 0 & 0 \\ 0 & 1 & 0 \\ 0 & 0 & 1 \end{bmatrix}$,

find the products AB, BA, and AC.

Solution Watch the interactive video to verify that

$$
AB = \begin{bmatrix} 6 & 10 & -2 \\ -20 & -13 & 23 \\ -32 & -22 & 36 \end{bmatrix}, BA = \begin{bmatrix} 1 & 2 & -2 \\ 9 & 20 & -22 \\ 5 & 6 & 8 \end{bmatrix}, \text{ and } AC = \begin{bmatrix} 1 & 1 & 2 \\ 2 & 4 & -3 \\ 3 & 6 & -5 \end{bmatrix}
$$

Figure 3 illustrates how a graphing utility can be used to perform matrix multiplication.

Using Technology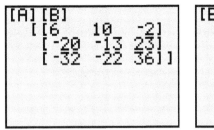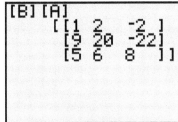

Figure 3 Using a graphing utility to compute matrices AB and BA

You Try It Work through this You Try It problem.

Work Exercises 11–20 in this textbook or in the My MathLab **Study Plan.**

In Example 3, notice that $AB \neq BA$. This implies that matrix multiplication is not commutative. Although matrix multiplication is not commutative in general, it is possible to find two matrices that commute under the operation of multiplication. For example, consider matrix A and matrix C from Example 3. We can easily verify that

$$AC = \begin{bmatrix} 1 & 1 & 2 \\ 2 & 4 & -3 \\ 3 & 6 & -5 \end{bmatrix} \begin{bmatrix} 1 & 0 & 0 \\ 0 & 1 & 0 \\ 0 & 0 & 1 \end{bmatrix} = \begin{bmatrix} 1 & 1 & 2 \\ 2 & 4 & -3 \\ 3 & 6 & -5 \end{bmatrix} = A \text{ and}$$

$$CA = \begin{bmatrix} 1 & 0 & 0 \\ 0 & 1 & 0 \\ 0 & 0 & 1 \end{bmatrix} \begin{bmatrix} 1 & 1 & 2 \\ 2 & 4 & -3 \\ 3 & 6 & -5 \end{bmatrix} = \begin{bmatrix} 1 & 1 & 2 \\ 2 & 4 & -3 \\ 3 & 6 & -5 \end{bmatrix} = A.$$

Notice that C is an **identity matrix**. We will typically use the notation I_n to denote an $n \times n$ identity matrix. For example,

$$I_2 = \begin{bmatrix} 1 & 0 \\ 0 & 1 \end{bmatrix}, I_3 = \begin{bmatrix} 1 & 0 & 0 \\ 0 & 1 & 0 \\ 0 & 0 & 1 \end{bmatrix}, I_4 = \begin{bmatrix} 1 & 0 & 0 & 0 \\ 0 & 1 & 0 & 0 \\ 0 & 0 & 1 & 0 \\ 0 & 0 & 0 & 1 \end{bmatrix}, \text{ and so on.}$$

Whenever an $m \times n$ matrix A is multiplied by the identity matrix I_n, the resultant matrix is always A. Thus, we have the following identity property for matrix multiplication. This property is extremely important to understand and is revisited in Section 8.2.

Identity Property for Matrix Multiplication

For any $m \times n$ matrix A, $I_m A = A$ and $AI_n = A$. If A is an $n \times n$ square matrix, then $AI_n = I_n A = A$.

Example 4 Multiplication of an Identity Matrix

If $A = \begin{bmatrix} 2 & -1 \\ 5 & 6 \\ 0 & -3 \end{bmatrix}$, find appropriate identity matrices I_m and I_n and verify that $I_m A = A$ and $AI_n = A$.

Solution A is a 3×2 matrix, so $m = 3$ and $n = 2$. Therefore,

$$I_m = I_3 = \begin{bmatrix} 1 & 0 & 0 \\ 0 & 1 & 0 \\ 0 & 0 & 1 \end{bmatrix} \text{ and } I_n = I_2 = \begin{bmatrix} 1 & 0 \\ 0 & 1 \end{bmatrix}. \text{ Thus,}$$

$$I_m A = I_3 A = \begin{bmatrix} 1 & 0 & 0 \\ 0 & 1 & 0 \\ 0 & 0 & 1 \end{bmatrix} \begin{bmatrix} 2 & -1 \\ 5 & 6 \\ 0 & -3 \end{bmatrix} = \begin{bmatrix} 2 & -1 \\ 5 & 6 \\ 0 & -3 \end{bmatrix} = A$$

$$\text{and } A I_n = A I_2 = \begin{bmatrix} 2 & -1 \\ 5 & 6 \\ 0 & -3 \end{bmatrix} \begin{bmatrix} 1 & 0 \\ 0 & 1 \end{bmatrix} = \begin{bmatrix} 2 & -1 \\ 5 & 6 \\ 0 & -3 \end{bmatrix} = A.$$

You Try It Work through this You Try It problem.

Work Exercises 21–23 in this textbook or in the MyMathLab Study Plan.

Summary of Matrix Operations

Assume matrix addition and matrix multiplication are defined for matrices A, B, and C and let k be a scalar, then

1. $A + B = B + A$ — Commutative Property for Matrix Addition

2. $(A + B) + C = A + (B + C)$ — Associative Property for Matrix Addition

3. $k(A + B) = kA + kB$ — Distributive Property of a Scalar with Matrix Addition

4. $(AB)C = A(BC)$ — Associative Property for Matrix Multiplication

5. $A(B + C) = AB + AC$ — Distributive Property

6. $(B + C)A = BA + CA$ — Distributive Property

7. $k(AB) = (kA)B = A(kB)$ — Associative Property for Scalar Multiplication

8. If A is an $m \times n$ matrix, then $I_m A = A$ and $A I_n = A$. — Identity Property

---

OBJECTIVE 5 APPLICATIONS OF MATRIX MULTIPLICATION

Matrices can be used as a convenient way to express data. For example, suppose that a certain sporting goods manufacturer has manufacturing plants in Atlanta, Boise, Columbus, and Detroit. The number of each type of ball produced (in thousands) during the month of June last year is represented by matrix A.

	Atlanta	Boise	Columbus	Detroit	
Basketballs	8	5	4.5	10	
Footballs	7	3	2	8	$= A$
Golf balls	10	0	8	15	

We can use this matrix to determine the number of each ball that was produced in each plant. For example, the Detroit plant produced 10,000 basketballs last June. We can deduce that the Boise plant does not produce golf balls (or perhaps the golf ball portion of the plant was shut down during June). Now, suppose that during the month of June, the selling price of a basketball was $25, footballs sold for $20, and golf balls sold for $1.50. We can calculate the total revenue at each plant by multiplying matrix A on the left by the 1×3 row matrix $[25 \quad 20 \quad 1.5] = B$, which represents the selling price of each ball. Thus, we can calculate the revenue matrix BA:

$$BA = [25 \quad 20 \quad 1.5] \begin{bmatrix} 8 & 5 & 4.5 & 10 \\ 7 & 3 & 2 & 8 \\ 10 & 0 & 8 & 15 \end{bmatrix} \begin{matrix} \text{Atlanta} & \text{Boise} & \text{Columbus} & \text{Detroit} \\ = [355 & 185 & 164.5 & 432.5] \end{matrix}$$

Therefore, we see that total revenue generated by each manufacturing plant in the month of June was

Atlanta: $355,000
Boise: $185,000
Columbus: $164,500
Detroit: $432,500

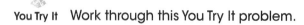

You Try It   Work through this You Try It problem.

Work Exercises 24–26 in this textbook or in the MyMathLab Study Plan.

8.1 Exercises

1. $A = \begin{bmatrix} 2 & -5 & 7 & 9 \\ 1 & 3 & -4 & 11 \\ 8 & 6 & 1 & 0 \end{bmatrix}$

   a. What is the size of matrix A?
   b. Determine whether A is a square matrix, row matrix, column matrix, or none of these.

2. $A = \begin{bmatrix} 1 & 0 & 5 \\ 2 & 1 & 0 \\ 0 & -1 & 1 \end{bmatrix}$

   a. What is the size of matrix A?
   b. Determine whether A is a square matrix, row matrix, column matrix, or none of these.

3. $A = [4 \quad -2 \quad \pi \quad \sqrt{3}]$

   a. What is the size of matrix A?
   b. Determine whether A is a square matrix, row matrix, column matrix, or none of these.

4. $A = \begin{bmatrix} 3.9 \\ -1.2 \\ 5.3 \\ -2.7 \end{bmatrix}$

   a. What is the size of matrix A?
   b. Determine whether A is a square matrix, row matrix, column matrix, or none of these.

In Exercises 5–10, use matrices A and B to find the indicated matrix:

$$A = \begin{bmatrix} 1 & 0 & -2 \\ 3 & 1 & 4 \\ 5 & -1 & -5 \end{bmatrix}, \quad B = \begin{bmatrix} 3 & 4 & -1 \\ 5 & 0 & 2 \\ 3 & 4 & -5 \end{bmatrix}$$

5. $A + B$

6. $A - B$

7. $B - A$

8. $2A$

9. $\frac{1}{2}B$

10. $2A - \frac{1}{2}B$

In Exercises 11–14, let $(m \times n)$ denote a matrix of size $m \times n$. Find the size of the following products, or state that the product is not defined.

11. $(2 \times 3)(3 \times 2)$

12. $(3 \times 4)(4 \times 5)$

13. $(2 \times 3)(2 \times 3)$

14. $(1 \times 4)(4 \times 1)$

In Exercises 15–20, use matrices $A, B, C,$ and D to find the indicated matrix:

$$A = \begin{bmatrix} -1 & 2 \\ 3 & 5 \end{bmatrix}, B = \begin{bmatrix} 1 & 3 & -1 \\ 4 & -2 & 0 \end{bmatrix}, C = \begin{bmatrix} 1 & 0 & -1 \\ 4 & 0 & 3 \\ 2 & -1 & -2 \end{bmatrix}, D = \begin{bmatrix} 1 & -2 \\ -3 & -4 \end{bmatrix}$$

15. AB

16. BC

17. $AB + BC$

18. A^2

19. ABC

20. $(A + D)B$

In Exercises 21–23, matrix A is given. Find appropriate identity matrices I_m and I_n such that $I_m A = A$ and $AI_n = A$.

21. $A = \begin{bmatrix} 5 & 1 & -1 \\ 3 & -2 & 4 \end{bmatrix}$

22. $A = \begin{bmatrix} 2 & -1 & 0 & 3 \\ -1 & 4 & -2 & 5 \\ 5 & -3 & 7 & 9 \end{bmatrix}$

23. $A = \begin{bmatrix} 3 & -1 \\ 9 & 6 \\ 2 & 7 \\ 0 & 4 \\ 1 & -3 \end{bmatrix}$

24. The youth from a local church are having a breakfast fund-raising event. They are planning on serving biscuits, pancakes, and waffles. The ingredients for one batch of each are as follows:

Biscuits: 3 cups baking mix, 1 egg, 1 cup milk, $\frac{1}{2}$ tablespoon cooking oil

Pancakes: 2 cups baking mix, 2 eggs, $\frac{2}{3}$ cup milk

Waffles: $2\frac{1}{2}$ cups baking mix, 1 egg, 1 cup milk, 2 tablespoons cooking oil

The ingredients can be summarized by matrix A.

$$\begin{array}{c} \\ \text{Biscuits} \\ \text{Pancakes} \\ \text{Waffles} \end{array} \begin{array}{cccc} \text{Mix} & \text{Eggs} & \text{Milk} & \text{Oil} \\ \begin{bmatrix} 3 & 1 & 1 & \frac{1}{2} \\ 2 & 2 & \frac{2}{3} & 0 \\ 2\frac{1}{2} & 1 & 1 & 2 \end{bmatrix} \end{array} = A$$

It was determined that they were going to need 16 batches of biscuits, 30 batches of pancakes, and 40 batches of waffles, which is summarized by matrix B.

$$\begin{array}{c} \quad\quad\quad\quad\text{Biscuits}\quad\text{Pancakes}\quad\text{Waffles} \\ \text{Batches}\quad \begin{bmatrix} 16 & 30 & 40 \end{bmatrix} = B \end{array}$$

a. Calculate matrix BA.
b. Interpret the meaning of each entry of matrix BA.

25. A lawn-mower manufacturer produces two types of lawn-mowers, a standard model and the self-propelled model. The production of each model requires assembly and quality assurance testing. The number of hours required for each model are given by the following matrix.

$$\begin{array}{c} \quad\quad\quad\quad\quad\text{Assembly}\quad\quad\text{Testing} \\ \begin{array}{c}\text{Standard}\\ \text{Self-propelled}\end{array} \begin{bmatrix} 2 & .5 \\ 2.5 & 1 \end{bmatrix} = A \end{array}$$

The company has three manufacturing plants located in New York, Los Angeles, and Beijing, China. The hourly labor rates (in dollars) for each plant is given by the following matrix.

$$\begin{array}{c} \quad\quad\quad\quad\text{New York}\quad\quad\text{Los Angeles}\quad\text{Beijing} \\ \text{Assembly Testing}\begin{bmatrix} \$12 & \$10 & \$4 \\ \$9 & \$8 & \$2 \end{bmatrix} = B \end{array}$$

a. Calculate matrix AB.
b. If ab_{ij} represents the entry of matrix AB in row i and column j, then interpret the meaning of the entries ab_{21}, ab_{12}, and ab_{23}.

26. A small clothing store has locations in Buffalo, Rochester, and Syracuse. The store sells only three items: sport coats, slacks, and shirts. During the first quarter, the Buffalo branch sold 100 sport coats, 120 pairs of slacks, and 250 shirts. The Rochester store sold 110 sport coats, 100 slacks, and 300 shirts. The Syracuse store sold 180 sport coats, 200 slacks, and 320 shirts. At each store, sport coats sell for $100, slacks sell for $50, and shirts sell for $35. Use matrix multiplication to get a matrix that shows the total revenue for each store. What was the total revenue of all three stores combined?

8.2 Inverses of Matrices and Matrix Equations

THINGS TO KNOW

Before working through this section, be sure that you are familiar with the following concepts:

VIDEO   ANIMATION   INTERACTIVE

You Try It
1. Solving a System of Three Equations in Three Variables Using Gauss-Jordan Elimination (Section 7.3)

You Try It
2. Scalar Multiplication of Matrices (Section 8.1)

You Try It
3. Multiplication of Matrices (Section 8.1)

You Try It
4. Applications of Matrix Multiplication (Section 8.1)

OBJECTIVES

1 Understanding the Definition of an Inverse Matrix

2 Finding the Inverse of a 2×2 Matrix Using a Formula

3 Finding the Inverse of an Invertible Square Matrix

4 Solving Systems of Equations Using an Inverse Matrix

OBJECTIVE 1   UNDERSTANDING THE DEFINITION OF AN INVERSE MATRIX

In Section 8.1 we introduced the identity matrix and the identity property for matrix multiplication. Recall, if $A = \begin{bmatrix} 4 & 6 \\ -1 & -2 \end{bmatrix}$, then

$$AI_2 = \begin{bmatrix} 4 & 6 \\ -1 & -2 \end{bmatrix} \begin{bmatrix} 1 & 0 \\ 0 & 1 \end{bmatrix} = \begin{bmatrix} 4 & 6 \\ -1 & -2 \end{bmatrix} = A \quad \text{and}$$

$$I_2 A = \begin{bmatrix} 1 & 0 \\ 0 & 1 \end{bmatrix} \begin{bmatrix} 4 & 6 \\ -1 & -2 \end{bmatrix} = \begin{bmatrix} 4 & 6 \\ -1 & -2 \end{bmatrix} = A.$$

Now, suppose there exists a 2×2 matrix B such that $AB = BA = I_2$. For example, if

$$A = \begin{bmatrix} 4 & 6 \\ -1 & -2 \end{bmatrix} \text{ and } B = \begin{bmatrix} 1 & 3 \\ -\dfrac{1}{2} & -2 \end{bmatrix}, \text{ then}$$

$$AB = \begin{bmatrix} 4 & 6 \\ -1 & -2 \end{bmatrix} \begin{bmatrix} 1 & 3 \\ -\dfrac{1}{2} & -2 \end{bmatrix} = \begin{bmatrix} 4 - 3 & 12 - 12 \\ -1 + 1 & -3 + 4 \end{bmatrix} = \begin{bmatrix} 1 & 0 \\ 0 & 1 \end{bmatrix} = I_2 \quad \text{and}$$

$$BA = \begin{bmatrix} 1 & 3 \\ -\dfrac{1}{2} & -2 \end{bmatrix} \begin{bmatrix} 4 & 6 \\ -1 & -2 \end{bmatrix} = \begin{bmatrix} 4 - 3 & 6 - 6 \\ -2 + 2 & -3 + 4 \end{bmatrix} = \begin{bmatrix} 1 & 0 \\ 0 & 1 \end{bmatrix} = I_2$$

We say that matrix B is the **multiplicative inverse** of matrix A. We typically use the notation A^{-1} to denote the multiplicative inverse of matrix A.

Definition   Multiplicative Inverse of a Matrix
Let A be an $n \times n$ square matrix. If there exists an $n \times n$ matrix A^{-1} such that $AA^{-1} = A^{-1}A = I_n$, then A^{-1} is called the **multiplicative inverse** of matrix A.

Note   For the remainder of this section, the term inverse implies multiplicative inverse.

My video summary   ◉ **Example 1** Verify the Inverse of a Square Matrix

Verify that $A = \begin{bmatrix} 1 & 1 & 2 \\ 2 & 4 & -3 \\ 3 & 6 & -5 \end{bmatrix}$ and $B = \begin{bmatrix} 2 & -17 & 11 \\ -1 & 11 & -7 \\ 0 & 3 & -2 \end{bmatrix}$ are inverse matrices.

Solution To verify that the two matrices are inverses, we must show that $AB = BA = I_3$.

$$AB = \begin{bmatrix} 1 & 1 & 2 \\ 2 & 4 & -3 \\ 3 & 6 & -5 \end{bmatrix} \begin{bmatrix} 2 & -17 & 11 \\ -1 & 11 & -7 \\ 0 & 3 & -2 \end{bmatrix} = \begin{bmatrix} 1 & 0 & 0 \\ 0 & 1 & 0 \\ 0 & 0 & 1 \end{bmatrix} = I_3$$

$$BA = \begin{bmatrix} 2 & -17 & 11 \\ -1 & 11 & -7 \\ 0 & 3 & -2 \end{bmatrix} \begin{bmatrix} 1 & 1 & 2 \\ 2 & 4 & -3 \\ 3 & 6 & -5 \end{bmatrix} = \begin{bmatrix} 1 & 0 & 0 \\ 0 & 1 & 0 \\ 0 & 0 & 1 \end{bmatrix} = I_3$$

Because $AB = BA = I_3$, we have verified that matrix A is the inverse of matrix B and vice versa. Watch the video to see each step of the multiplication process. ⬤

You Try It Work through this You Try It problem.

Work Exercises 1–4 in this textbook or in the MyMathLab Study Plan.

OBJECTIVE 2   FINDING THE INVERSE OF A 2 × 2 MATRIX USING A FORMULA

The definition of the inverse of a matrix indicates that a matrix must be square in order for it to have an inverse. However, not every square matrix has an inverse. If a square matrix has an inverse, we say that the matrix is **invertible**. If a square matrix does not have an inverse, we say that the matrix is **singular**. There is an easy way to determine whether a 2 × 2 matrix is invertible or singular. Every square matrix has a number associated with it called the **determinant**. (We learn more about determinants in Section 8.3.) The formula for the determinant of a 2 × 2 matrix is as follows.

Determinant of a 2 × 2 Matrix

Let $A = \begin{bmatrix} a & b \\ c & d \end{bmatrix}$. The determinant of matrix A is denoted as $|A|$ or $\begin{vmatrix} a & b \\ c & d \end{vmatrix}$ and

is defined by $|A| = \begin{vmatrix} a & b \\ c & d \end{vmatrix} = ad - cb$. If $|A| \neq 0$, then matrix A is invertible.

You can see that if $|A| \neq 0$, then matrix A has an inverse. For example, the

matrix $A = \begin{bmatrix} 4 & 6 \\ -1 & -2 \end{bmatrix}$ is invertible because $a = 4, b = 6, c = -1, d = -2$, and

$|A| = \begin{vmatrix} 4 & 6 \\ -1 & -2 \end{vmatrix} = (4)(-2) - (-1)(6) = -8 + 6 = -2 \neq 0$. Following is a formula

for determining the inverse of a 2 × 2 matrix.

Formula for Determining the Inverse of a 2 × 2 Matrix

My video summary

▶ Let $A = \begin{bmatrix} a & b \\ c & d \end{bmatrix}$. If $|A| \neq 0$, then A is invertible and $A^{-1} = \dfrac{1}{|A|} \begin{bmatrix} d & -b \\ -c & a \end{bmatrix}$. See

the video proof.

 *My video summary*

⊘ Example 2 Find the Inverse of a 2 × 2 Matrix

Determine whether the following matrices are invertible or singular. If the matrix is invertible, find its inverse and then verify by using matrix multiplication.

a. $A = \begin{bmatrix} 2 & 3 \\ -4 & -6 \end{bmatrix}$ **b.** $B = \begin{bmatrix} 2 & -1 \\ 4 & 3 \end{bmatrix}$

Solution

a. $|A| = \begin{vmatrix} 2 & 3 \\ -4 & -6 \end{vmatrix} = (2)(-6) - (-4)(3) = -12 + 12 = 0.$ The determinant is zero, so matrix A is singular.

b. $|B| = \begin{vmatrix} 2 & -1 \\ 4 & 3 \end{vmatrix} = (2)(3) - (4)(-1) = 6 + 4 = 10.$ The determinant is nonzero, so matrix B is invertible.

$$B^{-1} = \frac{1}{10}\begin{bmatrix} 3 & 1 \\ -4 & 2 \end{bmatrix} = \begin{bmatrix} \dfrac{3}{10} & \dfrac{1}{10} \\ -\dfrac{2}{5} & \dfrac{1}{5} \end{bmatrix}$$

We can verify that this is the inverse matrix by showing that $BB^{-1} = B^{-1}B = I_2$. Watch the video to see the solution worked out in detail.

Using Technology

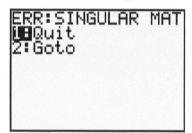

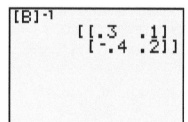

Using the matrix editor feature of a graphing utility, we can attempt to find the inverse of the matrices from Example 2. You can see in the first screenshot that an error occurs because matrix A is singular.

You Try It   Work through this You Try It problem.

Work Exercises 5–8 in this textbook or in the MyMathLab Study Plan.

OBJECTIVE 3   FINDING THE INVERSE OF AN INVERTIBLE SQUARE MATRIX

Before we learn how to find the inverse of any square invertible matrix, we start by showing an alternative way of finding the inverse of a 2 × 2 invertible matrix and then generalize this technique for any invertible $n \times n$ matrix. In Example 2, we saw that $B = \begin{bmatrix} 2 & -1 \\ 4 & 3 \end{bmatrix}$ was invertible.

Thus, there exists an inverse matrix B^{-1} such that $BB^{-1} = B^{-1}B = I_2$. If we let $B^{-1} = \begin{bmatrix} x & y \\ z & w \end{bmatrix}$, then

$$\underbrace{\begin{bmatrix} 2 & -1 \\ 4 & 3 \end{bmatrix}}_{B} \underbrace{\begin{bmatrix} x & y \\ z & w \end{bmatrix}}_{B^{-1}} = \underbrace{\begin{bmatrix} 1 & 0 \\ 0 & 1 \end{bmatrix}}_{I_2}.$$

Multiplying BB^{-1}, we get

$$\underbrace{\begin{bmatrix} 2x - z & 2y - w \\ 4x + 3z & 4y + 3w \end{bmatrix}}_{BB^{-1}} = \underbrace{\begin{bmatrix} 1 & 0 \\ 0 & 1 \end{bmatrix}}_{I_2}.$$

We can compare the corresponding entries of the previous two matrices to obtain the following two systems of equations:

$$\begin{array}{l} 2x - z = 1 \\ 4x + 3z = 0 \end{array} \quad \text{and} \quad \begin{array}{l} 2y - w = 0 \\ 4y + 3w = 1 \end{array}$$

These two systems correspond to the following two augmented matrices:

$$\left[\begin{array}{cc|c} 2 & -1 & 1 \\ 4 & 3 & 0 \end{array}\right] \quad \text{and} \quad \left[\begin{array}{cc|c} 2 & -1 & 0 \\ 4 & 3 & 1 \end{array}\right]$$

Because the coefficients of the left-hand side of the previous two matrices are exactly the same, we can combine the two augmented matrices into the following augmented matrix:

$$\left[\begin{array}{cc|c} 2 & -1 & 1 \\ 4 & 3 & 0 \end{array}\right] \left[\begin{array}{cc|c} 2 & -1 & 0 \\ 4 & 3 & 1 \end{array}\right]$$

$$\left[\begin{array}{cc|cc} 2 & -1 & 1 & 0 \\ 4 & 3 & 0 & 1 \end{array}\right]$$

To solve for each variable, we use Gauss-Jordan elimination to write the left-hand side of the augmented matrix in **reduced row-echelon form**. View these steps to see how the previous augmented matrix can be reduced to the following matrix:

$$\left[\begin{array}{cc|cc} 1 & 0 & \dfrac{3}{10} & \dfrac{1}{10} \\[2ex] 0 & 1 & -\dfrac{2}{5} & \dfrac{1}{5} \end{array}\right]$$

Therefore, we see that $x = \dfrac{3}{10}, y = \dfrac{1}{10}, z = -\dfrac{2}{5}$, and $w = \dfrac{1}{5}$. Because $B^{-1} = \begin{bmatrix} x & y \\ z & w \end{bmatrix}$,

then $B^{-1} = \begin{bmatrix} \dfrac{3}{10} & \dfrac{1}{10} \\[2ex] -\dfrac{2}{5} & \dfrac{1}{5} \end{bmatrix}$.

You can see that this is precisely what we obtained for the inverse in Example 2. Note that we started with the augmented matrix

$$\underbrace{\left[\begin{array}{cc|cc} 2 & -1 & 1 & 0 \\ 4 & 3 & 0 & 1 \end{array}\right]}_{B \qquad I_2},$$

then used Gauss-Jordan elimination to end up with the augmented matrix

$$\underbrace{\left[\begin{array}{cc|cc} 1 & 0 & \dfrac{3}{10} & \dfrac{1}{10} \\[2ex] 0 & 1 & -\dfrac{2}{5} & \dfrac{1}{5} \end{array}\right]}_{I_2 \qquad B^{-1}}.$$

This technique can be used to find the inverse of any invertible square matrix. The process is summarized as follows.

Steps for Finding the Multiplicative Inverse of an Invertible Square Matrix

If A is an $n \times n$ invertible matrix, then the multiplicative inverse of A can be obtained by following these steps:

Step 1. Form the augmented matrix $\left[A \,|\, I_n\right]$.

Step 2. Use Gauss-Jordan elimination to reduce A into the identity matrix I_n.

Step 3. The new augmented matrix is of the form $\left[I_n \,|\, A^{-1}\right]$, where A^{-1} is the multiplicative inverse of A.

 *My video summary*   ⊘ **Example 3** Find the Inverse of a 3 × 3 Matrix

If possible, determine the inverse of $A = \begin{bmatrix} 1 & 1 & 2 \\ 2 & 4 & -3 \\ 3 & 6 & -5 \end{bmatrix}$.

Solution To attempt to find the inverse of matrix A, we create the augmented matrix by attaching the identity matrix, I_3, to the right-hand side.

$$\left[\begin{array}{ccc|ccc} 1 & 1 & 2 & 1 & 0 & 0 \\ 2 & 4 & -3 & 0 & 1 & 0 \\ 3 & 6 & -5 & 0 & 0 & 1 \end{array}\right]$$
$$\underbrace{}_{A} \quad \underbrace{}_{I_3}$$

We now use Gauss-Jordan elimination to attempt to write the matrix on the left-hand side as I_3. Watch this video to see each step of the Gauss-Jordan elimination process. The reduced augmented matrix is

$$\left[\begin{array}{ccc|ccc} 1 & 0 & 0 & 2 & -17 & 11 \\ 0 & 1 & 0 & -1 & 11 & -7 \\ 0 & 0 & 1 & 0 & 3 & -2 \end{array}\right].$$
$$\underbrace{}_{I_3} \qquad \underbrace{}_{A^{-1}}$$

Therefore, $A^{-1} = \begin{bmatrix} 2 & -17 & 11 \\ -1 & 11 & -7 \\ 0 & 3 & -2 \end{bmatrix}$. You should verify that $AA^{-1} = A^{-1}A = I_3$. ●

You Try It Work through this You Try It problem.

Example 4 Example of a 3 × 3 Singular Matrix

If possible, determine the inverse of $A = \begin{bmatrix} 5 & 0 & -1 \\ 1 & -3 & -2 \\ 0 & 5 & 3 \end{bmatrix}$.

Solution To attempt to find the inverse of matrix A, we create the augmented matrix by attaching the identity matrix, I_3, to the right-hand side.

$$\underbrace{\begin{bmatrix} 5 & 0 & -1 \\ 1 & -3 & -2 \\ 0 & 5 & 3 \end{bmatrix}}_{A} \left| \underbrace{\begin{matrix} 1 & 0 & 0 \\ 0 & 1 & 0 \\ 0 & 0 & 1 \end{matrix}}_{I_3} \right.$$

We now use a series of row operations to try to obtain I_3 on the left-hand side:

$$\begin{bmatrix} 5 & 0 & -1 & | & 1 & 0 & 0 \\ 1 & -3 & -2 & | & 0 & 1 & 0 \\ 0 & 5 & 3 & | & 0 & 0 & 1 \end{bmatrix} \xrightarrow{R_1 \Leftrightarrow R_2} \begin{bmatrix} 1 & -3 & -2 & | & 0 & 1 & 0 \\ 5 & 0 & -1 & | & 1 & 0 & 0 \\ 0 & 5 & 3 & | & 0 & 0 & 1 \end{bmatrix} \xrightarrow{-5R_1 + R_2 \rightarrow \text{New } R_2}$$

$$\begin{bmatrix} 1 & -3 & -2 & | & 0 & 1 & 0 \\ 0 & 15 & 9 & | & 1 & -5 & 0 \\ 0 & 5 & 3 & | & 0 & 0 & 1 \end{bmatrix} \xrightarrow{\left(\frac{1}{15}\right) R_2 \rightarrow \text{New } R_2} \begin{bmatrix} 1 & -3 & -2 & | & 0 & 1 & 0 \\ 0 & 1 & \frac{3}{5} & | & \frac{1}{15} & -\frac{1}{3} & 0 \\ 0 & 5 & 3 & | & 0 & 0 & 1 \end{bmatrix} \xrightarrow{(-5)R_2 + R_3 \rightarrow \text{New } R_3}$$

$$\begin{bmatrix} 1 & -3 & -2 & | & 0 & 1 & 0 \\ 0 & 1 & \frac{3}{5} & | & \frac{1}{15} & -\frac{1}{3} & 0 \\ 0 & 0 & 0 & | & -\frac{1}{3} & \frac{5}{3} & 1 \end{bmatrix}$$

Because the last row of the matrix on the left-hand side is all zeros, there is no way to obtain the identity matrix. Therefore, matrix A does not have an inverse and thus is singular. In Section 8.3, we learn another way to determine whether a square matrix is invertible.

You Try It Work through this You Try It problem.

Work Exercises 9–14 in this textbook or in the MyMathLab Study Plan.

OBJECTIVE 4   SOLVING SYSTEMS OF EQUATIONS USING AN INVERSE MATRIX

My video summary   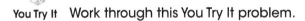 Suppose we are given the linear equation $ax = b$, where $a \neq 0$. If $a^{-1} = \dfrac{1}{a}$ (the multiplicative inverse of a), then we can solve for x as follows:

$$ax = b \qquad \text{Write the original linear equation.}$$

$$a^{-1}ax = a^{-1}b \qquad \text{Multiply both sides by the multiplicative inverse of } a.$$

$$\frac{1}{a}ax = a^{-1}b \qquad a^{-1} = \frac{1}{a}$$

$$x = a^{-1}b \qquad \frac{1}{a} \cdot a = 1$$

We can use this same technique to solve systems of n linear equations with n variables as follows.

Suppose we are given the linear system
$$a_{11}x_1 + a_{12}x_2 + a_{13}x_3 + \cdots \; a_{1n}x_n = b_1$$
$$a_{21}x_1 + a_{22}x_2 + a_{23}x_3 + \cdots \; a_{2n}x_n = b_2$$
$$\vdots \qquad \vdots \qquad \vdots \qquad \vdots \qquad \vdots$$
$$a_{n1}x_1 + a_{n2}x_2 + a_{n3}x_3 + \cdots \; a_{nn}x_n = b_n$$

This system is equivalent to the matrix equation

$$
\begin{bmatrix}
a_{11} & a_{12} & a_{13} & \cdots & a_{1n} \\
a_{21} & a_{22} & a_{23} & \cdots & a_{2n} \\
\vdots & \vdots & \vdots & & \vdots \\
a_{n1} & a_{n2} & a_{n3} & & a_{nn}
\end{bmatrix}
\begin{bmatrix}
x_1 \\ x_2 \\ \vdots \\ x_n
\end{bmatrix}
=
\begin{bmatrix}
b_1 \\ b_2 \\ \vdots \\ b_n
\end{bmatrix}.
$$

If we let $A = \begin{bmatrix} a_{11} & a_{12} & a_{13} & \cdots & a_{1n} \\ a_{21} & a_{22} & a_{23} & \cdots & a_{2n} \\ \vdots & \vdots & \vdots & & \vdots \\ a_{n1} & a_{n2} & a_{n3} & & a_{nn} \end{bmatrix}$, $X = \begin{bmatrix} x_1 \\ x_2 \\ \vdots \\ x_n \end{bmatrix}$, and $B = \begin{bmatrix} b_1 \\ b_2 \\ \vdots \\ b_n \end{bmatrix}$, then the previous

$\underbrace{\qquad\qquad\qquad\qquad}_{\text{Coefficient Matrix}}$ $\underbrace{\quad}_{\substack{\text{Variable} \\ \text{Matrix}}}$ $\underbrace{\quad}_{\substack{\text{Constant} \\ \text{Matrix}}}$

matrix equation can be written as $AX = B$. Matrix A is called the **coefficient matrix,** matrix X is called the **variable matrix,** and matrix B is called the **constant matrix.** If A is invertible, then we can determine the entries of matrix X by multiplying both sides of the matrix equation on the left by A^{-1}:

$AX = B$	Write the matrix equation.
$A^{-1}AX = A^{-1}B$	Multiply both sides of the equation on the left by A^{-1}.
$I_nX = A^{-1}B$	Definition of an inverse matrix ($A^{-1}A = I_n$)
$X = A^{-1}B$	Identity property for matrix multiplication ($I_nX = X$)

⚠ **Because matrix multiplication is** not commutative, **we must be careful when multiplying both sides of a matrix equation by another matrix. Notice that when we multiplied both sides of the previous equation by** A^{-1}, **we had to position** A^{-1} **on the left of each side.**

Note If $AX = B$, then $A^{-1}AX \neq BA^{-1}$.

My video summary ⊙ **Example 5** Solve a System of Equations Using an Inverse Matrix

Solve the following system of linear equations using an inverse matrix:

$$-3x + 2y = 4$$
$$5x - 4y = 9$$

Solution Let $A = \begin{bmatrix} -3 & 2 \\ 5 & -4 \end{bmatrix}$, $X = \begin{bmatrix} x \\ y \end{bmatrix}$, and $B = \begin{bmatrix} 4 \\ 9 \end{bmatrix}$. The linear system is equivalent to the matrix equation $AX = B$:

$$
\overset{A}{\begin{bmatrix} -3 & 2 \\ 5 & -4 \end{bmatrix}}
\overset{X}{\begin{bmatrix} x \\ y \end{bmatrix}}
=
\overset{B}{\begin{bmatrix} 4 \\ 9 \end{bmatrix}}
$$

We now calculate A^{-1} using the formula for finding the inverse of a 2×2 invertible matrix:

$$|A| = \begin{vmatrix} -3 & 2 \\ 5 & -4 \end{vmatrix} = (-3)(-4) - (5)(2) = 2, \text{ so } A^{-1} = \frac{1}{2}\begin{bmatrix} -4 & -2 \\ -5 & -3 \end{bmatrix} = \begin{bmatrix} -2 & -1 \\ -\dfrac{5}{2} & -\dfrac{3}{2} \end{bmatrix}$$

To solve for X, we must multiply both sides of the matrix equation on the left by A^{-1}.

$$\overbrace{\begin{bmatrix} -3 & 2 \\ 5 & -4 \end{bmatrix}}^{A}\overbrace{\begin{bmatrix} x \\ y \end{bmatrix}}^{X} = \overbrace{\begin{bmatrix} 4 \\ 9 \end{bmatrix}}^{B}$$

Write the matrix equation $AX = B$.

$$\overbrace{\begin{bmatrix} -2 & -1 \\ -\dfrac{5}{2} & -\dfrac{3}{2} \end{bmatrix}}^{A^{-1}}\overbrace{\begin{bmatrix} -3 & 2 \\ 5 & -4 \end{bmatrix}}^{A}\overbrace{\begin{bmatrix} x \\ y \end{bmatrix}}^{X} = \overbrace{\begin{bmatrix} -2 & -1 \\ -\dfrac{5}{2} & -\dfrac{3}{2} \end{bmatrix}}^{A^{-1}}\overbrace{\begin{bmatrix} 4 \\ 9 \end{bmatrix}}^{B}$$

Multiply both sides of the equation on the left by A^{-1}.

$$\underbrace{\begin{bmatrix} 1 & 0 \\ 0 & 1 \end{bmatrix}}_{I_2}\begin{bmatrix} x \\ y \end{bmatrix} = \begin{bmatrix} -17 \\ \dfrac{47}{2} \end{bmatrix}$$

Definition of an inverse matrix $(A^{-1}A = I_n)$

$$\begin{bmatrix} x \\ y \end{bmatrix} = \begin{bmatrix} -17 \\ \dfrac{47}{2} \end{bmatrix}$$

Identity property for matrix multiplication $(I_n X = X)$

So, $x = -17$ and $y = -\dfrac{47}{2}$.

Example 6 Solve a System of Equations Using an Inverse Matrix

Solve the following system of linear equations using an inverse matrix:

$$2x + 4y + z = 3$$
$$-x + y - z = 6$$
$$x + 4y \quad\;\; = 7$$

Solution The linear system can be written as the following matrix equation:

$$\underbrace{\begin{bmatrix} 2 & 4 & 1 \\ -1 & 1 & -1 \\ 1 & 4 & 0 \end{bmatrix}}_{A}\underbrace{\begin{bmatrix} x \\ y \\ z \end{bmatrix}}_{X} = \underbrace{\begin{bmatrix} 3 \\ 6 \\ 7 \end{bmatrix}}_{B}$$

The solution can be obtained by multiplying both sides of this matrix equation by A^{-1} to obtain $X = A^{-1}B$. To find the inverse matrix, we create the augmented matrix by attaching the identity matrix, I_3, to the right-hand side.

$$\left[\underbrace{\begin{array}{ccc} 2 & 4 & 1 \\ -1 & 1 & -1 \\ 1 & 4 & 0 \end{array}}_{A}\middle|\;\underbrace{\begin{array}{ccc} 1 & 0 & 0 \\ 0 & 1 & 0 \\ 0 & 0 & 1 \end{array}}_{I_3}\right]$$

After a series of row operations, we obtain the following augmented matrix:

$$\left[\begin{array}{ccc|ccc} 1 & 0 & 0 & -4 & -4 & 5 \\ 0 & 1 & 0 & 1 & 1 & -1 \\ 0 & 0 & 1 & 5 & 4 & -6 \end{array}\right]$$

$\underbrace{}_{I_3}$ $\underbrace{}_{A^{-1}}$

(View this row reduction process.)

Therefore, $A^{-1} = \begin{bmatrix} -4 & -4 & 5 \\ 1 & 1 & -1 \\ 5 & 4 & -6 \end{bmatrix}$.

The solution to the system of equations is

$$X = A^{-1}B$$

$$= \begin{bmatrix} -4 & -4 & 5 \\ 1 & 1 & -1 \\ 5 & 4 & -6 \end{bmatrix}\begin{bmatrix} 3 \\ 6 \\ 7 \end{bmatrix}$$

$$= \begin{bmatrix} -1 \\ 2 \\ -3 \end{bmatrix}$$

Thus, $x = -1, y = 2$, and $z = -3$.

Using Technology

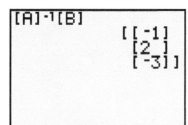

A graphing utility can be used to solve the linear system from Example 6. First, use the matrix editor feature to input the entries of the coefficient matrix A and the constant matrix B. Then, find the product $A^{-1}B$.

You Try It   Work through this You Try It problem.

Work Exercises 15–21 in this textbook or in the MyMathLab Study Plan.

8.2 Exercises

In Exercises 1–4, determine whether matrix B is the inverse of matrix A by finding the products AB and BA.

1. $A = \begin{bmatrix} -3 & 1 \\ -5 & 2 \end{bmatrix}, B = \begin{bmatrix} -2 & 1 \\ -5 & 3 \end{bmatrix}$

2. $A = \begin{bmatrix} 2 & -4 \\ 1 & -3 \end{bmatrix}, B = \begin{bmatrix} \dfrac{3}{2} & -2 \\ \dfrac{1}{2} & -1 \end{bmatrix}$

3. $A = \begin{bmatrix} -28 & -13 & 3 \\ 2 & 1 & 0 \\ -7 & -3 & 1 \end{bmatrix}, B = \begin{bmatrix} 1 & 4 & -3 \\ -2 & -7 & 6 \\ 1 & 7 & -2 \end{bmatrix}$

4. $A = \begin{bmatrix} -5 & 7 & 8 \\ -5 & 8 & 9 \\ -5 & 9 & 8 \end{bmatrix}, B = \begin{bmatrix} -\dfrac{17}{10} & \dfrac{8}{5} & -\dfrac{1}{10} \\ -\dfrac{1}{2} & 0 & \dfrac{1}{2} \\ -\dfrac{1}{2} & 1 & -\dfrac{1}{2} \end{bmatrix}$

In Exercises 5–8, determine whether the given 2×2 matrix is invertible or singular. If it is invertible, find the inverse matrix.

5. $\begin{bmatrix} 2 & 1 \\ 5 & 3 \end{bmatrix}$

6. $\begin{bmatrix} -5 & 4 \\ -2 & 2 \end{bmatrix}$

7. $\begin{bmatrix} -3 & 6 \\ 4 & -8 \end{bmatrix}$

8. $\begin{bmatrix} \dfrac{1}{3} & \dfrac{5}{4} \\ \dfrac{6}{5} & \dfrac{9}{2} \end{bmatrix}$

In Exercises 9–14, find the inverse of each matrix if possible.

9. $\begin{bmatrix} 1 & -1 \\ 2 & 0 \end{bmatrix}$

10. $\begin{bmatrix} 1 & -6 & -2 \\ 0 & -1 & 0 \\ 2 & -11 & -3 \end{bmatrix}$

11. $\begin{bmatrix} 1 & 3 & 2 \\ 2 & -1 & 4 \\ 3 & 2 & 6 \end{bmatrix}$

12. $\begin{bmatrix} 1 & 4 & -3 \\ -2 & -7 & 6 \\ 1 & 7 & -2 \end{bmatrix}$

13. $\begin{bmatrix} 1 & 1 & 3 \\ -1 & -1 & -2 \\ -1 & 3 & 5 \end{bmatrix}$

14. $\begin{bmatrix} 1 & 0 & -2 & 0 \\ 0 & 1 & 0 & -5 \\ -4 & 0 & 9 & 0 \\ 0 & 2 & 1 & -9 \end{bmatrix}$

In Exercises 15–21, solve each linear system using an inverse matrix.

15. $\begin{aligned} -x + y &= -5 \\ -2x + 3y &= -13 \end{aligned}$

16. $\begin{aligned} -3x + 4y &= 3 \\ x - 2y &= -\dfrac{4}{3} \end{aligned}$

17. $\begin{aligned} -\dfrac{3}{10}x - \dfrac{1}{10}y &= -3 \\ \dfrac{2}{5}x - 1\dfrac{1}{5}y &= -7 \end{aligned}$

18. $\begin{aligned} x + y + 3z &= 11 \\ -x - y - 2z &= -7 \\ x - 3y - 5z &= -13 \end{aligned}$

19. $\begin{aligned} x + 4y - 3z &= -5 \\ -2x - 7y + 6z &= 11 \\ x + 7y - 2z &= 1 \end{aligned}$

20. $\begin{aligned} x + y - 2z &= -4 \\ x + 3y + 4z &= 8 \\ x + 2y - z &= 2 \end{aligned}$

21. $\begin{aligned} x - y + 2z + 3w &= 2 \\ 2x - y + 6z + 5w &= 6 \\ 3x - y + 9z + 6w &= 9 \\ 2x - 2y + 4z + 7w &= 5 \end{aligned}$

8.3 Determinants and Cramer's Rule

THINGS TO KNOW

Before working through this section, be sure that you are familiar with the following concepts:

VIDEO   ANIMATION   INTERACTIVE

You Try It

1. Solving a System of Three Equations in Three Variables Using Gauss-Jordan Elimination (Section 7.3)

You Try It 2. Finding the Inverse of a 2 × 2 Matrix Using a Formula (Section 8.2)

 You Try It 3. Solving Systems of Equations Using an Inverse Matrix (Section 8.2)

OBJECTIVES

1 Using Cramer's Rule to Solve Linear Systems in Two Variables

2 Calculating the Determinant of an $n \times n$ Matrix by Expansion of Minors

3 Using Cramer's Rule to Solve Linear Systems in n Variables

OBJECTIVE 1 USING CRAMER'S RULE TO SOLVE LINEAR SYSTEMS IN TWO VARIABLES

In this section, we learn how determinants can be used to solve systems of linear equations. Let's start by considering the following system of linear equations:

$$ax + by = m$$

$$cx + dy = n$$

In Section 8.2, we saw that this linear system can be written in matrix form as

$$\underbrace{\begin{bmatrix} a & b \\ c & d \end{bmatrix}}_{A} \underbrace{\begin{bmatrix} x \\ y \end{bmatrix}}_{X} = \underbrace{\begin{bmatrix} m \\ n \end{bmatrix}}_{B}$$

We now define two new matrices, D_x and D_y. Matrix D_x is created by replacing the first column of the coefficient matrix by the entries of the constant matrix. So, $D_x = \begin{bmatrix} m & b \\ n & d \end{bmatrix}$. Similarly, matrix D_y is created by replacing the second column of the coefficient matrix by the entries of the constant matrix. Therefore, $D_y = \begin{bmatrix} a & m \\ c & n \end{bmatrix}$.

Recall that the determinant of the coefficient matrix is defined as $|A| = \begin{vmatrix} a & b \\ c & d \end{vmatrix} = ad - cb$. So, $|D_x| = \begin{vmatrix} m & b \\ n & d \end{vmatrix} = md - nb$ and $|D_y| = \begin{vmatrix} a & m \\ c & n \end{vmatrix} = an - cm$. If $|A| \neq 0$, then the linear system has a unique solution where x and y can be found by the following formula:

$$x = \frac{|D_x|}{|A|} \quad \text{and} \quad y = \frac{|D_y|}{|A|}$$

This technique of using determinants to solve a linear system is known as **Cramer's rule**.

Cramer's Rule for a System of Linear Equations in Two Variables

For the linear system of equations $\begin{matrix} ax + by = m \\ cx + dy = n \end{matrix}$, let $|A| = \begin{vmatrix} a & b \\ c & d \end{vmatrix}$, $|D_x| = \begin{vmatrix} m & b \\ n & d \end{vmatrix}$,

and $|D_y| = \begin{vmatrix} a & m \\ c & n \end{vmatrix}$. If $|A| \neq 0$, then there is a unique solution to this system

given by

$$x = \frac{|D_x|}{|A|} \quad \text{and} \quad y = \frac{|D_y|}{|A|}$$

My video summary ⊘ The proof of Cramer's rule is actually not too difficult. We can prove Cramer's rule for a system of linear equations in two variables by solving the linear system using the method of elimination that was discussed in Section 7.1. Watch this video proof of Cramer's rule.

My video summary ⊘ **Example 1** Use Cramer's Rule to Solve a Linear System in Two Variables

Use Cramer's rule to solve the following system:

$$\begin{aligned} 2x - 2y &= 4 \\ 3x + y &= -3 \end{aligned}$$

Solution For this particular system,

$$|A| = \begin{vmatrix} 2 & -2 \\ 3 & 1 \end{vmatrix} = (2)(1) - (3)(-2) = 8,$$

$$|D_x| = \begin{vmatrix} 4 & -2 \\ -3 & 1 \end{vmatrix} = (4)(1) - (-3)(-2) = -2, \quad \text{and}$$

$$|D_y| = \begin{vmatrix} 2 & 4 \\ 3 & -3 \end{vmatrix} = (2)(-3) - (3)(4) = -18.$$

Therefore, by Cramer's rule, the solution is

$$x = \frac{|D_x|}{|A|} = \frac{-2}{8} = -\frac{1}{4} \quad \text{and} \quad y = \frac{|D_y|}{|A|} = \frac{-18}{8} = -\frac{9}{4}$$

You Try It Work through this You Try It problem.

Work Exercises 1–4 in this textbook or in the MyMathLab Study Plan.

Cramer's rule can actually be used to solve larger systems of n equations with n unknowns. Before we learn how to extend the use of Cramer's rule for linear systems of more than two equations, we must first learn how to find the determinant of larger square matrices.

OBJECTIVE 2   CALCULATING THE DETERMINANT OF AN $n \times n$ MATRIX BY EXPANSION OF MINORS

Before we learn how to calculate the determinant of any square matrix, we first learn how to calculate the determinant of a 3×3 matrix. For every entry of a 3×3 matrix, there is a 2×2 matrix associated with it. This 2×2 matrix is obtained by deleting the row and the column in which this entry appears. The determinants of these 2×2 matrices are called **minors**. In general, if a_{ij} is the entry in the ith row and jth column of a 3×3 matrix, then matrix M_{ij} is obtained by deleting the ith row and the jth column of this matrix. The determinant $|M_{ij}|$ is the minor corresponding to entry a_{ij}. For example, consider the following 3×3 matrix:

$$\begin{bmatrix} 2 & -17 & 11 \\ -1 & 11 & -7 \\ 0 & 3 & -2 \end{bmatrix}$$

To find the minor for the entry $a_{11} = 2$, we delete the first row and the first column to obtain a 2×2 matrix.

$$\begin{bmatrix} 2 & 17 & 11 \\ -1 & 11 & -7 \\ 0 & 3 & -2 \end{bmatrix}$$ The 2×2 matrix associated with the entry $a_{11} = 2$ is

$$[M_{11}] = \begin{bmatrix} 11 & -7 \\ 3 & -2 \end{bmatrix}.$$

Therefore, the minor for $a_{11} = 2$ is $|M_{11}| = \begin{vmatrix} 11 & -7 \\ 3 & -2 \end{vmatrix} = -1$. Similarly, the minor for the entry $a_{12} = -17$ is found by deleting the first row and second column.

$$\begin{bmatrix} 2 & -17 & 11 \\ -1 & 11 & -7 \\ 0 & 3 & -2 \end{bmatrix}$$ The 2×2 matrix associated with the entry $a_{12} = -17$ is

$$[M_{12}] = \begin{bmatrix} -1 & -7 \\ 0 & -2 \end{bmatrix}.$$

Thus, the minor for the entry $a_{12} = -17$ is $|M_{12}| = \begin{vmatrix} -1 & -7 \\ 0 & -2 \end{vmatrix} = (-1)(-2) -$

$(0)(-7) = 2.$

To find the determinant of a 3×3 matrix, we use a technique called **expansion by minors**.

Determinant of a 3 × 3 Matrix Using Expansion by Minors

Let $A = \begin{bmatrix} a_{11} & a_{12} & a_{13} \\ a_{21} & a_{22} & a_{23} \\ a_{31} & a_{32} & a_{33} \end{bmatrix}$, then $|A| = a_{11}|M_{11}| - a_{12}|M_{12}| + a_{13}|M_{13}|$.

My video summary ⊘ **Example 2** Calculate the Determinant of a **3 × 3** Matrix

Calculate $|A|$ if $A = \begin{bmatrix} 2 & -17 & 11 \\ -1 & 11 & -7 \\ 0 & 3 & -2 \end{bmatrix}$.

Solution

$$|A| = a_{11}|M_{11}| - a_{12}|M_{12}| + a_{13}|M_{13}|$$

$$= 2\begin{vmatrix} 11 & -7 \\ 3 & -2 \end{vmatrix} - (-17)\begin{vmatrix} -1 & -7 \\ 0 & -2 \end{vmatrix} + 11\begin{vmatrix} -1 & 11 \\ 0 & 3 \end{vmatrix}$$

$$= 2(-1) + 17(2) + 11(-3)$$

$$= -1$$

Using Technology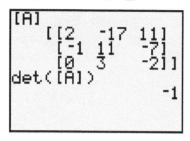

We can use a graphing utility to find the determinant of the matrix from Example 2. ●

In Example 2, we calculated the determinant by expanding the minors across the first row. Actually, we can find the determinant using the expansions of minors across any row or down any column. We have to be careful to properly alternate the signs of the coefficients of the minors. The coefficient, a_{ij}, of each minor must be preceded by a "+" or "−", which can be determined by the formula $(-1)^{i+j}$. If the sum $i + j$ is odd, then the sign preceding the minor is $(-1)^{i+j} = $ "−". Likewise, if the sum $i + j$ is even, the sign preceding the coefficient is $(-1)^{i+j} = $ "+". The signs can be summarized in the following matrix:

$$\begin{bmatrix} + & - & + \\ - & + & - \\ + & - & + \end{bmatrix}$$

My video summary   ◎ **Example 3** Calculate the Determinant of a **3 × 3** Matrix by Expansion of Minors Down the First Column

Find $|A|$ using the matrix from Example 2 by expanding the minors down the first column.

Solution Expanding the minors down the first column, we get

$$|A| = \begin{bmatrix} 2 & -17 & 11 \\ -1 & 11 & -7 \\ 0 & 3 & -2 \end{bmatrix}$$

$$= 2\begin{vmatrix} 11 & -7 \\ 3 & -2 \end{vmatrix} - (-1)\begin{vmatrix} -17 & 11 \\ 3 & -2 \end{vmatrix} + 0\begin{vmatrix} -17 & 11 \\ 11 & -7 \end{vmatrix}$$

$$= 2(-1) + 1(1) + 0$$

$$= -1$$

Notice that we obtained the same value as we did in Example 2. When calculating determinants, it is typically easier to expand by minors across rows or down columns that contain zeros.

In Example 4 of Section 8.2, we found that the matrix $A = \begin{bmatrix} 5 & 0 & -1 \\ 1 & -3 & -2 \\ 0 & 5 & 3 \end{bmatrix}$ did not have an inverse (i.e., the matrix is **singular**). The determinant of this matrix can be found by expanding the minors down the first column:

$$|A| = \begin{bmatrix} 5 & 0 & -1 \\ 1 & -3 & -2 \\ 0 & 5 & 3 \end{bmatrix} = 5\begin{vmatrix} -3 & -2 \\ 5 & 3 \end{vmatrix} - 1\begin{vmatrix} 0 & -1 \\ 5 & 3 \end{vmatrix} = 5(1) - 1(5) = 0$$

The fact that the determinant of this matrix is zero is no coincidence, as stated in the following theorem. ●

Theorem

A square matrix A is invertible if and only if $|A| \neq 0$. (If $|A| = 0$, then A is a singular matrix.)

You Try It   Work through this You Try It problem.

We can calculate the determinant of any $n \times n$ matrix in exactly the same way as with a 3×3 matrix by expansion of minors across any row or down any column. We have to make sure to alternate the signs of each coefficient properly. We define the determinant of any square matrix as follows.

Determinant of a Square Matrix

Let $A = \begin{bmatrix} a_{11} & a_{12} & \cdots & a_{1n} \\ a_{21} & a_{22} & \cdots & a_{2n} \\ \vdots & \vdots & \ddots & \vdots \\ a_{n1} & a_{n2} & \cdots & a_{nn} \end{bmatrix}$, then for each entry a_{ij} there is a corresponding

minor $|M_{ij}|$, which is the determinant of a $(n-1) \times (n-1)$ matrix. The determinant $|A|$ is equal to the sum of the products of the entries of any row or column and the corresponding minors. (Each coefficient must be preceded by $(-1)^{i+j}$.)

My video summary   ⊙ **Example 4 Calculate the Determinant of a 4 × 4 Matrix**

Calculate $|A|$ if $A = \begin{bmatrix} 8 & 0 & 1 & 10 \\ -1 & 0 & 0 & -5 \\ 5 & 2 & -2 & 4 \\ -9 & 0 & 0 & -12 \end{bmatrix}$.

Solution Because the second column only has one nonzero entry, we calculate $|A|$ by expanding down the second column. The first entry of column 2 is $a_{12} = 0$. Note that $i = 1$ and $j = 2$, so $i + j = 1 + 2 = 3$, which is an odd number. Therefore, $(-1)^{i+j} = (-1)^{1+2} = (-1)^3 = -1$. The coefficient of the first term must be preceded by a negative sign. The signs then alternate.

$$|A| = -0 \cdot |M_{12}| + 0 \cdot |M_{22}| - 2 \cdot |M_{32}| + 0 \cdot |M_{42}|$$

$$= -2 \cdot |M_{32}|$$

$$= -2 \cdot \begin{vmatrix} 8 & 1 & 10 \\ -1 & 0 & -5 \\ -9 & 0 & -12 \end{vmatrix} \qquad \text{Use expansion by minors down the second column.}$$

$$= -2 \cdot (-1) \begin{vmatrix} -1 & -5 \\ -9 & -12 \end{vmatrix}$$

$$= -2 \cdot (-1)(12 - 45)$$

$$= -66$$

Watch the video to see this solution worked out in detail.

You Try It Work through this You Try It problem.

Work Exercises 5–12 in this textbook or in the MyMathLab Study Plan.

OBJECTIVE 3 USING CRAMER'S RULE TO SOLVE LINEAR SYSTEMS IN n VARIABLES

Now that we know how to compute the determinant of an $n \times n$ matrix, we can extend the use of Cramer's rule to solve a system of linear equations in n variables.

Cramer's Rule for a System of Linear Equations in n Variables

Given the linear system

$$
\begin{aligned}
a_{11}x_1 + a_{12}x_2 + a_{13}x_3 + \cdots a_{1n}x_n &= b_1 \\
a_{21}x_1 + a_{22}x_2 + a_{23}x_3 + \cdots a_{2n}x_n &= b_2 \\
\vdots \qquad\qquad\qquad \vdots \\
a_{n1}x_1 + a_{n2}x_2 + a_{n3}x_3 + \cdots a_{nn}x_n &= b_n
\end{aligned}
$$
, or equivalently

$$
\underbrace{\begin{bmatrix}
a_{11} & a_{12} & a_{13} & \cdots & a_{1n} \\
a_{21} & a_{22} & a_{23} & \cdots & a_{2n} \\
\vdots & \vdots & \vdots & & \vdots \\
a_{n1} & a_{n2} & a_{n3} & & a_{nn}
\end{bmatrix}}_{\substack{\text{Coefficient} \\ \text{Matrix } A}}
\underbrace{\begin{bmatrix} x_1 \\ x_2 \\ \vdots \\ x_n \end{bmatrix}}_{\substack{\text{Variable} \\ \text{Matrix } X}}
=
\underbrace{\begin{bmatrix} b_1 \\ b_2 \\ \vdots \\ b_n \end{bmatrix}}_{\substack{\text{Constant} \\ \text{Matrix } B}},
$$

if $|D_{x_i}| = \underbrace{\begin{vmatrix}
a_{11} & a_{12} & \cdots & b_1 & \cdots & a_{1n} \\
a_{21} & a_{22} & \cdots & b_2 & \cdots & a_{2n} \\
\vdots & & \cdots & \vdots & & \vdots \\
a_{n1} & a_{n2} & \cdots & b_n & & a_{nn}
\end{vmatrix}}_{\substack{\text{Replace the } ith \text{ column of } A \text{ with the} \\ \text{coefficients of the constant matrix.}}}$ and if $|A| \neq 0,$

then there is a unique solution to the linear system given by $x_i = \dfrac{|D_{xi}|}{|A|}.$

My video summary ⊙ **Example 5** Use Cramer's Rule to Solve a Linear System in Three Variables

Use Cramer's rule to solve the system
$$
\begin{aligned}
x + y + z &= 2 \\
x - 2y + z &= 5. \\
2x + y - z &= -1
\end{aligned}
$$

Solution We must calculate the following four determinants:

$$
|A| = \begin{vmatrix} 1 & 1 & 1 \\ 1 & -2 & 1 \\ 2 & 1 & -1 \end{vmatrix}, \ |D_x| = \begin{vmatrix} 2 & 1 & 1 \\ 5 & -2 & 1 \\ -1 & 1 & -1 \end{vmatrix},
$$

$$
|D_y| = \begin{vmatrix} 1 & 2 & 1 \\ 1 & 5 & 1 \\ 2 & -1 & -1 \end{vmatrix}, \ \text{and} \ |D_z| = \begin{vmatrix} 1 & 1 & 2 \\ 1 & -2 & 5 \\ 2 & 1 & -1 \end{vmatrix}
$$

Work through the interactive video to verify that $|A| = 9$, $|D_x| = 9$, $|D_y| = -9$, and $|D_z| = 18$. Therefore,

$$x = \frac{|D_x|}{|A|} = \frac{9}{9} = 1, \quad y = \frac{|D_y|}{|A|} = \frac{-9}{9} = -1, \quad \text{and} \quad z = \frac{|D_z|}{|A|} = \frac{18}{9} = 2.$$

You Try It   Work through this You Try It problem.

Work Exercises 13–18 in this textbook or in the MyMathLab Study Plan.

8.3 Exercises

In Exercises 1–4, use Cramer's rule to solve each linear system.

1. $\begin{aligned} x + y &= 3 \\ -2x + y &= 0 \end{aligned}$
  2. $\begin{aligned} 2x - y &= 1 \\ -3x + 3y &= 0 \end{aligned}$
  3. $\begin{aligned} 4x - y &= -4 \\ -3x + 3y &= -6 \end{aligned}$
  4. $\begin{aligned} 2x - 3y &= 0 \\ 4x + 9y &= 5 \end{aligned}$

In Exercises 5–12, calculate $|A|$.

5. $A = \begin{bmatrix} 2 & 2 & 2 \\ -1 & 5 & -2 \\ 3 & 7 & 4 \end{bmatrix}$
  6. $A = \begin{bmatrix} 7 & -3 & 8 \\ 1 & -5 & -2 \\ 3 & 7 & 4 \end{bmatrix}$
  7. $A = \begin{bmatrix} 7 & 0 & 4 \\ -1 & 3 & -1 \\ 6 & 2 & 5 \end{bmatrix}$

8. $A = \begin{bmatrix} -1 & 0 & 2 \\ 4 & 0 & -1 \\ 3 & 0 & -4 \end{bmatrix}$
  9. $A = \begin{bmatrix} 2 & 0 & 0 \\ 0 & 3 & 0 \\ 0 & 0 & 1 \end{bmatrix}$
  10. $A = \begin{bmatrix} -5 & -1 & -2 \\ 8 & 10 & 0 \\ -7 & 5 & 0 \end{bmatrix}$

11. $A = \begin{bmatrix} 1 & 5 & 1 & 9 \\ 2 & -1 & 7 & 3 \\ 9 & 0 & 0 & 1 \\ 12 & -4 & -2 & 8 \end{bmatrix}$
  12. $A = \begin{bmatrix} 0 & -1 & 6 & 6 \\ 7 & -3 & 0 & 2 \\ 1 & 0 & 2 & 3 \\ 3 & 0 & 4 & 8 \end{bmatrix}$

In Exercises 13–18, use Cramer's rule to solve each linear system.

13. $\begin{aligned} x + y - z &= 4 \\ 5x - y + 3z &= 6 \\ x + y - 5z &= 8 \end{aligned}$
  14. $\begin{aligned} 3x + 3y + z &= 6 \\ 4x - y + z &= 8 \\ -x + y - 3z &= -2 \end{aligned}$
  15. $\begin{aligned} -2x + 5y + 5z &= -4 \\ 5x + 6y + 5z &= -23 \\ -3x + 4y + 5z &= -3 \end{aligned}$

16. $\begin{aligned} x \quad\quad - z &= -1 \\ 2y - z &= 8 \\ 2x + 2y \quad\quad &= 0 \end{aligned}$
  17. $\begin{aligned} 3x + 5z &= 0 \\ 2x + 4y &= -14 \\ -y + 2z &= -21 \end{aligned}$
  18. $\begin{aligned} -2w + 6x - 9y + z &= 57 \\ -w + x - y + z &= 11 \\ w + x + y + z &= 1 \\ 3w + 2x + 2y + z &= -2 \end{aligned}$

Sequences and Series; Counting and Probability

CHAPTER NINE CONTENTS

9.1 Introduction to Sequences and Series

THINGS TO KNOW

Before working through this section, be sure that you are familiar with the following concepts:

VIDEO   ANIMATION   INTERACTIVE

You Try It

1. Using the Vertical Line Test (Section 3.1)

OBJECTIVES

1 Writing the Terms of a Sequence

2 Writing the Terms of a Recursive Sequence

3 Writing the General Term for a Given Sequence

4 Computing Partial Sums of a Series

5 Determining the Sum of a Finite Series Written in Summation Notation

6 Writing a Series Using Summation Notation

OBJECTIVE 1   WRITING THE TERMS OF A SEQUENCE

Consider the function $f(n) = 2n - 1$, where n is a natural number. The graph of this function consists of infinitely many ordered pairs of the form $(n, 2n - 1)$, where $n \geq 1$. Therefore, the ordered pairs that lie on the graph of this function are $(1, 1), (2, 3), (3, 5), (4, 7)$, and so on. A portion of the graph of this function can be seen in Figure 1.

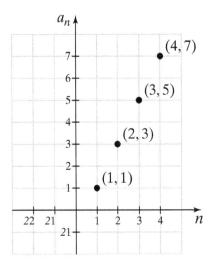

Figure 1 A portion of the graph of the function $f(n) = 2n - 1$, where n is a natural number

Clearly, the graph seen in Figure 1 is a function because the graph passes the vertical line test. Any function whose domain is the set of natural numbers is called an **infinite sequence**. Instead of using the conventional function notation $f(n)$ to name a sequence, we will use a subscript notation, such as a_n (read as "a sub n"), to name our sequences. We now formally define a sequence.

Definition   Sequence

A **finite sequence** is a function whose domain is the finite set $\{1, 2, 3, \ldots, n\}$, where n is a natural number.

An **infinite sequence** is a function whose domain is the set of all natural numbers.

The range values of a sequence are called the **terms** of the sequence.

We now rename the sequence $f(n) = 2n - 1$ as $a_n = 2n - 1$. The first four terms of this sequence are $a_1 = 1$, $a_2 = 3$, $a_3 = 5$, and $a_4 = 7$.

Using Technology

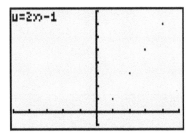

A graphing utility set to *sequence mode* can be used to sketch the graph of a sequence. The figure on the left shows a portion of the graph of the sequence $a_n = 2n - 1$.

Before we start to find the terms of a sequence it is important to introduce factorial notation.

Definition   The Factorial of a Non-Negative Integer

The factorial of a non-negative integer n, denoted as $n!$, is the product of all positive integers less than or equal to n. Thus, $n! = n(n - 1) \cdot \cdots \cdot 3 \cdot 2 \cdot 1$.

Note   By definition, we say that zero factorial is equal to 1 or $0! = 1$.

Some examples of factorial notation are

$$5! = 5 \cdot 4 \cdot 3 \cdot 2 \cdot 1 = 120 \text{ and } 8! = 8 \cdot 7 \cdot 6 \cdot 5 \cdot 4 \cdot 3 \cdot 2 \cdot 1 = 40{,}320.$$

As you can see, the value of $n!$ gets large rather quickly. In fact, the value of 13! is over 6 billion, which is a rough estimate of the population of Earth! Example 1c illustrates how factorial notation can be used to define a sequence.

 *My interactive video summary*

⟩ Example 1 Writing the Terms of a Sequence

Write the first four terms of each sequence whose nth term is given.

a. $a_n = 2n - 1$ **b.** $b_n = n^2 - 1$ **c.** $c_n = \dfrac{3^n}{(n-1)!}$ **d.** $d_n = (-1)^n 2^{n-1}$

Solution

a. To find the first four terms of the sequence, we evaluate $a_n = 2n - 1$ when n is 1, 2, 3, and 4.

$$a_1 = 2(1) - 1 = 2 - 1 = 1, a_2 = 2(2) - 1 = 4 - 1 = 3,$$
$$a_3 = 2(3) - 1 = 6 - 1 = 5, \text{ and } a_4 = 2(4) - 1 = 8 - 1 = 7$$

Therefore, the first four terms of the sequence $a_n = 2n - 1$ are $1, 3, 5,$ and 7.

b. Work through the interactive video to verify that the first four terms of the sequence $b_n = n^2 - 1$ are $0, 3, 8,$ and 15.

c. To find the first four terms of the sequence $c_n = \dfrac{3^n}{(n-1)!}$, we evaluate c_n when n is $1, 2, 3,$ and 4.

$$c_1 = \frac{3^1}{(1-1)!} = \frac{3}{0!} = \frac{3}{1} = 3, \qquad c_2 = \frac{3^2}{(2-1)!} = \frac{9}{1!} = \frac{9}{1} = 9,$$
$$c_3 = \frac{3^3}{(3-1)!} = \frac{27}{2!} = \frac{27}{2}, \text{ and } c_4 = \frac{3^4}{(4-1)!} = \frac{81}{3!} = \frac{81}{6} = \frac{27}{2}$$

Therefore, the first four terms of the sequence $c_n = \dfrac{3^n}{(n-1)!}$ are

$$3, 9, \frac{27}{2}, \text{ and } \frac{27}{2}.$$

d. Work through the interactive video to verify that the first four terms of the sequence $d_n = (-1)^n 2^{n-1}$ are $-1, 2, -4,$ and 8.

Note The sequence $d_n = (-1)^n 2^{n-1}$ is an example of an alternating sequence because the successive terms alternate in sign. ●

You Try It Work through this You Try It problem.

Work Exercises 1–10 in this textbook or in the MyMathLab Study Plan.

OBJECTIVE 2 WRITING THE TERMS OF A RECURSIVE SEQUENCE

Some sequences are defined recursively. A recursive sequence is a sequence in which each term is defined using one or more of its previous terms. Typically, the first term of a recursive sequence is given, followed by the formula for the nth term of the sequence. The following example illustrates two recursive sequences.

My interactive video summary

⊘ **Example 2** Writing the Terms of a Recursive Sequence

Write the first four terms of each of the following recursive sequences.

a. $a_1 = -3, a_n = 5a_{n-1} - 1$ for $n \geq 2$ **b.** $b_1 = 2, b_n = \dfrac{(-1)^{n-1}n}{b_{n-1}}$ for $n \geq 2$

Solution

a. The first four terms of this recursive sequence are $-3, -16, -81,$ and -406. Work through this interactive video to verify.

b. The first four terms of this recursive sequence are $2, -1, -3,$ and $\dfrac{4}{3}$. Work through this interactive video to verify.

Arguably the most famous recursively defined sequence is the Fibonacci sequence named after the 13th-century Italian mathematician Leonardo of Pisa, also known as Fibonacci. The Fibonacci sequence is defined in Example 3.

Example 3 Writing the Terms of the Fibonacci Sequence

The Fibonacci sequence is defined recursively by $a_n = a_{n-1} + a_{n-2}$, where $a_1 = 1$ and $a_2 = 1$. Write the first eight terms of the Fibonacci sequence.

Solution We are given that $a_1 = 1$ and $a_2 = 1$. We use the recursive formula $a_n = a_{n-1} + a_{n-2}$ to find the next six terms starting with $n = 3$.

$$a_3 = a_2 + a_1 = 1 + 1 = 2$$
$$a_4 = a_3 + a_2 = 2 + 1 = 3$$
$$a_5 = a_4 + a_3 = 3 + 2 = 5$$
$$a_6 = a_5 + a_4 = 5 + 3 = 8$$
$$a_7 = a_6 + a_5 = 8 + 5 = 13$$
$$a_8 = a_7 + a_6 = 13 + 8 = 21$$

You can see in Example 3 that each term of the Fibonacci sequence is the sum of the preceding two terms. We now write the first 12 terms of the Fibonacci sequence.

$$1, 1, 2, 3, 5, 8, 13, 21, 34, 55, 89, 144, \ldots$$

The numbers of this sequence are known as *Fibonacci numbers*. The Fibonacci sequence and Fibonacci numbers occur in many natural phenomena such as the spiral formation of seeds of various plants, the number of petals of a flower, and the formation of the branches of a tree. See Exercise 44.

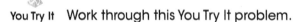
You Try It   Work through this You Try It problem.

Work Exercises 11–14 in this textbook or in the My Math**Lab Study Plan.**

OBJECTIVE 3   WRITING THE GENERAL TERM FOR A GIVEN SEQUENCE

Sometimes the first several terms of a sequence are given without listing the nth term. When this occurs, we must try to determine a pattern and use deductive reasoning to establish a rule that describes the general term, or nth term, of the sequence. Example 4 illustrates two such sequences.

My video summary   ⊘ **Example 4** Finding the General Term of a Sequence

Write a formula for the nth term of each infinite sequence, then use this formula to find the 8th term of the sequence.

a. $\dfrac{1}{1}, \dfrac{1}{2}, \dfrac{1}{3}, \dfrac{1}{4}, \dfrac{1}{5}, \cdots$

b. $-\dfrac{2}{1}, \dfrac{4}{2}, -\dfrac{8}{6}, \dfrac{16}{24}, -\dfrac{32}{120}, \cdots$

Solution

a. The nth term of the sequence is $a_n = \dfrac{1}{n}$. Thus, the 8th term of this sequence is

$$a_8 = \dfrac{1}{8}.$$

b. For this sequence, notice that the first term is negative and that terms alternate in sign. We can therefore represent the sign of each term as $(-1)^n$. Also, notice that the numerators are successive powers of 2. We now have the following pattern:

a_1	a_2	a_3	a_4	a_5
$\downarrow$	$\downarrow$	$\downarrow$	$\downarrow$	$\downarrow$
$-\dfrac{2}{1},$	$\dfrac{4}{2},$	$-\dfrac{8}{6},$	$\dfrac{16}{24},$	$-\dfrac{32}{120}, \cdots$
$\downarrow$	$\downarrow$	$\downarrow$	$\downarrow$	$\downarrow$
$\dfrac{(-1)^1 2^1}{1},$	$\dfrac{(-1)^2 2^2}{2},$	$\dfrac{(-1)^3 2^3}{6},$	$\dfrac{(-1)^4 2^4}{24},$	$\dfrac{(-1)^5 2^5}{120}, \cdots$

Finally, if we factor each successive denominator we get $1 = 1$, $2 = 2 \cdot 1$, $6 = 3 \cdot 2 \cdot 1$, $24 = 4 \cdot 3 \cdot 2 \cdot 1$, and $120 = 5 \cdot 4 \cdot 3 \cdot 2 \cdot 1$. This suggests that the denominator of the nth term can be represented by $n!$ Therefore, the nth term of the sequence is $a_n = \dfrac{(-1)^n 2^n}{n!}$. The 8th term of this sequence is

$$a_8 = \dfrac{(-1)^8 2^8}{8!} = \dfrac{256}{40{,}320} = \dfrac{2}{315}.$$

If you would like to see this solution worked out in detail, watch this video.   ⬤

You Try It   Work through this You Try It problem.

Work Exercises 15–22 in this textbook or in the My Math Lab **Study Plan**.

OBJECTIVE 4   COMPUTING PARTIAL SUMS OF A SERIES

Suppose that we wanted to find the sum of the first four terms of the sequence $a_n = 2n - 1$. From Example 1 we saw that the first four terms of this sequence were $a_1 = 1$, $a_2 = 3$, $a_3 = 5$, and $a_4 = 7$. Therefore, the sum of the first four terms is $a_1 + a_2 + a_3 + a_4 = 1 + 3 + 5 + 7 = 16$. The expression $1 + 3 + 5 + 7$ is called a **series**.

Definition Series

Let $a_1, a_2, a_3, \ldots$ be a sequence. The expression of the form $a_1 + a_2 + a_3 + \cdots + a_n$ is called a **finite series**.

The expression of the form $a_1 + a_2 + a_3 + \cdots + a_n + a_{n+1} + \cdots$ is called an **infinite series**.

The sum of the first n terms of a series is called the nth **partial sum** of the series and is denoted as S_n.

For the series $1 + 3 + 5 + 7 + 9 + \cdots + 2n - 1$, the first five partial sums are as follows:

$$S_1 = 1$$
$$S_2 = 1 + 3 = 4$$
$$S_3 = 1 + 3 + 5 = 9$$
$$S_4 = 1 + 3 + 5 + 7 = 16$$
$$S_5 = 1 + 3 + 5 + 7 + 9 = 25$$

It appears that $S_n = n^2$. In fact, it can be shown that for any positive integer n, the sum of the series $1 + 3 + 5 + 7 + 9 + \cdots + 2n - 1$ is equal to n^2. We will be able to prove this assertion in Section 9.5 using a method called *mathematical induction*.

Example 5 Computing Partial Sums of a Series

Given the general term of each sequence, find the indicated partial sum.

a. $a_n = \dfrac{1}{n}$, find S_3. **b.** $b_n = (-1)^n 2^{n-1}$, find S_5.

Solution

a. The first three terms are $a_1 = 1, a_2 = \dfrac{1}{2}$, and $a_3 = \dfrac{1}{3}$. Therefore, the partial sum, S_3,

is $S_3 = 1 + \dfrac{1}{2} + \dfrac{1}{3} = \dfrac{11}{6}$.

b. The first five terms are $b_1 = -1, b_2 = 2, b_3 = -4, b_4 = 8$, and $b_5 = -16$. Therefore, the partial sum, S_5, is $S_5 = -1 + 2 + (-4) + 8 + (-16) = -11$.

You Try It Work through this You Try It problem.

Work Exercises 23–28 in this textbook or in the MyMathLab Study Plan.

OBJECTIVE 5 DETERMINING THE SUM OF A FINITE SERIES WRITTEN IN SUMMATION NOTATION

 *My video summary*

Writing out an entire finite series of the form $a_1 + a_2 + a_3 + \cdots + a_n$ can be quite tedious, especially if n is fairly large. Fortunately, there is a convenient way to express a finite series using a short-hand notation called *summation notation* (also called *sigma notation*). This notation involves the use of the uppercase Greek letter sigma, which is written as Σ.

Definition   Summation Notation

If $a_1, a_2, a_3, \ldots$ is a sequence, then the finite series $a_1 + a_2 + a_3 + \cdots + a_n$ can

be written in **summation notation** as $\displaystyle\sum_{i=1}^{n} a_i$. The infinite series

$a_1 + a_2 + \cdots + a_n + a_{n+1} + \cdots$ can be written as $\displaystyle\sum_{i=1}^{\infty} a_i$.

The variable i is called the **index of summation**. The number 1 is the **lower limit of summation** and n is the **upper limit of summation**.

The lower limit of summation, $i = 1$, below the sigma tells us which term to start with. The upper limit of summation, n, that appears above the sigma tells us which term of the sequence will be the last term to add. There is nothing special about the letter i that is used to represent the index of summation. We will often use different letters such as j or k. Also, it is not necessary for the lower limit of summation to start at 1. In Examples 6b and 6c, the lower limits of summation are 2 and 0, respectively.

My interactive video summary

⊙ **Example 6** Determining the Sum of a Series Written in Summation Notation

Find the sum of each finite series.

a. $\displaystyle\sum_{i=1}^{5} i^2$ b. $\displaystyle\sum_{j=2}^{5} \frac{j-1}{j+1}$ c. $\displaystyle\sum_{k=0}^{6} \frac{1}{k!}$

(Round the sum to three decimal places.)

Solution

a. $\displaystyle\sum_{i=1}^{5} i^2 = 1^2 + 2^2 + 3^2 + 4^2 + 5^2 = 1 + 4 + 9 + 16 + 25 = 55$

b. $\displaystyle\sum_{j=2}^{5} \frac{j-1}{j+1} = \frac{2-1}{2+1} + \frac{3-1}{3+1} + \frac{4-1}{4+1} + \frac{5-1}{5+1}$

$$= \frac{1}{3} + \frac{2}{4} + \frac{3}{5} + \frac{4}{6} = \frac{21}{10}$$

c. $\displaystyle\sum_{k=0}^{6} \frac{1}{k!} = \frac{1}{0!} + \frac{1}{1!} + \frac{1}{2!} + \frac{1}{3!} + \frac{1}{4!} + \frac{1}{5!} + \frac{1}{6!}$

$$= 1 + 1 + \frac{1}{2} + \frac{1}{6} + \frac{1}{24} + \frac{1}{120} + \frac{1}{720} = \frac{1957}{720} \approx 2.718$$

You can work through this interactive video to see this solution worked out in detail.

Using Technology

```
sum(seq(1/n!,n,0
,6))
        2.718055556
```

Using a TI-83 Plus, we can calculate the sum obtained in Example 6c. Notice that this number is a good approximation of the number e. In fact, it can be shown using calculus that the exact value of e is $e = \displaystyle\sum_{n=0}^{\infty} \frac{1}{n!}$. Other irrational numbers, such as π, can be represented by a series. Again, using calculus, it can be shown that the exact value of π is $\pi = \displaystyle\sum_{n=0}^{\infty} \frac{4 \cdot (-1)^n}{2n + 1}$.

You Try It Work through this You Try It problem.

Work Exercises 29–35 in this textbook or in the MyMathLab Study Plan.

OBJECTIVE 6 WRITING A SERIES USING SUMMATION NOTATION

Given the first several terms of a series, it is important to be able to rewrite the series using summation notation as in Example 7.

Example 7 Writing a Series Using Summation Notation

Rewrite each series using summation notation. Use 1 as the lower limit of summation.

a. $2 + 4 + 6 + 8 + 10 + 12$ b. $1 + 2 + 6 + 24 + 120 + 720 + \cdots + 3{,}628{,}800$

Solution

a. This series is the sum of six terms. Therefore, the lower limit of summation is 1 and the upper limit of summation is 6. Each term is a successive multiple of 2. So, one possible series is $\displaystyle\sum_{i=1}^{6} 2i$.

b. Notice that $1 = 1!, 2 = 2!, 6 = 3!, 24 = 4!, 120 = 5!, 720 = 6!$, and $3{,}628{,}800 = 10!$ Thus, a possible series is $\displaystyle\sum_{n=1}^{10} n!$

 You Try It Work through this You Try It problem.

Work Exercises 36–43 in this textbook or in the MyMathLab Study Plan.

9.1 Exercises

In Exercises 1–10, write the first four terms of each sequence.

1. $a_n = 3n + 1$ 2. $a_n = 4^n$ 3. $a_n = \dfrac{4n}{n + 3}$ 4. $a_n = (-4)^n$

5. $a_n = 5(n + 2)!$ 6. $a_n = \dfrac{n^3}{(n + 1)!}$ 7. $a_n = (-1)^n(5n)$ 8. $a_n = \dfrac{3^n}{(-1)^{n+1} + 5}$

9. $a_n = \dfrac{(-1)^n}{(n + 5)(n + 6)}$ 10. $a_n = \dfrac{(-1)^n(3)^{2n+1}}{(2n + 1)!}$

In Exercises 11–14, write the first four terms of each recursive sequence.

11. $a_1 = 7, a_n = 3 + a_{n-1}$ for $n \geq 2$ 12. $a_1 = -1, a_n = n - a_{n-1}$ for $n \geq 2$

13. $a_1 = 6, a_n = \dfrac{a_{n-1}}{n^2}$ for $n \geq 2$ 14. $a_1 = -4, a_n = 1 - \dfrac{1}{a_{n-1}}$ for $n \geq 2$

In Exercises 15–22, write a formula for the general term, or nth term, for the given sequence. Then find the indicated term.

15. $-1, 1, 3, 5, 7, \ldots; a_{11}$.

16. $\dfrac{1}{5}, \dfrac{2}{6}, \dfrac{3}{7}, \dfrac{4}{8}, \dfrac{5}{9}, \ldots; a_8$.

17. $1 \cdot 6, 2 \cdot 7, 3 \cdot 8, 4 \cdot 9, \ldots; a_7$.

18. $-2, 4, -8, 16, \ldots; a_7$.

19. $\dfrac{2}{5}, \dfrac{2}{25}, \dfrac{2}{125}, \dfrac{2}{625}, \ldots; a_6$.

20. $-6, 12, -24, 48, -96, \ldots; a_9$.

21. $-6, 24, -120, 720, \ldots; a_5$.

22. $\dfrac{3}{2}, \dfrac{9}{6}, \dfrac{27}{24}, \dfrac{81}{120}, \ldots; a_5$.

In Exercises 23–25, the first several terms of a sequence are given. Find the indicated partial sum.

23. $2, 4, 6, 8, 10, \ldots; S_4$

24. $3, -6, 9, -12, 15, -18, \ldots; S_9$

25. $\dfrac{1}{2}, -\dfrac{1}{4}, \dfrac{1}{8}, -\dfrac{1}{16}, \ldots; S_5$

In Exercises 26–28, the general term of a sequence is given. Find the indicated partial sum.

26. $a_n = 3n + 8; S_6$

27. $a_n = (-1)^n \cdot (4n); S_6$

28. $a_1 = 4, a_n = a_{n-1} - 8$ for $n \geq 2; S_8$

In Exercises 29–35, find the sum of each series.

29. $\displaystyle\sum_{i=1}^{9} i$

30. $\displaystyle\sum_{i=1}^{6} (4i + 3)$

31. $\displaystyle\sum_{i=1}^{7} i(i + 2)$

32. $\displaystyle\sum_{i=1}^{21} 7$

33. $\displaystyle\sum_{j=0}^{5} (j + 4)^2$

34. $\displaystyle\sum_{k=2}^{7} \dfrac{k!}{(k - 2)!}$

35. $\displaystyle\sum_{j=0}^{5} (j - 2)^3$

In Exercises 36–43, rewrite each series using summation notation. Use 1 as the lower limit of summation.

36. $1 + 2 + 3 + \cdots + 29$

37. $5 + 10 + 15 + \cdots + 50$

38. $1^2 + 2^2 + 3^2 + \cdots + 11^2$

39. $\dfrac{4}{5} + \dfrac{5}{6} + \dfrac{6}{7} + \cdots + \dfrac{12}{13}$

40. $2 + (-4) + 8 + (-16) + \cdots + (-256)$

41. $-\dfrac{1}{9} + \dfrac{1}{18} - \dfrac{1}{27} + \cdots + \dfrac{1}{54}$

42. $5 + \dfrac{5^2}{2} + \dfrac{5^2}{3} + \cdots + \dfrac{5^n}{n}$

43. $1 + 7 + \dfrac{7^2}{2!} + \dfrac{7^3}{3!} + \dfrac{7^4}{4!} + \cdots + \dfrac{7^n}{n!}$

44. The figure below shows the progression of the branching of a tree during each stage of development. Notice that the number of branches formed during a given stage is a Fibonacci number. Assuming that this branching pattern continues, how many branches will form during the 10th stage of development?

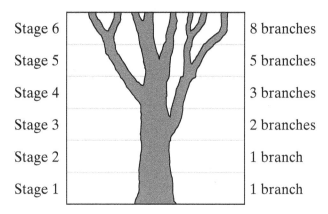

Stage 6 — 8 branches

Stage 5 — 5 branches

Stage 4 — 3 branches

Stage 3 — 2 branches

Stage 2 — 1 branch

Stage 1 — 1 branch

9.2 Arithmetic Sequences and Series

THINGS TO KNOW

Before working through this section, be sure that you are familiar with the following concepts:

VIDEO   ANIMATION   INTERACTIVE

 You Try It
1. Solving a System of Linear Equations Using the Substitution Method (Section 7.1)

 You Try It
2. Solving a System of Linear Equations Using the Elimination Method (Section 7.1)

 You Try It
3. Determining the Sum of a Finite Series Written in Summation Notation (Section 9.1)

OBJECTIVES

1 Determining If a Sequence Is Arithmetic

2 Finding the General Term or a Specific Term of an Arithmetic Sequence

3 Computing the nth Partial Sum of an Arithmetic Series

4 Applications of Arithmetic Sequences and Series

....

OBJECTIVE 1   DETERMINING IF A SEQUENCE IS ARITHMETIC

In this section, we will work exclusively with a specific type of sequence known as an **arithmetic sequence**. A sequence is arithmetic if the difference in any two successive terms is constant. For example, the sequence

$$5, 9, 13, 17, \ldots$$

is arithmetic because the difference of any two successive terms is 4. The first term of this sequence is $a_1 = 5$ and the common difference is $d = 4$. Notice that we can rewrite the terms of this sequence as $5, 5 + 4, 5 + 2(4), 5 + 3(4), \ldots$

In general, given an arithmetic sequence with a first term of a_1 and a common difference of d, the first n terms of the sequence are as follows:

$$a_1$$

$$a_2 = a_1 + d$$

$$a_3 = a_2 + d = \underbrace{(a_1 + d)}_{a_2} + d = a_1 + 2d$$

$$a_4 = a_3 + d = \underbrace{(a_1 + 2d)}_{a_3} + d = a_1 + 3d$$

$$\vdots$$

$$a_n = a_1 + (n - 1)d$$

Definition   Arithmetic Sequence

An **arithmetic sequence** is a sequence of the form $a_1, a_1 + d, a_1 + 2d, a_1 + 3d,$ $a_1 + 4d, \ldots$, where a_1 is the first term of the sequence and d is the common difference. The general term, or nth term, of an arithmetic sequence has the form $a_n = a_1 + (n - 1)d$.

 *My interactive video summary*

⊙ **Example 1 Determining If a Sequence Is Arithmetic**

For each of the following sequences, determine if it is arithmetic. If the sequence is arithmetic, find the common difference.

a. $1, 4, 7, 10, 13, \ldots$
b. $b_n = n^2 - n$

c. $a_n = -2n + 7$
d. $a_1 = 14, a_n = 3 + a_{n-1}$

Solution Watch this interactive video to verify that the sequences in parts a), c), and d) are arithmetic. The sequence in part b) is not arithmetic. Notice that the arithmetic sequence in part d) is a recursive sequence.

It is worth noting that every arithmetic sequence is a linear function whose domain is the natural numbers. A portion of the graphs of the arithmetic sequences from Example 1a and Example 1c are seen in Figure 2. Notice that the ordered pairs of each sequence are collinear.

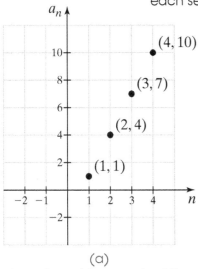

(a)
A portion of the graph of the sequence 1, 4, 7, 10, 13, . . .

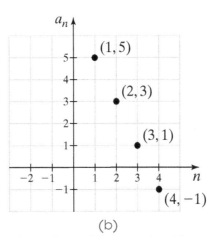

(b)
A portion of the graph of the sequence $a_n = -2n + 7$

Figure 2 The graph of every arithmetic sequence is represented by a set of ordered pairs that lies on a straight line.

9.2 Arithmetic Sequences and Series   9-11

Note When the common difference of an arithmetic sequence is positive, the terms of the sequence *increase* and the graph is represented by a set of ordered pairs that lie along a line with positive slope. When the common difference of an arithmetic sequence is negative, the terms of the sequence *decrease* and the graph is represented by a set of ordered pairs that lies along a line with negative slope.

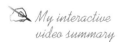
You Try It Work through this You Try It problem.

Work Exercises 1–6 in this textbook or in the MyMathLab Study Plan.

OBJECTIVE 2 FINDING THE GENERAL TERM OR A SPECIFIC TERM OF AN ARITHMETIC SEQUENCE

By the definition of an arithmetic sequence, the general term of an arithmetic sequence has the form $a_n = a_1 + (n - 1)d$. We can use this formula to find any term of an arithmetic sequence.

My interactive video summary

⊙ **Example 2** Finding the General Term of an Arithmetic Sequence

Find the general term of each arithmetic sequence, then find the indicated term of the sequence. (In part c, only a portion of the graph is given. Assume that the domain of this sequence is all natural numbers.)

a. $11, 17, 23, 29, 35, \ldots ; a_{50}$ b. $2, 0, -2, -4, -6, \ldots ; a_{90}$ c. Find a_{31}.

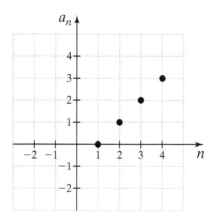

Solution

a. The first term of the sequence is $a_1 = 11$ and the common difference is $d = 6$. The general term is given by $a_n = 11 + (n - 1)(6) = 11 + 6n - 6 = 6n + 5$. Therefore, $a_{50} = 6(50) + 5 = 305$.

b. The first term of the sequence is $a_1 = 2$ and the common difference is $d = -2$. The general term is given by $a_n = 2 + (n - 1)(-2) = 2 - 2n + 2 = 4 - 2n$. Therefore, $a_{90} = 4 - 2(90) = -176$.

c. The first four terms of this sequence are $a_1 = 0, a_2 = 1, a_3 = 2$, and $a_4 = 3$. Thus, $a_1 = 0$ and the common difference is $d = 1$. The general term is given by $a_n = 0 + (n - 1)(1) = n - 1$. Thus, $a_{31} = 31 - 1 = 30$.

You may also watch this interactive video to see each of these solutions worked out in detail.

My interactive video summary

You Try It Work through this You Try It problem.

Work Exercises 7–12 in this textbook or in the MyMathLab Study Plan.

⊘ **Example 3** Finding a Specific Term of an Arithmetic Sequence

a. Given an arithmetic sequence with $d = -4$ and $a_3 = 14$, find a_{50}.

b. Given an arithmetic sequence with $a_4 = 12$ and $a_{15} = -10$, find a_{41}.

Solution

a. We are given that $d = -4$ and $a_3 = 14$. We can use this information to solve for a_1.

$$a_n = a_1 + (n - 1)d$$
  Use the formula for the general term of an arithmetic sequence.

$$a_3 = a_1 + (3 - 1)(-4)$$
  Substitute $n = 3$ and $d = -4$.

Simplifying, we get $a_3 = a_1 - 8$. We can now substitute $a_3 = 14$ to solve for a_1.

$$a_3 = a_1 - 8$$
  Start with the formula for a_3.

$$14 = a_1 - 8$$
  Substitute $a_3 = 14$.

$$22 = a_1$$
  Add 8 to both sides.

Using the formula $a_n = a_1 + (n - 1)d$ with $a_1 = 22$ and $d = -4$, we can simplify to get $a_n = 26 - 4n$. Therefore, $a_{50} = 26 - 4(50) = -174$. Watch this interactive video to see every step of this solution.

b. Using the fact that $a_n = a_1 + (n - 1)d$, we get $a_4 = a_1 + (4 - 1)d = 12$ and $a_{15} = a_1 + (15 - 1)d = -10$. This gives us the following system of linear equations:

$$a_1 + 3d = 12$$

$$a_1 + 14d = -10$$

My interactive video summary

⊘ Using the method of substitution or the method of elimination that were discussed in Section 7.1 to solve this system, we get $a_1 = 18$ and $d = -2$. (Watch this interactive video to see how to solve this system using either method.) Using the formula $a_n = a_1 + (n - 1)d$ with $a_1 = 18$ and $d = -2$, we can find the general term.

$$a_n = a_1 + (n - 1)d$$
  Use the formula for the general term of an arithmetic sequence.

$$= 18 + (n - 1)(-2)$$
  Substitute $a_1 = 18$ and $d = -2$.

$$= 18 - 2n + 2$$
  Use the distributive property.

$$= 20 - 2n$$
  Simplify.

The general term is $a_n = 20 - 2n$. Therefore, $a_{41} = 20 - 2(41) = -62$. Watch this interactive video to see this entire solution worked out in detail. ●

You Try It Work through this You Try It problem.

Work Exercises 13–18 in this textbook or in the MyMathLab Study Plan.

OBJECTIVE 3 COMPUTING THE nTH PARTIAL SUM OF AN ARITHMETIC SERIES

If $a_1, a_2, a_3, \ldots$ is an arithmetic sequence, then the expression $a_1 + a_2 + a_3 + \cdots + a_n + a_{n+1} + \cdots$ is called an **infinite arithmetic series** and can be written using summation notation as $\sum\limits_{i=1}^{\infty} a_i$. Recall that the sum of the first n terms of a series is called the **nth partial sum** of the series and is given by $S_n = a_1 + a_2 + a_3 + \cdots + a_n$. We can also represent the nth partial sum using summation notation as $S_n = \sum\limits_{i=1}^{n} a_i$.

As you can see, the nth partial sum is simply the sum of a finite arithmetic series. Fortunately, there is a convenient formula for computing the nth partial sum of an arithmetic series.

Formula for the nth Partial Sum of an Arithmetic Series

The sum of the first n terms of an arithmetic series is called the **nth partial sum** of the series and is given by $S_n = \sum\limits_{i=1}^{n} a_i = a_1 + a_2 + a_3 + \cdots + a_n$. This sum can be computed using the formula $S_n = \dfrac{n(a_1 + a_n)}{2}$.

 *My video summary* ⊙ Watch this video to see the derivation of this formula. (We will prove this formula again using mathematical induction in **Section 9.5**.)

Note The nth partial sum of an arithmetic series is simply the sum of a finite arithmetic series. An arithmetic series *must* be *finite* in order to compute the sum. This is not true for some other types of series. You will see how to find the sum of a special type of infinite series in **Section 9.3**.

Example 4 Finding the Sum of an Arithmetic Series

Find the sum of each arithmetic series.

a. $\sum\limits_{i=1}^{20}(2i - 11)$ 　　　　　　**b.** $-5 + (-1) + 3 + 7 + \cdots + 39$

Solution

a. We can use the formula $S_{20} = \dfrac{20(a_1 + a_{20})}{2}$ to compute the sum of the first 20 terms of this series.

$$a_1 = 2(1) - 11 = -9 \qquad \text{Substitute } i = 1 \text{ in the formula } 2i - 11 \text{ to find } a_1.$$

$$a_{20} = 2(20) - 11 = 29 \qquad \text{Substitute } i = 20 \text{ in the formula } 2i - 11 \text{ to find } a_{20}.$$

We now substitute $a_1 = -9$ and $a_{20} = 29$ into the formula $S_{20} = \dfrac{20(a_1 + a_{20})}{2}$.

$$S_{20} = \dfrac{20(a_1 + a_{20})}{2} \qquad \text{Use the formula for 20th partial sum of an arithmetic series.}$$

$$= \dfrac{20(-9 + 29)}{2} \qquad \text{Substitute } a_1 = -9 \text{ and } a_{20} = 29.$$

$$= 200 \qquad \text{Simplify.}$$

Therefore, $\sum_{i=1}^{20}(2i - 11) = 200$. You may also watch this interactive video to see this solution worked out in detail.

b. Work through the interactive video to verify that the sum of this arithmetic series is 204.

You Try It   Work through this You Try It problem.

Work Exercises 19–27 in this textbook or in the MyMathLab **Study Plan**.

OBJECTIVE 4   APPLICATIONS OF ARITHMETIC SEQUENCES AND SERIES

Example 5 Selling Newspaper Subscriptions

A local newspaper has hired teenagers to go door-to-door to try to solicit new subscribers. The teenagers receive $2 for selling the first subscription. For each additional subscription sold, the newspaper will pay the teenagers 10 cents more than what was paid for the previous subscription. How much will the teenagers get paid for selling the 100th subscription? How much money will the teenagers earn by selling 100 subscriptions?

Solution   The amount of money earned by selling one newspaper subscription can be represented by $a_1 = 2$. The money earned by selling the second subscription is $a_2 = 2.10$. The money earned by selling the third subscription is $a_3 = 2.20$. We see that the amount of money earned by selling n newspaper subscriptions is an arithmetic sequence with $a_1 = 2$ and $d = 0.10$. This sequence is defined by $a_n = 2 + (n - 1)(0.10) = 2 + (0.10)n - 0.10 = 0.10n + 1.90$.

The cash earned by selling the 100th subscription is the 100th term of this sequence, or $a_{100} = 0.10(100) + 1.90 = 11.90$. Therefore, the teenagers are paid $11.90 for selling the 100th subscription.

To find the total amount earned by selling 100 subscriptions, we must find the sum of the series $\sum_{i=1}^{100}[(0.10)i + 1.90]$.

Using the formula $S_n = \dfrac{n(a_1 + a_n)}{2}$ with $n = 100$, $a_1 = 2$, and $a_{100} = 11.90$, we get

$S_{100} = \dfrac{100(a_1 + a_{100})}{2} = \dfrac{100(2 + 11.90)}{2} = 50(13.90) = 695$. Thus, the teenagers will be paid $695 if they sell 100 subscriptions.

 *My video summary*   ⊙ **Example 6** Seats in a Theater

A large multiplex movie house has many theaters. The smallest theater has only 12 rows. There are six seats in the first row. Each row has two seats more than the previous row. How many total seats are there in this theater?

Solution   Try solving this word problem on your own. When you are done, watch this video to see if you are correct, then work through the following "You Try It" problem.

You Try It Work through this You Try It problem.

Work Exercises 28–35 in this textbook or in the MyMathLab Study Plan.

9.2 **Exercises**

In Exercises 1–6, determine if the sequence is arithmetic. If the sequence is arithmetic, find the common difference.

1. 8, 14, 20, 26, 32, . . .

2. 8, 11, 13, 16, 18, . . .

3. $a_n = \dfrac{3n + 1}{2}$

4. $a_n = n(n + 1)$

5. $a_1 = 8, a_n = 2 + a_{n-1}$

6. $a_1 = 5, a_n = 3a_{n-1} + 1$

In Exercises 7–12, find the general term of each arithmetic sequence then find the indicated term of the sequence. If the sequence is represented by a graph, assume that the domain of the sequence is all natural numbers.

7. 2, 7, 12, 17, . . . ; a_{10}

8. 5, 1, −3, −7, . . . ; a_{31}

9. $\dfrac{3}{2}, 3, \dfrac{9}{2}, 6, \dfrac{15}{2}, \ldots ;$ a_{50}

10. 5.0, 3.8, 2.6, 1.4, . . . ; a_{29}

11. Find a_{17}.

12. Find a_{11}.

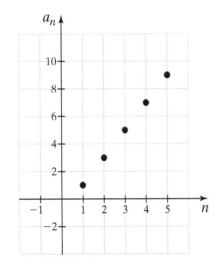

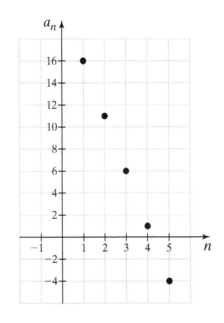

13. Given an arithmetic sequence with $d = 3$ and $a_8 = 5$, find a_{30}.

14. Given an arithmetic sequence with $d = -5$ and $a_7 = 11$, find a_{22}.

15. Given an arithmetic sequence with $a_5 = 4$ and $a_{22} = 55$, find a_{36}.

16. Given an arithmetic sequence with $a_6 = 4$ and $a_{20} = -52$, find a_{33}.

17. Given an arithmetic sequence with $a_{16} = 30$ and $a_{30} = 65$, find a_9.

18. Given an arithmetic sequence with $a_8 = -6$ and $a_{19} = -\dfrac{45}{2}$, find a_{34}.

In Exercises 19–27, find the indicated sum.

19. $\displaystyle\sum_{i=1}^{80} i$

20. $\displaystyle\sum_{j=1}^{10} (3j + 7)$

21. $7 + 10 + 13 + 16 + \cdots + 118$

22. $-14 + (-9) + (-4) + 1 + \cdots + 101$

23. $\displaystyle\sum_{i=3}^{14} (-7 - 9i)$

24. $1 + 11 + 21 + 31 + \cdots + a_{102}$

25. $6 + 14 + 22 + 30 + \cdots + (8n - 2)$

26. Find the sum of the first 100 odd integers.

27. Find the sum of the first 100 even positive integers.

28. A large multiplex movie house has many theaters. The largest theater has 40 rows. There are 12 seats in the first row. Each row has four seats more than the previous row. How many total seats are there in this theater?

29. A stack of logs has 47 logs on the bottom layer. Each subsequent layer has nine fewer logs than the previous layer. If the top layer has two logs, how many total logs are there in the pile?

30. A middle school mathematics teacher accepts a teaching position that pays $31,000 per year. Each year, the expected raise is $1,100. How much total money will this teacher earn teaching middle school mathematics over the first 12 years?

31. Suppose that you plan on taking a summer job selling magazine subscriptions. The magazine company will pay you $1 for selling the first subscription. For each additional subscription sold, the magazine company will pay you 15 cents more than what was paid for the previous subscription. How much will you earn by selling 200 magazine subscriptions?

32. Two companies have offered you a job. Alpha Company has offered you $35,000 per year with an annual raise of $2,000. Beta Company has offered you a $46,000 annual salary with an annual raise of $800 per year. Which company will pay you more over the first 10 years?

33. A city fund-raiser raffle is raffling off 25 cash prizes. First prize is $5,000. Each successive prize is $200 less than the preceding prize. What is the value of the 25th prize? What is the total amount of cash given out by this raffle?

34. Larry's Luxury Rental Car Company rents luxury cars for up to 18 days. The price is $300 for the first day, with the rental fee decreasing $7 for each additional day. How much will it cost to rent a luxury car for 18 days?

35. A ball thrown straight up in the air travels 48 inches in the first tenth of a second. In the next tenth of a second, the ball travels 44 inches. After each additional tenth of a second, the ball travels 4 inches less than it did during the preceding tenth of a second. How long will it take before the ball starts coming back down? What is the total distance that the ball has traveled when it has reached its maximum height?

9.3 Geometric Sequences and Series

THINGS TO KNOW

Before working through this section, be sure that you are familiar with the following concepts:

 You Try It   1. Using the Periodic Compound Interest Formula (Section 5.1)

You Try It   2. Solving a System of Linear Equations Using the Substitution Method (Section 7.1)

You Try It   3. Finding the General Term or a Specific Term of an Arithmetic Sequence (Section 9.2)

You Try It   4. Computing the nth Partial Sum of an Arithmetic Series (Section 9.2)

OBJECTIVES

1 Writing the Terms of a Geometric Sequence

2 Determining If a Sequence Is Geometric

3 Finding the General Term or a Specific Term of a Geometric Sequence

4 Computing the nth Partial Sum of a Geometric Series

5 Determining If an Infinite Geometric Series Converges or Diverges

6 Applications of Geometric Sequences and Series

OBJECTIVE 1   WRITING THE TERMS OF A GEOMETRIC SEQUENCE

Suppose that you have agreed to work for Donald Trump on a particular job for 21 days. Mr. Trump gives you two choices of payment. You can be paid $100 for the first day and an additional $50 per day for each subsequent day. Or, you can choose to be paid 1 penny for the first day with your pay doubling each subsequent day. Which method of payment would you choose? (We will revisit this question later in this section. See Example 7.) Notice that each payment method can be represented by a sequence:

Payment Method 1: 100, 150, 200, 250, 300, . . .

Payment Method 2: .01, .02, .04, .08, .16, . . .

The first method of payment is an **arithmetic sequence** with $a_1 = 100$ and $d = 50$. The second method of payment is an example of a **geometric sequence**. Each term of this geometric sequence can be obtained by multiplying the previous term by 2. The number 2 in this case is called the **common ratio**. We can obtain this common ratio by dividing any term of the sequence (except the first term) by the previous term. That is, $r = \dfrac{a_2}{a_1} = \dfrac{a_3}{a_2} = \cdots = \dfrac{a_{n+1}}{a_n}$. The first term of the payment method 2 sequence is $a_1 = .01$ and the common ratio is $r = 2$.

Notice that we can rewrite the terms of this sequence as $.01$, $(.01)(2)$, $(.01)(2^2)$, $(.01)(2^3)$, ... In general, given a geometric sequence with a first term of a_1 and a common ratio of r, the first n terms of the sequence are

$$a_1$$

$$a_2 = a_1 r$$

$$a_3 = a_2 r = \underbrace{(a_1 r)}_{a_2} r = a_1 r^2$$

$$a_4 = a_3 r = \underbrace{\left(a_1 r^2\right)}_{a_3} r = a_1 r^3$$

$$\vdots$$

$$a_n = a_1 r^{n-1}$$

Definition   Geometric Sequence

A **geometric sequence** is a sequence of the form $a_1, a_1 r, a_1 r^2, a_1 r^3, a_1 r^4, \ldots,$ where a_1 is the first term of the sequence and r is the common ratio such that $r = \dfrac{a_2}{a_1} = \dfrac{a_3}{a_2} = \cdots = \dfrac{a_{n+1}}{a_n}$ for all $n \geq 1$. The general term, or nth term, of a geometric sequence has the form $a_n = a_1 r^{n-1}$.

A portion of the two sequences representing the two payment methods are sketched in Figure 3. The sequence representing payment method 1 (Figure 3a) is arithmetic. The ordered pairs of this sequence are **collinear**. The sequence representing payment method 2 (Figure 3b) is geometric. Notice that the ordered pairs of this sequence do not lie along a common line but rather lie on an exponential curve.

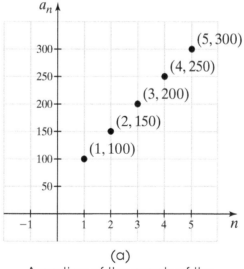

(a)
A portion of the graph of the
sequence 100, 150, 200, 250, 300, . . .

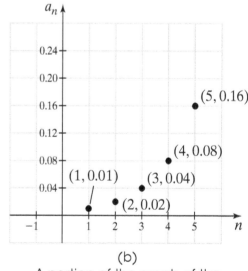

(b)
A portion of the graph of the
sequence .01, .02, .04, .08, .16, . . .

Figure 3 The graph of every arithmetic sequence is represented by a set of ordered pairs that lies on a straight line. The graph of a geometric sequence with $r > 0$ is represented by a set of ordered pairs that lie on an exponential curve.

If the first term and the common ratio of a geometric sequence are known, then we can determine the second term by multiplying the first term by the common ratio. The third term can be found by multiplying the second term by the common ratio. We continue this process to find the subsequent terms of the sequence.

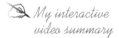
My interactive video summary

⊘ Example 1 Writing a Geometric Sequence

a. Write the first five terms of the geometric sequence having a first term of 2 and a common ratio of 3.

b. Write the first five terms of the geometric sequence such that $a_1 = -4$ and $a_n = -5a_{n-1}$ for $n \geq 2$.

Solution

a. We are given that the first term is $a_1 = 2$ and the common ratio is $r = 3$. The second term is $a_2 = 2 \cdot 3 = 6$. The third term is $a_3 = 6 \cdot 3 = 18$. The fourth term is $a_4 = 18 \cdot 3 = 54$. The fifth term is $a_5 = 54 \cdot 3 = 162$. Therefore, the first five terms of this sequence are 2, 6, 18, 54, and 162.

b. This sequence is defined recursively. The first term is $a_1 = -4$. To find a_2, substitute the value of 2 for n in the formula $a_n = -5a_{n-1}$ and simplify:

$$a_n = -5a_{n-1} \qquad \text{Given recursive formula}$$
$$a_2 = -5a_{2-1} \qquad \text{Substitute.}$$
$$= -5a_1 \qquad \text{Simplify.}$$
$$= -5(-4) \qquad \text{Substitute } a_1 = -4.$$
$$= 20 \qquad \text{Multiply.}$$

My interactive video summary

⊘ Therefore, $a_2 = 20$. We can follow this same procedure to find $a_3, a_4,$ and a_5. You should verify that $a_3 = -100, a_4 = 500,$ and $a_5 = -2500$ on your own, then watch this interactive video to see if you are correct. ●

You Try It Work through this You Try It problem.

Work Exercises 1–5 in this textbook or in the MyMathLab Study Plan.

OBJECTIVE 2   DETERMINING IF A SEQUENCE IS GEOMETRIC

To determine if a given sequence is geometric, we must check to see if each term of the sequence can be obtained by multiplying the previous term by a common ratio r. That is, we must check to see if there exists a constant value r such that

$$r = \frac{a_{n+1}}{a_n} \text{ for any } n \geq 1.$$

My interactive video summary

⊘ Example 2 Determining If a Sequence Is Geometric

For each of the following sequences, determine if it is geometric. If the sequence is geometric, find the common ratio.

a. $2, 4, 6, 8, 10, \ldots$ b. $\dfrac{2}{3}, \dfrac{4}{9}, \dfrac{8}{27}, \dfrac{16}{81}, \dfrac{32}{243}, \ldots$ c. $12, -6, 3, -\dfrac{3}{2}, \dfrac{3}{4}, \ldots$

Solution

a. For this sequence, $\frac{a_2}{a_1} = \frac{4}{2} = 2$ and $\frac{a_3}{a_2} = \frac{6}{4} = \frac{3}{2}$. Since $\frac{a_2}{a_1} \neq \frac{a_3}{a_2}$, there does not exist a common ratio. Hence, this sequence is not geometric. (Note that this sequence is an arithmetic sequence.)

b. For this sequence, $\frac{a_2}{a_1} = \frac{\frac{4}{9}}{\frac{2}{3}} = \frac{4}{9} \cdot \frac{3}{2} = \frac{2}{3}$. Note that each term of this sequence (other than the first term) can be obtained by multiplying the previous term by $\frac{2}{3}$. Therefore, this sequence is geometric with a common ratio of $\frac{2}{3}$.

c. Try to determine if this sequence is geometric. Work through the interactive video to see if you are correct.

You Try It   Work through this You Try It problem.

Work Exercises 6–10 in this textbook or in the MyMathLab Study Plan.

OBJECTIVE 3   FINDING THE GENERAL TERM OR A SPECIFIC TERM OF A GEOMETRIC SEQUENCE

By the definition of a geometric sequence, the general term, or nth term, of a geometric sequence has the form $a_n = a_1 r^{n-1}$. We can use this formula to find any term of a given geometric sequence.

Example 3 Finding the General Term of a Geometric Sequence

Find the general term of each geometric sequence.

a. $12, -6, 3, -\frac{3}{2}, \frac{3}{4}, \ldots$

b. $\frac{2}{3}, \frac{2}{9}, \frac{2}{27}, \frac{2}{81}, \frac{2}{243}, \ldots$

Solution

a. The first term of the sequence is $a_1 = 12$ and the common ratio is $r = \frac{a_2}{a_1} = \frac{-6}{12} = -\frac{1}{2}$. Therefore, $a_n = 12\left(-\frac{1}{2}\right)^{n-1}$.

b. The first term of the sequence is $a_1 = \frac{2}{3}$ and the common ratio is $r = \frac{a_2}{a_1} = \frac{\frac{2}{9}}{\frac{2}{3}} = \frac{1}{3}$. Therefore, $a_n = \left(\frac{2}{3}\right)\left(\frac{1}{3}\right)^{n-1}$.

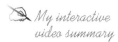
My interactive video summary

> **Example 4** Finding a Specific Term of a Geometric Sequence

a. Find the 7th term of the geometric sequence whose first term is 2 and whose common ratio is -3.

b. Given a geometric sequence such that $a_6 = 16$ and $a_9 = 2$, find a_{13}.

Solution

a. We can use the formula $a_n = a_1 r^{n-1}$ with $a_1 = 2$ and $r = -3$ to find the general term. The general term is $a_n = 2(-3)^{n-1}$. Therefore, $a_7 = 2(-3)^{7-1} = 2(-3)^6 = 2(729) = 1{,}458$.

b. Since $a_6 = 16$, we can substitute $n = 6$ into the formula $a_n = a_1 r^{n-1}$ to get $a_6 = a_1 r^5 = 16$. Similarly, we can substitute $n = 9$ into the formula $a_n = a_1 r^{n-1}$ to get $a_9 = a_1 r^8 = 2$. This gives the following two equations.

$$(1) \quad a_1 r^5 = 16$$

$$(2) \quad a_1 r^8 = 2$$

Divide both sides of equation (1) by r^5 to get $a_1 = \dfrac{16}{r^5}$. Now, substitute $a_1 = \dfrac{16}{r^5}$ into equation (2) and solve for a_1.

$$a_1 r^8 = 2 \qquad \text{Start with equation (2).}$$

$$\left(\frac{16}{r^5}\right) r^8 = 2 \qquad \text{Substitute } a_1 = \frac{16}{r^5}.$$

$$16 r^3 = 2 \qquad \frac{r^8}{r^5} = r^3$$

$$r^3 = \frac{1}{8} \qquad \text{Divide both sides by 16.}$$

$$r = \frac{1}{2} \qquad \text{Take the cube root of both sides.}$$

Now substitute $r = \dfrac{1}{2}$ into equation (1) to solve for a_1.

$$a_1 r^5 = 16 \qquad \text{Start with equation (1).}$$

$$a_1 \left(\frac{1}{2}\right)^5 = 16 \qquad \text{Substitute } r = \frac{1}{2}.$$

$$a_1 \left(\frac{1}{32}\right) = 16 \qquad \left(\frac{1}{2}\right)^5 = \frac{1}{32}$$

$$a_1 = 512 \qquad \text{Multiply both sides by 32.}$$

We can now use the formula $a_n = a_1 r^{n-1}$ with $a_1 = 512$ and $r = \dfrac{1}{2}$ to find the general term. The general term is $a_n = 512\left(\dfrac{1}{2}\right)^{n-1}$. Therefore, $a_{13} = 512\left(\dfrac{1}{2}\right)^{13-1} = 512\left(\dfrac{1}{2}\right)^{12} = \dfrac{512}{2^{12}} = \dfrac{512}{4{,}096} = \dfrac{1}{8}$.

My interactive video summary

⊙ You may wish to work through this interactive video to see these solutions worked out in detail.

You Try It Work through this You Try It problem.

Work Exercises 11–18 in this textbook or in the My MathLab **Study Plan.**

OBJECTIVE 4   COMPUTING THE nTH PARTIAL SUM OF A GEOMETRIC SERIES

If $a_1, a_1r, a_1r^2, a_1r^3, \ldots$ is a geometric sequence, then the expression $a_1 + a_1r + a_1r^2 + a_1r^3 + \cdots + a_1r^{n-1} + \cdots$ is called an **infinite geometric series** and can be written in summation notation as $\sum_{i=1}^{\infty} a_1r^{i-1}$. Recall that the sum of the first n terms of a series is called the **nth partial sum** of the series and is given by $S_n = a_1 + a_1r + a_1r^2 + a_1r^3 + \cdots + a_1r^{n-1}$. We can also represent the nth partial sum using summation notation as $S_n = \sum_{i=1}^{n} a_1r^{i-1}$. As you can see, this nth partial sum is simply the sum of a finite geometric series. Fortunately, there is a convenient formula for computing the nth partial sum of a geometric series.

Formula for the nth Partial Sum of a Geometric Series

The sum of the first n terms of a geometric series is called the **nth partial sum** of the series and is given by $S_n = \sum_{i=1}^{n} a_1r^{i-1} = a_1 + a_1r + a_1r^2 + a_1r^3 + \cdots + a_1r^{n-1}$.

This sum can be computed using the formula $S_n = \dfrac{a_1(1 - r^n)}{1 - r}$ for $r \neq 1$.

My video summary   ⊗ Watch this video to see the derivation of this formula. (We will also prove this formula using mathematical induction in Section 9.5.)

My interactive video summary   ⊗ **Example 5** Computing the nth Partial Sum of a Geometric Series

a. Find the sum of the series $\sum_{i=1}^{15} 5(-2)^{i-1}$.

b. Find the 7th partial sum of the geometric series $8 + 6 + \dfrac{9}{2} + \dfrac{27}{8} + \cdots$.

Solution

a. Using the formula $S_n = \dfrac{a_1(1 - r^n)}{1 - r}$ with $n = 15$, $a_1 = 5$, and $r = -2$, we get

$$S_{15} = \frac{5(1 - (-2)^{15})}{1 - (-2)} = \frac{5(1 - (-32{,}768))}{3} = \frac{5(32{,}769)}{3} = \frac{163{,}845}{3} = 54{,}615.$$

You may also watch this interactive video to see this solution worked out in detail.

b. Work through the interactive video to verify that the 7th partial sum of this series is $S_7 = \dfrac{14{,}197}{512}$.

You Try It   Work through this You Try It problem.

Work Exercises 19–23 in this textbook or in the MyMathLab **Study Plan**.

OBJECTIVE 5   DETERMINING IF AN INFINITE GEOMETRIC SERIES CONVERGES OR DIVERGES

My video summary

⊚ Consider the infinite geometric series $\sum_{n=1}^{\infty} a_1 r^{n-1} = a_1 + a_1 r + a_1 r^2 + \cdots + a_1 r^{n-1} + \cdots$.

Is it possible for a series of this form to have a finite sum? Is it possible to add infinitely many terms and get a finite sum? The answer is **yes**, it is possible, but it depends on the value of r. Before we determine the value(s) of r for which an infinite geometric series has a finite sum, let's look at an example. Consider the following geometric series

$$\frac{1}{2} + \frac{1}{3} + \frac{2}{9} + \frac{4}{27} + \frac{8}{81} + \cdots$$

Note that $a_1 = \frac{1}{2}$ and $r = \frac{2}{3}$. We can use the formula $S_n = \frac{a_1(1 - r^n)}{1 - r}$ to find the nth partial sum for any value of n of our choosing. The nth partial sums for $n = 5, 10, 20,$ and 40 are given in Table 1 as well as the value of r^n.

Table 1

n	$S_n = \dfrac{a_1(1 - r^n)}{1 - r} = \dfrac{\left(\frac{1}{2}\right)\left(1 - \left(\frac{2}{3}\right)^n\right)}{1 - \frac{2}{3}}$	$r^n = \left(\dfrac{2}{3}\right)^n$
5	1.3024691	.1316872
10	1.4739877	.0173415
20	1.4995489	.0003007
40	1.4999999	.0000001

Looking at Table 1, it appears that as n increases, the value of S_n is getting closer to $1.5 = \frac{3}{2}$. Also notice that as n increases, the value of $r^n = \left(\frac{2}{3}\right)^n$ is getting closer to 0. In fact, for any value of r between -1 and 1, the value of r^n will always approach 0 as n approaches infinity. We say, "For values of r between -1 and 1, r^n approaches zero as n approaches infinity" and write: For $|r| < 1$, $r^n \to 0$ as $n \to \infty$. Thus, if $|r| < 1$, $S_n = \frac{a_1(1 - r^n)}{1 - r} \approx \frac{a_1(1 - 0)}{1 - r} = \frac{a_1}{1 - r}$ for large values of n. Therefore, given an infinite geometric series with $|r| < 1$, the sum of the series is given by $S = \frac{a_1}{1 - r}$.

Note A formal proof of this formula requires calculus.

Formula for the Sum of an Infinite Geometric Series

Let $\sum_{n=1}^{\infty} a_1 r^{n-1} = a_1 + a_1 r + a_1 r^2 + a_1 r^3 + \cdots + a_1 r^{n-1} + \cdots$ be an infinite geometric series. If $|r| < 1$, then the sum of the series is given by $S = \frac{a_1}{1 - r}$.

Note that if $|r| < 1$, then the infinite geometric series has a finite sum and is said to **converge**. If $|r| \geq 1$, then the infinite geometric series does not have a finite sum and the series is said to **diverge**.

 *My interactive video summary*

⊘ **Example 6** Determining If an Infinite Geometric Series Converges or Diverges

Determine whether each of the following series converges or diverges. If the series converges, find the sum.

a. $\displaystyle\sum_{n=1}^{\infty} \frac{1}{2}\left(\frac{2}{3}\right)^{n-1}$

b. $3 - \dfrac{6}{5} + \dfrac{12}{25} - \dfrac{24}{125} + \cdots$

c. $12 + 18 + 27 + \dfrac{81}{2} + \dfrac{243}{4} + \cdots$

Solution

a. This is an infinite geometric series with $|r| = \left|\dfrac{2}{3}\right| < 1$. Since $|r| < 1$, the infinite series must converge and, thus, must have a finite sum. The sum of the series is

$$S = \frac{a_1}{1-r} = \frac{\dfrac{1}{2}}{1 - \dfrac{2}{3}} = \frac{\dfrac{1}{2}}{\dfrac{1}{3}} = \frac{1}{2} \cdot \frac{3}{1} = \frac{3}{2}.$$

b. For this infinite geometric series, the common ratio is $r = \dfrac{-\dfrac{6}{5}}{3} = -\dfrac{6}{5} \cdot \dfrac{1}{3} = -\dfrac{2}{5}.$

Since $|r| = \left|-\dfrac{2}{5}\right| = \dfrac{2}{5} < 1$, the infinite series converges. The sum is

$$S = \frac{a_1}{1-r} = \frac{3}{1 - \left(-\dfrac{2}{5}\right)} = \frac{3}{1 + \dfrac{2}{5}} = \frac{3}{\dfrac{7}{5}} = 3 \cdot \frac{5}{7} = \frac{15}{7}.$$

c. This infinite series diverges. Do you know why? Work through the interactive video to see why this series diverges.

You Try It Work through this You Try It problem.

Work Exercises 24–28 in this textbook or in the MyMathLab Study Plan.

OBJECTIVE 6 APPLICATIONS OF GEOMETRIC SEQUENCES AND SERIES

In Example 7, we revisit the question that was presented at the beginning of this section.

Example 7 Choosing a Payment Method

Suppose that you have agreed to work for Donald Trump on a particular job for 21 days. Mr. Trump gives you two choices of payment. You can be paid $100 for the first day and an additional $50 per day for each subsequent day. Or, you can choose to be paid 1 penny for the first day with your pay doubling each subsequent day. Which method of payment yields the most income?

Solution Each payment method can be represented by a sequence.

Payment method 1: 100, 150, 200, 250, . . .

Payment method 2: .01, .02, .04, .08, . . .

To find out which method of payment will yield the greatest income, we must find the sum of the first 21 terms of each sequence.

Payment method 1 is an arithmetic sequence with $a_1 = 100$ and $d = 50$. Using the formula for the general term of an arithmetic sequence we get $a_n = 100 + (n - 1)50 = 100 + 50n - 50 = 50n + 50$. Note that $a_{21} = 50(21) + 50 = 1,100$. Using the formula for the nth partial sum of an arithmetic series with $n = 21$, we get $S_{21} = \dfrac{n(a_1 + a_{21})}{2} = \dfrac{21(100 + 1,100)}{2} = \$12,600$.

Payment method 2 is a geometric sequence with $a_1 = .01$ and $r = 2$. Using the formula for the nth partial sum of a geometric series with $n = 21$, we get $S_{21} = = \dfrac{a_1(1 - r^{21})}{1 - r} = \dfrac{(.01)(1 - 2^{21})}{1 - 2} \approx \$20,971.51$. Clearly, payment method 2 is the better choice.

Example 8 Total Amount Given to a Local Charity

A local charity received \$8,500 in charitable contributions during the month of January. Because of a struggling economy, it is projected that contributions will decline each month to 95% of the previous month's contributions. What are the expected contributions for the month of October? What is the total expected contributions that this charity can expect at the end of the year?

Solution The monthly contributions can be represented by a geometric sequence with $a_1 = 8,500$ and $r = .95$. Thus, the contributions for the nth month is given by $a_n = 8,500(.95)^{n-1}$. The expected contributions for October, or when $n = 10$, are $a_{10} = 8,500(.95)^{10-1} \approx 5,357.12$. Thus, the contributions for the month of October are expected to be about \$5,357.12.

The total contributions for the year can be written as the following finite geometric series:

$$8,500 + (8,500)(.95) + (8,500)(.95)^2 + \cdots + (8,500)(.95)^{11} = \sum_{i=1}^{12} 8,500(.95)^{i-1}$$

Using the formula for the nth partial sum of a geometric series with $n = 12$, we get $S_{12} = \dfrac{a_1(1 - r^{12})}{1 - r} = \dfrac{8,500(1 - .95^{12})}{1 - .95} \approx 78,138.79$.

Therefore, the charity can expect about \$78,138.79 in donations for the year.

You Try It Work through this You Try It problem.

Work Exercises 29–33 in this textbook or in the My MathLab **Study Plan.**

My interactive video summary

⊘ Example 9 Expressing a Repeating Decimal as a Ratio of Two Integers

Every repeating decimal number is a rational number and can therefore be represented by the quotient of two integers. Write each of the following repeating decimal numbers as the quotient of two integers.

a. $.\overline{4}$ **b.** $.2\overline{13}$

Solution

a. We can rewrite $.\overline{4}$ as $.44444\ldots = \dfrac{4}{10} + \dfrac{4}{100} + \dfrac{4}{1,000} + \dfrac{4}{10,000} + \dfrac{4}{100,000} + \cdots$. This is an infinite geometric series with $a_1 = \dfrac{4}{10}$ and $r = \dfrac{1}{10}$. Because $|r| = \left|\dfrac{1}{10}\right| < 1$, we know that the series converges. Using the formula $S = \dfrac{a_1}{1 - r}$ we see that

$$.\overline{4} = \dfrac{\dfrac{4}{10}}{1 - \dfrac{1}{10}} = \dfrac{\dfrac{4}{10}}{\dfrac{9}{10}} = \dfrac{4}{9}.$$

b. Carefully work through the interactive video to see that $.2\overline{13} = \dfrac{211}{990}$.

You Try It Work through this You Try It problem.

Work Exercises 34–35 in this textbook or in the MyMathLab Study Plan.

ANNUITIES

You Try It

In Section 5.1 we derived a formula for periodic compound interest. This formula is used to determine the future value of a *one-time* investment. (Click on the You Try It icon to see a periodic compound interest practice exercise.) Suppose that instead of investing one lump sum, you wish to invest equal amounts of money at steady intervals. An investment of equal amounts deposited at equal time intervals is called an **annuity**. If these equal deposits are made at the end of a compound period, the annuity is called an **ordinary annuity**.

For example, suppose that you want to invest $\$P$ at the end of each payment period at an annual rate r, in decimal form. Then the interest rate per payment period is $i = \dfrac{r}{\text{number of payment periods per year}}$. We now summarize the total amount of the ordinary annuity after the first k payment periods.

End of 1st payment period: $\underbrace{P}_{\text{1st payment}}$

End of 2nd payment period: $\underbrace{P}_{\substack{\text{1st payment}}} + \underbrace{Pi}_{\substack{\text{Interest earned} \\ \text{on 1st payment}}} + \underbrace{P}_{\text{2nd payment}} = \underbrace{P(1 + i)}_{\substack{\text{Total amount} \\ \text{of 1st payment}}} + \underbrace{P}_{\text{2nd payment}}$

End of 3rd payment period:

$\underbrace{P(1 + i)}_{\substack{\text{Amount of} \\ \text{1st payment}}} + \underbrace{P(1 + i)i}_{\substack{\text{Interest earned} \\ \text{on amount of} \\ \text{1st payment}}} + \underbrace{P}_{\text{2nd payment}} + \underbrace{Pi}_{\substack{\text{Interest earned} \\ \text{on 2nd payment}}} + \underbrace{P}_{\text{3rd payment}} = \underbrace{P(1 + i)^2}_{\substack{\text{Total amount} \\ \text{of 1st payment}}} + \underbrace{P(1 + i)}_{\substack{\text{Total amount} \\ \text{of 2nd payment}}} + \underbrace{P}_{\text{3rd payment}}$

End of kth payment period: $\underbrace{P(1 + i)^{k-1}}_{\substack{\text{Total amount} \\ \text{of 1st payment}}} + \underbrace{P(1 + i)^{k-2}}_{\substack{\text{Total amount} \\ \text{of 2nd payment}}} + \cdots + \underbrace{P(1 + i)}_{\substack{\text{Total amount} \\ \text{of } (k-1)\text{st payment}}} + \underbrace{P}_{k\text{th payment}}$

The total amount of the ordinary annuity after k payment periods is $A = P + P(1 + i) + \cdots + P(1 + i)^{k-2} + P(1 + i)^{k-1}$.

This is a finite geometric series with $a_1 = P$ and a common ratio of $(1 + i)$. Thus, the amount of the annuity after the kth payment is

$$A = \frac{P(1 - (1 + i)^k)}{1 - (1 + i)} = \frac{P(1 - (1 + i)^k)}{-i} = \frac{P((1 + i)^k - 1)}{i}$$

Amount of an Ordinary Annuity after the kth Payment

The total amount of an ordinary annuity after the kth payment is given by the formula

$$A = \frac{P((1 + i)^k - 1)}{i}$$

where

A = Total amount of annuity after k payments

P = Deposit amount at the end of each payment period

i = Interest rate per payment period

 *My video summary*

⊘ **Example 10** Finding the Amount of an Ordinary Annuity

Chie and Ben decided to save for their newborn son Jack's college education. They decided to invest $200 every 3 months in an investment earning 8% interest compounded quarterly. How much is this investment worth after 18 years?

Solution This is an ordinary annuity with $P = \$200$ and $i = \dfrac{.08}{4} = .02$. What is k?

See if you can determine k and use the formula $A = \dfrac{P((1 + i)^k - 1)}{i}$ to determine

the total amount of this annuity. When you are done, watch this video to see if you are correct.

You Try It Work through this You Try It problem.

Work Exercises 36–38 in this textbook or in the MyMathLab Study Plan.

9.3 Exercises

In Exercises 1–5, write the first five terms of the geometric sequence with the given information.

1. The first term is 8 and the common ratio is 2.

2. The first term is 162 and the common ratio is $\dfrac{1}{3}$.

3. The first term is 25 and the common ratio is $-\dfrac{1}{5}$.

4. $a_n = 7a_{n-1}; a_1 = 3$

5. $a_n = -2a_{n-1}; a_1 = -4$

In Exercises 6–10, determine if the sequence is geometric. If the sequence is geometric, find the common ratio.

6. $4, 24, 144, 864, \ldots$

7. $-2, 2, -2, 2, \ldots$

8. $-3, 1, -1, -3, \ldots$

9. $2, -\dfrac{10}{3}, \dfrac{50}{9}, -\dfrac{250}{27}, \ldots$

10. $7.236, -5.7888, 4.63104, -3.704832, \ldots$

11. Determine the general term of the sequence $3, 6, 12, 24, \ldots$.

12. Determine the general term of the sequence $\dfrac{1}{2}, \dfrac{1}{8}, \dfrac{1}{32}, \dfrac{1}{128}, \ldots$.

13. Determine the general term of the sequence $\dfrac{1}{5}, -\dfrac{2}{15}, \dfrac{4}{45}, -\dfrac{8}{135}, \ldots$.

14. Find the 7th term of the geometric sequence whose first term is 5 and whose common ratio is 4.

15. Find the 6th term of the geometric sequence whose first term is 3,804 and whose common ratio is $-\dfrac{1}{4}$.

16. Find the 11th term of the geometric sequence $\$5,000, \$5,050, \$5,100.50, \ldots$.

17. Given a geometric sequence such that $a_4 = 108$ and $a_7 = 2916$, find a_{10}.

18. Given a geometric sequence such that $a_3 = 16$ and $a_8 = -\dfrac{1}{2}$, find a_{11}.

In Exercises 19–23, find the sum of each geometric series.

19. $\displaystyle\sum_{i=1}^{8} 7(-4)^{i-1}$

20. $\displaystyle\sum_{i=1}^{13} 2\left(\dfrac{3}{5}\right)^{i-1}$

21. $\displaystyle\sum_{i=1}^{10} 4(1.05)^{i-1}$

22. $2 + \dfrac{2}{3} + \dfrac{2}{9} + \dfrac{2}{27} + \cdots + \dfrac{2}{729}$

23. $1 - \dfrac{1}{2} + \dfrac{1}{4} - \dfrac{1}{8} + \cdots - \dfrac{1}{128}$

In Exercises 24–28, determine if each infinite geometric series converges or diverges. If the series converges, find the sum.

24. $-1 + \dfrac{1}{10} - \dfrac{1}{100} + \dfrac{1}{1000} - \cdots$

25. $343 + 49 + 7 + 1 + \cdots$

26. $\displaystyle\sum_{i=1}^{\infty} 7\left(\dfrac{1}{4}\right)^{i-1}$

27. $\displaystyle\sum_{i=1}^{\infty} \dfrac{1}{5}(3)^{i-1}$

28. $0.5 - 0.05 + 0.005 - 0.0005 + 0.00005 - \cdots$

29. Warren wanted to save money to purchase a new car. He started by saving $1 on the first of January. On the first of February, he saved $3. On the first of March he saved $9. So, on the first day of each month, he wanted to save three times as much as he did on the first day of the previous month. If Warren continues his savings pattern, how much will he need to save on the first day of September?

30. Suppose that you have accepted a job for 2 weeks that will pay $.07 for the first day, $.14 for the second day, $.28 for the third day, and so on. What will your total earnings be after 2 weeks?

31. Mary has accepted a teaching job that pays $25,000 for the first year. According to the Teacher's Union, Mary will get guaranteed salary increases of 4% per year. If Mary plans to teach for 30 years, what will be her total salary earnings?

32. A child is given an initial push on a rope swing. On the first swing, the rope swings through an arc of 12 feet. On each successive swing, the length of the arc is 80% of the previous length. After 10 swings, what is the total length the rope will have swung? When the child stops swinging, what is the total length the rope will have swung?

33. Randy dropped a rubber ball from his apartment window from a height of 50 feet. The ball always bounces $\dfrac{3}{5}$ of the distance fallen. How far does the ball travel once it is done bouncing?

34. Rewrite the number $.\overline{7}$ as the quotient of two integers.

35. Rewrite the number $.3\overline{25}$ as the quotient of two integers.

36. Kip contributes $200 every month to his 401(k). What will the value of Kip's 401(k) be in 10 years if the yearly rate of return is assumed to be 12% compounded monthly?

37. Mark and Lisa decide to invest $500 every 3 months in an education IRA to save for their son Beau's college education. What will the value of the IRA be after 10 years if the yearly assumed rate of return is 8% compounded quarterly?

38. Marv and Cindy decide to build a new home in 10 years. They will need $80,000 to purchase the lot of their dreams. How much should they save each month in an account that has an assumed yearly rate of return of 7% compounded monthly?

9.4 The Binomial Theorem

THINGS TO KNOW

Before working through this section, be sure that you are familiar with the following concepts:

VIDEO   ANIMATION   INTERACTIVE

You Try It

1. Multiplying Polynomials (Section R.4)

OBJECTIVES

1 Expanding Binomials Raised to a Power Using Pascal's Triangle

2 Evaluating Binomial Coefficients

3 Expanding Binomials Raised to a Power Using the Binomial Theorem

4 Finding a Particular Term or a Particular Coefficient of a Binomial Expansion

---

OBJECTIVE 1 EXPANDING BINOMIALS RAISED TO A POWER USING PASCAL'S TRIANGLE

My video summary

In this section, we will focus on expanding algebraic expressions of the form $(a + b)^n$, where n is an integer greater than or equal to zero. Because $(a + b)$ is a binomial, we call the expansion of $(a + b)^n$ a *binomial expansion*. Consider the expansion of $(a + b)^4$.

$$
\begin{aligned}
(a + b)^4 &= \underbrace{(a + b)(a + b)} \cdot \underbrace{(a + b)(a + b)} \\
&= (a^2 + 2ab + b^2)(a^2 + 2ab + b^2) \\
&= a^4 + 2a^3b + a^2b^2 + 2a^3b + 4a^2b^2 + 2ab^3 + a^2b^2 + 2ab^3 + b^4 \\
&= a^4 + 4a^3b + 6a^2b^2 + 4ab^3 + b^4
\end{aligned}
$$

Although the expansion of $(a + b)^4$ using the method above is not too complicated, it would not be desirable to use this method to expand $(a + b)^n$ for large values of n.

The goal in this section is to try to develop a method for expanding expressions of the form $(a + b)^n$ without actually performing all of the multiplication. We start by studying the expanded forms of $(a + b)^n$ for $n = 0, 1, 2, 3, 4$, and 5.

$$
\begin{aligned}
n = 0: (a + b)^0 &= 1 \\
n = 1: (a + b)^1 &= 1a + 1b \\
n = 2: (a + b)^2 &= 1a^2 + 2ab + 1b^2 \\
n = 3: (a + b)^3 &= 1a^3 + 3a^2b + 3ab^2 + 1b^3 \\
n = 4: (a + b)^4 &= 1a^4 + 4a^3b + 6a^2b^2 + 4ab^3 + 1b^4 \\
n = 5: (a + b)^5 &= 1a^5 + 5a^4b + 10a^3b^2 + 10a^2b^3 + 5ab^4 + 1b^5
\end{aligned}
$$

The coefficients of each expansion are highlighted in red. These coefficients are known as **binomial coefficients**. Before we determine a pattern for these coefficients, let's first observe the exponent pattern. Notice in each expansion of

$(a + b)^n$, there are always $n + 1$ terms. The sum of the exponents of each term is always equal to n. Also note that the first term is always a^n (or $a^n b^0$) and the last term is always b^n (or $a^0 b^n$). As we look at the terms of each expansion from left to right, the exponent of the first variable decreases by 1 and the exponent of the second variable increases by 1. Thus, the exponent pattern of the variables of each expansion is $a^n b^0, a^{n-1} b^1, a^{n-2} b^2, a^{n-3} b^3, \ldots, a^1 b^{n-1}, a^0 b^n$. The pattern for the binomial coefficients is less obvious. To see the pattern for the coefficients, we start by rewriting the six expansions of $(a + b)^n$ again, this time we only write the coefficients. See Figure 4.

$n = 0$:   1

$n = 1$:   1   1

$n = 2$:   1   2   1

$n = 3$:   1   3   3   1

$n = 4$:   1   4   6   4   1

$n = 5$:   1   5   10   10   5   1

Figure 4 The coefficients of the expansions of $(a + b)^n$, also called Pascal's triangle

Notice that the coefficients in Figure 4 form a "triangle." This triangle is known as Pascal's triangle, named after the French mathematician, **Blaise Pascal**. The first and last number of each row of Pascal's triangle is 1. Every other number is equal to the sum of the two numbers directly above it. We can now write the next row of Pascal's triangle, which is the row corresponding to $n = 6$.

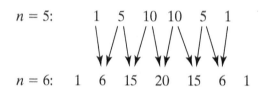

$n = 5$:   1   5   10   10   5   1

$n = 6$:   1   6   15   20   15   6   1

This new row of Pascal's triangle represents the coefficients of the expansion of $(a + b)^6$. Therefore,

$$(a + b)^6 = a^6 b^0 + 6a^5 b^1 + 15a^4 b^2 + 20a^3 b^3 + 15a^2 b^4 + 6a^1 b^5 + a^0 b^6$$
$$= a^6 + 6a^5 b + 15a^4 b^2 + 20a^3 b^3 + 15a^2 b^4 + 6ab^5 + b^6.$$

Notice the pattern of the exponents of each variable. The exponent of variable a starts with 6 and decreases by 1 for each successive term until it equals 0. The exponent of variable b starts with 0 and increases by 1 each term until it equals 6.

See if you can create Pascal's triangle for values of n up to 10. When you are done, view this version of Pascal's Triangle for $n = 0, 1, 2, \ldots 10$.

My interactive video summary

⊙ **Example 1** Using Pascal's Triangle to Expand a Binomial Raised to a Power

Use Pascal's triangle to expand each binomial.

a. $(x + 2)^4$ **b.** $(x - 3)^5$ **c.** $(2x - 3y)^3$

Solution

a. We start by looking at the row of Pascal's triangle corresponding with $n = 4$. We see that this row is 1 4 6 4 1. Using these coefficients and the exponent pattern we get

$$(x + 2)^4 = 1(x^4 \cdot 2^0) + 4(x^3 \cdot 2^1) + 6(x^2 \cdot 2^2) + 4(x \cdot 2^3) + 1(x^0 \cdot 2^4)$$

$$= x^4 + 8x^3 + 24x^2 + 32x + 16$$

b. The row of Pascal's triangle corresponding with $n = 5$ is 1 5 10 10 5 1. Using these coefficients and the exponent pattern we get

$$(x - 3)^5 = 1(x^5 \cdot (-3)^0) + 5(x^4 \cdot (-3)^1) + 10(x^3 \cdot (-3)^2)$$

$$+ 10(x^2 \cdot (-3)^3) + 5(x \cdot (-3)^4) + 1(x^0 \cdot (-3)^5)$$

$$= x^5 - 15x^4 + 90x^3 - 270x^2 + 405x - 243$$

c. See if you can use Pascal's triangle to show that $(2x - 3y)^3 = 8x^3 - 36x^2y + 54xy^2 - 27y^3$. Work through the interactive video to see the solution.

Note The terms of the expansion of the form $(a - b)^n$ will *always* alternate in sign with the sign of the first term being positive.

You Try It Work through this You Try It problem.

Work Exercises 1–6 in this textbook or in the MyMathLab Study Plan.

OBJECTIVE 2 EVALUATING BINOMIAL COEFFICIENTS

Although Pascal's triangle is useful for determining the binomial coefficients of $(a + b)^n$ for fairly small values of n, it is not that useful for large values of n. For example, to find the binomial coefficients of the expansion of $(a + b)^{50}$ using Pascal's triangle, we would need to write the first 51 rows of the triangle to determine the coefficients. Fortunately, there is a convenient formula for the binomial coefficients. This formula requires the use of factorials. Recall, $n! = n \cdot (n - 1) \cdot (n - 2) \cdot \cdots \cdot 3 \cdot 2 \cdot 1$ and $0! = 1$. To establish a formula for the binomial coefficients, let's take another look at the expansion for $(a + b)^6$.

$$(a + b)^6 = 1a^6b^0 + 6a^5b^1 + 15a^4b^2 + 20a^3b^3 + 15a^2b^4 + 6a^1b^5 + 1a^0b^6$$

Table 2 shows the relationship between the variable parts of the expansion of the form $a^{n-r}b^r$ and the corresponding coefficients. Click on any of the coefficients in Table 2 to verify that the formula using factorial notation is true.

Variables	Coefficient	Variables	Coefficient
a^6b^0	$1 = \dfrac{6!}{0! \cdot 6!}$	a^2b^4	$15 = \dfrac{6!}{4! \cdot 2!}$
a^5b^1	$6 = \dfrac{6!}{1! \cdot 5!}$	a^1b^5	$6 = \dfrac{6!}{5! \cdot 1!}$
a^4b^2	$15 = \dfrac{6!}{2! \cdot 4!}$	a^0b^6	$1 = \dfrac{6!}{6! \cdot 0!}$
a^3b^3	$20 = \dfrac{6!}{3! \cdot 3!}$		

Table 2

You can see in Table 2 that for each pair of variables of the form $a^{n-r}b^r$, the corresponding binomial coefficient is of the form $\dfrac{n!}{r! \cdot (n-r)!}$. We will use the shorthand notation $\dbinom{n}{r}$, read as "n choose r", to denote a binomial coefficient.

Formula for a Binomial Coefficient

For nonnegative integers n and r with $n \geq r$, the coefficient of the expansion of $(a+b)^n$ whose variable part is $a^{n-r}b^r$ is given by

$$\binom{n}{r} = \frac{n!}{r! \cdot (n-r)!}$$

Example 2 Evaluating Binomial Coefficients

Evaluate each of the following binomial coefficients.

a. $\dbinom{5}{3}$ b. $\dbinom{4}{1}$ c. $\dbinom{12}{8}$

Solution

a. $\dbinom{5}{3} = \dfrac{5!}{3!(5-3)!} = \dfrac{5!}{3! \cdot 2!} = \dfrac{5 \cdot 4 \cdot \cancel{3!}}{\cancel{3!} \cdot 2 \cdot 1} = \dfrac{20}{2} = 10$

b. $\dbinom{4}{1} = \dfrac{4!}{1!(4-1)!} = \dfrac{4!}{1! \cdot 3!} = \dfrac{4 \cdot \cancel{3!}}{1 \cdot \cancel{3!}} = \dfrac{4}{1} = 4$

c. $\dbinom{12}{8} = \dfrac{12!}{8!(12-8)!} = \dfrac{12!}{8! \cdot 4!} = \dfrac{12 \cdot 11 \cdot 10 \cdot 9 \cdot \cancel{8!}}{\cancel{8!} \cdot 4 \cdot 3 \cdot 2 \cdot 1} = \dfrac{11{,}880}{24} = 495$

Using Technology

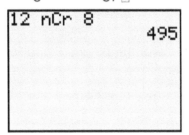

```
12 nCr 8
                495
```

A calculator can be used to compute binomial coefficients. Typically, the key $\boxed{nCr}$ is used. The figure on the left shows the computation of $\dbinom{12}{8}$ using a graphing utility.

You Try It Work through this You Try It problem.

Work Exercises 7–10 in this textbook or in the MyMathLab Study Plan.

OBJECTIVE 3   EXPANDING BINOMIALS RAISED TO A POWER USING
THE BINOMIAL THEOREM

Now that we know how to compute a binomial coefficient, we can state the
Binomial Theorem.

Binomial Theorem

If n is a positive integer then,

$$(a + b)^n = \binom{n}{0}a^n + \binom{n}{1}a^{n-1}b + \binom{n}{2}a^{n-2}b^2 + \cdots + \binom{n}{n}b^n$$

$$= \sum_{i=0}^{n} \binom{n}{i}a^{n-i}b^i$$

---

*My interactive
video summary*

⊙ **Example 3** Using the Binomial Theorem to Expand a Binomial
Raised to a Power

Use the Binomial Theorem to expand each binomial.

a. $(x - 1)^8$ **b.** $(\sqrt{x} + y^2)^5$

Solution

a. Work through the interactive video to verify that $(x - 1)^8 = x^8 - 8x^7 + 28x^6 - 56x^5 + 70x^4 - 56x^3 + 28x^2 - 8x + 1$ using the Binomial Theorem.

b. The expansion of $(\sqrt{x} + y^2)^5$ is as follows.

$$(\sqrt{x} + y^2)^5 = \binom{5}{0}(\sqrt{x})^5 + \binom{5}{1}(\sqrt{x})^4(y^2) + \binom{5}{2}(\sqrt{x})^3(y^2)^2 + \binom{5}{3}(\sqrt{x})^2(y^2)^3 + \binom{5}{4}(\sqrt{x})(y^2)^4 + \binom{5}{5}(y^2)^5$$

$$= 1 \cdot (\sqrt{x})^5 + 5 \cdot (\sqrt{x})^4(y^2) + 10 \cdot (\sqrt{x})^3(y^2)^2 + 10 \cdot (\sqrt{x})^2(y^2)^3 + 5 \cdot (\sqrt{x})(y^2)^4 + 1 \cdot (y^2)^5$$

$$= x^2\sqrt{x} + 5x^2y^2 + 10x\sqrt{x}y^4 + 10xy^6 + 5\sqrt{x}y^8 + y^{10}$$

You Try It   Work through this You Try It problem.

Work Exercises 11–17 in this textbook or in the MyMathLab Study Plan.

OBJECTIVE 4   FINDING A PARTICULAR TERM OR A PARTICULAR COEFFICIENT
OF A BINOMIAL EXPANSION

We may want to find a particular term of a binomial expansion. Fortunately, we
can use the Binomial Theorem to develop a formula for a particular term. We start
by writing out the first several terms of $(a + b)^n$ using the Binomial Theorem.

$$(a + b)^n = \binom{n}{0}a^n + \binom{n}{1}a^{n-1}b + \binom{n}{2}a^{n-2}b^2 + \binom{n}{3}a^{n-3}b^3 + \cdots + \binom{n}{n}b^n$$

The first term is $\binom{n}{0}a^n$. The second term is $\binom{n}{1}a^{n-1}b$. The third term is $\binom{n}{2}a^{n-2}b^2$.

Following this pattern, we can see that the formula for the $(r + 1)$st term (for $r \geq 0$)

is given by $\binom{n}{r}a^{n-r}b^r$.

> **Formula for the $(r + 1)$st Term of a Binomial Expansion**
>
> If n is a positive integer and if $r \geq 0$, then the $(r + 1)$st term of the expansion of $(a + b)^n$ is given by
>
> $$\binom{n}{r} a^{n-r} b^r = \frac{n!}{r! \cdot (n - r)!} a^{n-r} b^r$$

My video summary ⊙ **Example 4 Finding a Particular Term of a Binomial Expansion**

Find the third term of the expansion of $(2x - 3)^{10}$.

Solution Since we want to find the third term of this expansion, we will use the formula for the $(r + 1)$st term, which is equal to $\binom{n}{r} a^{n-r} b^r$ for $r = 2, n = 10, a = 2x$, and $b = -3$. Therefore, the third term is

$$\binom{10}{2} (2x)^{10-2}(-3)^2 = 45(256x^8)(9) = 103{,}680x^8$$

Watch this video to see every step of this solution.

My video summary ⊙ **Example 5 Finding a Particular Coefficient of a Binomial Expansion**

Find the coefficient of x^7 in the expansion of $(x + 4)^{11}$.

Solution The formula for the $(r + 1)$st term of the expansion of $(x + 4)^{11}$ is given by the formula $\binom{11}{r} x^{11-r} 4^r$. The term containing x^7 occurs when $11 - r = 7$. Solving this equation for r we get $r = 4$. Therefore, the term involving x^7 is $\binom{11}{4} x^7 4^4$.

Simplifying this expression we get $84{,}480x^7$. Thus, the coefficient of x^7 is $84{,}480$. Watch this video to see this solution worked out in detail.

You Try It Work through this You Try It problem.

Work Exercises 18–25 in this textbook or in the MyMathLab Study Plan.

9.4 Exercises

In Exercises 1–6, use Pascal's triangle to expand each binomial.

1. $(m + n)^6$ **2.** $(x + 5)^4$ **3.** $(x - y)^7$

4. $(x - 3)^5$ **5.** $(2x + 3y)^6$ **6.** $(3x^2 - 4y^3)^4$

In Exercises 7–10, evaluate each binomial coefficient.

7. $\binom{7}{1}$ **8.** $\binom{10}{4}$ **9.** $\binom{7}{7}$ **10.** $\binom{23}{3}$

In Exercises 11–17, use the Binomial Theorem to expand each binomial.

11. $(x + 2)^7$ **12.** $(x - 3)^6$ **13.** $(4x + 1)^5$

14. $(x + 3y)^4$ **15.** $(5x - 3y)^5$ **16.** $(x^4 + y^5)^6$

17. $(\sqrt{x} - \sqrt{2})^4$

18. Find the sixth term of the expansion of $(x + 4)^9$.

19. Find the fifth term of the expansion of $(a - b)^8$.

20. Find the third term of the expansion of $(3c - d)^7$.

21. Find the seventh term of the expansion of $(3x + 2)^{10}$.

22. Find the coefficient of x^5 in the expansion of $(x - 3)^{11}$.

23. Find the coefficient of x^4 in the expansion of $(4x + 1)^{12}$.

24. Find the coefficient of x^0 in the expansion of $\left(x^2 + \dfrac{1}{x} \right)^{12}$.

25. Find the coefficient of x^{10} in the expansion of $\left(x - \dfrac{3}{\sqrt{x}} \right)^{19}$.

9.5 Mathematical Induction

THINGS TO KNOW

Before working through this section, be sure that you are familiar with the following concepts:

VIDEO   ANIMATION   INTERACTIVE

You Try It

1. Computing the nth Partial Sum of an Arithmetic Series (Section 9.2)

2. Computing the nth Partial Sum of a Geometric Series (Section 9.3)

OBJECTIVES

1 Writing Mathematical Statements

2 Using the Principle of Mathematical Induction to Prove Statements

. . . .

OBJECTIVE 1   WRITING MATHEMATICAL STATEMENTS

In this section we will learn a technique that is often used in mathematics to prove certain mathematical statements that are declared to be true for all natural numbers. We will use the notation S_n to denote a mathematical statement. As a motivating example, consider the following mathematical statement

$$S_n: 1 + 2 + 3 + \cdots + n = \frac{n(n + 1)}{2} \text{ for all natural numbers } n$$

This statement suggests that the sum of the first n consecutive natural numbers is equal to the formula $\dfrac{n(n+1)}{2}$ for *all* natural numbers. We can verify that this statement is true for the first natural number, $n = 1$, by adding the "first one terms" on the left side and by substituting 1 for n on the right side.

$$S_1: \underbrace{1}_{\substack{\text{The sum of the} \\ \text{"first one terms"}}} = \underbrace{\dfrac{1(1+1)}{2}}_{\text{Substitute } n = 1}$$

The statement S_1 is true because the left- and right-hand sides of the equal sign are both equal to 1. In Example 1, we show that the statement $S_n: 1 + 2 + 3 + \cdots + n = \dfrac{n(n+1)}{2}$ is also true for the next four natural numbers.

Example 1 Verifying a Mathematical Statement for Specified Natural Numbers

Given the mathematical statement $S_n: 1 + 2 + 3 + \cdots + n = \dfrac{n(n+1)}{2}$, write the statements S_2, S_3, S_4, and S_5, and verify that each statement is true.

Solution

$$S_2: \underbrace{1 + 2}_{6} = \underbrace{\dfrac{2(2+1)}{2}}_{3}; \qquad S_3: \underbrace{1 + 2 + 3}_{6} = \underbrace{\dfrac{3(3+1)}{2}}_{6};$$

$$S_4: \underbrace{1 + 2 + 3 + 4}_{10} = \underbrace{\dfrac{4(4+1)}{2}}_{10}; \qquad S_5: \underbrace{1 + 2 + 3 + 4 + 5}_{15} = \underbrace{\dfrac{5(5+1)}{2}}_{15}$$

Because the left- and right-hand sides of the equal sign of each statement are equivalent, the statements S_2, S_3, S_4, and S_5 are true.

You Try It Work through this You Try It problem.

Work Exercises 1–4 in this textbook or in the My MathLab **Study Plan.**

In Example 1, we wrote mathematical statements for *specific* natural numbers. It is important that we can write mathematical statements for *arbitrary* natural numbers as in Example 2.

My video summary   ⊙ **Example 2** Writing a Mathematical Statement for Arbitrary Natural Numbers k and $k + 1$

Given the statement $S_n: 1 + 2 + 3 + \cdots + n = \dfrac{n(n+1)}{2}$, write the statements S_k and S_{k+1} for natural numbers k and $k + 1$.

Solution To find S_k, we replace n with k to get S_k: $1 + 2 + 3 + \cdots + k = \dfrac{k(k + 1)}{2}$.

Similarly, to find S_{k+1} we replace n with $k + 1$ to get S_{k+1}: $1 + 2 + 3 + \cdots + k +$

$(k + 1) = \dfrac{(k + 1)((k + 1) + 1)}{2} = \dfrac{(k + 1)(k + 2)}{2}$. Watch this video to see each

step of this solution.

You Try It Work through this You Try It problem.

Work Exercises 5–8 in this textbook or in the My MathLab **Study Plan.**

OBJECTIVE 2   USING THE PRINCIPLE OF MATHEMATICAL INDUCTION
TO PROVE STATEMENTS

So far in this section we have verified that the statement S_n: $1 + 2 + 3 + \cdots + n = \dfrac{n(n + 1)}{2}$ is true for the first five natural numbers, but how do we verify that this statement is true for *all* natural numbers? The answer is that we must use a technique called *mathematical induction*. This is how mathematical induction works: To prove that a mathematical statement S_n is true for *all* natural numbers, we must first show that the statement S_1 is true. Next, is the **inductive step**. We must prove that if S_k is true for any natural number k, then S_{k+1} is also true. Once we verify that S_{k+1} is true, it follows by this inductive step that S_{k+2} is true, S_{k+3} is true, and so on. Hence the statement is true for *all* natural numbers. We summarize this process by stating the principle of mathematical induction.

The Principle of Mathematical Induction

Let S_n be a mathematical statement involving the natural number n. The statement S_n is true for *all* natural numbers if the following two conditions are satisfied.

1. S_1 is true.

2. If, for any natural number k, S_k is true, then S_{k+1} is true.

Note The second condition is called the **inductive step**. We assume that S_k is true for some natural number k (this is called the **induction assumption**), then we must show that S_{k+1} is true.

Mathematical induction is like knocking down a sequence of dominoes. In order to knock down *all* of the dominoes, the first domino, S_1, must be knocked down. If we know that the kth domino, S_k, is knocked down and can verify that the $(k + 1)$st domino, S_{k+1}, will be knocked down, then we can conclude that *all* dominos will be knocked down. Note that both conditions must be true in order for all of the dominos to fall. Without knocking over the first domino, it is impossible to knock down *all* of the dominos. Similarly, if there is a domino that does not knock down a subsequent domino, then it is impossible to knock down *all* of the dominos. See Figure 5.

We can now use the principle of mathematical induction to prove that the statement S_n: $1 + 2 + 3 + \cdots + n = \dfrac{n(n + 1)}{2}$ is true for *all* natural numbers.

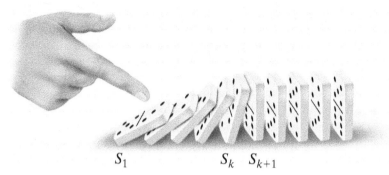

Figure 5 In order to knock down all dominos, the first domino, S_1, must fall. If it is assumed that domino S_k is knocked over and we can show that domino S_{k+1} will be knocked over, then we can conclude that *all* dominos will be knocked over.

 *My video summary*

⊙ **Example 3 Using the Principle of Mathematical Induction to Verify a Formula**

Prove that the mathematical statement $S_n: 1 + 2 + 3 + \cdots + n = \dfrac{n(n + 1)}{2}$ is true for all natural numbers n.

Solution

1. We need to show that S_1 is true.

    When we substitute $n = 1$ into the left- and right-hand side of this statement, we get $1 = \underbrace{\dfrac{1(1 + 1)}{2}}_{1}$. Thus, S_1 is true.

2. We assume that S_k is true for some natural number k. Thus, $1 + 2 + 3 + \cdots + k = \dfrac{k(k + 1)}{2}$. This is our induction assumption. Our goal is to show that S_{k+1} is true.

    Therefore, our goal is to verify that the following mathematical statement is true.

$$S_{k+1}: 1 + 2 + 3 + \cdots + k + (k + 1) = \frac{(k + 1)((k + 1) + 1)}{2} = \frac{(k + 1)(k + 2)}{2}$$

Start with the left-hand side of this equation and show that it is equal to the right-hand side.

$1 + 2 + 3 + \cdots + (k + 1)$	Start with the left-hand side of the statement S_{k+1}.
$= \underbrace{1 + 2 + 3 + \cdots + k} + (k + 1)$	The term preceding $k + 1$ is k.
$= \dfrac{k(k + 1)}{2} + (k + 1)$	Use the induction assumption.
$= \dfrac{k(k + 1) + 2(k + 1)}{2}$	Combine terms using a common denominator.
$= \dfrac{(k + 1)(k + 2)}{2}$	Factor.

Therefore, the statement S_{k+1}: $1 + 2 + 3 + \cdots + (k + 1) = \dfrac{(k + 1)(k + 2)}{2}$ is true. We have shown that S_1 is true. We have also shown that if the statement S_k is true, then the statement S_{k+1} is true. Therefore, by the principle of mathematical induction, S_n is true for all natural numbers. You may want to watch this video to see every step of this induction proof.

My video summary

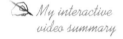
My interactive video summary

⊘ **Example 4** Using the Principle of Mathematical Induction to Verify a Formula

Prove that the mathematical statement S_n: $1^3 + 2^3 + 3^3 + \cdots + n^3 = \dfrac{n^2(n + 1)^2}{4}$ is true for all natural numbers n.

Solution Try following the same procedure as outlined in Example 3. When you are finished, watch this video to see if you are correct.

⊙ In Section 9.2 we studied arithmetic sequences and arithmetic series. Recall that if S_n is the nth partial sum of an arithmetic series, then this sum can be computed using the formula $S_n = \dfrac{n(a_1 + a_n)}{2}$. In Section 9.3 we studied geometric sequences and geometric series. If S_n is the nth partial sum of a geometric series, then this sum can be computed using the formula $S_n = \dfrac{a_1(1 - r^n)}{1 - r}$, where r is the common ratio. Watch this interactive video to see how to prove both of these formulas using the principle of mathematical induction.

So far, all of our examples involved verifying sum formulas. Mathematical induction is not limited to verifying only sum formulas. We can use the principle of mathematical induction to prove a variety of mathematical statements, as shown in Example 5.

Example 5 Using the Principle of Mathematical Induction to Verify a Mathematical Statement

Prove that the mathematical statement S_n: $2^n \geq 2n$ is true for all natural numbers n.

$$S_n\text{:} \; 2^n \geq 2n$$

Solution

1. We need to show that S_1 is true.

$$S_1\text{:} \; \underbrace{\frac{2^1}{2}} \geq \underbrace{\frac{2(1)}{2}}$$

Since 2 is greater than or equal to 2, we have verified that S_1 is true.

2. We assume that S_k is true for some natural number k. Thus, $2^k \geq 2k$ is our induction assumption. Our goal is to show that S_{k+1} is true. Therefore, our goal is to verify that the inequality $2^{k+1} \geq 2(k + 1)$ is true. To do this, start with the left-hand side of this inequality and verify that it is greater than or equal to the right-hand side.

2^{k+1}	Start with the left-hand side of the statement S_{k+1}.
$= 2^k \cdot 2$	Use the product rule of exponents.
$= 2^k + 2^k$	$2^k + 2^k = 2 \cdot 2^k$
$\geq 2k + 2k$	$2^k \geq 2k$ by the induction assumption.
$\geq 2k + 2$	$2k \geq 2$
$= 2(k + 1)$	Factor.

Thus, $2^{k+1} \geq 2(k + 1)$ and we have verified that S_{k+1} is true. Therefore, by the principle of mathematical induction, the statement S_n: $2^n \geq 2n$ is true for all natural numbers. ●

You Try It Work through this You Try It problem.

Work Exercises 9–16 in this textbook or in the MyMathLab Study Plan.

9.5 Exercises

In Exercises 1–4, write the statements S_1, S_2, and S_3, and determine if each statement is true.

1. S_n: $3 + 9 + 15 + \cdots + (6n - 3) = 3n^2$

2. S_n: $1^2 + 2^2 + 3^2 + \cdots + n^2 = \dfrac{n(n + 1)(2n + 1)}{6}$

3. S_n: $2^n > n^2$

4. S_n: 3 is a factor of $n(n^2 - 1)$.

In Exercises 5–8, write the statements S_k and S_{k+1}.

5. S_n: $6 + 12 + 18 + \cdots + 6n = 3n(n + 1)$

6. S_n: $5 + 11 + 17 + \cdots + (6n - 1) = n(3n + 2)$

7. S_n: $1^2 + 2^2 + 3^2 + \cdots + n^2 = \dfrac{n(n + 1)(2n + 1)}{6}$

8. S_n: $\dfrac{1}{1 \cdot 2} + \dfrac{1}{2 \cdot 3} + \dfrac{1}{3 \cdot 4} + \cdots + \dfrac{1}{n(n + 1)} = \dfrac{n}{n + 1}$

In Exercises 9–16, use the principle of mathematical induction to show that each mathematical statement is true for all natural numbers n.

9. S_n: $6 + 12 + 18 + \cdots + 6n = 3n(n + 1)$

10. S_n: $2 + 3 + 4 + \cdots + (n + 1) = \dfrac{1}{2}n(n + 3)$

11. S_n: $3 + 7 + 11 + \cdots + (4n - 1) = n(2n + 1)$

12. S_n: $1^2 + 2^2 + 3^2 + \cdots + n^2 = \dfrac{n(n + 1)(2n + 1)}{6}$

13. S_n: $3^0 + 3^1 + 3^2 + \cdots + 3^{n-1} = \dfrac{1}{2}(3^n - 1)$

14. S_n: $\dfrac{1}{1 \cdot 2} + \dfrac{1}{2 \cdot 3} + \dfrac{1}{3 \cdot 4} + \cdots + \dfrac{1}{n(n + 1)} = \dfrac{n}{n + 1}$

15. S_n: $n(n + 1) + 4$ is divisible by 2.

16. S_n: $3^n \geq 3n$

9.6 The Theory of Counting

OBJECTIVES

1 Using the Fundamental Counting Principle

2 Using Permutations

3 Using Combinations

Introduction to Section 9.6

Counting is a task we learn at an early age. It usually starts with the simple, "How old are you?" and we hold up two fingers. But counting gets more complicated as we mature and our lives get more complex. For example, when the Louisiana lottery began, people were asking, the question, "How many possible sets of numbers are there if you can choose six from the numbers 1 to 44?" To answer this question, it is necessary to learn simple counting skills. The answer is 7,059,052. See Example 12.

In addition to practical purposes, counting skills are a great exercise to improve analytical reasoning skills, which can be used for all sorts of problem-solving and decision-making situations encountered in many walks of life. We start by introducing the **fundamental counting principle.**

OBJECTIVE 1   USING THE FUNDAMENTAL COUNTING PRINCIPLE

An **event** is a specific part of an experiment or task that has zero or more countable outcomes. For example, suppose that you are getting dressed for a day at the beach and you have two pairs of shorts and three shirts that all coordinate. The task is to get dressed. Getting dressed consists of two events: choosing your shorts to wear and choosing your shirt to wear. Choosing your shorts is an event that has two possible outcomes. Choosing a shirt is an event that has three possible outcomes. To determine the total number of possible ways to dress for the beach, we use the fundamental counting principle.

Definition   The Fundamental Counting Principle

Suppose that an experiment or task is made up of two or more events. If the first event can occur in n_1 ways, the second event can occur in n_2 ways, and the third event can occur in n_3 ways and so on, then the number of possible outcomes of the experiment or task is $N = n_1 n_2 \ldots n_k$, where k is the total number of events.

The fundamental counting principle is sometimes referred to as the multiplication principle of counting because we simply multiply the number of possible outcomes of each event to determine the total number of possible outcomes of the experiment or task. In the example of getting dressed for the beach, the total number of possible outcomes is $N = n_1 n_2 = 2 \cdot 3 = 6$.

 *My video summary*

⊙ Example 1 Using the Fundamental Counting Principle

A woman wants to paint the exterior of her house using green for the shutters, peach for the siding, and off-white for the trim. She selects two shades of green, three shades of peach, and four shades of off-white that she likes. How many possible color schemes does this give her to choose from?

Solution We first recognize that the task of selecting a color scheme consists of three events.

Event 1: Choosing paint for the shutters; $n_1 = 2$ possible outcomes
Event 2: Choosing paint for the siding; $n_2 = 3$ possible outcomes
Event 3: Choosing paint for the trim; $n_3 = 4$ possible outcomes

By the fundamental counting principle, the total possible number of color schemes is $N = n_1 \, n_2 \, n_3 = 2 \cdot 3 \cdot 4 = 24$.

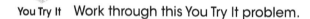

You Try It Work through this You Try It problem.

Work Exercises 1–3 in this textbook or in the MyMathLab Study Plan.

 *My video summary*

⊙ Example 2 Using the Fundamental Counting Principle

A student is to create a password for his MyMathLab homework account. He is informed that the password must consist of seven characters. How many possible ways can he create the password if the first three characters are letters followed by two digits and the last two characters can be letters or digits?

(Repeated letters are allowed and letters are not case sensitive. Also, repeated digits are allowed.)

Solution The letters are not case sensitive (there is no difference between a lowercase letter or an uppercase letter) and repetition of letters is allowed. Therefore, there are 26 possible letters to choose from for each character that can be a letter.

Because repetition of digits is allowed, there are 10 possible digits to choose from each time a digit can be chosen. We now consider the following seven events.

Event 1: Choose the first character, which must be a letter; $n_1 = 26$ possible letters
Event 2: Choose the second character, which must be
     a letter; $n_2 = 26$ possible letters
Event 3: Choose the third character, which must be a letter; $n_3 = 26$ possible letters
Event 4: Choose the fourth character, which must be a digit; $n_4 = 10$ possible digits
Event 5: Choose the fifth character, which must be a digit; $n_5 = 10$ possible digits
Event 6: Choose the sixth character, which can be a letter $n_6 = 26 + 10 = 36$
     or a digit;   possible characters
Event 7: Choose the seventh character, which can be a $n_7 = 26 + 10 = 36$
     letter or a digit;   possible characters

By the fundamental counting principle, there are

$$N = n_1 \cdot n_2 \cdot n_3 \cdot n_4 \cdot n_5 \cdot n_6 \cdot n_7$$

$$= 26 \cdot 26 \cdot 26 \cdot 10 \cdot 10 \cdot 36 \cdot 36$$

$$= 2{,}277{,}849{,}600 \text{ possible ways to create the password.}$$

You Try It   Work through this You Try It problem.

Work Exercises 4–6 in this textbook or in the MyMathLab **Study Plan.**

Sometimes the number of choices of a specific event is affected by the choice of a previous event. In this situation there will be more than one sequence of events to consider. Each sequence of events must be separated and counted. We call each sequence of events a **case,** and the technique is called **counting cases.** We illustrate this technique in Example 3 by slightly changing the password example seen previously.

Example 3 Using the Fundamental Counting Principle

A student is to create a password for his MyMathLab homework account. He is informed that the password must consist of seven characters. How many possible ways can he create the password if the first three characters are letters followed by two digits and the last two characters can be letters or digits?

(Repeated letters are not allowed and the letters are case sensitive. Also, repeated digits are allowed.)

Solution In this situation, repetition of letters is *not* allowed and the letters are case sensitive. The first three characters are letters so there are $26 \cdot 2 = 52$ distinct choices for the first character. There are 51 choices for the second character and there are 50 choices for the third character. The fourth and fifth characters are digits which can be repeated. Therefore there are 10 possible digits that can be used for the fourth and fifth characters. Note that the sixth and seventh characters can be either a letter or a digit. Since repetition of letters is not allowed, the number of possible outcomes for the seventh character is affected by the choice of the sixth character. So, we must consider two cases. The first case is to choose a letter as the sixth character. The second case is to choose a digit as the sixth character.

Case 1: The sixth character chosen is a letter:
In Case 1 we consider choosing a letter for the sixth character. There are only 49 possibilities because repetition of letters is not allowed. Then, there are 48 letters and 10 digits to choose from for the seventh character. Thus, the number of possibilities for the seventh character is $48 + 10 = 58$. Below is the number of outcomes for each of the seven events for this case.

$\dfrac{52}{n_1}$	$\dfrac{51}{n_2}$	$\dfrac{50}{n_3}$	$\dfrac{10}{n_4}$	$\dfrac{10}{n_5}$	$\dfrac{49}{n_6}$	$\dfrac{58}{n_7}$
choose a letter	*choose a letter*	*choose a letter*	*choose a digit*	*choose a digit*	*choose a letter*	*choose a letter or digit*

Using the Fundamental Counting Principle, we get

$$n_1 \cdot n_2 \cdot n_3 \cdot n_4 \cdot n_5 \cdot n_6 \cdot n_7$$

$$= 52 \cdot 51 \cdot 50 \cdot 10 \cdot 10 \cdot 49 \cdot 58$$

$$= 37{,}684{,}920{,}000 \text{ possible passwords for Case 1.}$$

Case 2: The sixth character chosen is a digit:
For Case 2 we consider the possibility of choosing a digit for the sixth character. Note that the number of choices for the first five characters is the same as in Case 1. For the sixth character, since repetition of digits is allowed, there are 10 possibilities. There are 49 letters and 10 digits to choose from for the seventh character. Therefore, there are $49 + 10 = 59$ possibilities for the seventh character.

(eText Screens 9.6-1–9.6-53)

$$\underbrace{\frac{52}{n_1}}_{\substack{\text{choose} \\ a \\ \text{letter}}} \cdot \underbrace{\frac{51}{n_2}}_{\substack{\text{choose} \\ a \\ \text{letter}}} \cdot \underbrace{\frac{50}{n_3}}_{\substack{\text{choose} \\ a \\ \text{letter}}} \cdot \underbrace{\frac{10}{n_4}}_{\substack{\text{choose} \\ a \\ \text{digit}}} \cdot \underbrace{\frac{10}{n_5}}_{\substack{\text{choose} \\ a \\ \text{digit}}} \cdot \underbrace{\frac{10}{n_6}}_{\substack{\text{choose} \\ a \\ \text{digit}}} \cdot \underbrace{\frac{59}{n_7}}_{\substack{\text{choose} \\ a \\ \text{letter or digit}}}$$

Using the Fundamental Counting Principle, we get

$$n_1 \cdot n_2 \cdot n_3 \cdot n_4 \cdot n_5 \cdot n_6 \cdot n_7$$

$$= 52 \cdot 51 \cdot 50 \cdot 10 \cdot 10 \cdot 10 \cdot 59$$

$$= 7{,}823{,}400{,}000 \text{ possible passwords for Case 2.}$$

Counting these two cases we get a total of $N = 37{,}684{,}920{,}000 + 7{,}823{,}400{,}000 = 45{,}508{,}320{,}000$ possible passwords.

In the previous example we added the possibilities from both cases. If there are many cases to consider, it may be easier to count all cases together and then subtract the cases that do not apply. This technique is called **all minus the exclusion**. In Example 4, we show two different methods for counting the number of outfits a tourist can choose for an upcoming weekend trip.

 *My video summary* ⊚ **Example 4** Using the Fundamental Counting Principle

Linda has five t-shirts of different colors, three pairs of shorts of different patterns, and four pairs of shoes of different colors. She is planning a weekend trip to Pensacola Beach, Florida. In how many ways can she combine these items if

a. the orange t-shirt clashes with the red shoes?

b. the red t-shirt clashes with the striped shorts?

Solution

a. Method I: Counting Cases

There are two non-clash cases for Linda. She can wear all t-shirts with all shorts and with the three non-red pairs of shoes. She can also wear her four non-orange t-shirts with all of her shorts and her red shoes.

Non-clash case 1: $\underline{\text{5 t-shirts}} \cdot \underline{\text{3 pairs of shorts}} \cdot \underline{\text{3 non-red pairs of shoes}} = 45$

Non-clash case 2: $\underline{\text{4 non-orange t-shirts}} \cdot \underline{\text{3 pairs of shorts}} \cdot \underline{\text{1 red pair of shoes}} = 12$

Therefore, we can count the cases to get $N = 45 + 12 = 57$ total possible non-clashing outfits.

Method II: All Minus the Exclusion

Using this method, we first determine the total number of possible ways for Linda to dress for the beach (ignoring clashes) and then subtract the number of outfits that clash and must be excluded.

Total possible outfits: $\underline{\text{5 t-shirts}} \cdot \underline{\text{3 pairs of shorts}} \cdot \underline{\text{4 pairs of shoes}} = 60$

Outfits that clash: $\underline{\text{1 orange t-shirt}} \cdot \underline{\text{3 pairs of shorts}} \cdot \underline{\text{1 red pairs of shoes}} = 3$

Now, subtract the outfits to exclude from the total possible outfits to get $N = 60 - 3 = 57$ total possible non-clashing outfits.

b. Method I: Counting Cases

Once again, there are two non-clash cases for Linda in this situation. She can wear all t-shirts with her two non-striped pairs of shorts and all pairs of shoes. She can also wear her four non-red t-shirts with the striped shorts and all pairs of shoes.

Non-clash case 1:   5 t-shirts · 2 non-striped pairs of shorts · 4 pairs of shoes = 40
Non-clash case 2:   4 non-red t-shirts · 1 striped pair of shorts   · 4 pairs of shoes = 16

Counting these two cases, we get $N = 40 + 16 = 56$ total possible non-clashing outfits.

Method II: All Minus the Exclusion

From part (a) we know that there are 60 total possible outfits. The outfits that clash are the ones with one red shirt, one pair of striped shorts, and any of the four pairs of shoes.

Total possible outfits: 5 t-shirts · 3 pairs of shorts   · 4 pairs of shoes = 60
Outfits that clash: 1 red t-shirt · 1 pair of striped shorts · 4 pairs of shoes = 4

 *My video summary*

⊙ Now, subtract the outfits to exclude from the total possible outfits to get $N = 60 - 4 = 56$ total possible non-clashing outfits. Watch this video to see the solution to this counting example worked out in detail. ●

You Try It   Work through this You Try It problem.

Work Exercises 7–12 in this textbook or in the My Math Lab **Study Plan.**

Often in life, a certain number of objects need to be chosen from a specific group or class. The objects in the class have various attributes. Determining the total possible outcomes of an event will depend on the desired attributes. To practice this skill, work through Examples 5 and 6 after considering the following demographic information collected from a teacher of mathematics at the local community college.

Total number of students in the class:	50
Number of boys:	35
Number of girls:	15
Number of freshmen:	22
Number of sophomores:	13
Number of juniors:	9
Number of seniors:	6
Number of education majors:	7
Number of non-education majors:	43

Example 5 Using the Fundamental Counting Principle

The instructor chooses a student to dim the lights and a student to shut the door (to use the overhead projector). In how many ways can she do this if a student may do both jobs and

a. both are girls?   **b.** one is a boy and one is a girl?

c. at least one is a boy?   **d.** the door closer is a boy?

e. neither are seniors?   **f.** neither are education majors?

Solution There are two events (or jobs) to consider: the number of possible students who can dim the lights and the number of possible students to shut the door. Once we determine the number of students who can complete each job, we use the fundamental counting principle to determine the total possible outcomes.

a. Because both jobs can be completed by the same girl, we know that there are 15 girls that can do each job.

Case: Girl/Girl   15 girls can dim the lights · 15 girls can shut the door = 225

Therefore, there are $N = 225$ possible ways for a girl to complete both jobs.

b. There are two cases for this situation.

Case 1: Boy/Girl   35 boys can dim the lights · 15 girls can shut the door = 525
Case 2: Girl/Boy   5 girls can dim the lights   · 35 boys can shut the door = 525

Adding the total outcomes from each case, we see that there are $N = 525 + 525 = 1050$ ways for a boy and a girl to complete the two jobs.

c. There are two methods that can be used to count how many outcomes there are if at least one boy completes a job.

Method I: Counting Cases

Count each case in which at least one boy does one of the jobs.

Case 1: Boy/Girl 35 boys can dim the lights · 15 girls can shut the door = 525
Case 2: Girl/Boy 15 girls can dim the lights · 35 boys can shut the door = 525
Case 3: Boy/Boy 35 boys can dim the lights · 35 boys can shut the door = 1225

Counting these three cases, we get $N = 525 + 525 + 1225 = 2275$ total possible ways for at least one boy to complete a job.

Method II: All Minus the Exclusion

The only case that we have to exclude is the case in which both jobs are completed by a girl. Therefore, we determine the total number of possible outcomes, then subtract the case in which both jobs are completed by a girl.

Total Possible Outcomes   50 students can dim the lights · 50 students can shut the door = 2500
Case: Girl/Girl   15 girls can dim the lights · 15 girls can shut the door   = 225

Therefore, we subtract 225 from 2500 to get $N = 2500 - 225 = 2275$ total possible ways for at least one boy to complete a job.

d. In this situation, any student can dim the lights but only a boy can shut the door.

Case: Any Student/Boy   50 students can dim the lights · 35 boys can shut the door = 1750

Thus, there are $N = 1750$ total possible ways for any student to dim the lights and a boy to shut the door.

e. There are 6 seniors in the class, so there are 44 non-seniors.

Case: Non-senior/Non-senior   44 non-seniors can dim the lights · 44 non-seniors can shut the door = 1936

Thus, there are $N = 1936$ ways that a non-senior can do both tasks.

f. There are 43 non-education majors.

Case: Non-ed major/Non-ed major   43 non-ed majors can dim the lights · 43 non-ed majors can shut the door = 1849

Thus, there are $N = 1849$ ways that a non-education major can do both tasks.

My interactive video summary

⊙ Example 6 Using the Fundamental Counting Principle

Using the same demographic information in Example 5, if the instructor chooses one student to dim the lights and one to shut the door, in how many ways can she do this if a student may **not** do both jobs and

a. both are girls?

b. one is a boy and one is a girl?

c. at least one is a boy?

d. the door closer is a boy?

e. neither are seniors?

f. neither are education majors?

Solution Try to determine the number of outcomes for (a)–(f) on your own; then check your answers or watch this interactive video to see a complete solution.

You Try It Work through this You Try It problem.

Work Exercises 13–17 in this textbook or in the MyMathLab Study Plan.

OBJECTIVE 2   USING PERMUTATIONS

Sometimes we are interested in counting the number of ways that objects can be arranged or ordered. For example, suppose that you purchased a new combination lock for your new mountain bike. The lock is a four-digit lock in which each digit is chosen from 0, 1, 2, 3, 4, 5, 6, 7, 8, and 9. Obviously, the precise order of the digits matters. This is an example of a **permutation**. (Perhaps we should call this lock a permutation lock.)

Definition   Permutation

A **permutation** is an ordered arrangement of n objects taken r at a time.

We will consider three counting techniques that involve permutations.

Permutations Involving Distinct Objects with Repetition

The first involves situations in which repetition is allowed.

Theorem   Permutations Involving Distinct Objects with Repetition Allowed

The number N of different arrangements of n objects using $r \leq n$ of them, in which

1. order matters,

2. an object can be repeated, and

3. the n objects are distinct,

is given by the formula $N = n^r$.

We have already seen a permutation of this type in Example 5a, where the selected students were distinct, a student was allowed to do both jobs, and the order of selection mattered because it determined the specific job. Now let's look at another example.

 *My video summary*

> **Example 7** Determining the Number of Permutations If Repetition Is Allowed

How many four-digit arrangements are possible for a bicycle permutation lock if repetition of digits is allowed?

Solution In any counting situation involving an arrangement, the order matters. Repetition of digits is allowed and there are ten distinct objects to choose from (the digits 0–9) Therefore, we use the permutation formula $N = n^r$ with $n = 10$ and $r = 4$. Thus, the number of total possibilities is
$N = n^r = 10^4 = 10,000.$

You Try It Work through this You Try It problem.

Work Exercises 18–20 in this textbook or in the MyMathLab Study Plan.

PERMUTATIONS INVOLVING DISTINCT OBJECTS WITHOUT REPETITION

In the previous example, repetition was allowed. Many counting problems involve situations where repetition is not allowed. Before we introduce the permutation formula for these situations, consider the permutation lock problem from Example 6, but this time suppose that each digit of the combination must be unique. In other words, the repetition of digits is not allowed. In this situation, there would be 10 possibilities for the first digit, 9 possibilities for the second digit, and so on. Thus, the total number of possibilities would be

$$\underline{10 \text{ possible digits}} \cdot \underline{9 \text{ possible digits}} \cdot \underline{8 \text{ possible digits}} \cdot \underline{7 \text{ possible digits}} = 5040$$

To illustrate how to produce the general formula for permutations of this nature, let's rewrite the left-hand side of the equation above. Before doing so, you may need to recall the definition of the **factorial of a non-negative number** that was introduced in Section 9.1.

$10 \cdot 9 \cdot 8 \cdot 7$	Start with the left-hand side of the equation above.
$= 10 \cdot 9 \cdot 8 \cdot 7 \cdot \dfrac{6 \cdot 5 \cdot 4 \cdot 3 \cdot 2 \cdot 1}{6 \cdot 5 \cdot 4 \cdot 3 \cdot 2 \cdot 1}$	Multiply by 1 in the form $1 = \dfrac{6 \cdot 5 \cdot 4 \cdot 3 \cdot 2 \cdot 1}{6 \cdot 5 \cdot 4 \cdot 3 \cdot 2 \cdot 1}$.
$= \dfrac{10!}{6!}$	Rewrite the numerator and denominator using factorial notation.
$= \dfrac{10!}{(10 - 4)!}$	Rewrite the denominator.

As you can see, the number of permutations of 10 digits taken 4 at a time without repetition is given by the formula $\dfrac{10!}{(10 - 4)!}$. We can generalize this permutation formula as $\dfrac{n!}{(n - r)!}$, where n is the number of distinct objects taken r at a time without repetition. This formula is restated in the following theorem.

Theorem Permutations Involving Distinct Objects with Repetition Not Allowed

The number N of different arrangements of n objects using $r \leq n$ of them, in which

1. order matters,

2. an object cannot be repeated, and

3. the n objects are distinct,

is given by the formula $N = {}_nP_r = \dfrac{n!}{(n - r)!}$.

The notation ${}_nP_r$ represents a permutation of n distinct objects taken r at a time where repetition is not allowed and order matters.

Many counting examples (and probability examples seen in Section 9.7) involve using a standard deck of playing cards. If you are unfamiliar with cards, read this explanation of a standard deck of **52** playing cards.

Example 8 Counting Cards Using Permutations

Four cards are drawn one at a time without replacement from a standard deck of 52 cards. In how many ways can they be drawn

a. when there are no restrictions as to what the cards are?

b. when all of the cards are face cards?

Solution

a. The cards are drawn one at a time, and therefore order matters. Thus, we count using a permutation. The cards are not replaced between draws, so repetition is not allowed. We are choosing from a deck of 52 distinct cards.

Therefore, we use the formula $N = {}_nP_r = \dfrac{n!}{(n-r)!}$ with $n = 52$ and $r = 4$.

$$N = {}_{52}P_4 = \frac{52!}{(52-4)!} = \frac{52!}{48!} = \frac{52 \cdot 51 \cdot 50 \cdot 49 \cdot 48!}{48!} = \frac{52 \cdot 51 \cdot 50 \cdot 49 \cdot \cancel{48!}}{\cancel{48!}} = 6{,}497{,}400$$

Therefore, there are $N = 6{,}497{,}400$ ways to draw four cards at random from a standard card deck.

b. As in the solution to part (a), the order matters, repetition is not allowed, and the cards are distinct. Thus we use the formula $N = {}_nP_r = \dfrac{n!}{(n-r)!}$. Again we are choosing $r = 4$ cards. However, in this case, $n = 12$ because there are exactly 12 face cards in a standard deck. (View this explanation of a standard deck of **52** playing cards if you need to see why.) Thus, we use the formula $N = {}_nP_r = \dfrac{n!}{(n-r)!}$ with $n = 12$ and $r = 4$.

$$N = {}_{12}P_4 = \frac{12!}{(12-4)!} = \frac{12!}{8!} = \frac{12 \cdot 11 \cdot 10 \cdot 9 \cdot 8!}{8!} = \frac{12 \cdot 11 \cdot 10 \cdot 9 \cdot \cancel{8!}}{\cancel{8!}} = 11{,}880$$

Therefore, there are $N = 11{,}880$ ways to draw four face cards.

My video summary

⊙ Example 9 Counting the Number of Competitors in a Race

There are nine competitors in a hurdle race.

a. In how many different ways can the nine competitors finish first, second, and third?

b. In how many different ways can all nine competitors finish the race?

Solution

a. The order in which the competitors finish the race matters, so we count using a permutation. Repetition is not allowed because a runner cannot finish in two places at once. Also, the nine competitors are all distinct. Thus, we use the formula $N = {}_nP_r = \dfrac{n!}{(n-r)!}$ with $n = 9$ and $r = 3$.

$$N = {}_9P_3 = \frac{9!}{(9-3)!} = \frac{9!}{6!} = \frac{9 \cdot 8 \cdot 7 \cdot 6!}{6!} = \frac{9 \cdot 8 \cdot 7 \cdot \cancel{6!}}{\cancel{6!}} = 504$$

There are $N = 504$ ways to arrange the nine competitors in first, second, and third place.

b. Again we use the permutation formula $N = {_nP_r} = \dfrac{n!}{(n-r)!}$, but this time we use $n = 9$ and $r = 9$ because we are interested in how the nine runners can finish in all nine places.

$$N = {_9P_9} = \frac{9!}{(9-9)!} = \frac{9!}{0!} = \frac{9!}{1} = 362{,}880$$

There are $N = 362{,}880$ ways to arrange the nine competitors in all nine places.

Note Example 9b is a special case of the formula $N = {_nP_r} = \dfrac{n!}{(n-r)!}$, where $n = r$. When permuting n distinct objects n at a time, we get

$N = {_nP_n} = \dfrac{n!}{(n-n)!} = \dfrac{n!}{0!} = \dfrac{n!}{1} = n!$. For example, there are $17!$ possible ways that 17 kindergarteners can be lined up to go to lunch, with 17 choices for who goes first, 16 choices for who goes next, and so on.

You Try It Work through this You Try It problem.

Work Exercises 21–27 in this textbook or in the MyMathLab Study Plan.

PERMUTATIONS INVOLVING SOME NON-DISTINCT OBJECTS

A third type of permutation involves a situation where some of the objects in a set of size n are not distinct.

Theorem Permutations Involving Some Objects That Are Not Distinct

The number N of different arrangements of n objects taken n at a time in which

1. order matters,

2. n_1 are of one kind, n_2 are of a second kind, . . . , and n_k are of a kth kind, and

3. some of the objects are not distinct,

is given by the formula $N = \dfrac{n!}{n_1!n_2!\ldots n_k!}$, where $n_1 + n_2 + \ldots + n_k = n$.

We illustrate the use of this theorem in Example 10.

My video summary ⊙ **Example 10** Arranging Letters in a Word That Has Repeated Letters
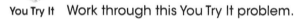

How many distinguishable letter codes can be formed from the word SUCCESSFUL if every letter is used?

Solution We are arranging the letters so order matters, so we count using a permutation. There are ten letters in the word "SUCCESSFUL" but there are only six different kinds of letters. There are $n_1 = 3$ "S's," $n_2 = 2$ "U's," $n_3 = 2$ "C's," $n_4 = 1$ "E," $n_5 = 1$ "F," and $n_6 = 1$ "L". Some of the letters are obviously not distinct. Therefore, we use the formula

$N = \dfrac{n!}{n_1!n_2!\ldots n_k!}$ with $n = 10, n_1 = 3, n_2 = 2, n_3 = 2, n_4 = 1, n_5 = 1,$ and $n_6 = 1$.

Therefore, the number of distinguishable arrangements is

$$N = \frac{n!}{n_1! n_2! n_3! n_4! n_5! n_6!} = \frac{10!}{3! 2! 2! 1! 1! 1!} = 151{,}200.$$

You Try It   Work through this You Try It problem.

Work Exercises 28–31 in this textbook or in the MyMathLab Study Plan.

OBJECTIVE 3   USING COMBINATIONS

Combinations are different from permutations. The difference is that order does not matter and there is never repetition of objects.

Definition   Combination

A **combination** is an unordered arrangement of n distinguishable objects taken r at a time without repetition.

Before we establish a formula for the number of combinations of n distinguishable objects taken r at a time, consider the following example, which illustrates the difference between a permutation and a combination.

Example 11   Choosing Books at the Library

Suppose that you went to the library and you chose the following four American novels off the shelf:

Tom Sawyer, Little Women, The Grapes of Wrath, and The Great Gatsby

a. Suppose that you place two of these books back on the shelf. How many arrangements are there?

b. How many groups of two books chosen from these four books can you check out to take home?

Solution

a. When arranging books on the shelf, the order matters. We could use the permutation formula $N = {_nP_r} = \dfrac{n!}{(n-r)!}$ with $n = 4$ and $r = 2$ because the objects are distinct and no repetition is allowed.

However, to illustrate a point, we list each possibility separately.

Possibilities of Arranging Two of Four Novels on a Shelf

1. *Tom Sawyer, Little Women*
2. *Tom Sawyer, The Grapes of Wrath*
3. *Tom Sawyer, The Great Gatsby*
4. *Little Women, Tom Sawyer*
5. *Little Women, The Grapes of Wrath*
6. *Little Women, The Great Gatsby*

7. *The Grapes of Wrath, Tom Sawyer*
8. *The Grapes of Wrath, Little Women*
9. *The Grapes of Wrath, The Great Gatsby*
10. *The Great Gatsby, Tom Sawyer*
11. *The Great Gatsby, Little Women*
12. *The Great Gatsby, The Grapes of Wrath*

We see that there are 12 ways to arrange these four classic novels on a shelf two at a time.

b. When checking out the books to take home, the order in which the books are checked out does not matter. So checking out *Tom Sawyer* and *Little Women* is no different than checking out *Little Women* and *Tom Sawyer*. Below are the possible ways to check out two of these four novels.

Possibilities of Checking Out Two of Four Novels from the Library

1. *Tom Sawyer, Little Women*

2. *Tom Sawyer, The Grapes of Wrath*

3. *Tom Sawyer, The Great Gatsby*

4. *Little Women, The Grapes of Wrath*

5. *Little Women, The Great Gatsby*

6. *The Grapes of Wrath, The Great Gatsby*

As you can see, there are six ways that we can check out two out of the four great American novels.

Notice in Example 11a that there are 12 arrangements of the four novels taken two at a time whereas in Example 11b there are only six groups of four novels taken two at a time. This is because in Example 11b we do not need to count the same pair of books twice when checking the books out. The process of counting used in Example 11b is a combination.

In Example 11b, for each pair of books chosen, there are exactly $2 = 2!$ ways to arrange them. In fact, for any collection of r objects (where $r \geq 1$) there are always $r!$ ways to arrange them. Therefore, to produce the formula to describe the combination of n distinct objects taken r at a time, we simply start with the formula for a permutation of n distinct objects taken r at a time, then divide by $r!$.

Theorem   Combinations Involving Objects That Are Distinct

The number N of different arrangements of n objects using $r \leq n$ of them, in which

1. order does not matter,

2. an object cannot be repeated, and

3. the n objects are distinct,

is given by the formula $N = {}_nC_r = \dfrac{n!}{(n - r)!\,r!}$.

The notation ${}_nC_r$ represents a combination of n distinct objects taken r at a time where repetition is not allowed and order does not matter.

My video summary   ⊙ **Example 12** Counting the Number of Possibilities in the Louisiana Lottery

The Louisiana lottery is a game in which six numbers are chosen from the numbers 1 to 44. Order does not matter and repetition of numbers is not allowed. How many possible sets of numbers are there in the Louisiana lottery?

Solution We use a combination to count the possible sets of numbers in the Louisiana lottery because the order does not matter. We must determine the number of combinations of the 44 numbers taken 6 at a time. Thus, we use the formula $N = {}_nC_r = \dfrac{n!}{(n-r)!\,r!}$, where $n = 44$ and $r = 6$.

$$N = {}_{44}C_6 = \frac{44!}{(44-6)!\cdot 6!} = \frac{44!}{38!\cdot 6!} = \frac{44\cdot 43\cdot 42\cdot 41\cdot 40\cdot 39\cdot \cancel{38!}}{\cancel{38!}\cdot 6!}$$

$$= \frac{44\cdot 43\cdot 42\cdot 41\cdot 40\cdot 39}{6\cdot 5\cdot 4\cdot 3\cdot 2\cdot 1} = 7{,}059{,}052$$

There are $N = 7{,}059{,}052$ possible sets of numbers in the Louisiana lottery.

You Try It Work through this You Try It problem.

Work Exercises 32–36 in this textbook or in the MyMathLab Study Plan.

⊙ **Example 13** Counting the Number of Poker Hands

A poker hand consists of 5 cards that are dealt from a typical standard deck of 52 cards.

a. How many total possible poker hands are there?

b. How many possible poker hands are there that consist of exactly two aces and exactly two kings? (The fifth card is a non-ace and non-king.)

c. How many possible poker hands are there that consist of exactly two queens and exactly three hearts?

Solution

a. The order of the five cards dealt in a poker hand does not matter, so we use a combination to count the total number of poker hands. To determine the number of possible poker hands, we must determine the number of combinations of the 52 cards taken 5 at a time.

Thus, we use the formula $N = {}_nC_r = \dfrac{n!}{(n-r)!\cdot r!}$, where $n = 52$ and $r = 5$.

$$N = {}_{52}C_5 = \frac{52!}{(52-5)!\cdot 5!} = \frac{52!}{47!\cdot 5!} = \frac{52\cdot 51\cdot 50\cdot 49\cdot 48\cdot \cancel{47!}}{\cancel{47!}\cdot 5!}$$

$$= \frac{52\cdot 51\cdot 50\cdot 49\cdot 48}{5\cdot 4\cdot 3\cdot 2\cdot 1} = 2{,}598{,}960$$

Therefore, there are $N = 2{,}598{,}960$ possible poker hands.

b. To determine the number of possible poker hands that consist of exactly two aces and exactly two kings, there are three different events that must be considered.

Event 1: Choose the two aces.

There are four aces from which two must be chosen. The order does not matter, so we count using a combination. The number of possible ways to choose two aces is ${}_4C_2 = 6$.

Event 2: Choose the two kings.

There are four kings from which two must be chosen. Again, the order does not matter, so we count using a combination. The number of possible ways to choose two kings is $_4C_2 = \mathbf{6}$.

Event 3: Choose one card that is a non-ace and non-king.

There are 44 remaining non-ace and non-king cards from which 1 must be chosen. The order does not matter, so we once again count using a combination. The number of possible ways to choose the remaining card is $_{44}C_1 = \mathbf{44}$.

Using the fundamental counting principle, we multiply the number of possible outcomes of each event.

$$N = \underbrace{_4C_2}_{\substack{\text{Number of ways} \\ \text{to choose an ace}}} \cdot \underbrace{_4C_2}_{\substack{\text{Number of ways} \\ \text{to choose a king}}} \cdot \underbrace{_{44}C_1}_{\substack{\text{Number of ways} \\ \text{to choose remaining card}}} = 6 \cdot 6 \cdot 44 = 1584$$

Thus, there are $N = 1584$ different poker hands that consist of exactly two aces and exactly two kings.

Note In the expression $_4C_2 \cdot {_4C_2} \cdot {_{44}C_1}$, the sum of the three values used for $n(4, 4, 44)$ is equal to 52. This can be used to check to see that you have considered all 52 cards in the deck. Also, the sum of the three values used for r $(2, 2, 1)$ is equal to 5. This can be used to check to see that you have chosen the correct number of cards.

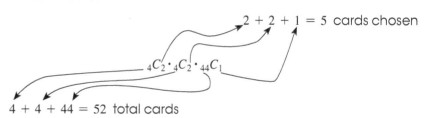

c. To determine the number of possible poker hands that consist of exactly two queens and exactly three hearts, there are two possible cases to consider. The first case is the case in which the poker hand contains the queen of hearts. The second case is the case in which the poker hand does not contain the queen of hearts.

Case 1: The hand contains the queen of hearts

For this case, there are four events to consider. Each event uses a combination because the order in which cards are chosen does not matter.

Event 1: Choose the queen of hearts.

There is one queen of hearts so there is $_1C_1 = \mathbf{1}$ way to choose the queen of hearts.

Event 2: Choose the second queen.

We must choose one out of the remaining three queens. Thus, there are $_3C_1 = \mathbf{3}$ ways to choose the remaining queen.

Event 3: Choose the remaining two hearts.

We must choose two out of the remaining 12 hearts. (One of the original 13 hearts, the queen of hearts, has already been accounted for in Event 1.) Thus, there are $_{12}C_2 = \mathbf{66}$ ways to choose the remaining two hearts.

Event 4: Choose the last card.

Out of the 36 non-queen and non-heart cards left, we must choose one of them to complete the five-card hand. Thus, there are $_{36}C_1 = 36$ ways to choose the last card.

Using the fundamental counting principle, we get the following:

Number of ways to choose the queen of hearts		Number of ways to choose the second queen		Number of ways to choose the remaining two hearts		Number of ways to choose the remaining card	
$_1C_1$	$\cdot$	$_3C_1$	$\cdot$	$_{12}C_2$	$\cdot$	$_{36}C_1$	$= 1 \cdot 3 \cdot 66 \cdot 36 = 7128$

Again, note **that the sum of the four values used for** n **adds up to** 52. $(1 + 3 + 12 + 36 = 52 \text{ total cards})$

The sum of the four values used for r adds up to 5. $(1 + 1 + 2 + 1 = 5 \text{ cards chosen})$

Case 2: The hand does not contain the queen of hearts

For this case, there are four events to consider. Each event uses a combination because the order in which cards are chosen does not matter.

Event 1: Consider the queen of hearts but do not choose it.

There is one queen of hearts so there is $_1C_0 = 1$ way not to choose the queen of hearts.

Event 2: Choose two queens out of the three queens that are not the queen of hearts. There are three queens that are not hearts in which two must be chosen, so there are $_3C_2 = 3$ ways to choose these two queens.

Event 3: Choose the three hearts.

We must choose three hearts out of the twelve non-queen hearts. Therefore, there are $_{12}C_3 = 220$ ways to choose the three hearts.

Event 4: Account for the cards remaining in the deck.

Out of the 36 non-queen and non-heart cards left, we do not choose any more cards because we already have a five-card hand. Thus, there are $_{36}C_0 = 1$ ways not to choose another card.

Using the fundamental counting principle, we get the following:

Number of ways to consider the queen of hearts but not choose it		Number of ways to choose two queens		Number of ways to choose the remaining three hearts		Number of ways to consider but not choose the remaining card	
$_1C_0$	$\cdot$	$_3C_2$	$\cdot$	$_{12}C_3$	$\cdot$	$_{36}C_0$	$= 1 \cdot 3 \cdot 220 \cdot 1 = 660$

Again, note that the sum of the four values used for n adds up to 52. $(1 + 3 + 12 + 36 = 52 \text{ total cards})$

The sum of the four values used for r adds up to 5. $(0 + 2 + 3 + 0 = 5$ cards chosen$)$

Adding the number of possible hands from each case, we get

$$N = \underbrace{{}_1C_1 \cdot {}_3C_1 \cdot {}_{12}C_2 \cdot {}_{36}C_1}_{7128} + \underbrace{{}_1C_0 \cdot {}_3C_2 \cdot {}_{12}C_3 \cdot {}_{36}C_0}_{660} = 7788$$

Therefore, there are $N = 7788$ possible hands that contain exactly two queens and exactly three hearts.

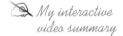 *My interactive video summary*

⊙ Watch this interactive video for a detailed explanation of parts (a)-(c).   ●

You Try It Work through this You Try It problem.

Work Exercises 37–45 in this textbook or in the My MathLab **Study Plan.**

Given a counting problem, it is crucial that we first identify if it is a permutation or a combination. Always try to decide whether or not the order matters. If the order does not matter, then the problem involves a combination. If the order matters, then the problem involves a permutation and then the type of permutation must be determined. Table 3 summarizes the fundamental counting principle and all of the counting techniques discussed in this section.

Table 3 The Fundamental Counting Principle and Counting Techniques

Fundamental Counting Principle	$N = n_1 n_2 \dots n_k$	Suppose that an experiment or task is made up of two or more events. If the first event can occur in n_1 ways, the second event can occur in n_2 ways, and the third event can occur in n_3 ways and so on, then the number of possible outcomes of the experiment or task is $N = n_1 n_2 \dots n_k$, where k is the total number of events.		
Counting Techniques	**Formula**	**Order Matters**	**Repetition Allowed**	**Distinct Objects**
Permutations	$N = n^r$	✔	✔	✔
	$N = {}_nP_r = \dfrac{n!}{(n-r)!}$	✔	✗	✔
	$N = \dfrac{n!}{n_1! \cdot n_2! \cdot \dots \cdot n_k!},$ $n_1 + n_2 + \dots + n_k = n$	✔	✗	✗
Combinations	$N = {}_nC_r = \dfrac{n!}{(n-r)! \cdot r!}$	✗	✗	✔

9.6 Exercises

1. A man has three pairs of slacks, four shirts, and five ties that all coordinate. How many different outfits of slacks, shirts, and ties can he put together?

2. The menu at Ben's Burgers is as follows:

Burgers	Sides	Drinks
Hamburger	Small fry	Water
Cheeseburger	Large fry	Coffee
Double cheeseburger	Small tater-tots	Tea
Double deluxe cheeseburger	Large tater-tots	Cola
	Salad	Root beer
		Sparkling cider

How many ways can a person order one burger, one side, and one drink?

3. You have narrowed your choice of a new car to either a four-door sedan or an SUV. Each car comes in white, black, tan, or green. Each model has the option of a gas engine, diesel engine, or a hybrid engine. Each model also has the option of a manual transmission or an automatic transmission. How many different choices do you still have to consider?

4. All automobile license plates in a certain state consist of two capital letters followed by five digits. (Repeated letters are allowed but repeated digits are not allowed.) How many different license plates can there be for this state?

5. A student is taking a five-question quiz. The first two questions are True/False questions. The last three are multiple choice in which she can choose the answer A, B, C, or D. In how many ways can she answer the entire test?

6. A student is to create a password for her MyMathLab homework account. She is informed that the password must consist of six characters. How many possible ways can she create the password if the first two characters are digits followed by three letters and one digit? (Repeated letters are allowed and the letters are case sensitive. Also, repeated digits are not allowed.)

7. Amy is planning a week-end trip to Atlanta. She plans to take six t-shirts all of different colors, five pairs of shorts all of different patterns, and two pairs of shoes of different styles. How many outfits can she mix and match if

   a. all pieces of clothing match?

   b. her pink t-shirt clashes with her red shorts and her orange polka-dot shorts, but everything else coordinates?

In Exercises 8–12, refer to the following menu. In each exercise, assume that an order consists of one main item, one side dish, and one choice of bread.

Main	Side Dishes	Bread
fried chicken	carrots	cornbread
BBQ steak	french fries	french bread
baked ham	potato salad	biscuits
	baked potato	
	dirty rice	

8. How many possible orders are there if you like everything on the menu?

9. How many possible orders are there if you don't like potato salad?

10. How many possible orders are there if you only like BBQ steak served with a baked potato and french bread, but you like all other combinations?

11. How many possible orders are there if you only like fried chicken, and you do not like baked potatoes or cornbread?

12. How many possible orders are there if you like everything except french fries and french bread?

In Exercises 13–17, refer to the following demographic information that has been collected by an instructor after surveying his class.

Total number of students in the class:	46
Number of boys:	19
Number of girls:	27
Number of freshmen:	7
Number of sophomores:	16
Number of juniors:	12
Number of seniors:	11
Number of science majors:	31
Number of non-science majors:	15

13. If the instructor chooses one student to pass out papers, one student to clean the chalkboard, and one student to empty the trash, then how many ways can he do this if he picks all girls and a student may do more than one job?

14. If the instructor chooses one student to pass out papers, one student to clean the chalkboard, and one student to empty the trash, then how many ways can he do this if at least one student is a girl and a student may not do more than one job?

15. If the instructor chooses one student to pass out papers, one student to clean the chalkboard, and one student to empty the trash, then how many ways can he do this if all three students are not the same sex and a student may not do more than one job?

16. If the instructor chooses one student to pass out papers, one student to clean the chalkboard, and one student to empty the trash, then how many ways can he do this if at least one student is a senior and a student may do more than one job?

17. If the instructor chooses one student to pass out papers, one student to clean the chalkboard, and one student to empty the trash, then how many ways can he do this if at most one student is a senior and a student may not do more than one job?

18. In how many ways can four cards be drawn from a standard deck if the card is replaced between draws?

19. How many six-digit passwords are possible if repetition of digits is allowed?

20. A football coach must decide who is going to receive the following awards at the end of season banquet:

    Most Valuable Player, Mr. "Hustle," Most Improved, Most Inspirational

    How many ways can the 56 players on the team be chosen to win these four awards if a player can win more than one award?

21. Four customers are heading to the checkout lines at a local grocery store. In how many ways can three of the four people arrive and stand in line at the express checkout line?

22. How many ways can the letters of the word ORBIT be arranged if all of the letters are used without repetition?

23. How many ways can eight DVDs be arranged on a shelf?

24. A typical game of SCRABBLE begins with each player randomly choosing seven letter tiles. How many ways can a player arrange five of the seven letter tiles assuming that each tile is a different letter?

25. Three students are chosen from a class of 35. One will get an A, one will get a C, and the other will get an F. In how many different ways can these three students be chosen?

26. Three cards are drawn one at a time without replacement from a standard deck of 52 cards. In how many ways can exactly three hearts be drawn?

27. A football coach must decide who is going to receive the following awards at the end of season banquet:

    Most Valuable Player, Mr. "Hustle," Most Improved, Most Inspirational

    How many ways can the 56 players on the team be chosen to win these awards if no player can win more than one award?

28. How many distinguishable letter codes can be formed from the word MISSISSIPPI if every letter is used?

29. A child has 5 red balls, 3 blue balls, and 2 green balls. All 10 balls are to be lined up on a shelf. How many total distinguishable arrangements of the 10 balls are there?

30. A binary number is a number that consists of only the digits 0 or 1. How many distinguishable binary numbers can be formed from the binary number 11011001?

31. In how many ways can the algebraic expression $x^5 y^3 z^2$ be written as a product without using any exponents?

SbS 32. How many five-member committees can be formed from a class of 30 students?

SbS 33. A kindergarten child is allowed to pick two toys from the goody box for demonstrating good classroom behavior. How many pairs of toys can she pick from a box of 25 toys?

SbS 34. A basketball conference consists of 12 teams. How many total in conference games must be played for each team to play each other exactly once?

SbS 35. A football coach must decide who are going to be the team captains for the last game of the season. There will be four team captains for the last game. How many distinct sets of team captains are possible if there are 46 players on the team?

SbS 36. How many ways can 7 cards be dealt from a standard deck of 52 cards?

SbS 37. A neighborhood contains 5 coffee shops, 3 pizza restaurants, and 2 sushi restaurants. Three of the businesses are to be chosen at random. How many possible outcomes are there?

SbS 38. A neighborhood contains 5 coffee shops, 3 pizza restaurants, and 2 sushi restaurants. Three of the businesses are to be chosen at random. Of the total number of possible outcomes, how many would consist only of coffee shops?

SbS 39. A neighborhood contains 5 coffee shops, 3 pizza restaurants, and 2 sushi restaurants. Three of the businesses are to be chosen at random. Of the total number of possible outcomes, how many would consist only of pizza restaurants?

SbS 40. Three cards are dealt without replacement from a standard deck of fifty-two. In how many ways can this happen if there are no restrictions?

SbS 41. Three cards are dealt without replacement from a standard deck of fifty-two. In how many ways can this happen if all three are hearts?

SbS 42. Three cards are dealt without replacement from a standard deck of fifty-two. In how many ways can this happen if two cards are hearts and one is a spade?

SbS 43. A gin rummy hand consists of 7 cards dealt without replacement from a standard deck of 52 cards. How many gin rummy hands are there that contain any 7 cards?

SbS 44. A gin rummy hand consists of 7 cards dealt without replacement from a standard deck of 52 cards. How many gin rummy hands are there that contain exactly two aces and exactly two tens?

SbS 45. A gin rummy hand consists of 7 cards dealt without replacement from a standard deck of 52 cards. How many gin rummy hands are there that contain exactly two jacks and exactly five hearts?

Brief Exercises

46. Four customers are heading to the checkout lines at a local grocery store. In how many ways can three of the four people arrive and stand in line at the express checkout line?

47. How many ways can the letters of the word ORBIT be arranged if all of the letters are used without repetition?

48. How many ways can eight DVDs be arranged on a shelf?

49. A typical game of SCRABBLE begins with each player randomly choosing seven letter tiles. How many ways can a player arrange five of the seven letter tiles assuming that each tile is a different letter?

50. Three students are chosen from a class of thirty-five. One will get an A, one will get a C, and the other will get an F. In how many different ways can these three students be chosen?

51. Three cards are drawn one at a time without replacement from a standard deck of fifty-two cards. In how many ways can exactly three hearts be drawn?

52. A football coach must decide who is going to receive the following awards at the end of season banquet:

Most Valuable Player, Mr. "Hustle," Most Improved, Most Inspirational

How many ways can the 56 players on the team be chosen to win these awards if no player can win more than one award?

53. How many five-member committees can be formed from a class of 30 students?

54. A kindergarten child is allowed to pick two toys from the goody box for demonstrating good classroom behavior. How many pairs of toys can she pick from a box of 25 toys?

55. A basketball conference consists of 12 teams. How many total in conference games must be played for each team to play each other exactly once?

56. A football coach must decide who are going to be the team captains for the last game of the season. There will be four team captains for the last game. How many distinct sets of team captains are possible if there are 46 players on the team?

57. How many ways can 7 cards be dealt from a standard deck of 52 cards?

58. A neighborhood contains 5 coffee shops, 3 pizza restaurants, and 2 sushi restaurants. Three of the businesses are to be chosen at random. How many possible outcomes are there?

59. A neighborhood contains 5 coffee shops, 3 pizza restaurants, and 2 sushi restaurants. Three of the businesses are to be chosen at random. Of the total number of possible outcomes, how many would consist only of coffee shops?

60. A neighborhood contains 5 coffee shops, 3 pizza restaurants, and 2 sushi restaurants. Three of the businesses are to be chosen at random. Of the total number of possible outcomes, how many would consist only of pizza restaurants?

61. Three cards are dealt without replacement from a standard deck of fifty-two. In how many ways can this happen if there are no restrictions?

62. Three cards are dealt without replacement from a standard deck of fifty-two. In how many ways can this happen if all three are hearts?

63. Three cards are dealt without replacement from a standard deck of fifty-two. In how many ways can this happen if two cards are hearts and one is a spade?

64. A gin rummy hand consists of 7 cards dealt without replacement from a standard deck of 52 cards. How many gin rummy hands are there that contain any 7 cards?

65. A gin rummy hand consists of 7 cards dealt without replacement from a standard deck of 52 cards. How many gin rummy hands are there that contain exactly two aces and exactly two tens?

66. A gin rummy hand consists of 7 cards dealt without replacement from a standard deck of 52 cards. How many gin rummy hands are there that contain exactly two jacks and exactly five hearts?

67. Four customers are heading to the checkout lines at a local grocery store. In how many ways can three of the four people arrive and stand in line at the express checkout line?

68. How many ways can the letters of the word ORBIT be arranged if all of the letters are used without repetition?

69. How many ways can eight DVDs be arranged on a shelf?

70. A typical game of SCRABBLE begins with each player randomly choosing seven letter tiles. How many ways can a player arrange five of the seven letter tiles assuming that each tile is a different letter?

71. Three students are chosen from a class of thirty-five. One will get an A, one will get a C, and the other will get an F. In how many different ways can these three students be chosen?

72. Three cards are drawn one at a time without replacement from a standard deck of fifty-two cards. In how many ways can exactly three hearts be drawn?

73. A football coach must decide who is going to receive the following awards at the end of season banquet:

    Most Valuable Player, Mr. "Hustle," Most Improved, Most Inspirational

    How many ways can the 56 players on the team be chosen to win these awards if no player can win more than one award?

74. How many five-member committees can be formed from a class of 30 students?

75. A kindergarten child is allowed to pick two toys from the goody box for demonstrating good classroom behavior. How many pairs of toys can she pick from a box of 25 toys?

76. A basketball conference consists of 12 teams. How many total in conference games must be played for each team to play each other exactly once?

77. A football coach must decide who are going to be the team captains for the last game of the season. There will be four team captains for the last game. How many distinct sets of team captains are possible if there are 46 players on the team?

78. How many ways can 7 cards be dealt from a standard deck of 52 cards?

79. A bag contains five red balls, three blue balls, and two green balls. Three balls are to be selected at random. How many possible outcomes are there?

80. A bag contains five red balls, three blue balls, and two green balls. Three balls are to be selected at random. Of the total number of possible outcomes, how many would only contain red balls?

81. A bag contains five red balls, three blue balls, and two green balls. Three balls are to be selected at random. Of the total number of possible outcomes, how many would only contain blue balls?

82. Three cards are dealt without replacement from a standard deck of fifty-two. In how many ways can this happen if there are no restrictions?

83. Three cards are dealt without replacement from a standard deck of fifty-two. In how many ways can this happen if all three are hearts?

84. Three cards are dealt without replacement from a standard deck of fifty-two. In how many ways can this happen if two cards are hearts and one is a spade?

85. A gin rummy hand consists of 7 cards dealt without replacement from a standard deck of 52 cards. How many gin rummy hands are there that contain any 7 cards?

86. A gin rummy hand consists of 7 cards dealt without replacement from a standard deck of 52 cards. How many gin rummy hands are there that contain exactly two aces and exactly two tens?

87. A gin rummy hand consists of 7 cards dealt without replacement from a standard deck of 52 cards. How many gin rummy hands are there that contain exactly two jacks and exactly five hearts?

9.7 An Introduction to Probability

THINGS TO KNOW

Before working through this section, be sure that you are familiar with the following concepts:

VIDEO ANIMATION INTERACTIVE

 You Try It 1. Using the Fundamental Counting Principle (Section 9.6)

 You Try It 2. Using Combinations (Section 9.6)

OBJECTIVES

1 Understanding Probability

2 Determining Probability Using the Additive Rule and Venn Diagrams

3 Using Combinations to Determine Probabilities

4 Using Conditional Probability

5 Understanding Odds

Introduction to Section 9.7

Probability has been of interest to man as far back as recorded history can determine. Games of chance were played by the Assyrians, the Sumerians, and the Egyptians. Though their games differed from the games we play today, the probability of certain outcomes was as predictable. The theory of probability was developed further during the 1600s by famous mathematicians such as Blaise Pascal and Pierre Fermat. During the late 1800s, Russian mathematicians became interested in probability, and this further enhanced the probability theory we use today.

But probability is not just for games. It has practical uses in any situation where a decision needs to be made in the face of uncertainty, provided that the uncertainties can be quantified. Life insurance companies rely heavily on probability to set accident and death rates for specific age groups and sexes. Also, probability arises in evaluating the data gathered by all of the different sciences—physics, chemistry, biology, psychology, sociology, and political science. Probability theory is a major branch of mathematics, with hundreds—if not thousands—of specialists. Research institutions often have groups of mathematicians called probabilists. Their work has applications in such diverse areas as data transmission, quantum theory, neurophysics, and the analysis of financial markets.

But whatever your reasons for studying probability, you will find that it is a fascinating topic. You will be surprised how many times you will notice probability or odds used after you have studied the subject.

OBJECTIVE 1   UNDERSTANDING PROBABILITY

Most of us have tossed a coin and asked a friend to call "heads or tails." If the coin is "fair," then the probability that the coin lands on heads is $\frac{1}{2} = 0.5 = 50\%$.

We can therefore think of probability as the likelihood that an event will occur. In the experiment of tossing a coin, the likelihood of the coin landing on heads is $\frac{1}{2} = 0.5 = 50\%$.

Definition   Probability

Probability is the mathematical estimate of the likelihood that an event or sequence of events will occur.

Probabilities are usually written as fractions or decimals, and sometimes they are written as percents, depending on what seems most appropriate for the situation. Probabilities range from 0 to 1, or from 0% to 100%.

Every experiment has a total number of possible outcomes. In the experiment of tossing a fair coin, there are two possible outcomes. The first outcome is the coin landing on heads and the second outcome is the coin landing on tails. The set of all possible outcomes of an experiment is known as the **sample space.**

Definition   Sample Space

The **sample space**, S, for an experiment is the set of all possible outcomes for the experiment. The number of possible outcomes of the sample space is denoted as $n(S)$.

 *My video summary*

⊘ **Example 1** Determining the Sample Space of an Experiment

State the number of elements in the sample space of the following experiments.

a. Of the 20,000 students on a college campus, one student is selected to be the mascot at the football game this weekend.

b. One card is to be selected at random from a standard deck of 52 playing cards that has all of the diamonds removed.

c. One marble is to be selected at random from a marble bag that contains two green marbles, nine red marbles, four white marbles, and five blue marbles.

d. Three marbles are to be selected at random from a marble bag that contains two green marbles, nine red marbles, four white marbles, and five blue marbles.

Solution

a. The number of elements of the sample space is $n(S) = 20,000$.

b. There are 13 diamonds in a standard deck of **52** playing cards. So there are $52 - 13 = 39$ cards from which exactly one is selected. Thus the size of the sample space of this experiment is $n(S) = {}_{39}C_1 = 39$.

c. There are a total of 20 marbles in the bag from which exactly one is selected. Therefore, the size of the sample space of this experiment is $n(S) = {}_{20}C_1 = 20$.

d. There are a total of 20 marbles in the bag in which three marbles are selected. Therefore, the size of the sample space of this experiment is $n(S) = {}_{20}C_3 = 1140$.

Note   In the solution to part c, above, we may have initially thought about considering counting only four possible outcomes for the sample space:

$$\{\text{green, red, white, blue}\}$$

This would suggest that the size of the sample space was 4, meaning that each color marble is equally likely to get chosen from the bag. This, of course, is not the case since the number of marbles of each color in the bag is not the same. Since we are counting the sample space in preparation for determining probability, we will see that we must consider all 20 marbles as if they are all distinct from one another. In the solution to part d, we also determined the size of the sample space by assuming that the marbles were distinct.

Thus, for part c, the list of total possible outcomes is really $\{\text{green1, green2}$, red1, red2, ... red9, white1, ... white4, blue1, ... blue5$\}$. This is a total of 20 or ${}_{20}C_1$. For part d, the total possible outcomes are much more difficult to list. But, if we tried to list every possible grouping of three marbles with the understanding that all marbles were distinct, then we would get a list that starts to look something like this:

$$\{\{\text{green1, green2, red1}\}, \{\text{green1, green2, red2}\}, \{\text{green1, green2, red3}\}, \dots\}$$

If we continued listing all outcomes, we would get a total of ${}_{20}C_3 = 1140$ groups of three marbles.

You Try It   Work through this You Try It problem.

Work Exercises 1–7 in this textbook or in the My MathLab **Study Plan.**

To compute the probability that an event E will occur, we must know the total number of possible ways that the event can occur and we must also know the size of the sample space of the experiment. The probability that the event will occur is the quotient of these two numeric values.

The Probability of an Event

Suppose that an experiment has a sample space S. The probability of event E occurring is denoted as $P(E)$ and is given by

$$P(E) = \frac{\text{the number of ways event } E \text{ can occur}}{\text{the total number of possible outcomes}} = \frac{n(E)}{n(S)}$$

When studying probability, there are several principles that we must be aware of before proceeding. Three such principles are outlined below.

Principles of Probability

1. The probability of a certain event (an event that is guaranteed to happen) is 1, written as $P(\text{certain } E) = 1$.

2. The probability of an impossible event is 0, written as $P(\text{impossible } E) = 0$.

3. The sum of the probabilities of all possible outcomes of an experiment is 1, written as $P(E_1) + P(E_2) + \ldots + P(E_n) = 1$, where $E_1, E_2, \ldots, E_n$ are the possible outcomes of the experiment.

We must be careful when considering principle 3. The sum of the probabilities of all possible outcomes of an experiment is equal to 1 only when each event of the experiment is mutually exclusive. If the events are not mutually exclusive, the sum will not be 1. See the note at the end of Example 3.

My video summary ◉ **Example 2** Determining Probabilities

A paint-ball gun ball hopper contains 11 yellow balls, 6 green balls, and 3 red balls. A ball is fired from the gun. Find the probabilities of each of the following events written as a fraction in lowest terms.

$$R: \text{ The ball is red}$$

$$G: \text{ The ball is green}$$

$$Y: \text{ The ball is yellow}$$

Solution The experiment is firing a ball from a gun. The size of the sample space is $n(S) = 11 + 6 + 3 = 20$. The probability that a certain ball is fired is the ratio of the number of balls of that color to the size of the sample space. Thus we get the following probabilities:

$$P(R) = \frac{n(R)}{n(S)} = \frac{3}{20}, \quad P(G) = \frac{n(G)}{n(S)} = \frac{6}{20} = \frac{3}{10}, \quad P(Y) = \frac{n(Y)}{n(S)} = \frac{11}{20}$$

The probability of firing a red ball is $\dfrac{3}{20}$.

The probability of firing a green ball is $\dfrac{6}{20} = \dfrac{3}{10}$.

The probability of firing a yellow ball is $\dfrac{11}{20}$.

Note In the previous example, the experiment was made up of exactly three events. The fired ball could be red, green, or yellow. Note that $P(R) + P(G) + P(Y) = \dfrac{3}{20} + \dfrac{6}{20} + \dfrac{11}{20} = \dfrac{20}{20} = 1$, which is exactly what we would expect from principle 3 because these events are mutually exclusive.

You Try It Work through this You Try It problem.

Work Exercises 8–12 in this textbook or in the MyMathLab Study Plan.

Examples 3 through 5 pertain to the following demographic information collected from a teacher of mathematics at the local community college.

Total number of students in the class:	50
Number of boys:	35
Number of girls:	15
Number of freshmen:	22
Number of sophomores:	13
Number of juniors:	9
Number of seniors:	6

Example 3 Determining Probabilities

The instructor chooses one student to dim the lights and the next one to shut the door (to use the overhead projector). If a student may do both jobs, determine the following probabilities rounded to the nearest tenth of a percent as needed.

a. P(both are girls)

b. P(one is a girl and one is a boy)

c. P(at least one is a boy)

d. P(the door closer is a boy)

e. P(neither are seniors)

Solution The experiment is choosing a student to dim the lights and a student to shut the door. Because a student can do both jobs, there are 50 students that can be chosen for each job. Thus, by the fundamental counting principle, the size of the sample space is $n(S) = 50 \cdot 50 = 2500$.

a. Because both jobs can be completed by the same girl, we know that there are 15 girls that can do each job. Therefore, the number of outcomes in which both students are girls is

$$n(\text{both are girls}) = \underline{15 \text{ girls can dim the lights}} \cdot \underline{15 \text{ girls can shut the door}} = 225$$

The probability that both jobs are done by a girl is

$$P(\text{both are girls}) = \frac{n(\text{both are girls})}{n(S)} = \frac{225}{2500} = \frac{9}{100} = 0.09.$$

Thus, there is a 9% probability that two girls will be chosen to do both jobs.

b. First determine n(one is a girl and one is a boy), the number of outcomes in which one student picked is a girl and one student picked is a boy. To do this, we must consider two cases.

Case 1: Boy-Girl $\underline{35 \text{ boys can dim the lights}} \cdot \underline{15 \text{ girls can shut the door}} = 525$

Case 2: Girl-Boy $\underline{15 \text{ girls can dim the lights}} \cdot \underline{35 \text{ boys can shut the door}} = 525$

Adding the total outcomes from each case we see that
n(one is a girl and one is a boy) $= 525 + 525 = 1050$. Thus,

$$P(\text{one is a girl and one is a boy}) = \frac{n(\text{one is a girl and one is a boy})}{n(S)} = \frac{1050}{2500} = \frac{21}{50} = 0.42.$$

Therefore, there is a 42% probability that one girl and one boy will be chosen to do the jobs.

c. To determine the number of outcomes in which at least one boy is chosen for a job, we can count the three separate cases in which at least one boy does one of the jobs.

Case 1: Boy-Girl   35 boys can dim the lights · 15 girls can shut the door = 525
Case 2: Girl-Boy   15 girls can dim the lights · 35 boys can shut the door = 525
Case 3: Boy-Boy   35 boys can dim the lights · 35 boys can shut the door = 1225

Adding the outcomes from each case we see that the total number of outcomes in which at least one boy is chosen for a job is n(at least one is a boy) = 525 + 525 + 1225 = 2275.

Therefore, P(at least one is a boy) $= \dfrac{n(\text{at least one is a boy})}{n(S)} = \dfrac{2275}{2500} = \dfrac{91}{100} = 0.91.$

So there is a 91% probability that at least one boy will be chosen for a job.

See Example 4 for an alternate way to compute this same probability.

d. There are 50 students (boys and girls) that can dim the lights and 35 boys that can shut the door. Therefore, the total number of outcomes in which a boy can shut the door is

n(the door closer is a boy) = 50 students can dim the lights · 35 boys can shut the door = 1750

Thus, P(the door closer is a boy) $= \dfrac{n(\text{the door closer is a boy})}{n(S)} = \dfrac{1750}{2500} = \dfrac{7}{10} = 0.7.$

Therefore, there is a 70% probability that the person chosen to close the door will be a boy.

e. There are 6 seniors in the class so there are 44 non-seniors. Therefore, the number of possible outcomes for which two non-seniors do both jobs is

n(neither are seniors) = 44 non-seniors can dim the lights · 44 non-seniors can shut the door = 1936

Thus, P(neither are seniors) $= \dfrac{n(\text{neither are seniors})}{n(S)} = \dfrac{1936}{2500} = \dfrac{484}{625} = 0.7744.$

Therefore, the probability that two non-seniors will be chosen to do both jobs is approximately 77.4%.

Note   The sum of the probabilities in parts (a)–(e) of Example 3 do not equal 1. This is because we did not determine the probabilities for all mutually exclusive events. For example, the girls who are counted in part (a) when determining P(both are girls) are counted again when determining P(one is a girl and one is a boy) in part (b).

It can sometimes be easier to determine the probability of the **complement** of an event rather than determining the probability of the event itself.

> **Definition**   Complement
>
> The **complement** of an event E is the set of all outcomes in the sample space that are not included in the outcomes of event E. The notation used for the complement of E is E'.

Some examples of the complement of an event are given below:

If A is the event that a tossed coin lands on heads, then A' is the event that the tossed coin lands on tails.

9.7 An Introduction to Probability   9-71

If B is the event that at least one girl is chosen for a two person committee, then B' is the event that no girls are chosen for the committee.

If C is the event that a heart is drawn from a deck of 52 cards, then C' is the event that a club, spade, or diamond is drawn.

The Complement Rule

Given the probability of an event E written as a fraction or a decimal, the probability of its complement can be found by subtracting the given probability from 1. That is, $P(E') = 1 - P(E)$. This equation can also be written as $P(E) = 1 - P(E')$.

Given the probability of an event E written as a percent, the probability of its complement can be found by subtracting the given probability from 100%. That is, $P(E') = 100\% - P(E)$. This equation can also be written as $P(E) = 100\% - P(E')$.

To illustrate the complement rule, consider Example 4, which is exactly the same exercise as Example 3c.

Example 4 Using the Compliment Rule to Determine Probability

Using the same demographic information given in Example 3, suppose that the instructor chooses one student to dim the lights and one to shut the door. If a student may do both jobs, determine the probability that at least one is a boy using the complement rule. Write the probability rounded correct to the nearest tenth of a percent as needed.

Solution The complement of the event "at least one is a boy" is the event "neither are boys" or "both are girls." From Example 3a, we see that P(both are girls) = 9%. Therefore, by the complement rule, P(at least one is a boy) = 100% − P(both are girls) = 100% − 9% = 91%. Thus, the probability that at least one boy is chosen for a job is 91%. This is exactly the same probability that was obtained in Example 3c.

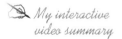
My interactive video summary

⊙ Example 5

Using the same demographic information given in Example 3, suppose that the instructor chooses one student to dim the lights and one to shut the door. If a student may **not** do both jobs, determine the following probabilities rounded correct to the nearest tenth of a percent as needed.

a. P(both are girls)

b. P(one is a girl and one is a boy)

c. P(at least one is a boy)

d. P(the door closer is a boy)

e. P(neither are seniors)

Solution Try to determine the probabilities for (a)–(e) on your own, then check your answers or watch this interactive video to see a complete solution.

You Try It Work through this You Try It problem.

Work Exercises 13–17 in this textbook or in the MyMathLab Study Plan.

OBJECTIVE 2 DETERMINING PROBABILITY USING THE ADDITIVE RULE
AND VENN DIAGRAMS

Often we are interested in determining the probability of the union or intersection of two events. For example, suppose that a single card is chosen at random from a standard deck of **52** playing cards. What is the probability of choosing a heart or an ace? This is an example of the probability of the union of two events. For this example we use the notation $P(H \text{ or } A) = P(H \cup A)$, where H is the event of drawing a heart and A is the event of drawing an ace. The union symbol, $\cup$, can be used interchangeably with the word *or*. We can determine $P(H \cup A)$ as follows:

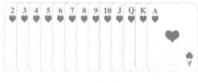

First, there are 13 hearts in a standard deck of 52 cards, so the probability of choosing a heart is $P(H) = \dfrac{13}{52}$.

Second, there are 4 aces in a standard deck of 52 cards, so the probability of choosing an ace is $P(A) = \dfrac{4}{52} = \dfrac{1}{13}$.

It might appear that we can simply add these two probabilities. However, note that we have accounted for the probability of choosing the ace of hearts *twice* in this process. Therefore, we must account for this "double counting" by subtracting the probability of choosing the ace of hearts, which is $P(H \text{ and } A) = P(H \cap A) = \dfrac{1}{52}$. Recall that the symbol $\cap$ can be used interchangeably with the word *and* which means "intersection." The notation $P(H \cap A)$ means "the probability of an ace and a heart."

Thus, the probability of choosing a heart or an ace is

$$P(H \cup A) = P(H) + P(A) - P(H \cap A) = \dfrac{13}{52} + \dfrac{4}{52} - \dfrac{1}{52} = \dfrac{16}{52} = \dfrac{4}{13}.$$ This formula is known as the **additive rule of probability**.

The Additive Rule of Probability

For any two events A and B, $P(A \cup B) = P(A) + P(B) - P(A \cap B)$.

The additive rule states that the probability of the union of two events is equal to the sum of the probabilities of the two events minus the probability of the intersection of the two events. Note that if events A and B are mutually exclusive, then $P(A \cup B) = P(A) + P(B)$.

 *My video summary*

⊙ **Example 6** Using the Additive Rule of Probability

Suppose $P(A) = 0.6$, $P(B) = 0.7$ and $P(A \cup B) = 0.9$. Find the following probability written as a decimal rounded to one decimal place as needed.

a. $P(A \cap B)$ **b.** $P((A \cap B)')$

Solution

a. We use the additive rule of probability to determine $P(A \cap B)$.

$P(A \cup B) = P(A) + P(B) - P(A \cap B)$ Write the additive rule of probability.

$P(A \cap B) = P(A) + P(B) - P(A \cup B)$ Solve for $P(A \cap B)$.

$$= 0.6 + 0.7 - 0.9$$ Substitute the given information.

$$= 0.4$$ Simplify.

b. We can use the complement rule to determine $P((A \cap B)')$.

$$P((A \cap B)') = 1 - P(A \cap B)$$ Write the formula representing the complement rule.

$$= 1 - 0.4$$ Substitute the information obtained from part (a).

$$= 0.6$$ Simplify.

 You Try It Work through this You Try It problem.

Work Exercises 18–23 in this textbook or in the My MathLab **Study Plan.**

A Venn diagram, named after British logician John Venn, can sometimes be used to determine probabilities.

Definition   Venn Diagram

A Venn diagram is a visual relationship between sets usually drawn using circles or other shapes. The Venn diagram below illustrates two sets A and B drawn within a sample space S. The shaded region in common to both sets is called the intersection of sets A and B and is denoted as $A \cap B$.

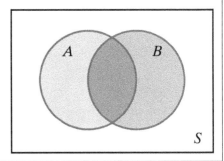

 *My video summary*

> **Example 7** Using a Venn Diagram to Determine Probabilities

In a building with 123 offices, 72 offices have computers in them. Forty-five offices have windows. Twenty-three offices have computers and windows. Construct a Venn diagram describing this information. If one office is chosen at random, determine the probability of the following events written as a fraction in lowest terms.

a. the office has a computer

b. the office has a window

c. the office has a computer and does not have a window

d. office does not have computer and does not have a window

Solution Watch this video to see how to construct the Venn diagram shown below.

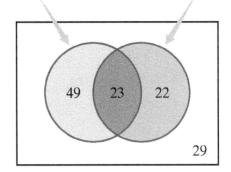

Office has a computer.   Office has a window.

Using the Venn diagram and using the fact that the size of the sample space is 123, we can determine each probability.

a. $P(\text{the office has a computer}) = \dfrac{49 + 23}{123} = \dfrac{72}{123} = \dfrac{24}{41}$

b. $P(\text{the office has a window}) = \dfrac{22 + 23}{123} = \dfrac{45}{123} = \dfrac{15}{41}$

c. $P(\text{the office has a computer and does not have a window}) = \dfrac{49}{123}$

d. $P(\text{the office does not have computer and does not have a window}) = \dfrac{29}{123}$

You Try It   Work through this You Try It problem.

Work Exercises 24–36 in this textbook or in the MyMathLab Study Plan.

OBJECTIVE 3   USING COMBINATIONS TO DETERMINE PROBABILITIES

In Section 9.6 we used the formula $N = {}_nC_r = \dfrac{n!}{(n - r)!r!}$ to count the number of different arrangements (combinations) of n objects taken r at a time when the following three criteria were satisfied:

1. order does not matter

2. an object cannot be repeated, and

3. the n objects are distinct

Combinations are often needed when determining certain probabilities.

My video summary   ⊙ **Example 8** Using Combinations to Determine Probabilities

A club has 50 members, and you and your best friend are both members. Four people will be selected from this club to be in a promotional video. Determine each of the following probabilities rounded correct to the nearest tenth of a percent as needed.

a. $P(\text{you are chosen})$

b. $P(\text{you and your best friend are chosen})$

c. $P(\text{neither you nor your best friend are chosen})$

d. $P(\text{at least one of you are chosen})$

Solution First we must determine the size of the sample space. The order in which the four people are chosen does not matter. The four people cannot be repeated and the 50 members are distinct. Therefore, we can determine the size of the sample space using the formula $n(S) = {}_nC_r = \dfrac{n!}{(n - r)!r!}$, where $n = 50$ and $r = 4$.

$$n(S) = {}_{50}C_4 = \dfrac{50!}{(50 - 4)!4!} = \dfrac{50!}{46!4!} = \dfrac{50 \cdot 49 \cdot 48 \cdot 47 \cdot 46!}{46! \cdot 4!} = \dfrac{50 \cdot 49 \cdot 48 \cdot 47}{4 \cdot 3 \cdot 2 \cdot 1} = 230{,}300$$

a. First determine the number of outcomes in which you can be chosen for the promotional video using the fundamental counting principle.

$$n(\text{you are chosen}) = \underbrace{{}_1C_1}_{\substack{\text{Number of ways} \\ \text{to choose you}}} \cdot \underbrace{{}_{49}C_3}_{\substack{\text{Number of ways to choose} \\ \text{the remaining three people}}}$$

Therefore, the probability that you will be chosen for the promotional video is

$$P(\text{you are chosen}) = \frac{n(\text{you are chosen})}{n(S)} = \frac{{}_1C_1 \cdot {}_{49}C_3}{{}_{50}C_4} = \frac{18{,}424}{230{,}300} = \frac{2}{25} = 0.08. \text{ Thus,}$$

there is an 8% chance that you will be chosen for the promotional video.

b. The number of ways that you and your best friend can be chosen for the promotional video is

$$n(\text{you and your best friend are chosen}) = \underbrace{{}_1C_1}_{\substack{\text{Number of ways} \\ \text{to choose you}}} \cdot \underbrace{{}_1C_1}_{\substack{\text{Number of ways to} \\ \text{choose your best friend}}} \cdot \underbrace{{}_{48}C_2}_{\substack{\text{Number of ways to choose} \\ \text{the remaining two people}}}$$

Therefore, the probability that you and your best friend will be chosen for the promotional video is

$$P(\text{you and your best friend are chosen}) = \frac{n(\text{you and your best friend are chosen})}{n(S)} = \frac{{}_1C_1 \cdot {}_1C_1 \cdot {}_{48}C_2}{{}_{50}C_4}$$

$$= \frac{1{,}128}{230{,}300} = \frac{6}{1225} \approx 0.0049$$

Thus, there is approximately a 0.5% probability that you and your best friend will be chosen for the video.

c. The number of ways that neither you nor your friend are chosen for the promotional video is

$$n(\text{neither you nor your best friend are chosen}) = \underbrace{{}_1C_0}_{\substack{\text{Number of ways} \\ \text{to consider but} \\ \text{not choose you}}} \cdot \underbrace{{}_1C_0}_{\substack{\text{Number of ways to} \\ \text{consider but not choose} \\ \text{your best friend}}} \cdot \underbrace{{}_{48}C_4}_{\substack{\text{Number of ways to choose} \\ \text{4 people out of the} \\ \text{remaining 48 people}}}$$

Therefore, the probability that neither you nor your best friend will be chosen for the video is

$$P(\text{neither you nor your best friend are chosen}) = \frac{n(\text{neither you nor your best friend are chosen})}{n(S)}$$

$$= \frac{{}_1C_0 \cdot {}_1C_0 \cdot {}_{48}C_4}{{}_{50}C_4} = \frac{194{,}580}{230{,}300} = \frac{207}{245} \approx 0.845$$

There is approximately an 84.5% probability that neither you nor your best friend will be chosen for the video.

d. To determine the probability that at least one of you is chosen, we can use the **complement rule**. Note that the **complement** of "at least one of you is chosen" is "neither of you nor your best friend is chosen." In part (c), we

determined that $P(\text{neither you nor your best friend are chosen}) = \frac{207}{245}$.

Therefore, by the complement rule we get

$P(\text{at least one of you are chosen}) = 1 - P(\text{neither you nor your best friend are chosen})$

$$= 1 - \frac{207}{245} = \frac{38}{245} \approx 0.155$$

Thus, the probability that at least one of you is chosen for the video is approximately 15.5%.

You Try It   Work through this You Try It problem.

Work Exercises 37–40 in this textbook or in the MyMathLab Study Plan.

 *My video summary*   ⊘ **Example 9** Using Combinations to Determine Probabilities

A box contains eight yellow, seven red, and six blue balls. Five balls are selected at random and the colors are noted. Determine each of the following probabilities written as a fraction in lowest terms.

a. $P(\text{all } 5 \text{ balls are yellow})$

b. $P(\text{exactly } 2 \text{ balls are red})$

c. $P(\text{at least one ball is blue})$

Solution Try to determine each probability on your own. Then check your answers or watch this video to see the entire worked out solution.

You Try It   Work through this You Try It problem.

Work Exercises 41–46 in this textbook or in the MyMathLab Study Plan.

OBJECTIVE 4   USING CONDITIONAL PROBABILITY

Sometimes a situation arises where we want to determine the probability of an event B happening given that a separate event A has already happened. This is called **conditional probability** because we determine the probability of an event only after some condition has been met.

Definition   Conditional Probability

The conditional probability of an event B in relation to an event A is the probability that event B will occur given that event A has already occurred. The notation used for conditional probability is $P(B|A)$, which is read as "the probability of B given A."

Conditional probability can be determined by two methods:

1. Reducing the sample space outright

2. Using a formula to guide the counting and reasoning

My video summary ▷ **Example 10** Determining Conditional Probability by Reducing the Sample Space

The Venn diagram below shows that in a building with 123 offices, 72 offices have computers in them. Forty-five offices have windows to the outside of the building. Twenty-three offices have computers and windows. If one office is chosen at random, determine each of the following probabilities written as a fraction in lowest terms.

a. $P(C|W)$ **b.** $P(W|C)$

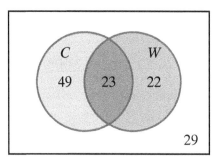

C: Offices that have a computer
W: Offices that have a window

Solution

a. To determine the probability that an office has a computer given that it has a window, we only consider the 45 offices that have a window. Thus, the sample space is reduced from 123 to 45. Of the offices that have a window, 23 of those offices have a computer. (This can be seen as the intersection of the two sets in the Venn diagram.) Thus, we get $P(C|W) = \dfrac{23}{45}$.

b. To determine the probability that an office has a window given that it has a computer, we only consider the 72 offices that have a computer. Thus, the sample space is reduced from 123 to 72. Of the offices that have a computer, 23 of those offices have a window. Thus, we get $P(W|C) = \dfrac{23}{72}$.

You Try It Work through this You Try It problem.

Work Exercises 47–48 in this textbook or in the MyMathLab **Study Plan**.

Note that in the previous example we determined the conditional probability by first determining the number of occurrences common to both events and then we divided that number by the number of occurrences of the given event. In the case of determining the conditional probability $P(C|W)$, we found that
$P(C|W) = \dfrac{n(C \cap W)}{n(W)} = \dfrac{23}{45}$. Thus, we can establish a formula for determining conditional probability $P(B|A)$ as follows.

$$\text{Multiply by 1 of the form } 1 = \frac{\dfrac{1}{n(S)}}{\dfrac{1}{n(S)}}$$

$$P(B|A) = \frac{n(A \cap B)}{n(A)} = \frac{n(A \cap B)}{n(A)} \cdot \frac{\dfrac{1}{n(S)}}{\dfrac{1}{n(S)}} = \frac{\dfrac{n(A \cap B)}{n(S)}}{\dfrac{n(A)}{n(S)}} = \frac{P(A \cap B)}{P(A)}$$

Conditional Probability Formula

The probability of an event B occurring given that event A has already occurred is given by the formula

$$P(B|A) = \frac{n(A \cap B)}{n(A)} = \frac{P(A \cap B)}{P(A)}.$$

Example 11 Determining Conditional Probability Using the Conditional Probability Formula

Suppose $P(A) = 0.6$, $P(B) = 0.7$, and $P(A \cup B) = 0.9$. Determine the following probabilities correct to two decimal places as needed.

a. $P(B|A)$ **b.** $P(A|B)$

Solution First, determine $P(A \cap B)$ using the additive rule of probability.

$P(A \cup B) = P(A) + P(B) - P(A \cap B)$	Write the additive rule of probability.
$P(A \cap B) = P(A) + P(B) - P(A \cup B)$	Solve for $P(A \cap B)$.
$\quad\quad = 0.6 + 0.7 - 0.9$	Substitute the given information.
$\quad\quad = 0.4$	Simplify.

We can now use the conditional probability formula to determine each conditional probability.

a. $P(B|A) = \dfrac{P(A \cap B)}{P(A)}$ Write the conditional probability formula.

$\quad\quad = \dfrac{0.4}{0.6}$ Substitute 0.4 for $P(A \cap B)$ and substitute 0.6 for $P(A)$.

$\quad\quad \approx 0.67$ Simplify.

b. $P(A|B) = \dfrac{P(A \cap B)}{P(B)}$ Write the conditional probability formula.

$\quad\quad = \dfrac{0.4}{0.7}$ Substitute 0.4 for $P(A \cap B)$ and substitute 0.7 for $P(B)$.

$\quad\quad \approx 0.57$ Simplify.

Using the conditional probability formula is not always the easiest way to determine conditional probability especially when the problem involves combinations. When combinations are involved, it is typically more straightforward to reduce the sample space as we see in the following example.

 *My video summary*   ⊘ **Example 12** Determining Conditional Probability

A box contains eight yellow, seven red, and six blue balls. Five balls are selected at random and the colors noted. Determine each of the following conditional probabilities written as a fraction reduced to lowest terms.

a. $P(\text{all yellow balls} \mid \text{all the same color})$ **b.** $P(\text{exactly 2 red balls} \mid \text{no blue balls})$

c. $P(\text{at least 1 blue ball} \mid \text{no red balls})$

Solution

a. First reduce the size of the sample space. That is, determine the number of possible outcomes of choosing five balls of the same color. We can determine the size of this reduced sample space by adding the number of outcomes of the three possibilities "choosing 5 yellow balls," "choosing 5 red balls," and "choosing 5 blue balls."

$$n(\text{all the same color}) = \overbrace{{}_8C_5}^{\substack{\text{Number of ways} \\ \text{to choose 5} \\ \text{yellow balls}}} + \overbrace{{}_7C_5}^{\substack{\text{Number of ways} \\ \text{to choose 5} \\ \text{red balls}}} + \overbrace{{}_6C_5}^{\substack{\text{Number of ways} \\ \text{to choose 5} \\ \text{blue balls}}}$$

The number of ways to choose all yellow balls is $n(\text{all yellow balls}) = {}_8C_5$. Therefore, the probability of choosing all yellow balls given that the balls are all of the same color is

$$P(\text{all yellow balls} \mid \text{all the same color}) = \frac{n(\text{all yellow balls})}{n(\text{all the same color})} = \frac{{}_8C_5}{{}_8C_5 + {}_7C_5 + {}_6C_5} = \frac{56}{56 + 21 + 6} = \frac{56}{83}$$

Try to determine the conditional probabilities for parts b and c on your own. Then check your answers or watch this video to see the entire worked out solution.

You Try It   Work through this You Try It problem.

Work Exercises 49–54 in this textbook or in the MyMathLab Study Plan.

OBJECTIVE 5   UNDERSTANDING ODDS

Sometimes probabilities are stated in terms of **odds**. For example, the probability of a tossed coin landing on heads is $P(\text{heads}) = \dfrac{1}{2}$. But the odds of a tossed coin landing on heads are 1:1, read as "one to one." We interpret the notation 1:1 to mean that there is 1 way for a head to occur to one way that a tail can occur. We treat the odds notation $a{:}b$ in much the same way as we treat a fraction in the sense that we always want to write the odds notation in lowest terms. For example, the odds notation 6:4 should be rewritten as 3:2.

Definition Odds

Let E be an event, E' be the complement of E, and S be the sample space.

The **odds of event E happening** is the ratio of the number of ways that event E can occur to the number of ways that event E cannot occur.
The odds of an event E happening can be calculated using one of the following methods:

$n(E): n(E')$, where $n(E') = n(S) - n(E)$

$P(E): P(E')$, where $P(E') = 1 - P(E)$ and when $0 < P(E) < 1$

$P(E): P(E')$, where $P(E') = 100\% - P(E)$ and when $0\% < P(E) < 100\%$

Example 13 Calculating Odds

A private school sold 2100 raffle tickets and you purchased 20. What are the odds of winning the raffle?

Solution Let E be the event of winning the raffle. Then the number of ways to win is $n(E) = 20$. The number of ways to not win is $n(E') = 2100 - 20 = 2080$. Therefore, the odds of winning the raffle are $n(E) : n(E') = 20 : 2080 = 1 : 104$.

 *My video summary*

⊙ Example 14 Calculating Odds

A paint-ball gun ball hopper contains 10 yellow balls, 6 green balls, and 3 red balls. A ball is fired from the gun. Find:

a. odds of firing a red ball   **b.** odds of firing a green ball

c. odds against firing a yellow ball

Solution

a. Let R be the event of firing a red ball. Then $n(R) = 3$ and $n(R') = 16$. Thus, the odds of firing a red ball are $n(R) : n(R') = 3 : 16$.

b. Let G be the event of firing a green ball. Then $n(G) = 6$ and $n(G') = 13$. Thus, the odds of firing a green ball are $n(G) : n(G') = 6 : 13$.

c. Let Y be the event of firing a yellow ball. Then $n(Y) = 10$ and $n(Y') = 9$. Thus, the odds against firing a yellow ball are $n(Y') : n(Y) = 9 : 10$.

You Try It Work through this You Try It problem.

Work Exercises 55–62 in this textbook or in the MyMath Lab **Study Plan.**

9.7 Exercises

In Exercises 1–7, determine the size of the sample space for each experiment described.

1. A man has three pairs of slacks, four shirts, and five ties that all coordinate. He wants to choose a pair of slacks, a shirt, and a tie to wear to work.

2. A group of 20,000 math students were selected to participate in a study to determine how many hours each week that they spend working on mathematics.

3. Four coins are tossed and the results of "Heads or Tails" from each coin is recorded.

4. A standard deck of 52 cards has all of the face cards removed. One card is drawn at random.

5. Three cards are selected without replacement from a standard deck of 52 cards.

6. A marble bag contains 7 orange marbles, 4 red marbles, 9 green marbles, and 12 white marbles. One marble is selected at random.

7. A marble bag contains 7 orange marbles, 4 red marbles, 9 green marbles, and 12 white marbles. Four marbles are selected at random without replacement.

8. A paint-ball gun ball hopper contains 9 yellow balls, 12 green balls, and 7 red balls. A ball is fired from the gun. What is the probability that the ball fired from the gun is red?

9. There are five cars in a driveway: one tan van, one green sedan, one blue four-wheel-drive Bronco, one red Mustang convertible, and one green VW Super Beetle. All of the keys were hanging on a rack on the wall. The rack fell and the keys landed in a pile on the floor. If you randomly grab one set of keys, determine the following probabilities:

   a. *P*(keys are for the Mustang)   b. *P*(keys are for a green car)

Roulette is a game played in many casinos in which a wheel containing numbers of various colors is spun. A ball is dropped and eventually lands on a number. Use the roulette wheel shown below to answer Exercises 10–12.

10. What is the probability that the ball lands on red?

11. What is the probability that the ball lands on an odd number?

12. What is the probability that the ball lands on green?

In Exercises 13–17, refer to the following demographic information that has been collected by an instructor after surveying her class.

Total number of students in the class:	46
Number of boys:	19
Number of girls:	27
Number of freshmen:	7
Number of sophomores:	16
Number of juniors:	12
Number of seniors:	11

13. The instructor randomly chooses one student to pass out papers, one student to clean the chalkboard, and one student to empty the trash. If a student may do more than one job, then what is the probability that the instructor will choose all girls to do the jobs?

14. The instructor randomly chooses one student to pass out papers, one student to clean the chalkboard, and one student to empty the trash. If a student may do more than one job, then what is the probability that the instructor will choose at least one girl to do the jobs?

15. The instructor randomly chooses one student to pass out papers, one student to clean the chalkboard, and one student to empty the trash. If a student may not do more than one job, then what is the probability that the instructor will choose all students of the same sex?

16. The instructor randomly chooses one student to pass out papers, one student to clean the chalkboard, and one student to empty the trash. If a student may do more than one job, then what is the probability that the instructor will choose at least one senior to do the jobs?

17. The instructor randomly chooses one student to pass out papers, one student to clean the chalkboard, and one student to empty the trash. If a student may not do more than one job, then what is the probability that the instructor will choose at most one senior to do the jobs?

For Exercises 18–21, suppose that $P(A) = 0.5$, $P(B) = 0.8$, and $P(A \cup B) = 0.9$.

18. Find $P(A \cap B)$.

19. Find $P(B')$.

20. Find $P(A')$.

21. Find $P((A \cap B)')$.

22. A single card is drawn from a standard deck of 52 cards. What is the probability that the card is black or a two?

23. A single card is drawn from a standard deck of 52 cards. What is the probability that the card is a face card or a club?

For Exercises 24–31, use the Venn diagram below to determine each probability.

24. $P(A)$

25. $P(B)$

26. $P(A \cap B)$

27. $P(A \cup B)$

28. $P(A')$

29. $P(B')$

30. $P(A' \cap B')$

31. $P(A' \cup B')$

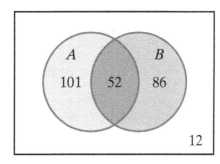

For Exercises 32–36, refer to the following information.

Two hundred sports fans were surveyed. It was found that 139 people were fans of the NFL (The National Football League), 64 were fans of MLB (Major League Baseball), and 35 were fans of both the NFL and MLB. One fan is selected at random. Create a Venn diagram and determine the indicated probabilities in Exercises 32–36.

32. What is the probability that the selected sports fan is a fan of the NFL?

33. What is the probability that the selected sports fan is a fan of MLB?

34. What is the probability that the selected sports fan is not a fan of the NFL?

35. What is the probability that the selected sports fan is a fan of MLB but not a fan of the NFL?

36. What is the probability that the selected sports fan is neither a fan of the NFL nor MLB?

SbS **37.** You and your best friend are part of a crowd of 60 people attending a magic show. The magician will choose five people to assist him during the show. What is the probability that you are chosen?

SbS **38.** You and your best friend are part of a crowd of 60 people attending a magic show. The magician will choose five people to assist him during the show. What is the probability that you and your best friend are chosen?

SbS **39.** You and your best friend are part of a crowd of 60 people attending a magic show. The magician will choose five people to assist him during the show. What is the probability that neither you nor your best friend is chosen?

SbS **40.** You and your best friend are part of a crowd of 60 people attending a magic show. The magician will choose five people to assist him during the show. What is the probability that at least one of you is chosen?

SbS **41.** A box contains 10 yellow, 5 red, and 7 blue balls. Four balls are selected at random and the colors are noted. Determine the probability that all four balls are yellow.

SbS **42.** A box contains 10 yellow, 5 red, and 7 blue balls. Four balls are selected at random and the colors are noted. Determine the probability that exactly three balls are blue.

SbS **43.** A box contains 10 yellow, 5 red, and 7 blue balls. Four balls are selected at random and the colors are noted. Determine the probability that none of the balls are yellow.

SbS **44.** A box contains 10 yellow, 5 red, and 7 blue balls. Four balls are selected at random and the colors are noted. Determine the probability that at least one ball is red.

SbS **45.** Seven cards are dealt from a standard deck of fifty-two cards. What is the probability that exactly one of the cards is a face card?

SbS **46.** Five cards are dealt from a standard deck of fifty-two cards. What is the probability that at least one of the cards is a face card?

The Venn diagram below shows that out of 1350 students surveyed who use an integrated eText in their College Algebra course, 994 students always watch a video before working a homework exercise. Eight hundred forty-five students always read the eText before working a homework exercise, and 720 students always watch a video and always read the eText before working a homework exercise. If one of these students is selected at random, determine the probabilities in Exercises 47 and 48.

47. $P(V \mid T)$

48. $P(T \mid V)$

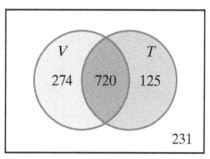

V: Always watch video before
    attempting homework
T: Always read eText before
    attempting homework

49. Suppose $P(A) = 0.4$, $P(B) = 0.6$, and $P(A \cup B) = 0.8$. Determine $P(B \mid A)$.

50. Suppose $P(A) = 0.4$, $P(B) = 0.6$, and $P(A \cup B) = 0.8$. Determine $P(A \mid B)$.

51. Two cards are drawn from a standard deck of fifty-two cards without replacement. What is the probability that the second card drawn is a diamond given that the first card drawn was a diamond?

52. A box contains 10 yellow, 5 red, and 7 blue balls. Five balls are selected at random and the colors are noted. Determine $P(\text{all yellow balls} \mid \text{all the same color})$.

53. A box contains 10 yellow, 5 red, and 7 blue balls. Six balls are selected at random and the colors are noted. Determine $P(\text{exactly 2 red balls} \mid \text{no blue balls})$.

54. A box contains 10 yellow, 5 red, and 7 blue balls. Five balls are selected at random and the colors are noted. Determine $P(\text{exactly 2 red balls} \mid \text{no blue balls})$.

55. A private school sold 1650 raffle tickets and you purchased 24. What are the odds of winning the raffle?

56. A marble bag contains 12 clear marbles, 8 red marbles, and 3 green marbles. One marble is chosen at random. Find the odds of choosing a clear marble.

57. A marble bag contains 12 clear marbles, 8 red marbles, and 3 green marbles. One marble is chosen at random. Find the odds of not choosing a clear marble.

58. Ten slips of paper are numbered consecutively 1–10 and placed in a hat. Two more slips of paper are numbered with a 5 and three more are numbered with a 6. Find the odds of drawing the number 5 from the hat.

59. Ten slips of paper are numbered consecutively 1–10 and placed in a hat. Two more slips of paper are numbered with a 5 and three more are numbered with a 4. Find the odds of drawing a number greater than 4.

Roulette is a game played in many casinos in which a wheel containing numbers of various colors is spun. A ball is dropped and eventually lands on a number. Use the roulette wheel shown below to answer Exercises 56–58.

60. What are the odds of the ball landing on red?

61. What are the odds of the ball landing on an odd number?

62. What are the odds against the ball landing on green?

Brief Exercises

63. You and your best friend are part of a crowd of 60 people attending a magic show. The magician will choose five people to assist him during the show. What is the probability that you are chosen?

64. You and your best friend are part of a crowd of 60 people attending a magic show. The magician will choose five people to assist him during the show. What is the probability that you and your best friend are chosen?

65. You and your best friend are part of a crowd of 60 people attending a magic show. The magician will choose five people to assist him during the show. What is the probability that neither you nor your best friend is chosen?

66. You and your best friend are part of a crowd of 60 people attending a magic show. The magician will choose five people to assist him during the show. What is the probability that at least one of you is chosen?

67. A box contains 10 yellow, 5 red, and 7 blue balls. Four balls are selected at random and the colors are noted. Determine the probability that all four balls are yellow.

68. A box contains 10 yellow, 5 red, and 7 blue balls. Four balls are selected at random and the colors are noted. Determine the probability that exactly three balls are blue.

69. A box contains 10 yellow, 5 red, and 7 blue balls. Four balls are selected at random and the colors are noted. Determine the probability that none of the balls are yellow.

70. A box contains 10 yellow, 5 red, and 7 blue balls. Four balls are selected at random and the colors are noted. Determine the probability that at least one ball is red.

71. Seven cards are dealt from a standard deck of fifty-two cards. What is the probability that exactly one of the cards is a face card?

72. Five cards are dealt from a standard deck of fifty-two cards. What is the probability that at least one of the cards is a face card?

73. A box contains 10 yellow, 5 red, and 7 blue balls. Five balls are selected at random and the colors are noted. Determine $P(\text{all yellow balls} \mid \text{all the same color})$.

74. A box contains 10 yellow, 5 red, and 7 blue balls. Six balls are selected at random and the colors are noted. Determine $P(\text{exactly 2 red balls} \mid \text{no blue balls})$.

75. A box contains 10 yellow, 5 red, and 7 blue balls. Five balls are selected at random and the colors are noted. Determine $P(\text{exactly 2 red balls} \mid \text{no blue balls})$.

76. You and your best friend are part of a crowd of 60 people attending a magic show. The magician will choose five people to assist him during the show. What is the probability that you are chosen?

77. You and your best friend are part of a crowd of 60 people attending a magic show. The magician will choose five people to assist him during the show. What is the probability that you and your best friend are chosen?

78. You and your best friend are part of a crowd of 60 people attending a magic show. The magician will choose five people to assist him during the show. What is the probability that neither you nor your best friend is chosen?

79. You and your best friend are part of a crowd of 60 people attending a magic show. The magician will choose five people to assist him during the show. What is the probability that at least one of you is chosen?

80. A box contains 10 yellow, 5 red, and 7 blue balls. Four balls are selected at random and the colors are noted. Determine the probability that all four balls are yellow.

81. A box contains 10 yellow, 5 red, and 7 blue balls. Four balls are selected at random and the colors are noted. Determine the probability that exactly three balls are blue.

82. A box contains 10 yellow, 5 red, and 7 blue balls. Four balls are selected at random and the colors are noted. Determine the probability that none of the balls are yellow.

83. A box contains 10 yellow, 5 red, and 7 blue balls. Four balls are selected at random and the colors are noted. Determine the probability that at least one ball is red.

84. Seven cards are dealt from a standard deck of fifty-two cards. What is the probability that exactly one of the cards is a face card?

85. Five cards are dealt from a standard deck of fifty-two cards. What is the probability that at least one of the cards is a face card?

86. A box contains 10 yellow, 5 red, and 7 blue balls. Five balls are selected at random and the colors are noted. Determine $P(\text{all yellow balls} \mid \text{all the same color})$.

87. A box contains 10 yellow, 5 red, and 7 blue balls. Six balls are selected at random and the colors are noted. Determine $P(\text{exactly 2 red balls} \mid \text{no blue balls})$.

88. A box contains 10 yellow, 5 red, and 7 blue balls. Five balls are selected at random and the colors are noted. Determine $P(\text{exactly 2 red balls} \mid \text{no blue balls})$.

Conic Section Proofs

APPENDIX A CONTENTS

A.1 Parabola Proofs

OBJECTIVES

1   Derivation of the Equation of a Parabola in Standard Form with a Vertical Axis of Symmetry

2   Derivation of the Equation of a Parabola in Standard Form with a Horizontal Axis of Symmetry

OBJECTIVE 1   DERIVATION OF THE EQUATION OF A PARABOLA IN STANDARD FORM WITH A VERTICAL AXIS OF SYMMETRY

To derive the equation of a parabola with a vertical axis of symmetry, start with the graph of a parabola with vertex $V(h, k)$, focus F, directrix D, which passes through the point $P(x, y)$ as shown in Figure 1.

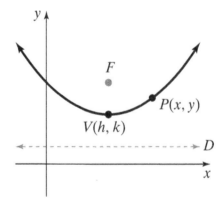

Figure 1

By the geometric definition of the parabola, the distance from any point that lies on the parabola to the focus must be equal to the distance from that point to the directrix. Therefore, the following equation must be true:

$$d(P, F) = d(P, D)$$

Because the vertex lies on the graph of the parabola, it implies that the distance from the vertex to the focus must be equal to the distance from the vertex to the directrix. Let $|p|$ be the distance from the vertex to the focus. Without loss of generality, suppose $p > 0$. Therefore, the coordinates of the focus must be $(h, k + p)$. Similarly, the equation of the directrix must be $y = k - p$. See Figure 2.

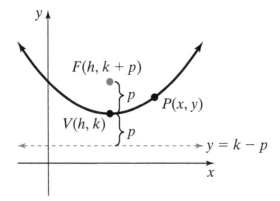

Figure 2

Consider the point $D(x, k - p)$ that lies on the directrix. By the geometric definition of the parabola, the distance from P to F must be the same as the distance from P to D. See Figure 3.

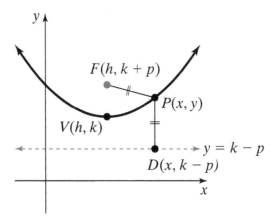

Figure 3

By the distance formula, we can rewrite the equation $d(P, F) = d(P, D)$ as

$$\sqrt{(x - h)^2 + (y - (k + p))^2} = \sqrt{(x - x)^2 + (y - (k - p))^2}$$

Squaring both sides of this equation and simplifying we get:

$$(x - h)^2 + (y - (k + p))^2 = (0)^2 + (y - (k - p))^2$$

$$(x - h)^2 + (y - k - p)(y - k - p) = (y - k + p)(y - k + p)$$

$$(x - h)^2 + y^2 - 2ky - 2py + 2pk + k^2 + p^2 = y^2 - 2ky + 2py - 2pk + k^2 + p^2$$

$$(x - h)^2 - 2py + 2pk = 2py - 2pk2$$

$$(x - h)^2 = 4py - 4pk$$

$$(x - h)^2 = 4p(y - k)$$

Therefore, the equation of a parabola with a vertical axis of symmetry (and a horizontal directrix) is $(x - h)^2 = 4p(y - k)$. If p is positive, the graph opens upward. If p is negative, the graph opens downward.

Equation of a Parabola in Standard Form with a Vertical Axis of Symmetry

The equation of a parabola with a vertical axis of symmetry is
$(x - h)^2 = 4p(y - k)$,

where

the vertex is $V(h, k)$,
$|p|$ = distance from the vertex to the focus = distance from the vertex to the directrix,
the focus is $F(h, k + p)$, and
the equation of the directrix is $y = k - p$.

The parabola opens *upward* if $p > 0$ or *downward* if $p < 0$.

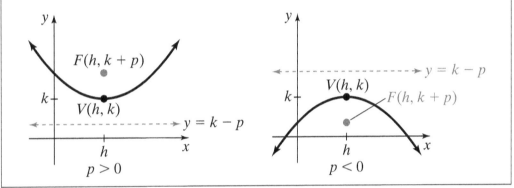

OBJECTIVE 2 DERIVATION OF THE EQUATION OF A PARABOLA IN STANDARD FORM WITH A HORIZONTAL AXIS OF SYMMETRY

To derive the equation of a parabola with a horizontal axis of symmetry, start with the graph of a parabola with vertex $V(h, k)$, focus F, directrix D, which passes through the point $P(x, y)$ as shown in Figure 4.

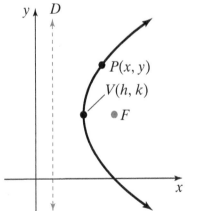

Figure 4

By the geometric definition of the parabola, the distance from any point that lies on the parabola to the focus must be equal to the distance from that point to the directrix. Therefore, the following equation must be true:

$$d(P, F) = d(P, D)$$

Because the vertex lies on the graph of the parabola, it implies that the distance from the vertex to the focus must be equal to the distance from the vertex to the directrix. Let $|p|$ be the distance from the vertex to the focus. Without loss of generality, suppose $p > 0$. Therefore, the coordinates of the focus must be $(h + p, k)$. Similarly, the equation of the directrix must be $x = h - p$. See Figure 5.

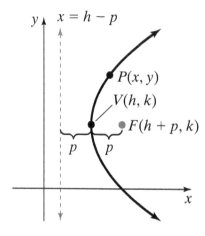

Figure 5

Consider the point $D(h - p, y)$ that lies on the directrix. By the geometric definition of the parabola, the distance from P to F must be the same as the distance from P to D. See Figure 6.

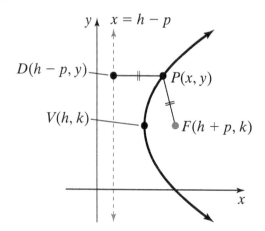

Figure 6

By the distance formula, we can rewrite the equation $d(P, F) = d(P, D)$ as

$$\sqrt{(x - (h + p))^2 + (y - k)^2} = \sqrt{(x - (h - p))^2 + (y - y)^2}$$

Squaring both sides of this equation and simplifying we get:

$$(x - (h + p))^2 + (y - k)^2 = (x - (h - p))^2 + (0)^2$$

$$(y - k)^2 + (x - h - p)(x - h - p) = (x - h + p)(x - h + p)$$

$$(y - k)^2 + x^2 - 2hx - 2px + 2ph + h^2 + p^2 = x^2 - 2hx + 2px - 2ph + h^2 + p^2$$

$$(y - k)^2 - 2px + 2ph = 2px - 2ph$$

$$(y - k)^2 = 4px - 4ph$$

$$(y - k)^2 = 4p(x - h)$$

Therefore, the equation of a parabola with a horizontal axis of symmetry (and a vertical directrix) is $(y - k)^2 = 4p(x - h)$. If p is positive, the graph opens to the right. If p is negative, the graph opens to the left.

Equation of a Parabola in Standard Form with a Horizontal Axis of Symmetry

The equation of a parabola with a horizontal axis of symmetry is
$(y - k)^2 = 4p(x - h),$

where

the vertex is $V(h, k)$,
$|p|$ = distance from the vertex to the focus = distance from the vertex to the directrix,
the focus is $F(h + p, k)$, and
the equation of the directrix is $x = h - p$.

The parabola opens *right* if $p > 0$ or *left* if $p < 0$.

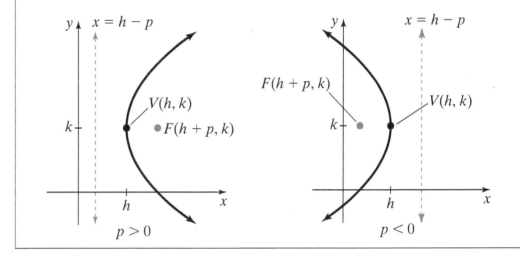

A.2 Ellipse Proofs

OBJECTIVES

1 Establishing Fact 1 for Ellipses
2 Establishing Fact 2 for Ellipses
3 Derivation of the Equation of an Ellipse in Standard Form with a Horizontal Major Axis

OBJECTIVE 1   ESTABLISHING FACT 1 FOR ELLIPSES

Fact 1 for Ellipses

Given the foci of an ellipse F_1 and F_2 and any point P that lies on the graph of the ellipse, the sum of the distances between P and the foci is equal to the length of the major axis or $d(F_1, P) + d(F_2, P) = 2a$.

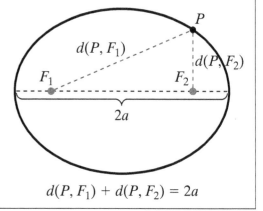

$$d(P, F_1) + d(P, F_2) = 2a$$

Proof

To establish Fact 1, note that for some $c > 0$, the foci are located along the major axis at points $F_1(h - c, k)$ and $F_2(h + c, k)$. Let $a > 0$ be the distance from the center of the ellipse to one of the vertices and let $b > 0$ be the distance from the center to an endpoint of the minor axis. See Figure 7. Because the length of the major axis is larger than the length of the minor axis, it follows that $a > b$

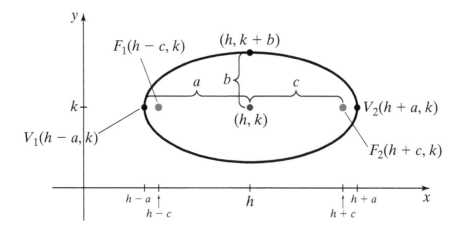

Figure 7

Consider the vertex $V_2(h + a, k)$. The distance from V_2 to F_2 is $(h + a) - (h + c) = a - c$ and the distance from V_2 to F_1 is $(h + a) - (h - c) = a + c$. See Figure 8.

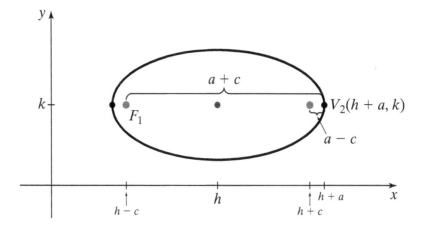

Figure 8

Thus, $d(V_2, F_1) + d(V_2, F_2) = (a + c) + (a - c) = 2a$. By the geometric definition of the ellipse, the sum of the distances from **any** point on the ellipse to the foci is a positive constant. Therefore, this positive constant must be equal to $2a$. This establishes Fact 1.

OBJECTIVE 2   ESTABLISHING FACT 2 FOR ELLIPSES

> **Fact 2 for Ellipses**
>
> Suppose $a > 0$ is the length from the center of an ellipse to a vertex, $b > 0$ is the length from the center to an endpoint of the minor axis, and $c > 0$ is the length from the center to a focus, then $b^2 = a^2 - c^2$ or equivalently, $c^2 = a^2 - b^2$.

Proof

Suppose that we label the coordinates of one of the endpoints of the minor axis as $(h, k + b)$. See Figure 9.

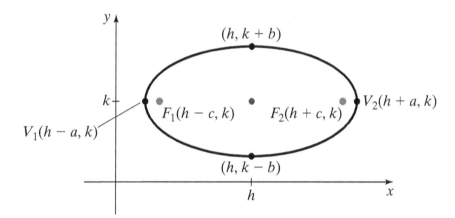

Figure 9

We can construct a triangle by connecting the point $(h, k + b)$ with the two foci as shown in Figure 10.

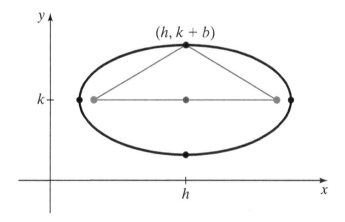

Figure 10

If we draw a line segment connecting the point $(h, k + b)$ with the center of the ellipse, we see that we have constructed two congruent right triangles. See Figure 11. The base of each right triangle has length c and the height of each right triangle has length b.

A.2   Ellipse Proofs   A-7

We know from Fact 1 that the sum of the distances between *any* point on the ellipse and the two foci is $2a$. Therefore, the sum of the distances between the point $(h, k + b)$ and the two foci is also equal to $2a$. Thus, the hypotenuse of each of the right triangles must be equal to a. Therefore, by the Pythagorean theorem we can establish the relationship $b^2 + c^2 = a^2$ or $b^2 = a^2 - c^2$. Thus, we have established Fact 2.

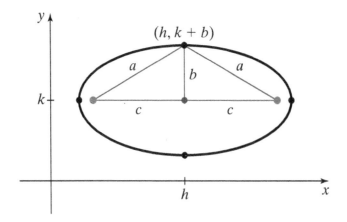

Figure 11

OBJECTIVE 3 DERIVATION OF THE EQUATION OF AN ELLIPSE IN STANDARD FORM WITH A HORIZONTAL MAJOR AXIS

Suppose we are given the ellipse with horizontal major axis centered at (h, k). See Figure 12. Let $c > 0, a > 0$, and $b > 0$ be the distance between the center and a focus, vertex, and endpoint of a minor axis respectively. Because the ellipse has a horizontal major axis, we know that $a > b$. By the definition of the ellipse and by Fact 1, we know that for any point P that lies on the graph of the ellipse, $d(P, F_1) + d(P, F_2) = 2a$.

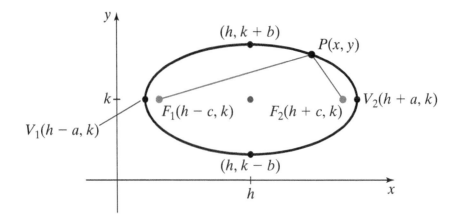

Figure 12

Using the distance formula, we can rewrite the equation $d(P, F_1) + d(P, F_2) = 2a$ as

$$\sqrt{(x - (h - c))^2 + (y - k)^2} + \sqrt{(x - (h + c))^2 + (y - k)^2} = 2a$$

Subtract a radical from both sides.

$$\sqrt{(x - (h - c))^2 + (y - k)^2} = 2a - \sqrt{(x - (h + c))^2 + (y - k)^2}$$

We now square both sides of the equation.

$$\left[\sqrt{(x-(h-c))^2+(y-k)^2}\right]^2 = \left[2a-\sqrt{(x-(h+c))^2+(y-k)^2}\right]^2$$

$$(x-(h-c))^2+(y-k)^2 = 4a^2-4a\sqrt{(x-(h+c))^2+(y-k)^2}+(x-(h+c))^2+(y-k)^2$$

Subtract $(y-k)^2$ from both sides and simplify.

$$(x-(h-c))^2 = 4a^2-4a\sqrt{(x-(h+c))^2+(y-k)^2}+(x-(h+c))^2$$

$$x^2-2xh+2cx-2ch+h^2+c^2 = 4a^2-4a\sqrt{(x-(h+c))^2+(y-k)^2}+x^2-2xh-2cx+2ch+h^2+c^2$$

Combine like terms and subtract $4a^2$ from both sides.

$$4cx-4ch-4a^2 = -4a\sqrt{(x-(h+c))^2+(y-k)^2}$$

Now divide both sides by -4.

$$-cx+ch+a^2 = a\sqrt{(x-(h+c))^2+(y-k)^2}$$

We now square both sides and simplify.

$$\left[-cx+ch+a^2\right]^2 = \left[a\sqrt{(x-(h+c))^2+(y-k)^2}\right]^2$$

$$(-cx+ch+a^2)(-cx+ch+a^2) = a^2[(x-(h+c))^2+(y-k)^2]$$

$$c^2x^2-2c^2xh-2a^2cx+2a^2ch+c^2h^2+a^4 = a^2[x^2-2xh-2cx+2ch+h^2+c^2+(y-k)^2]$$

$$c^2x^2-2c^2xh-2a^2cx+2a^2ch+c^2h^2+a^4 = a^2x^2-2a^2xh-2a^2cx+2a^2ch+a^2h^2+a^2c^2+a^2(y-k)^2$$

Collect the x and y terms on the left side.

$$-a^2x^2+2a^2xh-a^2h^2+c^2x^2-2c^2xh+c^2h^2-a^2(y-k)^2 = -a^4+a^2c^2$$

$$\underbrace{-a^2x^2+2a^2xh-a^2h^2}+\underbrace{c^2x^2-2c^2xh+c^2h^2}-a^2(y-k)^2 = -a^4+a^2c^2$$

Factor $-a^2$ out of the first three terms and factor c^2 out of the next three terms.

$$-a^2(x^2-2xh+h^2)+c^2(x^2-2xh+h^2)-a^2(y-k)^2 = -a^4+a^2c^2$$

$$\underbrace{-a^2(x-h)^2+c^2(x-h)^2}-a^2(y-k)^2 = -a^4+a^2c^2 \qquad \text{Factor.}$$

$$(-a^2+c^2)(x-h)^2-a^2(y-k)^2 = -a^2(a^2-c^2) \qquad \begin{array}{l}\text{Factor out }(x-h)^2\\\text{and }-a^2.\end{array}$$

$$-(a^2-c^2)(x-h)^2-a^2(y-k)^2 = -a^2(a^2-c^2) \qquad -a^2+c^2=-(a^2-c^2).$$

Recall that Fact 2 for ellipses is $b^2 = a^2 - c^2$. We can use this fact to substitute b^2 for $a^2 - c^2$.

$$-(a^2-c^2)(x-h)^2-a^2(y-k)^2 = -a^2(a^2-c^2) \qquad \text{Rewrite the equation from previous page.}$$

$$-b^2(x-h)^2-a^2(y-k)^2 = -a^2b^2 \qquad \text{Substitute } b^2=a^2-c^2.$$

$$\frac{-b^2(x-h)^2}{-a^2b^2}-\frac{a^2(y-k)^2}{-a^2b^2} = \frac{-a^2b^2}{-a^2b^2} \qquad \text{Divide both sides by } -a^2b^2.$$

$$\frac{(x-h)^2}{a^2}+\frac{(y-k)^2}{b^2} = 1.$$

Therefore, the standard equation of an ellipse centered at (h, k) with a horizontal major axis is $\dfrac{(x-h)^2}{a^2}+\dfrac{(y-k)^2}{b^2} = 1$, where $a > b$ and $b^2 = a^2 - c^2$. We can derive the standard equation of an ellipse centered at (h, k) with a vertical major axis in a similar fashion to obtain the equation $\dfrac{(x-h)^2}{b^2} + \dfrac{(y-k)^2}{a^2} = 1$, where $a > b$ and $b^2 = a^2 - c^2$.

Equation of an Ellipse in Standard Form with Center (h, k)

Horizontal Major Axis

$$\frac{(x-h)^2}{a^2} + \frac{(y-k)^2}{b^2} = 1$$

- $a > b > 0$
- Foci: $F_1(h - c, k)$ and $F_2(h + c, k)$
- Vertices: $V_1(h - a, k)$ and $V_2(h + a, k)$
- Endpoints of minor axis: $(h, k - b)$ and $(h, k + b)$
- $c^2 = a^2 - b^2$

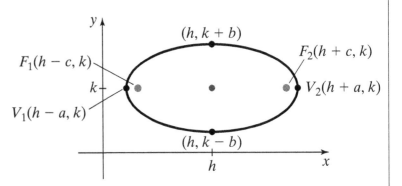

Equation of an Ellipse in Standard Form with Center (h, k)

Vertical Major Axis

$$\frac{(x-h)^2}{b^2} + \frac{(y-k)^2}{a^2} = 1$$

- $a > b > 0$
- Foci: $F_1(h, k - c)$ and $F_2(h, k + c)$
- Vertices: $V_1(h, k - a)$ and $V_2(h, k + a)$
- Endpoints of minor axis: $(h - b, k)$ and $(h + b, k)$.
- $c^2 = a^2 - b^2$

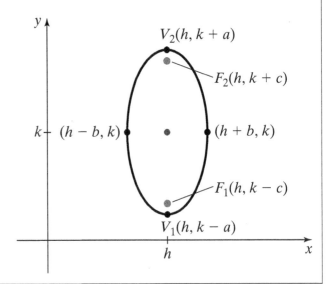

A.3 Hyperbola Proofs

OBJECTIVES

1 Establishing Fact 1 for Hyperbolas

2 Derivation of the Equation of a Hyperbola in Standard Form with a Horizontal Transverse Axis

OBJECTIVE 1   ESTABLISHING FACT 1 FOR HYPERBOLAS

Fact 1 for Hyperbolas

Given the foci of a hyperbola F_1 and F_2 and any point P that lies on the graph of the hyperbola, the difference of the distances between P and the foci is equal to $2a$. In other words, $|d(P, F_1) - d(P, F_2)| = 2a$. The constant $2a$ represents the length of the transverse axis.

Proof

Consider the hyperbola centered at (h, k) with vertices $V_1(h - a, k)$ and $V_2(h + a, k)$ and foci $F_1(h - c, k)$ and $F_2(h + c, k)$ as shown in Figure 13.

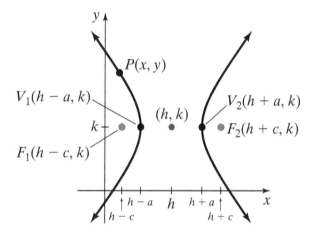

Figure 13

The distance between the center and either vertex is equal to a, so the distance between V_1 and V_2 is equal to $2a$. The distance between F_2 and V_2 is $|(h + c) - (h + a)| = |c - a| = c - a$ because $c > a$. Likewise, the distance between F_1 and V_1 is also equal to $c - a$. See Figure 14.

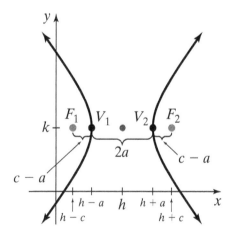

Figure 14

The distance between V_1 and F_2 is $d(V_1, F_2) = 2a + (c - a) = a + c$. The distance between V_1 and F_1 is $d(V_1, F_1) = c - a$. Therefore, $|d(V_1, F_2) - d(V_1, F_1)| = |a + c - (c - a)| = |a + c - c + a| = 2a$. Thus, by the geometric definition of the hyperbola, the difference between the distances from **any** point P on the hyperbola to the two foci must also be equal to $2a$. That is, $|d(P, F_1) - d(P, F_2)| = 2a$.

OBJECTIVE 2 DERIVATION OF THE EQUATION OF A HYPERBOLA IN STANDARD FORM WITH A HORIZONTAL TRANSVERSE AXIS

To derive the equation of the hyperbola with a horizontal transverse axis with center (h, k), use Figure 15. Without loss of generality, suppose that $d(P, F_2) > d(P, F_1)$. We will use Fact 1 for hyperbolas which states that $d(P, F_2) - d(P, F_1) = 2a$ to derive the equation.

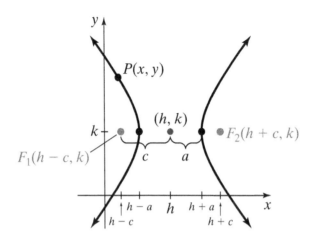

Figure 15

Using Fact 1 for hyperbolas and the distance formula, we get:

$$d(P, F_2) - d(P, F_1) = 2a$$

$$\sqrt{(x - (h + c))^2 + (y - k)^2} - \sqrt{(x - (h - c))^2 + (y - k)^2} = 2a$$

Now add a radical to both sides then square both sides of the equation.

$$\sqrt{(x - (h + c))^2 + (y - k)^2} = 2a + \sqrt{(x - (h - c))^2 + (y - k)^2}$$

$$\left[\sqrt{(x - (h + c))^2 + (y - k)^2}\right]^2 = \left[2a\sqrt{(x - (h - c))^2 + (y - k)^2}\right]^2$$

$$(x - (h + c))^2 + (y - k)^2 = 4a^2 + 4a\sqrt{(x - (h - c))^2 + (y - k)^2} + (x - (h - c))^2 + (y - k)^2$$

Subtract $(y - k)^2$ from both sides and simplify.

$$(x - (h + c))^2 = 4a^2 + 4a\sqrt{(x - (h - c))^2 + (y - k)^2} + (x - (h - c))^2$$

$$x^2 - 2hx - 2cx + 2ch + h^2 + c^2 = 4a^2 + 4a\sqrt{(x - (h - c))^2 + (y - k)^2}$$
$$+ x^2 - 2hx + 2cx - 2ch + h^2 + c^2$$

$$-4a^2 - 4cx + 4ch = 4a\sqrt{(x - (h - c))^2 + (y - k)^2}$$ Combine like terms and isolate the radical.

$$a^2 + cx - ch = -a\sqrt{(x - (h - c))^2 + (y - k)^2}$$ Divide both sides by -4.
$$a^2 + cx - ch = -a\sqrt{(x - (h - c))^2 + (y - k)^2}$$

Now square both sides and simplify.

$$[a^2 + cx - ch]^2 = \left[-a\sqrt{(x - (h - c))^2 + (y - k)^2}\right]^2$$

$$(a^2 + cx - ch)(a^2 + cx - ch) = a^2[(x - (h - c))^2 + (y - k)^2]$$

$$a^4 + 2a^2cx - 2a^2ch - 2c^2hx + c^2x^2 + c^2h^2 = a^2x^2 - 2a^2hx + 2a^2cx - 2a^2ch$$
$$+ a^2h^2 + a^2c^2 + a^2(y - k)^2$$

$$c^2x^2 - 2c^2hx + c^2h^2 - a^2x^2 + 2a^2hx - a^2h^2 - a^2(y - k)^2 = a^2c^2 - a^4$$

$\underbrace{c^2x^2 - 2c^2hx + c^2h^2}\ \underbrace{- a^2x^2 + 2a^2hx - a^2h^2}\ - a^2(y-k)^2 = a^2c^2 - a^4$

Factor c^2 out of the first three terms and factor $-a^2$ out of the next three terms.

$c^2(x^2 - 2xh + h^2) - a^2(x^2 - 2xh + h^2) - a^2(y-k)^2 = a^2c^2 - a^4$

$c^2(x-h)^2 - a^2(x-h)^2 - a^2(y-k)^2 = a^2c^2 - a^4$

$x^2 - 2xh + h^2 = (x-h)^2$

$(c^2 - a^2)(x-h)^2 - a^2(y-k)^2 = a^2(c^2 - a^2)$

Factor $(x-h)^2$ out of the first two terms.

$b^2(x-h)^2 - a^2(y-k)^2 = a^2b^2$

Let $b^2 = c^2 - a^2$.

$$\frac{b^2(x-h)^2}{a^2b^2} - \frac{a^2(y-k)^2}{a^2b^2} = \frac{a^2b^2}{a^2b^2}$$

Divide both sides by a^2b^2.

$$\frac{(x-h)^2}{a^2} - \frac{(y-k)^2}{b^2} = 1$$

Simplify.

Therefore, the standard equation of a hyperbola centered at (h, k) with a horizontal transverse axis is $\dfrac{(x-h)^2}{a^2} - \dfrac{(y-k)^2}{b^2} = 1$, where $b^2 = c^2 - a^2$. We can derive the standard equation of a hyprbola centered at (h, k) with a vertical transverse axis in a similar fashion to obtain the equation $\dfrac{(y-k)^2}{a^2} - \dfrac{(x-h)^2}{b^2} = 1$, where $b^2 = c^2 - a^2$.

Equation of a Hyperbola in Standard Form with Center (h, k)

Horizontal Transverse Axis

$$\frac{(x-h)^2}{a^2} - \frac{(y-k)^2}{b^2} = 1$$

- Foci: $F_1(h-c, k)$ and $F_2(h+c, k)$
- Vertices: $V_1(h-a, k)$ and $V_2(h+a, k)$
- Endpoints of conjugate axis: $(h, k-b)$ and $(h, k+b)$
- $b^2 = c^2 - a^2$

- Asymptotes: $y - k = \pm\dfrac{b}{a}(x-h)$

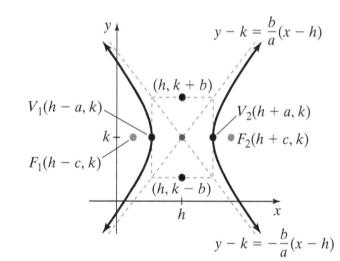

Equation of a Hyperbola in Standard Form with Center (h, k)

Vertical Transverse Axis

$$\frac{(y - k)^2}{a^2} - \frac{(x - h)^2}{b^2} = 1$$

- Foci: $F_1(h, k - c)$ and $F_2(h, k + c)$
- Vertices: $V_1(h, k - a)$ and $V_2(h, k + a)$
- Endpoints of conjugate axis: $(h - b, k)$ and $(h + b, k)$
- $b^2 = c^2 - a^2$
- Asymptotes: $y - k = \pm \dfrac{a}{b}(x - h)$

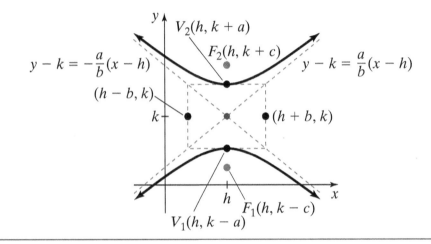

Glossary

Algebraic Expression An algebraic expression consists of one or more terms that may include variables, constants, and operating symbols such as + and −.

Argument The argument of a function is the variable, term, or expression on which a function operates.

Arithmetic Sequence An arithmetic sequence is a sequence of the form $a_1, a_1 + d, a_1 + 2d, a_1 + 3d,$ $a_1 + 4d, \ldots,$ where a_1 is the first term of the sequence and d is the common difference. The general term, or nth term, of an arithmetic sequence has the form $a_n = a_1 + (n − 1)d$.

Arithmetic Series An arithmetic series is of the form $a_1 + (a_1 + d) + (a_1 + 2d) + \cdots + (a_1 + (n − 1)d) + \cdots,$ where d is called the common difference.

Associative Property of Addition of Matrices If matrices $A, B,$ and C have the same size, then $(A + B) + C = A + (B + C)$. In other words, because we can only add matrices two at a time, it does not matter which two we add together first.

Axis of Symmetry An axis of symmetry is an invisible line that divides the graph into two equal halves. An ellipse has a horizontal axis of symmetry and a vertical axis of symmetry.

Binomial A binomial is a polynomial expression consisting of two terms.

Bisect Two line segments bisect each other if the intersection of the segments divides each segment into two parts of equal length.

Boundary Line A boundary line is a line that separates the ordered pair solutions to a linear inequality in two variables from the ordered pairs that are not solutions to the inequality.

Boundary Point A boundary point of a polynomial inequality $p < 0$ is a number for which $p = 0$.

Carrying Capacity The carrying capacity is the maximum population size that can be regularly sustained by an environment.

Coefficient A coefficient is a constant of an algebraic expression or equation that stands alone or is multiplied by a variable.

Coincide Two or more planes are said to coincide if each plane is the same representation of the other planes. Planes that coincide, therefore, have infinitely many points in common.

Collinear Ordered pairs (or points) in the rectangular coordinate system are said to be collinear if they all lie on the same line.

Column Matrix A column matrix is a matrix consisting of m rows and one column.

Combined Variation When a variable is related to more than one other variable, we call this combined variation.

Common Logarithmic Function For $x > 0$, the common logarithmic function is defined by $y = \log x$ if and only if $x = 10^y$

Commutative Property The commutative property is the algebraic property that states that a change in order produces the same result. Addition and multiplication are commutative because for any real numbers a and b, $a + b = b + a$ and $ab = ba$. Subtraction is *not* commutative.

Commutative Property of Matrices If matrices A and B have the same size, then $A + B = B + A$. In other words, the order in which we add matrices does not matter.

Complement The complement of an event E is the set of all outcomes in the sample space that are not included in the outcomes of event E. The notation used for the complement of E is E'.

Complement Rule Given the probability of an event E written as a fraction or a decimal, the probability of its complement can be found by subtracting the given probability from 1.

Completely Factored Form The polynomial function $f(x) = a_n x^n + a_{n−1} x^{n−1} + a_{n−2} x^{n−2} + \cdots + a_1 x + a_0,$ where $n \geq 1$ $a_n \neq 0$ is said to be in completely factored form if it is written in the form $f(x) = a_n(x − c_1)(x − c_2)(x − c_3) \cdots (x − c_n),$ where $c_1, c_2, c_3, \ldots c_n$ are complex numbers and not necessarily distinct. Note that in each factor of the form $(x − c)$, the coefficient of x is 1.

Complex Conjugate The complex conjugate of a complex number $a + bi$ is $a − bi$.

Complex Zero A complex zero is a zero of a polynomial $f(x)$ of the form $a + bi$, where a and b are real numbers.

Concentric Circles Concentric circles are circles that share a common center. When concentric circles intersect, their points of intersection form a hyperbola.

Constant A function f is constant on an interval (a, b) if, for any x_1 and x_2 chosen from the interval, $f(x_1) = f(x_2)$. The graph of a constant function is represented by a horizontal line segment.

Constant of Variation The constant of variation, or proportionality constant, is the constant ratio of two quantities that are directly related, or the constant product of two quantities that are inversely related.

Contradiction A contradiction (equation or inequality) is always false, regardless of the value of the variable, if any.

Coordinate Plane The coordinate plane is the plane represented by the rectangular coordinate system. It is also known as the Cartesian plane or the xy-plane.

Decreasing A function f is decreasing on an interval (a, b) if, for any x_1 and x_2 chosen from the interval with $x_1 < x_2$, then $f(x_1) > f(x_2)$.

Dependent System A system of linear equations is dependent if all solutions of one equation are also solutions of the other equations. A dependent system has infinitely many solutions.

Dependent Variable A dependent variable is the variable whose value is determined by the value assumed by an independent variable.

Difference of Two Squares If x and y are any two algebraic expressions, then $x^2 - y^2$ is called a difference of two squares.

Element An element of a cone is a straight line segment joining the outer edge of the base of the cone with the vertex of the cone. The set of all elements creates the surface of a cone.

Ellipse An ellipse is the set of all points in a plane, the sum of whose distances from two fixed points is a positive constant. The two fixed points, $F1$ and $F2$, are called the *foci*.

Equation of a Circle Every equation of a circle can be written in general form. However, not every equation of the form $Ax^2 + By^2 + Cx + Dy + E = 0$ has a graph that is a circle.

Equivalent System Two systems of linear equations are equivalent if and only if they have the same solution set.

Exponential Function An exponential function is of the form $f(x) = b^x$, where $b > 0$ and $b \neq 1$. The domain is the set of all real numbers, whereas the range is the set of all positive real numbers. The constant, b, is called the base of the exponential function.

Extraneous Solution An extraneous solution is a solution obtained through algebraic manipulations that is *not* a solution to the original equation.

Fact 1 for Ellipses The sum of the distances from any point on the ellipse to the two foci is equal to $2a$, the length of the major axis.

Fact 1 for Hyperbolas Given the foci of a hyperbola F_1 and F_2 and any point P that lies on the graph of the hyperbola, the difference of the distances between P and the foci is equal to $2a$. In other words, $|d(p, F_1) - d(p, F_2)| = 2a$ The constant $2a$ represents the length of the transverse axis.

Factor A factor is one of two or more expressions that are multiplied together to form a product.

Factorial of a Non-Negative Integer The factorial of a non-negative integer n, denoted as $n!$, is the product of all positive integers less than or equal to n.

Fundamental Counting Principle The Fundamental Counting Principle can be demonstrated in an experiment or task made up of two or more events. If the first event can occur in n_1 ways, the second event can occur in n_2 ways, and the third event can occur in n_3 ways and so on, then the number of possible outcomes of the experiment or task is $N = n_1 n_2 \ldots n_k$, where k is the total number of events.

Geometric Sequence A geometric sequence is a sequence of the form $a_1, a_1 r, a_1 r^2, a_1 r^3, a_1 r^4, \ldots$, where a_1 is the first term of the sequence and r is the common ratio such that $r = \dfrac{a_{n+1}}{a_n}$ for all $n \geq 1$. The general term, or nth term, of a geometric sequence has the form $a_n = a_1 r^{n-1}$.

Geometric Series If $a_1, a_1 r, a_1 r^2, a_1 r^3, \ldots$ is a geometric sequence, then $a_1 + a_1 r + a_1 r^2 + a_1 r^3 + \cdots + a_1 r^{n-1} + \cdots$ is called a geometric series.

Graph of an Equation The graph of an equation is the set of all ordered pairs (x, y) whose coordinates satisfy the equation.

Greatest Common Factor (GCF) The greatest common factor (GCF) is the largest number or algebraic expression that divides two or more algebraic expressions evenly.

Horizontal Asymptote A horizontal line $y = H$ is a horizontal asymptote of a function f if the values of $f(x)$ approach some fixed number H as the values of x approach ∞ or $-\infty$.

Horizontal Axis of Symmetry The horizontal axis of symmetry of a parabola is the invisible horizontal line that divides the parabola into two equal parts.

Horizontal Line Test Given the graph of a function f, if every horizontal line intersects the graph of f at most once, then f is a one-to-one function.

Horizontal Major Axis An ellipse has a horizontal major axis if the horizontal axis of symmetry passes through the two foci and is longer than the vertical axis of symmetry.

Identity An identity (equation or inequality) is always true, regardless of the value of the variable, if any.

Identity Matrix An identity matrix is a square matrix where all entries down the diagonal are 1's with zeros everywhere else.

Increasing A function f is increasing on an interval (a, b) if, for any x_1 and x_2 chosen from the interval with $x_1 < x_2$, then $f(x_1) < f(x_2)$.

Independent Variable An independent variable is the variable whose value determines the value of the other variables.

Index The index of a radical expression $\sqrt[n]{b}$ is the integer n.

Intercepts The intercepts of a graph are points where a graph crosses or touches a coordinate axis.

Inverse Function Let f be a one-to-one function with domain A and range B. Then f^{-1} is the inverse function of f with domain B and range A.

Irrational Number An irrational number is a real number that *cannot* be expressed as the quotient of two integers $\dfrac{p}{q}$. Irrational numbers have an infinite, nonrepeating decimal representation.

Irrational Solution An irrational solution is a real number solution to an equation that *cannot* be expressed as the quotient of two integers.

Irrational Zero The value $x = r$ is an irrational zero of f if $f(r) = 0$ and if r *cannot* be written as the quotient of two integers.

Joint Variation A special case of combined variation occurs when a variable is directly proportional to the product of two or more other variables. This is called joint variation.

Leading Coefficient The leading coefficient of a polynomial function is the constant that is multiplied by the variable of highest power.

Least Common Denominator (LCD) The least common denominator is the smallest multiple of all denominators in an algebraic expression or equation.

Linear Equation in n Variables A linear equation in n variables is an equation that can be written in the form $a_1 x_1 + a_2 x_2 + \cdots + a_n x_n = b$ for variables $x_1, x_2, \ldots x_n$ and real numbers $a_1 a_2, \ldots a_n, b$, where at least one of $a_1, a_2, \ldots a_n$ is nonzero

Linear Equation A linear equation in one variable is an equation that can be written in the form $ax + b = 0$, where a and b are real numbers and $a \neq 0$.

Linear Function A first-degree polynomial function of the form $f(x) = mx + b$ is called a linear function.

Linear Inequality in One Variable A linear inequality in one variable is an inequality that can be written in the form $ax + b < c$, where a, b, and c are real numbers and $a \neq 0$.

Logarithm Property of Equality The logarithm property of equality is if a logarithmic equation can be written as $\log_b u = \log_b v$, then $u = v$.

Method of Elimination The method of elimination is a method used to solve linear equations by adding a multiple of an equation to a multiple of another equation in the attempt to eliminate one of the variables.

Minor Axis The minor axis of an ellipse is the shorter of the two axes of symmetry.

Monomial Function A monomial function is a polynomial function consisting of a single term.

Mutually Exclusive Events Two or more events are mutually exclusive if they have no outcomes in common.

Natural Numbers The set of natural numbers, sometimes referred to as the *counting numbers,* is defined as $N = \{1, 2, 3, \ldots\}$.

Noncollinear Points Three points are said to be noncollinear if they do not all lie on the same line.

Non-Strict Inequality A non-strict inequality contains one or both of the following inequality symbols: $\leq$, $\geq$. The possibility of equality is included in a non-strict inequality.

Number e The number e, called the *natural base,* is an irrational number. The number e rounded to 6 decimal places is $e \approx 2.718282$.

One-to-One Function A function f is one-to-one if for any values $a \neq b$ in the domain of f, $f(a) \neq f(b)$.

Open Interval An open interval, denoted by the symbol (a, b), consists of all real numbers between a and b not including a and b.

Origin The origin is the point in the Cartesian plane where the x-axis and y-axis intersect. The origin has coordinates $(0, 0)$.

Parabola A parabola is the set of all points in a plane equidistant from a fixed point F and a fixed line D. The fixed point is called the focus, and the fixed line is called the directrix.

Parallelogram A parallelogram is a quadrilateral (four-sided polygon) in which opposite sides are parallel.

Perfect n**th Power** A perfect nth power is a rational number that can be written as the nth power of another rational number where n is a positive integer.

Periodic Compound Interest Formula Periodic compound interest can be calculated using the formula

$$A = P\left(1 + \frac{r}{n}\right)^{nt}$$

Point–Slope Form of the Equation of a Line Given the slope of a line m and a point on the line (x_1, y_1), the point-slope form of the equation of a line is given by $y - y_1 = m(x - x_1)$.

Polynomial Expression An algebraic expression in which the exponents of all variable factors are nonnegative integers is called a polynomial expression.

Polynomial Function The function $f(x) = a_n x^n + a_{n-1} x^{n-1} + a_{n-2} x^{n-2} + \cdots + a_1 x + a_0$ is a polynomial function of degree n, where n is a nonnegative integer. The numbers $a_0, a_1, a_2, \ldots, a_n$ are called the coefficients of the polynomial function. The number a_n is called the leading coefficient, and a_0 is called the constant coefficient.

Polynomial Inequality A polynomial inequality is an inequality that can be written in the form $a_n x^n + a_{n-1} x^{n-1} + \cdots + a_1 x + a_0 < 0$, where $a_0, a_1, \ldots, a_n$ are real numbers, and each exponent is an integer greater than or equal to 0.

Polynomial in Standard Form A polynomial in one variable is written in standard form when it is written in descending order with the term of highest power written first.

Present Value Formula Present value can be calculated using the formula $P = A\left(1 + \frac{r}{n}\right)^{-nt}$

Prime Polynomial A prime polynomial is a term that indicates that a polynomial cannot be factored using *integer* coefficients. It is possible for a prime polynomial to be factored using rational or irrational coefficients.

Quadratic Equation A quadratic equation is an equation that can be written in the form $ax^2 + bx + c = 0$, where a, b, and c are real numbers and $a \neq 0$.

Radical Conjugate The expression $\sqrt{5} - \sqrt{2}$ is called the radical conjugate of the expression $\sqrt{5} + \sqrt{2}$. The product of radical conjugates always yields an expression that does not contain a radical.

Radicand A radicand is the number, variable, or algebraic expression under the radical.

Range of a Parabola The range is the set of all values for the dependent variable. These are the second coordinates in the set of ordered pairs and are also known as *output values.*

Ratio A ratio is a comparison between two numbers.

Rational Equation A rational equation is an equation involving the quotient of two polynomial expressions where the polynomial in the denominator is non-constant.

Rational Expression A rational expression is the quotient of two polynomials.

Rational Function A rational function is of the form $f(x) = \frac{g(x)}{h(x)}$, where $g(x)$ and $h(x)$ are polynomial functions and $h(x) \neq 0$. The domain of a rational function is the set of all real numbers such that $h(x) \neq 0$.

Rational Inequality A rational inequality is an inequality that can be written in the form $\frac{p}{q} \geq 0$, where p and q are polynomials and $q \neq 0$.

Rational Number A rational number is a real number that can be expressed as the quotient of two integers $\frac{p}{q}$, where $q \neq 0$. All finite decimals and repeating decimals are rational numbers.

Rational Zero A rational zero is a real zero of a polynomial $f(x)$ of the form $\frac{p}{q}$, where p and q are integers and $q \neq 0$.

Rational Zeros Theorem Let f be a polynomial function of the form $f(x) = a_n x^n + a_{n-1} x^{n-1} + a_{n-2} x^{n-2} + \cdots + a_1 x + a_0$ of degree $n \geq 1$, where each coefficient is an integer. If $\frac{p}{q}$ is a rational zero of f (where $\frac{p}{q}$ is written in lowest terms), then p must be a factor of the constant coefficient, a_0, and q must be a factor of the leading coefficient a_n.

Real Number A real number can be either a rational or irrational number. Rational numbers include integers, whole numbers, and natural numbers. Irrational numbers include decimals that neither repeat nor terminate.

Real Zero   A real number $x = c$ is a real zero of a function f if $f(c) = 0$. Real zeros are also x-intercepts.

Reciprocal   A reciprocal is a number or algebraic expression related to another number or algebraic expression in such a way that when multiplied together their product is 1.

Recursive Sequence   A recursive sequence is a sequence in which each term is defined using one or more previous terms.

Reduced Row-Echelon Form   An augmented matrix is said to be in reduced row-echelon form when all zeros appear above and below the diagonal *and* 1's appear down the diagonal.

Right Circular Cone   A right circular cone is a cone that has a vertex directly above the center of a circular base.

Row Matrix   A row matrix is a matrix consisting of one row and n columns.

Row-Echelon Form   An augmented matrix is said to be in row-echelon form when all zeros appear under the diagonal *and* 1's appear down the diagonal.

Simple Interest   The formula for simple interest is $I = Prt$, where P = principal amount invested, r = interest rate, and t = time (in years).

Singular Matrix   A square matrix is singular if it does not have an inverse matrix.

Slope-Intercept Form of a Line   Given the slope of a line m and the y-intercept b, the slope-intercept form of the equation of a line is given by $y = mx + b$.

Square Function   We can sketch the square function $f(x) = x^2$ by setting up a table of values and plotting points. The square function assigns to each number in the domain the square of that number.

Square Matrix   A square matrix is a matrix that contains the same number of rows as columns.

Square Root Function   We can sketch the square root function $f(x) = \sqrt{x}$ by setting up a table of values and plotting points. The square root function assigns to each number in the domain the square root of that number.

Square Root Property   The solution to the quadratic equation $x^2 - c = 0$ or equivalently $x^2 = c$ is $x^2 = \pm\sqrt{c}$.

Standard Form   A polynomial in one variable is written in standard form when it is written in descending order with the term of highest power written first.

Standard Form Equation of a Line   The standard form of an equation of a line is given by $Ax + By = C$, where $A, B,$ and C are real numbers such that and are both not zero.

Strict Inequality   A strict inequality contains one or more of the following inequality symbols: $<, >, \neq$. There is no possibility of equality in a strict inequality.

Subset   A subset is a set contained within a set. A subset can be the set itself. The subset of an interval must contain points within the interval.

System of Linear Equations in Two Variables   A system of linear equations in two variables is the collection of two linear equations considered simultaneously. The solution to a system of equations in two variables is the set of all ordered pairs for which *both* equations are true.

Test Value of a Polynomial or Rational Inequality   After using the boundary points to divide the number line into intervals, a test value is a number chosen from each interval to test whether the inequality is positive or negative on the interval.

Test Value of a Polynomial Function   When sketching the graph of a polynomial function, a test value is an x-value chosen in such a way as to help complete the graph of a function.

Transverse Axis   The transverse axis of a hyperbola is the imaginary line segment whose endpoints are the vertices of the hyperbola.

Triangular Form   A system of three linear equations is said to be in triangular form when the leading coefficient of the first variable in the second equation is zero and the coefficients of the first two variables of the third equation are zero.

Turning Points   The turning points on the graph of a polynomial function are the points at which the graph changes from increasing to decreasing or from decreasing to increasing.

Vertex of a Parabola   The vertex of a parabola is the lowest point (if the parabola opens *up*) or the highest point (if the parabola opens *down*) of the graph of the parabola.

Vertical Asymptote   A vertical line $x = a$ is a vertical asymptote of a function f if the values of $f(x)$ approach ∞ or $-\infty$ as the values of x approach a.

Vertical Line Test   A graph in the Cartesian plane is the graph of a function if and only if no vertical line intersects the graph more than once.

Vertical Major Axis   An ellipse has a vertical major axis if the vertical axis of symmetry passes through the two foci and is longer than the horizontal axis of symmetry.

x-Intercept   An x-intercept is the x-coordinate of a point where a graph crosses or touches the x-axis. (The y-coordinate is 0.)

y-Intercept   A y-intercept is the y-coordinate of a point where a graph crosses or touches the y-axis. (The x-coordinate is 0.)

Zero   A zero of a function is a value of x for which $f(x) = 0$.

Answers

R.1 Exercises

1. natural, whole, integer, rational, real   **2.** integer, rational, real   **3.** rational, real   **4.** irrational, real   **5. a.** 1   **b.** 0, 1
c. $-17, 0, 1$ **d.** $-17, -\dfrac{25}{19}, 0, 0.331, 1$ **e.** $-\sqrt{5}, \dfrac{\pi}{2}$ **6. a.** none   **b.** none   **c.** -11 **d.** $-11, -3.\overline{2135}, -\dfrac{3}{9}, 21.1$
e. $\dfrac{\sqrt{7}}{2}$ **7. a.** open interval   **b.** $\{x \mid 0 < x < 3\}$ **c.** $(0, 3)$ **8. a.** closed infinite interval   **b.** $\left\{x \mid x \le \dfrac{1}{4}\right\}$
c. $\left(-\infty, \dfrac{1}{4}\right]$ **9. a.** half-open interval   **b.** $\left\{x \mid -\dfrac{3}{2} \le x < 2\right\}$ **c.** $\left[-\dfrac{3}{2}, 2\right)$ **10. a.** open infinite interval
b. $\{x \mid x > \sqrt{2}\}$ **c.** $(-\sqrt{2}, \infty)$ **11.** $\left\{x \mid -\dfrac{1}{2} \le x \le 5\right\}$,   **12.** $\left\{x \mid 0 < x < \dfrac{5}{2}\right\}$,
13. $\{x \mid x \le 3\}$,   **14.** $\{x \mid x > -1\}$,
15. $\left[-\dfrac{5}{2}, 1\right]$,   **16.** $[0, 3)$,   **17.** $\left[\dfrac{3}{4}, \infty\right)$,
18. $(-4, \infty)$,   **19. a.** $\{-1, 0, 1, 2, 4\}$ **b.** $\{0, 2\}$
20. a. $\left\{-1, -\dfrac{3}{4}, -\dfrac{1}{4}, \dfrac{1}{4}, \dfrac{3}{4}, 1\right\}$ **b.** $\left\{\dfrac{1}{4}\right\}$ **21. a.** $\{0, 1, 2, 3, 4, 5, 6\}$ **b.** $\varnothing$ **22.** $(-\infty, 3)$ **23.** $(-8, 3)$ **24.** $[0, 4]$
25. $(-\infty, 1) \cup (1, \infty)$ **26.** $[3, 0) \cup (0, \infty)$ **27.** $(-5, 1]$ **28.** 286 **29.** 2.2 **30.** 6 **31.** -3 **32.** $\pi - 2 \approx 1.14$
33. 11 **34.** 12 **35.** 22.5

R.2 Exercises

1. $m + 13$ **2.** $c + ab$ **3.** $10t + 11w$ **4.** $z \cdot 3$ **5.** nm **6.** $(v + 4)6$ **7.** $(4 + a) + 11$ **8.** $d + (a + 30)$
9. $zw + (a + y)$ **10.** $(45 \cdot y) \cdot z$ **11.** $(8 \cdot 2)(y + z)$ **12.** $132a + 48$ **13.** $20x + 20y$ **14.** $15a - 150 + 90b$
15. $5h + 10v - 30$ **16.** $32c$ **17.** $-2a$ **18.** $z(x - y)$ **19.** 4^3, base $= 4$, exponent $= 3$ **20.** w^4, base $= w$,
exponent $= 4$ **21.** $(-6y)^5$, base $= -6y$, exponent $= 5$ **22.** base $= 8$, exponent $= 3, 512$ **23.** base $= 5$, exponent $= 4$,
-625 **24.** base $= -4$, exponent $= 3, -64$ **25.** base $= 2z$, exponent $= 5, 32z^5$ **26.** 9 **27.** $-\dfrac{99}{7}$ **28.** $3{,}088$
29. $-\dfrac{3}{28}$ **30.** 116 **31.** $1{,}223$ **32.** $\dfrac{17}{8}$ **33.** -4 **34.** 16 **35.** -5 **36.** 9 **37.** -21 **38.** 126 **39.** 7
40. -23 **41.** $\dfrac{9}{2}$ **42.** $-17x$ **43.** $9z^2$ **44.** $-14y^5$ **45.** $-10a + 30$ **46.** $10k^2 + 10k$ **47.** $9t - 25$ **48.** $21r - 12$
49. $-11w + 59$ **50.** $52p^2 - 1$ **51.** $\dfrac{17x}{4} - \dfrac{1}{2}$

R.3 Exercises

1. 3^5 **2.** k^9 **3.** x^{26} **4.** $a^{12}b^6$ **5.** t^7 **6.** $\dfrac{1}{r^{35}}$ **7.** w^{48} **8.** n^8 **9.** m^3 **10.** 1 **11.** -1 **12.** $\dfrac{1}{4}$ **13.** $-\dfrac{1}{16}$
14. $\dfrac{1}{n^3}$ **15.** $\dfrac{5}{x^6}$ **16.** $-\dfrac{7}{y^2}$ **17.** t^2 **18.** $5y^9$ **19.** $\dfrac{b^4}{a^{11}}$ **20.** $-\dfrac{5q^4}{p^7w^5}$ **21.** b^7 **22.** $\dfrac{1}{a^{14}}$ **23.** $\dfrac{1}{27}$ **24.** z^{18}
25. $25x^4y^{10}$ **26.** $\dfrac{3y^{12}}{4x^5}$ **27.** 1 **28.** $\dfrac{1}{2a^8b^4c^6}$ **29.** $\dfrac{1}{q^{12}p^{16}w^{20}}$ **30.** $3x^4y^{15}$ **31.** 13 **32.** $\dfrac{1}{5}$ **33.** 6 **34.** -7

35. $7w$ **36.** $x + 1$ **37.** $4\sqrt{10}$ **38.** $11x\sqrt{3}$ **39.** $5\sqrt[3]{2}$ **40.** $-3x\sqrt[3]{x}$ **41.** $2z\sqrt[4]{2z}$ **42.** $6x^6y^{13}\sqrt{5y}$ **43.** x^3y
44. $\sqrt{30} - \sqrt{66}$ **45.** $-3\sqrt{14} - 36$ **46.** $19 + 8\sqrt{3}$ **47.** $8x\sqrt[3]{4} - 5x^2\sqrt[3]{x}$ **48.** $14\sqrt{5}$ **49.** $2\sqrt{3x}$ **50.** $7y\sqrt[3]{7x}$
51. $\dfrac{\sqrt{30}}{6}$ **52.** $\dfrac{\sqrt{7}}{7}$ **53.** $\dfrac{\sqrt[3]{28}}{2}$ **54.** $\dfrac{y\sqrt[3]{50x^2y^2}}{5x}$ **55.** $\dfrac{45 + 9\sqrt{2}}{23}$ **56.** $4 - \sqrt{15}$ **57.** $\dfrac{a - \sqrt{ab}}{a - b}$ **58.** 5 **59.** $\dfrac{1}{2}$
60. -3 **61.** 8 **62.** $\dfrac{1}{16}$ **63.** $\dfrac{1}{x^{1/3}}$ **64.** x **65.** $x^{1/12}$ **66.** xy^2 **67.** $x^{5/7}y^2$ **68.** $\dfrac{3x}{y^{5/4}}$ **69.** $\dfrac{x^2}{2y^{9/5}}$ **70.** $\dfrac{x^4}{y^2}$

R.4 Exercises

1. yes, 4, 5 **2.** no **3.** yes, 12, −1 **4.** no **5.** yes, 0, 0 **6.** no **7.** yes, 0 **8.** yes, 3 **9.** no **10.** yes, 7
11. $-x + 14$ **12.** $x^2 + 14x - 6$ **13.** $9x^3 - 15x^2 + 16x - 6$ **14.** $4x^5 + 7x^4 + 6x^3 + 8x^2 + 9x$ **15.** $-9x - 1$
16. $x^4 - x^2 + 7x$ **17.** $-35x^{10} + 10x^4$ **18.** $x^3 + 10x^2 + 15x - 18$ **19.** $x^2 + 5x - 36$ **20.** $3x^2 + 17x + 10$
21. $3x^2 + 31x + 56$ **22.** $36x^2 - 4$ **23.** $x^2 + 10x + 25$ **24.** $16x^2 - 72x + 81$ **25.** $64x^2 - y^2$ **26.** $x^3 - 18x^2 +$
$108x - 216$ **27.** $27x^3 + 54x^2 + 36x + 8$ **28.** $x^4 - 32x^2 + 256$ **29.** quotient $= 7x^2 - 26x + 79$, remainder $= -233$
30. quotient $= 9x^2 - 47$, remainder $= 2x + 244$ **31.** quotient $= 3x^2$, remainder $= -2x^2 + 7x + 6$ **32.** quotient $=$
$x^2 - x - 9$, remainder $= 18x + 88$

R.5 Exercises

1. $4(10x + 1)$ **2.** $w(x^4 + 1)$ **3.** $3y^7(1 + 2xy)$ **4.** $4x^2(x^2 - 5x + 4)$ **5.** $3wy(13w^8y^8 - 4wy + 4 + 5w)$
6. $(11 + 3a)(x + 7)$ **7.** $(8x + 1)(z + 1)$ **8.** $(9y - 7)(x^2 + 11)$ **9.** $(x + 3)(y + 5)$ **10.** $(a - 2)(b + 7)$
11. $(3x + 2)(3y + 4)$ **12.** $(x - 4)(3y - 2)$ **13.** $(x + 3)(5x + 4y)$ **14.** $(3x - 2)(x + y)$ **15.** $(x^2 + 4)(x + 9)$
16. $(x^2 - 7)(x - 1)$ **17.** $(x + 10)(x + 3)$ **18.** $(y + 7)(y - 2)$ **19.** $(w - 9)(w - 8)$ **20.** $(x - 8)(x + 6)$
21. prime **22.** $(x - 36)(x + 3)$ **23.** $(x + 4y)(x + 2y)$ **24.** $-(x - 6)(x + 4)$ **25.** $5(x + 5)(x - 3)$
26. $4w(x + 3)(x + 2)$ **27.** $5(x^2 - 6x - 5)$ **28.** $(3x + 5)(3x + 1)$ **29.** $(3y + 5)(4y + 3)$ **30.** $(x + 5)(2x - 5)$
31. $(2x - 3)(4x + 3)$ **32.** $(2y - 3)(5y - 4)$ **33.** $2(2x + 3)(4x + 3)$ **34.** $-(4x - 3)(5x + 4)$
35. $(x - 9)(x + 9)$ **36.** $(z + 5)^2$ **37.** $(x + 4)(x^2 - 4x + 16)$ **38.** $2(m - 12)^2$ **39.** $(8x - 7)(8x + 7)$
40. $(x - 7)(x^2 + 7x + 49)$ **41.** $3(6t - 7)(6t + 7)$ **42.** $(x + 8y - 5)(x + 8y + 5)$
43. $(5x + 6)(25x^2 - 30x + 36)$ **44.** $(2x - 7)(4x^2 - 16x + 19)$ **45.** $(w - 5)(w + 15)$ **46.** $(4x + 5)^2$
47. $x^{11}(x - 1)(x^2 + x + 1)$ **48.** $-(5x - 1)(5x + 1)(x^2 + 6)$ **49.** $y^4(x - 3)(x^2 + 3x + 9)$
50. $-(x^2 - x - 7)(x^2 + x + 7)$

R.6 Exercises

1. $1 - 3x$ **2.** $x - 9$ **3.** $\dfrac{9}{5}$ **4.** $x - 2$ **5.** -1 **6.** $-(x - 2)$ or $2 - x$ **7.** $\dfrac{x - 2}{x - 6}$ **8.** $\dfrac{x^2 + 5x + 25}{3}$
9. $\dfrac{x - 2}{2x^2 + 1}$ **10.** $\dfrac{1}{6x + 7}$ **11.** $\dfrac{3}{2(x - 1)}$ **12.** $-\dfrac{8}{7}$ **13.** $-\dfrac{5a}{3a + 1}$ **14.** $\dfrac{5(x + 2)}{2(x + 4)}$ **15.** $\dfrac{3a}{2(a - b)}$ **16.** $\dfrac{1}{6}$
17. $\dfrac{(x + 2)(x + 1)}{9}$ **18.** $\dfrac{x}{2}$ **19.** $\dfrac{7a^2}{a - b}$ **20.** $\dfrac{8}{(x + 6)(x + 7)}$ **21.** $\dfrac{1}{3}$ **22.** -1 **23.** $\dfrac{(x - 2)(x + 5)}{(x - 7)(5x + 1)}$
24. $\dfrac{40}{7x^3y}$ **25.** $\dfrac{(y - 4)(3x - 5)}{(y + 1)(2x - 3)}$ **26.** $\dfrac{5(3a + 2)}{a}$ **27.** $\dfrac{16}{x}$ **28.** $\dfrac{6x + 1}{x - 4}$ **29.** $\dfrac{(x - 2)(x + 2)}{3x + 1}$ **30.** $\dfrac{6x + 35}{49x^2}$
31. $\dfrac{9 - x}{x - 3}$ **32.** $\dfrac{2(x + 35)}{(x - 1)(x + 8)}$ **33.** $\dfrac{3x^2 - 4x - 21}{(x + 3)(x - 3)}$ **34.** $\dfrac{-11x + 4}{(x + 4)(x - 4)}$ **35.** $\dfrac{5x^2 - 144}{x(x - 6)(x + 6)}$ **36.** $\dfrac{7 - 39x}{12x^2y}$
37. $\dfrac{20a}{(a + b)(a - b)}$ **38.** $\dfrac{6(4z - 1)}{(z - 9)(z + 6)}$ **39.** $\dfrac{2(4x + 3y)}{(x - y)(x + y)}$ **40.** $\dfrac{2x^2 - 11x - 57}{(x + 9)(x + 3)(x - 3)}$
41. $\dfrac{x(5x - 34)}{(x + 6)(x - 2)(x - 6)}$ **42.** $\dfrac{-23y + 31}{(2y + 7)(y + 1)(y - 5)}$ **43.** $\dfrac{y + 45}{y(y - 3)^2}$ **44.** $\dfrac{(2x + 3)(x - 15)}{(x + 6)^2(2x - 9)}$ **45.** $\dfrac{7x + 1}{7x - 1}$
46. $\dfrac{x}{1 - 3x}$ **47.** $\dfrac{1}{(x - 6)(x + 5)}$ **48.** $\dfrac{6y + 5x}{5y - 6x}$ **49.** $\dfrac{x - 4}{4x}$ **50.** $\dfrac{6}{x + 7}$ **51.** $\dfrac{3}{x - 5}$ **52.** $\dfrac{x - 5}{x + 5}$
53. $\dfrac{x - 1}{x - 7}$ **54.** $\dfrac{6}{x - y}$

CHAPTER 1

1.1 Exercises
SCE-1. 18 **SCE-2.** 24 **SCE-3.** 30 **SCE-4.** $x^2 - 1$ **SCE-5.** $24x^2$ **SCE-6.** $2w^2 - 7w + 3$ **SCE-7.** $x^2 - 1$
SCE-8. $x^2 - x$

1. linear **2.** nonlinear **3.** linear **4.** nonlinear **5.** linear **6.** nonlinear **7.** linear **8.** $x = \dfrac{1}{3}$ **9.** $x = 0$

10. $x = -\dfrac{9}{10}$ **11.** $x = \dfrac{11}{10}$ **12.** $x = -\dfrac{17}{5}$ **13.** $x = 8$ **14.** $y = \dfrac{2}{29}$ **15.** $p = -\dfrac{19}{46}$ **16.** $x = -25$

17. $x = -6$ **18.** $a = -\dfrac{3}{4}$ **19.** $x = 2.925$ **20.** $k = 128.5$ **21.** $x = 1$ **22.** $x = -\dfrac{3}{7}$ **23.** $x = -\dfrac{24}{7}$ **24.** $\varnothing$

25. $w = -\dfrac{3}{4}$ **26.** $\varnothing$ **27.** $x = \dfrac{8}{5}$ **28.** $\varnothing$ **29.** $\varnothing$

1.2 Exercises

1. $2n + 10$ **2.** $3(n + 6) - 5$ **3.** $\dfrac{3}{4n}$ **4.** $2n = \dfrac{n}{4} - 3$ **5.** $\dfrac{4n}{7} - 3 = 2n - 8$ **6.** 7 and 19 **7.** 5 ft by 11 ft

8. 12, 14, 16 **9.** Steve: 81 tickets, Tom: 40 tickets **10.** 35 dimes, 41 quarters, and 5 silver dollars **11.** stock A:
64 shares, stock B: 32 shares, stock C: 16 shares **12.** 11.25 pounds of \$9.00/lb coffee and 18.75 pounds of \$5.00/lb coffee

13. 20 gallons of the 80% orange juice **14.** $8\dfrac{1}{3}$ L of pure water should be mixed with a 5-L solution of 80% acid

15. drain 2.25 L of coolant and fill up with 2.25 L of pure antifreeze **16.** \$8,000 at 4% simple annual interest and \$20,000 at
5% simple annual interest **17.** product A: \$21,000, product B: \$61,000 **18.** freight train: 30 mph, passenger train: 50 mph

19. 22 hours **20.** 30 mph **21.** 24 miles **22.** $\dfrac{5}{4}$ hours **23.** Joan: 60 minutes, Jane: 20 minutes **24.** 40 hrs

1.3 Exercises
SCE-1. $6\sqrt{2} - 2$ **SCE-2.** $9\sqrt{3} + 4$ **SCE-3.** $2\sqrt{5}$ **SCE-4.** $6\sqrt{3}$ **SCE-5.** $27 - 10\sqrt{2}$ **SCE-6.** $14 + 6\sqrt{5}$

1. i **2.** -1 **3.** -1 **4.** i **5.** $-i$ **6.** $-4 + 7i$ **7.** $10 - 11i$ **8.** -1 **9.** $3\sqrt{2}$ **10.** $6 + 8i$ **11.** $-1 - i$

12. $-10 + 5i$ **13.** $32 - 24i$ **14.** $1 - 2\sqrt{2}i$ **15.** 29 **16.** 2 **17.** $\dfrac{37}{4}$ **18.** 6 **19.** 1 **20.** $\dfrac{2}{25} - \dfrac{11}{25}i$

21. $\dfrac{2}{5} + \dfrac{1}{5}i$ **22.** $\dfrac{3}{4} + \dfrac{3}{4}i$ **23.** $\dfrac{12}{13} + \dfrac{5}{13}i$ **24.** $-7 + 6i$ **25.** $3 - 7i$ **26.** -6 **27.** -8 **28.** 4 **29.** $-2 - \sqrt{5}i$

30. $-\dfrac{1}{2} - \dfrac{3}{2}i$ **31.** $1 + \dfrac{\sqrt{2}}{2}i$

1.4 Exercises

SCE-1. 10 **SCE-2.** $-6 + 2\sqrt{11}$ **SCE-3.** $4 + 14i$ **SCE-4.** $6 - 2i\sqrt{29}$ **SCE-5.** $\dfrac{5}{3}$ **SCE-6.** $-3 + \sqrt{11}$

SCE-7. $\dfrac{2 + 7i}{3}$ **SCE-8.** $\dfrac{3 - i\sqrt{29}}{2}$ **SCE-9.** $(x + 4)(x + 6)$ **SCE-10.** $(x - 9)(x + 2)$ **SCE-11.** $(x - 7)(x - 3)$

SCE-12. $(2x + 1)(x + 4)$ **SCE-13.** $(3x - 5)(2x + 3)$

1. $x = 0$ or $x = 8$ **2.** $x = -2$ or $x = 3$ **3.** $x = -5$ or $x = -4$ **4.** $x = 4$ or $x = 6$ **5.** $x = -3$ or $x = \dfrac{1}{3}$

6. $x = -\dfrac{5}{6}$ or $x = 7$ **7.** $x = -4$ or $x = -\dfrac{1}{3}$ **8.** $x = -4$ or $x = \dfrac{3}{8}$ **9.** $m = -\dfrac{3}{4}$ or $m = \dfrac{5}{2}$ **10.** $z = -\dfrac{2}{7}$ or $z = \dfrac{3}{4}$

11. $x = \pm 8$ **12.** $x = \pm 8i$ **13.** $x = \pm 2\sqrt{6}$ **14.** $x = -5$ or $x = 1$ **15.** $x = -\dfrac{1}{2} \pm i$ **16.** 16 **17.** 25

18. $\dfrac{49}{4}$ **19.** $\dfrac{25}{36}$ **20.** $x = 4 \pm 3\sqrt{2}$ **21.** $x = -\dfrac{7}{2} \pm \dfrac{\sqrt{7}}{2}i$ **22.** $x = -\dfrac{3}{2} \pm \dfrac{\sqrt{7}}{2}i$ **23.** $x = -4 \pm \dfrac{\sqrt{165}}{3}$

24. $x = -\dfrac{5}{6} \pm \dfrac{\sqrt{119}}{6}i$ **25.** $x = -3$ or $x = \dfrac{1}{3}$ **26.** $x = 4 \pm 3\sqrt{2}$ **27.** $x = \dfrac{1}{8} \pm \dfrac{\sqrt{127}}{8}i$ **28.** $x = \dfrac{2}{3} \pm \dfrac{\sqrt{7}}{3}$

29. $x = \dfrac{1}{3}$ **30.** $x = -\dfrac{3}{10} \pm \dfrac{\sqrt{11}}{10}i$ **31.** $D = 0$: exactly one real solution   **32.** $D = 0$: exactly one real solution

33. $D = 41$: two real solutions   **34.** $D = -36$: two nonreal solutions

1.5 Exercises

1. $-\dfrac{4}{3}$ **2.** 11 or -12 **3.** $-\dfrac{5}{2} \pm \dfrac{\sqrt{313}}{2}$ **4.** 1, 3, and 5 or 3, 5, and 7   **5.** 4 seconds   **6.** 2 seconds   **7.** $\dfrac{10}{3}$ cm by 9 cm

8. 8 in. by 15 in.   **9.** $\dfrac{7}{2}$ ft or 3.5 ft   **10.** 25 mph   **11.** 5 mph   **12.** 35 mph   **13.** 120 min   **14.** approximately 26 hr 10 min   **15.** approximately 10 hr 43 min

1.6 Exercises

SCE-1. $2x + 4$ **SCE-2.** $18 - 7x$ **SCE-3.** $4x^2 - 4x + 1$ **SCE-4.** $128x + 144$ **SCE-5.** $y + 13 + 6\sqrt{y + 4}$
SCE-6. $4x + 9 - 4\sqrt{4x + 5}$

1. $\{0, i, -i\}$ **2.** $\{-5, 0, 3\}$ **3.** $\{-3, -1, 1, 3\}$ **4.** $\left\{-\sqrt{3}i, \sqrt{3}i, 1\right\}$ **5.** $\left\{-2, -\dfrac{3}{2}, 2\right\}$ **6.** $\{-1, 1\}$

7. $\left\{-2, 2, -\sqrt{2}, \sqrt{2}\right\}$ **8.** $\left\{0, \dfrac{4}{13}\right\}$ **9.** $\left\{\dfrac{1}{8}, 8\right\}$ **10.** $\{-1, 2\}$ **11.** $\{256\}$ **12.** $\left\{-2, \dfrac{3}{2}\right\}$ **13.** $\left\{-2, \dfrac{1}{2}\right\}$

14. $\{5\}$ **15.** $\{4\}$ **16.** $\{4\}$ **17.** $\{-2\}$ **18.** $\{5\}$ **19.** $\{-1\}$ **20.** $\left\{-\dfrac{7}{2}\right\}$ **21.** $\{-2, 8\}$

22. $\left\{-2, 2, -\sqrt{2}, \sqrt{2}\right\}$ **23.** $\left\{0, \dfrac{4}{13}\right\}$ **24.** $\left\{\dfrac{1}{8}, 8\right\}$ **25.** $\{-1, 2\}$ **26.** $\{256\}$ **27.** $\left\{-2, \dfrac{3}{2}\right\}$ **28.** $\left\{-2, \dfrac{1}{2}\right\}$

1.7 Exercises

1. $\left\{x \mid x > -\dfrac{5}{2}\right\}$ **2.** $\left\{y \mid y < \dfrac{1}{5}\right\}$ **3.** $\{a \mid a \geq -4\}$ **4.** $\left\{w \mid w > \dfrac{30}{7}\right\}$ **5.** $[-1, \infty)$,

6. $(-\infty, -3]$,   **7.** $\left(-\infty, -\dfrac{12}{5}\right)$,

8. $\left(-\infty, \dfrac{10}{7}\right)$,   **9.** $\left[\dfrac{3}{2}, \dfrac{7}{2}\right]$,

10. $\left(-\dfrac{3}{2}, 2\right)$,   **11.** $(2, 11]$,

12. $(-7, 5)$,   **13.** $\left[-\dfrac{11}{6}, -\dfrac{1}{2}\right)$,

14. $[-10, -9]$,   **15.** $(1, 4]$ **16.** $(-\infty, -3] \cup (4, \infty)$ **17.** $[4, 5]$

18. no solution   **19.** $(-\infty, \infty)$ **20.** 5 hrs   **21.** more than 320 miles   **22.** 92 or higher   **23.** $7'1''$ or taller
24. 28 feet to 68 feet   **25.** 20 feet to 30 feet

1.8 Exercises

1. $\{0, -2\}$ **2.** $\left\{-3, \dfrac{2}{3}\right\}$ **3.** $\left\{-\dfrac{13}{9}, \dfrac{7}{9}\right\}$ **4.** $\left\{1 - \sqrt{2}, 1, 1 + \sqrt{2}\right\}$ **5.** $\left\{-\dfrac{3}{7}\right\}$ **6.** $\varnothing$ **7.** $(-1, 5)$

8. $\left[-\dfrac{1}{5}, \dfrac{3}{5}\right]$ **9.** $\left(-2, \dfrac{7}{2}\right)$ **10.** $\left[\dfrac{11}{5}, \dfrac{14}{5}\right]$ **11.** $\left\{\dfrac{7}{3}\right\}$ **12.** $\varnothing$ **13.** $(-\infty, -9) \cup (1, \infty)$ **14.** $\left(-\infty, -\dfrac{12}{17}\right] \cup [2, \infty)$

15. $\left(-\infty, \dfrac{9}{4}\right] \cup \left[\dfrac{15}{4}, \infty\right)$ **16.** $\left(-\infty, \dfrac{1}{5}\right] \cup [1, \infty)$ **17.** $\left(-\infty, \dfrac{5}{7}\right) \cup \left(\dfrac{5}{7}, \infty\right)$ **18.** $(-\infty, \infty)$

1.9 Exercises

1. $(-\infty, -3] \cup [1, \infty)$ **2.** $\left[-\dfrac{2}{3}, 0\right]$ **3.** $[-4, 1] \cup [3, \infty)$ **4.** $(-\infty, 0) \cup (2, \infty)$ **5.** $\left(-\dfrac{3}{2} - \dfrac{\sqrt{93}}{2}, -\dfrac{3}{2} + \dfrac{\sqrt{93}}{2}\right)$

6. $[-1, 1]$ **7.** $\varnothing$ **8.** $(-4, 0) \cup (3, \infty)$ **9.** $(-\infty, 1]$ **10.** $\{-1\} \cup [0, \infty)$ **11.** $[-3, 1)$ **12.** $(-3, 2]$

13. $(-\infty, 0) \cup (1, \infty)$ **14.** $(-\infty, -3] \cup (-2, 3]$ **15.** $(-\infty, -8) \cup [2, 3)$ **16.** $(-\infty, -2) \cup (-2, 3)$

17. $(-1, 1]$ **18.** $\left(\dfrac{3}{2}, 8\right)$ **19.** $(-4, -2]$ **20.** $(-\infty, -2] \cup (0, 1] \cup (2, \infty)$ **21.** $\left(-3, -\dfrac{1}{2}\right] \cup (2, \infty)$

22. $\left(1, \dfrac{3}{2}\right]$ **23.** $(-\infty, -3] \cup [1, \infty)$ **24.** $\left[-\dfrac{2}{3}, 0\right]$ **25.** $[-4, 1] \cup [3, \infty)$ **26.** $(-\infty, 0) \cup (2, \infty)$

27. $\left(-\dfrac{3}{2} - \dfrac{\sqrt{93}}{2}, -\dfrac{3}{2} + \dfrac{\sqrt{93}}{2}\right)$ **28.** $[-1, 1]$ **29.** $\varnothing$ **30.** $(-4, 0) \cup (3, \infty)$ **31.** $(-\infty, 1]$ **32.** $\{-1\} \cup [0, \infty)$

33. $[-3, 1)$ **34.** $(-3, 2]$ **35.** $(-\infty, 0) \cup (1, \infty)$ **36.** $(-\infty, -3] \cup (-2, 3]$ **37.** $(-\infty, -8) \cup [2, 3)$

38. $(-\infty, -2) \cup (-2, 3)$ **39.** $(-1, 1]$ **40.** $\left(\dfrac{3}{2}, 8\right)$ **41.** $(-4, -2]$ **42.** $(-\infty, -2] \cup (0, 1] \cup (2, \infty)$

43. $\left(-3, -\dfrac{1}{2}\right] \cup (2, \infty)$ **44.** $\left(1, \dfrac{3}{2}\right]$

CHAPTER 2

2.1 Exercises

SCE-1. $\dfrac{5}{2}$ **SCE-2.** $-\dfrac{5}{6}$ **SCE-3.** $\dfrac{1}{70}$ **SCE-4.** $6\sqrt{3}$ **SCE-5.** 25 **SCE-6.** $3\sqrt{10}$ **SCE-7.** 5

1. 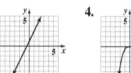   **a.** II   **b.** III   **c.** y-axis   **d.** I   **e.** x-axis   **2.** 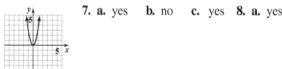   **a.** IV   **b.** II   **c.** I   **d.** y-axis

e. origin   **3.** 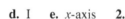   **4.**   **5.** 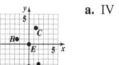   **6.**   **7. a.** yes   **b.** no   **c.** yes   **8. a.** yes

b. no   **c.** no   **9. a.** yes   **b.** yes   **c.** yes   **10.** $\left(-\dfrac{3}{2}, 1\right)$ **11.** $(3, -2)$ **12.** $\left(-\dfrac{3}{2}, \dfrac{3}{4}\right)$ **13.** $\left(\dfrac{a + c}{2}, \dfrac{b + d}{2}\right)$

14. $(-5, -2)$ **15.** yes   **16.** no   **17.** yes   **18.** 5 **19.** $\sqrt{74} \approx 8.602$ **20.** $5\sqrt{5} \approx 11.180$ **21.** $\sqrt{5} \approx 2.236$

22. yes   **23.** no   **24.** yes   **25.** 0 or 6

2.2 Exercises

SCE-1. $25, (x + 5)^2$ **SCE-2.** $\dfrac{25}{4}, \left(x - \dfrac{5}{2}\right)^2$ **SCE-3.** $16, (y - 4)^2$ **SCE-4.** $\dfrac{81}{4}, \left(y - \dfrac{9}{2}\right)^2$ **SCE-5.** $\dfrac{25}{36}, \left(x + \dfrac{5}{6}\right)^2$

SCE-6. 25 and 16 **SCE-7.** $\dfrac{25}{4}$ and 9 **SCE-8.** $\dfrac{49}{4}$ and $\dfrac{121}{4}$ **SCE-9.** $\dfrac{1}{9}$ and $\dfrac{25}{196}$

1. $x^2 + y^2 = 1$ **2.** $(x + 2)^2 + (y - 3)^2 = 16$ **3.** $(x - 1)^2 + (y + 4)^2 = 9$ **4.** $\left(x + \dfrac{1}{4}\right)^2 + \left(y + \dfrac{1}{3}\right)^2 = 4$

5. $x^2 + (y - 2)^2 = 25$ **6.** $(x + 4)^2 + (y - 7)^2 = 72$ **7.** $(x - 3)^2 + (y + 1)^2 = 13$

8. $(x - 4)^2 + \left(y + \dfrac{5}{2}\right)^2 = \dfrac{65}{4}$ **9.** $(x - 2)^2 + (y + 3)^2 = 9$ **10.** $(x + 4)^2 + (y - 1)^2 = 16$

11. $C(0,0), r = 1, x\text{-int} = \{\pm 1\}, y\text{-int} = \{\pm 1\},$ **12.** $C(0, 2), r = 2, x\text{-int} = \{0\}, y\text{-int} = \{0, 4\},$

13. $C(1, -5), r = 4, \text{no } x\text{-int}, y\text{-int} = \{-5 \pm \sqrt{15}\},$ **14.** $C(-2, -4), r = 6, x\text{-int} = \{-2 \pm 2\sqrt{5}\},$

$y\text{-int} = \{-4 \pm 4\sqrt{2}\},$ **15.** $C(4, -7), r = 2\sqrt{3}, \text{no } x\text{-int}, \text{no } y\text{-int},$ **16.** $C(-1, 3), r = 2\sqrt{5},$

$x\text{-int} = \{-1 \pm \sqrt{11}\}, y\text{-int} = \{3 \pm \sqrt{19}\},$ **17.** $C\left(\dfrac{1}{4}, -\dfrac{1}{2}\right), r = 2, x\text{-int} = \left\{\dfrac{1}{4} \pm \dfrac{\sqrt{15}}{2}\right\},$

$y\text{-int} = \left\{-\dfrac{1}{2} \pm \dfrac{3\sqrt{7}}{4}\right\},$ **18.** $C(-1, 2), r = 2, x\text{-int} = \{-1\}, y\text{-int} = \{2 \pm \sqrt{3}\},$

19. $C(5, -3), r = 4, x\text{-int} = \{5 \pm \sqrt{7}\}, \text{no } y\text{-int},$ **20.** $C(2, 4), r = 1, \text{no } x\text{-int}, \text{no } y\text{-int},$

21. $C(0, -1), r = 3, x\text{-int} = \{\pm 2\sqrt{2}\}, y\text{-int} = \{-4, 2\},$ **22.** $C(3, 6), r = 5\sqrt{2}, x\text{-int} = \{3 \pm \sqrt{14}\},$

$y\text{-int} = \{6 \pm \sqrt{41}\},$ 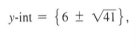   **23.** $C\left(\dfrac{3}{2}, \dfrac{1}{2}\right), r = \sqrt{3}, x\text{-int} = \left\{\dfrac{3}{2} \pm \dfrac{\sqrt{11}}{2}\right\}, y\text{-int} = \left\{\dfrac{1}{2} \pm \dfrac{\sqrt{3}}{2}\right\},$

24. $C\left(-\dfrac{1}{3}, \dfrac{1}{4}\right), r = \dfrac{3}{4}, x\text{-int} = \left\{-\dfrac{1}{3} \pm \dfrac{\sqrt{2}}{2}\right\}, y\text{-int} = \left\{\dfrac{1}{4} \pm \dfrac{\sqrt{65}}{12}\right\},$ **25.** $C(1, -2), r = 2, x\text{-int} = \{1\},$

$y\text{-int} = \{-2 \pm \sqrt{3}\},$ **26.** $C\left(\dfrac{1}{2}, -\dfrac{1}{4}\right), r = 1, x\text{-int} = \left\{\dfrac{1}{2} \pm \dfrac{\sqrt{15}}{4}\right\}, y\text{-int} = \left\{-\dfrac{1}{4} \pm \dfrac{\sqrt{3}}{2}\right\},$

27. $C\left(\dfrac{1}{4}, \dfrac{1}{3}\right), r = 2$, x-int $= \left\{\dfrac{1}{4} \pm \dfrac{\sqrt{35}}{3}\right\}$, y-int $= \left\{\dfrac{1}{3} \pm \dfrac{3\sqrt{7}}{4}\right\}$, **28.** $C\left(-\dfrac{1}{6}, -1\right), r = \sqrt{2}$,

x-int $= \left\{-\dfrac{7}{6}, \dfrac{5}{6}\right\}$, y-int $= \left\{-1 \pm \dfrac{\sqrt{71}}{6}\right\}$, **29.** $C(0, 0), r = 1$ **30.** $C(0, 2), r = 2$

31. $C(1, -5), r = 4$ **32.** $C(-2, -4), r = 6$ **33.** $C(4, -7), r = 2\sqrt{3}$ **34.** $C(-1, 3), r = 2\sqrt{5}$

35. $C\left(\dfrac{1}{4}, -\dfrac{1}{2}\right), r = 2$ **36.** x-int $= \{\pm 1\}$, y-int $= \{\pm 1\}$ **37.** x-int $= \{0\}$, y-int $= \{0, 4\}$

38. no x-int, y-int $= \left\{-5 \pm \sqrt{15}\right\}$ **39.** x-int $= \left\{-2 \pm 2\sqrt{5}\right\}$, y-int $= \left\{-4 \pm 4\sqrt{2}\right\}$ **40.** no x-int, no y-int

41. x-int $= \left\{-1 \pm \sqrt{11}\right\}$, y-int $= \left\{3 \pm \sqrt{19}\right\}$ **42.** x-int $= \left\{\dfrac{1}{4} \pm \dfrac{\sqrt{15}}{2}\right\}$, y-int $= \left\{-\dfrac{1}{2} \pm \dfrac{3\sqrt{7}}{4}\right\}$

43. 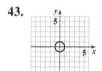 **44.** **45.** **46.** **47.** **48.**

49. **50.** $C(-1, 2), r = 2$ **51.** $C(5, -3), r = 4$ **52.** $C(2, 4), r = 1$ **53.** $C(0, -1), r = 3$

54. $C(3, 6), r = 5\sqrt{2}$ **55.** $C\left(\dfrac{3}{2}, \dfrac{1}{2}\right), r = \sqrt{3}$ **56.** $C\left(-\dfrac{1}{3}, \dfrac{1}{4}\right), r = \dfrac{3}{4}$ **57.** $C(1, -2), r = 2$ **58.** $C\left(\dfrac{1}{2}, -\dfrac{1}{4}\right), r = 1$

59. $C\left(\dfrac{1}{4}, \dfrac{1}{3}\right), r = 2$ **60.** $C\left(-\dfrac{1}{6}, -1\right), r = \sqrt{2}$ **61.** x-int $= \{-1\}$, y-int $= \left\{2 \pm \sqrt{3}\right\}$ **62.** x-int $= \left\{5 \pm \sqrt{7}\right\}$,

no y-int **63.** no x-int, no y-int **64.** x-int $= \left\{\pm 2\sqrt{2}\right\}$, y-int $= \{-4, 2\}$ **65.** x-int $= \left\{3 \pm \sqrt{14}\right\}$, y-int $= \left\{6 \pm \sqrt{41}\right\}$

66. x-int $= \left\{\dfrac{3}{2} \pm \dfrac{\sqrt{11}}{2}\right\}$, y-int $= \left\{\dfrac{1}{2} \pm \dfrac{\sqrt{3}}{2}\right\}$ **67.** x-int $= \left\{-\dfrac{1}{3} \pm \dfrac{\sqrt{2}}{2}\right\}$, y-int $= \left\{\dfrac{1}{4} \pm \dfrac{\sqrt{65}}{12}\right\}$

68. x-int $= \{1\}$, y-int $= \left\{-2 \pm \sqrt{3}\right\}$ **69.** x-int $= \left\{\dfrac{1}{2} \pm \dfrac{\sqrt{15}}{4}\right\}$, y-int $= \left\{-\dfrac{1}{4} \pm \dfrac{\sqrt{3}}{2}\right\}$ **70.** x-int $= \left\{\dfrac{1}{4} \pm \dfrac{\sqrt{35}}{3}\right\}$,

y-int $= \left\{\dfrac{1}{3} \pm \dfrac{3\sqrt{7}}{4}\right\}$ **71.** x-int $= \left\{-\dfrac{7}{6}, \dfrac{5}{6}\right\}$, y-int $= \left\{-1 \pm \dfrac{\sqrt{71}}{6}\right\}$ **72.** **73.**

74. **75.** **76.** 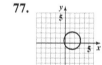 **77.** 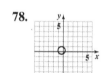 **78.** **79.**

80. 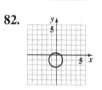 **81.** **82.**

2.3 Exercises

SCE-1. $y = 2x - 5$ **SCE-2.** $y = -\dfrac{3}{2}x + \dfrac{7}{2}$ **SCE-3.** $y = 2x + 7$ **SCE-4.** $y = -5x + 7$ **SCE-5.** $y = \dfrac{2}{7}x + \dfrac{41}{7}$

SCE-6. $y = -\dfrac{3}{5}x - \dfrac{17}{5}$

1. $-\dfrac{8}{7}$ **2.** $\dfrac{1}{2}$ **3.** $\dfrac{1}{2}$ **4.** $-\dfrac{1}{3}$ **5.** **6.** **7.** **8.**

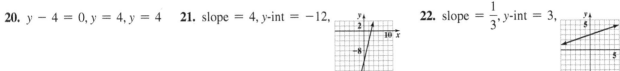

9. $y + 2 = \dfrac{1}{2}(x + 2)$ **10.** $y - 8 = -\dfrac{4}{3}(x - 3)$ **11.** $y - 3 = 3(x + 5)$ **12.** $y - 88 = \dfrac{5}{11}(x + 99)$

13. $y = x - 2$ **14.** $y = -\dfrac{1}{6}x + \dfrac{1}{2}$ **15.** $y = \dfrac{7}{9}x - \dfrac{8}{7}$ **16.** $y = 5$ **17.** $y - 2 = -1(x - 1), y = -x + 3, x + y = 3$

18. $y - 7 = -\dfrac{3}{2}(x + 5), y = -\dfrac{3}{2}x - \dfrac{1}{2}, 3x + 2y = -1$ **19.** $y - 1 = \dfrac{2}{15}\left(x + \dfrac{1}{2}\right), y = \dfrac{2}{15}x + \dfrac{16}{15}, 2x - 15y = -16$

20. $y - 4 = 0, y = 4, y = 4$ **21.** slope $= 4$, y-int $= -12$, **22.** slope $= \dfrac{1}{3}$, y-int $= 3$,

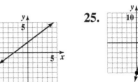

23. slope $= -\dfrac{5}{7}$, y-int $= \dfrac{12}{7}$, **24.** slope $= \dfrac{8}{11}$, y-int $= \dfrac{23}{11}$, **25.**

26. **27.** **28.** **29.** $y = -2$ **30.** $x = 5$ **31.** $x = 4$ **32.** $y = -3$

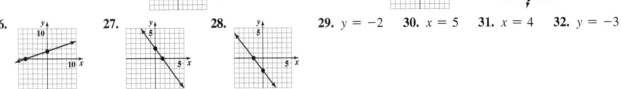

33. $y - 2 = -1(x - 1)$ **34.** $y - 7 = -\dfrac{3}{2}(x + 5)$ **35.** $y - 1 = \dfrac{2}{15}\left(x + \dfrac{1}{2}\right)$ **36.** $y - 4 = 0$ **37.** $y = -x + 3$

38. $y = -\dfrac{3}{2}x - \dfrac{1}{2}$ **39.** $y = \dfrac{2}{15}x + \dfrac{16}{15}$ **40.** $y = 4$ **41.** $x + y = 3$ **42.** $3x + 2y = -1$ **43.** $2x - 15y = -16$

44. $y = 4$ **45.** slope $= 4$, y-int $= -12$ **46.** slope $= \dfrac{1}{3}$, y-int $= 3$ **47.** slope $= -\dfrac{5}{7}$, y-int $= \dfrac{12}{7}$

48. slope $= \dfrac{8}{11}$, y-int $= \dfrac{23}{11}$ **49.** **50.** **51.** **52.**

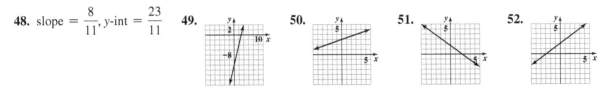

2.4 Exercises

SCE-1. $y = 2x - 5$ **SCE-2.** $y = -\dfrac{3}{2}x + \dfrac{7}{2}$ **SCE-3.** $y = 2x + 7$ **SCE-4.** $y = -5x + 7$ **SCE-5.** $y = \dfrac{2}{7}x + \dfrac{41}{7}$

SCE-6. $y = -\dfrac{3}{5}x - \dfrac{17}{5}$

1. perpendicular **2.** parallel **3.** parallel **4.** perpendicular **5.** neither **6.** neither **7.** $y - 3 = \dfrac{1}{4}(x + 2)$,

$y = \dfrac{1}{4}x + \dfrac{7}{2}, x - 4y = -14$ **8.** $y + 4 = \dfrac{3}{5}(x - 1), y = \dfrac{3}{5}x - \dfrac{23}{5}, 3x - 5y = 23$ **9.** $y + 5 = 3(x + 2), y = 3x + 1$,

$3x - y = -1$ **10.** $x = 11$ **11.** $y = -3$ **12.** $y - 3 = -4(x + 2), y = -4x - 5, 4x + y = -5$

13. $y + 4 = -\frac{5}{3}(x - 1)$, $y = -\frac{5}{3}x - \frac{7}{3}$, $5x + 3y = -7$ **14.** $y + 5 = -\frac{1}{3}(x + 2)$, $y = -\frac{1}{3}x - \frac{17}{3}$, $x + 3y = -17$

15. $y = -3$ **16.** $x = 11$ **17.** parallelogram and rhombus   **18.** parallelogram   **19.** neither

20. parallelogram and rhombus   **21.** $y - 3 = \frac{1}{4}(x + 2)$ **22.** $y + 4 = \frac{3}{5}(x - 1)$ **23.** $y + 5 = 3(x + 2)$

24. $y = \frac{1}{4}x + \frac{7}{2}$ **25.** $y = \frac{3}{5}x - \frac{23}{5}$ **26.** $y = 3x + 1$ **27.** $x - 4y = -14$ **28.** $3x - 5y = 23$ **29.** $3x - y = -1$

30. $y - 3 = -4(x + 2)$ **31.** $y + 4 = -\frac{5}{3}(x - 1)$ **32.** $y + 5 = -\frac{1}{3}(x + 2)$ **33.** $y = -4x - 5$

34. $y = -\frac{5}{3}x - \frac{7}{3}$ **35.** $y = -\frac{1}{3}x - \frac{17}{3}$ **36.** $4x + y = -5$ **37.** $5x + 3y = -7$ **38.** $x + 3y = -17$

CHAPTER 3

3.1 Exercises

SCE-1. $\left(\infty, \frac{2}{3}\right]$ **SCE-2.** $(\infty, -7] \cup [3, \infty)$ **SCE-3.** $[-4, 1] \cup [3, \infty)$ **SCE-4.** $(\infty, -3] \cup (1, \infty)$

SCE-5. $(-7, -3] \cup [3, \infty)$ **SCE-6.** $(-8, 2] \cup (3, \infty)$ **SCE-7.** $-18xh - 9h^2$ **SCE-8.** $\sqrt{x + h} - 2h - \sqrt{x}$

1. domain $= \{$Tim, Audrie, Sheryl, Jason$\}$, range $= \{$Blue, Green, Brown$\}$, function

2. domain $= \{$Ronald, William, George$\}$, range $= \{$Monica, Laura, Hillary, Nancy$\}$, not a function

3. domain $= \{-1, 0, 1, 2\}$, range $= \{2, 3, 5, 7\}$, not a function   **4.** domain $= \{-1, 0, 1, 2, 3\}$, range $= \{2, 3, 5, 7\}$, function

5. domain $= \{-3, -1, 1, 3\}$, range $= \{2\}$, function   **6.** domain $= \{-3, -1, 1, 3\}$, range $= \{-2, 2\}$, not a function

7. function   **8.** not a function   **9.** function   **10.** function   **11.** function   **12.** not a function

13. a. -13 **b.** $-\frac{5}{2}$ **c.** $3x - 19$ **14. a.** 5 **b.** 5 **c.** $x^2 - 2x + 2$ **15. a.** 2 **b.** $\frac{7}{8}$ **c.** $\frac{a}{a + 1}$ **16. a.** 36π

b. 288π **c.** $\frac{4}{3}\pi(r + 1)^3$ **17. a.** $2x - 3$ **b.** $2x + 2h - 5$ **18. a.** $22 - 7x$ **b.** $8 - 7x - 7h$ **19. a.** $x^2 - 1$

b. $x^2 + 2hx - 2x + h^2 - 2h$ **20. a.** $t^3 + 3t^2 + 2t$ **b.** $t^3 + 3t^2h + 3th^2 + h^3 - t - h$ **21. a.** -6

b. $\sqrt{x + h} - 2x - 2h$ **22.** 2 **23.** -7 **24.** $2x + h - 2$ **25.** $-18x - 9h$ **26.** $10x + 5h$

27. $3x^2 + 3xh + h^2 - 1$ **28.** $\frac{\sqrt{x + h} - 2h - \sqrt{x}}{h}$ **29.** function   **30.** function   **31.** function   **32.** not a function

33. function   **34.** function   **35.** not a function   **36.** not a function   **37.** function   **38.** polynomial function, $(-\infty, \infty)$

39. rational function, $(-\infty, 0) \cup (0, \infty)$ **40.** polynomial function, $(-\infty, \infty)$ **41.** root function, $(-\infty, 3]$

42. root function, $(-\infty, \infty)$ **43.** rational function, $\left(-\infty, -\frac{1}{2}\right) \cup \left(-\frac{1}{2}, \infty\right)$ **44.** root function, $(-\infty, 5) \cup (5, \infty)$

45. polynomial function, $(-\infty, \infty)$ **46.** rational function, $(-\infty, -7) \cup (-7, 6) \cup (6, \infty)$ **47.** polynomial function, $(-\infty, \infty)$

48. root function, $(-\infty, \infty)$ **49.** root function, $(-\infty, -1] \cup [1, \infty)$ **50.** rational function, $(-\infty-1) \cup (-1, 0) \cup$

$(0, 4) \cup (4, \infty)$ **51.** root function, $(-\infty, -5] \cup [3, \infty)$ **52.** polynomial function, $(-\infty, \infty)$

53. root function, $(-\infty, -3] \cup [3, \infty)$ **54.** root function, $(-\infty, 3) \cup (3, \infty)$ **55.** rational function, $(-\infty, \infty)$

56. root function, $(-4, 3] \cup (5, \infty)$ **57.** 2 **58.** -7 **59.** $2x + h - 2$ **60.** $-18x - 9h$ **61.** $10x + 5h$

62. $3x^2 + 3xh + h^2 - 1$ **63.** $\frac{\sqrt{x + h} - 2h - \sqrt{x}}{h}$

3.2 Exercises

SCE-1. $f(-x) = 3x^5 + 4x^4 - 3x^3 - 7x^2 - 21$ **SCE-2.** $f(-x) = x^4 - 7x^2 - 11$ **SCE-3.** $f(-x) = 3x^7 - 8x^3 + x$

SCE-4. $f(-x) = -\frac{2x^2 - 4}{x}$

1. x-int $= \frac{1}{2}$, y-int $= -1$ **2.** x-int $=$ none, y-int $= 3$ **3.** x-int $= \{-2, 3\}$, y-int $= -6$

4. x-int $= \left\{\frac{7}{4} - \frac{\sqrt{73}}{4}, \frac{7}{4} + \frac{\sqrt{73}}{4}\right\}$, y-int $= -3$ **5.** x-int $= \{-2, -1, 1\}$, y-int $= -2$ **6.** x-int $= \left\{-6, 0, \frac{1}{2}\right\}$, y-int $= 0$

7. domain $= [-5, 3]$, range $= [0, 4]$ **8.** domain $= (-4, 5]$, range $= (-5, 5]$ **9.** domain $= \left(-\frac{\pi}{2}, \frac{\pi}{2}\right)$,

range $= (-\infty, \infty)$ **10.** domain $= [-4, \infty)$, range $= (-\infty, 0]$ **11.** domain $= [-4, \infty)$, range $= (2, 5]$
12. domain $= (-\infty, -1) \cup (-1, 1) \cup (1, \infty)$, range $= (-\infty, 0] \cup (1, \infty)$ **13. a.** $(-\infty, -4) \cup (2, 5)$ **b.** $(-4, 2)$
c. none   **14. a.** $(0, 2)$ **b.** $(-4, 0) \cup (2, 4)$ **c.** $(4, \infty)$ **15. a.** $(-3, 0)$ **b.** $(-\infty, -3) \cup (4, \infty)$ **c.** $(0, 4)$
16. a. none   **b.** none   **c.** $(-1, 0) \cup (0, 1) \cup (1, 2) \cup (2, 3)$ **17. a.** $x = 6$ **b.** $f(6) = 0$ **c.** $x = 2$ **d.** $f(2) = 6$
18. a. none   **b.** none   **c.** $x = 0$ **d.** $f(0) = 3$ **19. a.** $x = -2$ **b.** $f(-2) = -1$ **c.** $x = 2$ **d.** $f(2) = 2$
20. a. none   **b.** none   **c.** none   **d.** none   **21. a.** $x = 0$ **b.** $f(0) = 0$ **c.** none   **d.** none   **22.** odd
23. neither   **24.** even   **25.** odd   **26.** even   **27.** neither   **28.** odd   **29.** odd   **30.** even   **31. a.** $(-8, 9]$
b. $[-5, 3]$ **c.** increasing: $(-8, -6) \cup (-2, 1) \cup (4, 7)$, decreasing: $(-6, -2) \cup (7, 9)$, constant: $(1, 4)$
d. $x = -2$, $f(-2) = -5$ **e.** $x = -6$ and $x = 7$, $f(-6) = 3$ and $f(7) = 2$ **f.** $x = \left\{ -\dfrac{15}{2}, -4, 5, 9 \right\}$ **g.** -3
h. $\left(-8, -\dfrac{15}{2} \right] \cup [-4, 5] \cup \{9\}$ **i.** two   **j.** -1 **32. a.** $(-\infty, \infty)$ **b.** $[-1, 7]$
c. increasing: $(-5, -4) \cup (0, 4) \cup (5, 6)$, decreasing: $(-6, -5) \cup (-4, 0) \cup (4, 5)$, constant $(-\infty, -6) \cup (6, \infty)$
d. $x = -5$, $x = 0$, and $x = 5$, $f(-5) = 5$, $f(0) = -1$, and $f(5) = 5$ **e.** $x = -4$ and $x = 4$, $f(-4) = 6$ and $f(4) = 6$
f. $x = \{-1, 1\}$ **g.** even   **h.** $(-\infty, -1) \cup (1, \infty)$ **i.** true   **j.** $(-\infty, -6] \cup [6, \infty)$ **33.** domain $= (-8, 9]$,
range $= [-5, 3]$ **34.** increasing: $(-8, -6) \cup (-2, 1) \cup (4, 7)$, decreasing: $(-6, -2) \cup (7, 9)$, constant: $(1, 4)$
35. a. $x = -2$, $f(-2) = -5$ **b.** $x = -6$ and $x = 7$, $f(-6) = 3$ and $f(7) = 2$ **36.** -3
37. a. $\left(-8, -\dfrac{15}{2} \right] \cup [-4, 5] \cup \{9\}$ **38. a.** one   **b.** -1 **39.** domain $= (-\infty, \infty)$, range $= [-1, 7]$
40. increasing: $(-5, -4) \cup (0, 4) \cup (5, 6)$, decreasing: $(-6, -5) \cup (-4, 0) \cup (4, 5)$, constant: $(-\infty, -6) \cup (6, \infty)$
41. a. $x = -5$, $x = 0$, and $x = 5$, $f(-5) = 5$, $f(0) = -1$, and $f(5) = 5$ **b.** $x = -4$ and $x = 4$, $f(-4) = 6$ and $f(4) = 6$
42. $x = \{-1, 1\}$ **43. a.** $(-\infty, -) \cup (1, \infty)$ **b.** false   **c.** $(-\infty, -6] \cup [6, \infty)$ **44.** even

3.3 Exercises

1. a, e,   **2.** b, c, e, f,   **3.** c, d, e, f,   **4.** a, e,   **5.** a, d, e,

6. b, c, d, e, f,   **7.** b, c, e, f,   **8.** none,   **9.** b,   **10. a.** $f(-2) = 1, f(0) = 1, f(2) = -3$

b.

c. $(-\infty, \infty)$

d. $\{-3, 1\}$

11. a. $f(-2) = -5, f(0) = -1, f(2) = 4$
b.

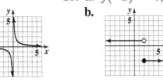

c. $(-\infty, \infty)$

d. $(-\infty, 1] \cup (3, \infty)$

12. a. $f(-2) = 4, f(0) = 0, f(2) = \dfrac{1}{2}$
b.

c. $(-\infty, \infty)$

d. $[0, \infty)$

13. a. $f(-2) = 3, f(0) = 0, f(2) = 1$
b.

c. $(-\infty, \infty)$
d. $[0, 2) \cup \{3\}$

14. a. $f(-2) = 4, f(0) = 2, f(2) = \sqrt{2}$
b.

c. $(-\infty, \infty)$
d. $(0, \infty)$

15. a. $f(-2) = -2, f(0) = 2, f(2) = 8$

b.

c. $[-3, \infty)$

d. $\{-3\} \cup \{-2\} \cup \{-1\} \cup [1, \infty)$

19. $f(x) = \begin{cases} 2x + 3 \text{ if } -3 \le x < -1 \\ x^2 \text{ if } -1 < x \le 2 \end{cases}$

21. a. NOK 1036

b. $T(x) = \begin{cases} 0 \text{ if } 0 \le x \le 26,300 \\ .28(x - 26,300) \text{ if } x > 26,300 \end{cases}$

c.

23. a. $\$1.47$

b. $P(x) = \begin{cases} 1.13 \text{ if } 0 < x \le 1 \\ 1.3 \text{ if } 1 < x \le 2 \\ 1.47 \text{ if } 2 < x \le 3 \\ 1.64 \text{ if } 3 < x \le 4 \\ 1.81 \text{ if } 4 < x \le 5 \end{cases}$

c.

25. a. $f(-2) = 1, f(0) = 1, f(2) = -3$

b.

27. a. $f(-2) = 4, f(0) = 0, f(2) = \dfrac{1}{2}$

b.

29. a. $f(-2) = 4, f(0) = 2, f(2) = \sqrt{2}$

b.

16. $f(x) = \begin{cases} 1 \text{ if } x < 0 \\ -1 \text{ if } x \ge 0 \end{cases}$

17. $f(x) = \begin{cases} x \text{ if } -4 \le x \le 2 \\ 3 \text{ if } x > 2 \end{cases}$

18. $f(x) = \begin{cases} |x| \text{ if } x < 1 \\ -x + 1 \text{ if } 1 \le x \le 4 \end{cases}$

20. $f(x) = \begin{cases} \dfrac{1}{x} \text{ if } x \le -\dfrac{1}{2} \\ \sqrt{x} \text{ if } 0 \le x < 4 \end{cases}$

22. a. $\$29.10$

b. $\$36.50$

c. $C(x) = \begin{cases} 9.5 + .07x \text{ if } 0 \le x \le 350 \\ 34 + .05(x - 350) \text{ if } x > 350 \end{cases}$

d.

24. a. $\$25.15$

b. $P(x) = \begin{cases} 25 \text{ if } x \le 500 \\ 25.05 \text{ if } 500 < x \le 501 \\ 25.1 \text{ if } 501 < x \le 502 \\ 25.15 \text{ if } 502 < x \le 503 \\ 25.2 \text{ if } 503 < x \le 504 \\ 25.25 \text{ if } 504 < x \le 505 \end{cases}$

c.
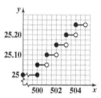

26. a. $f(-2) = -5, f(0) = -1, f(2) = 4$

b.

28. a. $f(-2) = 3, f(0) = 0, f(2) = 1$

b.

30. a. $f(-2) = -2, f(0) = 2, f(2) = 8$

b.

3.4 Exercises

1.

2.

3.

4.

5.

6.

7. (−1, 0) (1, 0) (−2, −1) (0, −1)

8. (−2, 4) (1, 3) (3, 1)

9.

10.

11.

12.

13.

14.

15. (1, 1) (3, 1) (2, 0)

16. (−4, 2) (−2, 0) (1, −1)

17.

18.

19.

20.

21.

22.

23. (1, 0) (3, 0) (0, −1)

24. (−3, 4) (0, 3) (2, 1)

25.

26.

27.

28.

29.

30.

31.

32.

33.

34.

35.

36.

37. (−2, 0) (−1, −1) (1, −1)

38. (−1, 1) (0, 0) (2, 0)

39.

40.

41.

42.

43.

44.

45.

46.

47.

48.

49.

50.

51. (−1, 3) (1, 3) (−2, 0)

52. $\left(-1, \frac{1}{2}\right)$ $\left(1, \frac{1}{2}\right)$ (−2, 0)

53.

54.

55.

56. $\left(-\frac{1}{2}, 1\right)$ $\left(\frac{1}{2}, 1\right)$ (−1, 0)

57. (−2, 1) (2, 1) (−4, 0)

58.

59.

60.

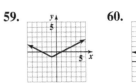

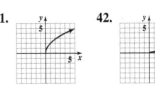

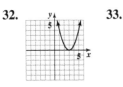

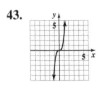

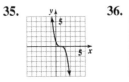
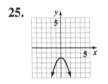

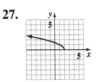
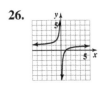
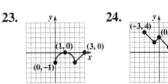
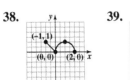
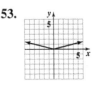
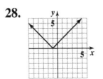

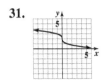
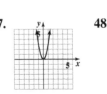

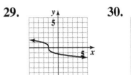
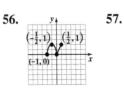
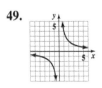
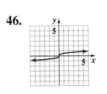

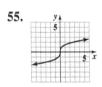

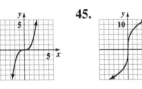
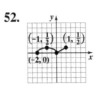
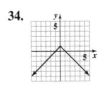

61. **62.** 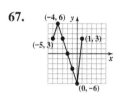 **63.** **64.** 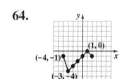 **65.** **66.**

67. **68.**

3.5 Exercises

SEC-1. $\dfrac{(2x + 1)(5x - 2)}{5x(x - 9)}$ **SEC-2.** $\dfrac{x(x - 5)}{(x - 2)(x + 2)}$ **SEC-3.** $\dfrac{(x - 7)^2(x + 7)}{(x - 3)(x + 2)(x + 8)}$ **SEC-4.** $\dfrac{7x + 29}{x + 3}$

SEC-5. $-\dfrac{2(3x + 4)}{x + 6}$ **SEC-6.** $\dfrac{3(x + 8)}{8x + 67}$

1. 9 **2.** −59 **3.** 1 **4.** $-\dfrac{1}{2}$ **5.** $\dfrac{4}{33}$ **6.** $-\dfrac{153}{11}$ **7.** undefined **8.** $7\sqrt{3}$ **9.** 186 **10.** $\dfrac{1}{36}$ **11. a.** 0

b. 8 **c.** −10 **d.** −1 **12. a.** −5 **b.** 0 **c.** 12 **d.** undefined **13. a.** 5 **b.** 1 **c.** 9 **d.** $\dfrac{2}{3}$ **14. a.** 4

b. 3 **c.** 4 **d.** $-\dfrac{1}{2}$ **15.** $(-\infty, 3)$, [number line: $-5\ -4\ -3\ -2\ -1\ 0\ 1\ 2\ 3\ 4\ 5$] **16.** $(-8, 3)$, [number line: $-10\ -8\ -6\ -4\ -2\ 0\ 2\ 4\ 6\ 8\ 10$]

17. $[0, 4]$, [number line: $-5\ -4\ -3\ -2\ -1\ 0\ 1\ 2\ 3\ 4\ 5$] **18.** $(-\infty, 1) \cup (1, \infty)$, [number line: $-5\ -4\ -3\ -2\ -1\ 0\ 1\ 2\ 3\ 4\ 5$]

19. $[-3, 0) \cup (0, \infty)$, [number line: $-5\ -4\ -3\ -2\ -1\ 0\ 1\ 2\ 3\ 4\ 5$] **20.** $(-5, 1]$, [number line: $-5\ -4\ -3\ -2\ -1\ 0\ 1\ 2\ 3\ 4\ 5$]

21. a. $x^2 + x + 1, (-\infty, \infty)$ **b.** $x^2 - x + 3, (-\infty, \infty)$ **c.** $x^3 - x^2 + 2x - 2, (-\infty, \infty)$

d. $\dfrac{x^2 + 2}{x - 1}, (-\infty, 1) \cup (1, \infty)$ **22. a.** $x^2 + x - 9, (-\infty, \infty)$ **b.** $-x^2 + x - 1, (-\infty, \infty)$

c. $x^3 - 5x^2 - 4x + 20, (-\infty, \infty)$ **d.** $\dfrac{x - 5}{x^2 - 4}, (-\infty, -2) \cup (-2, 2) \cup (2, \infty)$ **23. a.** $x^2 - 5x - 11, (-\infty, \infty)$

b. $-x^2 + 7x + 21, (-\infty, \infty)$ **c.** $x^3 - x^2 - 46x - 80, (-\infty, \infty)$ **d.** $\dfrac{x + 5}{x^2 - 6x - 16}, (-\infty, -2) \cup (-2, 8) \cup (8, \infty)$

24. a. $\sqrt{x} + x + 6, [0, \infty)$ **b.** $\sqrt{x} - x - 6, [0, \infty)$ **c.** $x\sqrt{x} + 6x, [0, \infty)$ **d.** $\dfrac{\sqrt{x}}{x + 6}, [0, \infty)$

25. a. $\dfrac{1}{x} + \sqrt{x - 1}, [1, \infty)$ **b.** $\dfrac{1}{x} - \sqrt{x - 1}, [1, \infty)$ **c.** $\dfrac{\sqrt{x - 1}}{x}, [1, \infty)$ **d.** $\dfrac{1}{x\sqrt{x - 1}}, (1, \infty)$

26. a. $\dfrac{7x^2 + 18x + 11}{3x^2 + 14x + 15}, (-\infty, -3) \cup \left(-3, -\dfrac{5}{3}\right) \cup \left(-\dfrac{5}{3}, \infty\right)$ **b.** $\dfrac{5x^2 + 8x - 1}{3x^2 + 14x + 15}, (-\infty, -3) \cup \left(-3, -\dfrac{5}{3}\right) \cup \left(-\dfrac{5}{3}, \infty\right)$

c. $\dfrac{2x^2 + 5x + 2}{3x^2 + 14x + 15}, (-\infty, -3) \cup \left(-3, -\dfrac{5}{3}\right) \cup \left(-\dfrac{5}{3}, \infty\right)$ **d.** $\dfrac{6x^2 + 13x + 5}{x^2 + 5x + 6}, (-\infty, -3) \cup (-3, -2) \cup \left(-2, -\dfrac{5}{3}\right) \cup$

$\left(-\dfrac{5}{3}, \infty\right)$ **27. a.** $\dfrac{2x^2 - 3x - 4}{x^3 - 3x^2 - 4x + 12}, (-\infty, -2) \cup (-2, 2) \cup (2, 3) \cup (3, \infty)$ **b.** $\dfrac{4 - 3x}{x^3 - 3x^2 - 4x + 12}, (-\infty, -2) \cup$

$(-2, 2) \cup (2, 3) \cup (3, \infty)$ **c.** $\dfrac{x}{x^3 - 3x^2 - 4x + 12}, (-\infty, -2) \cup (-2, 2) \cup (2, 3) \cup (3, \infty)$ **d.** $\dfrac{x^2 - 3x}{x^2 - 4}, (-\infty, -2) \cup$

$(-2, 2) \cup (2, 3) \cup (3, \infty)$ **28. a.** $\dfrac{2x^3 - 27x + 34}{x^4 - x^3 - 22x^2 + 16x + 96}, (-\infty, -4) \cup (-4, -2) \cup (-2, 3) \cup (3, 4) \cup (4, \infty)$

b. $\dfrac{-8x^2 - 5x + 94}{x^4 - x^3 - 22x^2 + 16x + 96}, (-\infty, -4) \cup (-4, -2) \cup (-2, 3) \cup (3, 4) \cup (4, \infty)$ **c.** $\dfrac{x + 5}{x^3 + 3x^2 - 10x - 24},$

$(-\infty, -4) \cup (-4, -2) \cup (-2, 3) \cup (3, 4) \cup (4, \infty)$ **d.** $\dfrac{x^3 - 4x^2 - 16x + 64}{x^3 + 4x^2 - 11x - 30},$

$(-\infty, -5) \cup (-5, -4) \cup (-4, -2) \cup (-2, 3) \cup (3, 4) \cup (4, \infty)$ **29. a.** $\sqrt{x + 2} + \sqrt{2 - x}, [-2, 2]$

b. $\sqrt{x + 2} - \sqrt{2 - x}, [-2, 2]$ **c.** $\sqrt{4 - x^2}, [-2, 2]$ **d.** $\dfrac{\sqrt{x + 2}}{\sqrt{2 - x}}, [-2, 2)$ **30. a.** $\sqrt{\dfrac{x + 1}{x + 2}} + \sqrt{x - 3}, [3, \infty)$

b. $\sqrt{\dfrac{x + 1}{x + 2}} - \sqrt{x - 3}, [3, \infty)$ **c.** $\sqrt{\dfrac{(x + 1)(x - 3)}{x + 2}}, [3, \infty)$ **d.** $\sqrt{\dfrac{x + 1}{(x + 2)(x - 3)}}, (3, \infty)$

31. a. $\sqrt[3]{x + 1} + \sqrt[3]{\dfrac{x - 1}{x + 2}}, (-\infty, -2) \cup (-2, \infty)$ **b.** $\sqrt[3]{x + 1} - \sqrt[3]{\dfrac{x^2 - 1}{x + 2}}, (-\infty, -2) \cup (-2, \infty)$

c. $\sqrt[3]{\dfrac{x^2 - 1}{x + 2}}, (-\infty, -2) \cup (-2, \infty)$ **d.** $\sqrt[3]{\dfrac{(x + 1)(x + 2)}{x - 1}}, (-\infty, -2) \cup (-2, 1) \cup (1, \infty)$

32. a. $\sqrt{7 - x} + \sqrt{x^2 - 2x - 15}, (-\infty, -3] \cup [5, 7]$ **b.** $\sqrt{7 - x} - \sqrt{x^2 - 2x - 15}, (-\infty, -3] \cup [5, 7]$

c. $\sqrt{-x^3 + 9x^2 + x - 105}, (-\infty, -3] \cup [5, 7]$ **d.** $\dfrac{\sqrt{7 - x}}{\sqrt{x^2 - 2x - 15}}, (-\infty, -3) \cup (5, 7]$ **33.** $\dfrac{6}{x + 1} + 1$

34. $\dfrac{2}{3x + 2}$ **35.** $3\sqrt{x + 3} + 1$ **36.** $\dfrac{2}{\sqrt{x + 3} + 1}$ **37.** $\sqrt{3x + 4}$ **38.** $\sqrt{\dfrac{3x + 5}{x + 1}}$ **39.** $9x + 4$ **40.** $\dfrac{2x + 2}{x + 3}$

41. $\sqrt{\sqrt{x + 3} + 3}$ **42.** $\dfrac{6}{\sqrt{x + 3} + 1} + 1$ or $\dfrac{7 + \sqrt{x + 3}}{\sqrt{x + 3} + 1}$ **43.** $\dfrac{2}{3\sqrt{x + 3} + 2}$ **44.** $\sqrt{\dfrac{4x + 10}{x + 1}}$ **45.** 7

46. 10 **47.** $\dfrac{2}{5}$ **48.** 1 **49.** 2 **50.** $\dfrac{\sqrt{14}}{2}$ **51.** -5 **52.** $\dfrac{10}{7}$ **53.** $\sqrt{5}$ **54.** 4 **55.** $\sqrt{6}$ **56.** $\dfrac{2}{11}$

57. a. -4 **b.** 0 **c.** 2 **d.** -1 **e.** -1 **f.** 0 **58. a.** 2 **b.** -2 **c.** 0 **d.** 1 **e.** 2 **f.** -1

59. $(-\infty, \infty), (-\infty, \infty)$ **60.** $(-\infty, \infty), (-\infty, \infty)$ **61.** $[0, \infty), (-\infty, \infty)$ **62.** $(-\infty, -2) \cup (-2, 2) \cup (2, \infty),$

$(-\infty, 0) \cup (0, \infty)$ **63.** $(-\infty, 1) \cup (1, 2) \cup (2, \infty), (-\infty, -1) \cup \left(-1, \dfrac{1}{2}\right) \cup \left(\dfrac{1}{2}, \infty\right)$ **64.** $(-\infty, 1) \cup (1, 2) \cup (2, \infty),$

$(-\infty, -3) \cup (-3, 3) \cup (3, \infty)$ **65.** $(-\infty, -3] \cup (2, \infty), [0, 4) \cup (4, \infty)$ **66.** $(-\infty, 0) \cup (0, 4], (-\infty, 2) \cup \left[\dfrac{9}{4}, \infty\right)$

67. a. $x^2 + x + 1$ **b.** $x^2 - x + 3$ **c.** $x^3 - x^2 + 2x - 2$ **d.** $\dfrac{x^2 + 2}{x - 1}$ **68. a.** $x^2 + x - 9$ **b.** $-x^2 + x - 1$

c. $x^3 - 5x^2 - 4x + 20$ **d.** $\dfrac{x - 5}{x^2 - 4}$ **69. a.** $x^2 - 5x - 11$ **b.** $-x^2 + 7x + 21$ **c.** $x^3 - x^2 - 46x - 80$

d. $\dfrac{x + 5}{x^2 - 6x - 16}$ **70. a.** $\sqrt{x} + x + 6$ **b.** $\sqrt{x} - x - 6$ **c.** $x\sqrt{x} + 6x$ **d.** $\dfrac{\sqrt{x}}{x + 6}$ **71. a.** $\dfrac{1}{x} + \sqrt{x - 1}$

b. $\dfrac{1}{x} - \sqrt{x - 1}$ **c.** $\dfrac{\sqrt{x - 1}}{x}$ **d.** $\dfrac{1}{x\sqrt{x - 1}}$ **72. a.** $\dfrac{7x^2 + 18x + 11}{3x^2 + 14x + 15}$ **b.** $\dfrac{5x^2 + 8x - 1}{3x^2 + 14x + 15}$ **c.** $\dfrac{2x^2 + 5x + 2}{3x^2 + 14x + 15}$

d. $\dfrac{6x^2 + 13x + 5}{x^2 + 5x + 6}$ **73. a.** $\dfrac{2x^2 - 3x - 4}{x^3 - 3x^2 - 4x + 12}$ **b.** $\dfrac{4 - 3x}{x^3 - 3x^2 - 4x + 12}$ **c.** $\dfrac{x}{x^3 - 3x^2 - 4x + 12}$

d. $\dfrac{x^2 - 3x}{x^2 - 4}$ **74. a.** $\dfrac{2x^3 - 27x + 34}{x^4 - 3x^2 - 22x^2 + 16x + 96}$ **b.** $\dfrac{-8x^2 - 5x + 94}{x^4 - x^3 - 22x^2 + 16x + 96}$ **c.** $\dfrac{x + 5}{x^3 + 3x^2 - 10x - 24}$

d. $\dfrac{x^3 - 4x^2 - 16x + 64}{x^3 + 4x^2 - 11x - 30}$ **75. a.** $\sqrt{x + 2} + \sqrt{2 - x}$ **b.** $\sqrt{x + 2} - \sqrt{2 - x}$ **c.** $\sqrt{4 - x^2}$ **d.** $\dfrac{\sqrt{x + 2}}{\sqrt{2 - x}}$

76. a. $\sqrt{\dfrac{x + 1}{x + 2}} + \sqrt{x - 3}$ **b.** $\sqrt{\dfrac{x + 1}{x + 2}} - \sqrt{x - 3}$ **c.** $\sqrt{\dfrac{(x + 1)(x - 3)}{x + 2}}$ **d.** $\sqrt{\dfrac{x + 1}{(x + 2)(x - 3)}}$

77. a. $\sqrt[3]{x + 1} + \sqrt[3]{\dfrac{x - 1}{x + 2}}$ **b.** $\sqrt[3]{x + 1} - \sqrt[3]{\dfrac{x - 1}{x + 2}}$ **c.** $\sqrt[3]{\dfrac{x^2 - 1}{x + 2}}$ **d.** $\sqrt[3]{\dfrac{(x + 1)(x + 2)}{x - 1}}$

78. a. $\sqrt{7 - x} + \sqrt{x^2 - 2x - 15}$ **b.** $\sqrt{7 - x} - \sqrt{x^2 - 2x - 15}$ **c.** $\sqrt{-x^3 + 9x^2 + x - 105}$

d. $\dfrac{\sqrt{7 - x}}{\sqrt{x^2 - 2x - 15}}$

3.6 Exercises

SEC-1. $\dfrac{-x + 13}{x + 5}$ **SEC-2.** $\dfrac{47x - 29}{6x + 40}$ **SEC-3.** $y = 4x + 20$ **SEC-4.** $y = \dfrac{7}{3}x - 3$ **SEC-5.** $y = \dfrac{7x + 1}{5x + 8}$

SEC-6. $y = \dfrac{-4x - 5}{8x - 4}$ **SEC-7.** $y = (6 - x)^3 + 3$ **SEC-8.** $y = -\sqrt{x - 9} + 4$

1. one-to-one **2.** not one-to-one **3.** one-to-one **4.** not one-to-one **5.** one-to-one **6.** one-to-one
7. not one-to-one **8.** not one-to-one **9.** one-to-one **10.** one-to-one **11.** not one-to-one **12.** not one-to-one
13. one-to-one **14.** one-to-one **15.** not one-to-one **16.** one-to-one **17.** one-to-one **18–24.** final result:
$(f \circ g)(x) = (g \circ f)(x) = x$ **25.** , domain: $(-\infty, 5]$, range: $[-1, \infty)$ **26.** , domain: $[2, \infty)$,

range: $[0, \infty)$ **27.** , domain: $(-\infty, \infty)$, range: $(-\infty, \infty)$ **28.** 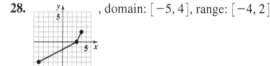 , domain: $[-5, 4]$, range: $[-4, 2]$

29. 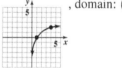 , domain: $[-4, 3]$, range: $[-4, 4]$ **30.** , domain: $(0, \infty)$, range: $(-\infty, \infty)$

31. a. $(-3, 4]$ **b.** $[-4, 3)$ **c.** -1 **d.** -1 **e.** -2 **f.** 0 **g.** -3 **h.** -4
32. $f^{-1}(x) = 3x + 15$, $\text{domain}\, f = \text{domain}\, f^{-1} = \text{range}\, f = \text{range}\, f^{-1} = (-\infty, \infty)$

33. $f^{-1}(x) = \dfrac{7x}{3} - 3$, $\text{domain}\, f = \text{domain}\, f^{-1} = \text{range}\, f = \text{range}\, f^{-1} = (-\infty, \infty)$

34. $f^{-1}(x) = \dfrac{x^3 + 3}{2}$, $\text{domain}\, f = \text{domain}\, f^{-1} = \text{range}\, f = \text{range}\, f^{-1} = (-\infty, \infty)$

35. $f^{-1}(x) = (1 - x)^5 - 4$, $\text{domain}\, f = \text{domain}\, f^{-1} = \text{range}\, f = \text{range}\, f^{-1} = (-\infty, \infty)$
36. $f^{-1}(x) = \sqrt{-x - 2}$, $\text{domain}\, f = \text{range}\, f^{-1} = [0, \infty)$, $\text{domain}\, f^{-1} = \text{range}\, f = (-\infty, -2]$
37. $f^{-1}(x) = -3 - \sqrt{x + 5}$, $\text{domain}\, f = \text{range}\, f^{-1} = (-\infty, -3]$, $\text{domain}\, f^{-1} = \text{range}\, f = [-5, \infty)$
38. $f^{-1}(x) = \dfrac{3}{x}$, $\text{domain}\, f = \text{domain}\, f^{-1} = \text{range}\, f = \text{range}\, f^{-1} = (-\infty, 0) \cup (0, \infty)$

39. $f^{-1}(x) = \dfrac{1}{2x + 1}$, $\text{domain}\, f = \text{range}\, f^{-1} = (-\infty, 0) \cup (0, \infty)$, $\text{domain}\, f^{-1} = \text{range}\, f = \left(-\infty, -\dfrac{1}{2}\right) \cup \left(-\dfrac{1}{2}, \infty\right)$

40. $f^{-1}(x) = \dfrac{7x + 1}{5x + 8}$, $\text{domain}\, f = \text{range}\, f^{-1} = \left(-\infty, \dfrac{7}{5}\right) \cup \left(\dfrac{7}{5}, \infty\right)$, $\text{domain}\, f^{-1} = \text{range}\, f = \left(-\infty, -\dfrac{8}{5}\right) \cup \left(-\dfrac{8}{5}, \infty\right)$

41. $f^{-1}(x) = \dfrac{-11x + 1}{13x + 4}$, $\text{domain}\, f = \text{range}\, f^{-1} = \left(-\infty, -\dfrac{11}{13}\right) \cup \left(-\dfrac{11}{13}, \infty\right)$, $\text{domain}\, f^{-1} = \text{range}\, f =$

$\left(-\infty, -\dfrac{4}{13}\right) \cup \left(-\dfrac{4}{13}, \infty\right)$ **42.** $f^{-1}(x) = \dfrac{-4x - 5}{8x - 4}$, $\text{domain}\, f = \text{range}\, f^{-1} = \left(-\infty, -\dfrac{1}{2}\right) \cup \left(-\dfrac{1}{2}, \infty\right)$, $\text{domain}\, f^{-1} =$

$\text{range}\, f = \left(-\infty, \dfrac{1}{2}\right) \cup \left(\dfrac{1}{2}, \infty\right)$ **43.** $f^{-1}(x) = -2 + \sqrt{5 + x}$, $\text{domain}\, f = \text{range}\, f^{-1} = [-2, \infty)$, $\text{domain}\, f^{-1} =$

$\text{range}\, f = [-5, \infty)$ **44.** $f^{-1}(x) = \dfrac{3}{2} - \dfrac{\sqrt{25 - 2x}}{2}$, $\text{domain}\, f = \text{range}\, f^{-1} = \left(-\infty, \dfrac{3}{2}\right]$, $\text{domain}\, f^{-1} =$

$\text{range}\, f = \left(-\infty, \dfrac{25}{2}\right]$

45.   **46.**   **47.** 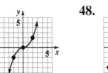   **48.**   **49.**   **50.**

51. a. $(-3, 4]$ **b.** $[-4, 3)$ **52.** -1 **53. a.** -1 **b.** -2 **c.** 0 **d.** -3 **e.** -4 **54.** $f^{-1}(x) = 3x + 15$

55. $f^{-1}(x) = \dfrac{7x}{3} - 3$ **56.** $f^{-1}(x) = \dfrac{x^3 + 3}{2}$ **57.** $f^{-1}(x) = (1 - x)^5 - 4$ **58.** $f^{-1}(x) = \sqrt{-x - 2}$

59. $f^{-1}(x) = -3 - \sqrt{x + 5}$ **60.** $f^{-1}(x) = \dfrac{3}{x}$ **61.** $f^{-1}(x) = \dfrac{1}{2x + 1}$ **62.** $f^{-1}(x) = \dfrac{7x + 1}{5x + 8}$

63. $f^{-1}(x) = \dfrac{-11x + 1}{13x + 4}$ **64.** $f^{-1}(x) = \dfrac{-4x - 5}{8x - 4}$ **65.** $f^{-1}(x) = -2 + \sqrt{5 + x}$

66. $f^{-1}(x) = \dfrac{3}{2} - \dfrac{\sqrt{25 - 2x}}{2}$

CHAPTER 4

4.1 Exercises

SCE-1. 25 **SCE-2.** $\dfrac{25}{4}$ **SCE-3.** $\dfrac{25}{36}$

1. up   **2.** down   **3.** up   **4.** down   **5. a.** $(2, -4)$ **b.** up   **c.** $x = 2$ **d.** $0, 4$ **e.** 0 **f.**

g. domain: $(-\infty, \infty)$, range: $[-4, \infty)$ **6. a.** $(-1, -9)$ **b.** down   **c.** $x = -1$ **d.** none   **e.** -10 **f.**

g. domain: $(-\infty, \infty)$, range: $(-\infty, -9]$ **7. a.** $(3, 2)$ **b.** down   **c.** $x = 3$ **d.** $2, 4$ **e.** -16 **f.**

g. domain: $(-\infty, \infty)$, range: $(-\infty, 2]$ **8. a.** $(-2, 2)$ **b.** up   **c.** $x = -2$ **d.** none   **e.** 4 **f.**

g. domain: $(-\infty, \infty)$, range: $[2, \infty)$ **9. a.** $(4, 2)$ **b.** down   **c.** $x = 4$ **d.** $4 - 2\sqrt{2}, 4 + 2\sqrt{2}$ **e.** -2

f.   **g.** domain: $(-\infty, \infty)$, range: $(-\infty, 2]$ **10. a.** $\left(-\dfrac{1}{3}, -4\right)$ **b.** up   **c.** $x = -\dfrac{1}{3}$ **d.** $-\dfrac{1}{3} - \dfrac{2\sqrt{3}}{3},$

$-\dfrac{1}{3} + \dfrac{2\sqrt{3}}{3}$ **e.** $-3\dfrac{2}{3}$ **f.**   **g.** domain: $(-\infty, \infty)$, range: $[-4, \infty)$ **11. a.** $(4, -16)$ **b.** up   **c.** $x = 4$

d. $0, 8$ **e.** 0 **f.**   **g.** domain: $(-\infty, \infty)$, range: $[-16, \infty)$ **12. a.** $(-2, 16)$ **b.** down   **c.** $x = -2$

d. $-6, 2$ **e.** 12 **f.** 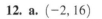   **g.** domain: $(-\infty, \infty)$, range: $(-\infty, 16]$ **13. a.** $(-1, -7)$ **b.** up   **c.** $x = -1$

d. $-1 - \dfrac{\sqrt{21}}{3}, -1 + \dfrac{\sqrt{21}}{3}$ **e.** -4 **f.**   **g.** domain: $(-\infty, \infty)$, range: $[-7, \infty)$ **14. a.** $\left(\dfrac{5}{4}, -\dfrac{49}{8}\right)$

b. up   **c.** $x = \dfrac{5}{4}$ **d.** $-\dfrac{1}{2}, 3$ **e.** -3 **f.**   **g.** domain: $(-\infty, \infty)$, range: $\left[-\dfrac{49}{8}, \infty\right)$

15. a. $(1, -5)$ **b.** down   **c.** $x = 1$ **d.** none   **e.** -6 **f.**   **g.** domain: $(-\infty, \infty)$, range: $(-\infty, -5]$

16. a. $(-6, -17)$ **b.** up   **c.** $x = -6$ **d.** $-6 - \sqrt{34}, -6 + \sqrt{34}$ **e.** 1 **f.**   **g.** domain: $(-\infty, \infty)$,

range: $[-17, \infty)$ **17. a.** $(-3, 8)$ **b.** down   **c.** $x = -3$ **d.** $-3 - 2\sqrt{6}, -3 + 2\sqrt{6}$ **e.** 5 **f.**

g. domain: $(-\infty, \infty)$, range: $(-\infty, 8]$ **18. a.** $\left(\dfrac{7}{6}, \dfrac{109}{12}\right)$ **b.** down   **c.** $x = \dfrac{7}{6}$ **d.** $\dfrac{7}{6} - \dfrac{\sqrt{109}}{6}, \dfrac{7}{6} + \dfrac{\sqrt{109}}{6}$ **e.** 5

f.   **g.** domain: $(-\infty, \infty)$, range: $\left(-\infty, \dfrac{109}{12}\right]$ **19. a.** $\left(-\dfrac{4}{3}, -\dfrac{25}{9}\right)$ **b.** up   **c.** $x = -\dfrac{4}{3}$ **d.** $-3, \dfrac{1}{3}$

e. -1 **f.**   **g.** domain: $(-\infty, \infty)$, range: $\left[-\dfrac{25}{9}, \infty\right)$ **20. a.** $(12, 35)$ **b.** down   **c.** $x = 12$

d. $12 - 2\sqrt{35}, 12 + 2\sqrt{35}$ **e.** -1 **f.**   **g.** domain: $(-\infty, \infty)$, range: $(-\infty, 35]$ **21. a.** $(4, -16)$

b. up   **c.** $x = 4$ **d.** $0, 8$ **e.** 0 **f.**   **g.** domain: $(-\infty, \infty)$, range: $[-16, \infty)$ **22. a.** $(-2, 12)$

b. down   **c.** $x = -2$ **d.** $-2 - 2\sqrt{3}, -2 + 2\sqrt{3}$ **e.** 8   **f.**   **g.** domain: $(-\infty, \infty)$, range: $(-\infty, 12]$

23. a. $(-1, -7)$ **b.** up   **c.** $x = -1$ **d.** $-1 - \dfrac{\sqrt{21}}{3}, -1 + \dfrac{\sqrt{21}}{3}$ **e.** -4 **f.**   **g.** domain: $(-\infty, \infty)$,

range: $[-7, \infty)$ **24. a.** $\left(\dfrac{5}{4}, -\dfrac{49}{8}\right)$ **b.** up   **c.** $x = \dfrac{5}{4}$ **d.** $-\dfrac{1}{2}, 3$ **e.** -3 **f.**   **g.** domain: $(-\infty, \infty)$,

range: $\left[-\dfrac{49}{8}, \infty\right)$ **25. a.** $(1, -5)$ **b.** down   **c.** $x = 1$ **d.** none   **e.** -6 **f.**   **g.** domain: $(-\infty, \infty)$,

range: $(-\infty, -5]$ **26. a.** $(-6, -17)$ **b.** up   **c.** $x = -6$ **d.** $-6 - \sqrt{34}, -6 + \sqrt{34}$ **e.** 1   **f.**

g. domain: $(-\infty, \infty)$, range: $[-17, \infty)$ **27. a.** $\left(-\dfrac{27}{2}, \dfrac{263}{4}\right)$ **b.** down   **c.** $x = -\dfrac{27}{2}$ **d.** $-\dfrac{27}{2} - \dfrac{\sqrt{789}}{2}$,

$-\dfrac{27}{2} + \dfrac{\sqrt{789}}{2}$ **e.** 5   **f.**   **g.** domain: $(-\infty, \infty)$, range: $\left(-\infty, \dfrac{263}{4}\right]$ **28. a.** $\left(\dfrac{7}{6}, \dfrac{109}{12}\right)$ **b.** down

c. $x = \dfrac{7}{6}$ **d.** $\dfrac{7}{6} - \dfrac{\sqrt{109}}{6}, \dfrac{7}{6} + \dfrac{\sqrt{109}}{6}$ **e.** 5   **f.**   **g.** domain: $(-\infty, \infty)$, range: $\left(-\infty, \dfrac{109}{12}\right]$

29. a. $\left(-\dfrac{4}{3}, -\dfrac{25}{9}\right)$ **b.** up   **c.** $x = -\dfrac{4}{3}$ **d.** $-3, \dfrac{1}{3}$ **e.** -1 **f.**   **g.** domain: $(-\infty, \infty)$, range:

$\left[-\dfrac{25}{9}, \infty\right)$ **30. a.** $(12, 35)$ **b.** down   **c.** $x = 12$ **d.** $12 - 2\sqrt{35}, 12 + 2\sqrt{35}$ **e.** -1 **f.**

g. domain: $(-\infty, \infty)$, range: $(-\infty, 35]$ **31. a.** positive   **b.** $(h, k) = (1, -2)$ **c.** 2   **d.** $f(x) = 2(x - 1)^2 - 2$

e. $f(x) = 2x^2 - 4x$ **32. a.** positive   **b.** $(h, k) = (3, 0)$ **c.** $\dfrac{2}{9}$ **d.** $f(x) = \dfrac{2}{9}(x - 3)^2$ **e.** $f(x) = \dfrac{2}{9}x^2 - \dfrac{4}{3}x + 2$

33. a. positive   **b.** $(h, k) = (2, 3)$ **c.** 3   **d.** $f(x) = 3(x - 2)^2 + 3$ **e.** $f(x) = 3x^2 - 12x + 15$

34. a. negative   **b.** $(h, k) = (-2, 3)$ **c.** $-\dfrac{3}{4}$ **d.** $f(x) = -\dfrac{3}{4}(x + 2)^2 + 3$ **e.** $f(x) = -\dfrac{3}{4}x^2 - 3x$ **35. a.** negative

b. $(h, k) = (1, 1)$ **c.** $-\dfrac{1}{3}$ **d.** $f(x) = -\dfrac{1}{3}(x - 1)^2 + 1$ **e.** $f(x) = -\dfrac{1}{3}x^2 + \dfrac{2}{3}x + \dfrac{2}{3}$ **36. a.** negative

b. $(h, k) = (-1, -2)$ **c.** $-\dfrac{1}{16}$ **d.** $f(x) = -\dfrac{1}{16}(x + 1)^2 - 2$ **e.** $f(x) = -\dfrac{1}{16}x^2 - \dfrac{x}{8} - \dfrac{33}{16}$ **37. a.** $(2, -4)$ **b.** up

c. $x = 2$ **38. a.** $(-1, -9)$ **b.** down   **c.** $x = -1$ **39. a.** $(3, 2)$ **b.** down   **c.** $x = 3$ **40. a.** $(-2, 2)$

b. up   **c.** $x = -2$ **41. a.** $(4, 2)$ **b.** down   **c.** $x = 4$ **42. a.** $\left(-\dfrac{1}{3}, -4\right)$ **b.** up   **c.** $x = -\dfrac{1}{3}$

43. a. $0, 4$ **b.** 0 **44. a.** none   **b.** -10 **45. a.** $2, 4$ **b.** -16 **46. a.** none   **b.** 4 **47. a.** $4 - 2\sqrt{2}, 4 + 2\sqrt{2}$

b. -2 **48. a.** $-\dfrac{1}{3} - \dfrac{2\sqrt{3}}{3}, -\dfrac{1}{3} + \dfrac{2\sqrt{3}}{3}$ **b.** $-3\dfrac{2}{3}$ **49. a.**   **b.** domain: $(-\infty, \infty)$, range: $[-4, \infty)$

50. a.    **b.** domain: $(-\infty, \infty)$, range: $(-\infty, -9]$ **51. a.**   **b.** domain: $(-\infty, \infty)$,

range: $(-\infty, -2]$ **52. a.**   **b.** domain: $(-\infty, \infty)$, range: $[2, \infty)$ **53. a.**

b. domain: $(-\infty, \infty)$, range: $(-\infty, 2]$ **54. a.**   **b.** domain: $(-\infty, \infty)$, range: $[-4, \infty)$ **55. a.** $(-1, -7)$

b. up   **c.** $x = -1$ **56. a.** $(1, -5)$ **b.** down   **c.** $x = 1$ **57. a.** $(-6, -17)$ **b.** up   **c.** $x = -6$ **58. a.** $\left(\dfrac{7}{6}, \dfrac{109}{12}\right)$

b. down   **c.** $x = \dfrac{7}{6}$ **59. a.** $(12, 35)$ **b.** down   **c.** $x = 12$ **60. a.** $-1 - \dfrac{\sqrt{21}}{3}, -1 + \dfrac{\sqrt{21}}{3}$ **b.** -4

61. a. none   **b.** -6 **62. a.** $-6 - \sqrt{34}, -6 + \sqrt{34}$ **b.** 1 **63. a.** $\dfrac{7}{6} - \dfrac{\sqrt{109}}{6}, \dfrac{7}{6} + \dfrac{\sqrt{109}}{6}$ **b.** 5

64. a. $12 - 2\sqrt{35}, 12 + 2\sqrt{35}$ **b.** -1 **65. a.**   **b.** domain: $(-\infty, \infty)$, range: $[-7, \infty)$

66. a.   **b.** domain: $(-\infty, \infty)$, range: $(-\infty, -5]$ **67. a.**   **b.** domain: $(-\infty, \infty)$,

range: $[-17, \infty)$ **68. a.**   **b.** domain: $(-\infty, \infty)$, range: $\left(-\infty, \dfrac{109}{12}\right]$ **69. a.**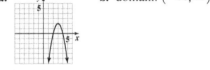

b. domain: $(-\infty, \infty)$, range: $(-\infty, 35]$ **70. a.** $(-1, -7)$ **b.** up   **c.** $x = -1$ **71. a.** $(1, -5)$ **b.** down   **c.** $x = 1$

72. a. $(-6, -17)$ **b.** up   **c.** $x = -6$ **73. a.** $\left(\dfrac{7}{6}, \dfrac{109}{12}\right)$ **b.** down   **c.** $x = \dfrac{7}{6}$ **74. a.** $(12, 35)$ **b.** down

c. $x = 12$ **75. a.** $-1 - \dfrac{\sqrt{21}}{3}, -1 + \dfrac{\sqrt{21}}{3}$ **b.** -4 **76. a.** none   **b.** -6 **77. a.** $-6 - \sqrt{34}, -6 + \sqrt{34}$

b. 1 **78. a.** $\dfrac{7}{6} - \dfrac{\sqrt{109}}{6}, \dfrac{7}{6} + \dfrac{\sqrt{109}}{6}$ **b.** 5 **79. a.** $12 - 2\sqrt{35}, 12 + 2\sqrt{35}$ **b.** -1 **80. a.**

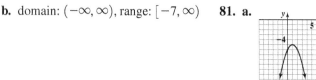

b. domain: $(-\infty, \infty)$, range: $[-7, \infty)$ **81. a.**   **b.** domain: $(-\infty, \infty)$, range: $(-\infty, -5]$

82. a.   **b.** domain: $(-\infty, \infty)$, range: $[-17, \infty)$ **83. a.**   **b.** domain: $(-\infty, \infty)$,

range: $\left(-\infty, \dfrac{109}{12}\right]$ **84. a.**   **b.** domain: $(-\infty, \infty)$, range: $(-\infty, 35]$ **85. a.** minimum   **b.** -16

86. a. maximum   **b.** 12 **87. a.** minimum   **b.** -7 **88. a.** minimum   **b.** $-\dfrac{49}{8}$ **89. a.** maximum   **b.** -5

90. a. minimum   **b.** -17 **91. a.** maximum   **b.** $\dfrac{91}{2}$ **92. a.** maximum   **b.** $\dfrac{109}{12}$ **93. a.** minimum

b. $-\dfrac{25}{9}$ **94. a.** maximum   **b.** 35 **95.** $f(x) = 2(x - 1)^2 - 2$ **96.** $f(x) = \dfrac{2}{9}(x - 3)^2$ **97.** $f(x) = 3(x - 2)^2 + 3$

98. $f(x) = -\dfrac{3}{4}(x + 2)^2 + 3$ **99.** $f(x) = -\dfrac{1}{3}(x - 1)^2 + 1$ **100.** $f(x) = -\dfrac{1}{16}(x + 1)^2 - 2$ **101.** $f(x) = 2x^2 - 4x$

102. $f(x) = \dfrac{2}{9}x^2 - \dfrac{4}{3}x + 2$ **103.** $f(x) = 3x^2 - 12x + 15$ **104.** $f(x) = -\dfrac{3}{4}x^2 - 3x$ **105.** $f(x) = -\dfrac{1}{3}x^2 + \dfrac{2}{3}x + \dfrac{2}{3}$

106. $f(x) = -\dfrac{1}{16}x^2 - \dfrac{x}{8} - \dfrac{33}{16}$

4.2 Exercises

1. 3.5 sec, 61.025 m   **2.** 3.75 sec, 78.9063 m   **3.** 3.5 sec, 201 ft   **4.** 87.8906 sec, 47.9453 ft   **5.** 104.775 ft, 90.7256 ft

6. a. $\dfrac{w_0 v_0}{g}$ **b.** $\dfrac{v_0^2}{2g} + h_0$ **c.** 128, 65   **7. a.** \$292.50   **b.** 1,000, \$150,000   **c.** \$150   **8. a.** \$71.60   **b.** 360

c. \$36   **9. a.** $R(x) = \dfrac{-x^2}{20} + 1{,}000x$ **b.** $P(x) = \dfrac{-x^2}{20} + 900x - 5{,}000$ **c.** 9,000, \$4,045,000   **d.** \$550

10. a. $R(x) = \dfrac{-x^2}{40} + 8{,}000x$ **b.** $P(x) = \dfrac{-x^2}{40} + 4{,}000x - 20{,}000$ **c.** 80,000, \$159,980,000   **d.** \$6,000

11. 40 scooters, \$400, \$6,500   **12.** 1 pitcher, \$12, \$7   **13.** \$275, \$226,875   **14.** 1500 trees, \$112,500   **15.** \$172.50
16. \$60   **17.** 40   **18.** \$15, \$1125   **19.** 40 trees   **20.** 5, 12,250   **21.** 450 ft by 900 ft, 405,000 ft^2 **22.** 781,250 ft^2
23. $833\dfrac{1}{3}$ ft   **24.** $\dfrac{48}{17}$ in.

4.3 Exercises

1. polynomial, $7, \frac{1}{3}, -8$ **2.** not polynomial   **3.** not polynomial   **4.** polynomial, $3, -\frac{5}{11}, 0$ **5.** polynomial, $5, -9, 1$

6. polynomial, $0, 12, 12$ **7.**   **8.**   **9.**   **10.**   **11.**

12.   **13.**   **14.**   **15.**   **16. a.** odd   **b.** negative

17. a. even   **b.** negative   **18. a.** odd   **b.** positive   **19. a.** even   **b.** positive   **20. a.** odd   **b.** negative

21. a. even   **b.** negative   **22.** $-2, 5$ **23.** $-4, -3, 4$ **24.** $-6, \frac{1}{2}$ **25.** $-3, 2, 3$ **26.** $x = 2$, multiplicity 2, touches, $x = -1$, multiplicity 3, crosses   **27.** $x = -1$, multiplicity 1, crosses, $x = -3$, multiplicity 2, touches   **28.** $x = 0$ and $x = 3$, multiplicity 1, crosses, $x = 4$, multiplicity 4, touches   **29.** $x = -\sqrt{2}$ and $x = \sqrt{2}$, multiplicity 2, touches   **30.** $x = 2$, multiplicity 4, touches, $x = -2$, multiplicity 3, crosses   **31.** $x = \frac{1 - \sqrt{13}}{2}$ and $x = \frac{1 + \sqrt{13}}{2}$, multiplicity 1, crosses, $x = 0$, multiplicity 2, touches   **32.** $x = -\sqrt{3}$ and $x = \sqrt{3}$, multiplicity 1, crosses, $x = -1$, multiplicity 3, crosses   **33.** $x = \sqrt{7}$, multiplicity 2, touches, $x = -\sqrt{7}$, multiplicity 5, crosses   **34.**

35.   **36.**   **37.**   **38.** 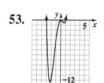   **39.**   **40.**

41.   **42.** 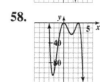   **43.**   **44.**   **45. a.** odd   **b.** positive   **c.** 0

d. $x = -2$, $x = 0$, and $x = 3$ all have odd multiplicity   **e.** ii   **46. a.** even   **b.** negative   **c.** 3   **d.** $x = -3$ has even multiplicity, and $x = -1$ and $x = 2$ have odd multiplicity   **e.** iv   **47. a.** even   **b.** positive   **c.** -3 **d.** $x = -3$ and $x = 1$ have even multiplicity, and $x = -1$ and $x = 3$ have odd multiplicity   **e.** ii   **48.**   **49.**

50.   **51.**   **52.**   **53.**   **54.**   **55.**

56.   **57.**   **58.**   **59.** ii   **60.** iv   **61.** ii

4.4 Exercises

SCE-1. quotient: $9x^2 - 47$, remainder: $2x + 244$

1. $f(x) = (x - 1)(x^2 - 2x - 1) - 6$ **2.** $f(x) = (x + 1)(x^2 + 4x - 7) + 6$ **3.** $f(x) = (x - 2)(6x^2 + 12x + 22) + 47$
4. $f(x) = (x + 3)(2x^2 - 7x + 21) - 59$ **5.** $f(x) = (x - 2)(x^3 + 3x^2 + 5x + 9) + 14$ **6.** $f(x) = (x + 1)(-2x^3 + 5x^2 - 6x + 13) - 11$ **7.** $f(x) = (x + 4)(-x^3 + 3x^2 - 12x + 43) - 172$ **8.** $f(x) = (x + 1)(3x^4 - 3x^3 + 3x^2 - 4x + 4) - 6$ **9.** $f(x) = (x - 1)(x^2 + x + 1)$ **10.** $f(x) = (x + 1)(x^4 - x^3 + x^2 - x + 1)$ **11.** -13 **12.** -7
13. 234 **14.** 101 **15.** -2 **16.** 0 **17.** factor **18.** not a factor **19.** factor **20.** factor **21.** not a factor
22. factor **23.** factor **24.** factor **25.** factor **26.** not a factor **27.** -2 and -1, $f(x) = (x - 1)(x + 1)(x + 2)$,

28. $3, f(x) = -(x - 3)^2(x + 1)$, **29.** -3 and -1, $f(x) = (x + 1)(x + 3)(2x - 1)$,

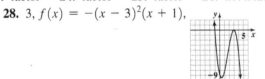

30. $-1, f(x) = \frac{1}{2}(x - 2)(x + 1)^2$, **31.** -2 and 1, $f(x) = -\frac{1}{3}(x - 1)(x + 2)(x + 3)$,

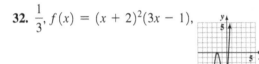

32. $\frac{1}{3}, f(x) = (x + 2)^2(3x - 1)$, **33.** $-\frac{1}{4}$ and 1, $f(x) = (x - 1)(x + 1)^2(4x + 1)$,

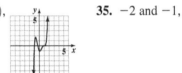

34. -1 and $-\frac{1}{3}, f(x) = (x - 1)^3(x + 1)(3x + 1)$, **35.** -2 and -1,

$f(x) = (x - 1)(x + 1)(x + 2)$ **36.** $3, f(x) = -(x - 3)^2(x + 1)$ **37.** -3 and $-1, f(x) = (x + 1)(x + 3)(2x - 1)$
38. $-1, f(x) = \frac{1}{2}(x - 2)(x + 1)^2$ **39.** -2 and $-1, f(x) = -\frac{1}{3}(x - 1)(x + 2)(x + 3)$ **40.** $\frac{1}{3}, f(x) = (x + 2)^2$
$(3x - 1)$ **41.** $-\frac{1}{4}$ and $1, f(x) = (x - 1)(x + 1)^2(4x + 1)$ **42.** -1 and $-\frac{1}{3}, f(x) = (x - 1)^3(x + 1)(3x + 1)$

4.5 Exercises

SCE-1. $f(-x) = 3x^5 + 4x^4 - 3x^3 - 7x^2 - 21$ **SCE-2.** $f(-x) = x^4 - 7x^2 - 11$ **SCE-3.** $f(-x) = 3x^7 - 8x^3 + x$

1. $\pm 1, \pm 2$ **2.** $\pm 1, \pm 2, \pm 4, \pm 8, \pm 16$ **3.** $\pm 1, \pm 2, \pm 3, \pm 4, \pm 6, \pm 12$ **4.** $\pm 1, \pm 2, \pm 4, \pm 8, \pm \frac{1}{3}, \pm \frac{2}{3}, \pm \frac{4}{3}, \pm \frac{8}{3}$ **5.** $\pm 1, \pm 2,$

$\pm 4, \pm \frac{1}{3}, \pm \frac{2}{3}, \pm \frac{4}{3}, \pm \frac{1}{9}, \pm \frac{2}{9}, \pm \frac{4}{9}$ **6.** $\pm 1, \pm 2, \pm 4, \pm 8, \pm 16, \pm \frac{1}{3}, \pm \frac{2}{3}, \pm \frac{4}{3}, \pm \frac{8}{3}, \pm \frac{16}{3}, \pm \frac{1}{9}, \pm \frac{2}{9}, \pm \frac{4}{9}, \pm \frac{8}{9}, \pm \frac{16}{9}, \pm \frac{1}{27},$

$\pm \frac{2}{27}, \pm \frac{4}{27}, \pm \frac{8}{27}, \pm \frac{16}{27}$ **7.** $5, 3,$ or 1 positive real zeros, 0 negative real zeros **8.** $4, 2,$ or 0 positive real zeros,

0 negative real zeros **9.** 1 positive real zero, 3 or 1 negative real zeros **10.** 0 positive real zeros, 0 negative real zeros

11. $-2, \frac{1}{4}, 3, f(x) = (x - 3)(x + 2)(4x - 1)$ **12.** $-\frac{7}{2}, -\sqrt{3}, \sqrt{3}, f(x) = (2x + 7)(x + \sqrt{3})(x - \sqrt{3})$

13. $\frac{1}{2}, -2 + 4i, -2 - 4i, f(x) = -(2x - 1)(x - (-2 + 4i))(x - (-2 - 4i))$ **14.** $-\frac{1}{2}, 1, \frac{5}{3},$

$f(x) = -(x - 1)^2(2x + 1)(3x - 5)$ **15.** $-\frac{1}{3}, 1, -1 + 5i, -1 - 5i, f(x) = (3x + 1)(x - 1)(x - (-1 + 5i))(x - (-1 - 5i))$

16. $-1, -\frac{5}{2}, \frac{-1 - \sqrt{5}}{2}, \frac{-1 + \sqrt{5}}{2}, f(x) = -(x + 1)^2(2x + 5)\left(x - \left(\frac{-1 - \sqrt{5}}{2}\right)\right)\left(x - \left(\frac{-1 + \sqrt{5}}{2}\right)\right)$

17. $\frac{1}{2}, -\frac{3}{2}, -2 + 2i, -2 - 2i, f(x) = (2x - 1)^2(2x + 3)(x - (-2 + 2i))(x - (-2 - 2i))$ **18.** $-\frac{1}{3}, \frac{1}{7}, \sqrt{2}, -\sqrt{2},$

$f(x) = (3x + 1)^2(7x - 1)(x - \sqrt{2})(x + \sqrt{2})$ **19.** $1, \frac{3}{5}, i, -i, f(x) = (x - 1)^3(5x - 3)(x - i)(x + i)$

20. $-\dfrac{1}{2} + \dfrac{\sqrt{3}}{2}i, -\dfrac{1}{2} - \dfrac{\sqrt{3}}{2}i, 1$ **21.** $i, -i, -1, 1$ **22.** $-1, 2, 3$ **23.** $\dfrac{1}{2} + \dfrac{\sqrt{7}}{2}i, \dfrac{1}{2} - \dfrac{\sqrt{7}}{2}i, -\dfrac{5}{2}, 1$

24. $-1 + 5i, -1 - 5i, -\dfrac{1}{3}, 1$ **25.** $-\dfrac{5}{2} - \dfrac{\sqrt{17}}{2}, -\dfrac{5}{2} + \dfrac{\sqrt{17}}{2}, -\dfrac{1}{6}, \dfrac{1}{2}$ **26.** answers may vary,

$f(x) = (x - (5 + i))(x - (5 - i))(x - 2)$ **27.** answers may vary, $f(x) = (x - 1)^2(x - (3 - 2i))(x - (3 + 2i))$

28. $f(x) = -\dfrac{1}{78}(x - 3i)(x + 3i)(x - (1 - 5i))(x - (1 + 5i))(x + 1)$ **29.** $f(x) = \dfrac{1}{4}(x - 5)(x - 2i)(x + 2i)$

30. $f(x) = -(x + 2)(x - 1)^2(x - (1 + i))(x - (1 - i))$ **31.** $f(x) = 2(x + 5)(x - 5)(x - (-1 - 2i))(x - (-1 + 2i))$

32. $f(1) = -2$ and $f(2) = 7$ **33.** $f(0) = -1$ and $f(1) = 10$ **34.** $f(-1) = -6$ and $f(0) = 1$ **35.** 1.33

36. -1 and 1   **37.** $-.20$ **38.**   **39.**   **40.**   **41.**

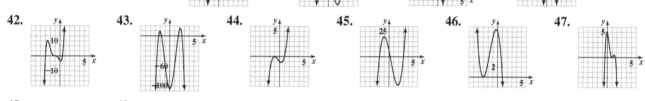

42.   **43.**   **44.**   **45.**   **46.**   **47.**

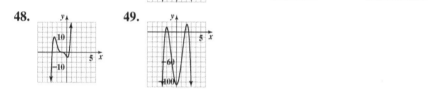

48.   **49.**

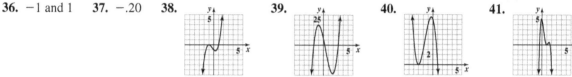

4.6 Exercises

1. a. $(-\infty, -1) \cup (-1, 1) \cup (1, \infty)$ **b.** 0   **c.** 0   **2. a.** $(-\infty, \infty)$ **b.** 0   **c.** 0   **3. a.** $(-\infty, 0) \cup (0, \infty)$

b. none   **c.** $1, -1$ **4. a.** $(-\infty, -2) \cup (-2, \infty)$ **b.** $-\dfrac{1}{2}$ **c.** $\dfrac{1}{3}$ **5. a.** $(-\infty, -1) \cup (-1, 0) \cup (0, \infty)$ **b.** none

c. $-4, 3$ **6. a.** $(-\infty, -1) \cup (-1, 5) \cup (5, \infty)$ **b.** $-\dfrac{6}{5}$ **c.** $-2, 1, 3$ **7.** $x = -1,$ **8.** $x = 2,$

9. $x = -1,$ **10.** $x = 2$ and $x = 4,$ **11.** $x = -2$ and $x = 1,$

12. $x = -2$ and $x = 1,$ **13.** $y = 0$ **14.** none   **15.** $y = \dfrac{3}{5}$ **16.** none   **17.** $y = -\dfrac{4}{7}$ **18.** $y = -9$

19. $y = 0$ **20.** $y = \dfrac{3}{11}$ **21.**   , $x = 0, y = -2$ **22.**   , $x = 0, y = 2$ **23.**   , $x = 1,$

$y = 0$ **24.**   , $x = 1, y = 0$ **25.**   , $x = -1, y = -3$ **26.**   , $x = -3, y = -2$

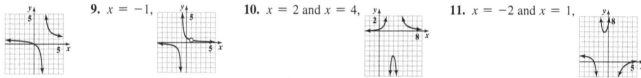
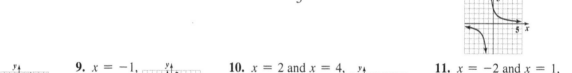

27. 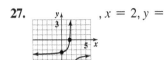 , $x = 2, y = -3$ **28.** , $x = -1, y = 2$ **29.** , $x = -1, y = 0$ **30.** ,

$x = 3, y = 0$ **31.** , $x = -1, y = -1$ **32.** , $x = 1, y = 3$ **33.** $(3, 6),$

34. $(-4, -5),$ **35.** $(1, 1),$, $x = 0, y = 0$ **36.** $\left(-2, -\dfrac{1}{5}\right),$, $x = 3, y = 0$

37. $\left(-2, \dfrac{1}{9}\right),$, $x = 1, y = 0$ **38.** $\left(1, \dfrac{1}{9}\right),$, $x = -2, y = 0$ **39.** $y = 2x$ **40.** $y = x - 3$

41. $y = x - 1$ **42.** $y = 2x$ **43.** , $x = 1, y = 2$ **44.** 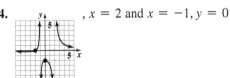 , $x = 2$ and $x = -1, y = 0$

45. , $y = 1$ **46.** , $x = -2, y = 2x - 9$ **47.** 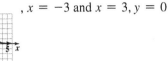 , $x = -3$ and $x = 3, y = 0$

48. , $x = -3$ and $x = 3, y = 0$ **49.** , $x = -1, y = x + 1$ **50.** , $x = 3, y = 2$

51. , $x = -2, y = x - 2$ **52.** , $x = 4$ and $x = -1, y = 1$ **53.**  , $x = -2, x = 1,$

and $x = 4, y = 2$ **54.** , $x = -3$ and $x = 3, y = 2x - 3$ **55.** $(-\infty, -1) \cup (-1, 1) \cup (1, \infty)$ **56.** $(-\infty, \infty)$

57. $(-\infty, 0) \cup (0, \infty)$ **58.** $(-\infty, -2) \cup (-2, \infty)$ **59.** $(-\infty, -1) \cup (-1, 0) \cup (0, \infty)$ **60.** $(-\infty, -1) \cup (-1, 5) \cup (5, \infty)$
61. $x = -1$ **62.** $x = 2$ **63.** $x = -1$ **64.** $x = 2$ and $x = 4$ **65.** $x = -2$ and $x = 1$ **66.** $x = -2$ and $x = 1$
67.   **68.**   **69.**   **70.**   **71.**   **72.**

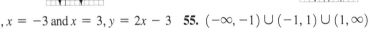

73. 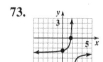 **74.** **75.** 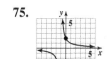 **76.** 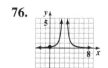 **77.** 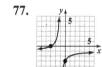 **78.**

79. **80.** **81.** **82.** **83.** **84.**

85. **86.** **87.** **88.** **89.** **90.**

91. **92.** **93.** **94.** **95.** **96.**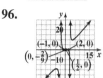

4.7 Exercises

1. 40 ft   **2.** 16 m   **3.** 900 gallons   **4.** 396.9 m   **5.** 9 years old   **6.** 3 ohms   **7.** about 79 kg   **8.** 18 lumens
9. \$5568   **10.** 5.2 horsepower   **11.** 45 kilograms per square meter   **12.** 47.75 revolutions per minute
13. 18.75 ft or less   **14.** 1638.4 watts

CHAPTER 5

5.1 Skill Check Exercises

SCE-1. 3^3 **SCE-2.** 2^{-3} **SCE-3.** $3^{1/4}$ **SCE-4.** 5^{-3x} **SCE-5.** $3^{x/3}$ **SCE-6.** 2^{1-x} **SCE-7.** $2^{7/4}$
SCE-8. $3^{4-x/5}$ **SCE-9.** 3^{x^2-2x} **SCE-10.** 2^{2x+x^2}

1. 0.303212   **2.** 5.805141   **3.** 20.085537   **4.** .818731   **5.** 1.395612   **6.** 88.426366   **7.** 32.725881
8. **9.** **10.** **11.** **12.** **13.**

14.   **15.** $f(x) = 10^x$ **16.** $f(x) = 2^x$ **17.** $f(x) = \left(\dfrac{2}{3}\right)^x$ **18.** $f(x) = \left(\dfrac{3}{4}\right)^x$ **19.** $f(x) = 8^x$

20. $f(x) = \left(\dfrac{1}{16}\right)^x$ **21.** $f(x) = 2.7^x$ **22.** , domain: $(-\infty, \infty)$, range: $(0, \infty)$, y-intercept: $\dfrac{1}{2}$, $y = 0$

23. , domain: $(-\infty, \infty)$, range: $(-1, \infty)$, y-intercept: 0, $y = -1$ **24.** , domain: $(-\infty, \infty)$, range:

$(-\infty, 0)$, y-intercept: -9, $y = 0$ **25.** , domain: $(-\infty, \infty)$, range: $(-\infty, -1)$, y-intercept: -3, $y = -1$

26. , domain: $(-\infty, \infty)$, range: $(0, \infty)$, y-intercept: $\dfrac{1}{2}$, $y = 0$ **27.** , domain: $(-\infty, \infty)$, range:

$(-3, \infty)$, y-intercept: -2, $y = -3$ **28.** , domain: $(-\infty, \infty)$, range: $(1, \infty)$, y-intercept: 2, $y = 1$

29. , domain: $(-\infty, \infty)$, range: $(-2, \infty)$, y-intercept: 1, $y = -2$ **30.** , domain: $(-\infty, \infty)$, range:

$(0, \infty)$, $y = 0$ **31.** , domain: $(-\infty, \infty)$, range: $(-1, \infty)$, $y = -1$ **32.** , domain: $(-\infty, \infty)$,

range: $(-\infty, 0)$, $y = 0$ **33.** , domain: $(-\infty, \infty)$, range: $(-\infty, -1)$, $y = -1$ **34.** , domain:

$(-\infty, \infty)$, range: $(-2, \infty)$, $y = -2$ **35.** 4 **36.** -1 **37.** $\dfrac{1}{4}$ **38.** 6 **39.** $-\dfrac{3}{5}$ **40.** $-\dfrac{7}{12}$ **41.** $\dfrac{1}{3}$ **42.** $-\dfrac{9}{5}$

43. $-1, 3$ **44.** $-1, 0, 2$ **45.** -2 **46.** $-\dfrac{1}{3}$ **47.** $\dfrac{1}{4}$ **48.** $-2, 2$ **49.** $-3, 4$ **50.** $-\dfrac{5}{2} - \dfrac{\sqrt{37}}{2}, -\dfrac{5}{2} + \dfrac{\sqrt{37}}{2}$

51. $-1, 1, 2$ **52.** $-1, 2, 3$ **53. a.** \$300,000   **b.** \$265,640   **c.** \$163,299   **54. a.** 130 lbs   **b.** 201 lbs   **c.** 235 lbs
55. a. 2,560 bacteria   **b.** 4,096 bacteria   **c.** 8,589,934,592 bacteria   **56. a.** 40 rabbits   **b.** 62 rabbits   **c.** 336 rabbits
d. 1595 rabbits, 1599 rabbits, 1600 rabbits   **57.** \$12,752.18   **58.** 7.35% compounded daily   **59.** compounded quarterly at
4.5%   **60.** $A(2) \approx$ \$6,798.89, $A(20) \approx$ \$20,942.06   **61.** \$55.23   **62.** \$332,570.82   **63.** \$5,862.47
64. \$469,068.04   **65.** \$2,000   **66.** \$10,202.05   **67.** a   **68. a.** 2,000 people   **b.** 3.5%   **c.** 6,808 people
69. a. $P(t) = 10e^{.25t}$ **b.** 4,034 bacteria   **70.** 5,488 people   **71. a.** 8 rabbits   **b.** 478,993 rabbits

72. , domain: $(-\infty, \infty)$, range: $(0, \infty)$, y-intercept: $\dfrac{1}{2}$, $y = 0$ **73.** , domain: $(-\infty, \infty)$, range:

$(-1, \infty)$, y-intercept: 0, $y = -1$ **74.** , domain: $(-\infty, \infty)$, range: $(-\infty, 0)$, y-intercept: -9, $y = 0$

75. , domain: $(-\infty, \infty)$, range: $(-\infty, -1)$, y-intercept: -3, $y = -1$ **76.** , domain: $(-\infty, \infty)$, range:

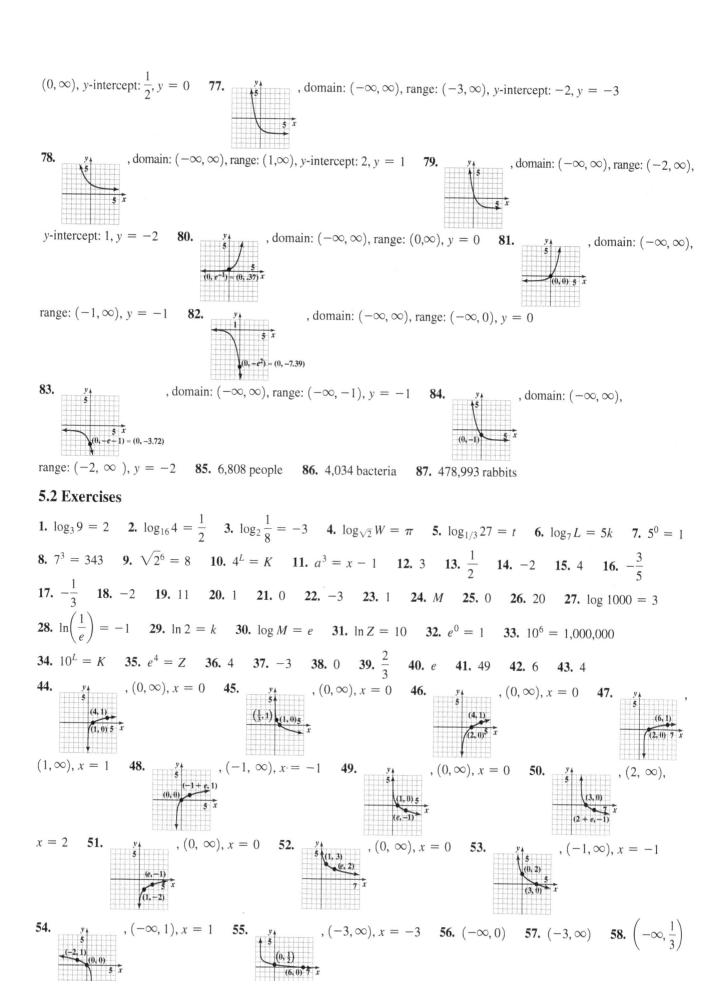

$(0, \infty)$, y-intercept: $\dfrac{1}{2}$, $y = 0$ **77.** , domain: $(-\infty, \infty)$, range: $(-3, \infty)$, y-intercept: -2, $y = -3$

78. , domain: $(-\infty, \infty)$, range: $(1, \infty)$, y-intercept: 2, $y = 1$ **79.** , domain: $(-\infty, \infty)$, range: $(-2, \infty)$,

y-intercept: 1, $y = -2$ **80.** , domain: $(-\infty, \infty)$, range: $(0, \infty)$, $y = 0$ **81.** , domain: $(-\infty, \infty)$,

range: $(-1, \infty)$, $y = -1$ **82.** , domain: $(-\infty, \infty)$, range: $(-\infty, 0)$, $y = 0$

83. , domain: $(-\infty, \infty)$, range: $(-\infty, -1)$, $y = -1$ **84.** , domain: $(-\infty, \infty)$,

range: $(-2, \infty)$, $y = -2$ **85.** 6,808 people **86.** 4,034 bacteria **87.** 478,993 rabbits

5.2 Exercises

1. $\log_3 9 = 2$ **2.** $\log_{16} 4 = \dfrac{1}{2}$ **3.** $\log_2 \dfrac{1}{8} = -3$ **4.** $\log_{\sqrt{2}} W = \pi$ **5.** $\log_{1/3} 27 = t$ **6.** $\log_7 L = 5k$ **7.** $5^0 = 1$

8. $7^3 = 343$ **9.** $\sqrt{2}^6 = 8$ **10.** $4^L = K$ **11.** $a^3 = x - 1$ **12.** 3 **13.** $\dfrac{1}{2}$ **14.** -2 **15.** 4 **16.** $-\dfrac{3}{5}$

17. $-\dfrac{1}{3}$ **18.** -2 **19.** 11 **20.** 1 **21.** 0 **22.** -3 **23.** 1 **24.** M **25.** 0 **26.** 20 **27.** $\log 1000 = 3$

28. $\ln\left(\dfrac{1}{e}\right) = -1$ **29.** $\ln 2 = k$ **30.** $\log M = e$ **31.** $\ln Z = 10$ **32.** $e^0 = 1$ **33.** $10^6 = 1{,}000{,}000$

34. $10^L = K$ **35.** $e^4 = Z$ **36.** 4 **37.** -3 **38.** 0 **39.** $\dfrac{2}{3}$ **40.** e **41.** 49 **42.** 6 **43.** 4

44. , $(0, \infty)$, $x = 0$ **45.** , $(0, \infty)$, $x = 0$ **46.** , $(0, \infty)$, $x = 0$ **47.** ,

$(1, \infty)$, $x = 1$ **48.** , $(-1, \infty)$, $x = -1$ **49.** , $(0, \infty)$, $x = 0$ **50.** , $(2, \infty)$,

$x = 2$ **51.** , $(0, \infty)$, $x = 0$ **52.** , $(0, \infty)$, $x = 0$ **53.** , $(-1, \infty)$, $x = -1$

54. , $(-\infty, 1)$, $x = 1$ **55.** , $(-3, \infty)$, $x = -3$ **56.** $(-\infty, 0)$ **57.** $(-3, \infty)$ **58.** $\left(-\infty, \dfrac{1}{3}\right)$

59. $(-\infty, -3) \cup (3, \infty)$ **60.** $(-\infty, -4) \cup (5, \infty)$ **61.** $(-\infty, -5) \cup (8, \infty)$ **62.** $(-10, -2) \cup (3, \infty)$

63.   **64.**   **65.**   **66.**   **67.**   **68.**

69.   **70.**   **71.**   **72.**   **73.**   **74.**

5.3 Exercises

1. $\log_4 x + \log_4 y$ **2.** $\log 9 - \log t$ **3.** $3\log_5 y$ **4.** $3 + \log_3 w$ **5.** $\ln 5 + 2$ **6.** $\frac{1}{4}\log_9 k$ **7.** $2 + \log P$

8. $5 - \ln r$ **9.** $6 + \log_{\sqrt{2}} x$ **10.** $\log_2 M - 5$ **11.** $2\log_7 x + 3\log_7 y$ **12.** $2\ln a + 3\ln b - 4\ln c$

13. $\frac{1}{2}\log x - 1 - 3\log y$ **14.** $2 + \log_3 (x^2 - 25) = 2 + \log_3 (x - 5) + \log_3 (x + 5)$ **15.** $1 + \frac{1}{2}\log_2 x + \frac{1}{2}\log_2 y$

16. $-\frac{1}{6} + \frac{5}{2}\log_5 x - \frac{4}{3}\log_5 y$ **17.** $\frac{1}{5} + \frac{1}{5}\ln z - \frac{1}{2}\ln (x - 1)$ **18.** $\frac{1}{2}\log_3 x + \frac{5}{4}\log_3 y - \frac{1}{2}$

19. $-\frac{4}{3}\ln (x + 1) - 2\ln (x - 1)$ **20.** $3 + 3\log x - \frac{9}{2}\log (x - 4) - 5\log (x + 4)$ **21.** $\log_b (AC)$ **22.** $\log_4 \left(\dfrac{M}{N}\right)$

23. $\log_8 \left(x^2 \sqrt[3]{y}\right)$ **24.** 2   **25.** 4   **26.** $\ln \left(x^{\frac{5}{12}}\right)$ **27.** $\log_5 (x^2 - 4)$ **28.** $\log_3 \dfrac{x + 1}{(x + 4)\sqrt{6x}}$ **29.** $\ln \dfrac{x - 1}{4x\sqrt{x + 5}}$

30. $\ln[7x^2(x - 3)^2]$ **31.** $\log_9 (x - 3)$ **32.** $\log \left(\dfrac{x + 3}{x + 1}\right)$ **33.** $\ln \left(\dfrac{1}{x - 1}\right)$ **34.** 5   **35.** 36   **36.** 1   **37.** $\dfrac{7}{8}$

38. 2   **39.** 6   **40.** 2.8362   **41.** $-.1147$ **42.** -2.6572 **43.** 4.7332   **44.** $\log_3 (xw^2)$ **45.** $\log_5 \left(\dfrac{1}{x^2}\right)$

46. $\log_2 (xyw)$ **47.** $\ln \left(\sqrt{x^{15}}\right)$ **48.** 5   **49.** $\dfrac{1}{20}$ **50.** 36   **51.** $\sqrt{5}$

5.4 Exercises

SCE-1. 1.2528   **SCE-2.** 5.1565   **SCE-3.** 1.6826   **SCE-4.** 1.8516   **SCE-5.** 738.1281   **SCE-6.** 6.0588
SCE-7. 11.5813   **SCE-8.** 72.5188

1. 1.4650   **2.** 12.7438   **3.** $-1, 3$ **4.** $\varnothing$ **5.** -13.3627 **6.** $\dfrac{1}{3}$ **7.** 4.4841   **8.** $-\dfrac{5}{6}$ **9.** .3334   **10.** .4833

11. .6931   **12.** .5988   **13.** $-.2055$ **14.** 2.6889   **15.** -3.1192 **16.** $\dfrac{6}{5}$ **17.** 7   **18.** -3 **19.** 2, 3   **20.** $\dfrac{15}{4}$

21. $-\dfrac{99}{5}$ **22.** $\dfrac{23}{9}$ **23.** $\dfrac{1}{3}$ **24.** 8   **25.** 6   **26.** -8 **27.** $\dfrac{19}{6}$ **28.** 10   **29.** -5 **30.** $\dfrac{7}{5}$ **31.** 8   **32.** $-1\dfrac{1}{3}$

33. 6   **34.** $\varnothing$ **35.** 2.0933

5.5 Exercises

1. about 4 years 8 months   **2.** about 7 years 9 months   **3.** 13.73 years   **4.** 17.33%   **5.** Bank A: 4.5%, Bank B: 4.6%
6. 21,086 people   **7.** 4,147 bacteria   **8.** 252 days   **9.** 200,119 people   **10.** 53 days   **11.** 64.84%   **12. a.** 53.03 grams
b. 17 years   **13. a.** 6,000 families   **b.** 2,000 families   **c.** 1988   **14. a.** 3,000 students   **b.** $B = 374$

c. $K = \ln \left(\dfrac{29}{374}\right)$ **d.** ≈ 3.63 days   **15. a.** 61,500   **b.** 40   **c.** $-\dfrac{\ln \left(\dfrac{43}{2}\right)}{6}$ **d.** 2009   **16.** just over 20 minutes
17. just before 11:52 P.M.   **18.** about 58.6°F   **19.** 1988   **20.** ≈ 3.63 days   **21.** 2009

CHAPTER 6

6.1 Exercises

1. vertex: $(0, 0)$, focus: $(0, 4)$, directrix: $y = -4$,   **2.** vertex: $(0, 0)$, focus: $(0, -2)$, directrix: $y = 2$,

   3. vertex: $(1, 4)$, focus: $(1, 1)$, directrix: $y = 7$,   **4.** vertex: $(-3, 1)$, focus: $\left(-3, \dfrac{5}{2}\right)$,

directrix: $y = -\dfrac{1}{2}$,   **5.** vertex: $(-2, -6)$, focus: $\left(-2, -\dfrac{19}{4}\right)$, directrix: $y = -\dfrac{29}{4}$,   **6.** vertex:

$(0, 0)$, focus: $(1, 0)$, directrix: $x = -1$,   **7.** vertex: $(0, 0)$, focus: $(-2, 0)$, directrix: $x = 2$,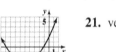

8. vertex: $(2, 5)$, focus: $(1, 5)$, directrix: $x = 3$, 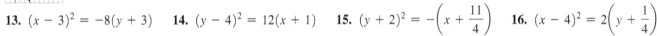   **9.** vertex: $(4, -3)$, focus: $(9, -3)$, directrix: $x = -1$,

   10. vertex: $(-3, -4)$, focus: $\left(-\dfrac{3}{4}, -4\right)$, directrix: $x = -\dfrac{21}{4}$,   **11.** $y^2 = 8x$ **12.** $x^2 = -2y$

13. $(x - 3)^2 = -8(y + 3)$ **14.** $(y - 4)^2 = 12(x + 1)$ **15.** $(y + 2)^2 = -\left(x + \dfrac{11}{4}\right)$ **16.** $(x - 4)^2 = 2\left(y + \dfrac{1}{4}\right)$

17. $(y - 4)^2 = 16(x + 3)$ **18.** $8(x + 2) = (y - 4)^2$ and $-8(x - 2) = (y - 4)^2$ **19.** $-2(y - 2) = (x - 2)^2$ and

$8\left(y + \dfrac{1}{2}\right) = (x - 2)^2$ **20.** vertex: $(-5, -3)$, focus: $\left(-5, -\dfrac{7}{4}\right)$, directrix: $y = -\dfrac{17}{4}$, 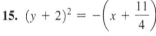   **21.** vertex: $(1, 6)$,

focus: $\left(-\dfrac{1}{2}, 6\right)$, directrix: $x = \dfrac{5}{2}$, 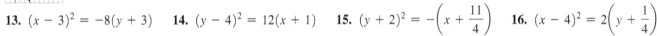   **22.** vertex: $(1, 2)$, focus: $\left(1, \dfrac{7}{4}\right)$, directrix: $y = \dfrac{9}{4}$,

23. vertex: $(-1, -3)$, focus: $\left(-\dfrac{5}{4}, -3\right)$, directrix: $x = -\dfrac{3}{4}$,   **24.** 10 cm   **25.** $\dfrac{81}{32}$ in.   **26.** $\dfrac{121}{16}$ ft

27. $-160(y - 40) = x^2$, no   **28.** 14.4 m   **29.**   **30.**   **31.**   **32.**

33. **34.** **35.** **36.** **37.** **38.**

39. **40.** **41.** **42.**

6.2 Exercises

1. center: $(0, 0)$, foci: $(2\sqrt{3}, 0)$ and $(-2\sqrt{3}, 0)$, vertices: $(4, 0)$ and $(-4, 0)$, 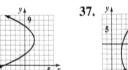 **2.** center: $(0, 0)$, foci: $(0, \sqrt{13})$

and $(0, -\sqrt{13})$, vertices: $(0, 5)$ and $(0, -5)$, 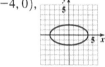 **3.** center: $(0, 0)$, foci: $(\sqrt{5}, 0)$ and $(-\sqrt{5}, 0)$, vertices: $(3, 0)$

and $(-3, 0)$, **4.** center: $(0, 0)$, foci: $(0, \sqrt{15})$ and $(0, -\sqrt{15})$, vertices: $(0, 2\sqrt{5})$ and $(0, -2\sqrt{5})$,

5. center: $(2, 4)$, foci: $(2 + \sqrt{11}, 4)$ and $(2 - \sqrt{11}, 4)$, vertices: $(8, 4)$ and $(-4, 4)$, **6.** center: $(1, -4)$,

foci: $(1, -4 + 2\sqrt{10})$ and $(1, -4 - 2\sqrt{10})$, vertices: $(1, 3)$ and $(1, -11)$, 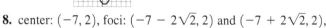 **7.** center: $(5, -1)$, foci: $(9, -1)$

and $(1, -1)$, vertices: $(0, -1)$ and $(10, -1)$, 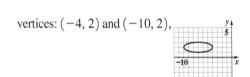 **8.** center: $(-7, 2)$, foci: $(-7 - 2\sqrt{2}, 2)$ and $(-7 + 2\sqrt{2}, 2)$,

vertices: $(-4, 2)$ and $(-10, 2)$, 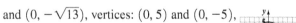 **9.** $\dfrac{(x + 4)^2}{16} + \dfrac{y^2}{36} = 1$ **10.** $\dfrac{(x - 3)^2}{16} + \dfrac{(y + 3)^2}{9} = 1$ **11.** $\dfrac{x^2}{49} + \dfrac{y^2}{24} = 1$

12. $\dfrac{(x - 3)^2}{36} + \dfrac{(y + 1)^2}{11} = 1$ **13.** $\dfrac{(x + 6)^2}{16} + \dfrac{(y - 5)^2}{25} = 1$ **14.** $\dfrac{(x - 4)^2}{100} + \dfrac{(y - 5)^2}{64} = 1$ **15.** $\dfrac{(x + 1)^2}{7} + \dfrac{(y - 4)^2}{16} = 1$

16. $\dfrac{(x - 2)^2}{20} + \dfrac{(y + 1)^2}{36} = 1$ **17.** $\dfrac{(x - 4)^2}{4} + \dfrac{(y - 1)^2}{53} = 1$ **18.** $\dfrac{(x - 1)^2}{1} + \dfrac{(y - 3)^2}{10} = 1$

19. $\dfrac{x^2}{16} + \dfrac{y^2}{12} = 1$ **20.** $\dfrac{x^2}{8} + \dfrac{y^2}{72} = 1$ **21.** $\dfrac{(x + 2)^2}{20} + \dfrac{(y - 1)^2}{16} = 1$, center: $(-2, 1)$, foci: $(-4, 1)$ and $(0, 1)$,

vertices: $(-2 - 2\sqrt{5}, 1)$ and $(-2 + 2\sqrt{5}, 1)$, **22.** $\dfrac{(x + 7)^2}{9} + \dfrac{(y + 3)^2}{16} = 1$, center: $(-7, -3)$, foci: $(-7, -3 - \sqrt{7})$ and $(-7, -3 + \sqrt{7})$, vertices: $(-7, -7)$ and $(-7, 1)$, **23.** $\dfrac{(x + 2)^2}{16} + \dfrac{(y - 1)^2}{1} = 1$,

center: $(-2, 1)$, foci: $(-2 - \sqrt{15}, 1)$ and $(-2 + \sqrt{15}, 1)$, vertices: $(-6, 1)$ and $(2, 1)$,

24. $\dfrac{x^2}{2} + \dfrac{(y + 10)^2}{100} = 1$, center: $(0, -10)$, foci: $(0, -10 - 7\sqrt{2})$ and $(0, -10 + 7\sqrt{2})$, vertices: $(0, -20)$ and $(0, 0)$,

 25. $\dfrac{x^2}{225} + \dfrac{y^2}{256} = 1$ **26.** $\dfrac{x^2}{81} + \dfrac{y^2}{441} = 1$ **27.** about 25.8 yds   **28.** 74.83 cm to the left and right of center

29. 44.80 in.   **30.** $\dfrac{x^2}{18{,}705{,}625} + \dfrac{y^2}{18{,}675{,}000} = 1$ **31.** about 39.69 ft   **32.**   **33.**

34.   **35.**   **36.**   **37.**   **38.**   **39.**

40.   **41.**   **42.**   **43.**

6.3 Exercises

1. center: $(0, 0)$, transverse axis: $y = 0$, vertices: $(-4, 0)$ and $(4, 0)$, foci: $(-5, 0)$ and $(5, 0)$, asymptotes: $y = \pm\dfrac{3}{4}x$,

2. center: $(0, 0)$, transverse axis: $x = 0$, vertices: $(0, -3)$ and $(0, 3)$, foci: $(0, -5)$ and $(0, 5)$, asymptotes: $y = \pm\dfrac{3}{4}x$,

3. center: $(2, 4)$, transverse axis: $x = 2$, vertices: $(2, -1)$ and $(2, 9)$, foci: $(2, 4 - \sqrt{61})$ and $(2, 4 + \sqrt{61})$, asymptotes:

$y - 4 = \pm\dfrac{5}{6}(x - 2)$,   **4.** center: $(-1, -3)$, transverse axis: $y = -3$, vertices: $(-4, -3)$ and $(2, -3)$,

foci: $(-1 - \sqrt{58}, -3)$ and $(-1 + \sqrt{58}, -3)$, asymptotes: $y + 3 = \pm\dfrac{7}{3}(x + 1)$,   **5.** center: $(0, 0)$, transverse axis: $y = 0$, vertices: $(-\sqrt{5}, 0)$ and $(\sqrt{5}, 0)$, foci: $(-5, 0)$ and $(5, 0)$, asymptotes: $y = \pm 2x$, **6.** center: $(1, 3)$,

transverse axis: $x = 1$, vertices: $(1, 3 - 2\sqrt{5})$ and $(1, 3 + 2\sqrt{5})$, foci: $(1, -3)$ and $(1, 9)$, asymptotes: $y - 3 = \pm\dfrac{\sqrt{5}}{2}(x - 1)$,

 7. $\dfrac{x^2}{9} - \dfrac{y^2}{16} = 1$, **8.** $\dfrac{y^2}{36} - \dfrac{x^2}{64} = 1$, **9.** $\dfrac{(x - 4)^2}{1} - \dfrac{(y + 4)^2}{3} = 1$,

10. $\dfrac{(y + 1)^2}{16} - \dfrac{(x + 6)^2}{48} = 1$, **11.** $\dfrac{(y - 6)^2}{4} - \dfrac{(x - 9)^2}{5} = 1$, **12.** $\dfrac{(x - 4)^2}{36} - \dfrac{(y + 3)^2}{49} = 1$,

13. $\dfrac{y^2}{20} - \dfrac{x^2}{16} = 1$, **14.** $\dfrac{(x - 2)^2}{16} - \dfrac{(y - 1)^2}{48} = 1$,

15. $\dfrac{(x - 2)^2}{4} - \dfrac{(y - 1)^2}{4} = 1$, center: $(2, 1)$, vertices: $(0, 1)$ and $(4, 1)$, foci: $(2 - 2\sqrt{2}, 1)$ and $(2 + 2\sqrt{2}, 1)$, endpoints of

conjugate axis: $(2, -1)$ and $(2, 3)$, asymptotes: $y - 1 = \pm(x - 2)$, **16.** $\dfrac{(y + 3)^2}{2} - \dfrac{(x + 4)^2}{2} = 1$,

center: $(-4, -3)$, vertices: $(-4, -3 - \sqrt{2})$ and $(-4, -3 + \sqrt{2})$, foci: $(-4, -5)$ and $(-4, -1)$, endpoints of conjugate axis:

$(-4 - \sqrt{2}, -3)$ and $(-4 + \sqrt{2}, -3)$, asymptotes: $y + 3 = \pm(x + 4)$, **17.** $\dfrac{(y - 6)^2}{9} - (x + 2)^2 = 1$,

center: $(-2, 6)$, vertices: $(-2, 3)$ and $(-2, 9)$, foci: $(-2, 6 - \sqrt{10})$ and $(-2, 6 + \sqrt{10})$, endpoints of conjugate axis: $(-3, 6)$

and $(-1, 6)$, asymptotes: $y - 6 = \pm 3(x + 2)$, **18.** $\dfrac{(x + 5)^2}{16} - (y - 2)^2 = 1$, center: $(-5, 2)$, vertices:

$(-9, 2)$ and $(-1, 2)$, foci: $(-5 - \sqrt{17}, 2)$ and $(-5 + \sqrt{17}, 2)$, endpoints of conjugate axis: $(-5, 1)$ and $(-5, 3)$, asymptotes:

$y - 2 = \pm\dfrac{1}{4}(x + 5)$, **19.** $\dfrac{(x - 6)^2}{25} - \dfrac{(y + 8)^2}{144} = 1$, center: $(6, -8)$, vertices: $(1, -8)$ and $(11, -8)$,

foci: $(-7, -8)$ and $(19, -8)$, endpoints of conjugate axis: $(6, -20)$ and $(6, 4)$, asymptotes: $y + 8 = \pm\dfrac{12}{5}(x - 6)$,

20. $\dfrac{(y-1)^2}{576} - \dfrac{(x+1)^2}{49} = 1$, center: $(-1, 1)$, vertices: $(-1, -23)$ and $(-1, 25)$, foci: $(-1, -24)$ and $(-1, 26)$, endpoints of

conjugate axis: $(-8, 1)$ and $(6, 1)$, asymptotes: $y - 1 = \pm\dfrac{24}{7}(x + 1)$,   **21.** $\dfrac{y^2}{49} - \dfrac{x^2}{15} = 1$

22. $\dfrac{x^2}{2500} - \dfrac{y^2}{7500} = 1$ **23. a.** $\dfrac{x^2}{900} - \dfrac{y^2}{2700} = 1$ **b.** 159m   **24.** $\dfrac{y^2}{324} - \dfrac{x^2}{1276} = 1$ **25.**   **26.**

27.   **28.**   **29.**   **30.**   **31.**   **32.**

33.   **34.**   **35.**   **36.**

CHAPTER 7

7.1 Exercises

1. $(-1, 3)$ yes, $(-3, 7)$ no   **2.** $\left(-\dfrac{1}{2}, \dfrac{1}{4}\right)$ no, $\left(-\dfrac{1}{4}, \dfrac{1}{2}\right)$ yes   **3.** $(.2, -.9)$ yes, $(-.3, .4)$ no   **4.** $(-2, 5)$ no, $(3, 5)$ yes

5. $(1, 3)$ **6.** $(-3, 2)$ **7.** $\left(-\dfrac{1}{2}, -4\right)$ **8.** $(-6, 6)$ **9.** $\left(0, -\dfrac{2}{3}\right)$ **10.** $\left(\dfrac{5}{2}, \dfrac{5}{4}\right)$ **11.** $(2, -7)$ **12.** $(6, 9)$

13. $(-8, 4)$ **14.** $(8, 8)$ **15.** no solution   **16.** $(2 + 4y, y)$ or $\left(x, \dfrac{x-2}{4}\right)$ **17.** $\left(-\dfrac{13}{16}, \dfrac{13}{2}\right)$ **18.** $\left(x, \dfrac{3}{4}x - \dfrac{1}{2}\right)$ or

$\left(\dfrac{2}{3} + \dfrac{4y}{3}, y\right)$ **19.** no solution   **20.** $\left(x, \dfrac{5x}{7} - 9\right)$ or $\left(\dfrac{63}{5} + \dfrac{7y}{5}, y\right)$ **21.** Ricky: 1,196 hits, Pedro: 1,477 hits
22. 93 two-bedroom and 107 three-bedroom   **23.** 301 children and 191 adults.   **24.** \$3,100 at 12% and \$2,800 at 7%
25. 394 beverages   **26.** 580 acres   **27.** plane: 550 mph, wind: 50 mph   **28.** 12 nickels and 34 quarters   **29.** about 42.5 lbs
of peanuts and about 16 lbs of cashews   **30.** \$19.52

7.2 Exercises

1. $(-1, 1, -2)$ yes, $(1, -1, 2)$ no   **2.** $(2, -1, 4)$ no, $(-2, 1, -4)$ yes   **3.** $(-1, 2, 3)$ **4.** $(1, -2, 1)$ **5.** $\left(\dfrac{1}{4}, -\dfrac{1}{2}, \dfrac{3}{8}\right)$

6. $(-2, 3, 1)$ **7.** $(-3, 0, 1)$ **8.** $\left(-1, \dfrac{2}{3}, \dfrac{1}{3}\right)$ **9.** $(29, 16, 3)$ **10.** $(19, -7, -15)$ **11.** $(5, -2, -4)$

12. $x + 3y - 2z = 4$ **13.** $2x - y + z = -6$ **14.** $(5 - z, 3 + z, z)$ **15.** $(2, z - 3, z)$ **16.** no solution
17. no solution   **18.** 20 long swordsmen, 20 spearmen, 35 crossbowmen   **19.** \$1.50 for each hot dog, \$2.50 for each hamburger,
\$3.75 for each chicken sandwich   **20.** 10 cheese pizzas, 15 pepperoni pizzas, 10 supreme pizzas   **21.** 520 adult tickets,
210 children tickets, 280 senior tickets   **22.** 249 free throws, 223 two-point field goals, 9 three-point field goals

23. $a = -\dfrac{1}{80}, b = \dfrac{37}{40}, c = -\dfrac{161}{16}$; 5.25 million

7.3 Exercises

1. $(-1, 2, 3)$ 2. $(1, -2, 1)$ 3. $(29, 16, 3)$ 4. $(19, -7, -15)$ 5. $\left(-\dfrac{13}{3}, \dfrac{47}{15}, \dfrac{6}{5}\right)$ 6. $(-2, 3, 1)$ 7. $(3, 1, 2)$

8. $(-1, -3, 5)$ 9. $\left(\dfrac{23}{21}, \dfrac{10}{7}, \dfrac{1}{21}\right)$ 10. $\left(\dfrac{1}{4}, -2, -\dfrac{9}{8}\right)$ 11. $(5, -2, -4)$ 12. $(-3, 0, 1)$ 13. $(-1, -3, 10)$

14. $\left(\dfrac{1}{2}, -1, 0\right)$ 15. $\left(-2, \dfrac{1}{2}, -3\right)$ 16. $\left(1, -3, \dfrac{1}{3}\right)$ 17. $(-2, 3, 1)$ 18. $(-1, -3, 10)$ 19. $(-3, 0, 1)$

20. $(1 + 2z, 2 - 3z, z)$ 21. $(-2 - z, 4 + 2z, z)$ 22. $\left(\dfrac{5}{2} - \dfrac{z}{2}, -\dfrac{1}{2} - \dfrac{z}{2}, z\right)$ 23. $(-7 + 2z, 3 - z, z)$ 24. no solution

25. no solution   26. $\left(\dfrac{21}{5}, -3, \dfrac{4}{5}\right)$ 27. inconsistent   28. $(-5 + 2z, 4 - z, z)$ 29. $(5 - 3y, y, 8 - y)$ 30. inconsistent

31. $(1 - 2y, y, -4)$ 32. $(-6 - 2z, 8 + 3z, z)$ 33. $(5 - z, 3 + z, z)$ 34. $(2y, y, -2)$ 35. $(-2 + 3z, 13 - 8z, z)$
36. $(2, -3 + z, z)$ 37. 10 cheese, 15 pepperoni, 10 supreme   38. 442 adults, 241 children, 266 senior citizens
39. 100 high performance, 80 ultra performance, 58 extreme performance   40. \$200 at 6%, \$1,400 at 8%, \$600 at 9%
41. run: 10 miles, swim: 5 miles, bike: 25 miles   42. $f(x) = 3x^2 + 3x$ 43. $f(x) = 3x^2 - 4x - 5$

7.4 Exercises

1. $\dfrac{-1}{x} + \dfrac{1}{x - 3}$ 2. $\dfrac{1}{x + 9} + \dfrac{3}{x + 3}$ 3. $\dfrac{-1}{x - 1} + \dfrac{2}{x - 2}$ 4. $\dfrac{\frac{1}{3}}{x - 2} + \dfrac{\frac{2}{3}}{x + 4}$ 5. $\dfrac{2}{x - 3} - \dfrac{1}{x - 5}$

6. $\dfrac{\frac{2}{3}}{x - 5} - \dfrac{\frac{4}{3}}{2x - 1}$ 7. $\dfrac{\frac{5}{12}}{x - 5} + \dfrac{\frac{7}{12}}{x + 7}$ 8. $\dfrac{2}{x - 2} - \dfrac{1}{x - 1}$ 9. $\dfrac{1}{x} - \dfrac{59}{x + 4} + \dfrac{66}{x + 5}$ 10. $\dfrac{\frac{1}{25}}{3x + 1} + \dfrac{\frac{7}{25}}{4x - 7}$

11. $\dfrac{12}{x} - \dfrac{3}{x - 1} - \dfrac{2}{x + 1}$ 12. $\dfrac{16}{x} - \dfrac{5}{x - 1} - \dfrac{4}{x + 1}$ 13. $\dfrac{\frac{5}{12}}{x - 5} + \dfrac{\frac{5}{8}}{x + 5} - \dfrac{\frac{1}{24}}{x + 1}$ 14. $\dfrac{-\frac{2}{21}}{x} + \dfrac{\frac{18}{91}}{x - 7} + \dfrac{\frac{4}{39}}{x + 6}$

15. $\dfrac{-7}{x - 5} - \dfrac{24}{(x - 5)^2}$ 16. $\dfrac{1}{x - 2} - \dfrac{2}{(x - 2)^2} - \dfrac{4}{(x - 2)^3}$ 17. $\dfrac{-2}{x - 4} + \dfrac{25}{(x - 4)^2} + \dfrac{2}{x + 5}$

18. $\dfrac{\frac{9}{25}}{x - 1} + \dfrac{\frac{1}{5}}{(x - 1)^2} + \dfrac{\frac{16}{25}}{x + 4}$ 19. $\dfrac{3}{x} - \dfrac{3}{x - 1} + \dfrac{4}{(x - 1)^2}$ 20. $\dfrac{-1}{x} + \dfrac{5}{x^2} + \dfrac{2}{x + 6}$

21. $\dfrac{\frac{1}{2}}{x - 2} + \dfrac{1}{(x - 2)^2} - \dfrac{\frac{1}{2}}{x + 2} + \dfrac{1}{(x + 2)^2}$ 22. $\dfrac{-\frac{7}{4}}{x + 3} + \dfrac{\frac{7}{4}}{x + 5} + \dfrac{\frac{9}{2}}{(x + 5)^2}$

23. $\dfrac{6}{x} - \dfrac{2}{x^2} - \dfrac{6}{x + 1} + \dfrac{3}{(x + 1)^2} - \dfrac{9}{(x + 1)^3}$ 24. $\dfrac{\frac{1}{2}}{x + 1} - \dfrac{\frac{1}{2}}{x - 1} + \dfrac{1}{(x - 1)^2}$ 25. $\dfrac{-\frac{6}{25}}{x} - \dfrac{\frac{1}{5}}{x^2} + \dfrac{\frac{7}{50}}{x - 5} + \dfrac{\frac{1}{10}}{x + 5}$

26. $\dfrac{\frac{13}{32}}{x + 4} + \dfrac{\frac{51}{32}}{x - 4} + \dfrac{\frac{25}{4}}{(x - 4)^2}$ 27. $\dfrac{1}{x} - \dfrac{x}{x^2 + 6}$ 28. $\dfrac{7}{x - 1} + \dfrac{-7x + 4}{x^2 + x + 1}$ 29. $\dfrac{2}{x - 6} + \dfrac{6x - 2}{x^2 + 5}$

30. $\dfrac{-\frac{4}{27}}{x - 3} + \dfrac{\frac{4}{27}x + \frac{8}{9}}{x^2 + 3x + 9}$ 31. $\dfrac{1}{x + 3} - \dfrac{1}{x - 3} + \dfrac{7}{x^2 + 8}$ 32. $\dfrac{\frac{1}{9}}{x} + \dfrac{\frac{7}{9}}{x^2} + \dfrac{-\frac{1}{9}x - \frac{7}{9}}{x^2 + 9}$

33. $\dfrac{3}{x + 3} + \dfrac{13}{(x + 3)^2} + \dfrac{-3x - 4}{x^2 + 4}$ 34. $\dfrac{4}{x + 3} + \dfrac{2x + 5}{x^2 + 4}$ 35. $\dfrac{\frac{3}{17}}{x + 3} + \dfrac{\frac{14}{17}x + \frac{45}{17}}{x^2 - 12x + 6}$ 36. $\dfrac{x + 1}{x^2 + 4} - \dfrac{4x - 1}{(x^2 + 4)^2}$

37. $\dfrac{5x - 6}{x^2 + 4x + 8} + \dfrac{6x + 11}{(x^2 + 4x + 8)^2}$ 38. $\dfrac{4x + 2}{(x^2 + 3)^2} + \dfrac{3x - 2}{(x^2 + 3)^3}$ 39. $\dfrac{-\frac{1}{9}}{x} + \dfrac{\frac{5}{9}x}{5x^2 + 3} + \dfrac{\frac{5}{3}x + 3}{(5x^2 + 3)^2}$

40. $\dfrac{3}{x + 2} + \dfrac{-2x + 2}{x^2 + 4} + \dfrac{5x + 3}{(x^2 + 4)^2}$ 41. $\dfrac{-2}{x} + \dfrac{1}{x^2} + \dfrac{2x + 2}{x^2 + 1} + \dfrac{2x}{(x^2 + 1)^2}$

7.5 Exercises

1. ; two real solutions   **2.** ; two real solutions   **3.** ; one real solution

4. ; two real solutions   **5.** ; four real solutions   **6.** ; two real solutions

7. ; one real solution   **8.** ; four real solutions   **9.** $(-5, 10), (4, 1)$ **10.** $(-6, -8), (8, 6)$

11. $(-1, 10)$ **12.** $(-10, -6), (8, 0)$ **13.** $(2, 0), (4, -2)$ **14.** $\left(-\dfrac{4}{5}, \dfrac{6}{5}\right), (0, 2)$ **15.** $(0, 2), (1, 1)$

16. $(-5, -6), (2, 15)$ **17.** $(-5, -1), (-1, -5), (1, 5), (5, 1)$ **18.** $(-\sqrt{3}, -\sqrt{3}), (-1, -3), (1, 3), (\sqrt{3}, \sqrt{3})$

19. $(-3, -4), (6, 5)$ **20.** $\left(\dfrac{1}{3}, \dfrac{29}{9}\right), \left(3, \dfrac{7}{3}\right)$ **21.** $(-8, -4), (-8, 4), (8, -4), (8, 4)$

22. $(-1, -2), (-1, 2), (1, -2), (1, 2)$ **23.** $(0, 3)$ **24.** $(-3, -\sqrt{7}), (-3, \sqrt{7}), (0, -4), (0, 4)$

25. $(-2, -4), (-2, 4), (2, -4), (2, 4)$ **26.** $\left(-\dfrac{1}{2}, -\dfrac{3}{2}\right), \left(-\dfrac{1}{2}, \dfrac{3}{2}\right), \left(\dfrac{1}{2}, -\dfrac{3}{2}\right), \left(\dfrac{1}{2}, \dfrac{3}{2}\right)$ **27.** $\left(-\sqrt{2}, -\dfrac{1}{\sqrt{2}}\right), \left(\sqrt{2}, \dfrac{1}{\sqrt{2}}\right)$

28. no real solutions   **29.** $(2, -8), (0, 0)$ **30.** $(3, -2)$ **31.** $(-6, 5)$ **32.** $(4, 2)$ **33.** 8 and 14   **34.** 17 and 8
35. 18 and 15   **36.** 94 or 49   **37.** 742   **38.** 6 ft by 13 ft, or 13 ft by 6 ft   **39.** 8 m and 13 m   **40.** 38.7 ft
41. 9 ft by 9 ft by 12 ft, or about 16.8 ft by 16.8 ft by 3.5 ft   **42.** 6 ft by 13 ft by 6 ft

7.6 Exercises

1. a. no   **b.** yes   **c.** no   **2. a.** yes   **b.** no   **c.** yes   **3. a.** no   **b.** no   **c.** yes   **4. a.** no   **b.** yes   **c.** no
5.   **6.**   **7.**   **8.**   **9.**   **10.**

11.   **12.**   **13.**   **14.**   **15.**   **16.**

17.   **18.**   **19.**   **20.**   **21.** 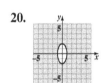   **22.**

23.   **24.**   **25. a.** yes   **b.** no   **c.** no   **26. a.** no   **b.** no   **c.** yes   **27. a.** yes   **b.** no

c. yes **28. a.** no **b.** no **c.** yes **29. a.** no **b.** no **c.** yes **30. a.** yes **b.** no **c.** yes **31.**

32.   **33.**   **34.**   **35.**   **36.**   **37.** ,

no solution **38.** 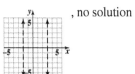 , no solution **39.**   **40.**   **41.**   **42.**

43.   **44.**   **45.**   **46.**   **47.**   **48.**

49.   **50.**   **51.**   **52.**   **53.**   **54.**

55.   **56.**   **57.**   **58.**

CHAPTER 8

8.1 Exercises

1. a. 3×4 **b.** none of these **2. a.** 3×3 **b.** square matrix **3. a.** 1×4 **b.** row matrix **4. a.** 4×1

b. column matrix **5.** $\begin{bmatrix} 4 & 4 & -3 \\ 8 & 1 & 6 \\ 8 & 3 & -10 \end{bmatrix}$ **6.** $\begin{bmatrix} -2 & -4 & -1 \\ -2 & 1 & 2 \\ 2 & -5 & 0 \end{bmatrix}$ **7.** $\begin{bmatrix} 2 & 4 & 1 \\ 2 & -1 & -2 \\ -2 & 5 & 0 \end{bmatrix}$ **8.** $\begin{bmatrix} 2 & 0 & -4 \\ 6 & 2 & 8 \\ 10 & -2 & -10 \end{bmatrix}$

9. $\begin{bmatrix} \frac{3}{2} & 2 & -\frac{1}{2} \\ \frac{5}{2} & 0 & 1 \\ \frac{3}{2} & 2 & -\frac{5}{2} \end{bmatrix}$ **10.** $\begin{bmatrix} \frac{1}{2} & -2 & -\frac{7}{2} \\ \frac{7}{2} & 2 & 7 \\ \frac{17}{2} & -4 & -\frac{15}{2} \end{bmatrix}$ **11.** 2×2 **12.** 3×5 **13.** not defined **14.** 1×1

15. $\begin{bmatrix} 7 & -7 & 1 \\ 23 & -1 & -3 \end{bmatrix}$ **16.** $\begin{bmatrix} 11 & 1 & 10 \\ -4 & 0 & -10 \end{bmatrix}$ **17.** $\begin{bmatrix} 18 & -6 & 11 \\ 19 & -1 & -13 \end{bmatrix}$ **18.** $\begin{bmatrix} 7 & 8 \\ 12 & 31 \end{bmatrix}$ **19.** $\begin{bmatrix} -19 & -1 & -30 \\ 13 & 3 & -20 \end{bmatrix}$

20. $\begin{bmatrix} 0 & 0 & 0 \\ 4 & -2 & 0 \end{bmatrix}$ **21.** $I_m = I_2 = \begin{bmatrix} 1 & 0 \\ 0 & 1 \end{bmatrix}$ and $I_n = I_3 = \begin{bmatrix} 1 & 0 & 0 \\ 0 & 1 & 0 \\ 0 & 0 & 1 \end{bmatrix}$ **22.** $I_m = I_3 = \begin{bmatrix} 1 & 0 & 0 \\ 0 & 1 & 0 \\ 0 & 0 & 1 \end{bmatrix}$ and

$$I_n = I_4 = \begin{bmatrix} 1 & 0 & 0 & 0 \\ 0 & 1 & 0 & 0 \\ 0 & 0 & 1 & 0 \\ 0 & 0 & 0 & 1 \end{bmatrix}$$
23. $I_m = I_5 = \begin{bmatrix} 1 & 0 & 0 & 0 & 0 \\ 0 & 1 & 0 & 0 & 0 \\ 0 & 0 & 1 & 0 & 0 \\ 0 & 0 & 0 & 1 & 0 \\ 0 & 0 & 0 & 0 & 1 \end{bmatrix}$ and $I_n = I_2 = \begin{bmatrix} 1 & 0 \\ 0 & 1 \end{bmatrix}$ **24. a.** $[208 \quad 116 \quad 76 \quad 88]$

b. 208 cups of baking mix, 116 eggs, 76 cups of milk, and 88 tablespoons of oil are needed to make the desired amount of food
25. a. $\begin{bmatrix} 28.5 & 24 & 9 \\ 39 & 33 & 12 \end{bmatrix}$ **b.** $ab_{21} = $ it costs \$39 to produce a self-propelled mower in New York, $ab_{12} = $ it costs \$24 to
produce a standard mower in Los Angeles, $ab_{23} = $ it costs \$12 to produce a self-propelled mower in Beijing
26. $[\text{Buffalo} \quad \text{Rochester} \quad \text{Syracuse}] = [24,750 \quad 26,500 \quad 39,200]$, \$90,450

8.2 Exercises

1. yes   **2.** yes   **3.** yes   **4.** yes   **5.** yes, $A^{-1} = \begin{bmatrix} 3 & -1 \\ -5 & 2 \end{bmatrix}$ **6.** yes, $A^{-1} = \begin{bmatrix} -1 & 2 \\ -1 & \frac{5}{2} \end{bmatrix}$ **7.** singular

8. singular   **9.** $\begin{bmatrix} 0 & \frac{1}{2} \\ -1 & \frac{1}{2} \end{bmatrix}$ **10.** $\begin{bmatrix} -3 & -4 & 2 \\ 0 & -1 & 0 \\ -2 & 1 & 1 \end{bmatrix}$ **11.** singular matrix   **12.** $\begin{bmatrix} -28 & -13 & 3 \\ 2 & 1 & 0 \\ -7 & -3 & 1 \end{bmatrix}$ **13.** $\begin{bmatrix} -\frac{1}{4} & -1 & -\frac{1}{4} \\ -\frac{7}{4} & -2 & \frac{1}{4} \\ 1 & 1 & 0 \end{bmatrix}$

14. $\begin{bmatrix} 9 & 0 & 2 & 0 \\ -20 & -9 & -5 & 5 \\ 4 & 0 & 1 & 0 \\ -4 & -2 & -1 & 1 \end{bmatrix}$ **15.** $x = 2, y = -3$ **16.** $x = -\frac{1}{3}, y = \frac{1}{2}$ **17.** $x = \frac{29}{4}, y = \frac{33}{4}$ **18.** $x = 1, y = -2, z = 4$
19. $x = 0, y = 1, z = 3$ **20.** $x = -10, y = 6, z = 0$ **21.** $x = 2, y = 3, z = 0, w = 1$

8.3 Exercises

1. $x = 1, y = 2$ **2.** $x = 1, y = 1$ **3.** $x = -2, y = -4$ **4.** $x = \frac{1}{2}, y = \frac{1}{3}$ **5.** 20   **6.** 164   **7.** 39   **8.** 0
9. 6   **10.** -220 **11.** 4,710   **12.** 42   **13.** $x = 2, y = 1, z = -1$ **14.** $x = 2, y = 0, z = 0$
15. $x = -3, y = 2, z = -4$ **16.** $x = -3, y = 3, z = -2$ **17.** $x = 35, y = -21, z = -21$
18. $w = -2, x = 4, y = -3, z = 2$

CHAPTER 9

9.1 Exercises

1. $4, 7, 10, 13$ **2.** $4, 16, 64, 256$ **3.** $1, \frac{8}{5}, 2, \frac{16}{7}$ **4.** $-4, 16, -64, 256$ **5.** $30, 120, 600, 3600$ **6.** $\frac{1}{2}, \frac{4}{3}, \frac{9}{8}, \frac{8}{15}$

7. $-5, 10, -15, 20$ **8.** $\frac{1}{2}, \frac{9}{4}, \frac{9}{2}, \frac{81}{4}$ **9.** $-\frac{1}{42}, \frac{1}{56}, -\frac{1}{72}, \frac{1}{90}$ **10.** $-\frac{9}{2}, \frac{81}{40}, -\frac{243}{560}, \frac{243}{4480}$ **11.** $7, 10, 13, 16$ **12.** $-1, 3, 0, 4$

13. $6, \frac{3}{2}, \frac{1}{6}, \frac{1}{96}$ **14.** $-4, \frac{5}{4}, \frac{1}{5}, -4$ **15.** $a_n = 2n - 3, a_{11} = 19$ **16.** $a_n = \frac{n}{n+4}, a_8 = \frac{8}{12}$ **17.** $a_n = n(n+5), a_7 = 7 \cdot 12$

18. $a_n = (-1)^n \cdot 2^n, a_7 = -128$ **19.** $a_n = \frac{2}{5^n}, a_6 = \frac{2}{15,625}$ **20.** $a_n = (-1)^n \cdot 6 \cdot 2^{n-1}, a_9 = -1536$

21. $a_n = (-1)^n (n+2)!, a_5 = -5040$ **22.** $a_n = \frac{3^n}{(n+1)!}, a_5 = \frac{243}{720}$ **23.** 20   **24.** 15   **25.** $\frac{11}{32}$ **26.** 111

27. 12   **28.** -192 **29.** 45   **30.** 102   **31.** 196   **32.** 147   **33.** 271   **34.** 112   **35.** 27   **36.** $\displaystyle\sum_{i=1}^{29} i$ **37.** $\displaystyle\sum_{i=1}^{10} 5i$

38. $\displaystyle\sum_{i=1}^{11} i^2$ **39.** $\displaystyle\sum_{i=1}^{9} \frac{i+3}{i+4}$ **40.** $\displaystyle\sum_{i=1}^{8} (-1)^{i+1} \cdot 2^i$ **41.** $\displaystyle\sum_{i=1}^{6} \frac{(-1)^i}{9i}$ **42.** $\displaystyle\sum_{i=1}^{n} \frac{5^i}{i}$ **43.** $\displaystyle\sum_{i=1}^{n+1} \frac{7^{i-1}}{(i-1)!}$ **44.** 55

9.2 Exercises

1. yes, 6 **2.** no **3.** yes, $\dfrac{3}{2}$ **4.** no **5.** yes, 2 **6.** no **7.** $a_n = 5n - 3, a_{10} = 47$ **8.** $a_n = 9 - 4n, a_{31} = -115$

9. $a_n = \dfrac{3}{2}n, a_{50} = 75$ **10.** $a_n = 6.2 - 1.2n, a_{29} = -28.6$ **11.** $a_n = 2n - 1, a_{17} = 33$ **12.** $a_n = 21 - 5n, a_{11} = -34$

13. 71 **14.** -64 **15.** 97 **16.** -104 **17.** $\dfrac{25}{2}$ **18.** -45 **19.** 3240 **20.** 235 **21.** 2375 **22.** 1044

23. -1002 **24.** 51,612 **25.** $4n^2 + 2n$ **26.** 10,000 **27.** 10,100 **28.** 3600 **29.** 147 **30.** \$444,600 **31.** \$3185
32. Beta Company **33.** \$200, \$65,000 **34.** \$4329 **35.** 1.3 sec, 312 in.

9.3 Exercises

1. 8, 16, 32, 64, 128 **2.** 162, 54, 18, 6, 2 **3.** 25, -5, 1, $-\dfrac{1}{5}, \dfrac{1}{25}$ **4.** 3, 21, 147, 1029, 7203 **5.** $-4, 8, -16, 32, -64$

6. yes, 6 **7.** yes, -1 **8.** no **9.** yes, $-\dfrac{5}{3}$ **10.** yes, $-.8$ **11.** $a_n = 3 \cdot 2^{n-1}$ **12.** $a_n = \dfrac{1}{2}\left(\dfrac{1}{4}\right)^{n-1}$ **13.** $a_n = \dfrac{1}{5}\left(-\dfrac{2}{3}\right)^{n-1}$

14. 20,480 **15.** $-\dfrac{951}{256}$ **16.** $\approx \$5,523.11$ **17.** 78,732 **18.** $\dfrac{1}{16}$ **19.** $-91,749$ **20.** ≈ 4.99 **21.** ≈ 50.31

22. $\dfrac{2186}{729}$ **23.** $\dfrac{85}{128}$ **24.** converges, $-\dfrac{10}{11}$ **25.** converges, $\dfrac{2401}{6}$ **26.** converges, $\dfrac{28}{3}$ **27.** diverges **28.** converges, $\dfrac{5}{11}$

29. \$6561 **30.** \$1146.81 **31.** $\approx \$1,402,123.44$ **32.** ≈ 53.56 ft, 60 ft **33.** 200 ft **34.** $\dfrac{7}{9}$ **35.** $\dfrac{161}{495}$

36. $\approx \$46,007.74$ **37.** $\approx \$30,200.99$ **38.** $\approx \$462.20$

9.4 Exercises

1. $m^6 + 6m^5n + 15m^4n^2 + 20m^3n^3 + 15m^2n^4 + 6mn^5 + n^6$ **2.** $x^4 + 20x^3 + 150x^2 + 500x + 625$ **3.** $x^7 - 7x^6y$
$+ 21x^5y^2 - 35x^4y^3 + 35x^3y^4 - 21x^2y^5 + 7xy^6 - y^7$ **4.** $x^5 - 15x^4 + 90x^3 - 270x^2 + 405x - 243$ **5.** $64x^6 + 576x^5y$
$+ 2160x^4y^2 + 4320x^3y^3 + 4860x^2y^4 + 2916xy^5 + 729y^6$ **6.** $81x^8 - 432x^6y^3 + 864x^4y^6 - 768x^2y^9 + 256y^{12}$ **7.** 7
8. 210 **9.** 1 **10.** 1771 **11.** $x^7 + 14x^6 + 84x^5 + 280x^4 + 560x^3 + 672x^2 + 448x + 128$ **12.** $x^6 - 18x^5 + 135x^4$
$- 540x^3 + 1215x^2 - 1458x + 729$ **13.** $1024x^5 + 1280x^4 + 640x^3 + 160x^2 + 20x + 1$ **14.** $x^4 + 12x^3y + 54x^2y^2$
$+ 108xy^3 + 81y^4$ **15.** $3125x^5 - 9375x^4y + 11250x^3y^2 - 6750x^2y^3 + 2025xy^4 - 243y^5$ **16.** $x^{24} + 6x^{20}y^5$

$+ 15x^{16}y^{10} + 20x^{12}y^{15} + 15x^8y^{20} + 6x^4y^{25} + y^{30}$ **17.** $x^2 - 4\sqrt{2}x^{\frac{3}{2}} + 12x - 8\sqrt{2}x^{\frac{1}{2}} + 4$ **18.** $129,024x^4$ **19.** $70a^4b^4$
20. $5103c^5d^2$ **21.** $1,088,640x^4$ **22.** 336,798 **23.** 126,720 **24.** 495 **25.** 19,779,228

9.5 Exercises

1. $S_1: 3 = 3(1)^2, S_2: 3 + 9 = 3(2)^2, S_3: 3 + 9 + 15 = 3(3)^2$, yes **2.** $S_1: 1^2 = \dfrac{1(2)(3)}{6}, S_2: 1^2 + 2^2 = \dfrac{2(3)(5)}{6}$,

$S_3: 1^2 + 2^2 + 3^2 = \dfrac{3(4)(7)}{6}$, yes **3.** $S_1: 2 > 1, S_2: 4 > 4, S_3: 8 > 9$, false **4.** S_1: 3 is not a factor of 0, S_2, 3 is a factor of 6,

S_3: 3 is a factor of 24, false **5.** $S_k: 6 + 12 + 18 + \cdots + 6k = 3k(k + 1), S_{k+1}: 6 + 12 + 18 + \cdots + 6k + 6(k + 1) =$
$3(k + 1)(k + 2)$ **6.** $S_k: 5 + 11 + 17 + \cdots + (6k - 1) = k(3k + 2), S_{k+1}: 5 + 11 + 17 + \cdots + (6k - 1)$

$+ (6k + 5) = (k + 1)(3k + 5)$ **7.** $S_k: 1^2 + 2^2 + 3^2 + \cdots + k^2 = \dfrac{k(k + 1)(2k + 1)}{6}, S_{k+1}: 1^2 + 2^2 + 3^2 + \cdots + k^2$

$+ (k + 1)^2 = \dfrac{(k + 1)(k + 2)(2k + 3)}{6}$ **8.** $S_k: \dfrac{1}{1 \cdot 2} + \dfrac{1}{2 \cdot 3} + \dfrac{1}{3 \cdot 4} + \cdots + \dfrac{1}{k(k + 1)} = \dfrac{k}{k + 1}, S_{k+1}: \dfrac{1}{1 \cdot 2} + \dfrac{1}{2 \cdot 3} + \dfrac{1}{3 \cdot 4}$

$+ \cdots + \dfrac{1}{k(k + 1)} + \dfrac{1}{(k + 1)(k + 2)} = \dfrac{k + 1}{k + 2}$ **9.** Show true for $S_1: 6 = 3 \cdot 1(1 + 1) = 6$ true. Assume S_k is true:

$S_k: 6 + 12 + 18 + \cdots + 6k = 3k(k + 1)$. We must show S_{k+1} is true. Start with the left-hand side of S_{k+1}. $6 + 12 + 18 + \cdots$
$+ 6k + 6(k + 1) = 3k(k + 1) + 6(k + 1) = 3k^2 + 3k + 6k + 6 = 3k^2 + 9k + 6 = 3(k + 1)(k + 2)$, so S_{k+1} is true.

10. Show true for S_1: $2 = \frac{1}{2} \cdot 1(1 + 3) = 2$ true. Assume S_k is true: S_k: $2 + 3 + 4 + \cdots + (k + 1) = \frac{1}{2}k(k + 3)$. We must

show S_{k+1} is true. Start with the left-hand side of S_{k+1}. $2 + 3 + 4 + \cdots + (k + 1) + (k + 2) = \frac{1}{2}k(k + 3) + (k + 2) =$

$\frac{1}{2}k^2 + \frac{3}{2}k + k + 2 = \frac{1}{2}k^2 + \frac{5}{2}k + 2 = \frac{1}{2}(k^2 + 5k + 4) = \frac{1}{2}(k + 1)(k + 4)$, so S_{k+1} is true. **11.** Show true for S_1:
$3 = 1(2 \cdot 1 + 1) = 3$ true. Assume S_k is true: S_k: $3 + 7 + 11 + \cdots + (4k - 1) = k(2k + 1)$. We must show S_{k+1} is true.
Start with the left-hand side of S_{k+1}. $3 + 7 + 11 + \cdots + (4k - 1) + (4(k + 1) - 1) = k(2k + 1) + (4(k + 1) - 1) =$
$2k^2 + k + 4k + 4 - 1 = 2k^2 + 5k + 3 = (k + 1)(2k + 3)$, so S_{k+1} is true. **12.** Show true for S_1:

$1^2 = \frac{1(1 + 1)(2 + 1)}{6} = 1$ true. Assume S_k is true: S_k: $1^2 + 2^2 + 3^2 + \cdots + k^2 = \frac{k(k + 1)(2k + 1)}{6}$. We must show S_{k+1}

is true. Start with the left-hand side of S_{k+1}. $1^2 + 2^2 + 3^2 + \cdots + k^2 + (k + 1)^2 = \frac{k(k + 1)(2k + 1)}{6} + (k + 1)^2 =$

$\frac{2k^3 + 3k^2 + k + 6k^2 + 12k + 6}{6} = \frac{2k^3 + 9k^2 + 13k + 6}{6} = \frac{(k + 1)(k + 2)(2k + 3)}{6}$, so S_{k+1} is true. **13.** Show true

for S_1: $3^0 = \frac{1}{2}(3^1 - 1) = \frac{1}{2}(2) = 1$ true. Assume S_k is true: S_k: $3^0 + 3^1 + 3^2 + \cdots + 3^{k-1} = \frac{1}{2}(3^k - 1)$. We must show S_{k+1}

is true. Start with the left-hand side of S_{k+1}. $3^0 + 3^1 + 3^2 + \cdots + 3^{k-1} + 3^k = \frac{1}{2}(3^k - 1) + 3^k = \frac{(3^k - 1) + 2 \cdot 3^k}{2} =$

$\frac{3 \cdot 3^k - 1}{2} = \frac{3^{k+1} - 1}{2} = \frac{1}{2}(3^{k+1} - 1)$, so S_{k+1} is true. **14.** Show true for S_1: $\frac{1}{1 \cdot 2} = \frac{1}{1 + 1} = \frac{1}{2}$ true. Assume S_k is true:

S_k: $\frac{1}{1 \cdot 2} + \frac{1}{2 \cdot 3} + \frac{1}{3 \cdot 4} + \cdots + \frac{1}{k(k + 1)} = \frac{k}{k + 1}$. We must show S_{k+1} is true. Start with the left-hand side of S_{k+1}. $\frac{1}{1 \cdot 2}$

$+ \frac{1}{2 \cdot 3} + \frac{1}{3 \cdot 4} + \cdots + \frac{1}{k(k + 1)} + \frac{1}{(k + 1)(k + 2)} = \frac{k}{k + 1} + \frac{1}{(k + 1)(k + 2)} = \frac{k(k + 2)}{(k + 1)(k + 2)} + \frac{1}{(k + 1)(k + 2)} =$

$\frac{k^2 + 2k + 1}{(k + 1)(k + 2)} = \frac{(k + 1)^2}{(k + 1)(k + 2)} = \frac{k + 1}{k + 2}$, so S_{k+1} is true. **15.** Show true for S_1: $1(1 + 1) + 4 = 6$, which is divisible

by 2. Assume S_k: $k(k + 1) + 4$ is divisible by 2. We must show that S_{k+1} is true. That is, we must show that $(k + 1)(k + 2) + 4$
is divisible by 2. $(k + 1)(k + 2) + 4 = k(k + 1) + 2(k + 1) + 4 = k(k + 1) + 4 + 2(k + 1)$ is divisible by 2 because
$k(k + 1) + 4$ is divisible by 2 by the induction assumption, $2(k + 1)$ is divisible by 2 since 2 is a factor, and the sum of two
numbers divisible by 2 is also divisible by 2, so S_{k+1} is true. **16.** Show true for S_1: $3^1 \geq 3 \cdot 1$ true. Assume S_k is true: S_k: $3^k \geq 3k$. We
must show S_{k+1} is true. Start with the left-hand side of S_{k+1}. $3^{k+1} = 3 \cdot 3^k = 3^k + 3^k + 3^k \geq 3k + 3k + 3k \geq 3k + 2 + 1 =$
$3k + 3 = 3(k + 1)$, so S_{k+1} is true.

9.6 Exercises
1. 60 **2.** 120 **3.** 48 **4.** 20,442,240 **5.** 256 **6.** 101,237,760 **7. a.** 60 **b.** 56 **8.** 45 **9.** 36 **10.** 31
11. 8 **12.** 24 **13.** 19,683 **14.** 85,266 **15.** 67,716 **16.** 54,461 **17.** 78,540 **18.** 7,311,616 **19.** 1,000,000
20. 9,834,496 **21.** 24 **22.** 120 **23.** 40,320 **24.** 2520 **25.** 39,270 **26.** 1716 **27.** 8,814,960 **28.** 34,650
29. 2520 **30.** 35 **31.** 2520 **32.** 142,506 **33.** 300 **34.** 66 **35.** 163,185 **36.** 133,784,560 **37.** 120 **38.** 10
39. 1 **40.** 22,100 **41.** 286 **42.** 1014 **43.** 133,784,560 **44.** 476,784 **45.** 55,836 **46.** 24 **47.** 120
48. 40,320 **49.** 2520 **50.** 39,270 **51.** 1716 **52.** 8,814,960 **53.** 142,506 **54.** 300 **55.** 66 **56.** 163,185
57. 133,784,560 **58.** 120 **59.** 10 **60.** 1 **61.** 22,100 **62.** 286 **63.** 1014 **64.** 133,784,560 **65.** 476,784
66. 55,836 **67.** 24 **68.** 120 **69.** 40,320 **70.** 2520 **71.** 39,270 **72.** 1716 **73.** 8,814,960 **74.** 142,506
75. 300 **76.** 66 **77.** 163,185 **78.** 133,784,560 **79.** 120 **80.** 10 **81.** 1 **82.** 22,100 **83.** 286 **84.** 1014
85. 133,784,560 **86.** 476,784 **87.** 55,836

9.7 Exercises

1. 60 **2.** 20,000 **3.** 16 **4.** 40 **5.** 22,100 **6.** 32 **7.** 863,040 **8.** $\frac{1}{4}$ **9. a.** $\frac{1}{5}$ **b.** $\frac{2}{5}$ **10.** $\frac{18}{37}$ **11.** $\frac{18}{37}$

12. $\frac{1}{37}$ **13.** $\approx 20.2\%$ **14.** $\approx 93.0\%$ **15.** $\approx 25.7\%$ **16.** $\approx 56.0\%$ **17.** $\approx 86.2\%$ **18.** 0.4 **19.** 0.2

20. 0.5　**21.** 0.6　**22.** $\frac{7}{13}$　**23.** $\frac{11}{26}$　**24.** $\frac{153}{251}$　**25.** $\frac{138}{251}$　**26.** $\frac{52}{251}$　**27.** $\frac{239}{251}$　**28.** $\frac{98}{251}$　**29.** $\frac{113}{251}$

30. $\frac{12}{251}$　**31.** $\frac{199}{251}$　**32.** $\frac{139}{200}$　**33.** $\frac{8}{25}$　**34.** $\frac{61}{200}$　**35.** $\frac{29}{200}$　**36.** $\frac{4}{25}$　**37.** $\approx 8.3\%$　**38.** $\approx 0.6\%$　**39.** $\approx 83.9\%$

40. $\approx 16.1\%$　**41.** $\frac{6}{209}$　**42.** $\frac{15}{209}$　**43.** $\frac{9}{133}$　**44.** $\frac{141}{209}$　**45.** $\approx 34.4\%$　**46.** $\approx 74.7\%$　**47.** $\frac{144}{169}$　**48.** $\frac{360}{497}$

49. 0.5　**50.** ≈ 0.33　**51.** $\frac{12}{51}$　**52.** $\frac{126}{137}$　**53.** $\frac{60}{143}$　**54.** $\frac{400}{1001}$　**55.** $4:271$　**56.** $12:11$　**57.** $11:12$　**58.** $1:4$

59. $8:7$　**60.** $18:19$　**61.** $18:19$　**62.** $36:1$　**63.** $\approx 8.3\%$　**64.** $\approx 0.6\%$　**65.** $\approx 83.9\%$　**66.** $\approx 16.1\%$　**67.** $\frac{6}{209}$

68. $\frac{15}{209}$　**69.** $\frac{9}{133}$　**70.** $\frac{141}{209}$　**71.** $\approx 34.4\%$　**72.** $\approx 74.7\%$　**73.** $\frac{126}{137}$　**74.** $\frac{60}{143}$　**75.** $\frac{400}{1001}$　**76.** $\approx 8.3\%$

77. $\approx 0.6\%$　**78.** $\approx 83.9\%$　**79.** $\approx 16.1\%$　**80.** $\frac{6}{209}$　**81.** $\frac{15}{209}$　**82.** $\frac{9}{133}$　**83.** $\frac{141}{209}$　**84.** $\approx 34.4\%$　**85.** $\approx 74.7\%$

86. $\frac{126}{137}$　**87.** $\frac{60}{143}$　**88.** $\frac{400}{1001}$

Index